PUTTING DOWN ROOTS:
FOUNDATIONS OF BOTANY AT CAROLINA

UNC botanist John N. Couch and UNC botany research assistant Anne (Whitlock) Nielsen observing specimens from the collection of antibiotic fungi (in background), Davie Hall, Department of Botany, University of North Carolina, Chapel Hill, March 1962. (Photographic Archives #22605, North Carolina Collection, Wilson Special Collections Library, University of North Carolina, Chapel Hill).

PUTTING DOWN ROOTS: FOUNDATIONS OF BOTANY AT CAROLINA

With a Concluding Chapter on the History of the Department of Botany

By William R. Burk

BRIT PRESS

Fort Worth Botanic Garden | Botanical Research Institute of Texas
Center for Botanical Exploration & Discovery
Fort Worth, Texas

Putting Down Roots: Foundations of Botany at Carolina

First edition, 2023

Printed in Korea
ISBN: 978-1-889878-71-3

Published by BRIT Press, Fort Worth Botanic Garden | Botanical Research Institute of Texas, Fort Worth, Texas, U.S.A.

Cover design: Linda Jacobson
Book design, layout, and production: Ricki Herdendorf
Compilation of indexes: Kelly Burch

Front Cover
UNC botanist Henry R. Totten (right foreground) conducting a botany field trip, April 1958. He and his students are examining the exotic princess tree, *Paulownia tomentosa*, which is an invasive non-native tree. (Photographic Archives #15696, North Carolina Collection, Wilson Special Collections Library, University of North Carolina, Chapel Hill). Cover design: Linda Jacobson, Keeper, North Carolina Collection Gallery, Wilson Special Collections Library, University of North Carolina, Chapel Hill.

Please cite this book as follows:
Burk, W.R. 2023. Putting Down Roots: Foundations of Botany at Carolina. Sida, Bot. Misc. 62. Botanical Research Institute of Texas, Fort Worth, Texas, U.S.A.

Distribution of copies by:
BRIT Press, Fort Worth Botanic Garden | Botanical Research Institute of Texas
1700 University Drive
Fort Worth, Texas 76107-3400, U.S.A.
Telephone: 1.817.332.4441
Fax: 1.817.332.4112
Email: press@brit.org
Websites: britpress.org, shop.brit.org

BRIT PRESS

Botanical Miscellany, No. 62
ISSN: 0883-1475
20 Feb 2023

Preface

Around 1992 I became interested in the history of botany as an academic discipline in North America. Emanuel D. Rudolph and Ronald L. Stuckey, professors of botany at The Ohio State University, were scholars in this field, and their work caught my attention. The spark that ignited my pursuit of a history of botany at the University of North Carolina at Chapel Hill was a document that I found in the archives of the UNC Herbarium in 1996. The item was a single sheet that listed the names of thirteen men who taught botany at the university. The list accompanied a letter of October 12, 1927, which was probably written by UNC botanist William C. Coker[a] or his research assistant, Alma Holland. The recipient of the correspondence was University of Pennsylvania botanist J. Harshberger.

After reading the document, I wanted to know more about the early years of botanical instruction at UNC–Chapel Hill and the professors who taught the subject before Coker's arrival there in 1902. No written history on this topic exists. I began in earnest to identify sources that would provide insight into those early years of instruction. My investigations commenced with extensive research in UNC's Wilson Special Collections Library, where I examined countless letters, documents, diaries, and publications in the North Carolina Collection, the Southern Historical Collection, and the UNC Archives. Concurrently, I visited a number of other institutions, including the University of Michigan, the St. Clair (Michigan) Public Library, Johns Hopkins University, Hampton University, the Friends Library at Swarthmore College, Cornell University, Duke University, Guilford College, the University of North Carolina at Greensboro, the North Carolina State University, and the North Carolina State Archives in Raleigh.

I began piecing together the bits of information I identified during eighteen years of research. The emerging picture consisted of eight men who laid the foundation for the instruction of botany at Carolina. Each person constituted a one-man department. That academic hierarchy influenced my decision to devote each chapter to a particular professor, in chronological order. The role of botanical instruction in the Normal School and Summer School is also included. The volume concludes with a chapter on the history of the Department of Botany. From 1908 to 1982 four chairpersons guided the department, which earned national recognition for its contributions in research and botanical education.

The formative period of botanical instruction at UNC spanned more than eighty years, in which at least twelve educators taught botany in some form until a formal botanical program was established in 1908. That was the year that Departments of Botany and Zoology were created from the division of the original Department of Biology. The departments shared their quarters in a new building, Davie Hall.

---

[a] Coker's interest in the history of botany developed when he studied at the Rheinische Friedrich-Wilhelms Universität. At Carolina in the early 1920s, Coker established a course titled History of Botany (later called Studies in the History of Botany). Alma Holland subsequently taught the course from 1935 to 1944.

This book examines each professor who instructed in botany from 1819 to 1902, including their biographical experience, pedagogical style, scientific instruction, and contributions to science. I also explain the academic influences that each of these educators brought to Carolina. I am pleased to note that this book provides considerable background on aspects of UNC's nineteenth-century history not covered elsewhere. The concluding chapter chronicles the history of the Department of Botany until 1982, when it merged with the Department of Zoology to form the Department of Biology.

William R. Burk

Acknowledgments

While writing this book, I received assistance from numerous people to whom I am very grateful. Librarians and staff members at the institutions that are cited in the text provided guidance in accessing archival collections. I thank the following librarians at the Wilson Special Collections Library, UNC, who introduced me to useful reference tools and resources when I initiated my research: Alice R. Cotten,[b] John White, and Susan Ballinger. Subsequently, other librarians assisted me, and I particularly thank Jason Tomberlin, Matt Turi, Susan Bales, and Taylor (de Klerk) Butler.

In addition, I am grateful for the assistance of the following individuals at the University of North Carolina at Chapel Hill: Sarah Parnham of the UNC General Alumni Association; Ron Strauss, Executive Vice Chancellor and Provost; Cecelia Moore, former University Historian; Michael Shelor, Library Technical Assistant; Linda Jacobson, Keeper, North Carolina Collection Gallery, who designed the cover of this book; Marie Fholer, Kenlyn R. Merritt, Logan Brackett, Dani Shirilla, and Lori Shamblin of the Department of Biology, who provided historical information; Daniel K. Pennington, Administrative Support Specialist, Office of Human Resources and Records, who provided various dates of employment of faculty members; and David Romito, Science Librarian, who provided guidance in word processing and formatting and answered various reference inquiries. In the Biology Photo Lab, Susan Whitfield and Brian Nalley assisted me with numerous aspects of preparing the illustrations. Susan patiently and expertly reviewed and improved each digital image.

As my investigations progressed, I shared my discoveries with UNC botanical researcher Rogers McVaugh. Through our frequent conversations, I received his encouragement and constructive suggestions. Regarding historical information that I uncovered, McVaugh repeatedly advised me, "Do not leave it out." I heeded his advice.

Several individuals provided editorial assistance and commentary for which I am sincerely indebted. Ruth Homrighaus of Ruthless Editing edited the majority of the text. Miriam Downey and Laura M. Cotterman edited several chapters that I wrote at the end of the

[b] In her response to a reference inquiry that I emailed to Alice in January 2001, she concluded: "Keep those questions coming, as each one teaches me something." Her support and assistance inspired and encouraged me throughout my research and writing.

project. I gratefully acknowledge the following individuals, who each reviewed a chapter: Robert Dirig of Cornell University and, at the University of North Carolina at Chapel Hill, Janis Holder and Susan Ballinger of [the] Wilson Special Collections Library; Cecelia Moore; Robert K. Peet of the Department of Biology; and Carol Ann McCormick and Alan S. Weakley of the UNC Herbarium. Miriam Kennard, former UNC Geological Sciences Librarian, reviewed a passage in chapter 15. I extend special thanks to John Nelson, curator of the A.C. Moore Herbarium, University of South Carolina, who reviewed my manuscript and provided numerous comments. The following individuals provided assistance in citing several bibliographical entries: Douglas Brown, public services unit manager, Division of Archives and Records, North Carolina Department of Natural and Cultural Resources; and James Stimpert, senior reference archivist, Special Collections, University Archives, Johns Hopkins Sheridan Libraries. Kelly Burch proofread an advance copy of the manuscript with great attention to details. I am very grateful for the expert help of Kanchi N. Gandhi, Senior Nomenclature Registrar and Bibliographer in the Harvard University Herbaria & Libraries, who edited numerous plant names in Appendix 5.1. Eddie Herdendorf, Professor Emeritus of Earth Sciences at The Ohio State University, edited and enhanced the frontispiece portrait of chapter 7 and also provided constructive comments with various aspects of the manuscript.

Ricki Herdendorf of EcoSphere Associates patiently and exactingly formatted the book and assisted in the preparation of the indexes to the volume. I am very grateful for her careful attention to details and her keen power of identifying problems. Kelly Burch prepared the final version of the indexes. Candace H. Haigler provided background on the research investigations of R. Malcolm Brown, Jr. I am deeply grateful to Barney L. Lipscomb, director of BRIT Press and Library, for his guidance and assistance throughout the preparation of this book.

With deep gratitude, I acknowledge an Archie K. Davis Fellowship that I received in 2004. The funds helped to defray travel costs to several out-of-state archival collections.

I dedicate this volume to several people who enriched my life and inspired in me a pursuit of botanical knowledge.

My parents Oliver G. and Lillian M. Burk,

and

Botanists Ronald L. Stuckey, Rogers McVaugh, and Emanuel D. Rudolph.

TABLE OF CONTENTS

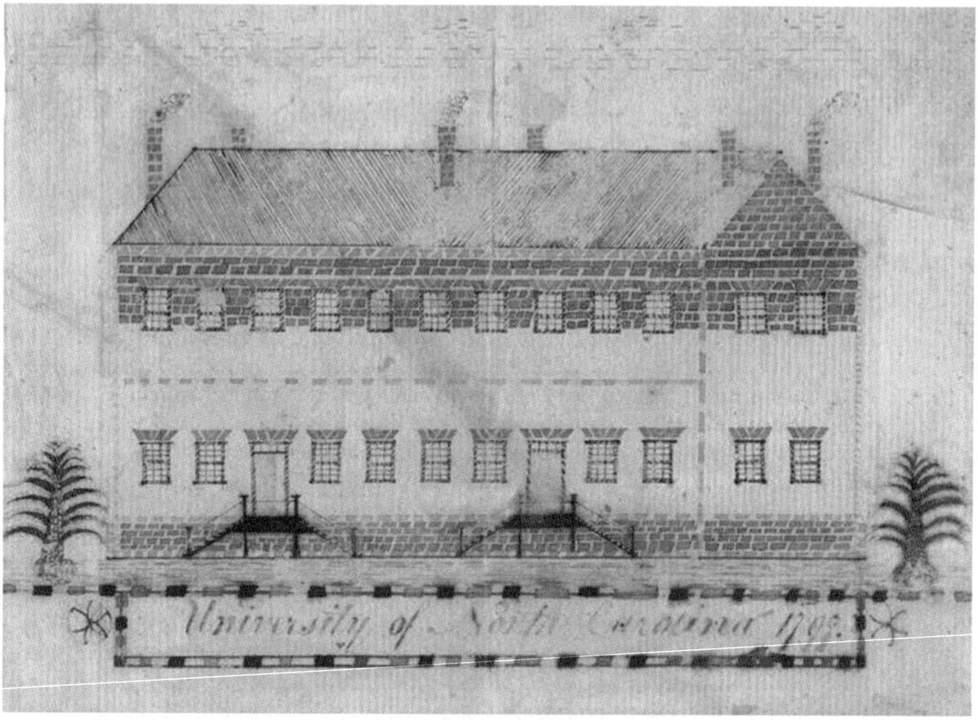
University of

Chapter 1.

Early Influences

Introduction

Soon after the North Carolina General Assembly chartered the University of North Carolina in December 1789, the roots of botanical instruction at the University of North Carolina were developing.[1] The Board of Trustees established a committee in 1792 to propose a plan of education for the institution. Chaired by Presbyterian minister and educator Rev. Samuel Eusebius McCorkle and composed of five other people, the group developed a plan for educating the students. The committee presented its proposal at a meeting on December 5 of that year. The suggested curriculum included "Information in Botany to which should be added a competent knowledge in the theory and practice of Agriculture but suited to the climate and Soils of the State" (BOT 1: 88–89). The committee clearly intended that botanical knowledge would be applied in an agricultural context, reflecting the agrarian nature, and needs, of the state. A few years later, on February 6, 1795, the trustees adopted the university's first curriculum, which was issued in the "Laws and Regulations for the University of North Carolina" (Connor, 1953, 1: 375–379). It differed slightly from the one that McCorkle's committee had developed. Among the listed duties of the president and professors was a third item: "That until a Professor shall be appointed for the purpose, the President or next officer shall deliver a lecture once in the week on the Principles of Agriculture, Botany, Zoology, Mineralogy [or] Commerce" (BOT 1: 163).

From 1819 to 1902 a series of eight men gave lectures or courses on botany in a one-man department under the following professorships: Chemistry, Mineralogy, and Geology; Chemistry Applied to Agriculture and the Arts; Agriculture; Natural History; Geology and Natural History; and Biology (Table 1.1). Based on extant archival documents, the first known instance of botany and its applications being taught was in 1819 in Denison Olmsted's class on chemistry. The educational background and training of these professors greatly influenced their approach and methods for teaching at UNC. Two institutions in particular had a significant impact on the early instruction in botany as well as other sciences at Chapel Hill: Yale College and Cornell University. The influence of other institutions on the instruction of botany is discussed in the following chapters.

[1] Numerous publications provide historical background on the University of North Carolina at Chapel Hill (Parker and Harrelson, 1992). UNC professor and president Kemp P. Battle (1907–1912) prepared a two-volume treatise that chronicles the institution's history from 1789 to 1912.

Frontispiece Credits: Early buildings on the University of North Carolina at Chapel Hill campus. (above) Old East Building, 1797, in which the Natural History Museum was displayed on the third floor, ca. 1875–ca. 1891. From an ink drawing by John Pettigrew. (below) Smith Hall, 1892, which served as the University Library, 1852–1907. (Courtesy of North Carolina Collection Photographic Archives, Wilson Special Collections Library, University of North Carolina, Chapel Hill).

TABLE 1.1. A CHRONOLOGICAL LIST OF MILESTONES IN THE DEVELOPMENT OF BOTANY AT THE UNIVERSITY OF NORTH CAROLINA, 1789 TO 1904.

1789	North Carolina General Assembly ratifies Charter for the University.
1792	Committee established by trustees includes the subject botany in a plan of education.
1795	University opens its doors and offers occasional lectures in botany.
1817	Denison Olmsted accepts newly established Professorship in Chemistry, Mineralogy, and Geology.
1817	Elisha Mitchell elected Professor of Mathematics and Natural Philosophy.
1818	Mitchell begins teaching at UNC; studies the plants of Chapel Hill and surrounding region.
1819	Olmsted begins teaching; lectures include Linnean system of plant classification, the effect of light on plants, and plant products and their use by man.
1825	Olmsted resigns his chair to accept a faculty post at Yale College; Mitchell fills Olmsted's chair in Chemistry, Mineralogy, and Geology.
1840	Mitchell publishes *Natural History*, a syllabus used for his lectures on botany, mineralogy, and zoology; his botanical lecture focuses on the identification of plants.
1857	Mitchell dies on June 27, falling from a precipice on Black Mountain while verifying its highest peak.
1857–1866	John Kimberly serves as Professor of Chemistry Applied to Agriculture and the Arts; his instruction includes at least six lectures on plants with an agricultural emphasis.
1861–1865	University remains open during the Civil War.
1868 (July 17)– January 2, 1869	University suspends operation due to actions of the Reconstruction government of North Carolina.
1869–1870	George Dixon fills newly created professorship of Agriculture; charged with teaching agriculture, chemistry, and natural history.
1871 (February 1)– September 6, 1875	University closed during Reconstruction.
1875–1876	John Kimberly serves as Professor of Agriculture; instruction probably includes vegetable physiology.
1876–1877	William Henry Smith, Professor of Natural History, teaches botany; organizes the University Museum. (Smith was the first professor at UNC who held a doctorate degree when hired.)
1877–1881	Frederic William Simonds, Professor of Natural History, teaches first semblance of a course in botany.
1877–1884	Normal School includes special lectures (but not courses) in botany and natural history.
1878	"Herbarium of local flora" first mentioned in the UNC catalog of 1877–1878 (p. 20).
1881–1891	Joseph Austin Holmes, Professor of Geology and Natural History, expands curriculum in botany; develops laboratory for botany and zoology; gives first course in horticulture; supervises University Museum.

TABLE 1.1. (CONTINUED)

1883	Elisha Mitchell Scientific Society established, promotes a forum for scientific exchange through meetings and its *Journal*.
1891	Department of Biology and Department of Geology formed by the division of the Department of Geology and Natural History.
1891–1902	Henry Van Peters Wilson, Professor of Biology, instruction includes botanical topics; laboratory work includes the use of microscopes to study plant cells and tissue; teaches zoology the remainder of his life.
1894–1904	Summer School includes courses in botany, biology, nature studies, and marine biology; in the first session Dixie Lee Bryant, a professor at the Normal and Industrial School (now UNC Greensboro), teaches a course in botany, distinguishing her as the first woman to teach that subject at UNC; Austin Craig Apgar (1896, 1898) and George Francis Atkinson (1901) give courses in botany.

The Influence of Yale College on Natural History and Chemistry Instruction at UNC

For nearly forty years (1819–1857), instruction in chemistry, geology, and mineralogy, as well as botany and agriculture, at UNC was vested in the hands of first Denison Olmsted and then Elisha Mitchell. The two men had been classmates at Yale College (now Yale University) from 1809 to 1813 and later served as tutors there. Their intellectual development and their moral and religious character were shaped by this academic setting, particularly by three notable professors: Timothy Dwight, Jeremiah Day, and Benjamin Silliman (Fig. 1.1).

Fig. 1.1. Early Yale professors who promoted science. (left) Timothy Dwight (image 4208); (center) Jeremiah Day (image 35450); (right) Benjamin Silliman (image 695).

(Images of Yale Individuals, [RU 684], Manuscripts and Archives, Yale University Library).

In the early nineteenth century, scientific exploration was in its infancy in the country. Yale was an important center from which scientific enlightenment was disseminated (Fulton and Thomson, 1947). In 1809 the scheme of instruction was structured around a four-year program. Freshmen and sophomores studied primarily the classics (Greek and Latin), as well as geography, mathematics, English grammar, and history. Juniors took trigonometry, navigation, surveying, fluxions (*i.e.*, calculus), science (astronomy, natural philosophy, and chemistry), as well as more classics and grammar. In their senior year, students took additional science courses, including more chemistry from Silliman and natural philosophy taught by Jeremiah Day, who later (1817) became Yale's president (Schoepflin, 1977).

Timothy Dwight was elected president of the college in 1795, served as professor of divinity, and taught rhetoric, ethics, logic, and metaphysics. He had a "strong dominating character, active mind, and untiring energy" (Baldwin, S.E., 1901–1903). Olmsted and Mitchell were among a long lineage of students who heard his four-year-long series of lectures on divinity given on Sundays. Dwight's philosophy of life, in which he favored discipline through persuasion, rather than punishment, influenced their moral thinking (Schoepflin, 1977; Dowling, 1999). He fostered the study of natural sciences while maintaining a balance with religious doctrine by holding that "God's divine plan could be appreciated even more fully as scholars mastered the intricacies of the emerging scientific fields" (Ellsworth, 1999). He trained the inquisitive minds of students to make careful observations and to organize "their store of knowledge so as to be readily recalled" (Schoepflin, 1977).

Jeremiah Day was elected to the chair of mathematics and natural philosophy in 1803. He had received his college education at Yale, where he later served as a tutor (Baldwin, E.,1831). Upon Dwight's death, Day was elected Yale's next president in 1817. He continued Dwight's efforts to promote science in the curriculum and persevered in acquiring the Gibbs Cabinet of minerals, which is discussed below (Kelley, 1974).

Benjamin Silliman was appointed professor of chemistry and natural history in 1802. The November 1812 *Catalogue of the Faculty and Students of Yale College* notes his title as "Prof. of Chem[istry] and Mineralogy." He spent his first two years studying chemistry to prepare for teaching that discipline and presented his first lectures in 1804. His chemistry course included over 100 lectures as well as demonstrations (Schoepflin, 1977). According to his biographers (Fulton and Thomson, 1947), he made two significant scientific contributions. He was a leader in advancing science, most notably its application to industry and in assessing the country's natural resources. As an educator, he inspired and mentored numerous students of science. He also popularized science, promoting its application to mining, agriculture, and industry (Warren, 1999). In 1818 Silliman founded the *American Journal of Science and Arts*, which covered practical and theoretical science. The journal became the preeminent science periodical in America and won Silliman international acclaim (Greene, 1975). Olmsted and Mitchell would later publish articles in "Silliman's journal," as it was commonly known.

While Olmsted and Mitchell were upperclassmen, educational resources in geology were greatly augmented with the acquisition of a collection of nearly 10,000 mineralogical specimens. Around 1809 to 1810, Silliman brokered a deal with Colonel George Gibbs, a man of great wealth and the owner of the magnificent collection, to display the specimens at Yale, where a suitable exhibition hall was built for them in South Middle Hall, later called Connecticut Hall. Following the arrangement of the specimens for display, the Gibbs Cabinet, as it was called, was officially opened for viewing in June 1812. In 1820 the collection (Fig. 1.2 [left]) was moved to a building called "the Cabinet" (Fig. 1.2 [right]), later known as the Philosophical Building. Yale purchased the collection in 1825 (Baldwin, E., 1831; Fulton and Thomson, 1947; Hartt, 1926). The cabinet became an important illustrative aid in Silliman's geology and mineralogy instruction, as it allowed students to examine specimens firsthand.

It also attracted visitors to the campus. Silliman subsequently promoted geological surveys through his students, such as Olmsted, Amos Eaton, and Edward Hitchcock (Fulton and Thomson, 1947).

Fig. 1.2. The Gibbs Collection, Yale University. (left) Gibbs Cabinet (image 8719). (right) The Cabinet built at Yale College in 1819 and razed in 1890 (image 8718).

(Yale University Buildings and Grounds Photographs, [RU 703], Manuscripts and Archives, Yale University Library).

In contrast to mineralogy, geology, and chemistry, whose instruction was aided by excellent collections and a well-equipped laboratory, botanical instruction was yet to be offered at Yale in the first decade of the nineteenth century. Following the establishment of Yale's Medical Institution in 1810, the earliest course of lectures in botany was added to the curriculum. Within the medical school, Eli Ives, M.D. (Fig. 1.3), professor of *materia medica* and botany (later professor of the theory and practice of physic), first taught botany in 1813 (Baldwin, E., 1831; Dexter, 1911). Although Eneas Munson, who most likely played a role in establishing the medical school, was appointed emeritus professor of *materia medica*, that title was honorary (Quen, 1976). He probably did not formally teach in the newly established school.

During their postgraduate years at Yale, Olmsted and Mitchell may have associated with a number of individuals who held botanical interests. Among these men was Melines Conklin Leavenworth, who enrolled at Yale in 1813 and graduated in 1817, at which time he was deemed "one of the rising stars of the botanical firmament" (McVaugh, 1947). Subsequently, he entered the Medical Institution there, graduating in 1818. Prior to Leavenworth's arrival in New Haven, Yale had neither a botanical garden nor a conservatory. According to lawyer and historian Ebenezer Baldwin[2] (1831), Leavenworth was "engaged" to collect indigenous plants for a garden[3] at the institution. The garden was financed by the "then Professor of Botany," Eli Ives. In writing about Ives, Henry Bronson (1867) noted that the garden was situated on the eastern side of the campus and also contained some non-native species. Yale historian Franklin Bowditch Dexter (1911) noted that the garden was near Sheffield Hall. Due to a lack of maintenance and to expenses, the garden eventually fell into neglect.

[2] Ebenezer Baldwin (1790–1837), born in New Haven, Connecticut, studied law in Albany, New York, and attained distinction in that profession. For health reasons, he returned to his hometown, writing a history of Yale (Dexter, 1912).

[3] Physician Philo G. Rockwell ([1867]) noted that Leavenworth was placed in charge of the garden, but he did not mention that he developed it.

Fig. 1.3. Eli Ives, Professor of *Materia Medica* and Botany at Yale College.
(Images of Yale Individuals, [RU 684: image 930]. Manuscripts and Archives, Yale University Library).

Other opportunities for botanical investigations were also developing. Leavenworth collaborated with Ives and William Tully[4] in compiling an inventory of nearly 1,200 plants in the vicinity of Yale. They later published a *Catalogue of the Phenogamous Plants and of the Ferns Growing without Cultivation within Five Miles of Yale College* ([Ives et al.], 1831).[5] In recognizing Leavenworth's passion for botanical studies, the faculty chose him, along with Dr. James Gates Percival (a poet), to collect plants, primarily those from the South (McVaugh, 1947).

Another possible associate of Olmsted and Mitchell was famed naturalist Amos Eaton, who arrived at Yale in the spring of 1816 to study chemistry and mineralogy under Silliman. In his spare time, he also studied botany under the tutelage of Ives (McAllister, 1941). In fact, Eaton had proposed that Ives collaborate with him to publish *A Practical Digest of American Botany*. The proposal was not carried out, however.

4 A graduate of the 1806 Yale class, William Tully (1785–1859) developed avid interests in botany and chemistry. In 1810, he studied under Eli Ives, particularly focusing on botany. He practiced medicine a number of years and was appointed a professor of *materia medica* and therapeutics at Yale in 1829 (Dexter, 1912; Mead, 1928).

5 In 1811, Eli Ives (1811) had compiled an earlier list of plants found in the vicinity of New Haven, Connecticut.

In addition to these intellectual influences were campus societies that furthered and advanced the students' scholarly interests, as well as their ethical character. Among these organizations was the Moral Society, which the students themselves had organized. Its aim was to suppress vice, promote the morality of the Bible, and abstain from profanity and the playing of cards. The society also sponsored religious debates (Schoepflin, 1977). Another influential organization was the Connecticut Academy of Arts and Sciences, founded in 1799 to promote "useful knowledge" and the exchange of ideas (Anonymous, 1810). According to the academy's Constitution (CAAS, 1810), its object was "to cultivate every Art and Science which may tend to advance the Interest and Happiness of a free and virtuous People." In the early years, the academy held meetings six times a year and was associated with Yale College. Timothy Dwight served as the first president from 1799 to 1817 and was succeeded by Jeremiah Day from 1817 to 1836 (Ellsworth, 1999). Although Olmsted and Mitchell were not elected into membership until the spring of 1817, they most likely attended meetings with their academic mentors before then.

The Influence of Cornell University on Botanical Instruction at UNC

Three Cornell-trained men—Frederic William Simonds, Joseph Austin Holmes, and George Francis Atkinson—taught natural history at UNC. These men were instrumental in establishing and developing laboratory practice and a system of lectures as part of their pedagogical methodology, which was influenced to a great extent by their experiences at Cornell.

Opening its doors on October 7, 1868, Cornell University was founded on several primary principles: that agriculture and mechanic arts would be prominent subjects; that liberal arts would include courses in history, political science, and modern literature; and that students would have great latitude in selecting courses as well as greater freedom of expression and conduct than at other contemporary institutions (Becker, 1943). The "Cornell idea" took root and fostered an invigorating academic climate, stimulating the inquiring minds of future scientists and thinkers.

A botanical department was in place when Cornell opened. From 1871 to 1885, the period during which Simonds, Holmes, and Atkinson attended Cornell, botanical instruction at the institution continued to develop. Botanists Albert Nelson Prentiss (1836–1896) and William Russel Dudley (1849–1911) (Fig. 1.4) were the earliest teachers. Prentiss was appointed chair and professor of botany, arboriculture, and horticulture when Cornell opened in 1868, remaining there the rest of his life. His botanical teaching emphasized systematics, for which he used a system of lectures. His "easy illuminating style in his lectures and his course attracted students" (Fitzpatrick, 1926). Dudley was an instructor of botany from 1872 to 1876 and became an assistant professor of botany in 1876, a position he held until 1892, when he accepted a professorship in botany at Stanford University (Jordan, 1911). He was deemed "one of the most successful teachers of botany this country has ever produced," in large part due to his human and personal techniques of instruction (Branner, 1913). According to a student at Cornell, his scientific lectures were thorough and were "always delivered with that easy graceful command of language so characteristic of [him]" (Anonymous, 1880).

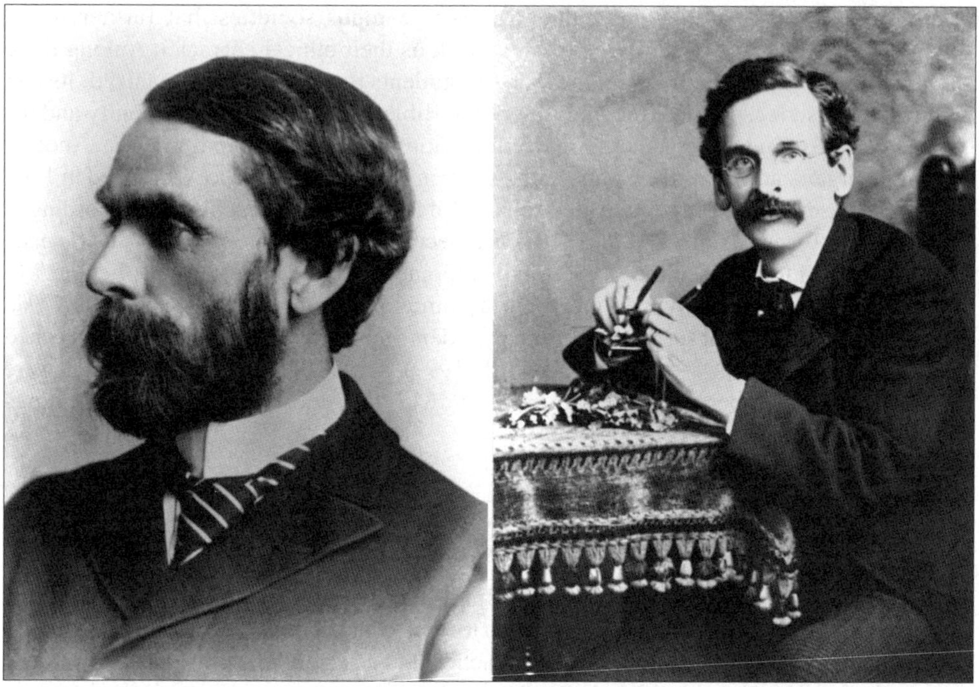

Fig. 1.4. Early Cornell botany faculty members. (left) Albert Nelson Prentiss, ca. mid-1870s; (right) William Russel Dudley, ca. late 1870s.

(Courtesy of Hunt Institute for Botanical Documentation, Carnegie Mellon University, Pittsburgh, Pennsylvania; copied from W.J. Beal Collection, Michigan State University).

During Cornell's early years, courses on structural botany, systematic and applied botany, and vegetable physiology were offered. Several additional botanical courses were later given: fungi in 1873, and mosses and algae and ferns in 1877 (Hewett, 1976). By 1881 courses on the Gramineae and Compositae and on plant culture were offered.

Besides lectures in the above-mentioned courses, the educational setting featured laboratory work, field studies, and an active Natural History Society, in addition to several specialized instructional tools. In the botanical laboratory, students identified plants, examined the anatomy of vegetable fabric, prepared an herbarium, and observed living and dried plant specimens. In the field, students learned the local plants and made a study of species within a locality (CU-R, no. 1).

The Natural History Society was the first departmental club at Cornell. Established in 1869, the society aimed "to work up the Natural History and Geology of Ithaca, and to obtain specimens, &c., from other parts of the United States" (Anonymous, 1869). The organization furthered a scholarly environment in which students contributed the results of their research and shared in the exchange of ideas. Faculty and students presented talks, which were followed by remarks and constructive comments from the listeners. All aspects of natural history were covered at the meetings, and botany was well represented. Botanical presentations covered such topics as wild crabapples, autumn leaves, *Saxifraga azoides*, fungi, ferns, pitcher plants, the use of vegetable hairs, wildflowers of Niagara, shade trees, plants of Ithaca, the longevity of trees, the fertilization of *Yucca* flowers, and the flora of the driftless region (NHSR, 1876–1881; Wright, 1954).

Several educational aids also enriched the botanical lessons (CU-R, no. 1, 2; CU-RC). Students had access to an herbarium. The plant collections included the Horace Mann (Jr.) Herbarium, containing about 12,500 specimens purchased from the Mann estate by Cornell President Andrew Dickson White; a developing collection of plant specimens from the Ithaca region; and a set of 300 wood samples.

Other instructional tools included three-dimensional plant models that were produced by Louis Auzoux[6] of Paris and Robert Brendel[7] of Breslau, Germany (now Wroclaw, Poland), who were prominent plant modellers of the nineteenth century. Auzoux's botanical models were known as *modèles clastiques* (from the Greek *klastos*, broken into pieces, referring to the fact that they could be dismantled and reassembled by students). Typically made of papier mâché (Tuckwell, 1869), the models represented "the more difficult orders of plants in gross anatomy of the flowers and fruits" (Paine, 1894). The Brendel models of algae, liverworts, mosses, fungi, ferns, and higher plants were usually also made of papier mâché, although other materials, including wood, rattan, pulp cane, glass beads, feathers, and gelatin, were also employed (Reiling, 2003) to create various textures. The models emphasized the microscopic structures of plant organs. In describing the *Brendel'schen Modellen*, Alexander Tschirch (1885), lecturer of botany at the University of Berlin, noted they were true to nature and based on current morphological research.

Plant models were an integral component of botanical teaching in the latter half of the nineteenth century, particularly in the study of systematics, and Cornell's botanists made frequent use of them. Through the enlargement of minute plant parts, particularly of flowers, the models offered an easily observable way to study plants. The models could be used year round, whereas living specimens were available only in the growing season. A number of the models (Fig. 1.5) survive at Cornell, where instructors still use them in courses on introductory botany.

An additional teaching aid was scientific wall charts. By the fall of 1874, the School of Botany had acquired a variety of educational charts (CU-RC). Among them was a set of nine sheets titled "Prof. Henslow's Botanical Diagrams," which were based on illustrations by the Reverend John Stevens Henslow[8] and were drawn by botanical artist Walter Hood Fitch

6 Louis Thomas Jérome Auzoux (1797–1880) studied medicine in Paris and received his medical degree in 1822. He dedicated his life to popularizing human anatomy, physiology, and hygiene by creating models (Lautour, 1948; Mulder, 1994).

7 Robert Brendel (ca. 1821–1898) made models of plants and other items in a factory that he established in Breslau in 1866. He became interested in model making through the encouragement of the noted botanist Geheimat Ferdinand Julius Cohn, who was a professor at the University of Breslau and father of bacteriology (Brendel, 1895). Brendel's son Reinhold Brendel (ca. 1861–1927) later assisted in the venture, and in about 1902 he moved the business to Grunewald, Germany.

8 John Stevens Henslow (1796–1861) entered St. John's College, Cambridge, in 1814, graduating in 1818. He subsequently became a fellow of the Linnean Society. At Cambridge, he was appointed chair of mineralogy in 1822 and then Regius Professor of Botany in 1825. He was also ordained a priest in 1824. Henslow became a mentor to Charles Darwin. From 1837 until his death, Henslow served as rector of Hitcham parish in Suffolk (Walters and Stow, 2001).

"for the Committee of Council on Education: Department of Science and Art, Phaenogamous Plants." Printed by Day & Son in 1857, they were early examples of printed wall charts, which were widely used in teaching during the late Victorian period (Walters and Stow, 2001). Measuring 40 x 30 inches, each chart contained enlarged colored diagrams of flowers, inflorescences, leaves, stems, roots, and seeds. According to Eliza Ann Youmans (1873a), a nineteenth-century author of botanical textbooks, the diagrams were highly acclaimed "for their scientific accuracy, their completeness of illustration, their judicious selection of typical specimens, and their skillful arrangement for the purpose of education." Because of their expense, however, Youmans encouraged her publisher (D. Appleton and Company) to prepare a revised set of Henslow's diagrams for the American classroom. The effort resulted in six sheets, along with a separately published thirty-four-page key to facilitate their use (Youmans, 1873b). Although Henslow's original set of charts does not seem to have survived at Cornell, a complete set is housed at the Hunt Institute for Botanical Documentation in Pittsburgh (HIBD).

Fig. 1.5. Nineteenth-century plant models. (left) Capsule of moss [*Brachythecium rutabulum*], x $^{1}/_{10}$, probably by Robert Brendel. (center) Pea flower, x $^{14}/_{100}$, artisan not given. (right) Longitudinal section of antheridial head of *Marchantia polymorpha* L. (original x 35), by R. Brendel.

(Courtesy of the Department of Plant Biology, Cornell University. Photographs by Kent E. Loeffler, December 2004; with special assistance from Kathie Hodge).

Cornell would become one of fourteen pioneering botanical departments that established modern curricula in America during the period 1870 to 1900 (Ford, 1962). Its students were aware of the superb quality of their botanical department. In 1877 a reporter of the *Cornell Era*, a student weekly journal of commentary and opinion, noted: "The facilities for the study of Botany at Cornell are not equaled by any other University in the country, excepting Harvard" (Anonymous, 1877). In writing for the *Cornell Review* (a student-run literary magazine) in 1880, another student praised "the splendid facilities, which [the] University affords for pursuing the study of advanced botany" (Anonymous, 1880). The instructive effect of Cornell on Simonds, Holmes, and Atkinson is discussed in the context of their tenure at UNC in subsequent chapters of this book.

Educational Enhancements at UNC

A number of experiences enhanced and enriched the students' academic training and scholarly pursuits in the first century of UNC. The museum and the Elisha Mitchell Scientific Society provided particular enhancement in the sciences, including botany. The library, Normal School, and laboratory also offered opportunities to strengthen the students' learning environment. The library and museum had their beginnings when the campus first opened its doors to students in 1795, the student laboratory experience and Normal School emerged in the late 1870s, and the Elisha Mitchell Scientific Society was founded in 1883.

The Library[9]

Three library collections developed early in the history of the university: the University Library and the libraries of the university's literary societies (the Dialectic and the Philanthropic, or the Di and Phi). The university collection at that time was not diverse. It emphasized the classics and theology, reflecting a curriculum strong on ancient languages, mathematics, natural philosophy, chemistry, and moral philosophy (York, 1982) and weak on general literature. Student use of the library was infrequent, due in part to the fact that the favored method of instruction at the university was based on recitations and textbooks and not on outside reading and library research.

The early library's primary means of growth was from donations. From 1795 to 1797, the library received 133 books (comprising fifty-six titles) as donations (Connor, 1953, 2: 290–291). Among the books was *An Introduction to Botany* (1794, 5th edition, corrected) (Fig. 1.6) by James Lee. The first known botany book to be acquired by the University Library, it was the earliest (first published in 1760) botanical textbook in the English language, based chiefly on Carl Linnaeus's (1751) *Philosophia Botanica*. Richard Bennehan, Esquire presented the book, along with thirty-one others, in July 1796 (Ashe, 1796). A successful merchant, planter, and builder as well as a pioneer in the tobacco industry and in education, Bennehan was a proponent for establishing the university at Chapel Hill (Sanders, 1979) and served the university as a trustee from 1799 to 1804. UNC President Joseph Caldwell purchased and acquired nearly 979 books for the library during his book-buying expedition to Europe, which lasted from early May 1824 to mid-March 1825. During the first century of the library its collection grew modestly: ca. 869 books in 1802; 1,900 in 1836; 3,500 in 1849; 3,600 in 1854; 6,540 in 1869; 8,394 in 1875; 9,000 (and 2,000 pamphlets) in 1884; 24,400 (includes the Di and Phi Libraries' books) in 1894; 26,000 (and 10,000 pamphlets) in 1895; and 40,272 in 1902.

Concurrently with the development of the University Library,[10] the Dialectic and Philanthropic Societies each collected books and managed libraries for their members. They amassed collections in a competitive spirit, often duplicating books within each library. According to UNC Associate Professor of Mathematics James Lee Love (1944), students used only the society book collections primarily during the period 1875–1886. He also noted that "there was no 'research' in science or literature . . . carried on by students." History, biography,

[9] Adapted from "The University of North Carolina at Chapel Hill Library and the Sciences, 1795–1902," by William R. Burk (*North Carolina Libraries* 67: 2–16, 2009).

[10] The early library was located in the first president's house, then in a room in South Building, and next in its own building, Smith Hall, where it remained until 1907.

government, and English literature, rather than science, were the subject strengths of the collections, according to Love. In 1884 a student writer also confirmed the bleak assessment of the use of whatever scientific books the society libraries owned (Anonymous, 1884a). He noted that "the books on Scientific subjects" were seldom removed from the shelves and that many students considered the library as a pleasant place, where they could pass away the time.

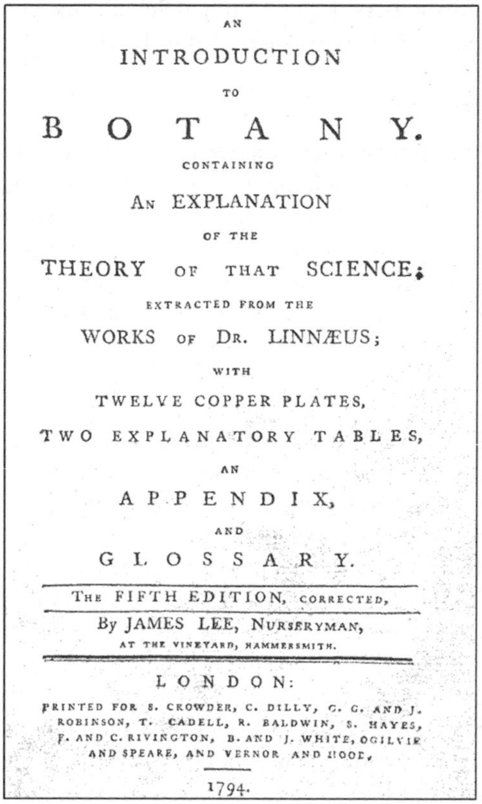
AN
INTRODUCTION
TO
BOTANY.
CONTAINING
AN EXPLANATION
OF THE
THEORY OF THAT SCIENCE;
EXTRACTED FROM THE
WORKS OF DR. LINNÆUS;
WITH
TWELVE COPPER PLATES,
TWO EXPLANATORY TABLES,
AN
APPENDIX,
AND
GLOSSARY.
THE FIFTH EDITION, CORRECTED,
By JAMES LEE, NURSERYMAN,
AT THE VINEYARD, HAMMERSMITH.
LONDON:
PRINTED FOR S. CROWDER, C. DILLY, G. G. AND J. ROBINSON, T. CADELL, R. BALDWIN, S. HAYES, F. AND C. RIVINGTON, B. AND J. WHITE, OGILVIE AND SPEARE, AND VERNOR AND HOOD,
1794.

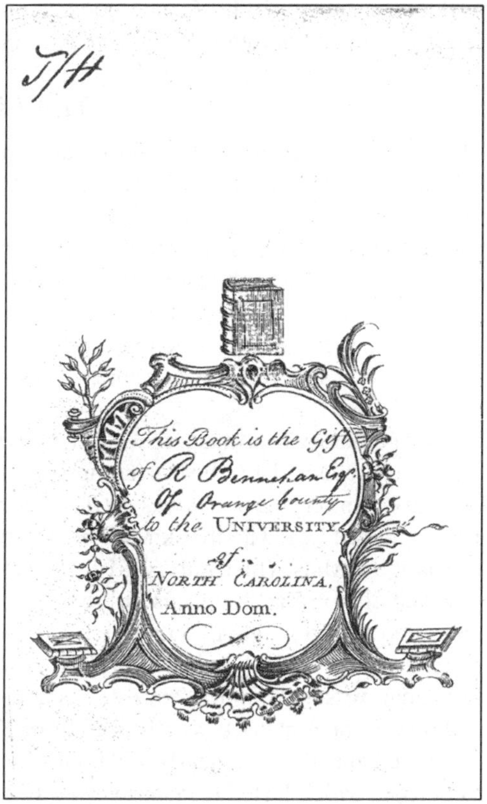

Fig. 1.6. (left) Title page and (right) book plate from UNC's copy of *An Introduction to Botany*, by James Lee. 5th ed., corrected. (1794).

(Courtesy of North Carolina Collection, Wilson Special Collections Library, University of North Carolina, Chapel Hill).

In the 1880s the university focused its attention on organizing the three campus libraries under one roof. The Di and Phi Societies agreed to move their libraries into Smith Hall (Fig. 1.7), which already housed the University Library, with the provision that they were to be shelved as distinct collections. This arrangement proved to be inefficient, so the three collections were unified in 1891 into one collection. About the same time, the faculty, particularly the scientists, were developing stronger interests in research. It became important for them to have access to collections containing current books for research and education. Teaching, too, increasingly required students to consult literature outside the classroom, and the library became the center of the university.

Fig. 1.7. Smith Hall, ca. 1860.

(Courtesy of North Carolina Collection Photographic Archives, Wilson Special Collections Library, University of North Carolina, Chapel Hill).

By the last decade of the nineteenth century, the library had become "the heart of the University," according to UNC President Edwin Alderman (BOT 9: 571). The students were now active users of the library. Professors of most departments now required that their students "consult and use" the library books in their daily assignments (BOT 9: 139). They were proud of the library resources and appreciative of the improvements made inside and outside of Smith Hall. The importance of the library was further recognized by the trustees, who authorized the establishment of a librarian (BOT 9: 271) and selected Benjamin Wyche for the position in 1894 (BOT 9: 359).

Louis Round Wilson became university librarian in 1901. Under his guidance, the library developed into a prominent research collection. Work on restructuring the library began as soon as the Summer School of 1901 concluded. In 1907 the library was moved into the newly erected Carnegie Library (now Hill Hall), a turning point in its development as an eminent research library.

During the last half of the nineteenth century the library acquired several collections particularly strong in science and medicine. Nearly six months after the death of Elisha Mitchell on June 27, 1857, the university bought his library of 1,869 volumes, which was evenly divided between belles lettres and science. A discussion of this library and its acquisition is noted further in the chapter on Mitchell.

Several noteworthy acquisitions in the last quarter of the nineteenth century strengthened the University Library's scientific and medical holdings. The first occurred after Doctor Francis Jones Smith died on April 17, 1877; his sister, Mary Ruffin Smith, donated his medical library of 200 volumes to UNC, according to his desire.

In the 1880s the university's science and math professors increasingly voiced their concern about the lack of modern books in their disciplines. Being more involved in research, they were developing modern laboratories by acquiring necessary space and new apparatus. At the same time, they acquired the literature needed to carry out their research.

In the 1890s three significant medical and scientific book collections (Fig. 1.8) were acquired: the Thomas F. Wood Collection in 1892, the Elisha Mitchell Scientific Society Collection in 1894, and the Francis W. Potter Collection in 1899. Thomas Fanning Wood was a well-known physician in the latter half of the nineteenth century in Wilmington, North Carolina. He was also an accomplished botanist of that region (Burk, 2006). He had developed an impressive personal library consisting of 1,200 volumes and 1,500 pamphlets covering his areas of professional interest. Known as "the Wood Collection," the books were shelved separately as a special unit of the University Library. Newly purchased "iron frame cases" would eventually provide shelving for the collection as well as for several thousand other volumes (Anonymous, 1893). In later years, the collection was distributed among the university's various library collections.

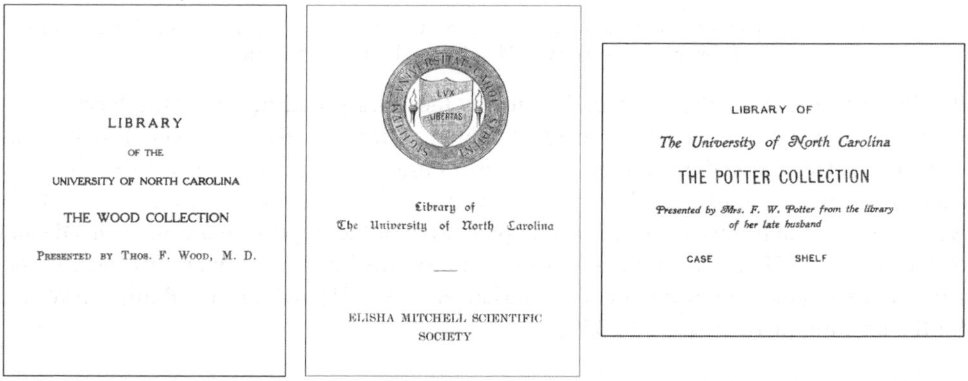

Fig. 1.8. Book plates from three medical and scientific book collections donated to UNC in the 1890s. (left) The Wood Collection, from UNC's copy of *American Woods and Useful Plants*, by William Darlington, revised with additions by George Thurber (1879), (Courtesy of the Science Library Annex, University of North Carolina, Chapel Hill). (center) Elisha Mitchell Scientific Society, from UNC's copy of the *Journal of the Franklin Institute, devoted to Science and the Mechanic Arts*, volume 130 (1890), (Courtesy of the University of North Carolina Library, Chapel Hill). (right) The Potter Collection, from UNC's copy of *The Writings of Hippocrates and Galen*, by John Redman Coxx (1846), (Courtesy of the Rare Book Collection, Health Sciences Library, University of North Carolina, Chapel Hill).

The transfer in 1894 of the library of the Elisha Mitchell Scientific Society to Smith Hall was another important event. When the Elisha Mitchell Scientific Society was established in the fall of 1883, one of its objectives was to publish a journal. As a means of building up its own library, the society journal was offered in exchange to other scientific associations. This practice of trading journals and books issued by learned societies and academies of science in this country was coming into vogue by the end of the nineteenth century. These exchanges were mutually beneficial to the trading partners. The volume count of books and pamphlets of the society's collection was 11,000 in March 1894, around the time it was moved into the University Library (Anonymous, 1894). The society's collection greatly augmented the scientific libraries being developed in the laboratories of the biology, chemistry, and geology departments.

Around 1899 the University Library received a third significant collection of books. Upon the death of Francis Whitford Potter, a physician who lived in Wilmington, North Carolina, his widow donated his private library and a collection of medical instruments to the university (Battle, 1907–1912, 2: 568). Known as "the Potter Collection," the donation consisted of 206 volumes.

Due to the need for current publications and for these to be located close to the research setting, the laboratories of biology, chemistry, and geology/mineralogy were convenient and logical sites for the development of departmental library collections. The formative period of the biology library was the last quarter of the nineteenth century, but the kernel of it formed when Professor Henry Van Peters Wilson constituted a one-man department of biology beginning in the fall of 1891. Trained at the Johns Hopkins University in research techniques styled on the German model, Wilson believed that the consultation of literature was second only to the subject of research. While a student at Hopkins, Wilson had used the biology library, which had been established in the Biological Laboratory in 1883. The collection was open for "the freest use of all books in it" (Anonymous, 1884b). For the "real" student, Wilson (1899) argued, textbooks and compendia would not suffice. He stated, "The works and figures of previous students must be critically studied, if an actual advance is to be made." As early as 1893, the literature in the UNC Biology Laboratory included seminal periodicals.

By the end of the nineteenth century, the Biology Library in the New East Building held 110 titles of books, consisting of approximately 140 volumes. Of these, twenty-two titles (twenty-six volumes) were on botany or horticulture, accounting for one-fifth of the titles housed in the Biology Library.

During the last decades of the nineteenth century, the UNC library increased its collection of scientific and medical literature, and smaller libraries developed by and located in the scientific departments grew. By 1901 the departmental libraries in law, medicine, chemistry, pharmacy, biology, and geology were constantly used (BOT 10: 346). In the twentieth century the departmental science libraries would flourish and develop into preeminent collections supporting modern scientific research and education in the natural and physical sciences. The university's libraries were a pivotal factor in the emergence of UNC as a major national university (Holley, 1998). Table 1.2 reviews the milestones of the University Library with an emphasis on the science collection.

Table 1.2. University Library Chronology, 1796–1902, with Emphasis on Its Science Collection.

1796	University Library acquires its first botany book, *An Introduction to Botany* (1794), by James Lee.
1799	Archibald DeBow Murphey, first known tutor/librarian.
1802	Total number of volumes, ca. 869.
1816	Total number of volumes, ca. 975.
1824–1825	President Joseph Caldwell acquires 1,039 volumes in Europe.
1836	Total number of volumes, 1,900.
1849	Total number of volumes, 3,500.
1852	Smith Hall completed as library building.
1853	Books moved to Smith Hall.
1854	Total number of volumes, 3,600 volumes.
1858	Elisha Mitchell library (1,897 volumes) purchased.
1868–69	Suspended because of Reconstruction.
1869	Total number of volumes, 6,540.
1869–70	Fisk Brewer, librarian, acquires 300 books.
1871–75	Closes because of Reconstruction.
1875	Total number of volumes, 8,394.
1877	Francis Jones Smith medical library (about 200 volumes) donated.
1881–83	Henry Horace Williams, librarian, sets up residency in Smith Hall.
1884	Total number of volumes, 9,000; pamphlets, 2,000.
1886	Joseph A. Holmes book collection approved for purchase.
1886	Dialectic and Philanthropic Societies libraries (about 15,000 volumes) moved to Smith Hall, but maintained as distinct collections.
1890s	Library becomes "heart of the University."[11]
1891	Dialectic and Philanthropic Societies libraries combined with University Library.
1891	Department of Biology library formed.
1892	Duplicate books sold to raise money to acquire new books.
1892	Thomas Fanning Wood medical and botanical collection (1,200 volumes and 1,000 pamphlets) donated.
1893	Dialectic and Philanthropic Societies donate their libraries; official name of the Library became "The Library of the University of North Carolina, Endowed by the Dialectic and Philanthropic Societies."
1894	Elisha Mitchell Scientific Society Library (about 11,000 books and pamphlets) moved to Smith Hall.
1894	Benjamin Wyche becomes "first regular full-time University librarian."[12]
1894	Total number of volumes, 24,400.
1895	Total number of volumes, 26,000; pamphlets, 10,000.

[11] Edwin Alderman, *Minutes*, Board of Trustees of UNC—CH, vol. 9 (April 1891–January 1898): 571.

[12] Anonymous, "Librarian Wyche," *Alumni Review* 18 (1930): 277.

TABLE 1.2. (CONTINUED)

1898	Smith Hall entrance moved to eastern end of building.
1899	Francis Whitford Potter medical library (206 volumes) donated.
1899	Henry Weil gives $1,000 to purchase library equipment and books.
1901	Louis Round Wilson becomes librarian.
1902	Total number of volumes, 40,272.

The University Museum

From the beginning of the university, the trustees supported the acquisition of natural objects and worldly curiosities to form a museum for the enlightenment of the students and the public. On July 14, 1795 they recommended that David Ker, who was UNC's "Presiding Professor" and first executive officer, purchase a blank book in which museum donations would be recorded (BOT 1: 186–187). A week later, Charles W. Harris (1795), who was a tutor (and later professor) of mathematics at UNC, wrote to his uncle, Dr. Charles Harris, about the endeavor of developing a museum. "We have at length determined to collect a Museum at this place. The trustees unanimously came into the idea and have agreed to use all their influence individually to procure curiosities."[13] Harris also noted that the blank book for recording the names of donors and their donations had been procured. The sole specimen in the nascent museum was an ostrich egg.

According to an inventory of donations listed in the journal of faculty meetings (UNC Faculty 2: 103–104), the collection had developed with various specimens by 1799. In addition to the ostrich egg, the following were listed: pieces of cloth made of bark brought from Otaheite by Captain Cooke; the tooth of a young mammoth from the banks of the Ohio; a sea leaf; a vial containing an eel; a piece of asbestos; a pine limb and a piece of petrified resin; the incisors of a beaver; a porcupine skin, which was subsequently eaten by rats; a petrified beechnut; and a shell bracelet found in an Indian grave near Nashville. An anonymous (1802) source enumerated nearly forty items that the university received from at least eight people around 1801 to 1802. The breadth of the new acquisitions was truly diverse, including not only natural objects but also such items as medals of Bonaparte in bronze denoting him at the Battle of Marengo; a number of specimens of paper money emitted by the Continental Congress, and by the State of North Carolina; and Indian pipes. UNC President Joseph Caldwell supported the acquisitions of natural objects as teaching aids.

The importance of recording and acknowledging donations was further codified by the trustees. At its meeting of January 23, 1815, the board resolved that the "Museum Keeper" present a report of all articles that were donated along with the names of donors to be done annually "on the aforesaid 22nd of June" (BOT 4: 93). The keeper would then hand over the donor list to the Committee of Visitation, who would publish it (as well as a list of books and their donors that the librarian would compile) in at least one newspaper in North Carolina.

Around early 1824 General Calvin Jones, a physician of Wake County and a longtime trustee of the university, offered the university his "Museum of Artificial and Natural Curiosities"

[13] The ink of Charles W. Harris's letter is severely faded; however, historians H.M. Wagstaff (1916) and R.D.W. Connor, 1953 (1: 417–419) independently transcribed and published it.

(BOT 5: 35). The diverse collection included small botanical specimens, mastodon teeth, and the bones of other animals of prehistoric times, according to his biographer Marshall D. Haywood ([1919]). It is further noted that Jones commanded a "varied store of knowledge in state-craft, medicine, surgery, science, history, botany, and polite literature."

Upon learning about Jones's intended donation, the faculty met and concurred in accepting his museum. On January 28, 1824 UNC President Joseph Caldwell (1824) informed Jones about the faculty's opinion and prospects for obtaining adequate space for displaying it. Although there were then no rooms available to house the collection, the faculty expressed the hope that it would "be in their power to have such structures erected, as will be adapted to them, and that this may probably be effected in the course of the ensuing spring." The trustees (BOT 5: 35) were grateful for the offer and agreed to provide a suitable room to display Jones's collection when appropriate funds became available.

Caldwell (1824) particularly noted that a natural history collection would serve a useful educational role. He wrote:

> The department of Natural History is one which we wish to have illustrated by such collections as are usually made in a Museum. Our opportunity of directing the eye with examination upon a work of nature is of more value for making us acquainted with the object, than any description which it is possible to give. We think that if something interesting and respectable can be instituted, pertaining to this department, and exhibited in such a manner as to make impression upon visitors, of the importance of the object, and of the justice done to the munificence of the Donor, it may prove the occasion of exciting others to exert themselves for augmenting it, when they shall find their opportunities put it into their power.

Some insight into the background on the Jones collection is revealed by various archival records. He and several other noted physicians of the state established the North Carolina Medical Society in 1799, which was active at least until the end of 1804. The organization developed a botanical garden and natural history museum. Jones had developed his collection privately, but obligingly transferred it to the North Carolina Museum in Raleigh around 1818 (J. Marling and Company, 1818). In addition to Haywood's ([1919]) statement that the collection may have been the same as the one amassed by the society, Caldwell's letter mentioned above clearly confirms that the Jones collection had been "a portion of the Museum in Raleigh, which upon division among the Partners, had become your separate property."

From 1852 to 1882 state geologists W.C. Kerr and Ebenezer Emmons donated numerous mineralogical specimens that they either collected themselves or procured from other sources. Following the reopening of the university in 1875 there was a renewed growth of the university museum, including botanical collections, a number of the plants represented by specimens having economic importance. The increase of the museum's botanical specimens occurred during the tenures of the natural history professors (William H. Smith, Frederic W. Simonds, and Joseph A. Holmes). From 1876 to 1891 collections of "native woods, local plants, and various seeds and grains of the world" were acquired through donations and collections made by professors and students. Professor Smith played a pivotal role in consolidating and labeling the diverse natural history collection. Under his guidance the third-floor room in Old East that formerly served as the Philanthropic Society's library was renovated to house the natural history museum.

In publicizing the museum and to encourage donations, Smith (1876) prepared a "Circular to the Friends of the University of North Carolina in Regard to Specimens of Natural History." The noteworthy array of botanical items served as an important instructive tool. The museum also attracted visitors to the campus. Holmes took particular care in maintaining the museum and adding specimens. When UNC President Battle presented his annual report to the trustees on February 11, 1891, he mentioned that a new chapel was needed and then "Gerrard Hall could be filled up for a museum which is greatly needed for the large and valuable collection we have on hand" (BOT 8: 531). He realized, however, that finances were not available for that undertaking unless "some liberal soul" came forward. Instead, a number of the botanical and zoological specimens were probably moved to the upper floor of the New East Building, where the newly created Department of Biology, called the Biological Laboratory, was originally located in 1891. The museum collections were displayed in cases in the main laboratory. The UNC catalog of 1897–1898 (p. [115]) mentioned that exhibitions included the marine fauna of the East Coast, bird skins, bird eggs, insects, and flowering plants. In that academic year Thomas Gilbert Pearson was hired as an assistant curator of the Biological Museum to manage the collection.

By the end of the nineteenth century, various collections had developed on campus and contained valuable scientific specimens and historical articles and relics. Administrators such as UNC President Kemp P. Battle recognized the importance of preserving the collections, using them as articles for instruction, and letting the public examine them. In his annual report of January 23, 1899, Battle suggested that the university "sadly need[ed] a scientific and historical museum in a fire-proof building under a salaried curator" (BOT 10: 51).

A document by Neil Fulghum ([2008]), former Keeper of UNC's North Carolina Collection Gallery, as well as later chapters of this book provide additional historical background on the museum, including its development, significant acquisitions, and role in the education of science students.

The Laboratory

Scientist Thomas H. Huxley introduced the laboratory experience to standard methodology for biology education around 1872. Based on the German model, the laboratory method provided a practical and hands-on learning opportunity and would change biological teaching (Ferry, 1920). Botanist Charles E. Bessey was instrumental in establishing teaching laboratories for botany in America in the early 1870s. Among the first institutions to develop botanical teaching laboratories in the United States were University of Illinois, 1869; Cornell, 1872; Harvard, 1872; and Iowa State University, 1873 (Ford, 1962). Historian of botany Emanuel D. Rudolph (1996) traced the history of this learning method in the United States.

At UNC laboratory practice in botany was established in the 1880s under the tutelage of Joseph A. Holmes; however, classroom experiments and demonstrations that reinforced scientific principles were developing as early as 1819. Yale-educated scientists Denison Olmsted and Elisha Mitchell did not restrict their classes to lectures and recitations. They enriched their instruction with classroom demonstrations in chemistry. Mitchell was well known for the experiments and demonstrations that he presented before his students. In chemistry he enlivened his lectures with experiments on such topics as oxygen, electromagnetism, and nitric acid. He also demonstrated new scientific devices at public presentations. Mitchell clearly used field trips in geology and mineralogy as a laboratory experience. It is certainly likely that he pointed out distinctive plants while on class excursions into the countryside.

Under John Kimberly's instruction (1857–1866) in analytical chemistry, which touched on botanical aspects, students gained hands-on experience in performing chemical analyses themselves. Although Kimberly presented at least six lectures on plants, it is unknown if laboratory work on plants was carried out. His educational background included training at the Lawrence Scientific School, Harvard University, where he studied under the influential scientists Louis Agassiz and Eben N. Horsford, and at the University of Heidelberg in Germany, where he came under the influence of German theory and laboratory methods in agricultural chemistry. When George Dixon taught agriculture from 1869 to 1870, his students most likely collected plants and prepared a herbarium for individual use. His students may have had access to a microscope in their studies.

The first attempt in botanical laboratory teaching at UNC occurred when William H. Smith used a laboratory setting (1876–1877) in which his students could observe the growth of plants. This activity coincided with the mission of the Agricultural Experiment Fertilizer Control Station established at UNC in early 1877. University records note that Smith's instruction focused on morphology and histology of plants as well as structural, physiological, and systematic botany. In addition, his students were required to collect plants that they would later identify. Following the departure of Smith in 1877, Cornell-trained scientist Frederic W. Simonds took charge of teaching natural history. Among his courses of instruction were Physiological Botany and Vegetable Physiology, an advanced course in botany in which microscopes were used. His students collected, identified, and preserved plants that were added to the herbarium and became the property of the University Museum. During his tenure (1877–1881), the botanical laboratory experience entailed the analysis (identification) of plants collected in the field. Among his students' required reference texts was Julius Sachs's (1875) *Text-book of Botany,* which became a model for future American textbooks. The book was influential in introducing American students to the new botany, which included topics beyond the usual descriptive botany, such as histology, cytology, ecology, embryology, and genetics.

Separate botany instruction started to take form at UNC under Simonds's tenure, but botany was first taught as distinct courses by Joseph A. Holmes, who had gained broad botanical training at Cornell. Historian of botany Charles E. Ford (1962) notes that the emergence of distinct botanical courses derived particularly from natural history resulted from "the culmination of institutional and scientific developments of the post-Civil War years." From 1881 to 1891 Holmes taught botany classes that had a distinct focus on practical agriculture. At UNC students could now enroll in different botanical courses: (1) Elements of Botany; (2) Advanced Botany, which included Systematic and Applied Botany and Agricultural Botany; and (3) Horticulture. Holmes was a strong proponent of adopting laboratory practice as a learning tool, and he procured botanical charts and models (Fig. 1.9) as teaching aids.

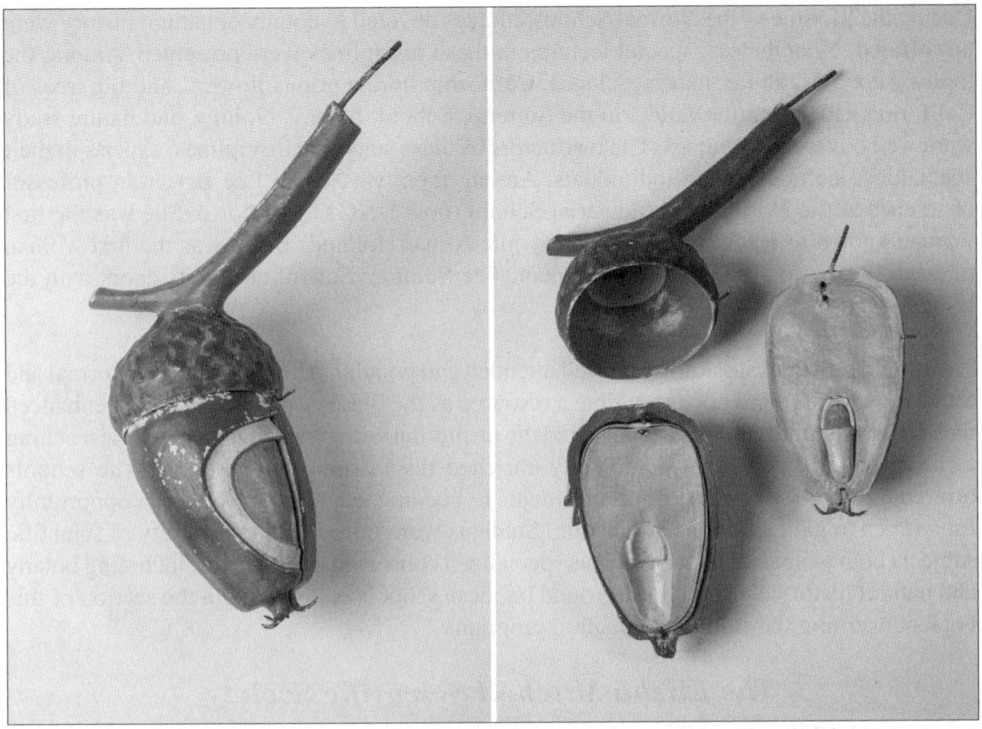

Fig. 1.9. Model of an acorn used in botanical instruction at UNC in the early twentieth century, x $^{1}/_{10}$. (left) Intact view. (right) Partially disassembled view. [UNC students employed the model pictured here in the early twentieth century, but it is indicative of models that Holmes's classes used.]

(Ck. 1341 North Carolina Collection Gallery, Wilson Special Collections Library, University of North Carolina, Chapel Hill).

Around 1891 the Department of Geology and Natural History split into two departments: Biology and Geology. Holmes now taught geology and Professor of Biology Henry Van Peters Wilson taught botany on occasion, though clearly the instruction was secondary to zoology, his area of expertise. Concerning what little botany Wilson taught, his instruction went beyond descriptive aspects of the discipline and required students to use microscopes in studying plant cells and tissue. The new botany continued to be applied but would not gain momentum until William C. Coker was hired as associate botany professor in 1902.

Normal School and Summer School

The Normal School was established at UNC to train North Carolina teachers. Even though the school was established for white males associated with the university, it admitted North Carolina teachers as well as those intending to become teachers, including females. The enterprise aimed to teach the most current and best methods of class management, to inspire interest, to provide knowledge, to develop the minds of students, and to teach subjects that were to be covered in public schools (Highsmith, 1923). Sessions were held during summer vacations of the university. The school closed in 1884 but reopened as the Summer School in 1894.

During the lifetime of the Normal School, courses devoted to botany or natural history were not offered. Nonetheless, special lectures in these disciplines were presented. Among the topics were agriculture, natural science, California flora, curious flowers, and big trees of California and Yosemite Valley. In the Summer School, botany, biology, and nature study were well covered as courses. The instructors of these scientific disciplines, experts in their specialties, included seven individuals. Among them was Dixie Lee Bryant, a professor of science at the Normal and Industrial School (now UNC Greensboro). She was the first woman known to teach botany at UNC while Alma (Holland) Beers was the first woman to teach botany in the Botany Department. The Summer School closed its doors with the last class taught in 1904.

The Summer School sessions were well attended and popular. Although both the Normal and Summer Schools were not as tangible a resource as the library and museum, they enhanced the academic environment and connected the institution with other educators whose teaching techniques and knowledge indubitably enriched the learning environment. The schools offered training for North Carolina residents to become teachers as well as an opportunity for women to gain additional education. Students were introduced to a variety of scientific subjects both as special lectures and as specialized courses in the sciences, including botany and natural history. Detailed background on these schools is provided in the section of this book concerning the summer education programs.

The Elisha Mitchell Scientific Society

By the mid-1880s the science departments at the University of North Carolina were developing research interests and a nascent professionalization in science was taking hold. A growing environment of scientific inquiry fostered a need for a forum to share and promote scientific discoveries. Four young professors of science with a zeal for uncovering scientific truths met in September 1883 to consider organizing a learned scientific society on campus. Joseph A. Holmes (geology/natural history professor) and three of his science colleagues—chemistry professor Francis P. Venable, mathematics professor Ralph H. Graves, and physics professor Joshua W. Gore—convened a meeting on September 24, 1883 to discuss this issue at the home of Holmes and Venable (Venable, 1924). They invited several other scholars (Kemp P. Battle, Thomas W. Harris, J. Manning, William B. Phillips, and Emile A. de Schweinitz) to join them. The assembled men chose the name "the Elisha Mitchell Scientific Society" for the newly formed association, to honor the accomplishments of that versatile scientist. They elected Venable as president and Gore as secretary and treasurer. The group sent out a three-page "Proposal for the Mitchell Scientific Society" (Table 1.3) to prospective members. The Dabney Papers, which are housed in the Southern Historical Collection at UNC, contain a facsimile of the promotional flyer. On October 1, 1883, a number of members approved the organization's constitution and elected Holmes as the resident vice-president. It was also recommended that President Battle notify the students of their eligibility to become associate members. The establishment of the society was among the early catalysts promoting the professionalization of science at the university. The society was the first of its kind to be connected with a university in the southern states, and its official organ, the *Journal of the Elisha Mitchell Scientific Society*, was a medium through which members could communicate their scientific discoveries.

The objectives of the fledgling organization are described in the promotional flyer noted above. At monthly meetings members and invited guests, such as students, presented papers and shared their discoveries and observations with the attendees. The meetings provided an informal, though influential, environment where students and professors shared a spirit of scientific inquiry. In its early years, the society admitted only men; however, women were permitted to attend the meetings. UNC sophomore J. Bryan Grimes (1884) noted that "a great many very pretty young ladies were present" at the March 8, 1884 meeting. The scientific atmosphere inspired the young scholars and enriched their education. A newspaper reporter expressed great hope that the organization would cultivate "a corps of educated scientists; men who know how to look at nature, and what to see and what is worth developing, and who can tell our people what to do with our resources, and how to do it" (Alumnus, [1886]). Among the presentations made from 1883 to 1899, at least twenty-eight concerned botanical topics (Table 1.4). Presenters of botanical topics included two UNC students of natural history: Leander W. Lynch and W.W. Ashe, the latter becoming a noted botanist.

The *Journal of the Elisha Mitchell Scientific Society* began publication in 1884. After a century of service, the society disbanded in 1984, but its journal continued under the title *Journal of the North Carolina Academy of Science*, carrying on the mission of the Mitchell Society. The early history of the society is recorded in the minutes of the society's proceedings (EMSS). Several published accounts provide a thorough review of the society's history and development as well as background on its journal (Venable, 1924; Henderson, 1934; Wiggins, 1963; Corgan, 1984; Webster, 2002).

TABLE 1.3. PROPOSAL FOR THE MITCHELL SCIENTIFIC SOCIETY.[14]

More thorough scientific training and increased interest in scientific work is a clearly felt want of the State. The building up of a true spirit of scientific research cannot be effected all at once, but it is believed that the formation of a Scientific Society, thus bringing about a union of strength and effort, will aid greatly in increasing the goal of those already at work and arousing the interest of others. We have thought it well then to send out this plan of organization of such a society, hoping to enlist in the cause those to whom this paper is sent.

No more appropriate name could be shown than that of the "Mitchell Society." If the society flourishes it will be one of the few memorials left of that great worker at the scene of his labors. It is a name that breathes a spirit of indefatigable zeal, patience, and perseverance—a spirit that must be emulated if this undertaking is to be carried out to success.

The objects of the foundation of the Society are–

1st. The cultivation of an interest in Natural History and general scientific subjects.

2nd. The encouragement of individual workers who, isolated as they are at present, with no one of congenial pursuits to turn to for advice or encouragement, are apt to become disheartened or indifferent and give up this work.

3rd. Increased knowledge of the State, and its resources.

4th. Building up of smaller local societies.

5th. Collection of specimens.

[14] A transcription of the facsimile housed in the Dabney Papers, Southern Historical Collection, Wilson Special Collections Library, University of North Carolina, Chapel Hill.

Table 1.3. (Continued)

The plan of working is as follows—

1. A small fee is to be paid by each member (say $2.00) for the annual publication of a Journal to contain papers from members. These papers are to be on original researches, new observations, or facts collected—as for instance, reports of experiments carried out in any branch of science, observations on storms, strange growths, animal or plant life histories, collection of facts as to minerals, climatology, flora, and generally interesting statistics.
2. The publication of names of members, occupations, addresses, and scientific work accomplished.
3. At the centre of the organization—the University—a series of lectures, popularizing science so far as possible–making its branches attractive and explaining methods of observation.

Lastly as to the Constitution or Government.

1. Members—any one duly recommended by members of the Society, and paying the annual fee can become a member.

2. A President and certain Vice-presidents will be chosen by the members at large, and hence these will be positions of honor to be held in high esteem.

3. A Secretary, Treasurer, and Executive Committee at the University will transact all necessary routine business.

It is hoped that this plan will meet with your approval, that you will allow us to enroll your name as a member, and that you will aid us in the work by original papers or observations.

K.P. Battle	J.W. Gore
T.W. Harris	J.A. Holmes
J.S. Manning	F.P. Venable
E.A. de Schweinitz	R.H. Graves
J.M. Manning	W.B. Phillips

Table 1.4. Botany Promoted by the Elisha Mitchell Scientific Society; A List of Talks Presented at the Society's Meetings and Recorded in Its Minutes, 1883–1899.[15]

p. 15. November 10, 1883. Insectivorous Plants. By J.A. Holmes.

p. 20. April 12, 1884. Flowering Plants. By Charles Phillips.

p. 25. October 18, 1884. Analysis of Yaupon Leaf. By F.P. Venable.

p. 25. October 18, 1884. Transpiration of Water by Plants. By F.P. Venable.

p. 25. October 18, 1884. Flora of Angola Bay. By J.A. Holmes.

p. 27. March 7, 1885. A Botanical Note. By [M.E.] Hyams.

p. 27. March 7, 1885. Occurrence of Malic and Citric Acids in Pea-nuts. By E.A. de Schweinitz.

p. 28. April 25, 1885. Peculiar Animal and Plant Life in Geology. By J.A. Holmes.

p. 28. April 25, 1885. Additions to Catalogue of North Carolina Plants. By [M.E.] Hyams.

p. 28. April 25, 1885. Distribution of Rhododendrons in N.C. By J.A. Holmes.

p. 28. April 25, 1885. The Manufacture of Pine Fibre. By [C.] Phillips & [F.P.] Venable.

[15] (EMSS, volume 1).

TABLE 1.4. (CONTINUED)

p. 29.	May 24, 1885. Biography of Dr. [Moses Ashley] Curtis. By Thomas F. Wood.
p. 29.	February 17, 1886. Biographical Sketch of Dr. L.D. Von Schweinitz. [unsigned].
p. 37.	March 17, 1886. Sugar-beet Culture in Kentucky. By J[ames] L[ewis]. Howe [Ph.D., M.D., Polytechnic Society, Louisville, KY; corresponding member].
p. 39.	May 19, 1886. Note on the *Shortia*. By M.E. Hyams (Statesville).
p. 47.	April 12, 1887. Flowering plants in the Chapel Hill Section. By [Leander Williams] Lynch. Mr. Lynch gave a talk on the flowering plants of Chapel Hill. He reported the dates of flowering of up-wards of fifty species of plants in this section of the country observed by him during the present season. The observations will be continued.
p. 51.	October 11, 1887. Rare Specimen of Cotton Plant (Seeds without Lint). By J.A. Holmes.
p. 53.	January 11, 1888. Study of Local Flora. By Gerald McCarthy. (By title only.)
p. 55.	May 8, 1888. N.C. Desmids. By W.L. Poteat (from Wake Forest).
p. 56.	December 5, 1888. Chemical Examination of Some Species of the Genus *Ilex*. By F.P. Venable.
p. 57.	February 12, 1889. Natural History of the Cereals. By Gerald McCarthy.
p. 57.	February 12, 1889. Some Sources for Sugar Proposed at the Close of Last Century. By H.L. Harris.
p. 59.	May 7, 1890. Distribution of Boracic Acid in Plants. By J.S. Callison.
p. 60.	October 14, 1890. A Raining Tree in the Campus of the University. By P. Dalrymple.
p. 61.	April 21, 1891. A Brief Sketch of the Pea-nut Plant. By Gaston Battle.
p. 70.	January 29, 1894. Absorption of Insoluble Material by Plants. By F.P. Venable.
p. 70.	March 20, 1894. The Pine Barrens of Paint Rock, the Tree Shrubs of Kings Mountain & the Dwarf Oaks of Harnett County. By Collier Cobb.
p. 71.	October 10, 1894. The Long-leaf Pine in Its Struggle for Existence. By W.W. Ashe (read by F.P. Venable).
p. 71.	October 10, 1894. The Stunted Oaks Following Long-leaf Pine in the Burned-over District. By Collier Cobb.
p. 71.	December 11, 1894. Two Cycads from the Triassic of Hornsboro, S.C. By Collier Cobb.
p. 72.	September 15, 1895. Experiments in Restocking Burned-over Areas with Long-leaf Pine. By J.A. Holmes.
p. 99.	April 13, 1897. Natural History of the Wilmington Region. By H.V.P. Wilson.
p. 100.	November 23, 1897. The Influence of High Altitudes upon the Growth of Plants. (Specimens from the Rockies shown). By W.W. Ashe.
p. 102.	March 21, 1899. The Relation between Forestry and Geology in North Carolina. By Collier Cobb.

References Cited

(References that conclude with page numbers within parentheses indicate sources from where quotes and pertinent information were obtained.)

Alumnus. [1886]. Practical and progressive. The attention that is given to the study of the sciences at the university, Chapel Hill, p. 21 in Kemp P. Battle, *Scrapbook of Material Relating to the University of North Carolina, 1869–1893*, vol. 8. (Newspaper clipping without complete bibliographical data. North Carolina Collection, Wilson Special Collections Library, University of North Carolina, Chapel Hill)

Anonymous. 1802. North-Carolina University. *North-Carolina Minerva* 7(348): [1], Tuesday, December 7, 1802.

Anonymous. 1810. Preface. *Memoirs of the Connecticut Academy of Arts and Sciences* 1(1): [iii].

Anonymous. 1869. Natural History Society. *Cornell Era* 1(14): 6, March 13, 1869.

Anonymous. 1877. The course in botany. *Cornell Era* 10(6): 4.

Anonymous. 1880. The Botanical Department. *Cornell Review* 7(7): 288.

Anonymous. 1884a. Our libraries. *University Magazine* (University of North Carolina, Chapel Hill), new series [4], 4: 117–118.

Anonymous. 1884b. The Biological Laboratory of the Johns Hopkins University. *Science* 3: 354.

Anonymous. 1893. Library improvements. *The Tar Heel* (University of North Carolina, Chapel Hill) 2(4): [4], October 19, 1893.

Anonymous. 1894. Library notes. *The Tar Heel* (University of North Carolina, Chapel Hill) 2(18): [2], March 23, 1894.

Ashe, Samuel. 1796. Letter of July 13, 1796 to Richard Bennehan. (Box 3, folder 63, Cameron Family Papers #133, Southern Historical Collection, subseries 1.1, Wilson Special Collections Library, University of North Carolina, Chapel Hill)

Baldwin, Ebenezer. 1831. Annals of Yale College in *New Haven, Connecticut, from Its Foundation, to the Year 1831. With an Appendix, Containing Statistical Tables, and Exhibiting the Present Condition of the Institution*. Hezekiah Howe, New Haven. (pp. 155, 222, 239–253, 263–264)

Baldwin, Simeon Eben. 1901–1903. The first century of the Connecticut Academy of Arts and Sciences; an historical address delivered before the academy at its centennial anniversary, on October 11th, 1899. *Transactions of the Connecticut Academy of Arts and Sciences* 11(1): [xviii]–xlv.

Battle, Kemp P. 1907–1912. *History of the University of North Carolina*. Printed by Edwards & Broughton, Raleigh. 2 vols.

Becker, Carl. L. 1943. *Cornell University: Founders and the founding*. Cornell University Press, Ithaca, NY. (p. 118)

BOT. Board of Trustees, University of North Carolina, Chapel Hill. *Minutes of the Board of Trustees.* (University Archives #40001, Wilson Special Collections Library, University of North Carolina, Chapel Hill) (Vol. 1, 1789–1798; vol. 4, July 1811–December 1822; vol. 5, December 1823–December 1840; vol. 8, June 1883–February 1891; vol. 9, April 1891–January 1898; vol. 10, May 1898–April 1904)

Branner, John Casper. 1913. William Russel Dudley, pp. [7]–10 in Anonymous, *Dudley Memorial Volume, Containing a Paper by William Russel Dudley and Appreciations and Contributions in His Memory by Friends and Colleagues.* Published by the University, Stanford University, California.

Brendel, R. 1895. *Verzeichnis der botanischen Modelle von R. Brendel.* Druck von H. Klokaw, Berlin.

Bronson, Henry. 1867. *Biographical Notice of Prof. Eli Ives, M.D.* "Republished from the *Proceedings of the Connecticut Medical Society* [1863]." Tuttle, Morehouse & Taylor, New Haven, CT. (portrait opposite cover page) (p. 6)

Burk, William R. 2006. Thomas Fanning Wood (1841–1892) and his botanical contributions. *Huntia* 12: 113–148.

Burk, William R. 2009. The University of North Carolina Library and the sciences, 1795–1902. *North Carolina Libraries* 67: 2–16.

CAAS. 1810. Connecticut Academy of Arts and Sciences. *Constitution of the Connecticut Academy of Arts and Sciences.* [no statement of publisher or place of publication]. (Microopaque, Readex Microprint, New York; courtesy of American Antiquarian Society, 1974. *Early American Imprints*, second series, no. 19843)

Caldwell, Joseph. 1824. Letter of January 28, 1824 to Calvin Jones. (Folder 1, Calvin Jones Papers #127-z, Southern Historical Collection, Wilson Special Collections Library, University of North Carolina, Chapel Hill)

Connor, R.D.W. 1953. *A Documentary History of the University of North Carolina 1776–1799.* Edited with a preface by Louis R. Wilson and Hugh T. Lefler. University of North Carolina Press, Chapel Hill. 2 vols.

Corgan, James X. 1984. The *Journal of the Elisha Mitchell Scientific Society* and the history of science in the South. *Journal of the Elisha Mitchell Scientific Society* 100: 74–81.

CU-R. *The Cornell University Register.* At the University, Ithaca, NY. No. 1. 1868–1869. (pp. 35, 77–78); no. 2. 1871–1872. (p. 69)

CU-RC. The *Cornell University Register and Catalogue, 1874–75.* 2nd ed. At the University, Ithaca, NY. (p. 88)

Dabney Papers. (Box 5, folder 68, Charles William Dabney Papers #1412, Southern Historical Collection, Wilson Special Collections Library, University of North Carolina, Chapel Hill)

Dexter, Franklin Bowditch. 1911. Eli Ives, pp. 358–362 in *Biographical Sketches of the Graduates of Yale College with Annals of the College History*, vol. 5. Henry Holt and Company, New York.

Dexter, Franklin Bowditch. 1912. Ebenezer Baldwin, pp. 172–173; William Tully, pp. 68–73 in *Biographical Sketches of the Graduates of Yale College with Annals of the College History*, vol. 6. Yale University Press, New Haven, CT.

Dowling, William C. 1999. Dwight, Timothy (14 May 1752–11 Jan. 1817), pp. 192–194 in John A. Garraty and Mark C. Carnes (general editors), *American National Biography*, vol. 7. Oxford University Press, New York, Oxford.

Ellsworth, Mary Ellen. 1999. *A History of the Connecticut Academy of Arts and Sciences 1799–1999.* (*Transactions, The Connecticut Academy of Arts and Sciences*, vol. 55). Connecticut Academy of Arts and Sciences, New Haven, CT. (pp. 46, 220)

EMSS. Elisha Mitchell Scientific Society. *Records of the Elisha Mitchell Scientific Society*. (University Archives #40183, Wilson Special Collections Library, University of North Carolina, Chapel Hill). (Box 1 contains two volumes of minutes of the society's proceedings; volume 1 covers 1883 to 1940)

Ferry, Lily M. 1920. *An Historical Survey of the Teaching of Botany*. M.S. Thesis, Ogden Graduate School of Science, University of Chicago. Chicago. (pp. 12–13)

Fitzpatrick, Harry M. 1926. "An Historical Sketch of the Early Days of Botany in Cornell." Typed manuscript. (p. 9). (Copies at Bailey Hortorium and Herbarium, Cornell University)

Ford, Charles Edward. 1962. *Botany in the American College Curriculum 1642–1914*. Doctoral Dissertation, Graduate Board of Washington University, Department of Education, Washington University, Saint Louis, MO. (pp. 46–47)

Fulghum, Neil. [2008]. "The North Carolina Collection Gallery; Including the mission statement of NCC Gallery & an early history of museums at the university." Unpublished document in the files of the North Carolina Collection Gallery, Wilson Special Collections Library, University of North Carolina, Chapel Hill.

Fulton, John F. and Elizabeth H. Thomson. 1947. *Benjamin Silliman 1779–1864, Pathfinder in American Science*. Henry Schuman, New York. (pp. vii, 82–89, 139–145, 147)

Greene, John C. 1975. Silliman, Benjamin, pp. 432–434 in Charles Coulston Gillispie (editor in chief), *Dictionary of Scientific Biography*, vol. 12. Charles Scribner's Sons, New York.

Grimes, J. Bryan. 1884. Letter of March 9, 1884 to Miss Mannie. (Box 1, folder b, J. Bryan Grimes Papers, Collection 54, East Carolina Manuscript Collection, J.Y. Joyner Library, East Carolina University, Greenville, NC)

Harris, Charles. 1795. Letter of July 21, 1795 to Dr. Charles Harris. (Box 1, folder 8, University Papers #40005, University Archives, Wilson Special Collections Library, University of North Carolina, Chapel Hill)

Hartt, Rollin Lynde. 1926. The story of Yale's great museum. *Scientific Monthly* 22: 473–493. (pp. 475–476)

Haywood, Marshall DeLancey. [1919]. *Calvin Jones, Physician, Soldier and Freemason (1775–1846); Being an account of his career in North Carolina and Tennessee*. Reprint issued by James W. Jones, Bolivar, TN. (p. 6)

Henderson, Archibald. 1934. The Elisha Mitchell Scientific Society: Its history and achievements. *Journal of the Elisha Mitchell Scientific Society* 50: 1–13.

Hewett, Waterman Thomas. 1976. [Chapter] XV. Natural Science, p. 583–621 in *The History of Cornell University in the Twenty-five Years of Its Existence, 1868–1893*, pp. [357]–704 in John H. Selkreg (editor), *Landmarks of Tompkins County, New York, including a History of Cornell University by W. T. Hew[e]tt*. D. Mason and Company, Syracuse, NY, 1894; reproduction by Unigraphic, Inc., 1976. (Botany, pp. 587–596)

HIBD. Hunt Institute for Botanical Documentation. *Prof. Henslow's Botanical Diagrams*. For description and images, search for accession 6866 in the catalog of the HIBD art collection, the first database listed at <http://huntbot.andrew.cmu.edu/HIBD/Departments/Databases.shtml>

Highsmith, Edwin M. 1923. The state normal school idea in North Carolina. *Meredith College Quarterly Bulletin*, series 17, no. 1, pp. 5–38. (p. 21)

Holley, Edward G. 1998. *The Library, Philanthropy & UNC's Emergence as a Major American University*. Hanes Foundation, Rare Book Collection, Academic Affairs Library, The University of North Carolina at Chapel Hill. (pp. 8–9, 18–19)

Ives, Eli. 1811. [Vegetable productions found in New-Haven], pp. 29–34 in Timothy Dwight, *A Statistical Account of the City of New-Haven*, pp. [i]–xi, 83,[1] in *A Statistical Account of the Towns and Parishes in the State of Connecticut*. "Published by the Connecticut Academy of Arts and Sciences, Vol. I, no. I." Printed and Sold by Walter and Steele, New-Haven, CT. (*Early American Imprints*, Second Series, no. 22603)

[Ives, Eli, William Tully, and Melines C. Leavenworth]. 1831. *Catalogue of the Phenogamous Plants, and of the Ferns, Found within Five Miles of Yale College*, pp. 264–302 in Ebenezer Baldwin, *Annals of Yale College, in New Haven, Connecticut, from Its Foundation, to the Year 1831, with an Appendix, Containing Statistical Tables, and Exhibiting the Present Condition of the Institution. September 25, 1818*. Hezekiah Howe, New Haven, CT.

J. Marling and Company. 1818. North Carolina Museum. *Raleigh Minerva* 20(1173): [1], Friday, September 25, 1818.

Jordan, David Starr. 1911. William Russell [*i.e.*, Russel] Dudley. *Science*, n.s. 34: 142–145.

Kelley, Brooks Mather. 1974. *Yale, a History*. Yale University Press, New Haven, CT. (p. 149)

Lautour, A.-M. 1948. Auzoux (Louis-Thomas-Jérome), column 808 in M. Prevost and Roman D'Amat, *Dictionnaire de Biographie Française*, tome 4. Librairie Letouzey et Ané, Paris.

Lee, James. 1794. *An Introduction to Botany: Containing an explanation of the theory of that science*. Extracted from the works of Dr. Linnaeus, 5th ed., corrected. Printed for S. Crowder [et al.], London.

Linnaeus, Carl. 1751. *Philosophia Botanica: In qua explanatur fundamenta botanica*. Apud Godofr. Kiesewetter, Stockholmiae.

Love, James L. 1944. James L. Love to Dr. Charles E. Rush, Chapel Hill, November 2, 1944. "The University of North Carolina Library in the period, 1875–1884." (North Carolina Collection, Wilson Special Collections Library, University of North Carolina, Chapel Hill)

McAllister, Ethel M. 1941. *Amos Eaton, Scientist and Educator 1776–1842*. University of Pennsylvania Press, Philadelphia. (pp. 163–165)

McVaugh, Rogers. 1947. The travels and botanical collections of Dr. Melines Conklin Leavenworth. *Field & Laboratory* 15: 57–70. (p. 57)

Mead, Kate Campbell. 1928. Tully, William (1785–1859), pp. 1230–1232 in Howard A. Kelly and Walter L. Burrage, *Dictionary of American Medical Biography; Lives of eminent physicians of the United States and Canada, from the earliest times*. D. Appleton and Company, New York.

Mulder, Willem J. 1994. L.Th.J. Auzoux's models used for the dissemination of anatomic knowledge, pp. 129–136 in K.S. Grooss and B. Baljet (editors), *Actes du 6e Colloque des Conservateurs des Musées d'Histoire des Sciences Médicales: 2–5 Septembre 1992*. Leiden, Pays-Bas, The Netherlands. Fondation Mérieux, Lyon, France.

NHSR. *Natural History Society Records. Proceedings [of Meetings], 1876–1881*. (Archives #37-6-584, Carl A. Kroch Library, Rare and Manuscript Collections, Cornell University)

Paine, Emily Florence. 1894. The women's colleges of the United States.—No. 6. A girl's life and work at Mount Holyoke. *Delineator*, October 1894.

Parker, David B. and Edward L. Harrelson. 1992. *The University of North Carolina at Chapel Hill, a Historical Bibliography*. University of North Carolina at Chapel Hill, Bicentennial Policy Committee, Chapel Hill.

Quen, Jacques M. 1976. Dr. Eneas Munson (1734–1826). *Journal of the History of Medicine and Allied Sciences* 31: 307–319.

Reiling, Henri. 2003. Beter dan le natuur, pp. 221–235 in Brand, Jan and Alex de Vries (editors), *Neo. Centraal Museum, Utrecht*. 415 pp. ("Botanische modellen: Robert en Reinhold Brendel," pp. 227, 230). Translated online <http://64.233.161.104/search?q=cache:Yfn0_K1-eLUJ:home.wanadoo.nl/here/neo.html+henri+reiling+better+than+nature&hl=en> (Accessed December 4, 2004)

Rockwell, P.G. [1867]. Biographical sketch of Melines C. Leavenworth, M.D. *Medical Communications with the Proceedings of the 1866 Annual Convention of the Connecticut Medical Society* 1866: [269]–272.

Rudolph, Emanuel D. 1996. History of the botanical teaching laboratory in the United States. *American Journal of Botany* 83: 661–671.

Sachs, Julius. 1875. *Text-book of Botany, Morphological and Physiological*. Translated and annotated by Alfred W. Bennett, assisted by W.T. Thiselton Dyer. At the Clarendon Press, Oxford, [England].

Sanders, Charles Richard. 1979. Bennehan, Richard, pp. 136–137 in William S. Powell (editor), *Dictionary of North Carolina Biography*, vol. 1. University of North Carolina Press, Chapel Hill.

Schoepflin, Gary Lee. 1977. *Denison Olmsted (1791–1859): Scientist, teacher, Christian: A biographical study of the connection of science with religion in antebellum America*. Ph.D. Oregon State University, 1977. Facsimile by microfilm xerography. University Microfilms International, Ann Arbor, MI. (pp. 45–46)

Smith, William H. 1876. *Circular to the Friends of the University of North Carolina in Regard to Specimens of Natural History*. Raleigh News Job Office and Book Bindery, Raleigh.

Tschirch, A. 1885. *Erläuterungen zu den botanischen Modellen von Robert Brendel*. Druck von Gebr. Unger (Th. Grimm), Berlin.

Tuckwell, W. 1869. Science-teaching in schools. *Nature* 1(1): 18–20.

UNC Faculty. University of North Carolina, Chapel Hill. *Minutes of Faculty Meetings*. (University Archives #40106, Wilson Special Collections Library, University of North Carolina, Chapel Hill) (Vol. 2, 1814–1821)

Venable, Francis P. 1924. Historical sketch of the Elisha Mitchell Scientific Society. *Journal of the Elisha Mitchell Scientific Society* 39: 117–122.

Wagstaff, H.M. 1916. *The Harris Letters*. Seeman Printery, Durham, NC. (pp. 19–21)

Walters, S.M. and E.A. Stow. 2001. *Darwin's Mentor, John Stevens Henslow 1796–1861*. Cambridge University Press, Cambridge, England. (pp. 240–241)

Warren, Leonard. 1999. Silliman, Benjamin (8 Aug. 1779–24 Nov. 1864), pp. 933–934 in John A. Garraty and Mark C. Carnes (general editors), *American National Biography*, vol. 19. Oxford University Press, New York, Oxford.

Webster, Wm. David. 2002. Elisha Mitchell and the Elisha Mitchell Scientific Society, North Carolina Academy of Science, and *Journal of the Elisha Mitchell Scientific Society*. *Journal of the North Carolina Academy of Science* 118: 1–11.

Wiggins, Minnie Marguerite. 1963. *The Journal of the Elisha Mitchell Scientific Society, 1883–1963*. Thesis (M.S.L.S.), University of North Carolina, Chapel Hill.

Wilson, Henry Van Peters. 1899. Biology at the University of North Carolina. *Gulf Fauna and Flora Bulletin* 1: 23–25.

Wright, Albert Hazen. 1954. *Pre-Cornell and Early Cornell V. Natural History Society of Cornell University 1869–1899. The History of Natural Science in Cornell University for the First Thirty Years*. (*Studies in History*, no. 19). [The Author], Ithaca, NY. (pp. 587–596)

York, Maurice C. 1982. The Dialectic and Philanthropic Societies' contributions to the Library of the University of North Carolina, 1886–1906. *North Carolina Historical Review* 59: 327–353. (p. 350)

Youmans, Eliza A. 1873a. *Second Book of Botany. A Practical Guide to the Observation and Study of Plants*. D. Appleton & Company, NY. ("Henslow's Botanical Charts," pp. [7]–8)

Youmans, Eliza A. 1873b. *A Key to Henslow's Botanical Charts*. D. Appleton and Company, NY.

Denison Olmsted

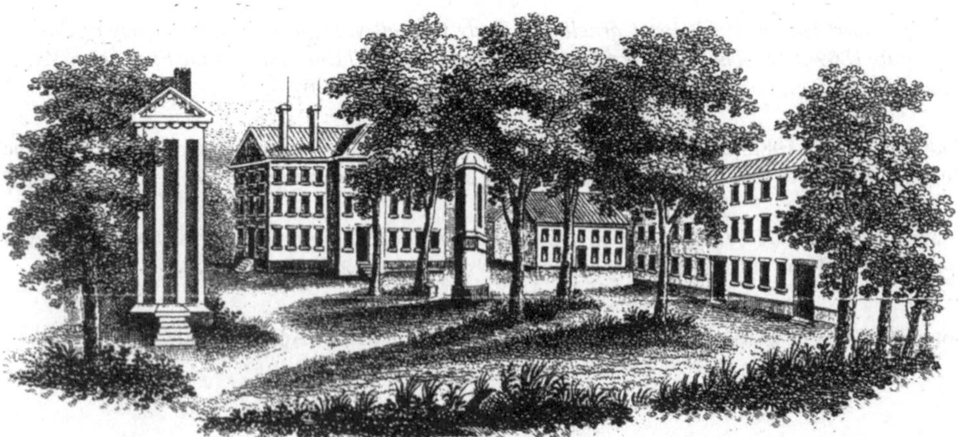

Chapter 2.

Denison Olmsted (1791–1859)

Professor of Chemistry, Mineralogy, and Geology,[16] (1817/1818–December 1825) (resident professorship began 1819)

Education and Appointment

Born June 18, 1791, in East Hartford, Connecticut, Denison Olmsted was the son of Nathaniel and Eunice (Kingsbury) Olmsted. His father, a farmer, died when Olmsted was one year old. Young Olmsted showed serious interests in English literature and mathematics. He prepared for college under the instruction of Mr. James Morris at Litchfield South Farms, and Rev. Noah Porter in Farmington, both in Connecticut. In 1809, he entered Yale College, now Yale University, where he was greatly influenced by Dr. Timothy Dwight, who helped shape his character and mind (Woolsey, 1859). Among scientific courses, he took mineralogy in his senior year from Benjamin Silliman (Johnson, 1985), who later served as his scientific mentor at Yale. Probably due to financial problems, he left Yale in April 1813 and taught at the Union School for boys in New London, Connecticut. On September 1, 1813, Olmsted attended graduation ceremonies at Yale, where he received a Bachelor of Arts degree with high honors. He delivered an oration "On the Causes of Intellectual Greatness" (Yale-S1). From 1815 to 1817, he was a tutor at Yale, during which time he also studied theology under Dwight. In completing his Master of Arts degree in 1816,[17] he gave an oration on the state of education in Connecticut (Lyman, 1859; Yale-S2).

In addition to the influence of his professors at Yale, which is discussed in Chapter 1 of this book, Olmsted was active in campus societies. According to the minutes of the proceedings of the Moral Society (Yale-MS), Olmsted was elected into the organization in April 1810, becoming a debator and orator as well as its librarian in 1812. Joining the Connecticut Academy of Arts and Sciences in 1817, Olmsted became an active member in later years as

[16] Olmsted's academic title is variously cited as "Professor of Chemistry and Mineralogy" and as "Professor of Chemistry, Mineralogy, and Geology." UNC catalogs of 1819 and 1821–1823 cite the former title; whereas UNC catalogs of 1824 and 1825, as well as the trustees minutes of the December 25, 1818 meeting (BOT 4: 125) and a broadside (UNC, 1819b), give the latter title. Kemp Battle (1907–1912, 1: 417) mentions that Olmsted added Geology to his title when he headed the State Geological Survey.

[17] The Master of Arts degree conferred on Olmsted was not an academic degree, but rather one awarded automatically on application to alumni three years after receiving the undergraduate degree. Applicants of the Master of Arts "were supposed to present 'a written synopsis either of logick or natural philosophy or metaphysicks as also a Common place on some Divinity thesis' and solve several problems posed by the rector; and at commencement they had to demonstrate their advanced knowledge in response to 'Quaestiones' similar in nature to the B.A." (Kelley, 1974).

Frontispiece Credits: (above) Denison Olmsted, ca. late 1830s (image 10595) (Images of Yale Individuals, [RU 684], Manuscripts and Archives, Yale University Library). (below) Scene of UNC campus, ca. 1847; although dated about 1847/1848, this view shows the campus buildings that stood when Olmsted left UNC in 1825; left to right: Old East, South Building, Person Hall, Old West. (From Commencement, Ball Programs, North Carolina Collection, Wilson Special Collections Library, University of North Carolina, Chapel Hill, VC378 UL v. 1, 1814–1888).

a presenter of an annual meeting on the history of chemical science in 1826, its recording secretary in 1829, and its president from 1848 to 1850. A number of publications provide biographical background on Olmsted (Anonymous, 1852; Anonymous, 1859; Bursey, 1982; Goodrum, 1999; Jones et al., 1859; Kingsley, 1879; Lyman, 1859; Parker, 1991; Schoepflin, 1981; W[heeler], 1859), and several archival collections house relevant manuscripts on him (Bowdoin, NCSA, UNC-P, Yale).

In 1817, Olmsted abandoned his studies in theology and was hired as the chair of chemistry and mineralogy, a newly created professorship at the University of North Carolina. The events leading to his appointment at UNC are described by Gary L. Schoepflin (1981) and Maurice M. Bursey (1982). The trustees established the professorship, hoping it "would not only enhance the fame and add to the growing reputation of [the] infant Establishment, but would also operate in a way highly beneficial to the community at large, and particularly to the agriculturalist of our Country" (BOT 4: 146). During this period, it was customary for university faculty to carry out duties associated with the functioning of the campus. In December 1820, Olmsted was appointed the university bursar, with responsibilities for collecting tuition, room rent, and other college dues (Olmsted, 1820a), a post he held until the close of 1824 (Olmsted, 1824a). He was compensated for this extra work by receiving 2% of the total monies collected for room and tuition. His payment averaged about $100 each session.

Olmsted accepted the professorship with the proviso that he would remain at Yale one additional year (with full salary—$1,000) to work under Benjamin Silliman, a chemist and leading promoter of science at that time, to gain additional expertise in the theory and practice of chemistry. This extra year of studies prepared him for the new professorship at Chapel Hill. Olmsted (1817b) also had assumed that the trustees would procure scientific apparatus to which he could adapt his lectures.

While gaining a practical knowledge of chemistry (as well as geology and mineralogy) at Yale, Olmsted (1817a) wrote a letter to Elisha Mitchell, his classmate at Yale and then Professor of Mathematics and Natural Philosophy at UNC. Olmsted inquired about housing and general aspects of life in Chapel Hill. He also asked Mitchell: "Won't [it] be best for me to study Botany with Dr. Ives this summer?" From this inquiry we can assume that Olmsted was probably anticipating some involvement with botanical instruction at Chapel Hill and that he felt his knowledge of botany was lacking or weak. Olmsted was also thinking about his lectureship in chemistry at UNC. In this same letter, he wrote as follows:

> What I have thought of (and I lay awake last night a great while thinking of it) is this. To begin with the present Junior class next January and hear them twice a week in 'Conversations on Chemistry' (tis an excellent little comprende) performing all the experiments which are alluded to there and that I can do with little but apparatus. This course would probably continue till near the close of the session. This plan would afford me leisure to be preparing my lectures and chemical text etc. (Many of which I shall be obliged to manufacture myself for the next year). By that time I hope to be able to commence a regular course of lectures to the Senior class twice a week for the greater part of the year, and in the mean time to hear the Junior Class recite 'Conversations on Chemistry.' The present Sophomore class therefore may recite this book next year and hear a course of lectures the year after.

According to historian Kevin Cherry (1993), Olmsted had been contemplating that instruction in the sciences should be given to sophomores and seniors and not simply to juniors, as was then the common practice. Wasting little time, the university approved Olmsted's ideas for expanding the scientific curriculum. At their December 19, 1818, meeting, the Board of Trustees passed this plan, as noted in a broadside listing the university's courses of instruction as of June 28, 1819 (UNC, 1819b) (Fig. 2.1). Seniors were now offered instruction in several scientific disciplines.

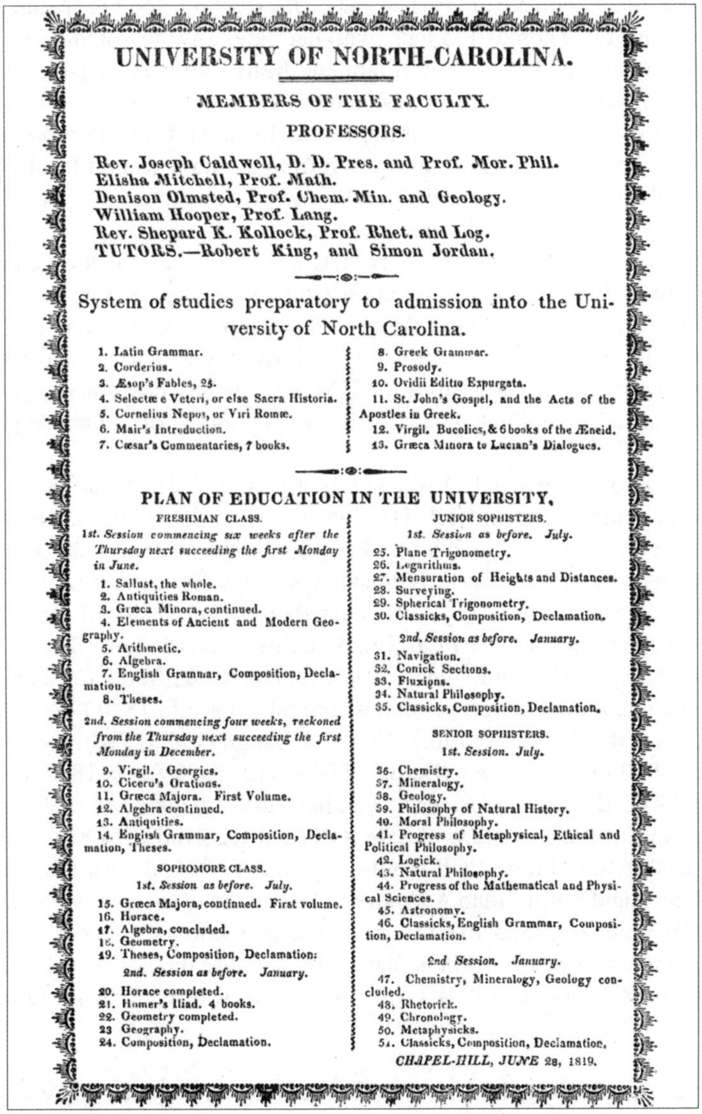

UNIVERSITY OF NORTH-CAROLINA.

MEMBERS OF THE FACULTY.

PROFESSORS.

Rev. Joseph Caldwell, D. D. Pres. and Prof. Mor. Phil.
Elisha Mitchell, Prof. Math.
Denison Olmsted, Prof. Chem. Min. and Geology.
William Hooper, Prof. Lang.
Rev. Shepard K. Kollock, Prof. Rhet. and Log.
TUTORS.—Robert King, and Simon Jordan.

System of studies preparatory to admission into the University of North Carolina.

1. Latin Grammar.
2. Corderius.
3. Æsop's Fables, 25.
4. Selectæ e Veteri, or else Sacra Historia.
5. Cornelius Nepos, or Viri Romæ.
6. Mair's Introduction.
7. Cæsar's Commentaries, 7 books.
8. Greek Grammar.
9. Prosody.
10. Ovidii Editio Expurgata.
11. St. John's Gospel, and the Acts of the Apostles in Greek.
12. Virgil. Bucolics, & 6 books of the Æneid.
13. Græca Minora to Lucian's Dialogues.

PLAN OF EDUCATION IN THE UNIVERSITY.

FRESHMAN CLASS.

1st. Session commencing six weeks after the Thursday next succeeding the first Monday in June.

1. Sallust, the whole.
2. Antiquities Roman.
3. Græca Minora, continued.
4. Elements of Ancient and Modern Geography.
5. Arithmetic.
6. Algebra.
7. English Grammar, Composition, Declamation.
8. Theses.

2nd. Session commencing four weeks, reckoned from the Thursday next succeeding the first Monday in December.

9. Virgil. Georgics.
10. Cicero's Orations.
11. Græca Majora. First Volume.
12. Algebra continued.
13. Antiquities.
14. English Grammar, Composition, Declamation, Theses.

SOPHOMORE CLASS.

1st. Session as before. July.

15. Græca Majora, continued. First volume.
16. Horace.
17. Algebra, concluded.
18. Geometry.
19. Theses, Composition, Declamation:

2nd. Session as before. January.

20. Horace completed.
21. Homer's Iliad. 4 books.
22. Geometry completed.
23 Geography.
24. Composition, Declamation.

JUNIOR SOPHISTERS.

1st. Session as before. July.

25. Plane Trigonometry.
26. Logarithms.
27. Mensuration of Heights and Distances.
28. Surveying.
29. Spherical Trigonometry.
30. Classicks, Composition, Declamation.

2nd. Session as before. January.

31. Navigation.
32. Conick Sections.
33. Fluxions.
34. Natural Philosophy.
35. Classicks, Composition, Declamation.

SENIOR SOPHISTERS.

1st. Session. July.

36. Chemistry.
37. Mineralogy.
38. Geology.
39. Philosophy of Natural History.
40. Moral Philosophy.
41. Progress of Metaphysical, Ethical and Political Philosophy.
42. Logick.
43. Natural Philosophy.
44. Progress of the Mathematical and Physical Sciences.
45. Astronomy.
46. Classicks, English Grammar, Composition, Declamation.

2nd. Session. January.

47. Chemistry, Mineralogy, Geology concluded.
48. Rhetorick.
49. Chronology.
50. Metaphysicks.
51. Classicks, Composition, Declamation.

CHAPEL-HILL, JUNE 28, 1819.

Fig. 2.1. [Broadside]. University of North-Carolina: Members of the Faculty; Plan of Education in the University, June 28, 1819.

(Courtesy of North Carolina Collection, Wilson Special Collections Library, University of North Carolina, Chapel Hill).

Living and Teaching in Chapel Hill

When leaving New Haven in early November 1818, Olmsted confided with Silliman, expressing "his sense of the advantages which he had enjoyed in the course of preparatory labor and instruction through which he had passed, without which he said that he should not have dared to enter upon the duties of his station" (Fisher, 1866). En route to North Carolina, Olmsted (1818) stopped in New York, then Philadelphia, where he met several noted scientists. He also attended a lecture on chemistry given by lawyer, chemist, and educator Thomas Cooper; met chemist Robert Hare, who showed Olmsted his chemical laboratory; and conversed with Dr. Thomas P. Jones, who had held a professorship in chemistry and natural philosophy at the College of William and Mary and was well-known for his skill in making chemical apparatus. Olmsted became so enthralled with the sights of interest in the city that he missed the steamboat to Baltimore, his next destination on his itinerary to North Carolina. The delay, however, allowed him to attend a second lecture given by Dr. Hare. He probably arrived in Chapel Hill at the end of November.

Unlike the culturally rich town of New Haven, Chapel Hill was at that time a village in the wilderness of Orange County, containing "at least thirteen residences, four stores, two 'hotels,' and a blacksmith shop, in addition to the grammar school and four university buildings" (Vickers, 1985). Its main street was dirt; it was dusty in the summer and often became a muddy mire after rainfalls. When visiting the UNC campus as a possible prospective student, David L. Swain (1822) noted that Chapel Hill "is upon the whole a handsome place."

Olmsted moved into the original president's house (Fig. 2.2) on campus, which was in the vicinity of the current Swain Hall. The house was available because President Joseph Caldwell preferred living in a dwelling on Franklin Street. When Elisha Mitchell married in 1819 he and his wife moved into the Olmsted residence. Soon after, Olmsted moved into the widow Jane Puckett's house (Fig. 2.3), located on the north side of Franklin Street and east of Hillsborough Street, now 501 East Franklin Street and probably one of the two oldest extant houses in Chapel Hill. Although the house and lot were offered for sale in April 1819, the terms of sale stipulated that possession would take place in 1820 (Anonymous, 1819). Olmsted purchased the dwelling in June 1820 (McVaugh et al., 1996). He paid $1,300 for it and invested $900 in renovations (Battle, 1907–1912, 1: 299). In 1822, he successfully convinced the UNC Board of Trustees to purchase the structure and make it available to faculty. According to the indenture of sale, Olmsted received $2,400 for the house and lot (UNC-Papers). After Olmsted left North Carolina in 1825, the trustees authorized that Professor of Greek and Latin Ethan A. Andrews be permitted to occupy the house rent free (BOT 5: 53). Whether Andrews lived in the house is unknown; however, James Phillips, the newly appointed Professor of Mathematics, moved into the house in 1826, living there until his death in 1867 (Bryant, 1999). Professor Phillips was the father of famed university historian Cornelia Phillips Spencer.

Because of the historical significance of the house, a number of people have described aspects of its history. Lucy Phillips Russell (1931), granddaughter of James Phillips, provides a brief history of the house and its occupants up to 1931. Architectural historian M. Ruth Little (2006) gives background on and provides an illustration of the house. Researcher Elaine Vaudreuil (1992) summarizes the history of the house and provides a description and analysis of the structure. Botanical historian Burk (2011) offers background on the house as well as notes on its occupants, including its current ones.

Fig. 2.2. Olmsted's Residence, University of North Carolina at Chapel Hill. "Caldwell, Mitchell house," painted by Caroline Lee Whiting Hentz.

(Folder 9, Cornelia Phillips Spencer Papers #P-683/58, Southern Historical Collection, Wilson Special Collections Library, University of North Carolina, Chapel Hill).

On Christmas Day, 1818, the trustees of the university set Olmsted's annual salary at $1,240 (BOT 4: 195). In the same year, rhetoric and logic were added to the curriculum (Battle, 1907–1912, 1: 255–257). Teaching responsibilities were vested in the hands of five professors: Rev. Joseph Caldwell (D.D.), President and Professor of Moral Philosophy; Elisha Mitchell (A.M.), Professor of Mathematics; Denison Olmsted (A.M.), Professor of Chemistry, Mineralogy and Geology; William Hooper (A.M.), Professor of Language; and Rev. Shepard K. Kollock (A.M.), Professor of Rhetoric and Logic. Tutors Robert Rufus King (A.M.) and Simon Jordan (A.B.) assisted the faculty. The UNC catalog of October 1819 (UNC, 1819a) (Fig. 2.4) records that the student body numbered 118 men in Olmsted's first year of residency at Chapel Hill. The first session commenced "six weeks after the Thursday next succeeding the first Monday in June" and the second session commenced "four weeks, reckoned from the Thursday next succeeding the first Monday in December."

According to the *Laws of the University of North-Carolina* as revised in 1813 (Laws, 1822), student expenses were $55 each session: board $30, tuition $10, room rent $1, library fee $1, servant hire $1, washing and mending $5, wood and candles $4, and bed hire $3. (In 1802, yearly expenses were about $100 [Polk, W., 1802]). The university trustees were concerned that parents might give their sons more money than necessary to meet their expenses. "To prevent the evils that may result from improper credit being given to the students[18] by shop keepers and others in the village, the trustees recommend[ed] in the strongest terms, that [they] should forbid all purchasing on credit, even necessaries, without the approbation of the president" (Laws, 1822).

Fig. 2.3. The Widow Puckett house at 501 East Franklin Street, Chapel Hill, North Carolina, ca. 1890.

(Courtesy of North Carolina Collection Photographic Archives, Wilson Special Collections Library, University of North Carolina, Chapel Hill).

In laying the foundations for teaching chemistry at UNC, Olmsted would not only develop lectures and write a textbook on the subject but would also require a laboratory for lecturing and demonstrating scientific principles before the students. Hitherto, the campus did not have a chemical laboratory. In a letter of March 11, 1817, to Elisha Mitchell, Olmsted (1817a) mentioned that about $700 would be required to procure apparatus and supplies for the undertaking.

UNC President Joseph Caldwell (1818) provided the leadership in establishing the laboratory. Writing to UNC trustee Colonel William Polk on April 1, 1818, he noted his earlier correspondence with State Treasurer and UNC trustee John Haywood, in which he informed him and the trustees about the possible need of funds to build a laboratory. After writing to Haywood, Caldwell developed an alternative plan to building a facility. Instead, he recommended the renovation of Steward's Hall (Fig. 2.5), a two-storied wooden building

[18] In the early years of the university, students were considerably younger than currently. When visiting the UNC campus in April 1822 as a possible prospective student, David L. Swain (1822) wrote to his father: "The usual age at which young men graduate at the University is 19. Some as early as 17."

that was situated on Cameron Avenue, in an area now directly north of Carr Hall. Serving as the student commons and as a ballroom at commencement, the structure has a noteworthy history (DocSouth; Henderson, 1949; Link, 1941; Powell, 1972). The hall was dismantled around 1848.

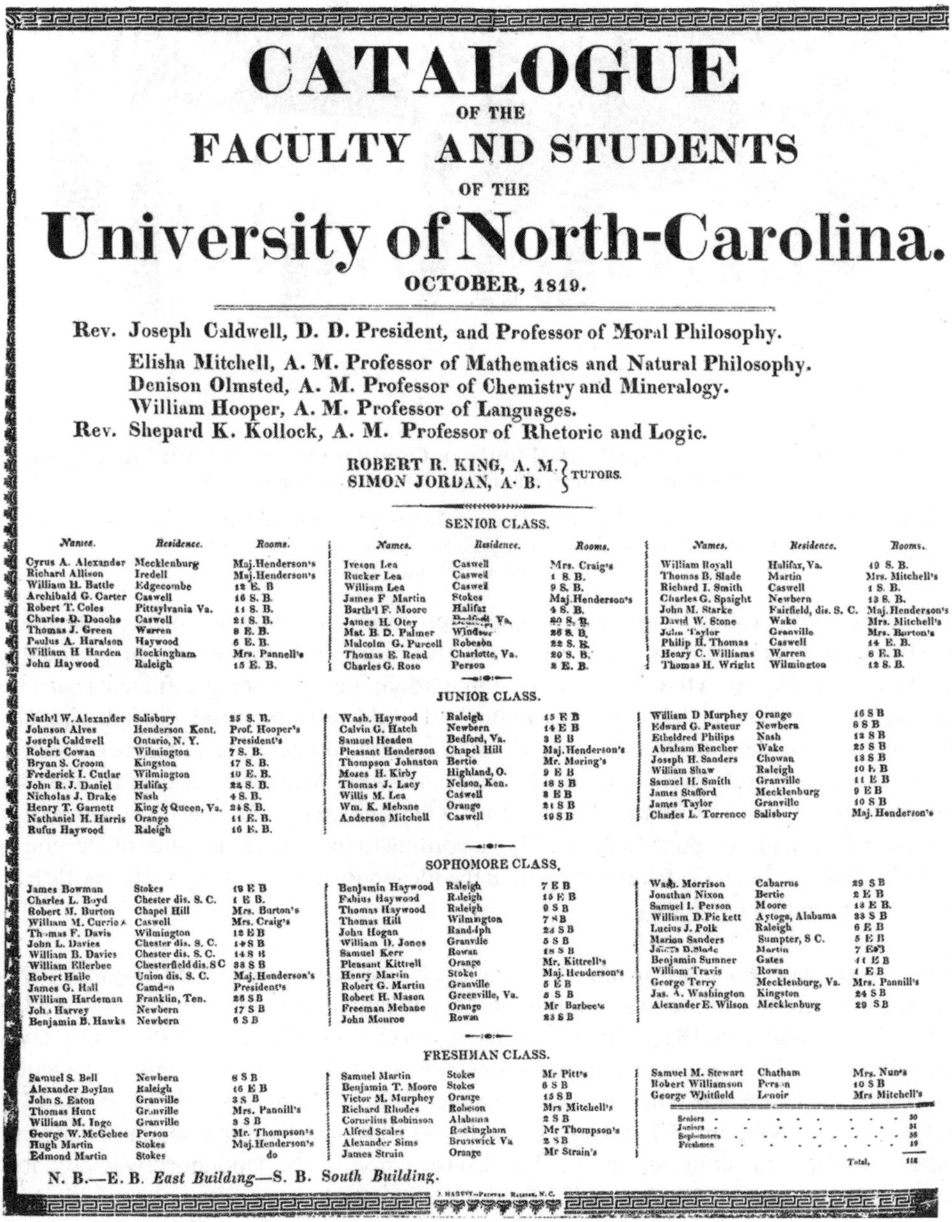

CATALOGUE
OF THE
FACULTY AND STUDENTS
OF THE
University of North-Carolina.
OCTOBER, 1819.

Rev. Joseph Caldwell, D. D. President, and Professor of Moral Philosophy.
Elisha Mitchell, A. M. Professor of Mathematics and Natural Philosophy.
Denison Olmsted, A. M. Professor of Chemistry and Mineralogy.
William Hooper, A. M. Professor of Languages.
Rev. Shepard K. Kollock, A. M. Professor of Rhetoric and Logic.

ROBERT R. KING, A. M. } TUTORS.
SIMON JORDAN, A. B. }

SENIOR CLASS.

Names.	Residence.	Rooms.
Cyrus A. Alexander	Mecklenburg	Maj.Henderson's
Richard Allison	Iredell	Maj.Henderson's
William H. Battle	Edgecombe	12 E. B
Archibald G. Carter	Caswell	16 S. B.
Robert T. Coles	Pittsylvania Va.	11 S. B.
Charles D. Donoho	Caswell	21 S. B.
Thomas J. Green	Warren	8 E. B.
Paulus A. Haralson	Haywood	6 E. B.
William H Harden	Rockingham	Mrs. Pannell's
John Haywood	Raleigh	15 E. B.
Iveson Lea	Caswell	Mrs. Craig's
Rucker Lea	Caswell	1 S. B.
William Lea	Caswell	9 S. B.
James F Martin	Stokes	Maj.Henderson's
Barth'l F. Moore	Halifax	4 S. B.
James H. Otey	Bedford, Va.	20 S. B.
Mat. B. D. Palmer	Windsor	26 S. B.
Malcolm G. Purcell	Robeson	22 S. B.
Thomas E. Read	Charlotte, Va.	20 S. B.
Charles G. Rose	Person	2 E. B.
William Royall	Halifax, Va.	19 S. B.
Thomas B. Slade	Martin	Mrs. Mitchell's
Richard I. Smith	Caswell	1 S. B.
Charles G. Spaight	Newbern	13 S. B.
John M. Starke	Fairfield, dis. S. C.	Maj.Henderson's
David W. Stone	Wake	Mrs. Mitchell's
John Taylor	Granville	Mrs. Burton's
Philip H. Thomas	Caswell	14 E. B.
Henry C. Williams	Warren	8 E. B.
Thomas H. Wright	Wilmington	12 S. B.

JUNIOR CLASS.

Names.	Residence.	Rooms.
Nath'l W. Alexander	Salisbury	25 S. B.
Johnson Alves	Henderson Kent.	Prof. Hooper's
Joseph Caldwell	Ontario, N. Y.	President's
Robert Cowan	Wilmington	7 S. B.
Bryan S. Croom	Kingston	17 S. B.
Frederick I. Cutlar	Wilmington	10 E. B.
John R. J. Daniel	Halifax	22 S. B.
Nicholas J. Drake	Nash	4 S. B.
Henry T. Garnett	King & Queen, Va.	24 S. B.
Nathaniel H. Harris	Orange	11 E. B.
Rufus Haywood	Raleigh	16 E. B.
Wash. Haywood	Raleigh	15 E B
Calvin G. Hatch	Newbern	14 E B
Samuel Headen	Bedford, Va.	3 E B
Pleasant Henderson	Chapel Hill	Maj. Henderson's
Thompson Johnston	Bertie	Mr. Moring's
Moses H. Kirby	Highland, O.	9 E B
Thomas J. Lacy	Nelson, Ken.	18 S B
Willis M. Lea	Caswell	3 E B
Wm. K. Mebane	Orange	21 S B
Anderson Mitchell	Caswell	19 S B
William D Murphey	Orange	16 S B
Edward G. Pasteur	Newbern	8 S B
Etheldred Philips	Nash	12 S B
Abraham Rencher	Wake	25 S B
Joseph H. Sanders	Chowan	13 S B
William Shaw	Raleigh	10 E B
Samuel H. Smith	Granville	11 E B
James Stafford	Mecklenburg	9 E B
James Taylor	Granville	10 S B
Charles L. Torrence	Salisbury	Maj. Henderson's

SOPHOMORE CLASS.

Names.	Residence.	Rooms.
James Bowman	Stokes	19 E B
Charles L. Boyd	Chester dis. S. C.	1 E B.
Robert M. Burton	Chapel Hill	Mrs. Burton's
William M. Currie	Caswell	Mrs. Craig's
Thomas F. Davis	Wilmington	12 E B
John L. Davies	Chester dis. S. C.	14 S B
William B. Davies	Chester dis. S. C.	14 S B
William Ellerbee	Chesterfield dis. S C	33 S B
Robert Haile	Union dis. S. C.	Maj. Henderson's
James G. Hall	Camden	President's
William Hardeman	Franklin, Ten.	26 S B
John Harvey	Newbern	17 S B
Benjamin B. Hawks	Newbern	6 S B
Benjamin Haywood	Raleigh	7 E B
Fabius Haywood	Raleigh	15 E B
Thomas Haywood	Raleigh	9 S B
Thomas Hill	Wilmington	7 S B
John Hogan	Randolph	24 S B
William D. Jones	Granville	5 S B
Samuel Kerr	Rowan	18 S B
Pleasant Kittrell	Orange	Mr. Kittrell's
Henry Martin	Stokes	Maj. Henderson's
Robert G. Martin	Granville	5 E B
Robert H. Mason	Greenville, Va.	5 S B
Freeman Mebane	Orange	Mr Barbee's
John Monroe	Rowan	23 S B
Wash. Morrison	Cabarrus	29 S B
Jonathan Nixon	Bertie	2 E B
Samuel I. Person	Moore	13 E B.
William D. Pickett	Aytoga, Alabama	33 S B
Lucius J. Polk	Raleigh	6 E B
Marion Sanders	Sumpter, S C.	5 E B
J[illegible] D. Slade	Martin	7 E B
Benjamin Sumner	Gates	11 E B
William Travis	Rowan	1 E B
George Terry	Mecklenburg, Va.	Mrs. Pannill's
Jas. A. Washington	Kingston	24 S B
Alexander E. Wilson	Mecklenburg	29 S B

FRESHMAN CLASS.

Names.	Residence.	Rooms.
Samuel S. Bell	Newbern	8 S B
Alexander Boylan	Raleigh	16 E B
John S. Eaton	Granville	3 S B
Thomas Hunt	Granville	Mrs. Pannill's
William M. Inge	Granville	3 S B
George W. McGehee	Person	Mr. Thompson's
Hugh Martin	Stokes	Maj. Henderson's
Edmond Martin	Stokes	do
Samuel Martin	Stokes	Mr Pitt's
Benjamin T. Moore	Stokes	6 S B
Victor M. Murphey	Orange	15 S B
Richard Rhodes	Robeson	Mrs Mitchell's
Cornelius Robinson	Alabama	2 S B
Alfred Scales	Rockingham	Mr Thompson's
Alexander Sims	Brunswick Va	2 S B
James Strain	Orange	Mr Strain's
Samuel M. Stewart	Chatham	Mrs. Nun's
Robert Williamson	Person	10 S B
George Whitfield	Lenoir	Mrs Mitchell's

Seniors	30
Juniors	31
Sophomores	38
Freshmen	19
Total,	118

N. B.—E. B. *East Building*—S. B. *South Building*.

J. HARVEY—Printer, Raleigh, N. C.

Fig. 2.4. [Broadside]. Catalogue of the Faculty and Students of the University of North-Carolina, October 1819.

(Courtesy of North Carolina Collection, Wilson Special Collections Library, University of North Carolina, Chapel Hill).

Fig. 2.5. Silhouette of Steward's Hall (left) and Main (now South Building) (right), University of North Carolina, Chapel Hill, ca. 1815.

(Made by Frances Jones—Mrs. William—Hooper. Louis and Mildred Graves Papers #4010/3, Southern Historical Collection, Wilson Special Collections Library, University of North Carolina, Chapel Hill).

In his letter to Polk, Caldwell noted that immediately following the enlargement of Steward's Hall by about 16 feet, students were allowed to also board in the village. He suggested that a partition be built across the middle of the dining room, thus providing sufficient space for boarders and creating the lab in the remaining half. Caldwell's proposal was subsequently taken up by the trustees on December 16, 1818, when they appointed a committee consisting of President Caldwell, Archibald D. Murphey, and John Winslow "to enquire into the expediency of building a [chemical] Laboratory at the University" (BOT 4: 175). Three days later, Murphey made a report (BOT 4: 177). According to the trustees minutes of December 22, 1818 (BOT 4: 191), the faculty supported the idea to lop off a portion of Steward's Hall and convert it into a laboratory.

Even before plans for constructing the laboratory were completed, John Haywood was directed to travel to Europe to procure in part chemical apparatus in the summer of 1818 (UNC-B: 184; Williams, 1818). Soon after arriving in Chapel Hill in late 1818, Olmsted was also reimbursed for acquiring appropriate apparatus (UNC-B: 188; Williams, 1819 [entry a]). Former UNC steward William Barbee obtained lumber "for the use of a Laboratory" in Steward's Hall (Williams, 1819 [entry b]; UNC-B: 188), and Olmsted obtained materials and also assisted in setting up the facility. Construction of the laboratory was probably completed by summer 1819, about when Olmsted taught his first chemistry class. Additional procurements followed: $351.33 in 1820 for paying for work and materials (UNC-B: 192; Williams, 1820), $100 in 1821 to purchase apparatus (UNC-B: 200), $400 in 1822 for apparatus (Manly, 1822; UNC-B: 200), $111.22 in 1823 for various expenditures (UNC-B: 217), $136.01 in 1824 (Manly, 1824; UNC-B: 222), and $103.78 in 1825 (UNC-B: 230). In the chemical laboratory, Olmsted performed demonstrations before his classes (although the students did not get handson experience) and carried out experiments in pursuit of his own scientific investigations. After Olmsted resigned his professorship in 1825, the university developed a new chemistry laboratory on the first floor in the middle of the south side of South Building. Chapter 3 provides background on the development of the new laboratory.

As noted in the university's plan of education (Fig. 2.1), Olmsted's instruction in chemistry, mineralogy, and geology was confined to seniors (then called "senior sophisters"). He taught the course Chemistry in the first session of the school year, typically continuing it into the second session, which concluded with the courses Mineralogy and Geology. One can gain insight into Olmsted's early methods of teaching chemistry by examining his *Outlines of the Lectures on Chemistry, Mineralogy, & Geology, Delivered at the University of North-Carolina, for the Use of the Students*, published in 1819 (Olmsted, 1819c) (Fig. 2.6). In this forty-four-page book, his course of chemistry lectures was divided into two main sections, one dealing with general principles, the other treating the chemical properties and relations of specific substances. In the first portion of the course, he covered the observed action of light upon plants. Throughout the text, Olmsted gave bibliographical citations for students' additional consultation and reading.

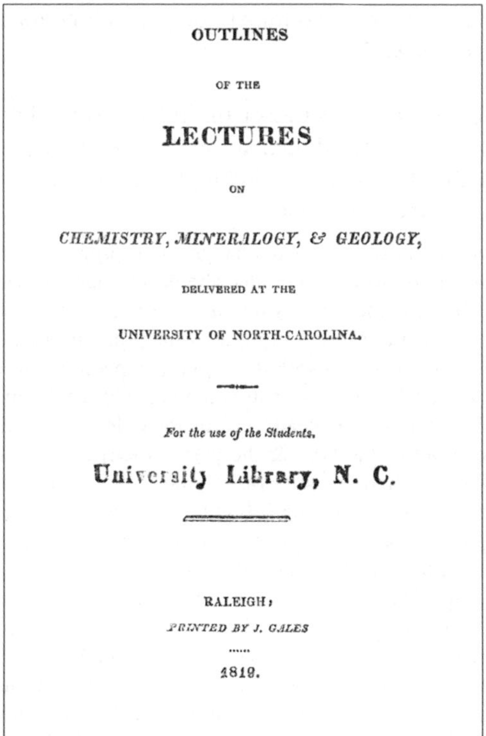

OUTLINES

OF THE

LECTURES

ON

CHEMISTRY, MINERALOGY, & GEOLOGY,

DELIVERED AT THE

UNIVERSITY OF NORTH-CAROLINA.

For the use of the Students.

RALEIGH:

PRINTED BY J. GALES

1819.

Fig. 2.6. Title page from *Outlines of the Lectures on Chemistry, Mineralogy, & Geology,* by Denison Olmsted (1819c).

(Courtesy of North Carolina Collection, Wilson Special Collections Library, University of North Carolina, Chapel Hill).

Course notes taken by his students provide additional insight into Olmsted's teaching, including his instruction of botanical topics. Several student notebooks from Olmsted's lectures on chemistry survive in the Southern Historical Collection (Bryan, 1823; Graham, 1823–1824; Moore, 1820; Polk, L. J., 1821). Of these notebooks, the one by Augustus Moore provides the most complete coverage of the lectures. Augustus Moore was a student from Edenton, North Carolina, who later in life studied law and served as a judge (Wheeler, 1884). The flyleaf of his notebook is inscribed: "Notes on a Course of Lectures delivered at the University Laboratory of N. Carolina by Professor D. Olmstead [*sic*], began 13th Sept. 1820"; thus the notes were taken in Olmsted's second year of teaching at UNC. Here, Moore also cited several textbooks that Olmsted recommended: the Greenfield edition of Marcet's *Conversations on Chemistry* (1818), the Edinburgh edition of Dr. Murray's *Elements of Chemistry* (1817), Parkes's *Chemical Catechism* (1818), and Black's three-volume work, *Lectures of Chemistry* (1806–1807). Olmsted's lectures reflected current knowledge of scientific concepts. He gave a "Demonstration used in the Lecture on cold" that Moore witnessed. Although Olmsted performed demonstrations for lectures, the apparatus in chemistry appeared incomplete according to prospective UNC student D. Wilkins (1820). He "found there was no philosophical apparatus & that the chemical laboratory was so very deficient" that a student would not acquire the necessary exhibit of experiments to gain sufficient knowledge in either natural philosopy or chemistry. Wilkins also noted that the faculty were "very haughty and impervious in their conduct, without the essential quallifications [*sic*] of politeness."

Apparently the content of Olmsted's lecture on the influence of light on plants was gleaned in great part from Black's treatise, including notes in its appendix by John Robison, Professor of Natural Philosophy at the University of Edinburgh. On the topic of light and plants, Moore (1820, pp. 18–21) recorded the following:

> Light combines with various boddies & changes their properties 1st with Vegetables. Their properties are found to be materially altered by their growing in ye [notation henceforth transcribed as 'the' in this quotation] dark, they loose all their colour, their smell, pungency & all of their properties which identify their species: plants so circumstanced have a tendency to seek the Light. If a plant be placed in a dark room near the side of it, & then be made a small hole in the side of the room so as to let the light into it, the plant will turn its head towards the hole as if seeking after light—& it will also proceede out of the hole & the part wh[ich] is out, will turn to a green colour & have all of the characteristics of its kind. The exterior leaves of a cabbage are not fit for the table but are noxious & highly disagreeable to the taste & to health; & it is only such parts of Cellery as grow under the ground, & are shielded from the influence of the light, that are fit to be eaten, the other parts being rendered totally unfit by light.—so also Potatoes, wh. Grow uncovered become green & are totally poisonous; all these effects are produced by the agency of Light. Prof. Robertson [*i.e.*, Robison] of Edinburgh, having some business which caused him to go down into the drain of a coal work underground, while there he laid his hand on a plant which when a light was brought to it he discovered had a copious deep indented white foliage; he had never seen anything like to it, nor could any of the Colliers or the workmen tell what it was; he caused the sod to be carried up to the day light—& set in the ground & carefully watered it; in a few days he observed that the plant had languished & died & that a new vegetation was beginning with little sproutings from the same stem, & that this new growth was of a green colour, & after about a week he discovered the plant to be common Tansy; for it had then become green & when

> bruised had the same smell as Tansy; he repeated the experiment on several other pieces & all proved to be common Tansy; yet when the plant was bruised in the coal work, it had no smell, or aromatic fragrance. Prof. Olmstead [*sic*] thinks these experiments prove fully that the green colour, the pungency & in short all the distinguishing properties of vegetables are dependent for their existence on Light. It should have been mentioned that, it was afterwards, found out, that the sods containing, in these coal works, those white plants, had been originally taken from a Garden, where a great deal of Tansy was still growing; among the grasses. If light be applied to a plant by a Lamp, it will grow as if it were in the open air. Vegetation is most fruitful & fragrant & abundant in the Equatorial regions where there is most light.

Concluding topics in the course were vegetable chemistry, animal chemistry, and agricultural chemistry. The text for these lectures is not covered in Olmsted's *Outlines*; however, Moore recorded in detail the lectures on vegetable chemistry. I have transcribed these notes, which constitute Appendix 2.1. From these selections of lectures, we learn that Olmsted taught the Linnean sexual system of plant classification, plant physiology, and the uses of plant products by man.

In April 1819, Olmsted (1819a) shared his thoughts about the Chapel Hill weather and vegetation, the university, and other topics with James L. Kingsley, his former Yale professor of languages and ecclesiastical history. Olmsted was amazed that peach trees blossomed in early March and that garden peas and ivy of temple were growing in April. He commented: "A week past the therm.[ometer] has been on different days, when at the maximum, from 65° to 78°. For a month past it has been called sickly in this region, but this fact convinces me of the general healthiness of the country, for such a quantum of sickness as has prevailed here would be little regarded in Connecticut." In the same letter, Olmsted provided insight about the curriculum and the nature of the students at UNC. In discussing a number of similarities between UNC and Yale, the governance at UNC, and President Caldwell's approach to the discipline of students, he wrote:

> I have been a little disappointed to find a college in N. Carolina in so many respects the same thing as it is in N. England. The regular succession of prayers, recitations, speaking, and especially the "unwell Sir" & "I didn't hear the bell Sir" savor so strongly of old Yale that I am rather pleased than otherwise to take in the whole together. The students would bear a better comparison with those of Yale in point of application than in morals. The requisites for admission have lately been increased and the college studies augmented, so that they do not materially differ from those of your printed scheme. —But some of them are still only on paper—Dr. C. has been obliged to contend with the limited views of the country and the deficiencies of preparatory schools—His own views are by no means limited. We think very well of him as a scholar—such have been his efforts to raise the institution that he has occasionally supplied the place of every instructor through all the classes. He has thus become entirely familiar with everything taught in college, but has lacked opportunity to make himself as much master of his own profession as his talents and industry would otherwise have enabled him to do. I wish I could see you and talk over with you the peculiarities of our system of government—it differs more from yours than the system of instruction. I cannot write many particulars for fear of accidents. We are amused too [*sic*] see with what cool indifference Dr. C. will receive from an offender an inflated harangue which would have drawn down upon his head from a certain other Dr. of honored memory thunder more terrible than those of the Vatican. But in doing so he entirely humours the feelings of the country.

> 'Mr. C. you will please to state the reason why you assaulted your fellow student.' C. states some idle reason and adds 'It may be my lot to suffer censure or perhaps to be cast off from your favour; but I shall ever congratulate myself on having persevered in the path of honour and agreeably to the dictates of conscious rectitude.' 'Well, Mr. C. have you any other reasons to state.' C. is sent away and the conclusion of his defence proves to be borrowed from the speech of "a man of honour," who was on his trial for killing his neighbor in a duel and acquitted, probably for the extraordinary merit of his speech.—The faculty have no check upon their authority, except the right of public expulsion—you will be apt to suppose they make public dismission take its place without often troubling their betters with a case of discipline. You will see that such matters as these are *inter nos et amicos*.

At the end of May, when Olmsted (1820b) concluded teaching his first class of seniors at UNC, he reported having given sixty-two lectures, of which twenty-eight were based on his *Outlines*. After nine months of instruction, he was "encouraged by the proficiency & attention of [the] class." He was particularly pleased with the respect with which new ideas were received on the campus. Realizing that southern colleges operated on a more slim basis than those in New England, he was prepared for the likelihood of "ups & downs."

Interest in Agriculture

George Washington Jeffreys (1822), a clergyman and farmer of Person County, North Carolina, noted in 1822 that North Carolina was nearly entirely an agricultural state and was in need of agricultural reform. He had earlier, beginning in 1817, written a series of columns on agricultural improvement in the *Raleigh Register*, writing under the pseudonym "Agricola" (Elliott, R.N., 1955). Olmsted also realized the need of agricultural improvement in the state. He emphasized the importance of thinking as well as observation and inquiry. Olmsted (1824c) believed farmers not only needed to learn about new methods of farming but also should become enlightened thinkers, which would provide them with a means of conducting business more intelligently and skillfully.

Recognizing that many of his pupils would follow agricultural pursuits, Olmsted devoted a portion of his course on chemistry to agricultural topics. According to UNC catalogs of 1824 and 1825, "Philosophy of Agriculture" was taught to seniors in the second session. Olmsted (1824c) was most likely the instructor.[19] The course is also listed in the 1826 catalog, but it may not have been taught. Olmsted recommended two publications to students preparing for livelihoods in agriculture: Davy's *Agricultural Chemistry* (1821) and the *American Farmer* (a periodical "containing original essays and selections on agriculture, horticulture, rural and domestic economy, and internal improvements"). The former would provide philosophical principles, the latter practical advice. Olmsted's early instruction on agricultural philosophy no doubt inspired his students, including William K. Mebane and William D. Murphey who presented a debate at the UNC commencement exercises held on June 7, 1821. Their topic was: "Has the art of husbandry been advanced more by the philosophical Agriculturist or practical Farmer?" (UNC, 1821). Olmsted's coverage of agricultural topics in his classes garnered attention in the state. In a transmittal to the North Carolina General Assembly's meeting of November 20, 1822, Governor Gabriel Holmes (1822) wrote:

[19] The UNC catalog of 1824 lists a course on the philosophy of agriculture offered to seniors in the second session.

> I would not have you suppose that this subject [*i.e.*, agriculture] is altogether neglected in our University. We have there a professor of chemistry and mineralogy, which bear an intimate relationship to agriculture, and it gives me much pleasure to state, that I have been informed that the gentleman who has the charge of that department of instruction, takes a lively interest in the improvement of the agriculture of the county, and devotes a part of his course of lectures to that subject alone, and loses no opportunity of imparting to his pupils every article of knowledge which will be of service in the business of life.

In December 1818, the Agricultural Society of North-Carolina was established in Raleigh to promote "Agriculture and Rural Affairs" (Gales, 1819). Three years later, on December 14, Olmsted (1825f) addressed the society about the preparation of mortar, a topic falling within the purview of rural economy. Through his experimentation, he had tested the making of lime-mortar for building purposes. By 1822, he had gained the respect of farming advocates. In speaking to the Agricultural Society of North-Carolina on December 16, 1822, Smithfield, North Carolina, physician Robert H. Helme[20] ([1823]) recommended that a professorship of agriculture should be established at UNC. He further noted that if the society had ample funds, he would "recommend the appointment of a skillful and scientific agriculturist, assisted by Professor Olmsted, of our University, as a geologist, to make an agricultural and geological tour of the state."

Outside his classroom teaching and public speaking, Olmsted performed scientific investigations that he hoped would have practical applications for agriculture. He began research on the applications of obtaining gas for illumination from cotton seed (Olmsted, 1824b). He summarized the technique used in making his discovery in a local newspaper as follows (Olmsted, 1824e):

> Ex. 1. Stick a pin into a single seed, and hold it with the tongs near hot coals, or in the flame of a candle. The only vapour being set on fire as it issues from the seed, will burn with a white and pure flame.
>
> Ex. 2. Put five or six seeds into a tobacco-pipe, and close the orifice with a clay stopper. When this is quite dry, place the bowl of the pipe among hot coals, and speedily raise it to a *low* red heat. The gas will issue at the stem, when it may be set on fire.—The stopper may be made of brick-dust, beaten up with water sufficient to make it into a paste. If a little quick-lime be added, it will dry the sooner.

Through his experimentation, Olmsted determined that one pound of cotton seed could produce 9.425 cubic feet of illuminating gas. In fact, the amount of cotton seed available could produce sufficient gas each year to supply the needs of seven cities the size of London. Nothing of importance developed from this research, however.

The originality of his research on this work came into question by the editor of the *Baltimore Patriot*. Results of similar work on cotton seeds as a source of illumination had been published in the *Patriot* on June 14, 1823, and republished in other newspapers, including those in the

[20] Robert H. Helme was a member of the North Carolina House of Commons, representing Johnston County in 1818, 1823, and 1824. A strong advocate for agricultural improvement in the state, he was a charter officer of the Agricultural Society of North-Carolina (Gales, 1819). He considered slavery as one of the principal causes for the lack of progress in agricultural development, believing that agriculture would flourish more successfully when freemen (rather than enslaved people) performed the manual labor (Helme, [1823]). After a short illness, Helme died on February 23, 1830 in Smithfield, North Carolina (Anonymous, 1830).

South. In replying to the editor's accusation of plagarism, Olmsted noted that he had not seen the other researchers' reports, because he had been traveling in the western region of the state on a geological survey. Furthermore, he noted that the other experiments were based on extracting oil from cotton seeds and then using it to manufacture gas. Olmsted's studies focused on using the cotton seed itself, not its oil, as the source of gas light (Olmsted, 1824d).

Geological and Mineralogical Investigations

Robert Helme's recommendation that he and Olmsted make a scientific tour of the state resonated with Olmsted's interest in the geological and mineralogical features of North Carolina. While broadening his education in chemistry at Yale from 1817 to 1818 to prepare for teaching at UNC, Olmsted (1817b) was also eager to learn more about geology and mineralogy. He observed at that time how students avidly traversed the countryside, collecting specimens of minerals, which were developed into a cabinet. Olmsted anticipated that he would promote "a spirit of mineralogical & geological inquiry among the hilly & mountainous counties of North Carolina." He planned to develop a cabinet with a wide variety of specimens. He even considered the possibility that he would discover mineralogical deposits of economic significance.

Six months after he began teaching at UNC, Olmsted gave the university a small collection of minerals that he had collected and received from friends while he resided in Connecticut (Olmsted, 1819b). He wished that this collection "might contribute to diffuse the knowledge of mineralogy, and to elucidate the hitherto unexplored mineral formations of this part of our country." The specimens were housed in the apparatus room. From subsequent geological excursions in North Carolina, he added specimens to this collection. He also sent specimens to Benjamin Silliman at Yale (Olmsted, 1822). In 1828, Olmsted reflected on his North Carolina geological studies in a letter to educator Parker Cleaveland: "When I went to N.C. in the year 1818, I soon saw that the field for geological researches was much more ample ... than had been generally supposed. I determined therefore to complete my vacations in geological rambles, but soon found the expenses, travelling greater than I could well afford" (Olmsted, 1828). His subsequent interest in the state's geology led to his proposal in June 1821 to the Board of Internal Improvements that a North Carolina geological and mineralogical survey be undertaken (Olmsted, 1821). In his communiqué to the board, Olmsted proposed to carry out a geological study as a means to identify and develop North Carolina's native resources, such as stone for building; deposits of gypsum, salt, coal, marl, potter's clay, and various fossils employed in agriculture; deposits of ochre, bole, reddle, and plumbago used in painting; metallic ores, including iron, lead, antimony, and zinc; and mineral springs. He suggested that he might make excursions when his university responsibilities were least pressing (*i.e.*, during lulls at the end of the year and during the summer). He recommended that he be given a stipend of $100 to cover the cost of a servant and horse and any traveling charges he incurred.

Olmsted's proposal was turned down. His letter supporting the formation of a geological survey was then published and presented to the legislature (Stuckey, 1965). His suggestion was declined by the 1821 legislature. Olmsted (1827a) thought that the legislators "had never heard of the name of geology before & formed no conception either of the object itself or of the benefit that would accrue from it to the State." In 1822, however, the North Carolina

General Assembly authorized the establishment of a Board of Agriculture (NC Laws, 1823), which first met in Raleigh in late 1823 (BOA, 1823). The board was established to "promote Agriculture and Family Domestic Manufactures within this State" (Anonymous, 1823).

A number of letters that supported the cause of a state geological survey, appeared in the local newspapers. Of particular note is a series of eight letters that were published in the *Raleigh Register* between March 21 and May 9, 1823. Written under the pen name of "Walter Raleigh" (1823), the letters were addressed to Charles Fisher, who was then secretary of the Rowan Agricultural Society. Fisher also was a North Carolina congressman and served as Speaker of the House of Commons. The anonymous writer noted that he had relied on a friend for geological information, and it is certainly possible that the friend was Olmsted himself. The writer believed that the survey would "ascertain what useful minerals we have in this State, which either are or may be employed in agriculture, or any of the arts of life or which may become sources of profit to the State as articles of commerce" (Raleigh, 1823: Letter II). He further recommended that the results of the survey be published in a style understood by the public, thereby assisting farmers in putting its discoveries into practice. The majority of the remaining letters enumerated the minerals that would benefit North Carolina agriculture and the mechanic arts. These included iron, gypsum, salt, limestone, coal, plumbago (graphite), freestone and marble, paints and dyestuffs, slates, and metallic ores.

Hearing from these and other advocates of agricultural improvements most likely influenced the legislature. On December 31, 1823, the state legislators in both the Senate and the House of Commons passed bills "directing a geological and mineralogical survey be made of the State of North-Carolina" (Journals, 1824: p. 109). The bill was duly enacted, giving the Board of Agriculture consent to employ a person to conduct a survey (NC Laws, 1824) (Fig. 2.7). The legislature appropriated "$250 a year, for 4 years" for this purpose. Olmsted was appointed to carry out the survey, becoming the first person in the Union to conduct a geological survey. North Carolina geologist Timothy Clark (1998) later stated that Olmsted was the "creator of [the] first geological survey in the United States." But some historians of geology place Olmsted's fame in a different light. Scientist and historian of geology George P. Merrill (1924) wrote, "credit must be given to Massachusetts for the first geological survey made at the expense of the state." Stuckey (1965) remarked that the "Olmsted Survey ... can hardly qualify as a state geological survey as later established and conducted by the several states including North Carolina." In a letter written to North Carolina Governor William A. Graham, Olmsted (1846) himself called his geological work more of a "beginning" than something "entitled to the name of 'Survey.'" Michael S. Smith (1999), associate professor of geology at the University of North Carolina at Wilmington, placed Olmsted's (and Mitchell's) surveys in historical perspective, noting that "their reports form the basis of study for the study of the geological framework of North Carolina."

In the same letter, Olmsted (1846) mentioned that when he first moved to North Carolina, mineralogists and geologists thought the state had no useful minerals except for some gold around Cabarrus. He contemplated making tours to assess the mineralogical assets of the state in a Jersey wagon but was told that he "ought as well mineralize among the waves of the ocean." Not long after moving to Chapel Hill, however, he realized that the mineralogical wealth of North Carolina was underrated. His subsequent excursions throughout the borders of the state confirmed his convictions. He "rode hastily over the state, intending, by this general and rapid examination of the whole, to determine where the points of greatest interest lay, and afterwards to give these a more special and scientific examination," particularly regarding their economic potential. Olmsted was able to do some of the survey work without

CHAPTER XIV.

An Act directing a geological and mineralogical survey to be made of the State of North-Carolina.

> Board of Agriculture to employ a person of skill to make the survey.

Be it enacted by the General Assembly of the State of North-Carolina, and it is hereby enacted by the authority of the same, That it is hereby made the duty of the Board of Agriculture of North-Carolina to employ some person of competent skill and science, to commence and carry on a geological and mineralogical survey of the various regions of this State; and that the person or persons so employed shall, at stated periods, furnish to the Board true and correct accounts of the results of said surveys and investigations, which shall annually be published by the Board aforesaid, for the benefit of the public, as provided by the sixth section of the act of the last General Assembly, entitled "An act to promote Agriculture and family Domestic Manufactures within this State.

> $250 a year, for 4 years, appropriated out of the agricultural fund for this purpose.

II. *And be it further enacted,* That for the purpose of carrying the intention of the foregoing section into effect, a sum not exceeding two hundred and fifty dollars, be, and the same is hereby annually appropriated for four successive years, out of the unexpended balance of the agricultural fund, as created and set apart by the above recited act; and that the Treasurer of the State is hereby directed to pay the same to the order of the Board of Agriculture of North-Carolina.

ACTS

PASSED BY

THE GENERAL ASSEMBLY

OF

THE STATE OF NORTH-CAROLINA.

At its Session,

COMMENCING ON THE 17TH OF NOVEMBER, 1823.

RALEIGH:

PRINTED BY J. GALES & SON—STATE PRINTERS.

1824.

Fig. 2.7. An Act directing a geological and mineralogical survey to be made of the State of North-Carolina.

(*Acts Passed by the General Assembly of the State of North-Carolina ... 17th of November, 1823.* Raleigh, 1824. p. 17).

leaving Chapel Hill. In addition to soliciting specimens of minerals and rocks from the state's citizens, Olmsted asked "those who inhabit the lower districts of the State ... to make collections of organic remains, such as fossils, bones, teeth, shells and petrified wood" (BOA, 1825). He provided details for the method of collecting specimens, noting the kinds to be selected, how to document the samples (including making a record of the place found and assigning a number to each), and how to prepare them for shipment to him.

Although this fact is unconfirmed, Royster et al. (1910) reported that Elisha Mitchell assisted Olmsted on the first survey. It encompassed a reconnaissance of the plumbago and mineral deposits of Wake County; the freestone and coal of Chatham and Orange counties; and the slate formations of Person, Orange, Randolph, and Montgomery counties. (The section of the report dealing with freestone and coal of Orange and Chatham counties was republished in 1855 [Olmsted, 1855].) Olmsted also reported on the gold mines of the state (including those of Anson, Cabarrus, and Montgomery counties), later publishing a journal article on the state's gold deposits (Olmsted, 1825d). At meetings of the North Carolina Board of Agriculture held in Raleigh in late fall 1824, Olmsted read the results of his investigations on the first survey.

After making his first geological tour, Olmsted (1825e, p. 20) described in his resulting geological report (mentioned below) the process of boring, which was being used elsewhere to determine the extent and quality of coal deposits. It would be sixty years before the state took his suggestion and began boring, however. In 1884, the North Carolina Board of Agriculture promoted explorations that included drilling first of the Deep River Coalfield, then of the Dan River Coalfield (McGehee, 1885). These explorations are the focus of a report written by geologist Henry Martyn Chance (1885), who directed those efforts.

In the proceedings of the December 10, 1824 meeting of the Board of Agriculture of North Carolina, mention is made that Olmsted was to begin his second survey in "a few days hence" (BOA, 1825). On this survey, miner and mineralogist Charles E. Rothe, who conducted reconnaissance on the state's gold mines, assisted Olmsted. Olmsted traveled to the coastal plain, studying marl, limestone, sand, and iron beds. In his second report, he mentioned the agricultural possibilities of the eastern part of the state and provided additional text from his first survey (Olmsted, 1827b). While making the surveys, Olmsted (1828) covered "a space of nearly 50,000 square miles." Originally hoping the surveys would promote science, he later noted to Parker Cleaveland that his "observations were not scientific—the objects were economical and the information collected was such as suited that purpose well enough" (Olmsted, 1828).

Olmsted completed his geological field studies during his summer and winter vacations of 1824 and 1825, and he published two reports on his investigations (Olmsted 1825e, 1827b).[21] (Before the first report was published, he read the first half of it on November 30, 1824,

[21] These publications are usually cited as individually paged documents. The reports that I consulted in the UNC North Carolina Collection, however, are combined with other miscellaneous publications and issued by the Board of Agriculture of North Carolina. Of the four known volumes of collected papers, three (volumes 1, 2, and 4) have their own cover pages, each entitled *Papers on Agricultural Subjects*, and volume three is entitled *Professor Mitchell's Report [etc.]*. The paging of the first two volumes is continuous and that of the last two independent.

the second half on December 5, at a meeting of the North Carolina Board of Agriculture [Anonymous, 1824b; BOA, 1825].) In the introduction to his first published report, Olmsted mentioned that his observations "assume[d] the form rather of a STATISTICAL MEMOIR, on the USEFUL minerals which our State embraces, than of a scientific Geological Survey" (Olmsted, 1825e). He continued by noting that the board wanted him to include in his report other observations that deal with "Agriculture and Rural Economy." The second report includes additional observations from his first survey. In the winter of 1823, the Board of Agriculture constituted a Committee of Selection and Publication to choose papers on agricultural subjects and rural economy from its archives and other sources. The Committee consisted of Daniel M. Forney (Beattie's Ford, Lincoln County), Rev. John Mushat (Statesville), James W. Clark (Tarborough), and Denison Olmsted (University of North Carolina) (Anonymous, 1824a). The publications chosen by the committee, including the geological reports of Olmsted and later of Elisha Mitchell, were collected into bound volumes and distributed "among the good people of North-Carolina, by means of the County Societies" (BOA, 1825). Fifteen hundred copies of each volume were printed (Anonymous, 1824a).

Accounts of Olmsted's geological discoveries were serially published as twenty-six articles in the *Raleigh Register* (Table 2.1). Joseph Gales, the editor of the *Register*, was also the printer of Olmsted's published reports. The first three articles, published in 1825, provide an overview of Olmsted's observations of the eastern section of the state, information that was included in his second report issued in 1827. Published from June to November 1826, the concluding twenty-three articles covered facts of Olmsted's investigations of the western region of the state that were published in his first and second reports. It is noteworthy that the results of Olmsted's investigations that were published in his second report were preceded in publication by the newspaper articles.

The establishment of a state geological survey and Olmsted's subsequent geological explorations gained the respect of people not only in North Carolina but also in adjacent states. An editorial essay in the April 8, 1825, *Raleigh Register and North-Carolina State Gazette* praised Olmsted's work: "More than ever do we applaud our Legislature for sanctioning a Geological Survey, which we trust will be continued until every part of our state shall have been explored by Professor Olmsted, the gentleman appointed for the purpose, and who is very eminently qualified to accomplish this desirable object" (Anonymous, 1825a). In another editorial, it was noted that Olmsted "had done more to develop our boundless stores of Mineral Wealth, than any other individual" (Anonymous, 1825b). Several notices of Olmsted's geological reports were published. Among these notifications were two abstracts (Anonymous, 1833; B[oué], 1829) that were published in foreign scientific periodicals and two notices that were published in American sources (Anonymous, 1835; Niles, 1825). In addition to these items, two book reviews appeared in scholarly journals. Massachusetts geologist Edward Hitchcock (1828), in reviewing Olmsted's reports, wrote: "This geological survey of North Carolina was authorized by an act of its legislature, and to this day it remains a conspicuous and solitary instance, in which any of our state governments have undertaken thoroughly to develope their mineral resources."

TABLE 2.1. LIST OF CITATIONS OF ARTICLES IN THE *RALEIGH REGISTER* [SEMI-WEEKLY], PROVIDING RESULTS OF DENISON OLMSTED'S GEOLOGICAL REPORTS, CHRONOLOGICALLY ARRANGED.

Newspaper citations include volume, (issue), [inferred page], and date. Citations from Olmsted's reports provide pages from which the newspaper text is derived. Concerning the newspaper articles, there is a decreasing amount of editing and cutting of text for each article with the advance of time, the last nine being virtually verbatim.

A. Citations relating to the eastern region of North Carolina with corresponding pages of Olmsted's second *Report* (1827).

Newspaper	Olmsted Report
2(208): [3], Tuesday, November 15, 1825	88–95
3(212): [3], Tuesday, November 29, 1825	95–98
3(217): [3], Friday, December 16, 1825	102–104

B. Citations relating to the western region of North Carolina with corresponding pages of Olmsted's first *Report* (1825).

Newspaper	Olmsted Report
3(266): [3], Tuesday, June 6, 1826	4–12
3(267): [3], Friday, June 9, 1826	12–15
3(269): [3], Friday, June 16, 1826	15–18
3(271): [3], Friday, June 23, 1826	18–20
3(273): [3], Friday, June 30, 1826	21–22
3(275): [3], Friday, July 7, 1826	23–26
3(277): [3], Friday, July 14, 1826	26–29
3(279): [3], Friday, July 21, 1826	30–32
3(281): [3], Friday, July 28, 1826	33–34
3(283): [3], Friday, August 4, 1826	34–40

C. Citations relating to the western region of North Carolina with corresponding pages of Olmsted's second *Report* (1827).

Newspaper	Olmsted Report
3(287): [3], Friday, August 18, 1826	105–110
3(289): [3], Friday, August 25, 1826	110–113
3(291): [3], Friday, September 1, 1826	113–115
3(293): [3], Friday, September 8, 1826	118–120
3(295): [3], Friday, September 15, 1826	120–121
3(297): [3], Friday, September 22, 1826	122–123
3(299): [3], Friday, September 29, 1826	123–125
3(303): [3], Friday, October 13, 1826	125–127
3(305): [3], Friday, October 20, 1826	127–128
3(307): [2], Friday, October 27, 1826	129–132
3(309): [2], Friday, November 3, 1826	132–136
3(310): [2], Tuesday, November 7, 1826	136–137
4(315): [2], Friday, November 24, 1826	138–141

Another review appeared in the first volume of the *Southern Review*, providing a holistic view of Olmsted's surveys ([Elliott, S., Sr.], 1828). Because the review is unsigned, debate has arisen about the identity of the author. There are two men from whose pen it may have come. In her master's thesis on the *Southern Review*, Elizabeth Steele Bearden (1925) named English-born American scientist and educator Thomas Cooper as the author. Two of her sources, however, believe the reviewer was the noted South Carolina botanist and naturalist

Stephen Elliott Sr. (Rembert, 2004). There is evidence that Elliott wrote it in a letter from Columbia, South Carolina clergyman Rev. Peter J. Shand (1869) to Rev. Whitefoord Smith of Charleston, South Carolina. Shand consulted volumes of the *Southern Review* from the private library of Robert Woodward Barnwell. Most of the reviews in this collection were annotated with the reviewer's name; the review of Olmsted's reports is attributed to Stephen Elliott Sr., who co-edited the *Southern Review* with Hugh S. Legaré. Barnwell believed that the annotations were made by Bishop Stephen Elliott, the son of Stephen Elliott Sr. and later co-editor of the journal with Legaré.

According to Joseph A. Holmes (1889), the collections of rocks and minerals that Olmsted made on his geological excursions were housed at the university but later disappeared. Surviving, however, is a map in which Olmsted recorded eight geological divisions of the state in colored inks. Believed to be the first geological map in the country, it is housed in the State Archives of North Carolina in Raleigh (Anonymous, 1999; Clark, 1999).

Olmsted also conducted analytical studies of water. According to an 1838 advertisement in the *Western Carolinian*, in 1824 he analyzed the water of Catawba Springs (Lincoln County, North Carolina), then noted for its therapeutic virtues. He determined that the water contained several minerals: sulphuretted hydrogen, sulphate of lime, sulphate of magnesia, and muriate of lime (Hampton, 1838). William S. Simonton, who owned the springs in 1831, claimed that Olmsted had recommended "its use for all complaints of liver or debility" (Sherrill, 1937).

During his tenure at Chapel Hill, Olmsted published fifteen essays and papers. Nearly half of these appeared in the *American Journal of Science and Arts*, established in 1818 by his academic mentor Benjamin Silliman. UNC botanist Alma Holland (1934), in her landmark bibliography of the publications of UNC scientists between 1795 to 1934, listed twelve of Olmsted's papers. University of North Carolina chemist Maurice Bursey (1982) mentioned an additional publication, which appeared in the August 27, 1819, issue of the Raleigh, North Carolina, newspaper the *Star* and was the first research article by a chemistry professor at UNC. Bursey reproduced the article, which focused on "thermometrical observations," in his book on UNC chemists, commenting that it "was not a great portent of interest in chemistry." Two additional publications appeared as editorials in periodicals. One explained Olmsted's (1824b) research on cotton seed as a source of illumination, the other his (1824c) philosophical view of agriculture.

Departure

After seven years in Chapel Hill, Olmsted halted his teaching, geological fieldwork, and research on illuminating gas at UNC when he was tapped to become Professor of Mathematics and Natural Philosophy (later Professor of Natural Philosophy and Astronomy) at his alma mater in New Haven, Connecticut. His election at Yale took place at its September 14, 1825, commencement; he filled the place of Professor of Mathematics and Natural Philosophy Matthew Rice Dutton, who had died on July 17, 1825 (Anonymous, 1825b). In a letter of October 9, 1825 to Governor Hutchins G. Burton, who then served as the president of the Board of Trustees at UNC, Olmsted (1825a) stated his intention of resigning. He also mentioned that Elisha Mitchell, then Professor of Mathematics at UNC, had asked him to announce his intention of offering himself to the Board as a candidate for the place. Olmsted went on to say that Mitchell was qualified "to the discharge of the duties of this department." In another letter to Governor Burton, dated December 8, 1825, Olmsted (1825c) stated that when he submitted his letter of resignation, he did not know whether he would leave at the close of the school session or not. With numerous affairs to complete and the advance of the cold season, he asked that his resignation be placed before the Board of Trustees at its first meeting (Olmsted, 1825b; 1825c). The Committee of Appointment recommended approval of his resignation (and also of the request of Elisha Mitchell to transfer from his chair of mathematics to Olmsted's vacated chair), which was duly accepted by the Board of Trustees at their December 13, 1825, meeting (BOT 5: 47).

Olmsted (1824–1841) wrote in his private journal that he left Chapel Hill on December 28, 1825, a cold and rainy day. His ensuing northward journey, which took nearly two weeks, began with a private carriage to Raleigh, Tarborro, and Norfolk. He then boarded a steamboat destined for Baltimore but detoured to Annapolis. Subsequently, he traveled by stage coach to Baltimore, Philadelphia, and New York. The final leg of his trip to New Haven was aboard the steamboat *Connecticut*. Olmsted related his fears and apprehensions about the journey, mentioning the notoriety of highway thieves on the road to Philadelphia and the treacherous, life-threatening gale he had encountered on the boat to New Haven. He was relieved to have arrived safely, but he lamented the fact that his family remained in North Carolina until spring, when they would join him in Connecticut.

At Yale College, Olmsted (Fig. 2.8) devoted the remaining thirty-four years of his life to teaching and research with singular dedication (Woolsey, 1859). In 1836, his professorship was divided, and he became Professor of Natural Philosophy and Astronomy. Most of Olmsted's publications, including the textbooks *An Introduction to Natural Philosophy* (1831), *A Compendium of Natural Philosophy* (1833), *Rudiments of Natural Philosophy and Astronomy* (1844), *A Compendium of Astronomy* (1839a), and *Introduction to Astronomy* (1839b), are enumerated by Schoepflin (1981). Olmsted was particularly renowned for a number of contributions in astronomy, including investigations of shooting stars, the aurora, and the zodiacal light (Goodrum, 1999). Concerning his instruction in astronomy, the State Archives of North Carolina in Raleigh contains a student notebook, "Lectures on Astronomy, by Professor Olmsted, Yale College, Oct., 1838" (Olmsted, 1838). In 1835, Olmsted and his Yale colleague Elias Loomis were the first Americans to view the return of Halley's comet.

Following a brief period of illness, Denison Olmsted (Fig. 2.9) died at age sixty-seven at his residence in New Haven, Connecticut, on May 13, 1859. The cause of his death was neuralgia of the stomach (W[heeler], 1859). He was survived by his children, five sons (Francis Allyn, John Howard, Alexander Fisher, Denison, and Lucius Duncan) and two daughters (Eliza and Cornelia) by his first wife, Eliza Allyn, who died eleven years after their marriage in 1818, and one daughter (Julia Mason) by his second wife, Julia Mason, whom he had married in 1831. He was a third cousin, once removed, of renowned landscape architect Frederick Law Olmsted (Olmsted and Ward, 1912). In professional organizations, he was an active member of the American Academy for the Advancement of Science and was elected to the Connecticut Academy of Arts and Sciences. He was a member of the short-lived (1819–1828) American Geological Society (Merrill, 1906), in which he gave presentations. He is buried in the Grove Street Cemetery in New Haven, Connecticut (Fig. 2.10).

Fig. 2.8. Denison Olmsted, 1833. Portrait painting by Nathaniel Jocelyn.
(Courtesy of Yale University Art Gallery, #1833.3).

Fig. 2.9. Denison Olmsted, ca. late 1850s.

(Images of Yale Individuals, [RU 684: image 10494], Manuscripts and Archives, Yale University Library).

Memorials and Legacy

Olmsted's memory and accomplishments have been commemorated in North Carolina and at Yale University. The North Carolina General Assembly issued a joint resolution (Resolution 19; Senate Joint Resolution 1169) on July 14, 1999, honoring him and marking the 175th anniversary of the state's first geological survey (Fig. 2.11). In celebrating the anniversary, the North Carolina Geological Survey issued a poster entitled "America's First State Geologic Map," which Timothy W. Clark (1999) designed. It includes a facsimile of Olmsted's original map as well as some historical background on his surveys. At Yale University, a stained glass window on the south wall of Battell Chapel is dedicated to Olmsted. He is also remembered as an inventor. In 1835, he designed and patented a fireplace stove (Fig. 2.12) for heating study chambers, but its utility extended to heating family apartments (Leggett, 1874). Smith & Sherman (1836) of New Haven, Connecticut, manufactured the stove, which was made primarily of sheet iron. It was distinguished by an apparatus termed a "radiator," which was used instead of the typical transverse stove pipe to distribute heat in a room.

Fig. 2.10. Tombstone of Denison Olmsted, Grove Street Cemetery, New Haven, CT.
(Courtesy of Susan Williams).

As a teacher of chemistry, mineralogy, and geology (including topics in agriculture and botany) at UNC, Denison Olmsted was a pioneer whose academic torch was passed on to Elisha Mitchell, his classmate from Yale and colleague at Chapel Hill. According to historical records, Olmsted was the first faculty member to instruct on a botanical topic at Carolina. Under his guidance, the first chemical laboratory was established at UNC. In 1819, he presented his first lectures, which were enhanced with scientific demonstrations in the classroom. At UNC, he also served as bursar. Olmsted was the first person in the United States to conduct a geological survey at state expense, writing two survey reports on North Carolina. He also made a geological map, probably the first in the country. He promoted agricultural interests in the Chapel Hill/Raleigh region and actively served on the Board of Agriculture of North Carolina. Outside classroom teaching and geological pursuits, he conducted experiments and published on practical aspects of using cotton seed as a source for lighting. As an educator and promoter of science, Olmsted furthered the causes of science and agriculture in the state.

GENERAL ASSEMBLY OF NORTH CAROLINA
SESSION 1999
RATIFIED BILL

RESOLUTION 19
SENATE JOINT RESOLUTION 1169

A JOINT RESOLUTION HONORING THE MEMORY OF DR. DENISON OLMSTED, THE FIRST NORTH CAROLINA STATE GEOLOGIST ON THE ONE HUNDRED SEVENTY-FIFTH ANNIVERSARY OF THE NORTH CAROLINA GEOLOGICAL SURVEY.

Whereas, the North Carolina Geological Survey is the nation's oldest geological survey, having been established by the General Assembly on December 31, 1823; and

Whereas, Dr. Denison Olmsted, as the first state geologist, produced the first geologic map of a state in the nation in 1825, thereby producing the first state geological survey; and

Whereas, the North Carolina Geological Survey serves the citizens of North Carolina by examining the geology and mineral resources of the State; and

Whereas, the North Carolina Geological Survey administers cooperative geologic and topographic mapping programs; and

Whereas, the North Carolina Geological Survey's mission is to provide unbiased and technically accurate applied earth science scientific information to address societal needs through geologic maps, mineral resources and geochemical information, topographic maps, digital products, and earth science education materials; and

Whereas, a strong state geological survey will be essential for continued prosperity of the State and its citizens into the twenty-first century; and

Whereas, geology and the other earth sciences are fundamental to the safety, health, and welfare of North Carolinians and to the State's economy; and

Whereas, knowledge of the State's geology and geologic processes is essential for the proper understanding, protection, and management of the environment, and for mitigation of natural hazards; and

Whereas, the earth sciences are integral to finding, developing, and conserving the mineral, energy, and water resources needed for the continuing prosperity of North Carolina; Now, therefore,

Be it resolved by the Senate, the House of Representatives concurring:

Section 1. The General Assembly honors the memory of Dr. Denison Olmsted, and calls upon all North Carolinians to observe, celebrate, and participate in events commemorating the 175th anniversary of the first geological survey of a state authorized by a state legislature.

Section 2. This resolution is effective upon ratification.

In the General Assembly read three times and ratified this the 14th day of July, 1999.

Dennis A. Wicker
President of the Senate

James B. Black
Speaker of the House of Representatives

Fig. 2.11. A Joint Resolution Honoring the Memory of Dr. Denison Olmsted.

(Resolution 19, Senate Joint Resolution 1169—General Assembly of North Carolina, Session 1999, Ratified Bill).

In delivering a memorial tribute on Olmsted before his Yale colleagues, President Theodore Dwight Woolsey (1859) wrote: "A life like Professor Olmsted's is full of encouragement to those who are entering life, for it shows that a man with a steady will and sound principles makes himself what he is, that it is not patronage, nor favoring circumstances, which determine his position and usefulness, but the character which he has acquired, and the abilities which he has improved by his own painstaking."

Fig. 2.12. Diagram of Olmsted's "Fire place stove."

(From United States Patent Office, *Index of Patents Issued from the United States Patent Office from 1790 to 1873, Inclusive–Continued,* vol. 3. Government Printing Office, Washington, D.C. 1874, page 1479).

References Cited

(References that conclude with page numbers within parentheses indicate sources from where quotes and pertinent information were obtained.)

Anonymous. 1819. House and lot for sale at Chapel Hill. *Star and North-Carolina State Gazette* (Raleigh) 11(17): [3], Friday, April 23, 1819.

Anonymous. 1823. Board of Agriculture. *Raleigh Register, and North-Carolina State Gazette* [Semi-Weekly] 1(4): [3], Friday, November 28, 1823.

Anonymous. 1824a. Agricultural notice. *Raleigh Register, and North-Carolina State Gazette* [Semi-Weekly] 1(52): [2], Friday, May 14, 1824.

Anonymous. 1824b. [The Agricultural Board]. *Raleigh Register, and North-Carolina State Gazette* [Semi-Weekly] 2(110): [3], Friday, December 3, 1824.

Anonymous. 1825a. [The geological resources of our state]. *Raleigh Register, and North-Carolina State Gazette* [Semi-Weekly] 2(146): [3], Friday, April 8, 1825.

Anonymous. 1825b. [Annual commencement of Yale College]. *Raleigh Register, and North-Carolina State Gazette* [Semi-Weekly] 2(194): [3], Tuesday, September 27, 1825.

Anonymous. 1830. Died. *Raleigh Register, and North-Carolina Gazette* [*Weekly*] 29(1899): [1], Thursday, March 4, 1830.

Anonymous. 1833. [Abstract of Denison Olmsted's geological reports.] *Neuesjahrbuch für Mineralogie, Geognosie, Geologie und Petrefaktenkunde* 1833: 501–502. [Abstract is based on Hitchcock's (1828) review.]

Anonymous. 1835. [Notice on Denison Olmsted's geological reports]. *Transactions of the Geological Society of Pennsylvania* 1: 173–174.

Anonymous. 1852. Professor Olmsted. *North Carolina University Magazine* 1(8): 339–343. ("From the Ohio Teacher.")

Anonymous. 1859. Death of Professor Olmsted—outline of his life. *New York Times* 8(2387): 4, Saturday, May 14, 1859.

Anonymous. 1999. The first state geologic map. *GeoTimes* 44(4): 11.

Battle, Kemp P. 1907–1912. *History of the University of North Carolina.* Printed by Edwards & Broughton, Raleigh. 2 vols. (Vol. 1, pp. 255–257, 299, 417)

Bearden, Elizabeth Steele. 1925. *The Southern Review.* Submitted in partial fulfillment of the requirements for the degree of Master of Arts in the Faculty of Philosophy, Columbia University.

Black, Joseph. 1806–1807. *Lectures on the Elements of Chemistry. Published from His Manuscripts by John Robison.* Printed for M. Carey by B. Graves, Philadelphia.

BOA. Board of Agriculture of North Carolina. 1823. *Proceedings of the Board of Agriculture of North-Carolina. At their session in 1823.* Published by Order of the Board [by] J. Gales & Son, State Printers, Raleigh. (pp. 5–6)

BOA. Board of Agriculture of North Carolina. 1825. *Proceedings of the Board of Agriculture of North-Carolina. At their session in 1824.* Published by Order of the Board [by] J. Gales & Son, State Printers, Raleigh. (p. 7)

BOT. Board of Trustees, University of North Carolina, Chapel Hill. *Minutes of the Board of Trustees.* (University Archives #40001, Wilson Special Collections Library, University of North Carolina, Chapel Hill) (Vol. 4, July 1811–December 1822; vol. 5, December 1823–December 1840)

B[oué], A. 1829. [Abstract of Denison Olmsted's geological reports]. *Bulletin des Sciences Naturelles et de Géologie* 17: 38–40.

Bowdoin. George J. Mitchell Department of Special Collections & Archives, Bowdoin College Library, Bowdoin College.

Bryan, James W. 1823. "From Professor Olmstead's [*sic*] Lectures, October 8th, 1823." Pp. [76]–[81]. (Volume 4, folder 595, Bryan Family Papers #96, Southern Historical Collection, Series 3.1, Wilson Special Collections Library, University of North Carolina, Chapel Hill)

Bryant, Bernard Lee, Jr. 1999. *Occupants and Structures of Franklin Street, Chapel Hill, North Carolina at 5-year Intervals, 1793–1998*. Chapel Hill Historical Society, Chapel Hill, NC. (pp. 21–22)

Burk, William R. 2011. Puckett house has distinguished history. *Chapel Hill News*, Section A, Wednesday, January 5, 2011, p. 5A.

Bursey, Maurice M. 1982. *Carolina Chemists; Sketches from Chapel Hill, Using the Records Prepared by Horace D. Crockford; With recollections by Arthur Roe*. Department of Chemistry, University of North Carolina, Chapel Hill. (pp. [3]–8)

Caldwell, Joseph. 1818. Letter of April 1, 1818 to Col. William Polk. (Box 2, folder 59, University Papers #40005, University Archives, Wilson Special Collections Library, University of North Carolina, Chapel Hill)

Chance, H.M. 1885. *Report on an Exploration of the Coalfields of North Carolina, Made for the State Board of Agriculture*. Published by Order of the Board. P.M. Hale, State Printer and Binder, Raleigh.

Cherry, Thomas Kevin. 1993. *Science at the University of North Carolina 1795–1853*. M.A. Thesis. Department of History, University of North Carolina, Chapel Hill.

Clark, T. Wyche. 1998. A history of early geologic research in the Deep River Triassic Basin, North Carolina. *Southeastern Geology* 38: 65–76. (caption of figure 2, p. 66)

Clark, T. Wyche. 1999. America's First State Geologic Map. (North Carolina Geological Survey Geologic Map 7). North Carolina Geological Survey, Raleigh. 1 poster (25 × 39 inches). (includes portrait of Olmsted)

Davy, Humphry, Sir. 1821. *Elements of Agricultural Chemistry in a Course of Lectures for the Board of Agriculture*. B. Warner, M. Cary & Son, and Bennett & Walton, Philadelphia.

DocSouth. Documenting the American South. Steward's Hall. <http://docsouth.unc.edu/getBio.phtml?type=place&id=name0001086&name=Steward's%20Hall> (Accessed September 4, 2006)

Elliott, Robert Neal, Jr. 1955. The *Raleigh Register 1799–1863*. (*The James Sprunt Studies in History and Political Science*, vol. 36). University of North Carolina Press, Chapel Hill.

[Elliott, Stephen, Sr.]. 1828. [Review]. Art. IX.—*Report on the Geology of North-Carolina, conducted under the direction of the Board of Agriculture*. By Denison Olmsted, Professor of Chemistry and Mineralogy in the University of North-Carolina. Part I. Nov. 1824.—Part II. Nov. 1825. Raleigh. Gales & Son. *Southern Review* 1: 235–261.

Fisher, George P. 1866. *Life of Benjamin Silliman, M.D., LL.D., Late Professor of Chemistry, Mineralogy, and Geology in Yale College. Chiefly from His Manuscript Reminiscences, Diaries, and Correspondence*. Sampson Low, Son, and Marston, London. (vol. 1, p. 300)

Gales, Joseph. 1819. [Circular], *The Agricultural Society of North-Carolina, June 1819*. (Box 4, folder 253, Hayes Collection: Johnston Family Series #324, Southern Historical Collection, Wilson Special Collections Library, University of North Carolina, Chapel Hill)

Goodrum, Matthew R. 1999. Olmsted, Denison, pp. 696–697 in John A. Garraty and Mark C. Carnes (general editors), *American National Biography*, vol. 16. Oxford University Press, New York.

Graham, William A. 1823–1824. "Notes on the Science of Chymistry. Taken from the Lectures of Denison Olmsted Esq. Professor in University of N. Car. in 1823–24." Notebook containing 100 pages of notes. (Volume 3, William A. Graham Papers #285, Southern Historical Collection, Wilson Special Collections Library, University of North Carolina, Chapel Hill)

Hampton, Jos. W. 1838. The Catawba Springs. *Western Carolinian* (Salisbury, NC) 19(1): [4], Friday, June 15, 1838.

Helme, R.H. [1823]. *An Address Delivered before the Agricultural Society of North-Carolina, at Raleigh, December 16, 1822*. Printed by J. Gales & Son, Raleigh. (p. 7). (Also published as "An Address," in the *Raleigh Register, and North-Carolina Gazette* [*Weekly*] 26(1256): [1–2], Friday, May 30, 1823 and in the *Star & North-Carolina State Gazette* (Raleigh) 14(24): [1–2], Friday, June 13, 1823.)

Henderson, Archibald. 1949. *The Campus of the First State University*. (*The University of North Carolina Sesquicentennial Publications*). University of North Carolina Press, Chapel Hill. (pp. 45–52)

[Hitchcock, Edward]. 1828. Notice of the *Report on the Geology of North Carolina*, conducted under the direction of the Board of Agriculture; by Denison Olmsted. *American Journal of Science and Arts* 14: 230–251. (p. 230)

Holland, Alma. 1934. Publications in the field of science from the University of North Carolina (1795–1934). *Journal of the Elisha Mitchell Scientific* Society 50: 303–415.

Holmes, [Gabriel]. 1822. [Communication] to the Honourable the General Assembly of the State of North-Carolina. *Journal of the House of Commons*, [North Carolina]. 1822: 103–110.

Holmes, J.A. 1889. Historical notes concerning the North Carolina Geological Survey. *Journal of the Elisha Mitchell Scientific Society* 6: 5–18.

Jeffreys, George W. 1822. On deep ploughing and the cultivation of Indian corn. *American Farmer* 3: 39–40.

Johnson, Markes E. 1985. Early survey work and the roots of geological education in the Carolinas. *Earth Sciences History* 4: 3–9.

Jones, Geo. Wm., Truman A. Post and Robert A. Stiles. 1859. Death of Professor Olmsted. *New Haven Morning News* (New Haven, CT), May 14, 1859.

Journals. 1824. *Journals of the Senate and House of Commons, of the General Assembly of the State of North-Carolina, at Its Session in 1823*. Printed by J. Gales & Son—State Printer, Raleigh. (Senate confirmation, p. 109; House of Commons confirmation, p. 230)

Kelley, Brooks Mather. 1974. *Yale, a History*. (*Yale Scene. University Series,* 3). Yale University Press, New Haven, CT. (p. 21)

Kingsley, William L. (editor). 1879. Denison Olmsted, pp. 232–249 and portrait in William L. Kingsley, *Yale College: A sketch of its history, with notices of its several departments, instructors, and benefactors together with some account of student life and amusements by various authors*, vol. 2. Henry Holt and Company, New York.

Laws. 1822. *The Laws of the University of North-Carolina as Revised in 1813*. Printed by Dennis Heartt, Hillsborough, NC. (pp. [23]–24)

Leggett, M.D. (compiler). 1874. *The United States Patent Office. Index of Patents Issued from the United States Patent Office from 1790 to 1873, Inclusive—Continued*, vol. 3. G[overnment] P[rinting] O[ffice], Washington, [DC]. (p. 1479: "Stove or furnace for burning anthracite coal: D. Olmsted/New Haven, Conn./Oct. 14, 1835.")

Link, Arthur Stanley. 1941. *A History of the Buildings at the University of North Carolina*. A thesis submitted to the Faculty of the University of North Carolina in partial fulfillment of the requirements for the degree of Bachelor of Arts, with Honors, in the Department of History, Chapel Hill. (pp. 11–12)

Little, M. Ruth. 2006. *The Town and Gown Architecture of Chapel Hill, North Carolina, 1795–1975*. The Preservation Society of Chapel Hill, Chapel Hill, NC. (Widow Puckett House, p. 151)

Lyman, C.S., Rev. 1859. Biographical sketch of Professor Denison Olmsted. *American Journal of Science and Arts* 78: 109–118.

Manly, Charles. 1822. "Trustees of the University of N. Carolina. Fr. Chas. Manly Treas: from 20th Nov. 1821 to 20th Nov. 1822." (Box 2, folder 76, University Papers #40005, University Archives, Wilson Special Collections Library, University of North Carolina, Chapel Hill) (entry of March 22, [1822])

Manly, Charles. 1824. "Trustees of the University [of] N. C. fr. Charles Manly Treasurer of the Board. Account current from 20th Nov. 1823 to 20th Nov. 1824." (Box 2, folder 88, University Papers #40005, University Archives, Wilson Special Collections Library, University of North Carolina, Chapel Hill) (entry of October 25, 1824)

Marcet, Mrs. (Jane Haldimand). 1818. *Conversations on Chemistry, in Which the Elements of That Science Are Familiarly Explained and Illustrated by Experiments: To which are added, notes and observations*. 4th English edition, revised, corrected, and considerably enlarged. Denio and Phelps, Greenfield, MA.

McGehee, Montford. 1885. *Report of the Board of Agriculture*. Ashe & Gatling, State Printers and Binders, Raleigh. (Issued as Document no. 16, Executive and Legislative Documents of the State of North Carolina. Session 1885; bound with other documents by P.M. Hale, State Printer and Binder, Raleigh)

McVaugh, Rogers, Michael R. McVaugh and Mary Ayers. 1996. *Chapel Hill and Elisha Mitchell the Botanist*. The Botanical Garden Foundation, Chapel Hill, NC. (pp. 65–66, 104)

Merrill, George P. 1906. Contributions to the History of American Geology. Government Printing Office, Washington, DC. ("No. 135. From the *Report of the United States National Museum for 1904*, pages 189–734 [*i.e.*, 733], with 37 plates.")

Merrill, George P. 1924. *The First One Hundred Years of American Geology*. Yale University Press, New Haven, CT. (p. 94)

Moore, Augustus. 1820. "Notes on a Course of Lectures delivered at the University of N. Carolina by Professor D. Olmstead [*sic*] began 13th September 1820 … ." Pp. 1–103, [1]. (Box 1, volume 4, Miscellaneous Students' Notebooks #3286.4, Southern Historical Collection, Wilson Special Collections Library, University of North Carolina, Chapel Hill)

Murray, John. 1817. *Elements of Chemistry*. 4th ed. Printed for F. Pillans, Edinburgh; and for Longman, Hurst, Rees, Orme, & Brown, London.

NC Laws. 1823. *The Laws of North-Carolina, Enacted in the Year 1822*. Printed at Bell & Lawrence, Raleigh. (p. 19)

NC Laws. 1824. *Acts Passed by the General Assembly of the State of North-Carolina, at Its Session, Commencing on the 17th of November, 1823*. Printed by J. Gales & Son—State Printers, Raleigh. (p. 17)

Niles, H. (editor). 1825. [Notice on Denison Olmsted's geological reports]. *Niles' Weekly Register* 29 (744): 247.

Olmsted, Denison. 1817a. Letter of March 11, 1817 to Elisha Mitchell. (Box 2, folder 55, University Papers #40005, University Archives, Wilson Special Collections Library, University of North Carolina, Chapel Hill). (This letter is probably wrongly dated by one year as various passages would indicate. The date should be March 11, 1818, according to Gary Lee Schoepflin [1981].)

Olmsted, Denison. 1817b. Letter of September 12, 1817 to J. Haywood. (Box 9, folder 81, Ernest Haywood Papers #1290, Southern Historical Collection, Wilson Special Collections Library, University of North Carolina, Chapel Hill)

Olmsted, Denison. 1818. Letter of November 7, 1818 to Benjamin Silliman. (Reel 4, Film 6463, Series II, *Silliman Family Papers, General Correspondence, 1796–1885*, Microforms Center, Memorial Library, University of Wisconsin, Madison)

Olmsted, Denison. 1819a. Letter of April 18, 1819 to [James L.] Kingsley. (Denison Olmsted Papers, Group Number 604, Box no. 98, Manuscripts and Archives, Yale University, New Haven, CT)

Olmsted, Denison. 1819b. Letter of June 7, 1819 to Board of Trustees of the University of North Carolina. (Box 2, folder 63, University Papers #40005, University Archives, Wilson Special Collections Library, University of North Carolina, Chapel Hill)

Olmsted, Denison. 1819c. *Outlines of the Lectures on Chemistry, Mineralogy, & Geology, Delivered at the University of North-Carolina, for the Use of the Students*. Printed by J. Gales, Raleigh.

Olmsted, Denison. 1820a. Letter of December 22, 1820 to Robert Williams, Treasurer of the University of North Carolina. (Box 2, folder 68, University Papers #40005, University Archives, Wilson Special Collections Library, University of North Carolina, Chapel Hill)

Olmsted, Denison. 1820b. Letter of August 1820 to Benjamin Silliman. (Reel 4, Film 6463, Series II, *Silliman Family Papers, General Correspondence, 1796–1885*, Microforms Center, Memorial Library, University of Wisconsin, Madison)

Olmsted, Denison. 1821. Communication from Professor Olmsted to the honorable Board of Internal Improvements, pp. 65–67 in *Annual Report of the Board of Public Improvements of North-Carolina to the General Assembly, November 26, 1821*. . . . Printed by J. Gales, Raleigh.

Olmsted, Denison. 1822. Letter of May 5, 1822 to Benjamin Silliman. (Reel 4, Film 6463, Series II, *Silliman Family Papers, General Correspondence, 1796–1885*, Microforms Center, Memorial Library, University of Wisconsin, Madison)

Olmsted, Denison. 1824a. Letter of November 22, 1824 to Ch. Manly. (Box 2, folder 88, University Papers #40005, University Archives, Wilson Special Collections Library, University of North Carolina, Chapel Hill)

Olmsted, Denison. 1824b. Notice of illuminating gas from cotton seed. *American Journal of Science and Arts* 8: 294–298.

Olmsted, Denison. 1824c. The utility of thought, and the necessity of investigation, to the progress of agricultural improvements. *American Farmer* 6: 318.

Olmsted, Denison. 1824d. For the *Register*. *Raleigh Register, and North-Carolina State Gazette* [Semi-Weekly] 1(53): [3], Tuesday, May 18, 1824.

Olmsted, Denison. 1824e. Gas lights. *Raleigh Register, and North-Carolina State Gazette* [Semi-Weekly] 1(52): [3], Friday, May 14, 1824.

Olmsted, Denison. 1824–1841. "Journal" (includes lectures on Chemistry delivered at the University of North Carolina in 1824). (Denison Olmsted Papers, #MS583, Manuscripts and Archives, Yale University Library, New Haven, CT)

Olmsted, Denison. 1825a. Letter of October 9, 1825 to Governor Burton. (Box 3, folder 94, University Papers #40005, University Archives, Wilson Special Collections Library, University of North Carolina, Chapel Hill)

Olmsted, Denison. 1825b. Letter of December 8, 1825 to Governor Burton. (Box 3, folder 94, University Papers #40005, University Archives, Wilson Special Collections Library, University of North Carolina, Chapel Hill)

Olmsted, Denison. 1825c. Letter of December 8, 1825 to Governor Burton. (Box 3, folder 94, University Papers #40005, University Archives, Wilson Special Collections Library, University of North Carolina, Chapel Hill). (Second, of two letters, dated December 8, 1825, from Olmsted to Burton)

Olmsted, Denison. 1825d. On the gold mines of North Carolina. *American Journal of Science and Arts* 9: 5–15.

Olmsted, Denison. 1825e. *Report on the Geology of North-Carolina, Conducted under the Direction of the Board of Agriculture*. Part I. Pp. [1]–44 (pp. 43–44 comprise an index to the *Report*. At end of text is date, November 10, 1824.) Special note: The Board of Agriculture of North Carolina published Olmsted's report, along with four other papers, in a volume entitled, *Papers on Agricultural Subjects and Professor Olmsted's Report on the Geology of Part of the Western Counties of North-Carolina*. It was printed in 1825 by J. Gales & Son, Raleigh.

Olmsted, Denison. 1825f. Remarks on the Preparation of Mortar. Addressed to the Agricultural Society of North-Carolina. Read in Society and ordered to be printed, December 14, 1821. Reprinted by order of the Board of Agriculture June, 1825, [J. Gales & Son, Raleigh.] Pp. [59]–72. (Published in *Papers on Agricultural Subjects*; *cf.* previous citation)

Olmsted, Denison. 1827a. Letter of December 11, 1827 to Parker Cleaveland. (Special Collections and Archives, Bowdoin College, Brunswick, ME)

Olmsted, Denison. 1827b. *Report on the Geology of North-Carolina, Conducted under the Direction of the Board of Agriculture*. Report II. Pp. [85]–142. ("*Report, &c.*," pp. [87]–104; "Observations on the Geology of the Western Section of the State, continued from Part I," pp. 105–[141]); "Index to Part II," p. [142]. Report is dated November, 1825, *cf.* title page.) Special note: The Board of Agriculture of North Carolina published Olmsted's second report, along with one other paper as [No. 2], *Papers on Agricultural Subjects and Professor Olmsted's Report on the Geology of North-Carolina*, Part II. It was printed by J. Gales & Son, Raleigh, in 1827, with pages [85]–155, which continued the paging of the first volume of *Papers on Agricultural Subjects*.

Olmsted, Denison. 1828. Letter of January 24,1828 to Parker Cleaveland. (Special Collections and Archives, Bowdoin College, Brunswick, ME)

Olmsted, Denison. 1831. *An Introduction to Natural Philosophy: Designed as a text-book for the use of students in Yale College*. Published by Hezekiah Howe, New Haven, CT. [At least 33 additional printings (including new editions) were issued, 1832–1895.]

Olmsted, Denison. 1833. *A Compendium of Natural Philosophy: Adapted to the use of the general reader and of schools and academies*. Babcock, New Haven, CT. [At least 22 additional printings were issued, 1837–1870.]

Olmsted, Denison. 1838. "Lectures on Astronomy by Professor Olmsted. Yale College. Oct., 1838." (Polk Papers, #75.10, State Archives of North Carolina)

Olmsted, Denison. 1839a. *A Compendium of Astronomy: Containing the elements of the science, familiarity explained and illustrated, with the latest discoveries. Adapted to the use of schools and academies, and to the general reader*. Collins, Keese & Co., New York. [At least 22 additional printings were issued, 1840–1869.]

Olmsted, Denison. 1839b. *Introduction to Astronomy: Designed as a text-book for the students of Yale College*. Collins, Keese & Co., New York. [At least 28 additional printings (including new editions) were issued, 1841–1890.]

Olmsted, Denison. 1844. *Rudiments of Natural Philosophy and Astronomy: Designed for the younger classes in academies, and for common schools*. S. Babcock, New Haven, CT. [At least 11 additional printings (including new editions) were issued, 1845–1864.]

Olmsted, Denison. 1846. Letter of December 11, 1846 to Governor William A. Graham. (William A. Graham Papers, #61.3, State Archives of North Carolina)

Olmsted, Denison. 1855. On the sandstone and coal of North Carolina of the age of the Richmond coal basins. *American Journal of Science and Arts*, 2nd series 19:430–433. (Republished from Olmsted's 1824 [*i.e.*, 1825] *Report on the Geology of North-Carolina)*

Olmsted, Henry King and Rev. Geo. K. Ward. 1912. *Genealogy of the Olmsted Family in America: Embracing the descendants of James and Richard Olmsted and covering a period of nearly three centuries 1632–1912*. A.T. De La Mare Printing and Publishing Company, New York.

Parker, John M., III. 1991. Olmsted, Denison, pp. 395–396 in William S. Powell (editor), *Dictionary of North Carolina Biography*, vol. 4. University of North Carolina, Chapel Hill.

Parkes, Samuel. 1818. *The Chemical Catechism, with Notes, Illustrations, and Experiments*. Printed and Published by Collins and Co., New York.

Polk, Lucius J. 1821. "Notes on the Lectures of Chemistry Delivered by Professor Olmstead [*sic*] at the University of North Carolina, September 18th, 1821." Pp. [31]–[88]. (Volume 7, folder 110, Polk and Yeatman Family Papers #606, Southern Historical Collection, Subseries 4.1, Wilson Special Collections Library, University of North Carolina, Chapel Hill)

Polk, William. 1802. [Circular for the Parents of Students] Including an Abstract of the Laws of the University. Raleigh, December 15, 1802. (Reel #1, Leonidas Polk Papers #M-2965, Southern Historical Collection, University of North Carolina, Chapel Hill)

Powell, William S. 1972. *The First State University: A pictorial history of the University of North Carolina*. University of North Carolina Press, Chapel Hill. (Steward's Hall, p. 15)

Raleigh, Walter [pen name]. 1823. Agriculture of North-Carolina. To Charles Fisher, Esq. Secretary of the Rowan Agricultural Society. (A series of eight letters published in the *Raleigh Register* of 1823. Letter I. 26(1226): [3], Friday, March 21; II. 26(1227): [3], Friday, March 28; III. 26(1228): [3], Friday, April 4; IV. 26(1229): [3], Friday, April 11; V. 26(1230): [3], Friday, April 18; VI. 26(1231): [3], Friday, April 25; VII. 26(1232): [3], Friday, May 2; VIII. 26(1233): [3], Friday, May 9.)

Rembert, David. 2004. Electronic mail of February 16, 2004 to William R. Burk.

Royster, James F., J.G.deR. Hamilton and L.R. Wilson (Faculty Committee on Publication of the *Record*). 1910. Scientific investigation by the faculty of the University of North Carolina previous to the restoration of the university in 1875, pp. 1–60. (*The University Record*, vol. 86). University Press, Chapel Hill, NC. (Scientific work of Elisha Mitchell, pp. 8–10)

Russell, Lucy Phillips. 1931. An old house; history of a celebrated dwelling on East Franklin Street. *Chapel Hill Weekly* (Chapel Hill, NC) 9(10): 6–7, Friday, May 8, 1931.

SANC. State Archives of North Carolina, Raleigh.

Schoepflin, Gary Lee. 1981. *Denison Olmsted (1791–1859): Scientist, teacher, Christian: A biographical study of the connection of science with religion in antebellum America*. Ph.D. (Oregon State University, 1977). Facsimile by microfilm xerography. University Microfilms International, Ann Arbor, MI.

Shand, Peter J. 1869. Letter of July 26, 1869 to Rev. Whitefoord Smith. (Archives, Charleston Library Society, Charleston, SC)

Sherrill, William L. 1937. *Annals of Lincoln County, North Carolina. Containing Interesting and Authentic Facts of Lincoln County History through the Years 1749 to 1937.* The Observer Printing House, Inc., Charlotte, NC. (p. 103)

Smith, Michael S. 1999. The conflict between "practical utility" and geology: Denison Olmsted, Elisha Mitchell and the 1823 to 1828 geologic surveys of North Carolina. *Southeastern Geology* 38: 145–154. (p. 153)

Smith & Sherman. 1836. Olmsted's Patent Stove Manufactured by Smith & Sherman, New Haven, CT. Promotional brochure. (The Connecticut Historical Society, New Haven, Connecticut)

Stuckey, Jasper Leonidas. 1965. *North Carolina: Its geology and mineral resources.* Department of Conservation and Development, Raleigh. (pp. 224–229)

Swain, David L. 1822. Letter of April 24, 1822 to George Swain. (Epistolary Correspondence of David L. Swain, vol. 5, Needham B. Cobb Loan, North Carolina Collection, Wilson Special Collections Library, University of North Carolina, Chapel Hill) (Volume spine reads: Epistolary Correspondence, David L. Swain, no. 2, 1819–1831)

UNC. 1819a. [University of North Carolina]. [Broadside]. *Catalogue of the Faculty and Students of the University of North Carolina. October, 1819. Chapel Hill.* (North Carolina Collection, Wilson Special Collections Library, University of North Carolina, Chapel Hill)

UNC. 1819b. University of North-Carolina. [Broadside]. *Members of the Faculty. System of Studies Preparatory to Admission into the University of North Carolina. Plan of Education in the University. June 28, 1819.* (North Carolina Collection, Wilson Special Collections Library, University of North Carolina, Chapel Hill)

UNC. 1821. [Broadside]. *Scheme of the Exercises at the Public Commencement of the University of North-Carolina, June 7th, 1821.* J. Gales, Printer, Raleigh. (Folder 1821, Announcements, bulletins, etc., North Carolina Collection, Wilson Special Collections Library, University of North Carolina, Chapel Hill)

UNC-B. University of North Carolina, Business and Finance: Finance Division, Series 1: Historical Financial Book, vol. S-5 (1789–1859). (University Archives #40095, Wilson Special Collections Library, University of North Carolina, Chapel Hill) The following entries are cited in the text: June 30, 1818–p. 184; November 23, 1818–p. 188 (D. Olmsted), May 14, 1819–p. 188 (W. Barbee); January 21, 1820–p. 192; December 6, 1821–p. 200; March 22, 1822–p. 200; September 23, 1823–p. 217; November 2, 1824–p. 222; December 14, 1825–p. 230; and June 25, 1827–p. 242.

UNC-P. University of North Carolina, University Papers #40005, University Archives, Wilson Special Collections Library, University of North Carolina at Chapel Hill.

UNC Papers. [Indenture between Denison Olmsted and the University of North Carolina, made September 21, 1822]. (Box 2, folder 76, University Papers #40005, University Archives, Wilson Special Collections Library, University of North Carolina, Chapel Hill)

Vaudreuil, Elaine. 1992. "The Widow Puckett House: An Inventory Prepared by Elaine Vaudreuil, for PLAN 201, Historic Preservation Planning, Myrick Howard, Fall 1992." [no statement of publication or publisher]. [Original copy filed in the State Historic Preservation Office, North Carolina Department of Cultural Resources; copy in author's files]

Vickers, James. 1985. *Chapel Hill, An Illustrated History*. James Vickers (text), Thomas Scism (illustrations), and Dixon Qualls (color photography). Barclay Publishers, Chapel Hill. (p. 23)

W[heeler], J.H. 1859. The late Professor Olmsted. *Boston Evening Transcript* 30(8911): [1], Monday, May 16, 1859.

Wheeler, J.H. 1884. *Reminiscences and Memoirs of North Carolina and Eminent North Carolinians*. Columbus Printing Works, Columbus, OH. (p. 126)

Wilkins, D. 1820. Letter of April 2, 1820 to Thomas W. Jones. (Folder 2, Thomas Williamson Jones Papers #3684-z, Southern Historical Collection, Wilson Special Collections Library, University of North Carolina, Chapel Hill)

Williams, Robert. 1818. "Robert Williams, Treasurer of the University of North Carolina in account with the Board of Trustees &c: from the 20th of November AD 1817, unto the 20th of November AD 1818." (Box 2, folder 61, University Papers #40005, University Archives, Wilson Special Collections Library, University of North Carolina, Chapel Hill) (entry of June 30, 1818)

Williams, Robert. 1819. "Robert Williams, Treasurer of the University of North Carolina in account with the Board of Trustees &c: from the 20th November AD 1818, unto the 20th of November AD 1819." (Box 2, folder 64, University Papers #40005, University Archives, Wilson Special Collections Library, University of North Carolina, Chapel Hill) (entry "a", November 23, 1818; entry "b", May 14, 1819)

Williams, Robert. 1820. "Robert Williams, Treasurer of the University of North Carolina, in account with the Board of Trustees &c: from the 20th November AD 1819; unto the 20th November AD 1820." (Box 2, folder 67, University Papers #40005, University Archives, Wilson Special Collections Library, University of North Carolina, Chapel Hill) (entry of January 21, 1820)

Woolsey, T.D. 1859. Discourse commemorative of Professor Denison Olmsted, LL.D. *New Englander* 17(67): [575]–600.

Yale. Yale University Library, Manuscripts and Archives, New Haven, CT.

Yale-MS. 1797–1819. *Journal of the Meetings, Proceedings, and Transactions of the Moral Society* (Apr. 6 1797–July 22, 1819). (Records of the Moral Society, Yale College, Record unit 56, YRG 40, Box 58, Yale University Library, Manuscripts and Archives) (pp. 134, 156–158)

Yale-S1. 1813. [Broadside]. *Scheme of the Exercices at the Public Commencement of Yale-College, September 1st, 1813*. O. Steele, Printer, New-Haven. (Yale College Orations, Exams, etc., Yale University Library, Manuscripts and Archives)

Yale-S2. 1816. [Broadside]. *Scheme of the Exercices at the Public Commencement of Yale-College, September 11, 1816*. (Yale College Orations, Exams, etc., Yale University Library, Manuscripts and Archives)

APPENDIX 2.1.

TRANSCRIPTION OF SELECTED BOTANICAL NOTES FROM LECTURES OF DENISON OLMSTED'S CHEMISTRY COURSE TAKEN BY AUGUSTUS MOORE[22] IN 1820.

Vegetable Chemistry

Organised boddies are such as have life & consequently organs & vessels to support that life. Unorganised boddies have not life. It is the province of the botanist to ascertain the existence of the plants, to name them & to arrange them systematically; of the Physiologist to attend to the nature of their growth & vegetation, he considers them only while they are alive; of the Chemist to examine all the successive changes which it undergoes after it is dead & while it is decaying, & to find out what principles it receives from the earth, water & atmosphere, which promotes its growth, [& to resolve it into its proximate principles].

Organization of a plant is as follows: 1) root, 2) trunk, 3) leaves, 4) flowers, 5) fruit & seed. The office of the roots is twofold, to support the tree in its natural position & to afford nutriment for the growth of the tree. The root is nothing but the trunk extended in various directions under ground, by the roots grapling hold of the soil they prevent the tree's falling down & absorbing various nutritious substances from the earth, they help to support vegetation. If plants be abstracted from the ground & their branches be placed in the ground & their roots in the air, the plants will still continue to grow; their roots becoming soon covered with foliage & their branches turned into roots. <u>The Trunk</u> consists of two parts 1) the bark divided into the epidermis, parenchyma, & cortical layers. The office of the epidermis is to serve as a coat of mail, to shield the plant from external injury & sometimes it serves to support the upright position of the plant, in which case, its texture is generally silicious, as in Hay, Flags & bull rushes. The Parenchyma is the green skin just under the epidermis. This is the storehouse of provisions for the plant; through it the juices for the nourishment of the plant descend after having passed up the tree in the alburnum. The cortical layers are situated next to the wood of the tree; one of their layers is converted annually into the wood of the tree; by this means, the age of many trees can be ascertained with considerable accuracy. 2) <u>Wood</u>, this is the second part of the trunk, & is divided into the Alburnum & the heart wood, the vital functions of the plant are carried on in the alburnum, through this, the <u>common vessels</u> ascending from the roots carry the sap up the tree, which during its passage is elaborated. In the heart wood, the vital functions have ceased, & it is in fact dead–if the tree be in a very flourishing condition-then will be but little heart wood, but if it is pining away, then is much of it.–The branches of a tree are in every respect, composed of the same substances as the Trunk. <u>The Leaves</u>: As the roots were merely the trunk of the [tree] extended and divided, so the leaves are only the bark extended on a large surface; this is done for the production of chemical changes which are essential to vegetation. The epidermis, parenchyma & cortical layers can all be traced on the leaves. The leaves decompose carbonic acid and give out oxygen in the light, but during the absence of light, oxygen is consumed & carbonic acid formed (on this subject there is a great diversity of sentiment; see Dr. Murray vol. 2nd page 298). On account of the plants absorbing carbon, they serve to keep the atmosphere pure & its elements in [word illegible] proportion; as the absorption of Carbon is necessary for the growth of plants & this takes place only in the Light,—it—follows that light is essential to perfect vegetation. As there is a great preponderance of sunshine & other light during the season of vegitation [*sic*]—more carbon is absorbed by plants than is given out. The foregoing parts relate to the support & growth of plants, but the following. <u>The Flowers & The Seeds</u> to the reproduction of a plant

[22] Moore, Augustus. 1820. [Student Notebook]. [The Author, Chapel Hill]. 1 volume. (Transcription covers pages 87–102; Southern Historical Collection #3286.4, Wilson Special Collections Library, University of North Carolina, Chapel Hill).

Appendix 2.1. (Continued)

of the same species. The Sexual System of Linnaeus was derived from the adaptation of the stamina to the Pistils. It is necessary that a communication should take place between the stamenes & Pistil for a reproduction of the Plants: In some plants both the stamen & pistil are found; and again the stamenes are found in one, & the Pistil in another. The stamen is the male & the Pistil the Female: Linneus derived the order of his System from the number of Pistils which a Plant has; those having only one Pistil belonging to the first order, which he called Mononigenia [*i.e.*, monogynia]; & the various classes under each order he formed according to the number of stamena; which Plants have; those having only one stamen, being of 1st class, monandria. Thus a Plant of only one stamen & one pistil would be called Monandria-monagenia [*i.e.*, monogynia] &c. Most astonishing care & provision is manifested in the preservation of seeds, & the sort of care is wonderfully adapted to the particular exigencies of each Plant. Sometimes the seed is covered in a hard shell as Hickrinuts, Wallnuts; sometimes in a hard shell, covered over with a pulpy substance, as Peaches; sometimes scarcely any covering as Indian corn, wheat.

Vegetable Principles are divided into Ultimate & Proximate; the Ultimate Principles are the simple Elements of which all vegetables are composed; i.e. carbon, oxygen & hydrogen, or expressed scientifically Carbon & the Elements of Water. The Proximate Principles are the products of vegetation, or the ultimate principles variously combined—of the Prox. Prin., Dr. Murray enumerates 35 divided into 15 species at page 319, vol. 2nd. Mr. Olmsted, however lectures on only 8. When we speak of the decomposition of a Plant, we mean a resolution of it into its Proximate Principles; the most important of which are 1. Acids, 2. Sugar, 3. Starch, 4. Gluten, 5. Gums & Resins, 6. Oils, 7. Tannin & 8. Colouring matter.

1. Vegetable Acids; the composition of these acids, is Carbon & the elements of water; Carbon & Hydrogen forming a compound base of which oxygen is the acidifyer: the greatest difference among the acids, is produced by the different proportions in which their elements are united; hence they can be easily converted into each other, by altering the natural proportion of their elements. The chief Veg[etable] Ac[ids] are 1st Oxalic, 2nd Tartaric, 3rd Malic, 4th Citric, & 5th Acetic. (note Dr. Murray describes 10 of them). Oxalic acid (Murray 2. vol., page 376) contains more oxygen than any other vegetable acid, having 70 per cent. When more oxygen is added to this acid, Carbonic acid is formed by the union of carbon & oxygen, & water is also formed by the union of the hydrogen & oxygen. This acid exists in the juice of the leaves of the sorrel, combined with a portion of Potash, not sufficient to neutralise it; hence it is called salt of sorrel. This acid can be prepared from sugar in the following manner & this is the most usual way of forming it—one part of refined sugar in the form of powder is put with 3 parts of nitrous acid into a retort; a strong mutual action is exerted & nitrous acid vapours are disengaged; when this has ceased, more acid is added & a moderate [heat] applyed; the liquor on cooling affords a large quantity of oxalic acid crystallised in slender prisms. The acid in this case is formed not only from the oxygenation of the sugar by the nitric acid, but also from more complicated changes of composition not well understood. Oxalic acid has a great affinity for lime, hence it is a test for that substance; & in fact it is the best test known, forming with it a most insoluable precipitate. Oxalic acid is also used for removing stains, as of ink &c.; it dissolves the staining matter, which is insoluable in water, or, common fluids.

Tartaric Acid. Murray 2. vol., 380. This contains nearly as much oxygen as the last, having 63 percent, it is remarkably sour, & it is this acid which is so prevelent [*sic*] in grapes & Tamarinds & other like fruits from this source (of grapes) it is frequently found deposited in wine casks, particularly in those where there is much vegitable [*sic*] matter; it is generally united with Potash forming a Tartrate of Potash, which is commonly called Tartar; in this state is the crude Tartar of the Shops. It is purifyed by boiling it in water with the addition of a small quantity of fine clay, which attracts the colouring; by evaporation it is obtained

Appendix 2.1. (Continued)

crystallised forming the purifyed Tartar, or Cream of Tartar of the Shops; this substance is the salts of Tartar. Tartaric acid is concrete, & may be made into powders; hence it is used with sub-carbonate of soda, to form soda powders.

Malic & Citric Acids. Malic acid exists—sour apples & unripe fruit generally; it is citric acid which exists in Lemons, both these acids are united in Strawberries, Currents [*sic*] & other fruits; & it is this, which imparts to them that delicious flavour, which is so refreshing in a hot sultry day. Malic acid is distinguished from the other vegitable acids, by its not being concrete, it cannot be crystallized. The common way of preparing citric acid, (& this way will serve for most other vegitable acids) is as follows, Dr. Murray, 2 vol. 375. The expressed juice of the lemon after filtration, is neutralized by carbonate of lime, a citrate of lime is formed, (the carbonic acid being disengaged from the Lime, escapes) which is insoluable in water; it is washed to carry off any remaining mucilage or other impurity, & is then decomposed by adding half of its weight of sulphuric acid, diluted with six parts of water, the action being promoted by a gentle heat; the sulphuric acid unites with the lime forming a sulphate of lime, which being insoluable, is precipitated; the citric acid remains dissolved & by evaporation may be obtained in crystalline form.

Acetic Acid; acid of vinegar, more peculiarly the product of Fermentation; to which it is referred. 2nd Proximate Principle Sugar. This substance is from cane, maple, Beets, sweet potatoes & other fruits as Water Mellons [*sic*] &c. It is manufactured in the West Indies in the following manner—Murray 2. 327. The juice of the cane is expressed by pressing the cane between two cylinders of iron & boiled, a quantity of Lime being added, the operation of which is to neutralize a portion of oxalic in the juice, which would prevent the concretion of sugar, & which also makes a thick scum rise to the surface; the juice after the proper decretion is allowed to remain at rest till it becomes clear, when it is let off below & boiled down in another vessel, (or, suffered to evaporate) when on cooling the sugar separates in a concrete state; a sweet liquor remaining, molasses, which is drained off through holes perforated in the bottom of the vessel. This is brown sugar, but to make loose sugar as the process is thus continued. The Sugar is dissolved in water & this solution is boiled in lime water with the addition of a quantity of blood; the lime neutralizing any acid which may still be present; & the blood coagulating & the coagulum entangling the impurities. The liquor after being sufficiently concentrated is drawn off & run into inverted conical moulds of clay, where it becomes concrete; the base of the inverted cone is then covered with clay, which is moistened with water & this trickling slowly through the mass removes any saccharine matter according to Dr. Murray and according to Conver. Chem. 2nd page 121 it removes all the colouring matter present. When sugar is exposed to heat, it is blackened, swells up & burns to a coal, the products of the combustion are carbonic acid & water—the action of nitric acid on sugar, is that it affords the sugar oxygen, & thereby the oxalic acid is formed. (see page 92). Sugar is made in Switzerland from milk- Honey is another vegitable product, similar in manner [*i.e.*, many] respects to sugar; altogether except its fluidity & viscosity. Sugar is perhaps the most nutritious, perhaps of vegitable substances; it is also antiputrescent; meat can be kept sound by curing it = as sweet meats-as fruits as preserves & West India fruit.

3. Principle, Starch, or, Fecula. Murray 2. vol., p. 323. This makes up a great part of rice & all esculent seeds—it exists in Potatoes, Wheat, Rye—it is thus manufactured from wheat—the grain being cleaned is bruised & suffered to remain in water for some time; the starch is then pressed out of the swollen wheat—it is diffused through the water & subsides when it is dryed.—It differs in composition from sugar in having [notes not finished for this lecture].

4. Principal Gluten-Flour is made up of Starch & Gluten & they can be separated from each other by washing, because the starch is dissolved & the gluten remains in a viscous elastic state-of all kinds of flour, wheat flour contains the most gluten, it amounting to ½

Appendix 2.1. (Continued)

part of the flour in a dry state, but in a humid state it is from 1/3 to 1/5 of the weight. It is gluten which enables flour to ferment, & as wheat flour contains a great quantity of gluten, it makes the best light-bread-& the reason why rice is unsuitable for bread is because it contains scarcely any Gluten.

5. Principle Gums & Resins. The greatest distinction between a gum & resin is that a Gum is soluable in water, but, not in Alkohol & a Resin is insoluable in water, but, soluable in Alkohol. A gum may be precipitated from its solution in water, by pouring in Alkohol, & a resin may be precipitated from it–solution in Alkohol by water. Gums exude from certain trees as the Peach tree, Plum tree, Percimon [*sic*] tree &—sometimes the Gums become so abundant in Trees as to injure the fruit, in this case, it may be got out, cutting slits in the Tree longitudinally—Resins also exude from some plants. Gum arabic is very nutritional (Con. on Chem. 2. pg. 119) it forms the chief nourishment of the natives of those parts (where it is obtained) who procure it in great quantities, by incisions which they make in the Tree. Caravans passing through the great African desert have been supported on this gum for weeks. Ships' companies have been saved from starving by this example of a prize of the *Snap Dragon* during the late war [*i.e.*, Spanish American War], coming to New Bern. It is also frequently taken medicinally & is very nourishing. A solution of gums for paste, is much better, & nicer than the stuff made of wheat flour.

[6.] Resins are of various kinds, including turpentine, copal, rosin. The manufacture of the Spirits of turpentine is done by distilling turpentine. Rosin purified becomes Resin-a varnish is some Resin dissolved either in oil, or, Alkohol. Lac is a resinous substance containing a portion of extract, vegitable gluten & a matter analagous to wax. Sealing wax is lac coloured with the red oxide of mercury.

6. [*i.e.*, 7]. Principle Vegitable Oils. Murray 2. 341 page. All oils are divided into two sorts, fixed & volatile oils. The difference between a fixed and volatile oil is that the former is decomposed by heat-& the later [*i.e.*, latter] is volatized by it without decomposition. Fixed oils are generally contained in the seeds & fruits of those vegitables of which they are the products, they are obtained by expression-hence sometimes called expressed oils. The purest of these oils are those expressed from the fruit of the olive, or seed of the almond-others less pure are extracted from linseed, hempseed & the seeds of other plants. Exported to a warm atmosphere, expressed or, fixed oils, gradually acquire a sharp taste & smell & become thick—this change, named Rancidity, is caused by the absorption of oxygen—hence it takes place sooner when the oil is exposed to oxygen gas. Frying oils, as those expressed with the aid of heat are called, do not become rancid; but by absorbing oxygen, are rather converted into a resinous kind of mater [*i.e.*, matter]—by these drying oils, the absorption of oxygen, under favourable circumstances, is so rapid as to produce spontaneous combustion. Castor oil is made from Palma Christie seed & is used as a cathartic-*i.e.*- Volatile, or, Essential oil, is contained in the fruits, flowers, leaves, wood, or, bark of many vegitables—it exists sometimes in distinct vesicles & then may be obtained by expression but the usual method is to subject the vegitable [*sic*] matter containing it, to distillation with water, the oil is volatized with aqueous paper & is easily condensed—being usually lighter, it rises to the surface. The reason why certain plants possess such fragrancy is the quantity of volatile oil which they contain. These oils are sometimes prepared in Families, I believe, by merely soaking the Plants in water & distilling it—the water comes over impregnated with the oil. Insects of every kind have a great aversion to these oils, so that they are employed with success in Museums for the preservation of stuffed birds & other species of animals; they are also of the utmost importance in protecting the flowers, fruits & other parts of the plant in which they reside; Hence Cedar is not ever worm eaten & it is said that some gate posts of Constantinople of this wood lasted 1100 years & were perfectly sound when destroyed.

APPENDIX 2.1. (CONTINUED)

These oils are also used to remove grease spots; this, they do, because they dissolve the substance as causing the spot. Volatile oils are much more soluable in Alkohol than in water, some of them unite with it in every proportion, others in limited quantities. Cordials are prepared by dissolving some volatile oil (or the part of the plant which contains it) in some intoxicating spirits & then distilling it—the product will partake both of the delightful fragrancy of the oil & of the strength of the spirits.

7. [*i.e.*, 8.]. Principle Tannin. This is the astringent principle in certain vegitable substances as in oak bark, cherry-gall nutts. Tannin strongly unites with gelatine (Murray 2. 366) & forms an insoluable compound, which on drying becomes hard & brittle. On the formation of combination depends the art of Tanning, for leather is composed of Tannin & Gelatine, it differs from the skin of an animal, in not being soluable in water, not liable to putrefaction, not permeable to water & not subject to the spontaneous change, which the skin would soon undergo. The process of tanning is this, the skin of an animal when freed from the hair, epidermis & cellular fibre, (which is done in the mannipulations [*sic*] of the art, principally by the action of lime), consists principally of Gelatine. By immersion in the tan-liquor (which is an infusion of some bark containing Tannin) the combination of the Tannin with the organized gelatine, which forms the animal fibre, is slowly established, & the compound consisting of Tannin & Gelatine, is Leather. To render the leather of equal consistence throughout the action of the Tan liquor must be gradual—hence the Tanning is performed by successive immersions of the Skin in liquors of different strength—a portion of the extractive matter of the infusion appears also to enter into the combination—the skin increases in weight from 1/5 to 1/3.

Some gentleman in Connecticut has discovered that the whole of the chesnut tree contains Tannin & that the whole of it can be applyed to the process of Tanning—this discovery promises to be of much importance in that art; and by it—all of the tree can now be used, whereas formerly only the bark employed. Murray 2. 364. Tannin exerts a strong action on the metalic oxides & particularly on the oxide of iron.

Elisha Mitchell

Chapter 3.

Elisha Mitchell (1793–1857)

Professor of Mathematics (1818–1825)
Professor of Chemistry, Mineralogy, and Geology (1826/1827–1856/1857)

Early Life, Education, and University Appointment

Born August 19, 1793 in Washington, Connecticut, Elisha Mitchell was the eldest son of Abner and Phoebe (Eliot) Mitchell. His father was a farmer and his mother was a descendant of John Eliot, noted "Apostle to the Indians." As a child, Elisha Mitchell was an avid reader and sought intellectual enrichment. In preparing for college, he received training from Reverend Azel Backus, D.D., who ran a classical school in Bethlehem, Connecticut. Backus was a man of genial disposition, good common sense, keen wit, noteworthy attainments in science, and deference to the will of God (Phillips, 1858a). This educational experience had a lasting effect on Mitchell and no doubt helped form his character as an adult and educator.

Mitchell entered Yale College in 1809 and earned a Bachelor of Arts degree with distinction in 1813. At the Yale commencement exercises held on September 1, 1813, he gave an oration "On Superstition" in the morning and participated in a dialogue with several fellow students on the topic "The Intriguing Politician" in the afternoon (Yale-S1). He next became a teacher at the Union Hall Academy, a school for boys, in Jamaica, Long Island, New York. The principal of the school was Lewis E.A. Eigenbrodt, a noted educator. After nearly a year of teaching, Mitchell returned to Yale, where he became a graduate student and served as a tutor (Yale-College, 1814). Beginning around March 1, 1815, he became director of the Union Academy, a school for girls, in New London, Connecticut. On April 22, 1815, Mitchell (1813–1816) noted in his diary that his students were "composed of young ladies from the first families in the city" and had "almost universally fine dispositions and good minds and [had] given him very little trouble." (One of his students, Maria Sybil North, would become his wife.) In 1816 he was appointed a tutor at Yale and also received his M.A. degree. At the September 1816 commencement exercises he delivered an oration "On Some of the Probable Effects of the Downfall of the Turkish Empire" (Yale-S2).

In 1817 the professorship in mathematics at the University of North Carolina opened up. Joseph Caldwell, who had held the post, was selected as president of the institution (for the second time) because Roland Chapman had resigned the presidency. The Reverend Sereno

Frontispiece Credits: (above) Elisha Mitchell, 1851; portrait painting by Nathaniel Jocelyn; owned by The Dialectic and Philanthropic Societies of the University of North Carolina, Chapel Hill. (Courtesy of North Carolina Collection Photographic Archives, Wilson Special Collections Library, University of North Carolina, Chapel Hill). (below) UNC campus, ca. 1847; engraving from *The Illustrated Hand-book, a New Guide for Travelers through the United States of America* by J. Calvin Smith. Sherman & Smith, New York (1847). (Courtesy of North Carolina Collection, Wilson Special Collections Library, University of North Carolina, Chapel Hill.)

E. Dwight, chaplain of the U.S. Senate, recommended Mitchell for the professorship of mathematics (he also recommended Mitchell's Yale classmate Denison Olmsted for the professorship in chemistry, mineralogy and geology) to William Gaston, a congressman from North Carolina and UNC trustee. It is noteworthy that Reverend Dwight was the son of Timothy Dwight, who was president of and a professor at Yale while Mitchell studied there.

After several unsuccessful attempts to fill the mathematics chair, the Committee of Appointment announced to the Board of Trustees at its December 5, 1817 meeting that "they [had] engaged and appointed to the professorship of Mathematics . . . Mr. E. Mitchell of Connecticut" (BOT 4: 144–145). The committee believed Mitchell to be well qualified. Subsequently, the trustees set his annual salary at $1,240 (BOT 4: 195).

Upon accepting his appointment at UNC but before moving to North Carolina, Mitchell wasted little time in further preparing for the ministry. He studied at the Andover (Massachusetts) Theological Seminary and then received his licensure from the Western Association of New Haven (Connecticut) County on December 30, 1817 (Dexter, 1912).

When Mitchell (1825a) left his home state of Connecticut to accept a professorship at UNC, he hoped to escape ideological arguments that filled every village there. He "hoped to be so far secluded from the contentions of the day as to be at liberty to pass [his] life in peace." Mitchell, however, would not always obtain that peace. In Chapel Hill, he would find himself defending against accusations that the faculty was too Presbyterian. On another front, he and John Stark Ravenscroft, first bishop of the Episcopal Diocese of North Carolina, would exchange eighteen letters concerning their different religious views and interpretations.

Mitchell's extensive and dedicated service to the university and to the state formed a foundation for his fame and popularity. Numerous biographies and obituaries on Mitchell have been published. Appendix 3.1 provides a bibliography of these biographical sources.

Getting Established in Chapel Hill

Soon after his arrival in Chapel Hill in January 1818, Mitchell (1818) wrote a letter to Maria North, his fiancée who had been a student at Union Academy in New London, Connecticut. Mitchell described the village, which then consisted of about fifteen houses, and its residents. His letter also included a diagram[23] (Fig. 3.1) depicting UNC buildings and several village

[23] Mitchell drew "two 3/4-inch squares and filled them in with a house plan of the lower and upper stories of his four-room house. A diagonal line across the sheet of paper represents the main road, now called Franklin Street, through a 'Village of about 12 or 16 Houses.' Two roads intersect this diagonal line, one labeled 'Road to Raleigh,' the other unidentified but now known as Columbia Street. Below the diagonal line, Mitchell has drawn six buildings labeled A through G. A legend near the right margin identifies these structures as follows: A—Old College [Old East]; B—New College [South Building] 'Much the largest with a cupola'; C—Chapel [Person Hall]; D and E—'My house with its kitchen apart together with the yard and gardens enclosed in one square'; F—Dr. Caldwell; G—Mr. Hooper. Mitchell's map is bordered north and south by two lines parallel to the text of his letter. Underneath the legend Mitchell has written 'The plan is very crude. I had no instruments to draw with and therefore took no pains. The whole span between the two parallel lines is intended for a plot of the village over which I have written wherever there was room—I acknowledge that the whole is very slovenly—'" ([Lindemann], 2006).

houses that were adjacent to the college grounds. He drew rough floor plans of the house where he roomed. Captain James Hogg, a veteran Revolutionary War officer and his wife and her niece lived in the house that was situated in the same area intersected by Franklin and Columbia Streets and adjoining the UNC campus (Battle, 1974, 1: 271–272). Mitchell (1819a) wrote about his plan to furnish his future residence (the first UNC president's house) following his marriage at the end of the year. He selected a furniture craftsman in Farmington, Connecticut, to make several pieces of furniture, including a mahogany table, a secretary, a mahogany dining set, and a bureau, all totaling $250.[24]

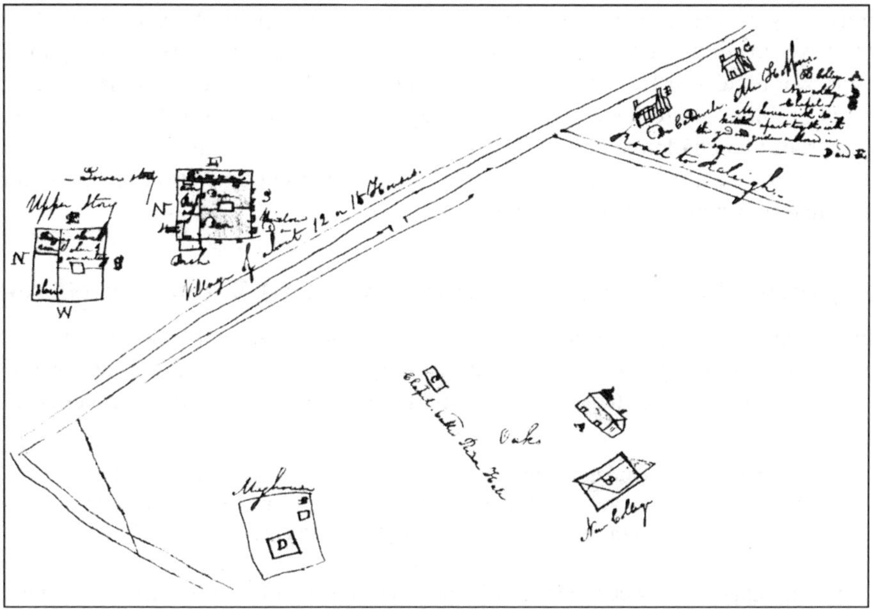

Fig. 3.1. Diagram of the UNC campus and adjacent borders of Chapel Hill, 1818. Extracted from Elisha Mitchell's letter to Maria North, February 11, 1818.

(Box 1, folder 1, Elisha Mitchell Papers #518, Southern Historical Collection, Wilson Special Collections Library, University of North Carolina, Chapel Hill).

By early 1830 Mitchell (1818–1842, p. 253) described his living situation: "We are very much in the woods here—a village of little more than 30 dwelling houses." He noted that there were four small stores that stocked only a few goods. When Mitchell began teaching, there were nearly 110 students, whose annual college expenses were about $100.

When Mitchell (1819c) wrote to Maria North about the plans for his trip to Connecticut for their marriage, he cautioned her about their subsequent travels to North Carolina. He noted that the weather would be cold and stormy and the travelling dangerous and disagreeable. The marriage ceremony took place on Friday, November 19, 1819. The newlyweds later

[24] Mitchell (1818–1842, p. 262) noted the cost of several articles of the furniture in his account book. Several pieces of the Mitchell family's furniture are known to have survived in Chapel Hill and Raleigh. These include: Victorian armchair (on loan to the North Carolina State Capitol), mahogany sofa (in renovated lounge of Toy Hall, UNC), mahogany sofa and mahogany desk and bookcase (North Carolina Collection, UNC), and sofa (in Octagon Room of the Horace Williams House, Preservation Society of Chapel Hill).

embarked on their journey to Chapel Hill, an 815-mile trip that Maria Mitchell (1820) recounted to her mother in a letter. The couple arrived in Chapel Hill on December 29, 1819 (Mitchell, 1818–1842, p. 425). They "began to board"[25] with Denison Olmsted's family in the first university president's house, paying $144[26] per year (Mitchell, 1818–1842, p. 425). Olmsted's servants were to provide laundry services. The Olmsted family vacated the dwelling around June 1820 and moved into the Widow Puckett house on East Franklin Street. The Mitchell family remained in the president's house until the end of 1857, following Elisha Mitchell's death that June.[27]

The university trustees were responsible for authorizing the repairs of the dwelling. Mitchell, however, would continually advocate for basic upkeep of the house. Workers began a long-term series of repairs in 1833, which were completed in 1836. Among the more significant structural problems were fences in ruins, decaying support posts for the front piazza, and a leaking roof that challenged him in finding a place in his bedroom where he and his wife "could keep dry of a rainy night" (Mitchell, 1836c). Upon completion of the repairs the cost for materials and labor was $249.67.

Religious Background, Preaching, and Views on Slavery

While in Connecticut Mitchell associated with the Congregational Church. Commencing in his senior year at Yale, he reflected on the subject of religion and the fate of his soul. Excerpts from his personal diary provide insight into his thoughts on these topics (Mitchell, 1813–1816).

> Let me reflect that if I am not going forward, I am travelling backward. . . . I have been troubled with vain and worldly thoughts to day. I must pray more and live more above the world and frequent the company of the thoughtful and religious. . . . (July 18, 1813)
>
> The day approaches when I am to take upon me the vows of the Lord. How cold and stupid am I. May the Lord have mercy upon me and not suffer me to eat and drink damnation to my soul. . . . Oh Holy Spirit sanctify me. (June 30, 1814)
>
> It is now time to look about me and to see what is the state of my soul–what are my views, prospects, and resolutions. And firstly if I have any degree of the Christian temper that degree is certainly very small. (July 7, 1816)

At UNC Mitchell frequently presided over the morning and evening prayers that students were required to attend in the college chapel, and he preached there on alternate Sundays. He also gave sermons in the village chapel on numerous nights. He was subsequently ordained in gospel ministry by the Orange Presbytery on September 2, 1821 (Anonymous, 1821). On September 23, 1849, the Presbyterian Church dedicated a new house of worship in Chapel Hill that replaced the village chapel. Mitchell (1849a) celebrated the occasion by giving

[25] Although Elisha Mitchell uses the word "board," he and his wife undoubtedly resided and dined in the Olmsted home.

[26] Cornelia P. Spencer (1884a) noted an annual rent of $288 and Kemp Battle (1974, 1: 252), who cited Spencer's account of the Mitchells' early years in Chapel Hill, also noted the same annual rent.

[27] Following Mitchell's death in June 1857, the Executive Committee of the Board of Trustees resolved that the Mitchell family be permitted to remain in the house until the end of the year. It also paid Mitchell's wife, Maria, his remaining half-year's salary (Manly, 1857).

a brief history of public worship in the village. Between 1795 and 1828 the only place of public worship in the area was in the campus chapel. With the erection of a small wooden structure in 1829, which was completed mostly at the expense of the faculty, services were offered Sunday and Thursday nights in the village.

Overall, Mitchell's contemporaries considered his sermons sound, but his manner of presentation was dry. Former President Kemp P. Battle (1915) thought Mitchell's delivery lacked eloquence and inspiration. He further noted that his "delivery was tame and awkward. His eyes were fixed on his manuscript, and he never raised his voice, but his sermons were always sound and sensible." Even Cornelia Spencer (1871), who praised Mitchell's accomplishments in science and academia, held a similar assessment of his dry manner of delivery.

Among the students commenting on the shortcomings of the sermons was William Sidney Mullins (1840–1841). After attending a sermon that Mitchell gave in the college chapel on July 25, 1841, Mullins recorded in his diary that the sermon "was characterized with the Doctor's usual dryness and coldness of style." Mullins, nonetheless, thought the sermon provided much instruction and contained a number of interesting passages. A sermon delivered on September 19, 1841, however, was "one of his most rambling discursive flights."

Mitchell's deep religious beliefs touched on his arguments against intemperance. He delivered a sermon on temperance at UNC on March 13, 1831. He noted the evils of alcohol and responded to the critics of temperance societies. Rebuffing critics who thought the societies violated biblical truth by forbidding the use of wine, he claimed that modern wines were adulterated by French brandy and therefore were different from those of biblical times (Watson, 1984). His students were so impressed by the sermon that they asked him to publish it (Mitchell, 1831a).

Mitchell (1850a) also wrote another religious tract, and it concerned Palestine. The topic was the focus of his Sunday junior recitations. Drawn heavily from Robinson and Smith's (1846) book titled *Biblical Researches*, the pamphlet provides a brief history and description of Palestine. UNC student Thomas M. Garrett (1849–1850) found Mitchell's recitation of August 19, 1849 "very interesting" and mentioned using the pamphlet.

According to historian Elgiva Watson (1984), Mitchell "justified his support of slavery in traditional southern terms with an underlying religious theme." He wouldn't give up his slaves unless he was persuaded that he had to forego his interests and all of his worldly property (Mitchell, [1848]). Mitchell owned enslaved individuals who worked in his home, and he directed enslaved individuals who worked in the college with such endeavors as building rock walls, cutting wood on university land, and general maintenance.

Correspondence furthers a perspective on his management of servants. In a letter to president Swain, Mitchell (1838a) wrote, "Will you look at Harrison's servant and see whether he is better than Abram whom we have had–and whether it may be best to take him and send Abel to the wood pile. Or could I send my Rascal Mark to the Rail Road and take one of them?"

In another letter to Swain, who was then absent from campus, Mitchell (1840a) wrote about hiring servants, including possibilities of hiring servants for university work, including cutting wood from university land for college use. He was especially concerned about having a backup person in case other servants became ill. He also inquired about the going rate to pay servants' board. Mitchell noted that one servant, who had been receiving $54 for board the past year, was "clamouring for 6 or 12 dollars additional." Another servant complained about the quality and quantity of the food.

Student Activities and Demeanor

The minutes of UNC faculty meetings are replete with incidents of student infractions, including intoxication, fights, insubordination, and misconduct. Mitchell, as well as his colleagues, were responsible for visiting the rooms of students. Mitchell (1828d) noted that in carrying out this duty the professors were "exposed to the petty tricks, and occult insulting behaviors of those who at any time may be unwilling to permit such opportunities to escape them, of directing capricious indignities upon the person of a professor. One of the chief inconveniences as yet experienced, is a drenching with water clean or foul, as they pass the steps or walk the passages." The professors spent considerable time in resolving student misconduct. In dealing with the students Mitchell was known for his "leniency in punishment" (Battle, 1915).

Boredom was a common complaint of the students. Freshman Rufus Patterson's (1847a) letter of October 12, 1847 to his father reveals the experience of boredom on campus. He wrote, "College life is very dull, as there is nothing in the world going on here, & the regular routine of College duties comes as natural to me as eating my meals." Despite the prevalence of misconduct and boredom, students enjoyed a number of activities. Chief among them was hunting on the campus grounds and surrounding woodlands. Historian Kemp Battle (1974, 1: 591) notes that students hunted opossums and raccoons. Some conservation-minded students, such as Rufus Patterson (1847b), were concerned about the extinction of the fauna. He noted that opossums "will not be here long if the students hunt them as they have been doing lately." Students also sought other wildlife, such as rabbits, and went fishing. Regarding rabbit hunting, UNC student William S. Mullins (1840–1841) describes, in his diary, a hunt that he and several of his classmates made on January 21, 1841: "Our course lay through Prof. Green's woods, across the road at King's mill, up the stream to the pond, and thence to College through the woods. During all this route we start[l]ed no rabbits, but we had considerable fun."

Bird sporting also provided a pastime for students. This sporting was popular in the 1870s and no doubt also occurred during Mitchell's tenure at UNC. An anonymous (1878) writer noted that "the chief amusement Christmas [1878] to those who remain here will be: . . . running owls (of which there are a great number) out of the trees in the Campus." Another anonymous reporter (1879) mentioned that turkeys and partridges were abundant in Chapel Hill during the hunting season.

Passenger pigeons, now extinct, were the focus of a favorite sport for the students. During their migration, these birds would form large flocks that would darken the skies as they flew overhead. In an entry in his personal diary on February 18, 1858, sophomore Henry Francis Jones (1857–1858) wrote: "I never saw the like of pigeons in my life as has been passing over here to day. Immense droves ha[ve] been going over all day. Good many of the students ha[ve] been shooting them all day." Hunting campus wildlife was a popular sport for the students, but university officials banned the practice in late 1876 (BOT 16: 29–30).

Students also enjoyed the beauty of the natural environment. They went on walks to favorite local destinations, such as Mt. Bolus, Piney Prospect, Morgan Creek, and Rock Quarry. On June 29, 1849, Thomas Miles Garrett (1849–1850) joined several of his classmates for a stroll, first walking through the village and observing "the transcendent splendor of the setting sun and the hues with which he fringes the clouds which hung about his setting– the dark cloud which rested in massive blackness in the south and south-west, the colour which tinged its western part, varying yet almost by no perceptible difference of hue." The group walked on and passed a bridge and a mill and came into view of a hill "which rose before us in wild, native grandeur." The students ascended the hill and delighted in a view that stretched before them for nearly twenty miles. They ended their stroll and returned to campus, where they sat in the grove and engaged in a conversation.

A variety of games and pastimes provided another form of diversion for the students, including baseball, bandy, ball bouncing, marbles, and skating. In the early years of the university, students often passed the time by hitting a ball against the walls of campus buildings. The trustees proposed a ban against this activity in December 1817, just before Mitchell's arrival on campus (BOT 4: 151). This amusement, however, continued among the students.

Nearly all the students played bandy (also called shinny). Players wielded curved sticks to strike hard wooden balls (Battle, 1974, 1: 589). The game, however, was very dangerous, particularly if a wooden ball was used, as was the case at UNC. President Joseph Caldwell (1830) reported to the trustees his and the faculty's concern about the danger of the sport. He noted that frequently students were knocked "lifeless" for a period of time. In 1830 Caldwell suggested that a resolution be considered to forbid any kind of ball, except ones made of yarn and covered with leather. Other common pastimes were playing cards and backgammon.

Early Teaching—Mathematics

A note in Mitchell's (1818–1847, p. 195) diary shows his agenda for January 12, 1818. It reflected a long, busy day. "Plan of Studies for the Session. From the time of rising in the Morning till 9 o'clock the Lessons of the Day. From 9 to 11 Vince's *Fluxions* [Fig. 3.2].[28] From 12 to 2. Letters. Private. Recitations, Miscellanies. The lessons of the Day. From 2 to 4. Optics. From Prayers to bed Time 1/2 an hour on Euclid, Locke. Language, Poetry. Religion &c. Tuesday Evening 2 o'clock Hutton[29] Page 399. Example 13, [volume I]." On the remainder of the page Mitchell recorded his calculations for the selected example.

[28] Samuel Vince, *A Treatise on Fluxions*. 5th edition, corrected and enlarged throughout. Printed by J. Smith, Cambridge, 1818. Mitchell's copy of the book survives in the University of North Carolina at Chapel Hill Library.

[29] Charles Hutton, *A Course of Mathematics in Two Volumes; for the use of academies, as well as private tuition*. From the 5th and 6th London editions, revised and corrected by Robert Adrain. S. Campbell, &c., New-York, 1818.

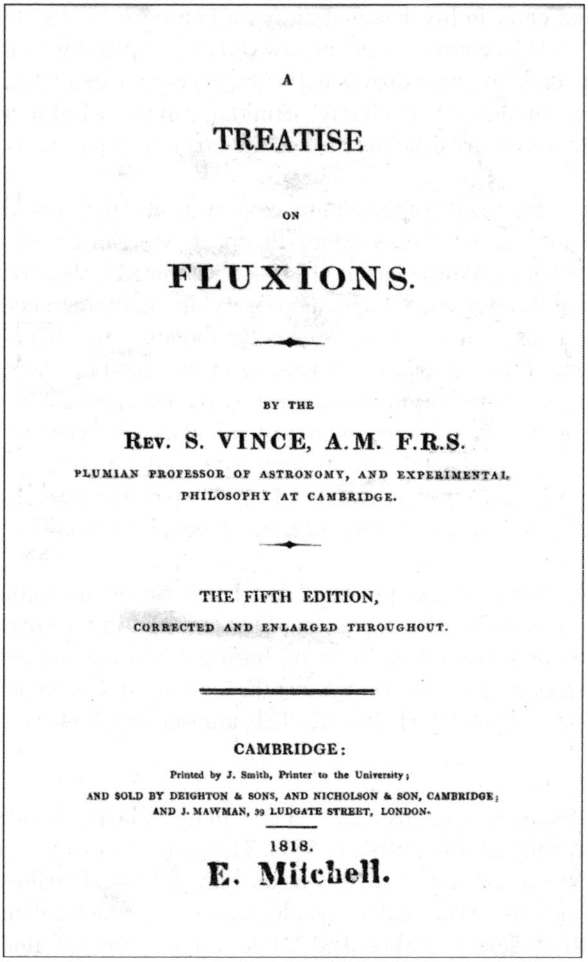

A

TREATISE

ON

FLUXIONS.

BY THE

Rev. S. VINCE, A.M. F.R.S.

PLUMIAN PROFESSOR OF ASTRONOMY, AND EXPERIMENTAL PHILOSOPHY AT CAMBRIDGE.

THE FIFTH EDITION,

CORRECTED AND ENLARGED THROUGHOUT.

CAMBRIDGE:

Printed by J. Smith, Printer to the University;

AND SOLD BY DEIGHTON & SONS, AND NICHOLSON & SON, CAMBRIDGE; AND J. MAWMAN, 39 LUDGATE STREET, LONDON.

1818.

E. Mitchell.

Fig. 3.2. Title page from *A Treatise on Fluxions*, 5th edition, by Samuel Vince (1818). (Note Mitchell's ownership stamp at the bottom of the image).

(Courtesy of Kenan Science Library Annex, University of North Carolina, Chapel Hill).

Schooled in current theory and leading practices in science at Yale, Elisha Mitchell began teaching with an eye toward expanding science at UNC. As professor of mathematics, he taught juniors. In his first year of instruction, he introduced fluxions, now known as differential and integral calculus, into the course of studies. After Mitchell's second week of teaching, Iveson L. Brookes (1818a), a junior from Rockingham County, North Carolina, wrote to his father, Jonathan Brookes, about the course change.

> Our new Professor of Mathematics has arrived, late from the North. Since his arrival the number of our recitations has been doubled and the length of the lessons but little diminished; thus these studies being at best difficult, are become [*sic*] somewhat oppressive; at least are so to me inasmuch as they are not very entertaining to me and are so extensive as to require close application. I am now debar[r]ed from the opportunity measurably of attending to reading, composing and some other sources of literary improvement which to me are more entertaining and the knowledge or possession of

> which is to me far more desirable than the mear [*sic*] knowledge of the properties of Angles and Triangles in which the branch of Mathematics that constitutes our present study, principally abounds. Our new Professor appears to be quite an accomplished sholar [*sic*] and is pleasant in his address & agreeable in his manners; but the class complains heavily of the long lessons he appoints. But we flatter ourselves that when he becomes better acquainted with the customs here he will be more d[i]sposed to comply with the former method of studying at this place.

Brookes's complaint about Mitchell's rigorous style of teaching and his introduction of new theory was tempered when he wrote to his father on June 6, 1818 (Brookes, 1818b). Upon being examined at the end of the session, his junior class had been highly praised by the trustees and professors for its outstanding performance, the best of any preceding junior class.

Although the expanded curriculum of 1818 included natural philosophy (physics), the course was most likely not taught while Mitchell served as mathematics professor. When Denison Olmsted resigned the chair of chemistry, mineralogy, and geology in late 1825, Mitchell (1825b) wrote to North Carolina Governor Burton to express his desire to relinquish his current post and fill Olmsted's vacancy. Mitchell considered himself qualified for the position and had been engaged in chemical experiments in the laboratory to gain practical experience during the fall months in 1825. Mitchell also recommended that the professorship include responsibilities in teaching natural history. He reminded Burton that nearly five or six years earlier a course in the Philosophy of Natural History had been proposed for the scheme of instruction at UNC, but there was no one at the university then who was prepared to teach it. Since that time, he had devoted his leisure time to studying it. He suggested that the trustees restructure the course into "The Professorship of Chemistry and Natural History (in which mineralogy and geology would of course be included)." The Committee of Appointment (Haywood, 1825) recommended Mitchell for chair in chemistry but did not add natural history to its charge. The latter could be considered whenever a Revision of the Plan of Education would take place. On December 13, 1825, the Board of Trustees approved Mitchell's appointment as professor of chemistry, mineralogy, and geology (BOT 5: 47).

Teaching Physical Sciences

Teaching Chemistry

As the newly appointed professor of chemistry, mineralogy, and geology, Mitchell taught in the chemistry laboratory in Steward's Hall, where Olmsted had also instructed his classes. In the 1830s new space for instruction in chemistry was created in South Building. Historians Kemp Battle (1974, 1: 281–282) and Thomas Kevin Cherry (1993) describe how the space was developed in South Building for the chemistry laboratory. A floor was built across the second level of a two-story hall in the middle of the south side of the building. The area on the first level became the new lab, where Mitchell presented lectures and conducted experiments for his chemistry classes. A room was built above the laboratory, where the University Library and a lecture room shared the space collectively.

In a letter of September 2, 1835 that Mitchell (1835a) wrote to Duncan Cameron, president of the Executive Committee of the Board of Trustees, he reported that "The Laboratory was transferred from the Steward's Hall to the room you were in when we had the pleasure of your company here last session by the Direction of Dr. Caldwell." In the same letter Mitchell noted that around $50–$75 was needed "to put the room into decent condition."

The laboratory subsequently underwent repairs during the summer vacation of 1837 (Mitchell, 1837a). Mitchell (1837b) recommended an allocation of $100 per year for the operation and upkeep of the lab, which the trustees approved. Mitchell's semi-annual reports as bursar (see below) reveal that the trustees provided $100 per year as well as additional funds when needed. For example, the chemistry department was allocated $1,542 in 1827, $498 in 1848, and $390 in 1851 (UNC-A).

Chemistry was offered to seniors. According to university catalogs, Mitchell taught the subject in the first session and into the beginning of the second session, which concluded with mineralogy and geology. He followed this plan of instruction from 1826 to mid-1831, after which he taught chemistry and mineralogy in the first session and chemistry and geology in the second, using this scheme until 1854. It is noteworthy that the UNC catalog of 1837–1838 announced that "occasional Lectures on the other departments of Natural History and on Technology" would be taught. A significant change in Mitchell's teaching took place in 1855/1856 when he expanded the curriculum. He began teaching chemistry and mineralogy to juniors in the first and second sessions and chemistry and geology to seniors, also in both sessions.

Mitchell's instruction entailed a combination of lectures and recitations, the common practice of his time. In his early years of teaching chemistry, he probably followed Olmsted's plan of instruction. By the early 1840s, however, "he published a short system of Chemistry & Geology, which [was] used by the students as text books" (W., 1844). According to the UNC catalog of 1846–1847 (p. 24), he assigned his students the *Text-book on Chemistry, for Use of Schools and Colleges* by John William Draper (1846).

Draper's textbook was considered a thorough and systematic introduction to the subject. It followed the author's outline of lectures that he delivered in his college course at the University of New York. According to the author's preface to the volume, the book was to serve "as a manual, arranged in such divisions as practice has shown to be suitable for daily instruction."

The book apparently appealed to some of Mitchell's students. In his diary entry of August 7, 1850, UNC senior Thomas Garrett (1849–1850) was "delighted with the text-book adopted." He considered the author's style as being "forcible, mainly from its perspicuity." He further noted the book's conciseness and plainness and concluded, "I must betake myself to it with diligence."

Shortly before his death in June 1857, Mitchell (1856a) had nearly accomplished the publication of his own chemistry textbook, *Elements of Chemistry* (Fig. 3.3). The only known extant printing of the volume is a copy bound with several of his publications held in the North Carolina Collection at UNC. President Kemp Battle (1974, 1: 681) stated that Mitchell had prepared a second edition of a chemical manual prior to his death, but it was never published. The version of his textbook held at UNC is most likely the edition to which Battle refers. The first edition may have been the text of the early 1840s mentioned above.

Lacking a title page and table of contents, the second edition consists of two major divisions: "Elements of Chemistry" (pp. [13]–258) and "Organic Chemistry" (pp. [259]–268). Clearly, additional text followed, or was to be added, as the last sentence of the last page stops in mid-sentence. Topics are presented under headings "Sec.," *i.e.*, Section, and are continuously numbered throughout the text.

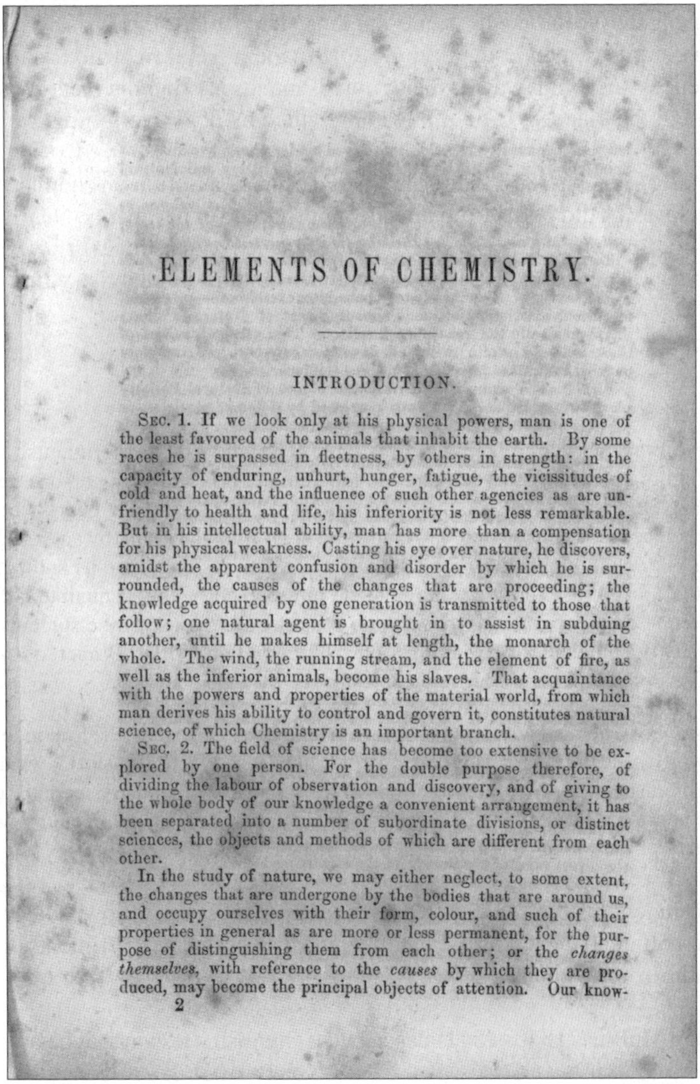

ELEMENTS OF CHEMISTRY.

INTRODUCTION.

SEC. 1. If we look only at his physical powers, man is one of the least favoured of the animals that inhabit the earth. By some races he is surpassed in fleetness, by others in strength: in the capacity of enduring, unhurt, hunger, fatigue, the vicissitudes of cold and heat, and the influence of such other agencies as are unfriendly to health and life, his inferiority is not less remarkable. But in his intellectual ability, man has more than a compensation for his physical weakness. Casting his eye over nature, he discovers, amidst the apparent confusion and disorder by which he is surrounded, the causes of the changes that are proceeding; the knowledge acquired by one generation is transmitted to those that follow; one natural agent is brought in to assist in subduing another, until he makes himself at length, the monarch of the whole. The wind, the running stream, and the element of fire, as well as the inferior animals, become his slaves. That acquaintance with the powers and properties of the material world, from which man derives his ability to control and govern it, constitutes natural science, of which Chemistry is an important branch.

SEC. 2. The field of science has become too extensive to be explored by one person. For the double purpose therefore, of dividing the labour of observation and discovery, and of giving to the whole body of our knowledge a convenient arrangement, it has been separated into a number of subordinate divisions, or distinct sciences, the objects and methods of which are different from each other.

In the study of nature, we may either neglect, to some extent, the changes that are undergone by the bodies that are around us, and occupy ourselves with their form, colour, and such of their properties in general as are more or less permanent, for the purpose of distinguishing them from each other; or the *changes themselves*, with reference to the *causes* by which they are produced, may become the principal objects of attention. Our know-

2

Fig. 3.3. Introduction to *Elements of Chemistry* [an advance proof copy by Elisha Mitchell] ([1856]a).

(Volume 7, Cobb Loan Collection, North Carolina Collection, Wilson Special Collections Library, University of North Carolina, Chapel Hill).

Mitchell's textbook begins with a history of chemistry and chemical nomenclature and then covers numerous topics, including specific gravity, the thermometer, the steam engine, vaporization, galvanism, crystallography, and various chemical elements. The book concludes with discussions on the theory of vegetation, organic analysis, classification of organic compounds and neutral organic compounds.

Students had been using advance copies of some of the text for their class. An anonymous (1856) student reported that the book was to consist of about 300 octavo pages of text. Commenting on the volume, the same student stated that "the Dr's. style is not very captivating we confess, nor is there any effort made at rhetorical flourish or beautifully rounded sentences—nothing was further from his intention—but the facts are stated clearly, concisely, and in the business kind of a way in which he does everything."

The UNC catalog of 1827–1828 (page 12) gives a brief description on chemistry instruction, noting that lectures would be accompanied by appropriate experiments. Nearly two decades later, the *Catalogue of the Trustees, Faculty, and Students, of the University of North Carolina 1846/1847* (page 24) briefly described the course of study in chemistry that Mitchell had developed.

> Department of Chemistry,
> Mineralogy and Geology
> Rev. Elisha Mitchell, Professor
>
> On these branches, three lectures, illustrated by appropriate experiments, and the exhibition of specimens, are delivered to the Senior Class and such persons pursuing a Partial Course, as choose to apply themselves to these studies, in every week. The lecture is succeeded after an interval of an hour, or an hour and a half, by an examination extending through one hour, upon the facts and doctrines that were its subject. With these, the Student is expected to have made himself thoroughly acquainted, through the medium, either of certain printed notes prepared for the use of the students of the University or some other book of Elements. The text book of Dr. Draper, of the New York University will be used hereafter.[30]
>
> The history of Chemistry, its Nomenclature and General Doctrines, the Imponderables, under which head the phenomena and laws of Light, Heat, Electricity, and Galvanism are treated of at such length as their relations to the science of Chemistry seem to require, and the non-metallic elements, if any part remain unfinished, are completed—with the Chemistry of the Metals and the Organized bodies.

The course of student instruction entailed attendance at a lecture, followed later in the day by a recitation pertaining to the lesson. Mitchell would also perform experiments or demonstrate a new apparatus or instrument.

William Sidney Mullins (1840–1841) described a chemistry lesson that he attended on July 21, 1841.

> This class convened in the Labrotary [*sic*] at half past eight and heard Dr. Mitchell's opening Lecture on the Science of Chemistry. Misses Margaret and Ellen Mitchell, and Miss Whitaker were also present, and perhaps drew off a little attention that should have been given to the Lecture.

[30] In later years Silliman's (1830–1831) *Elements of Chemistry* was also a recommended textbook for Mitchell's chemistry class.

> ... The Lecture embraced nearly an hour and a half and he gave us a lesson to prepare for Recitation at eleven, so that the intermediate time is generally pretty well employed. By this contrivance, he manages to make the Seniors recite three times a day notwithstanding their exemption from ante-breakfast recitation. Indeed when we recite to him twice on the same day, (as is the case on Wednesday) we go to the Labrotary [*sic*] four times a day, and spend at least an hour more there than any other class in College devotes to the Recitation room. This circumstance is however by no means displeasing to Classes generally.

As an educational enrichment (and probably to break the tedium of lectures), Mitchell performed experiments and demonstrated newly acquired apparatus before his class. Mitchell (1828c) believed that chemistry was a branch of science "which the student never will acquire if it is not illustrated by experiments." William S. Mullins (1840–1841) reported on two experiments given in different classes:

> October 6, 1841. In return for unmerciful bores inflicted on us in the Labrotary [*sic*] for the last week or two, the Dr. gave us this morning some very interesting experiments on Electro-Magnetism ... .
>
> October 27, 1841. Oxygen was the subject of our morning lecture and the experiments were quite brilliant. The Ladies of Dr. Mitchell's family and Miss Hall were present, and animated the scene of course.[31]

Well known for injecting humor and amusement into the classroom, Mitchell liked to tell jokes. During a lecture on October 10, 1841, Mitchell showed his audience, including several young ladies of the village, newly acquired galvanic and electro-magnetic instruments. Mitchell quipped "that we could pay no higher compliment to a young lady than to call her an Electro-Magnet" (Dusenbery, 1841–1842). On the evening of May 31, 1842, Mitchell gathered the seniors, as well as visitors, together in the grove behind South Building. He gave students, who volunteered, nitrous oxide ("exhilarating gas"), now called laughing gas (Dusenbery, 1841–1842). Some students became pugnacious, one said he felt "glorious," and another was dazed and walked about aimlessly.

A number of students took offense at Mitchell's style. One such individual was senior Ruffin Wirt Tomlinson (Sanders, 1953), who recorded his thoughts on Mitchell in a diary entry dated July 21, 1841. "I made my first recitation to Professor Mitchell to day at eleven o'clock upon Chemistry. He gave us a lecture between eight and nine. He is one of the most vulgar old men I ever saw, he brings in vulgarity upon every occasion he possibl[y] can—his experiments were fine. In the evening we had another lecture and recitation to Professor Mitchell." William S. Mullins (1840–1841) also noted Mitchell's inappropriate manner. At a recitation conducted on February 7, 1841 he wrote, "as usual Prof. Mitchell's remarks took a sexual turn. It is quite surprising how he delights to lead our attention to such subjects, and is, I think, rather ill taste in him."

Occasionally the students would direct jokes on Mitchell, or Old Mike, as he was known. One such incident took place in Mitchell's chemistry class around 1835, as recalled by UNC

[31] Mitchell permitted women, including his daughters, to attend his lectures. Their presence sometimes created distractions to the students. UNC senior Ruffin Tomlinson noted that ten ladies (Mrs. Cabia and her schoolgirls) attended Mitchell's chemistry class on February 23, 1842 (Sanders, 1953).

alumnus (Class of 1835) William P. Webb (1889). The students were studying chemistry with only Mitchell's "mss. Lectures" and had assembled under the grand oak trees near South Building. One student read the lecture while his classmates listened. Because Mitchell often repeated the words "from time to time" in his own lectures, the class decided to mock Mitchell's usage of these words.

> All of us agreed that when we met him in the class-room, each man called to recite should repeat the words 'from time to time.' That was done, several times, and the Dr. saw the point. He never said a word, but when we entered the Lecture-room the next day (on time) we found him (in loco) & the long table before him covered with books and he was covered with smiles! The roll was called; and he began to read authorities, ancient & modern, -lay and clerical, prose & poetical, classical & scientific,—using the words 'from time to time,' and after exhausting the hour for recitation, dismissed the class, with one of his bland smiles. We retired with a loud 'hurrah' for Old Mike.

Mitchell also expressed his sense of humor outside of the classroom. In answering a letter of inquiry from North Carolina agriculturist Frank M. Parker, Mitchell ([1852]) referred to his old age, stating that he could improve his condition if he acquired "a few hundred pounds of youth."

When Mitchell (1826) assumed responsibility for teaching chemistry, he also took on the long-term duty of purchasing materials for the laboratory. In 1826 he spent nearly $700 for laboratory apparatus. On various occasions he informed the trustees about the need for apparatus. For instance, in 1836 Mitchell (1836d) noted that apparatus would contribute to the students' education. Among the most pressing needs were new and improved devices for demonstrating electromagnetism and the polarization of light.

He used electromagnetic devices for classroom demonstrations (as noted above) and as an "instructive and entertaining exhibit" for the public. Using an electromagnetic device purchased around 1835 from Boston, he performed experiments at commencement exercises in June 1835. An anonymous (1835) newspaper reporter described Mitchell's demonstrations.

> He produced powerful and most rapid motion by magnetism alone, and demonstrated the practicability of its application to the useful arts, but seemed to think that the rust of copper and zinc (materials indispensable to the excitation of the magnetic influence in such a degree as to be useful) would hinder it from coming into successful competition with steam.

Mitchell also gave demonstrations to regional learned organizations—for example, a lecture on galvanic electricity before the Hillsborough Literary Association in July 1842 (Anonymous, 1842). He illustrated his presentation with various experiments.

Outside the classroom Mitchell conducted private chemical investigations in a large wooden building[32] (Fig. 3.4) located across from the present-day Carolina Inn on Columbia Street and to the south of current Phillips Hall. His chemical laboratory was on the second floor. The ground floor contained the campus carpenters' quarters and the attic the observatory (Cobb, 1896; Henderson, 1949). The building was razed in the beginning of the twentieth century (Henderson, 1949).

[32] The building, known as the second observatory, replaced the first one destroyed in 1838 by fire (Cobb, 1896).

Fig. 3.4. Astronomical Observatory (second), University of North Carolina, Chapel Hill; also housing the campus carpenters' workshop and Elisha Mitchell's private chemical laboratory, ca. 1895.

(Courtesy of North Carolina Collection Photographic Archives, Wilson Special Collections Library, University of North Carolina, Chapel Hill).

Mitchell's (1828c) outlook on research resonates with the present-day academic experience. He wrote, "It is absolutely necessary ... for a Professor of Chemistry to pass his life in his laboratory whether he be pursuing his studies for his own improvement or preparing instruction for others. Experiments and processes which are described in the compass of a page or two will occupy the whole of a long day—(hardly leaving time for the operator time to take his meals) and they must be watched over all the while and accelerated or retarded as occasion shall require or mischief will ensue."

Mitchell (1829b, 1830) published at least two papers in chemistry. His paper of 1829 concerns the controversy over the law of definite proportions that had divided chemists in the early decades of the nineteenth century. According to historian of science Seymour H. Mauskopf (1984), Mitchell was well versed with some of the pertinent chemical literature and its writers, such as C.L. Berthollet and John Dalton; however, his paper of 1829 "shows his limitations as a 'chemist'–an inability to mount a sustained theoretical argument." In the discipline of chemistry, nonetheless, Mitchell served well in his role as an up-to-date and well-read teacher.

Teaching Geology

Mitchell's (1842a) volume on the *Elements of Geology* (Fig. 3.5) served as a textbook for his students. The book was based on his lectures given at UNC and provided an understanding into his coverage of the subject. He decided to publish a textbook as a more "satisfactory" means of conveying information to the students. In the classroom he used plates and illustrations as teaching aids; however, these were not included in the volume. A folded map depicting the position and breadth of the rock formations in North Carolina was included as a frontispiece. Geologist Joseph Holmes (1887) noted that Mitchell's geological map of the state was the "first published" of its kind and remained uniquely so until 1875, when another appeared. As already noted, Mitchell expanded coverage of geology and mineralogy in the curriculum of the 1854/1855 academic year.

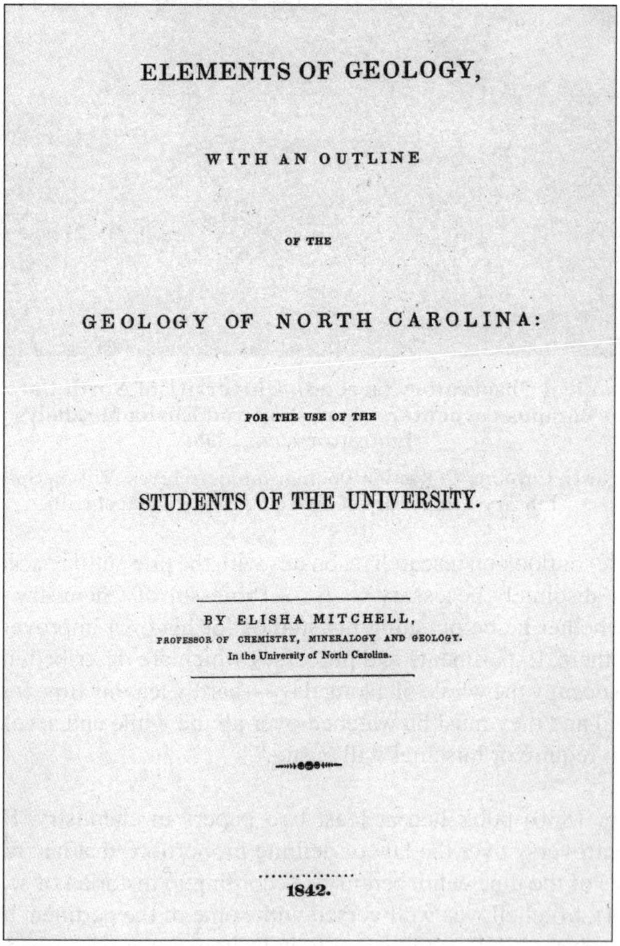

ELEMENTS OF GEOLOGY,

WITH AN OUTLINE

OF THE

GEOLOGY OF NORTH CAROLINA:

FOR THE USE OF THE

STUDENTS OF THE UNIVERSITY.

BY ELISHA MITCHELL,
PROFESSOR OF CHEMISTRY, MINERALOGY AND GEOLOGY.
In the University of North Carolina.

1842.

Fig. 3.5. Title page of *Elements of Geology* by Elisha Mitchell (1842a).

(Courtesy of North Carolina Collection, Wilson Special Collections Library, University of North Carolina, Chapel Hill).

Mitchell began his lectures on geology by defining the discipline as "the Science which treats of the composition and structure of the exterior crust of the earth; the changes that are now proceeding in it; its condition in the most ancient times; and the causes by which its existing form and the present distribution and position of the materials of which it is composed, have been produced." Topics covered by his lecture include the following: the earth as a mass, primary classification of rocks, mineral beds or strata, composition and structure of the rocks, Werner's theory, granite, gneiss, mica slate, simple and primitive rocks, transition rocks, clay slate, sandstones, secondary and tertiary rocks, influence of the rocks upon the fertility of soil, metallic veins and beds of valuable minerals, secondary strata of England, fossil salt and gypsum, theoretical geology, Hutton's theory, causes of geological changes, volcanoes, earthquakes, minerals, history of the earth, mineral geography, and geology of North Carolina. In his presentation of scientific information, it is clear that he "reveal[ed] his unwillingness to constrain geology by a literal interpretation of the Biblical story of creation" (Wheeler, 1984).

Geological field excursions enhanced the students' learning experience. UNC student Ruffin Wirt Tomlinson (Sanders, 1953) recorded his observations of what appears to have been a geological field trip led by Mitchell that took place on Wednesday, September 1, 1841.

> ... took us about 3 m[i]les from College, we went up and down hills, valleys, thickets, fences, bryers [*sic*], cornfields, by fish traps. I never had such a walk–he tried to outrun us every one and get back to College first. Jack [Hayes] and myself run the old gentleman clear down. The students all complained very much of being tired.

Teaching Mineralogy

In teaching mineralogy Mitchell ([1840]b) used a booklet titled *Natural History* (Fig. 3.6) that he had written to serve as a student textbook. The UNC libraries hold several copies of the pamphlet that appear to be different printings but with the text unchanged. I examined a copy that belonged to a student who graduated in 1847 and another belonging to a student who graduated in 1856.

In the booklet, Mitchell introduced natural history, stating that it "ascertains the different substances that exist naturally either within or upon the surface of the earth, and classifies and describes them by their properties or characters." His syllabus on the topic is divided into three sections: Botany (pages 2–8) (Appendix 3.2), Zoology (pages 8–18), and Mineralogy (pages 19–27). His coverage of these disciplines was decidedly descriptive.

Mitchell's introduction to the discipline of mineralogy in the same booklet began as follows: "Minerals are those bodies destitute of organization, which exist naturally either within the earth or at its surface. Their different kinds or species it is the object of the science of Mineralogy to ascertain, classify, and describe."

Lecture notes taken by two students in a mineralogy class reveal Mitchell's coverage of the discipline in the classroom. Several surviving pages from the notebook of UNC student David McMichen Lees ([1828]) provide additional insight into Mitchell's approach to mineralogy.

> Mineralogy prepares us for Geology. Classification is not by any means as important in Mineralogy as in Botany and Zoology. Why?
>
> Geographical Mineralogy which communicates a knowledge of the manner in which they are distributed over the surface of the earth and the countries in which they are to be found & Economical Mineralogy which points out the use to which the minerals may be applied in domestick economy and the arts. Why classification is more important in Botany & Zoology than Mineralogy? Because the number of distinct animal and vegetable species amount to some thousand. The number of distinct species of minerals

do not reach 300 including every subspecies and variety not to 800. The difficulties which we have to encounter in Botany and Zoology and to enable us to overcome which especially our scheme of classification was invented to present themselves in the science of Mineralogy.

Fig. 3.6. Cover page of *Natural History* by Elisha Mitchell ([1840]b).

(Courtesy of North Carolina Collection, Wilson Special Collections Library, University of North Carolina, Chapel Hill).

Several pages of notes concern the meaning of terms used to describe the "character" of minerals. Terms include fracture, splintering, frangibility, hardness, flexibility and elasticity, transparency, touch, smell (when breathed on), streak, magnetism, and crystallization. In Mitchell's textbook twenty characters are presented. In several instances Mitchell refers to researchers and their work.

UNC student John T. Jones (1835) kept a notebook on Mitchell's lectures on mineralogy. The class was introduced to the discipline, thusly:

Mineralogy is divided into 2 great classes—1st simple minerals which have all their integrant particles alike; Example Quarts [*sic*], carbonate of Lime.

2nd compound minerals; which contain two or more kinds of integr[a]nt particles. Example granite which contains three: viz. Felspar, Quarts [*sic*], and miche [*sic*].

Simple minerals belong to the science of Mineralogy; compound ones to that of Geology.

Lectures covered minerals distinguished by their properties (*i.e.,* characters), structure, and classification. Mitchell liked to provide tangible examples for his students. He noted, for example, that the statue of Washington in Raleigh was made of "carrara marble" from Tuscany.

Just as new, modern apparatus was essential in teaching and demonstrating the sciences, an extensive collection of minerals was an important aid in teaching and illustrating mineralogy. Such a collection "would lead to a better understanding among the students of the mineral resources of their region and would, therefore, lead to their better use of this resource" (Cherry, 1993). In studying mineralogy Mitchell (1828c) considered it essential to have specimens at hand as it would be "of little use to read without them." Mitchell (1835b), in fact, deemed that mineralogy "can be taught only by means of specimens."

As mentioned in the previous chapter, Denison Olmsted gave UNC a collection of minerals from Connecticut after his arrival in Chapel Hill. During excursions throughout North Carolina, Olmsted and Mitchell collected specimens for teaching. Joseph Caldwell bought a small mineralogical collection when he sailed to Europe in 1824/1825 to purchase books and apparatus. According to Mitchell (1835b), the minerals that Caldwell purchased in Switzerland cost $15. Mitchell built up his own collection, which the university purchased upon his death. He also inquired about collections offered for sale through advertisements in scientific journals and catalogs. For example, in 1835 Mitchell (1835d) wrote to Messrs. Mather and Hall in Albany, New York, about purchasing a number of specimens. Mitchell noted that the specimens he was ordering were "on my own account." But he stated that the trustees of the university had allotted $1,000 for the purchase of a cabinet of minerals, and he was particularly interested in acquiring foreign specimens.

Within several years of writing to Messrs. Mather and Hall, Mitchell fulfilled his desire to find a major cabinet of minerals for the university. Historian of science Thomas Kevin Cherry (1993) described the acquisition of the so-called Vienna Cabinet in his master's thesis. University senior Ruffin Wirt Tomlinson (Sanders, 1953) recorded in his diary that the "University Cabinet of Minerals" arrived in early February 1842. "They were purchased in Vienna by John Randolph Clay Charge d'affairs to St. Petersburg. It is a most splendid collection; they cost between 12 and 15 hundred dollars." He further noted that "Dr. Mitchel[l] is the best pleased I ever saw a man. He showed me several specimens of minerals as soon as the boxes were opened." Mitchell (1842c) spent nearly a week unpacking and arranging the mineral specimens. The cabinet actually contained over 2,000 fine specimens of minerals, including some noteworthy crystals, from all corners of the world (Anonymous, 1877). This resource became a useful tool in illustrating courses in mineralogy and geology.

Mitchell took his mineralogical classes on field trips. Senior William S. Mullins (1840–1841) wrote about a "Mineralogical Tour" that took place in an afternoon in August 1841. "The ramble would have been delightful, but for the heat which was intense, and as it was we were all tolerably well pleased. I derived some important information from the Dr. as we went along."

Geological and Mineralogical Investigations

In addition to his scientific training at Yale, Mitchell gained an understanding of geology and mineralogy through reading and through excursions within the state. His personal library contained over ninety titles of books on geology and mineralogy. Among them were such classics as Charles Lyell's (1839) *Elements of Geology,* Edward Hitchcock's (1840) *Elementary Geology*, James Dwight Dana's (1844) *A System of Mineralogy*, and Parker Cleaveland's (1822) *An Elementary Treatise on Mineralogy and Geology*. One of Mitchell's earliest tours took place in the summer of 1818 when he joined UNC President Joseph Caldwell on a survey of "The Narrows" on the Yadkin River (Turner, 1971). They aimed to ascertain the fall of the river and whether it would be feasible to build a canal along this section of it. Mitchell also answered inquiries about geological specimens. For example, in 1823 David L. Swain (1823) sent him a specimen of limestone or marble from a quarry about twelve miles south of Asheville, North Carolina, as well as a specimen of ore discovered in the same village when a well was being dug. In the same letter Swain invited Mitchell, and also Olmsted, for a visit to Asheville. He believed that many discoveries could be made in the vegetable and mineral kingdoms.

Mitchell continued the state geological survey when Denison Olmsted removed to Yale College. The Board of Agriculture of North Carolina appointed him "Geologist of the State" around late December 1825 (Anonymous, 1825). He considered the opportunity to pursue the survey as a way to further his knowledge of the state's geology, to share the knowledge with his classes, and to collect specimens for purposes of illustration (Mitchell, 1828a).

Mitchell changed the original focus of the survey from an emphasis on practical utility to "a more systematic investigation of the geology of the state" (Smith, 1999) that would provide a broad view of the state's geology. He intended to identify and locate the major formations and their boundaries as well as to roughly ascertain their age and origin. He embarked on his first leg of the geological survey during the winter recess of 1826. He traveled to the state's Low Country (now called Coastal Plain). His resulting report to the Board of Agriculture addressed the aims that he had proposed (Mitchell, 1828a). His next major geological excursion, carried out in the summer of 1827, was made in the western and southwestern regions of the state. He investigated the gold mines of Mecklenberg, Montgomery, and Rowan Counties (Smith, 1999) and subsequently wrote his second (and final) report (Mitchell, 1829a).

During his geological tours, Mitchell wrote a series of letters to his wife in which he chronicled his travels and also recorded his observations on the geology of the places he visited. He used this recorded information to prepare reports to the board or even articles for scientific periodicals. This fact is substantiated by a letter that Mitchell (1842b) wrote to his wife on June 27, 1842.

> In the older time you may recollect—if your memory reaches so far back—I was accustomed when out on a geological expedition to note down all that I did, heard or saw and send the same home under the form of a letter to my wife, which letter became thence forth a memorandum of such facts as I wished to preserve and precluded the necessity of any other kind of journal and which also you were at liberty to read to such extent as to yourself might seem good and proper.

Kemp Battle (1905) transcribed the letters with numerous annotations and published them as a diary of Mitchell's excursions. By de-emphasizing the economic aspects of the surveys, Mitchell lost the financial backing of the Board of Agriculture. Kemp Battle believed that economic hardship following the Panic of 1825, however, contributed to the discontinuance of the surveys. Mitchell wrote to his wife on December 28, 1827, "The Geological Survey dies a natural death at the end of this year. There is no one who takes any interest in the business, nor, in the present state of the Treasury, did I find there was any the least prospect of succeeding in any application to the legislature and I therefore gave it up at once" (Battle, 1905).

Geologist Walter H. Wheeler (1984) assessed Mitchell's contributions to geology. Through his geological surveys, Mitchell gained an understanding of the general character and origin of the Coastal Plain. Additionally, he developed a good knowledge of the outcrops of the state. Regarding Piedmont geology, and specifically the gold region of the state, he refuted the diluvian theory regarding the deposition of gold.

Besides Mitchell's two survey reports, he published several other geological papers, two of which also reflected results of his surveys. One concerns the character and origin of the Low Country in North Carolina (Mitchell, 1828b) and the other the geology of the gold region of the state (Mitchell, 1829c). He published a short paper on Leibnitz's protogaea (Mitchell, 1831b) and a number of articles in the state's newspapers, including several letters in the *Raleigh Register* (Mitchell, 1853). In one article he disagreed with state geologist Ebenezer Emmons's (1852) report on the coal deposits on the Deep River.[33] Mitchell noted that there was probably only a four-foot seam with an incline of about 18 degrees to the surface of the earth. Later excavations backed his theory.

Teaching Botany

The UNC catalog of 1837–1838 (page 12) provides the first documentation that Mitchell's lectures included botany and zoology. Mitchell's ([1840]b) textbook on natural history, mentioned above, also corroborates the fact that he gave lectures on botany and zoology. Through the successive years of teaching these disciplines, he did not revise the text, however. With such an unaltered text, Mitchell's lectures on botany as well as those on zoology and mineralogy probably "did not evolve" with updated information derived from advances made in those scientific fields (McVaugh et al., 1996).

Mitchell's ([1840]b) objective in teaching botany was "principally to enable us to determine the particular species to which a given vegetable belongs." He explained two systems of plant classification, that of Carl Linnaeus and of Antoine Laurent de Jussieu. The former system used the number and arrangement of flower parts (stamens and pistils) in classification, whereas the latter grouped plants according to natural character states into 100 families. In their book on Mitchell as a botanist, Rogers McVaugh, Michael McVaugh, and Mary Ayers (1996) describe Mitchell's botanical knowledge as outdated: "still back in the eighteenth century." Mitchell treated the discipline as merely plant identification and followed the Linnean system that had been virtually replaced by the natural system of Jussieu (and

[33] Emmons estimated that the Deep River coal field would yield about 272,000 tons of coal per square mile. He noted that the coal field covers 43 square miles at the lowest estimate to be taken.

also of A.P. de Candolle). According to the same authors, Mitchell lacked an awareness of progress in various areas of botany, such as plant anatomy, nutrition, growth, cell theory, plant sexuality, the alteration of generations, and reproductive biology. With his narrow view of botany his students probably were unaware of recent botanical advances.

Mitchell's classroom coverage of botany (Appendix 3.2), and also of zoology, consisted of one lecture each, whereas coverage in mineralogy was more extensive. William S. Mullins (1840–1841) reflected on a lesson in botany given on July 21, 1841.

> In the afternoon we recited on Natural History, and the lesson embraced the entire department of Botany, as treated in the Text-Book [*Natural History*] used here, which is one of the Professor's own productions. The general outlines of the study only are given and these in a very concise manner. He showed us the general principles by reference to flowers and gave just so much information as will easily enable anyone to prosecute the study of Botany without farther assistance.

Such a cursory and confined coverage of botany, therefore, would be in keeping with Mitchell's ([1840]b) concluding remarks in presenting his lesson: "As part of a liberal education, the study of Botany is to be defended chiefly on the ground that it affords an excellent exercise to the mental powers, generates habits of accurate observation and furnishes a source of elegant and innocent recreation and amusement."

Sometimes Mitchell interjected botanical topics while lecturing on other disciplines. While conducting a recitation of a junior class on Sunday evening, January 24, 1841, he read an extensive account of the palm tree and handed out samples of dates to the students, thereby giving them the experience of tasting and learning about the fruit's properties (Mullins, 1840–1841).

Mitchell may have begun to broaden the coverage of his botanical lessons just before his death in 1857. In his forthcoming textbook on the *Elements of Chemistry* (Mitchell, [1856]a), he covered the theory of vegetation, the growth of plants, and the chemical composition of plant substances.

> Page 260. Theory of Vegetation. All plants spring from seeds. A seed consists of two or three principal parts: a small plantule, or miniature plant, already formed; and one or two masses of vegetable matter, called cotyledons, and destined to the nourishment of the plantule, till it shall have struck its roots into the soil, and raised its stem and leaves, to where the latter can feel the influence, and receive the action of the air, and of the sun, producing both heat and light, and the plant be thus prepared to sustain itself.
>
> Page 261. Growth of Plants. There could be no sound theory of vegetation before the rise of the modern chemistry. It is now well understood, that the principal food of plants and trees is carbonic acid, derived partly from the atmosphere, and partly from water which holds this gas dissolved in it. Some require more or less of ammonia, which, when present in small quantity in the air, probably favours the growth of them all. Carbonic acid is everywhere, in the atmosphere, in rain water, as it sinks into the ground, and afterwards issues from it in springs; it is perpetually produced in immense quantities, by the slow decay of vegetable matter, by combustion, and the respiration of animals; so that the supply corresponds very well to the demand for it. The minute fibrils of the roots of vegetables absorb the water thus charged with carbonic acid, and it is carried up into the leaves. The leaves themselves collect the same acid from the

atmosphere; and in the leaf, under the influence of the sun's rays, by a chemical agency of which we [page 262] know absolutely nothing, controlled and rendered effective, apparently, by the power of life, the carbonic acid is decomposed, its oxygen given out, whilst its *carbon* is retained for the use of the plant. In what particular condition or connexion it may be in the first instance, we do not know. The leaves are therefore to the vegetable, what the intestines and lungs, the digestive and respiratory systems together, are to the animal. With these facts before us, we can very well explain the increase in size and weight of Van Helmont's willow.[34]

Page 262. The grove around the buildings of the University furnishes an experiment of the same kind, on a larger scale. The soil is composed essentially of silica and alumina, with a small quantity only of magnesia, lime, potassa, phosphoric acid, and other substances. It contains very little carbon under any form. The leaves with which the trees are clothed in summer, are mostly swept from the ground after they have fallen by the winds of winter; some are burnt at the opening of the spring, but in neither case do they furnish carbon to the soil. Yet the amount of the vegetable matter, rich in carbon, that is brought into existence, year by year, in the grove, including both the annual increase of the trunks and limbs of the trees, and their foliage and fruit, is by no means inconsiderable. The carbon must evidently be derived from the atmosphere; either directly, or indirectly, when it has been dissolved out of the air by rain water as it falls.

The actions of the leaves that have been stated, are neither uniform nor constant. Light is necessary to the continuance of the process, and the brighter the light, the more rapid are the changes. The luxuriance of the vegetation in the intertropical regions, may be supposed to depend upon the influence of the light, hardly less than of the heat, that is poured down upon them in the solar beam. At night, the changes that took place during the day, are reversed; oxygen is absorbed and carbonic acid given out. When the sky is constantly covered with clouds, the formation of vegetable matter is less rapid, but on the whole, the absorption of carbonic acid is much greater than its emission.

In the textbook, Mitchell also describes the chemistry of plants, noting its vastness and numerous compounds. He covers organic analysis and how organic substances are identified. He next discusses neutral organic compounds, the starch group and their composition. He includes starch, gum, cane sugar, grape sugar, and lignine. As mentioned previously, the text of the chemistry manual ends in mid-sentence; clearly, additional coverage was intended.

Botanical Investigations

As noted in chapter one of this book, Mitchell probably associated with several botanists while at Yale and most likely gained a background on that subject. Much that is known about his botanical endeavors outside of the classroom is elucidated in *Chapel Hill and Elisha Mitchell the Botanist* (McVaugh et al., 1996). Mitchell (1818–1847) recorded many of his botanical activities as notes in a ledger book now housed in UNC's Southern Historical Collection. McVaugh and co-authors transcribed the notes dealing with botany, thereby

[34] "Van Helmont planted a willow which weighed five pounds, in an earthen vessel, filled with two hundred pounds of soil, previously dried in an oven, and moistened with rain water. This vessel he sunk into the earth, and he watered his willow sometimes with rain, and sometimes with distilled water. After five years, it weighed one hundred and sixty-nine and a quarter pounds, and the earth in which it was planted, when again dried, was found to have lost only two ounces of its original weight. Here, it was said, was an increase of one hundred and sixty-four pounds, and yet the only food of the willow was pure water; therefore it follows, that pure water is sufficient to afford nourishment to plants" (Mitchell, [1856]a, p. 261).

gaining insight into Mitchell's botanical studies, which were particularly assiduous during the period 1820–1822 and focused on plant collecting and identification. According to an account by Mitchell's daughter Ellen, he would include his daughters on botanical and geological field outings around Chapel Hill. Upon returning home, they verified and recorded the collections in notebooks (Chamberlain, 1945). Mitchell's personal library contained the majority of published floras dealing with North America (Table 3.1), and he therefore had access to the relevant literature up to that period of time.

Table 3.1. A List of Titles of Botanical Books in the Library of Elisha Mitchell Arranged by Author(s) With Brief Bibliographical Citations.

Acharius, Erik. 1810. *Lichenographia Universalis*. J.F. Danckwerts, Gottingen.

Barton, William P.C. 1818. *Compendium Florae Philadelphiae*. M. Carey & Son, Philadelphia. 2 volumes.

Barton, William P.C. 1821–1823. *A Flora of North America*. M. Carey & Sons, Philadelphia.

Bigelow, Jacob. 1814. *Florula Bostoniensis: A collection of plants of Boston and its vicinity*. Cummings and Hilliard, Boston.

Bigelow, Jacob. 1817–1820. *American Medical Botany*. Cummings and Hilliard, Boston. 3 volumes. Volume 1 only.

Bridel, Samuel-Elisée de. 1797–1803. *Muscologia, Recentiorum seu Analysis*. C.G. Ettinger, Gotha. 2 volumes.

Brisseau-Mirbel, Ch.F. 1815. *Élémens de Physiologie Végétale*. Magirnel, Paris.

Browne, Daniel Jay. 1857. *The Trees of America*. Harper and Brothers, New York.

Croom, H.B. 1837. *A Catalogue of Plants, Native or Naturalized, in the Vicinity of New Bern, North Carolina*. G.P. Scott and Co., Printers, New York.

Curtis, William. *Botanical Magazine*. London. 7 volumes. (various volumes and years.)

Darby, John. 1855. *Botany of the Southern States*. A.S. Barnes & Co., Cincinnati.

DeCandolle, Augustin Pyramus. 1818. *Regni Vegetabilis Systema Naturae*. Treuttel et Würtz, Paris. 2 volumes.

DeCandolle, Augustin Pyramus. 1823–1849. *Prodromus Systematis Naturalis. Regni Vegetalis*. Treuttel et Würtz, Paris. At least 12 volumes.

DeLessert, Jules Paul Benjamin. 1820–1846. *Icones Selectae Plantarum*. [Apud Fortin Masson et Sociorum], Parisiis. 5 volumes. Volume 1 only.

Don, George. 1831–1838. *A General History of the Dichlamydeous Plants*. J.G. and F. Rivington, London. 4 volumes.

Gray, Asa. 1842. *The Botanical Text-book*. 1st edition. Wiley and Putnam, New York.

Griffith, William. 1847–1854. *Icones Plantarum Asiaticarum*. Bishop's College Press (volume 1), Calcutta; C.A. Serrao (volumes 2–4), Calcutta.

Jussieu, Adrien de. 1855. *La Botanique*. Langlois et Leclercq, Paris.

Lindley, John. 1830. *An Outline of the First Principles of Botany*. Longman and Co., London.

Main, James. 1835. *Illustrations of Vegetable Physiology*. Orr and Smith, London.

Muhlenberg, Gotthilf Henry Ernest. 1817. *Descriptio Uberior Graminum*. Impensis Solomon W. Conrad, Philadelphiae.

Nuttall, Thomas. 1818. *The Genera of North American Plants*. Printed for the Author by D. Heartt, Philadelphia. 2 volumes.

TABLE 3.1. (CONTINUED)

Persoon, Christiaan Hendrik. 1818. *Traité sur les Champignons Comestibles*. Chez Belin-Leprieur, Paris.

Pursh, Frederick T. 1816. *Flora Americae Septrionalis*. 2nd edition. Printed for James Black and Son, London.

Roget, Peter Mark. 1840. *Animal and Vegetable Physiology*. William Pickering, London.

Smith, James Edward. 1800–1804. *Flora Brittanica*. Typos J. Davis, London. 3 volumes.

Smith, James Edward. 1822. *A Grammar of Botany*. American edition. James V. Seaman, New York.

Sprengel, Kurt. 1807. *An Introduction to the Study of Cryptogamous Plants*. Printed for J. White, London.

Sumner, George. 1820. *A Compendium of Physiological and Systematic Botany*. Oliver D. Cook, Hartford.

Torrey, John. 1824. *A Flora of the Northern and Middle Sections of the United States*. Printed and sold by T. and J. Swords, New York.

Torrey, John. 1843. *A Flora of the State of New York*. Carroll and Cook, Albany. (Published for the New York State Natural History Survey).

Torrey, John and Asa Gray. 1838–1843. *A Flora of North America*. Wiley and Putnam, New-York. 2 volumes.

Willdenow, Karl Ludwig. 1811. *The Principles of Botany and of Vegetable Physiology*. English translation, new edition. Printed at the University Press, Edinburgh.

In the nineteenth century three notable botanists in North Carolina furthered the knowledge of the state's flora: Hardy Bryan Croom, Lewis David von Schweinitz, and Moses Ashley Curtis (Fig. 3.7). Mitchell interacted with each of these men on various levels. A native of New Bern, North Carolina, Croom is considered the most serious student of North Carolina botany before Curtis arrived in North Carolina (McVaugh et al., 1996). There are no known extant records to confirm that Croom and Mitchell ever met. Croom received his A.B. in 1817 from UNC and therefore preceded Mitchell's professorship there. If Croom did attend graduation ceremonies at Chapel Hill in 1820, when he was awarded an A.M.,[35] he could have met Mitchell. It is more likely that the two men corresponded with each other while Croom (1833) was completing an article on gama grass for publication in the *Harbinger*, a campus newspaper supervised by university faculty members. Mitchell (1833) probably oversaw the publication at that time as he gave a lengthy commentary of his own in the same issue and praised the grass's economic potential (McVaugh et al., 1996).

> The individual who offers these few remarks to the readers of the *Harbinger*, is not much given to enthusiasm on any subject; but in regard to the Gama Grass, is sanguine in the expectation and belief that a vast addition will be made by it to the numbers, wealth and comforts of the population of North-Carolina.

Croom and his family were drowned in a shipwreck off the Outer Banks of North Carolina near Ocracoke in 1837, thus ending his promising botanical pursuits.

[35] In the early days of the university, the Master of Arts degree (Artium Magister, *i.e.*, A.M.) could be obtained by alumni who graduated and petitioned for one after pursuing a learned profession for three years (Battle, 1974, 1: 781).

Fig. 3.7. North Carolina botanists who were contemporaries of Elisha Mitchell: (left) Hardy Bryan Croom (image RC 12849) (Courtesy of State Archives of Florida). (center) Lewis David von Schweinitz, ca. 1835 (from *A Memoir of the Late Lewis David von Schweinitz, P.D.*, by Walter R. Johnson [1835], North Carolina Collection, Wilson Special Collections Library, University of North Carolina, Chapel Hill). (right) Moses Ashley Curtis, ca. 1870 (Courtesy of North Carolina Collection Photographic Archives, Wilson Special Collections Library, University of North Carolina, Chapel Hill).

In summer of 1819, about 18 months after starting his professorship at UNC, Mitchell visited Lewis David von Schweinitz, a Moravian theologian and botanist who lived in Salem, North Carolina. Moravian records provide an entry for Mitchell's arrival in June 1819: "Professor Mitchell, from the University at Chapel Hill, came to stay for some time so that he might gather more knowledge of botany from Br. von Schweinitz" (Fries, 1947). During this visit, Mitchell also accompanied Schweinitz on a botanical excursion in Salisbury, North Carolina (Mitchell, 1819a). Schweinitz was clearly impressed with Mitchell's role as a professor and as a budding botanist. He expressed these sentiments to his botanist friend John LeConte, writing, "My friend Prof. Mitchell at Chapel Hill (where they have got together as smart & zealous a badge of Yankee professors as we would wish who are evidently inspiring a scientific spirit into the hitherto inanimate Carolinian youth) has made astonishing progress in botanical labors" (Schweinitz, 1820). In a letter to Schweinitz, Mitchell (1819b) noted his desire to return to Salem during another vacation. Mitchell's association and correspondence with Schweinitz coincided with his incipient interest in identifying plants in the vicinity of Chapel Hill. There are six known extant letters that Mitchell (1819b, 1820, 1821a, 1821b, 1821c, 1821d) wrote to Schweinitz. Some of these letters accompanied plant specimens that he sent to Schweinitz. For instance, in 1821 he sent Schweinitz over 225 plants, some to identify, others to donate to Schweinitz's collection (Mitchell, 1821d).

In 1821, when Schweinitz moved to Bethlehem, Pennsylvania, to assume a new position in the Moravian Church, Mitchell's activities in botany waned. He considered Schweinitz's removal to Pennsylvania as the loss of "a much valued friend," which was also like the separation of brothers (Mitchell, 1821b). He also predicted that his advances in botanical studies would suffer. After settling into his job in Pennsylvania, Schweinitz (1822) asked Mitchell whether he was continuing his botanical pursuits or if mineralogical and geological investigations had distracted that pastime. He also inquired if Mitchell would collect some specimens for him. "I want badly to procure of your means some particular plants vz. the beautiful new anemone & the *Carex* a little Salemward of Hillsborough. Perhaps you might even get me a few Wilmingtonians." Mitchell's friends showed compassion over the loss of his botanical mentor. Lawyer and plantsman of Orange County, North Carolina, Archibald D. Murphey (1822) offered consoling words. "I regret with you the Loss of Mr. Schweinitz. He was an ornament to North Carolina, and a Man whom I loved with great affection. I'm glad you have agreed to become the Lord Paramount of our Botany. If you will devote your attention to the Subject, you will soon occupy any place you please in this or any other Department of Science." Murphey also noted that the last act of his public duties in the General Assembly was to recommend to the Board of Internal Improvements to employ Denison Olmsted to carry out a geological and mineralogical survey of North Carolina. Even after retiring from public service, Murphey would make a recommendation that a Botanical Survey[36] be made if Mitchell would agree to undertake the work.

In addition to losing contact with his in-state botanical friend, Mitchell's academic responsibilities increased and required considerable time. By 1824 his teaching and administrative responsibilities as well as preaching increasingly consumed time that could have been directed to pursuing botany. Of particular note, in 1826 he acquired the responsibility for giving lessons in chemistry, mineralogy, and geology when he relinquished his post as mathematics professor to assume the professorship in those subjects. His geological excursions in the state, however, provided an opportunity to observe the natural world and collect plants. Concerning these trips, his colleague, Charles Phillips (1858b), noted how Mitchell would fill "his herbarium with many specimens of the plants he met."

Mitchell also interacted with Moses Ashley Curtis, an Episcopalian preacher in Hillsborough, North Carolina, and noted botanist. Curtis moved to Hillsborough, North Carolina, where he became the pastor at St. Matthews Church in 1841. He also conducted extensive botanical investigations and authored a number of significant publications based on his pioneering work. Curtis had wished to establish a network of scientists through correspondence to advance an understanding of natural history in the state (Berkeley and Berkely, 1986). Curtis and Mitchell developed an association and visited together in Chapel Hill around April 1836 (Berkeley and Berkeley, 1986). This association formed after Mitchell's active period of botanical pursuits, but nevertheless he shared his botanical expertise and provided plant specimens with Curtis (McVaugh et al., 1996). Mitchell gave Curtis a volume compiled by botanist John Lyon and containing published sources to the names and descriptions of North American plants (Lyon, 1845). Curtis highly respected Mitchell's contributions to the local flora and honored his friend by naming a sedge, *Carex mitchelliana,* for him in 1834.

It is likely that Mitchell developed a personal herbarium of the local flora. According to his own notes, Mitchell used a vasculum when collecting plants (Phillips, 1858b). In correspondence with Schweinitz, Mitchell (1821c) mentioned specimens that had been wafered, implying that they were specimens to be mounted on herbarium paper. Soon after his marriage Mitchell (1820) intended to recruit his wife in assisting him with his botanical pursuits. He told Schweinitz that he was giving "her first lessons in botany. She says she will not like it and I tell her in return she must then give up all pretensions to good taste and that I'll whip her if she don't. I think I shall make a proselyte of her." Apparently his wife developed some botanical skills as she later assisted in drying and preserving specimens

[36] A botanical survey came to fruition over four decades later when Mitchell's botanical friend, Moses Ashley Curtis (1867), prepared a catalog of the plants of North Carolina. The publication represented *Part III, Botany,* of the *Geological and Natural History Survey of North Carolina.*

(Mitchell, 1821a). A number of plant specimens, representing twenty-eight species, and presumably collected by Mitchell, are held in Schweinitz's collection, which now resides at The Academy of Natural Sciences of Drexel University (McVaugh et al., 1996). When Mitchell's books and scientific apparatus were sold after his death, there was no mention of his herbarium. Its fate is unknown. Also, there are no known botanical publications by Mitchell, except for a number of articles, all of minor importance, that were published in the *Harbinger* (Table 3.2).

TABLE 3.2. A LIST OF ARTICLE TITLES WITH CITATIONS WRITTEN BY ELISHA MITCHELL AND PUBLISHED IN THE *HARBINGER*.[37]

1833

Rural economy. 1(1): 3, Tuesday, August 27.
Remarks [on an article concerning gama grass]. 1(2): [5], Tuesday, September 3.
Rural economy. Notice of the new *American Orchardist*. 1(3): 11, Tuesday, September 10.
Rural economy. Kendrick's *Orchardist*—Apples. 1(4): 15, Tuesday, September 17.
Mathematics. 1(7): 26–27, Tuesday, October 8.
Rural economy. The pear. 1(10): 39, Tuesday, October 29.
Luminous meteors. 1(13): 51, Thursday, November 19. (unsigned)
The pines. 1(15): 59, Thursday, December 5.
The long-leaved pine. 1(16): 63, Thursday, December 12.
Products of the pine (from Michaux). 1(17): 67, Thursday, December 19.

1834

Rural economy. [*Southern Agriculturist*]. 1(20): 79, Thursday, January 16.
Products of the pine—in conclusion. 1(21): 83, Thursday, January 23.
Counties of North Carolina. 1(22): 87, Thursday, January 30.
Counties of North Carolina (continued). 1(25): 98–98, Thursday, February 20.
Counties of North Carolina (concluded). 1(26): 103, Thursday, February 27.
Natural history. 1(27): 107, Thursday, March 6.
How to ride an alligator. 1(30): 119, Thursday, March 27.
Engraving on steel. 1(31): 123, Thursday, April 3.
Rural economy. The vine. 1(32): 127, Thursday, April 10.
Methods of producing fire. 1(35): 139, Thursday, May 1.
The *Watchman*. 1(36): 143, Thursday, May 8.
The *Watchman*. 1(38): 158, Thursday, May 29.
Rural economy. Red and white clover. 1(40): 160, Thursday, June 12. (unsigned)
Slavery amongst the insects. 1(44): 175, Saturday, July 12.

Elisha Mitchell's botanical knowledge was further applied in the areas of agriculture and horticulture. In an article entitled "Agricultural Speculations," he wrote that agriculture was not mere theory for him but was based on his engagement "in the business of Agriculture" when he lived in Connecticut (Mitchell, 1825c). Mitchell also drew heavily from his reading of scientific and agricultural literature and was fond of observing and reflecting on the subject. According to historian Melton A. McLaurin (2003), Mitchell "was one of the state's earliest and most ardent advocates of scientific agriculture."

[37] The *Harbinger* was a weekly newspaper supervised by UNC professors. Published in Chapel Hill by Isaac C. Patridge, the serial was issued from August 1833 to August 1834. The contributions signed with "N" are undoubtedly from the pen of Elisha Mitchell. A few unsigned articles were most likely also by Mitchell.

Agricultural Interests

Mitchell's major contributions in agriculture were in connection with North Carolina agricultural societies, for which he published papers and gave presentations. He also corresponded with various members of the North-Carolina Agricultural Society as well as inquiring citizens. Mitchell's (1825c) earliest known published paper on agriculture was based on an essay that he had sent to James Mebane, a North Carolina legislator and longtime UNC trustee, in 1822. Two years later he read the essay before the North Carolina Board of Agriculture at its meeting held in Raleigh in December 1824. The board immediately gave its approval to publish the talk (Anonymous, 1824). The resulting article demonstrates Mitchell's progressive view of the field. He mentions how countries such as France had sent naturalists to foreign lands to make scientific discoveries, especially with an eye toward potential agricultural products. He noted that French naturalists sent seeds of plants and trees back to France, where they were planted for the country's advancement. Mitchell asked, "Is it not possible for us to draw from other countries some vegetables which may be cultivated with advantage in our own?" He thought that exploration for potential plants should focus on areas that lie in the same latitude though he further explained that such introductions are not always successful. For example, he knew that the olive tree grows in the same latitude, yet it does not grow well in North Carolina. It grows best in calcareous soil, and North Carolina's soil is siliceous. The extremes of hot and cold also place stress on the tree.

UNC botanist William C. Coker (1946) considered this publication an extremely insightful contribution for its time. "[Mitchell] foresaw many of the problems and suggested improvements in agriculture in a way that has proved to be almost prophetic. His suggestions for the improvement of soil, use of adequate cover crops, introduction of foreign plants, and improvement by selection of our own crops are truly modern in their attitude."

Mitchell shared information on economically important plants with readers of newspapers, particularly those of Raleigh, North Carolina. In 1853 Mitchell (1853) wrote a series of eight newspaper articles titled "Agriculture and the Geological Survey" that were published in the *Raleigh Register*. In the first article he praised the wealth of agricultural literature, particularly treatises on farming and agricultural journals, that was available to North Carolinians. The next three articles focused on the application of chemistry in farming and on soil nutrition, areas of research that were being considered in the state. He mentioned that the UNC trustees had created a professorship of agricultural chemistry and a person to fill it. The fifth article concerned his response to a criticism printed in the *Raleigh Register* by State Geologist Ebenezer Emmons. The concluding three articles regarded the geology of the state. He also wrote a noteworthy article on the grapevine borer (Mitchell, 1854).

Mitchell's broad knowledge of scientific and agricultural topics gained the respect of citizens and scholars. A prominent person had been slated to give the annual address before the North-Carolina Agricultural Society at the State Fair in Raleigh in October 1856. He, however, cancelled due to circumstances, and Mitchell was approached to fill in. Thomas Robbin (1856), a member of the society, considered Mitchell because of his "fondness for agriculture & horticulture & a devoted attachment to the interests of North Carolina in those elements of her prosperity." Although the invitation to speak was given within only a few weeks of the talk, Mitchell agreed. He reviewed the history of farming, beginning with Egypt, then Palestine (the Holy Land), Greece, and Italy. His talk then advanced to comparatively modern times and focused on English agricultural pioneer Jethro Tull (1829) of Great Britain, who had published a treatise on horse-hoeing. Next Mitchell introduced the topic of chemistry and its impact on agriculture, which composed nearly half of his presentation. Mitchell's talk was published (1856b).

Richmond M. Pearson (1856), an American jurist and farm owner, of Richmond Hills, Asheville, North Carolina, read the paper and wrote to Mitchell to express his pleasure with the instructive value of the talk as well as its overall information. In this letter, Pearson also related his own experience in preparing the soil on his farm. With his practical knowledge, he wished to inform Mitchell of another way in which soil could be conserved. He turned over the soil with a light plow, followed by a subsoil plow without turning up the ground. This technique allowed the roots to grow deeper and retain moisture, thus preventing soil erosion.

Horticultural Interests

Mitchell (Fig. 3.8) also held an interest in horticulture. His extensive geological excursions in the state, his travels to other states, and his correspondence provided opportunities to acquire plants for cultivation and to visit gardens and nurseries. His keen sense of observing the flora aided him in his discoveries of plants, particularly those that were edible. In a letter of August 28, 1825, Joseph Saunders (1825) of Edenton, North Carolina, notified Mitchell that he would send him some slips of grapevines. On a geological survey in the summer of 1845 Mitchell (1845a) rediscovered a type of blackberry growing in the vicinity of Greensboro, North Carolina. On previous occasions he had tried growing it in Chapel Hill but without success, so he wished to obtain new stock and to grow it again. On an excursion to the western part of the state Mitchell (1829a) observed and described a cranberry plant from Ashe County, which he found growing "almost exclusively on the east side of New River in the glades and swamps lying between that stream and the Blue Ridge." He astutely attributed its lush growth, which was mostly on the northwestern declivity of the Blue Ridge, to the character of the soil and to the elevation and moisture of the substrate.

Mitchell kept current with the plant trade and new introductions by acquiring nursery catalogs. American landscape designer and horticulturalist Andrew J. Downing (1847), who was also the proprietor of a nursery at Highland Gardens in Newburgh, New York, sent Mitchell his nursery catalog. Mitchell had mentioned seeing an ever-bearing mulberry. Downing pointed out that the tree was probably not a new variety.

On a trip to New York in the summer of 1836, Mitchell stopped in Philadelphia, where he visited two noted gardens: Carr's Botanical Garden (also known as Bartram's Garden) and Landreth's Garden. In a letter to his daughter Ellen, Mitchell (1836a) described his observations at Carr's.

> The gardens are chiefly on some knolls rather sandy and gravelly that rise on the west side of the river [*i.e.*, Schuylkill River]. It is a very pleasant spot and I cannot tell you all I saw—ever so many kinds of cactus, the plant that yields the milk that makes India rubber. Carr's son cut off a leaf and I tried on the spot. Oranges and lemons in flower and fruit, the agave too much cultivated in Mexico, the large palm, etc., etc. There were very many fine things, and yet the gardens did not meet my expectations after all. The family treated me with great politeness. On my way back I stopped at Landreth's garden, a rival establishment and saw more exotics.

Fig. 3.8. Elisha Mitchell, 1850s.

(Courtesy of North Carolina Collection Photographic Archives, Wilson Special Collections Library, University of North Carolina, Chapel Hill).

Physical Geography Investigations

During his geological excursions in the western part of North Carolina, Mitchell's scientific curiosity no doubt piqued an interest in the region's physical geography. From reading about plant hunters André Michaux and son François André, Mitchell was aware that the [Appalachians] attained their greatest altitude in North Carolina ([Phillips], 1858b). André had noticed that trees and plants growing on the high elevations of the mountains of Yancey County were not encountered again until Canada. The father and son team heeded "the well-known law which determines the height to which we must ascend from any place in order to secure the mean temperature that prevails in some more northerly zone" ([Phillips], 1858b). In his second (and last) geological report as North Carolina geologist, Mitchell (1829a) mentioned that the Black Mountain was probably higher than Grandfather Mountain and represented the highest peak between the White Mountains of New Hampshire and the Gulf of Mexico.

In 1835 Mitchell's investigations of the mountains and their measurements began in earnest. He used barometric pressure readings and mathematical calculations to measure the mountains. Regarding his first visit of 1835 to the Black Mountain range, he recorded in his memorandum book the following: "Objects of attention–Geology; Botany; Height of the Mountains; Positions by Trigonometry; Woods, as the Fir, Spruce, Magnolia, Birch; Fish, especially Trout; Springs; Biography, &c." (Mitchell, 1818–1847). He published his findings in a local newspaper (Mitchell, 1835c) and then published these as a scientific paper (Mitchell, 1839). Historian S. Kent Schwarzkopf (1985) discusses in depth the 1835 trip as well as four subsequent trips that Mitchell took in 1838, 1844, 1856, and 1857.

Particularly noteworthy was Mitchell's trip of 1844 to determine the highest peak of the range. Eleven years later Senator Thomas Clingman contested that Mitchell had not measured the highest peak, but that he (Clingman) had. Evidence supports the fact that Mitchell had actually climbed Mt. Gibbes, which was not the highest peak. Other evidence shows that Mitchell had indeed climbed the highest peak in 1835 and measured it at 6,476 feet above sea level (208 feet less than the now known altitude of 6,684 feet above sea level).

Clingman and Mitchell wrote a number of letters in support of their own claims to measuring the highest peak that were published in the Asheville, North Carolina, newspapers. Other relevant publications include a letter by Clingman ([1856]) and one by Mitchell (1849b). Resolved to settle the controversy, Mitchell returned to the Black Mountain in June 1857 to verify his measurements and to prove Clingman wrong. On June 27 he set out alone most likely to contact some of his previous guides. En route he experienced a fatal accident when he fell off the edge of a waterfall into a pool of water (Fig. 3.9), a nearly forty-foot plummet. His misfortune happened in the evening, as indicated by his stopped pocket watch (Fig. 3.10), and he subsequently drowned. According to Asheville citizen James Buxton (1857), Mitchell's body was "in a complete state of preservation, with only a bruise on his forehead" when searchers found him in the icy water on July 7, 1857. Mitchell's body was transferred to Asheville, where interment took place in the churchyard of the First Presbyterian Church. Several days later his wife Maria gave permission to rebury his remains on the top of the highest peak of the Black Mountains destined to be called Mount Mitchell.[38] His body was exhumed in mid-June 1858 and reburied on the peak. Historians S. Kent Schwarzkopf (1985) and Timothy Silver (2003) provide details on the search for his body and its subsequent burial atop the peak now known as Mount Mitchell.

The gravesite was first marked by a pile of rocks and a plaque. Zebulon Baird Vance, a friend of the Mitchell family and governor of North Carolina during the Civil War and again from 1877 to 1879, led an effort to erect a granite monument by placing announcements in the Asheville newspaper *Spectator* and seeking contributions (Vance, 1858). Pledges were not sufficient, and the effort was unsuccessful. Three decades later, in 1888, the gravesite was replaced by a monument of white bronze (Fig. 3.11), but vandals destroyed that around 1915. Currently Mitchell's grave (Fig. 3.12) is enclosed within a fence and marked with a plaque.

Nearly 135 years after Mitchell's fatal fall on Mt. Mitchell, mathematician Perrin Wright (1992) conducted long-distance surveying from the area of Yeates Knob, whereby he was able to determine the elevations of the peaks of the Black Mountains. If Mitchell had used

[38] Mathematician and historian Archibald Henderson (1957) noted that a "triangulation marker, placed by the United States Coast and Geodetic Survey on Mount Mitchell in 1876, first officially gave it that name."

this surveying technique, "he could have identified the high peak in 1835 beyond question thereby avoiding the controversy that led to his death."

Fig. 3.9. Mitchell Falls, ca. 1858.

(Courtesy of North Carolina Collection #D734-5, Pack Memorial Library, Buncombe County Public Libraries, Asheville, North Carolina).

Fig. 3.10. Pocket watch belonging to Elisha Mitchell.

(Courtesy of North Carolina Collection Gallery, Wilson Special Collections Library, University of North Carolina, Chapel Hill).

Fig. 3.11. Mitchell's Monument on top of Mount Mitchell, North Carolina, ca. 1910.

(Post card No. 962, published by International Post Card Co., New York; made in Germany. From the collection of William R. Burk).

Fig. 3.12. Elisha Mitchell's grave on top of Mount Mitchell, North Carolina, 1953.
(Courtesy of North Carolina Collection Photographic Archives, Wilson Special Collections Library, University of North Carolina, Chapel Hill).

Engagement Beyond Teaching

Bursar

In addition to his teaching discussed above, Mitchell assumed other responsibilities; chief among them were bursar, preacher, and director of library. In his early years at UNC Mitchell may have been involved with the supervision of buildings. Around 1818, Professor of Ancient Languages William Hooper ([1818]) asked the Committee of Appointment to relieve him of the office of supervisor of "publick buildings." He put forward Mitchell's name to replace him, noting that Mitchell "has no family, [and] it will be more convenient for him to attend to this business than myself."

Denison Olmsted had served as the receiver of college dues (room rent, tuition, etc.) after his arrival at UNC in 1818, but he resigned that post in 1824. Professor of Languages Ethan Allen Andrews took over that responsibility in January 1825 (BOT 5: 29). In 1827 the trustees (BOT 5: 96–99) created the position of Superintendant to oversee not only the financial matters of the institute but also its property. Thomas H. Taylor was appointed to the post, but this arrangement was deemed unsatisfactory by the board (BOT 5: 117). The trustees later resolved that a member of the faculty should fill the position (BOT 5: 146), and Mitchell subsequently filled the post.

As receiver of college dues, he collected money that the students brought with them and disbursed the same to pay for their board, tuition, fees, room rent, servant hire, washing, and other essential expenses. He also paid the faculty their salaries as well as paid for building repairs and related work. At the end of each session, he sent the university treasurer a report noting the amounts collected and disbursed. He also prepared circulars (Fig. 3.13) that were sent to the parents or guardians of students at the end of each session. Special note was made of the number of absences each student may have made from prayers and recitations. In addition to his faculty salary, he was compensated for this additional, time-consuming work. He initially received about $200 per year and as much as $795 in 1839. He received additional compensation for this work, but it "was ordered to be so regulated that his total salary should not exceed that of the President" (Battle, 1974, 1: 652).

The Executive Committee of the Board of Trustees (BOT 15: 40) appointed Mitchell bursar on July 4, 1836. As bursar, Mitchell's added responsibilities now included basic work and repairs on campus buildings and structures, construction of rock walls, and other similar endeavors.

Mitchell became burdened by the amount and complexity of the work. In his bursar's report of November 29, 1838, to Charles Manly, treasurer of the university, Mitchell (1838b) quipped that this work "is of as complicated and vexatious a character as is done in N. Car[olina]." He held this duty until his death in 1857, when Professor of Greek Language and Literature Manuel Fetter became the new bursar (BOT 15: 210). The post was abolished in 1912.

Even prior to serving as bursar, the strain of being overworked contributed to his thoughts of leaving his post at UNC. Around 1826 he had considered, but turned down, "a professorship in a respectable College in one of our northern cities with a salary of 2100 dollars" (Mitchell, 1828c). In 1829 the University of Alabama considered him for the professorship in natural history, but John Fielding Wallis won the seat (Sellers, 1953).

Campus Improvement and Landscaping

Mitchell was instrumental in improving the campus landscape. As early as 1826, he assumed a role in improving the campus grounds with such work as trimming the vegetation (Mitchell, 1826). Under his direction and with University President David L. Swain's ardent support, work began in 1838 to enclose the campus proper with rock walls to replace the deteriorating rail fences on its perimeter. Following a technique already well known in New England, of not using mortar or other cementing materials, Mitchell (1818–1842, p. 184) and his workers, probably enslaved people whom he owned or hired from other enslavers, stacked and interlocked rocks into place.

The construction of the rock walls was a daunting undertaking. Writing to Charles Manly, the university's secretary and treasurer of the Board of Trustees, Mitchell (1838c) explained the work that had been done.

> In hauling the rock for the wall that has run back of College my own cattle were principally employed. But they having given out from the extreme heat I was obliged to hire a waggon [*sic*] for three days and a half and I may mention also in this place that the <u>hauling</u> of the rock and <u>building</u> of the wall having been done altogether by <u>my</u> hands. I expect to get a like amount of work out of the College servants in the next summer vacation.

Chapel Hill, Dec 9th 1824

SIR,

IN conducting the education of the youth committed to our care, we, as the Faculty of the University, and as individuals, are accustomed to look around with solicitude, for the most efficacious means of attaining our object. The students of a college, are most of them at a distance from home. With many, the period of entering within its walls, is the first at which they have been left to choose, or at least to continue their own associations and friendships. Amidst the dangers to which they become exposed, the only means of preventing evil consequences, or of recovery to such as have incurred them, must be found in previous habits, and if these should fail, in such motives only as may be addressed immediately to the understanding and the heart. No means, then, should be neglected, by which such motives may be fortified, that the utmost assurance may be given to the prospects and anxious wishes of parents, in the education of their children. Among these means, we have ever been convinced that parental influence and authority, and the estimation of relatives and friends, are of the first importance. The same is to be said of guardians, on whom the paternal care and responsibility devolve. Through a reference to these, we can ever address ourselves in the language of counsel and remonstrance, with the greatest prospect of success. But parents, for the most part, have but little opportunity of knowing the course which their sons are taking in the university, or the dangers with which their habits are threatened: and too often, for want of this information, and of the seasonable and salutary interposition which would be the natural consequence, a youth passes on, regardless of his opportunities, and of every effort of the Faculty, until it ceases to be in our power, consistent with our duty to the institution, to permit his longer continuance in the enjoyment of its privileges.

We propose therefore to address a letter to you, at the close of every session, while your son, or the son of your friend, now under your guardianship and parental direction, shall continue in the college, containing statements from which you may form a judgment of his standing and progress during the past session. This to many will be valuable, by preventing or putting an end to needless anxieties, if not by imparting the richest enjoyment which a parent can feel. It will also be conferring one of the highest rewards upon merit in a youth, the knowledge that he renders his parents happy. From much experience in colleges the Faculty have found it to be a general rule, that the fidelity of a student in the prosecution of knowledge, and in most cases too, the correctness of his morals, are to be

Fig. 3.13. Letter to the parents of students at the close of every session, December 9, 1824.

(Courtesy of North Carolina Collection, Wilson Special Collections Library, University of North Carolina, Chapel Hill).

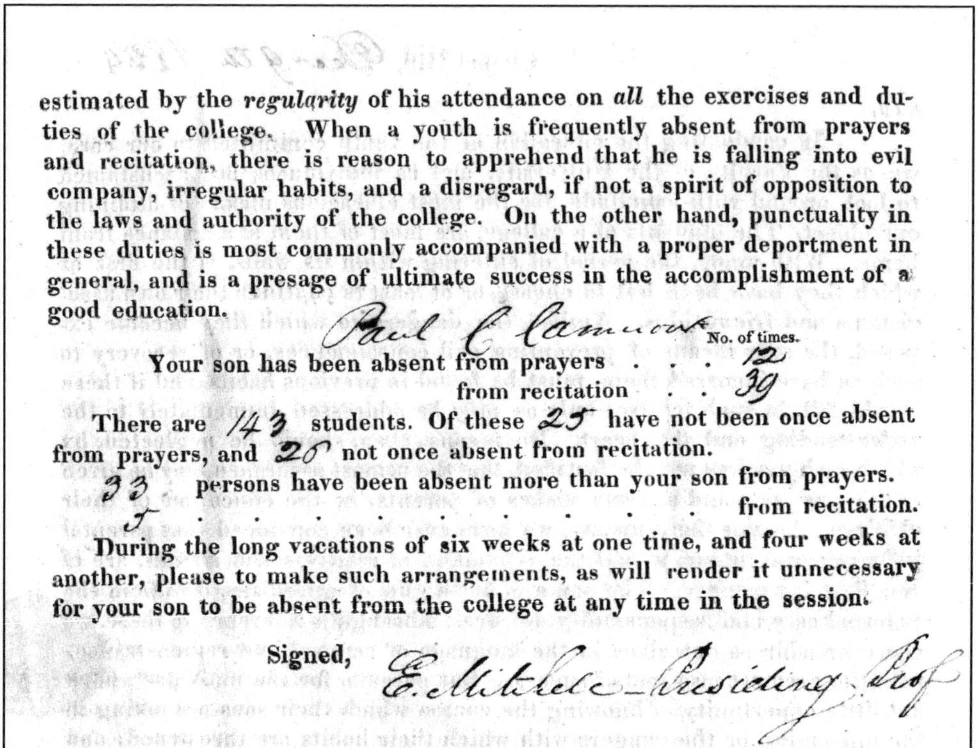

estimated by the *regularity* of his attendance on *all* the exercises and duties of the college. When a youth is frequently absent from prayers and recitation, there is reason to apprehend that he is falling into evil company, irregular habits, and a disregard, if not a spirit of opposition to the laws and authority of the college. On the other hand, punctuality in these duties is most commonly accompanied with a proper deportment in general, and is a presage of ultimate success in the accomplishment of a good education.

Paul C Cameron

No. of times.

Your son has been absent from prayers . . . 12

from recitation . . 39

There are 143 students. Of these 25 have not been once absent from prayers, and 26 not once absent from recitation.

33 persons have been absent more than your son from prayers.

5 . . . . from recitation.

During the long vacations of six weeks at one time, and four weeks at another, please to make such arrangements, as will render it unnecessary for your son to be absent from the college at any time in the session.

Signed, E. Mitchell—Presiding Prof

Fig. 3.13 (Continued). Letter to the parents of students at the close of every session, December 9, 1824.

In June 1841, a reporter (Anonymous, 1841) for the newspaper *Raleigh Register* noted that "the 'campus'–or grounds immediately adjoining the buildings–will soon be enclosed with a neat and substantial stone wall." Much of the campus had been enclosed by 1844, and Mitchell received $500 as compensation for his work and supervision. An additional section (40 rods, approximately 660 feet) of new wall was completed by 1852 under the direction of Captain John Brooks Tenny[#], an Orange County, North Carolina, farmer (UNC Trustees, 1852).

These stone fences served several functions for the university landscape. From a practical perspective, they were supposed to keep out the village livestock (cows and pigs) that freely roamed the region, according to the custom of the times. In his history of the university, President Kemp P. Battle (1974, 1: 471) regarded this free-range grazing as "evil." Even after the walls were completed, however, the trustees continued to deal with the livestock pasturing on the grounds (BOT 16: 273). Aesthetically, the walls enhanced the picturesque setting of the campus. Soon the villagers adopted the practice of erecting similar fences around their yards. With age, the rock fences took on a special character as mosses, lichens, and the elements softened their ruggedness. Having been repaired and filled in with mortar, some of the walls survive today. They continue to be a signature feature of the university campus.

The construction of the rock walls ushered in the beautification and improvement of the grounds that would prosper and flourish through the late 1850s. Plantings and landscaping contributed to those efforts. President Swain (1842) provided strong leadership in improving the grove and campus grounds. In addition to his encouragement for enclosing the campus with rock walls, Swain was enthusiastic to plant various species of trees and grass plots. According to UNC student David A. Barnes (1840), workers were rendering the college grounds more inviting: "ornamenting the campus by planting out trees" in February 1840. By June 1841, an anonymous writer (1841) was impressed with the changes, noting: "Trees have been set out in various parts of the College area, and the venerable old grove has been so grubbed and trimmed and thinned, as to lose much of its rude forest garb, while increased vigor has been given to the growth that remains." In early 1844, renowned landscaper and architect Alexander Jackson Davis was a guest at a meeting of the trustees where he presented his plans for landscaping the campus. He proposed several campus gardens, including plots meant for experimental agriculture, but this proposal came to naught (Allcott, 1986).

Efforts to hire a landscape gardener began in 1845. Nearly two years lapsed until the position was filled by John Loader, who soon established a nursery in the rear of the campus proper, which establishment is discussed below. There, a stock of plants and shrubs was grown for subsequent transplanting to the campus grounds. With the campus landscaper on board and with an eagerness to continue the campus improvements, Swain asked Mitchell for his suggestions for furthering the beautification of the campus grounds. Although Mitchell (1847) was overwhelmed with college duties of preparing circulars and bursar accounts, he sent Swain a letter on his views to improve the campus in early December 1847. His letter included a plan (Fig. 3.14) of the village of Chapel Hill and the grounds around the university proper (known as "the grove"). To maintain and enhance the vigor and health of the trees in the grove, he noted that the soil should be amended with potash. The removal of leaves each fall contributed to a deficiency of this mineral, but he thought that the then common practice of spreading ashes from the students' rooms would help alleviate the problem. He

[#] The surname is also variously spelled Tenney or Tinney.

recommended the grounds should be coultered.[39] He also thought that it would be futile to grow grass under the trees in the grove because the tree roots would "rob" the grass of all moisture. Mitchell suggested that shrubs and plants be considered in the open spaces as noted on his plan. He also recommended such plants as sunflowers, Jack beans, and hollyhocks.

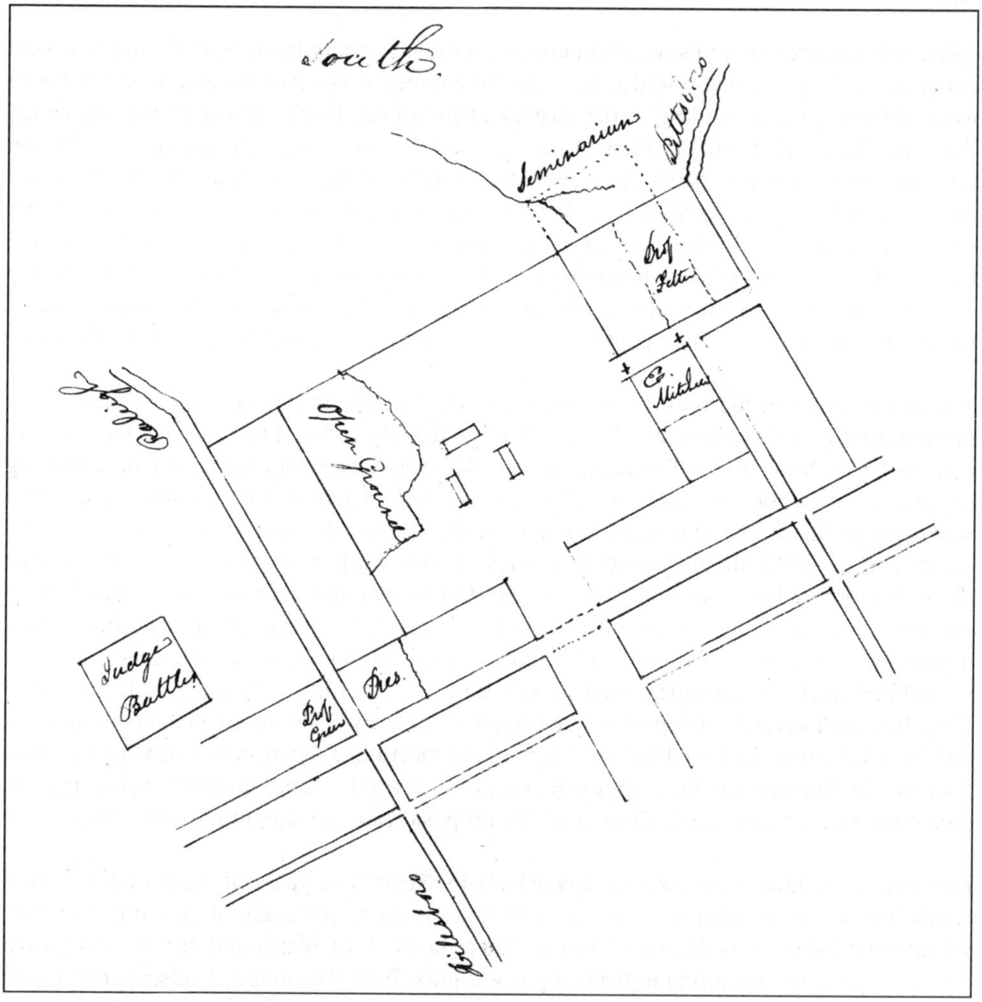

Fig. 3.14. Drawing of UNC campus showing open grounds and proposed site of a seminarium (plant nursery) with natural springs by Elisha Mitchell (1847).

(From letter of Elisha Mitchell to David L. Swain, December 11, 1847. Miscellaneous Papers, Volume 21.4, p. 70, State Archives of North Carolina).

Because of the extensive area of the grounds needing planting, he put forward the idea of setting up a nursery, which he called a "seminarium," where plants could be raised. Two places were considered for the nursery: behind Mr. Loader's house or on land behind

[39] Coultering plows use a knife or sharp disc attached to the beam of a plow to make vertical cuts into the surface of the soil. This permits a clean separation and effective covering of the soil and materials being turned under. <http://www.merriam-webster.com/dictionary/coulter> (Accessed February 19, 2012).

Professor Fetter's house. Mitchell favored the latter location, which had "two copious and perennial springs," and it became the site of the nursery, where he desired to grow shrubs and trees from abroad. (Today, Sitterson Hall occupies in part the area on the UNC campus.) He acquired seeds from a number of foreign correspondents in countries such as Argentina, Mexico, Spain, and the area around the Dead Sea. Among the correspondents was Daniel Moreau Barringer (1853), a foreign minister in Madrid and a former student of Mitchell. He sent two boxes filled with acorns of the cork oak to a friend in Raleigh from whom Mitchell was to request specimens. Mitchell also desired to obtain herbaceous plants. On August 27, 1850 he wrote to William A. Graham, head of the U.S. Department of the Interior and acting supervisor of War and Navy, requesting living sod specimens of buffalo grass[40] (Mitchell, 1850b). Graham (1850) passed along this request to a clerk in the Agricultural Bureau, who promised to procure the grass the following year. In another transaction, Graham arranged to have a package of seeds from the Mediterranean that Lieutenant Porter of the U.S. Navy presented and that would be forwarded to Mitchell.

In 1849 Mitchell (1849c) noted that improvements around the college buildings in 1847 were mainly at the expense of Swain and himself, but that an appropriation had been made for the years 1848 to 1849. The trustees allotted $1,000 annually for landscaping: $400 for Loader's salary; and $600 for other expenses, such as wages of workers, purchases of plants and shrubs, and the use of horse and wagon. In 1848, a faculty committee composed of Elisha Mitchell, William Mercer Green, and Charles Phillips (1848) submitted an account of expenditures (Fig. 3.15) for landscaping in 1848. The account showed that $991 was spent, including the purchase of oats for a lot as well as other seeds. By the close of 1849, Mitchell (1849c) thought that a "good deal" of work had been accomplished and that a look of grace and beauty to the approaches of the buildings and to the walks had been created. When Loader resigned at the end of 1851, Thomas Paxton[41] became his successor. Both men trained a number of enslaved individuals who learned horticultural skills and transferred plants onto campus.

Director of Library

Following his first year at UNC, Mitchell began buying books in earnest for the library, and he could be considered the director of the library at that time. Others who assisted in the management of the library were Joseph Hubbard Saunders, tutor and librarian (1824); William Hayes Owen, tutor and librarian (1836–1843); and Ashbel Green Brown, tutor (later adjunct professor) and librarian (1844–1856).

A glimpse into Mitchell's (1818–1842, p. 29) purchase of books is gained from passages in one of his private journals held in the UNC Southern Historical Collection. He noted that the nearest bookstore was twenty-eight miles distant, in Raleigh. He thought it impractical to send for every item needed as it seemed "a matter of much vexation and expense." He often was unable to acquire what was needed. He preferred to purchase books for the

[40] Mitchell (1850b) noted that "buffaloe grass" grew east of the Rockies and didn't produce seeds. He was interested in inoculating plugs of the grass, probably on the UNC campus. He believed that the area so treated would soon be covered with grass.

[41] Paxton's horticultural accomplishments were admired. Cornelia Spencer (1884b) noted that "everything that he put in the ground grew and flourished forthwith."

university from large companies with an extensive inventory. One of these establishments was Collins & Co., a New York book dealer. Although Mitchell stopped buying books for the university in 1827, when the student library fee was abolished, he continued buying books for himself (Burk, 2009; McVaugh, M., 1984).

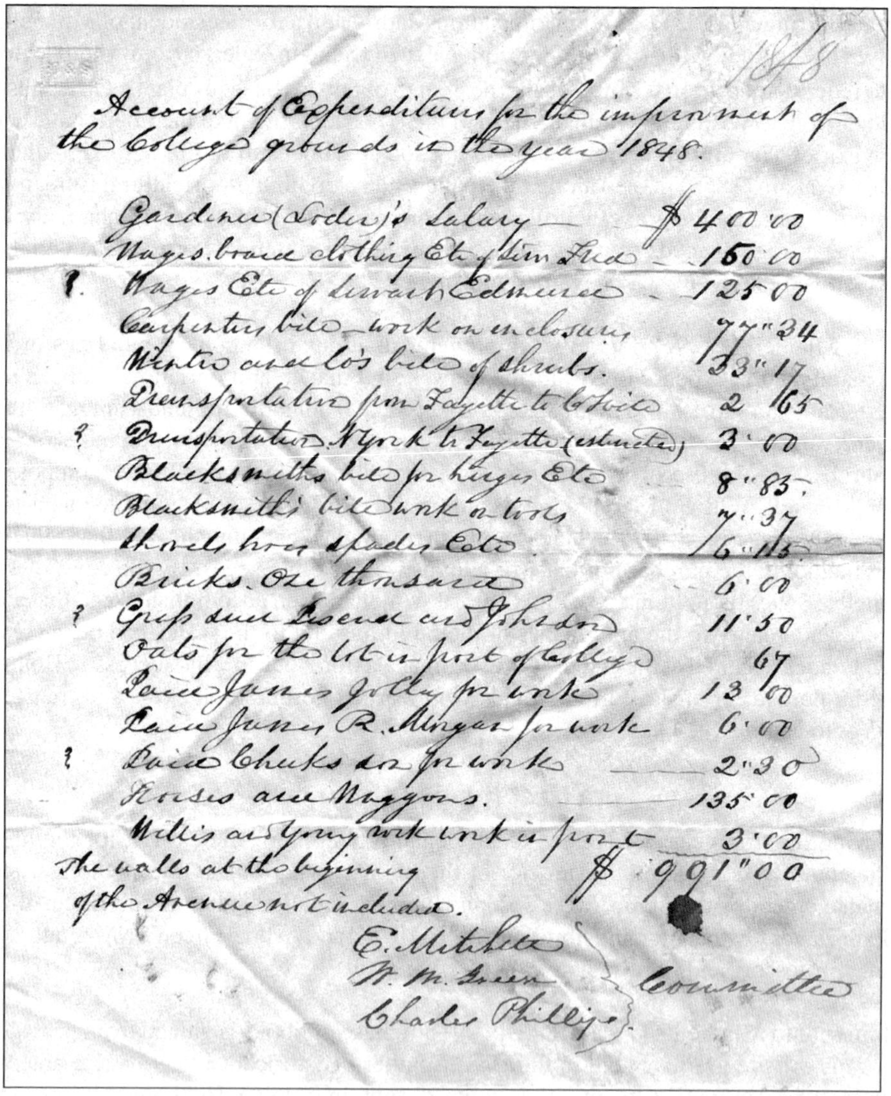
1848

Account of Expenditures for the improvement of the College grounds in the year 1848.

Gardener (Loder)'s Salary — $400.00
Wages, board clothing Etc of Wm Fred — 160.00
? Wages Etc of Stewart Edmund — 125.00
Carpenters bills, work on enclosures 77.24
Winter and Co's bill of shrubs. 33.17
Transportation from Fayette to C Hill 2.65
? Transportation N York to Fayette (estimated) 3.00
Blacksmiths bills for hinges Etc 8.85
Blacksmith's bill work on tools 7.37
Shovels hoes spades Etc 6.15
Bricks. One thousand 6.00
? Grass seed Bissend and Johnson 11.50
Oats for the lot in front of College .67
Paid James Jolly for work 13.00
Paid James R. Morgan for work 6.00
? Paid Chuks ... for work — 2.30
Horses and Waggons. — 135.00
Willis and Young work in front 3.00
the walls at the beginning of the Arrangement included. $991.00

E. Mitchell
W. M. Green } Committee
Charles Phillips

Fig. 3.15. Account of expenditures for the improvement of the college grounds in the year 1848.

(Box 5, folder 174, University Papers #40005, University Archives, Wilson Special Collections Library, University of North Carolina, Chapel Hill).

Nearly six months after Mitchell's death, a committee met in Chapel Hill to audit and settle his estate (Battle, W.H., 1857). The trustees later resolved "to purchase the whole of the Library of the late Dr. Mitchell & to pay his Administrator the sum of Three Thousand

Five Hundred Dollars in full for the same" (BOT 6: 216–217). According to Fisk P. Brewer ([1870]), who served as the librarian from 1869 to 1870, there were 1,897 books. Mitchell's library was practically evenly divided between belles lettres/theology and science (chemistry, geology, mathematics, mineralogy, botany, and zoology).

The holdings of the collection were itemized in a fifteen-page "Catalogue" (Anonymous, 1857). Based on the inventory of Mitchell's library, there were thirty-three titles on basic botany, all concerning systematics and descriptive aspects, except for three on physiology. Four titles were on cryptogams and one on mosses with the remainder on higher plants. Table 3.1 provides a list of the authors, titles, and bibliographical data of the botanical books found in Mitchell's library.

In 1987 and 1990, UNC Professor of History Michael McVaugh (1987, 1990) republished the catalog of Mitchell's library and added annotations, including information on the current locations of volumes in the campus libraries. Today nearly three-fourths of the books from Mitchell's library survive and are housed in the university's libraries. It is noteworthy that a large number of books that Mitchell had purchased for the library turned up in his own library. Before the University Library had its own building in 1852, the faculty customarily removed books from its shelves, especially those in their fields of specialty, to their study quarters (Burk, 2009). Taken together, the purchase of Mitchell's library and the collection of books purchased by UNC President and Professor of Moral Philosophy Joseph Caldwell during a book buying trip to Europe in 1824–1825, "had given North Carolina one of the two or three best scientific libraries in the [antebellum] United States" (McVaugh, M., 1984).

Acting President of the University

On two occasions Mitchell served as acting president of the university. The first instance occurred when Joseph Caldwell made a trip to Europe to buy books and apparatus. The faculty met on May 12, 1824 and agreed that "The Rev. Prof. Mitchell discharge the duties of President" during Caldwell's absence (UNC Faculty 3: 39). This responsibility covered the period of time from early May 1824 to mid-March 1825. The second instance occurred following President Caldwell's death on January 27, 1835. Three days later the executive committee of the Board of Trustees called a meeting and resolved that Mitchell would be authorized to act as president pro tempore until the new president took office (Cameron and Graham, 1835). He held this post until David Lowry Swain, who was selected as the new head of the university, began his tenure on January 1, 1836.

Swain would serve as the president of UNC until his death in 1868 and would work with Mitchell for the next two decades. Their association was a productive one concerning the development of the university; however, there were occasional misunderstandings. Soon after Swain took office Mitchell ([1836]b) expressed his thoughts on being overextended and conveyed his sentiments to President Swain. He also revealed his concern about Swain's view of him: "You have represented me as 'thin-skinned,' irascible, suspicious, uncharitable, simple (not in the good sense of the word) in allowing the gibes and jokes leveled at me for a series of years to influence me to the adoption of certain rules for the regulation of my conduct and finally as a man the effect of an observation upon whose mind was to be calculated–not upon the peculiarities of his own character."

Around 1846 further tension between the two men was revealed when Swain thought that Mitchell had neglected some of his duties during a vacation period. In his private journal Mitchell (1818–1847, pp. [20–24]) wrote a couple of drafts of a letter to Swain, revealing some tension between them. He drafted his comments prior to a visit of the trustees to the campus and stated his opinions about the possibilities of their contention. Mitchell wished they would dispel "any unkindness of feeling there may now be between us." Mitchell was not certain about the causes of Swain's feeling toward him, but reassured Swain that he labored hard and continuously in promoting the welfare of the college.

Surveyor of Roads

Mitchell (Fig. 3.16) also held the title of engineer regarding a project in surveying a major road in North Carolina. In an effort to promote internal improvements, the State of North Carolina focused attention on building roads. One such effort was a turnpike road from Raleigh west to Buncombe Turnpike Road, and ending at the Georgia state line in Cherokee County. An eastern branch was proposed from the eastern side of the Yadkin River to Fayetteville (Graham, 1846). The North Carolina General Assembly (1845) ratified an act on January 8, 1845 to commence the survey for the road. Governor William A. Graham appointed Mitchell as the survey engineer. Mitchell carried out the work and wrote an eighteen-page report which was published in a message that Graham (1846) presented to the General Assembly. He was paid $400 for the work (Mitchell, 1845b).

Fig. 3.16. Elisha Mitchell, ca. 1850. Photograph enlarged from a daguerreotype made in Asheville, North Carolina.

(Courtesy of North Carolina Collection Photographic Archives, Wilson Special Collections Library, University of North Carolina, Chapel Hill).

Honors, Memberships, and Commendations

Mitchell's scientific investigations earned him recognition from learned institutions and scholars. According to the *Register of Officers and Students of The University of Alabama from 1831–1901* (Palmer, 1901), Mitchell was awarded the D.D. in 1838 from that university.

During his lifetime, Mitchell furthered his scientific contacts and learning through memberships in scholarly associations. The Connecticut Academy of Arts and Sciences elected him a member on April 22, 1817 (CAAS). He probably became inactive in the organization after leaving Connecticut: there are no surviving archival records to prove otherwise. Nearly fourteen years after Mitchell assumed his faculty appointment at UNC, the Academy of Natural Sciences of Philadelphia (now of Drexel University) elected him a corresponding member on August 28, 1832. Although the Academy was devoted to all facets of natural history, geology was a primary focus when Mitchell became a member. Next, the Boston Society of Natural History elected him a corresponding member on July 19, 1844. Upon the founding of the American Association for the Advancement of Science in 1848, Mitchell joined; but he let his membership lapse in 1854.

The memory of Elisha Mitchell is respected throughout North Carolina and in scientific circles. His longtime professorship at UNC and his numerous scientific excursions provided a foundation for connecting with students, scholars, and citizens. His fame in and contributions to education and science have been memorialized in various ways. As already noted Mount Mitchell immortalizes his name. In 1861 Mitchell County, North Carolina, was created from portions of Yancey, Watauga, Caldwell, Burke, and McDowell Counties. In 1883 a group of UNC faculty members established the Elisha Mitchell Scientific Society to promote science on the campus and in the state. The organization first published its official organ, the *Journal of the Elisha Mitchell Scientific Society*, in 1884. The society was disbanded in 1993, but the journal continues to be published; however, it is now known as the *Journal of the North Carolina Academy of Science*. Geologist Joseph Hyde Pratt (1899) designated a new variety of the mineral chromite as Mitchellite; it occurs in the western part of North Carolina. Mount Mitchell State Park, the first state park in North Carolina, was established in 1915. At UNC the Department of Geology (now Geological Sciences) moved into its new building, Mitchell Hall, in 1964 (Fig. 3.17).

In 1856 the Princeton University Natural History Museum was founded and later named the Elisha Mitchell Museum of Geology and Archaeology. The reference to Elisha Mitchell was subsequently dropped when an anonymous donor of the museum withdrew monetary support. The museum closed its doors in September 2000 (Brennan, 2006).

Other tributes include a memorial marble slab (Fig. 3.18) in UNC's Memorial Hall and in the Presbyterian Church on Franklin Street in Chapel Hill. Writer Robert Dellinger (2003) based his book *Mitchell's Peak, a Novel of Discovery* on the explorations made by Mitchell in the Black Mountains. In their play *Mountains of Hope*, playwrights William Gregg and Perry Deane Young (2004) dramatize Mitchell's life as well as his explorations and discoveries.

Fig. 3.17. Mitchell Hall, University of North Carolina, Chapel Hill, 2012, home to the Department of Geological Sciences.

(Courtesy of Brian Nalley, Photo Lab, Department of Biology, University of North Carolina, Chapel Hill).

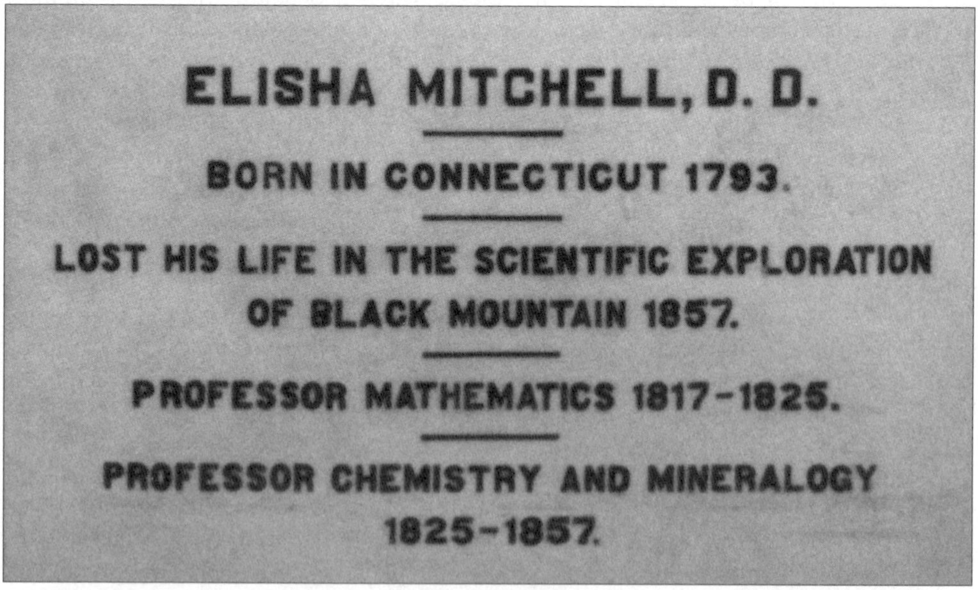

Fig. 3.18. Commemorative tablet, Memorial Hall, University of North Carolina, Chapel Hill.

(Courtesy of North Carolina Collection Photographic Archives, Wilson Special Collections Library, University of North Carolina, Chapel Hill).

Legacy

Elisha Mitchell's scientific pursuits and investigations were diverse, including the fields of agriculture, astronomy, botany, chemistry, geology, mathematics, and physical geography. Although his interests were spread over numerous disciplines, he focused more extensively on chemistry, geology, and physical geography than the other subjects. UNC botanist William C. Coker (1946) considered Mitchell "a naturalist in the good old sense of that word." As a Presbyterian preacher, he was a creationist, but he heeded a uniformitarian view (Moore, 1974). In his *Elements of Geology,* Mitchell (1842a) provided a window into his view: "A knowledge of the present will assist in explaining the past."

In the classroom Mitchell presented information supported by his educational background and extensive reading. When teaching mathematics, he introduced calculus into the course. He avidly pursued botanical studies in his early years at UNC, making collections of plants and identifying them. But in the classroom, his botanical instruction was not based on the most current knowledge. In chemistry instruction, he performed experiments and demonstrations to reinforce scientific principles, but the students did not perform experiments. Mitchell enhanced students' learning experiences in geology and mineralogy with field trips and by exhibiting specimens in the laboratory. To aid his instruction he wrote textbooks on chemistry and geology as well as a handbook on natural history. He was known to interject humor into his lectures.

In addition to teaching mathematics, chemistry, geology, mineralogy, botany, and zoology, Mitchell's time-consuming responsibilities at UNC involved the bursarship, superintendence of the grounds and buildings, preaching in the chapel, directorship of the library, and disciplinarian. Mitchell also served as town commissioner and justice of the peace in the village of Chapel Hill, and he preached in the local Presbyterian Church as well as the campus chapel.

During Mitchell's tenure at UNC from 1818 to 1857, student enrollment quadrupled from nearly 110 to 431 students; annual room rent increased fivefold, \$2 to \$10; yearly tuition fees grew from \$30 to \$50; and the number of faculty members grew from five to eleven.

Elisha and Maria Mitchell had seven children: Mary Phoebe, Margaret Eliot, Ellen Hannah, Eliza North, Charles Andrews,[42] Matthew Henry, and Henry Eliot. The last two mentioned sons died as infants and were buried in a small graveyard behind the president's house, where the Mitchell family lived for nearly forty years. Richard Henry Ashe, son of Mary Phoebe (Richard) Ashe and Richard Mitchell Grant, son of Eliza North (Swepson) Grant, were also buried there. In the center of the graveyard, which measured about ten feet square,

[42] Following Mitchell's death, President Swain recommended that Charles be offered a position as assistant tutor in chemistry at a salary of \$700 per year (Summerell, 1857). This was considered a "munificent offer for a boy of 19." Accepting a position as an officer of instruction, Charles taught until his father's position was filled by William Joseph Martin in 1858 (Bursey, 1982). In her diary entry of August 12, 1866, historian Cornelia Spencer (1866) expressed her thoughts about Charles, who had died on July 10, 1866 in Alabama. "His habits were those of an opium-eater. Dr. Mitchell's only son. I never admired Charles Mitchell as a boy or as [a] young man. There was always a certain repulsive something in the expression of his face–handsome as it was. Something brutal. But I always thought he would be an able & respectable man."

was a four-foot tombstone on which the four names were engraved (Bartholomew, 1955). On February 9, 1961 the bodies were exhumed and reburied in the Old Chapel Hill Cemetery to make room for facilities to accommodate a growing television and communications program at UNC (Ivey, 1961).

Historian Kemp P. Battle (1905) stated that Mitchell was among "the most conspicuous men" in North Carolina. "As a teacher he was most interesting, abounding in illustrations, often humorous, which illuminated the subject. As a preacher he was able and logical but lacking in grace of delivery. As a disciplinarian he was active in detection and prevention, but mild in punishment. As a neighbor he was boundless in charity, ready with purse and wise counsels to aid all who needed help."

References Cited

(References that conclude with page numbers within parentheses indicate sources from where quotes and pertinent information were obtained.)

Allcott, John V. 1986. *The Campus at Chapel Hill: Two hundred years of architecture*. Chapel Hill Historical Society, Chapel Hill. (p. 31; fig. 37)

Anonymous. 1821. [Elisha Mitchell ordained]. *Hillsborough Recorder* (Hillsborough, NC) 2(84): [3], Wednesday, September 19, 1821.

Anonymous. 1824. [Elisha Mitchell's agricultural paper]. *Raleigh Register, and North-Carolina State Gazette* [Semi-Weekly] 2(111): [3], Tuesday, December 7, 1824.

Anonymous. 1825. [Appointment of State Geologist]. *Raleigh Register, and North-Carolina Gazette* 25 (1363): [3], Friday, December 28 [*i.e.*, 23], 1825.

Anonymous. 1835. Our University. *Hillsborough Recorder* (Hillsborough, NC) 22(1127): [3], Thursday, June 9, 1835.

Anonymous. 1841. Our University. *Raleigh Register, and North-Carolina Gazette* 42(47): p. [3], Tuesday, June 8, 1841.

Anonymous. 1842. Hillsborough Literary Association. *Hillsborough Recorder* (Hillsborough, NC) 22(1134): [3], Thursday, July 28, 1842.

Anonymous. 1856. Chemistry at the university. *North-Carolina University Magazine* 5: 139–140.

Anonymous. 1857. *Catalogue of Books and Instruments Belonging to Dr. E. Mitchell, Late Professor in the University of North Carolina, for Sale at His Late Residence*. "Gazette" Office, Chapel Hill.

Anonymous. 1877. Progress of industrial education, pp. 326–362 in *Report of the Commissioner of Agriculture of the Operations of the Department for the Year 1876*. Government Printing Office, Washington, D.C. (p. 345)

Anonymous. 1878. [University items]. *Weekly Ledger* (Chapel Hill, NC) 2(9): 2, Saturday, December 21, 1878.

Anonymous. 1879. [Notice]. *Chapel Hill Ledger* 3: [2], Saturday, September 20, 1879.

Barnes, David Alexander. 1840. Diary. (Box 1, folder 16 [volume 6], Southern Historical Collection #3484, Wilson Special Collections Library, University of North Carolina, Chapel Hill) (entry of February 13, 1840)

Barringer, Daniel Moreau. 1853. Letter of February 15, 1853 to Elisha Mitchell. (Box 1, folder 6, Elisha Mitchell Papers #518, Southern Historical Collection, Wilson Special Collections Library, University of North Carolina, Chapel Hill)

Bartholomew, Robert. 1955. Monument to Mitchell descendants. *Durham Morning Herald*, Sunday, July 3, 1955, Section IV, p. 5.

Battle, Kemp P. ([editor]). 1905. *Diary of a Geological Tour by Dr. Elisha Mitchell in 1827 and 1828 with Introduction and Notes by Dr. Kemp P. Battle, LLD*. (*James Sprunt Historical Monograph,* no. 6). Published by the University, Chapel Hill. (pp. 5, 7)

Battle, Kemp P. 1915. Elisha Mitchell, D.D. *Journal of the Elisha Mitchell Scientific Society* 30: 157–164. (p. 159)

Battle, Kemp P. 1974. *History of the University of North Carolina*. Reprint edition. The Reprint Company, Publishers, Spartanburg, SC. 2 vols. (Vol. 1, pp. 252, 271–272, 281–282, 471, 589, 591, 652, 681, 781)

Battle, Will H. 1857. [Report of the Committee on Elisha Mitchell's accounts, December 15, 1857]. (Box 5, folder 205, University Papers #40005, University Archives, Wilson Special Collections Library, University of North Carolina, Chapel Hill)

Berkeley, Edmund and Dorothy Smith Berkeley. 1986. *A Yankee Botanist in the Carolinas: The Reverend Moses Ashley Curtis, D.D. (1808–1872)*. J. Cramer, Berlin. (pp. 53–54)

BOT. Board of Trustees, University of North Carolina, Chapel Hill. *Minutes of the Board of Trustees*. (University Archives #40001, Wilson Special Collections Library, University of North Carolina, Chapel Hill) (Vol. 4, July 1811–December 1822; vol. 5, December 1822–December 1840; vol. 6, December 1841–June 1868; vol. 15, 1835–1873; vol. 16, 1875–1881)

Brennan, Daniel. 2006. Electronic mail to William R. Burk, November 10, 2006.

Brewer, Fisk P. [1870]. *The Library of the University of North Carolina*. [no statement of publisher, Chapel Hill.]

Brookes, Iveson Lewis. 1818a. Letter of February 13, 1818 to Jonathan Brookes. (Box 1, folder 2, Iveson Lewis Brookes Papers #3249, Southern Historical Collection, Wilson Special Collections Library, University of North Carolina, Chapel Hill)

Brookes, Iveson Lewis. 1818b. Letter of June 6, 1818 to Jonathan Brookes. (Box 1, folder 2, Iveson Lewis Brookes Papers #3249, Southern Historical Collection, Wilson Special Collections Library, University of North Carolina, Chapel Hill)

Burk, William R. 2009. The University of North Carolina at Chapel Hill Library and the sciences, 1795–1902. *North Carolina Libraries* 67: 2–16.

Bursey, Maurice M. 1982. *Carolina Chemists; Sketches from Chapel Hill, using the records prepared by Horace D. Crockford with recollections by Arthur Roe*. The Department of Chemistry, University of North Carolina, Chapel Hill. (pp. 9, 13)

Buxton, James. 1857. Letter of July 12, 1857 to Ralph Buxton. (Box 2, folder 24, Ralph Potts Buxton Papers #116, Southern Historical Collection, Wilson Special Collections Library, University of North Carolina, Chapel Hill)

CAAS. Connecticut Academy of Arts and Sciences. *Record Book, 1799–1858*. (Group 1373, series no. 1, box 8, folder 121, Manuscripts and Archives, Yale University Library)

Caldwell, Joseph. 1830. Letter [of December 20, 1830] to the President and members of the UNC Board of Trustees. (Box 3, folder 123, University Papers #40005, University Archives, Wilson Special Collections Library, University of North Carolina, Chapel Hill)

Cameron, [Duncan] (chairman) and William A. Graham (secretary). 1835. [Resolution of the UNC Executive Committee, January 30, 1835.] (Box 1, folder 3, Elisha Mitchell Papers #518, Southern Historical Collection, Wilson Special Collections Library, University of North Carolina, Chapel Hill)

Chamberlain, Hope Summerrell. 1945. "Life Story of Elisha Mitchell, D.D., 1793–1857: Professor in the University of North Carolina from 1818 until his death in 1857." [The Author]. (Typescript in the North Carolina Collection, Wilson Special Collections Library, University of North Carolina, Chapel Hill)

Cherry, Thomas Kevin. 1993. *Science at the University of North Carolina 1795–1853*. A thesis submitted to the faculty of the University of North Carolina in partial fulfillment of the requirements for the degree of Masters [*sic*] of Arts in the Department of History, Chapel Hill. (pp. 71, 78, 82–83)

Cleaveland, Parker. 1822. *An Elementary Treatise on Mineralogy and Geology*. Cummings and Hilliard, Boston.

Clingman, T.L. [1856]. *Measurements of the Black Mountain: Letter of T.L. Clingman*. G. S. Gideon, Print., Washington, D.C. (p. 36)

Cobb, Collier. 1896. Some beginnings in science. *Appleton's Popular Science Monthly* 49: 763–771. (p. 767)

Coker, William C. 1946. Letter of November 18, 1946 to Robert D.W. Connor. (Robert Digges Wimberly Connor Papers #2427, Southern Historical Collection, Wilson Special Collections Library, University of North Carolina, Chapel Hill)

C[room], H[ardy] B[ryan]. 1833. Gama grass. *Harbinger* 1(2): [1], Tuesday, September 3, 1833.

Curtis, M[oses] A[shley]. 1867. *Geological and Natural History Survey of North Carolina. Part III, Botany; Containing a catalogue of the indigenous and naturalized plants of the state*. Printed at N.C. Institution for the Deaf and Dumb and the Blind, Raleigh.

Dana, James Dwight. 1844. *A System of Mineralogy*. Wiley and Putnam, New York, London.

Dellinger, Robert. 2003. *Mitchell's Peak, a Novel of Discovery*. Caleb's Press, High Point, NC.

Dexter, Franklin Bowditch. 1912. Elisha Mitchell, pp. 586–589 in *Biographical Sketches of the Graduates of Yale College with Annals of the College History*, vol. 6. Yale University Press, New Haven.

Downing, A.J. 1847. Letter of January 26, 1847 to Elisha Mitchell. (Box 1, folder 6, Elisha Mitchell Papers #518, Southern Historical Collection, Wilson Special Collections Library, University of North Carolina, Chapel Hill)

Draper, John William. 1846. *Text-book on Chemistry, for Use of Schools and Colleges*. Harper Brothers, Publishers, New York.

Dusenbery, James Laurence. 1841–1842. Diary (typed version). (Southern Historical Collection #2561-z, Wilson Special Collections Library, University of North Carolina, Chapel Hill) (pp. [111], [151–152])

Emmons, Ebenezer. 1852. *Report of Professor Emmons, on the Geological Survey of North Carolina*. Seaton Gales, Printer to the Legislature, Raleigh.

Fries, Adelaide L. (editor). 1947. *Records of the Moravians in North Carolina*, vol. 7 (1809–1822). (*Publications of the State Department of Archives and History*). State Department of Archives and History, Raleigh. (p. 3402)

Garrett, Thomas Miles. 1849–1850. "Diary of Thomas Miles Garrett at the University of North Carolina 1849–[1850] copied from original placed by Mrs. E.M. Denbo." (Typescript) (Southern Historical Collection #1171-z, Wilson Special Collections Library, University of North Carolina, Chapel Hill) (pp. 34, 96–97, 280)

Graham, William A. 1846. *Message from His Excellency, Gov. Graham, Communicating Report of Professor Mitchell, on the Turnpike from Raleigh, West.* W.R. Gales, Printer to the Legislature, Raleigh.

Graham, William A. 1850. Letter of October 23, 1850 to Elisha Mitchell. (Box 1, folder 6, Elisha Mitchell Papers #518, Southern Historical Collection, Wilson Special Collections Library, University of North Carolina, Chapel Hill)

Gregg, William and Perry Deane Young. 2004. *Mountain of Hope: The story of Elisha Mitchell and the betrayal that led to his death on Mount Mitchell: A play in two acts*. Don Congdon Associates, New York.

Haywood, John (for the Committee of Appointment). 1825. [Recommendation of appointment of the chair of chemistry], December 13, 1825. (Box 3, folder 95, University Papers #40005, University Archives, Wilson Special Collections Library, University of North Carolina, Chapel Hill)

Henderson, Archibald. 1949. *The Campus of the First State University*. (*University of North Carolina Sesquicentennial Publications*). University of North Carolina Press, Chapel Hill. (pp. 107, 363)

Henderson, Archibald. 1957. Elisha Mitchell: Scientist and man. *Journal of the Elisha Mitchell Scientific Society* 73: 204–212. (p. 212)

Hitchcock, Edward. 1840. *Elementary Geology*. J.S. & C. Adams, Amherst, MA.

Holmes, Joseph A. 1887. A sketch of Professor Washington Caruthers Kerr, M.A., D.Ph. *Journal of the Elisha Mitchell Scientific Society* 4: [1]–24. (p. 8)

Hooper, William. [1818]. Letter to The Honorable Committee of Appointment. (Box 2, folder 62, University Papers #40005, Wilson Special Collections Library, University of North Carolina, Chapel Hill)

Hutton, Charles. 1818. *A Course of Mathematics in Two Volumes; For the use of academies, as well as private tuition*. From the 5th and 6th London editions, revised and corrected by Robert Adrain. S. Campbell, &c., New-York.

Ivey, Pete. 1961. University to relocate old campus cemetery. *Greensboro Daily News* 99(22): A5, Wednesday, February 8, 1961.

Jones, Henry Francis. 1857–1858. Diary. (Southern Historical Collection #3019-z, Wilson Special Collections Library, University of North Carolina, Chapel Hill)

Jones, John T. 1835. ["Notebook on Mineralogy"]. (Box 23, folder 276 [volume 68], Jones-Patterson Papers #578, Southern Historical Collection, Wilson Special Collections Library, University of North Carolina, Chapel Hill)

Lees, David McMichen. [1828]. ["Notes on mineralogy," pp. 5–12]. (Southern Historical Collection #3705-z, Wilson Special Collections Library, University of North Carolina, Chapel Hill) (p. 5)

[Lindemann, Erika] (editor). 2006. True and Candid Compositions website. <http://docsouth.unc.edu/true/mss01-18/mss01-18.html> (Accessed November 23, 2015)

Lyell, Charles. 1839. *Elements of Geology*. James Kay, Jun. and Brother, Philadelphia; C.H. Kay & Co., Pittsburgh.

Lyon, John. 1845. ["Manuscript volume listing names and descriptions of North American plants and compiled by John Lyon"]. (Box 8, folder 106, Moses Ashley Curtis Papers #199, Southern Historical Collection, Wilson Special Collections Library, University of North Carolina at Chapel Hill)

Manly, Charles (secretary). 1857. "Proceedings of the UNC Executive Committee of the Board of Trustees, July 11, 1857." (Box 5, folder 202, University Papers #40005, University Archives, University of North Carolina, Chapel Hill)

Mauskopf, Seymour H. 1984. Elisha Mitchell and European science: The case of chemistry. *Journal of the Elisha Mitchell Scientific Society* 100: 57–60.

McLaurin, Melton A. 2003. *The North Carolina State Fair, the First 150 Years*. Illustrations selected by Paul Blankinship. Office of Archives and History, North Carolina Department of Cultural Resources, Raleigh; North Carolina State Fair Division, Department of Agriculture and Consumer Services, Raleigh. (p. 144)

McVaugh, Michael. 1984. Elisha Mitchell's library as an index to his scientific interests. *Journal of the Elisha Mitchell Scientific Society* 100: 50–56.

McVaugh, Michael. 1987, 1990. Elisha Mitchell's books and the University of North Carolina Library, [and] Part 2. *Bookmark* 55: 27–54; 56: 31–70.

McVaugh, Rogers, Michael R. McVaugh, and Mary Ayers. 1996. *Chapel Hill and Elisha Mitchell the Botanist*. Botanical Garden Foundation, Chapel Hill. (pp. 12–17)

Mitchell, Elisha. 1813–1816. Diary. (Box 1, folder 11, Elisha Mitchell Papers #518, Southern Historical Collection, Wilson Special Collections Library, University of North Carolina, Chapel Hill)

Mitchell, Elisha. 1818. Letter of February 11, 1818 to Maria North. (Box 1, folder 1, Elisha Mitchell Papers #518, Southern Historical Collection, Wilson Special Collections Library, University of North Carolina, Chapel Hill)

Mitchell, Elisha. 1818–1842. Journal, Letter Book, Account Book. (Folder 14 [volume 4], Elisha Mitchell Papers #518, Southern Historical Collection, Wilson Special Collections Library, University of North Carolina, Chapel Hill) (pp. 184, 253, 425)

Mitchell, Elisha. 1818–1847. Private Notebook. (Folder 12 [volume 2], Elisha Mitchell Papers #518, Southern Historical Collection, Wilson Special Collections Library, University of North Carolina, Chapel Hill)

Mitchell, Elisha. 1819a. Letter of July 29, 1819 to Maria North. (Miscellaneous papers, vol. 21.3, p. 91, State Archives of North Carolina)

Mitchell, Elisha. 1819b. Letter of September 29, 1819 to Lewis David von Schweinitz. (Collection No. 438, Ewell Sale Stewart Library, Archives & Manuscripts Collection, The Academy of Natural Sciences of Drexel University)

Mitchell, Elisha. 1819c. Letter of September 20, 1819 to Maria North. (Miscellaneous papers, volume 21.3, p. 92, State Archives of North Carolina)

Mitchell, Elisha. 1820. Letter of February 18, 1820 to Lewis David von Schweinitz. (Collection F221:8, Moravian Archives, Winston-Salem, North Carolina)

Mitchell, Elisha. 1821a. Letter of June 1821 to Lewis David von Schweinitz. (Collection No. 438, Ewell Sale Stewart Library, Archives & Manuscripts Collection, The Academy of Natural Sciences of Drexel University)

Mitchell, Elisha. 1821b. Letter of October 5, 1821 to Lewis David von Schweinitz. (Collection No. 438, Ewell Sale Stewart Library, Archives & Manuscripts Collection, The Academy of Natural Sciences of Drexel University)

Mitchell, Elisha. 1821c. Letter of October 20, 1821 to Lewis David von Schweinitz. (Collection No. 438, Ewell Sale Stewart Library, Archives & Manuscripts Collection, The Academy of Natural Sciences of Drexel University)

Mitchell, Elisha. 1821d. Letter of November 8, 1821 to Lewis David von Schweinitz. (Collection No. 438, Ewell Sale Stewart Library, Archives & Manuscripts Collection, The Academy of Natural Sciences of Drexel University)

Mitchell, Elisha. 1825a. Excerpt of a Letter of February 8, 1825 from Elisha Mitchell to John Ravenscroft. <http://docsouth.unc.edu/true/mss02-08/mss02-08.html> (Accessed October 25, 2006)

Mitchell, Elisha. 1825b. Letter of December 5, 1825 to Governor Burton. (Box 3, folder 94, University Papers #40005, University Archives, Wilson Special Collections Library, University of North Carolina, Chapel Hill)

Mitchell, Elisha. 1825c. Agricultural speculations, pp. 49–58 in *Papers on Agricultural Subjects and Professor Olmsted's Report on the Geology of Part of the Western Counties of North-Carolina.* Published by Order of the Board of Agriculture; Printed by J. Gale, Raleigh. (pp. 49–50)

Mitchell, Elisha. 1826. Letter of December 15, 1826 to Charles Manly, Treasurer of the University of North Carolina. (Box 3, folder 105, University Papers #40005, University Archives, Wilson Special Collections Library, University of North Carolina, Chapel Hill)

Mitchell, Elisha. 1828a. *Report on the Geology of North-Carolina. Part III. Conducted under the direction of the Board of Agriculture*, pp. [1]–27 in [*Papers on Agricultural Subjects*], [No. 3]: Professor Mitchell's Report on the Geology of North-Carolina: with C.R. Rothe's Remarks on the Gold Country: and A Report on the Subject of Manufacturing Establishments, and the Raising of Sheep. Published by Order of the Board of Agriculture; Printed by J. Gales & Son, Raleigh. (p. 4) (The date on the title page reads November 1827, but the year of publication is 1828.)

Mitchell, Elisha. 1828b. On the character and origin of the Low Country of North Carolina. *American Journal of Science and Arts* 13: 336–347.

Mitchell, Elisha. 1828c. Letter of April 7, 1828 to Governor Iredell. (Box 3, folder 112, University Papers #40005, University Archives, Wilson Special Collections Library, University of North Carolina, Chapel Hill)

Mitchell, Elisha. 1828d. Letter of December 15, 1828 to UNC Board of Trustees. (Box 3, folder 115, University Papers #40005, University Archives, Wilson Special Collections Library, University of North Carolina, Chapel Hill)

Mitchell, Elisha. 1829a. *Geological Report of Professor Mitchell*, pp. [101]–108 in *Papers on Agricultural Subjects*, [No. 4]: on the Silk Worm and Making Silk; on the Vine and Making Wine, and Professor Mitchell's Report on the Geology of North-Carolina. Published by Order of the Board of Agriculture; Printed by J. Gales & Son, Raleigh. (p. 107)

Mitchell, Elisha. 1829b. On the effect of quantity of matter in modifying the force of chemical attraction. *American Journal of Science and Arts* 16: 234–242.

Mitchell, Elisha. 1829c. On the geology of the gold region of North Carolina. *American Journal of Science and Arts* 16: 1–19.

Mitchell, Elisha. 1830. On a substitute for Welther's tube of safety [and notice on the preparation of the substance commonly called phosphoret of lime]. *American Journal of Science and Arts* 19: 345–350.

Mitchell, Elisha. 1831a. *Arguments for Temperance, a Sermon Addressed to the Students of the University of North Carolina, March 13, 1831, and Published by Their Request.* J. Gales & Son, Raleigh.

Mitchell, Elisha. 1831b. Analysis of the protogaea of Leibnitz. *American Journal of Science and Arts* 20: 56–64.

Mitchell, Elisha. 1833. Remarks [on an article concerning gama grass]. *Harbinger* 1(2): [5], Tuesday, September 3, 1833.

Mitchell, Elisha. 1835a. Letter of September 2, 1835 to Duncan Cameron. (Box 4, folder 138, University Papers #40005, Wilson Special Collections Library, University of North Carolina, Chapel Hill)

Mitchell, Elisha. 1835b. Letter of December 28, 1835 to Duncan Cameron. (Box 4, folder 139, University Papers #40005, Wilson Special Collections Library, University of North Carolina, Chapel Hill)

[Mitchell, Elisha]. 1835c. The mountains of North Carolina. *Raleigh Register, and North Carolina Gazette* 36(51): [1]. Tuesday, November 3, 1835. (The column is signed "N."—a pen name that Elisha Mitchell sometimes used when authoring articles in newspapers.)

Mitchell, Elisha. 1835d. Letter of October 29, 1835 to Mather and Hall. (George P. Merrill Collection, Record Unit 7177, Box 8, folder 12, Smithsonian Institution)

Mitchell, Elisha. 1836a. Letter of July 4, 1836 to Ellen Mitchell. (Box 1, folder 4, Elisha Mitchell Papers #518, Southern Historical Collection, Wilson Special Collections Library, University of North Carolina, Chapel Hill)

Mitchell, Elisha.[1836]b. Letter [ca. 1836] to David L. Swain. (David Lowry Swain Papers, Private Collections, 84.14, North Carolina State Archives, Raleigh) (Typescript copy of records at the Southern Historical Collection, Wilson Special Collections Library, University of North Carolina, Chapel Hill)

Mitchell, Elisha. 1836c. Letter of December 19, 1836 to Charles Manly, Treasurer of the University. (Box 4, folder 141, University Papers #40005, University Archives, Wilson Special Collections Library, University of North Carolina, Chapel Hill)

Mitchell, Elisha. 1836d. Letter of September 1836 to the Executive Committee of the Board of Trustees. <http://docsouth.unc.edu/unc/unc04-27/unc04-27 p01.html> (Accessed October 25, 2006)

Mitchell, Elisha. 1837a. Letter of July 31, 1837 to Dr. E. Strudwick. (Transcription of letter by Dr. H.A. Royster) (Box 1, folder 4, Elisha Mitchell Papers #518, Southern Historical Collection, Wilson Special Collections Library, University of North Carolina, Chapel Hill)

Mitchell, Elisha. 1837b. Letter [of June 16, 1837] to Charles Manly. (Box 4, folder 143, University Papers #40005, University Archives, Wilson Special Collections Library, University of North Carolina, Chapel Hill)

Mitchell, Elisha. 1838a. Letter of January 1, 1838 to David L. Swain. (David Lowry Swain Papers, Private Collections, Box 84.1, Folder Correspondence 1825–1840, State Archives of North Carolina)

Mitchell, Elisha. 1838b. Letter of November 29, 1838 to Charles Manly. (Box 4, folder 149, University Papers #40005, Wilson Special Collections Library, University of North Carolina, Chapel Hill)

Mitchell, Elisha. 1838c. Letter of December 15, 1838 to Charles Manly. (Box 4, folder 150, University Papers #40005, University Archives, Wilson Special Collections Library, University of North Carolina, Chapel Hill)

Mitchell, Elisha. 1839. Notice of the height of mountains in North Carolina. *American Journal of Science and* Arts 35: 377–380.

Mitchell, Elisha. 1840a. Letter of December 28, 1840 to David L. Swain. (Box 1, folder 8, David L. Swain Papers #706, Southern Historical Collection, Wilson Special Collections Library, University of North Carolina)

[Mitchell, Elisha]. [1840]b. *Natural History*. [The Author, Chapel Hill]. (pp. 2, 8)

Mitchell, Elisha. 1842a. *Elements of Geology, with an Outline of the Geology of North Carolina: For the use of the students of the university*. [The Author, Chapel Hill].

Mitchell, Elisha. 1842b. Letter of June 27, 1842 to "My Dear Wife" [*i.e.,* Maria Mitchell]. (Miscellaneous Papers, volume 21.4, p. 39, State Archives of North Carolina)

Mitchell, Elisha. 1842c. Letter of February 16, 1842 to William A. Graham. (Box 3, folder 39, William A. Graham Papers #285, Southern Historical Collection, Wilson Special Collections Library, University of North Carolina, Chapel Hill)

Mitchell, Elisha. 1845a. Letter of February 16, 1842 to Maria Mitchell. (Miscellaneous Papers 21.4, p. 61, State Archives of North Carolina)

Mitchell, Elisha. 1845b. Receipt and letter [of August 28, 1845] to William A. Graham. (William A. Graham Papers, 61.3, State Archives of North Carolina)

Mitchell, Elisha. 1847. Letter of December 11, 1847 to David L. Swain. (Miscellaneous Papers, volume 21.4, p. 70, State Archives of North Carolina)

Mitchell, Elisha. [1848]. *The Other Leaf of the Book of Nature and the Word of God*. [no statement of publisher or place.] (Year of publication is misprinted as 1948)

[Mitchell, Elisha]. 1849a. [Brief history on places of public worship in Chapel Hill, pp. 8–13]. (Chapel Hill Presbyterian Church Records #1978-z, Southern Historical Collection, Wilson Special Collections Library, University of North Carolina, Chapel Hill)

Mitchell, Elisha. 1849b. To Hon. T.L. Clingman, but originally published in the New-York *Albion*, pp. 192–198 in Charles Lanman, *Letters from the Alleghany Mountains*. Geo. P. Putnam, New-York.

Mitchell, Elisha. 1849c. Letter of December 27, 1849 to Charles Manly. (Box 5, folder 177, University Papers #40005, University Archives, Wilson Special Collections Library, University of North Carolina, Chapel Hill)

Mitchell, Elisha. 1850a. *Statistics, Facts, and Dates, for the Sunday Recitations of the Junior Class in the University*. R. Craighead, Printer, New York.

Mitchell, Elisha. 1850b. [Letter of August 27, 1850 to William A. Graham], pp. 373–374 in J.G. de Roulhouc Hamilton (editor), *The Papers of William Alexander Graham*, vol. 3. State Department of Archives and History, Raleigh.

Mitchell, Elisha. [1852]. Letter of June 21, [1852] to Frank M. Parker. (Box 1, folder 6, Elisha Mitchell Papers #518, Southern Historical Collection, Wilson Special Collections Library, University of North Carolina, Chapel Hill)

Mitchell, Elisha. 1853. Agriculture and the Geological Survey, I–VIII. *Semi-Weekly Raleigh Register*. No. I. 53(39): [3], Wednesday, May 11, 1853; No. II. 53(40): [2], Saturday, May 14, 1853; No. III. 53(42): [2], Saturday, May 21, 1853; No. IV. 53(44): [2], Saturday, May 28, 1853; No. V. 53(49): [2], Wednesday, June 15, 1853; No. VI. 53(51): [2], Wednesday, June 22, 1853; No. VII. 53(69): [2], Saturday, July 23, 1853; No. VIII. 53(75): [2], Wednesday, August 14, 1853.

Mitchell, Elisha. 1854. The Grape-vine Borer: A communication on the grape-vine. [no statement of publisher or place]. (An expanded publication of an article published in the *Semi-Weekly Raleigh Register*, vol. 55(28): [2], Wednesday, April 5, 1854.)

Mitchell, Elisha. [1856]a. [*Elements of Chemistry*]. (This is an advance printing of Mitchell's chemistry textbook. It is bound with a collection of his publications in a volume titled "Dr. Mitchell's Works" and filed in the Cobb Loan Collection, volume 7, North Carolina Collection, Wilson Special Collections Library, University of North Carolina, Chapel Hill.)

Mitchell, Elisha. 1856b. Annual address delivered before the N.C. Agricultural Society, at the Fair held at Raleigh, October, 1856, by Prof. E. Mitchell, of the University. *North-Carolina Arator* 2: 644–648; [Concluded] 2: 670–675.

Mitchell, Elisha, William M. Green and Charles Phillips (Committee). 1848. "Account of Expenditures for the Improvement of the College Grounds in the Year 1848." (Box 5, folder 174, University Papers #40005, University Archives, Wilson Special Collections Library, University of North Carolina, Chapel Hill)

Mitchell, Maria. 1820. Letter of January 1, 1820 to Mrs. North. (Box 1, folder 1, Elisha Mitchell Papers #518, Southern Historical Collection, Wilson Special Collections Library, University of North Carolina, Chapel Hill)

Moore, Ellen J. 1974. Mitchell, Elisha, pp. 420–421 in Charles Coulston Gillispie (editor in chief), vol. 9, *Dictionary of Scientific Biography*, Charles Scribner's Sons, New York.

Mullins, William Sidney. 1840–1841. Diary (typescript). (Southern Historical Collection #531-z, Wilson Special Collections Library, University of North Carolina, Chapel Hill) (folder 3 [volume 1], pp. 101, 105; folder 4 [volume 2], pp. 71, 124; folder 5 [volume 3], pp. 40, 42, 43, 84, 95, 109)

Murphey, Archibald De Bow. 1822. Letter of January 22, 1822 to Elisha Mitchell. (Box 1, folder 2, Elisha Mitchell Papers #518, Southern Historical Collection, Wilson Special Collections Library, University of North Carolina, Chapel Hill)

North Carolina General Assembly. 1845. *Laws of the State of North Carolina, Passed by the General Assembly, at the Session of 1844–45*. Thomas J. Lemay, Raleigh.

Palmer, Thomas Waverly. 1901. *Register of Officers and Students of The University of Alabama from 1831–1901*. University of Alabama, Tuscaloosa. (p.[449])

Patterson, Rufus. 1847a. Letter of October 12, 1847 to his father, General Samuel F. Patterson. (Box 2, folder 18, Jones and Patterson Papers #578, Southern Historical Collection, Wilson Special Collections Library, University of North Carolina, Chapel Hill)

Patterson, Rufus. 1847b. Letter of October 26, 1847 to his father, General Samuel F. Patterson. (Box 2, folder 18, Jones and Patterson Papers #578, Southern Historical Collection, Wilson Special Collections Library, University of North Carolina, Chapel Hill)

Pearson, Richard M. 1856. Letter of October 29, 1856 to Elisha Mitchell. (Box 1, folder 7, Elisha Mitchell Papers #518, Southern Historical Collection, Wilson Special Collections Library, University of North Carolina, Chapel Hill)

Phillips, Charles. 1858a. Memoir, pp. [5]–12 in *A Memoir of the Rev. Elisha Mitchell, D.D., Late Professor of Chemistry, Mineralogy & Geology in the University of North Carolina. Together with the Tributes of Respect to His Memory, by Various Public Meetings and Literary Associations, and the Addresses Delivered at the Re-interment of His Remains, by Rt. Rev. James H. Otey, D.D., Bishop of Tennessee, and Hon. David L. Swain, LL.D., President of the University*. Published by J.M. Henderson, Chapel Hill.

[Phillips, Charles]. 1858b. Dr. Mitchell's investigations among the mountains of Yancey County. *North Carolina University Magazine* 7: 293–318.

Pratt, J.H. 1899. On the occurrence, origin and chemical composition of chromite. *American Journal of Science*, 4th series (whole number 157): 281–286.

Robbin, Thomas. 1856. Letter of September 20, 1856 to Elisha Mitchell. (Box 1, folder 7, Elisha Mitchell Papers #518, Southern Historical Collection, Wilson Special Collections Library, University of North Carolina, Chapel Hill)

Robinson, Edward and E. Smith. 1846. *Biblical Researches in Palestine, Mount Sinai and Arabia Petraea: A journal of travels in the year 1838 by E. Robinson and E. Smith; Undertaken in reference to biblical geography; drawn up from the original diaries with historical illus. by Edward Robinson*. Crocker & Brewster, Boston.

Sanders, John L. (editor). 1953. The journal of Ruffin Wirt Tomlinson, The University of North Carolina 1841–1842. *North Carolina Historical* Review 30: [86]–114; [233]–[260]. (pp. 96, 103, 244–245, 248)

Saunders, Joseph. 1825. Letter of August 28, 1825 to Elisha Mitchell. (Box 1, folder 2, Elisha Mitchell Papers #518, Southern Historical Collection, Wilson Special Collections Library, University of North Carolina, Chapel Hill)

Schwarzkopf, S. Kent. 1985. *A History of Mount Mitchell and the Black Mountains: Exploration, development, and preservation*. Division of Archives and History, North Carolina Department of Cultural Resources, Division of Parks and Recreation, North Carolina Department of Natural Resources and Community Development, Raleigh.

Schweinitz, Lewis David von. 1820. Letter of June 25, 1820 to John LeConte. (Herbarium Library, University of Michigan, Ann Arbor)

Schweinitz, Lewis David von. 1822. Letter of November 15, 1822 to Elisha Mitchell. (Box 1, folder 1, Elisha Mitchell Papers #518, Southern Historical Collection, Wilson Special Collections Library, University of North Carolina, Chapel Hill)

Sellers, James B. 1953. *History of the University of Alabama*. University of Alabama Press, University, AL. (pp. 43, 47)

Silliman, Benjamin. 1830–1831. *Elements of Chemistry: In the order of the lectures given in Yale College*. In two volumes. Printed and published by Hezekiah Howe, New Haven, CT.

Silver, Timothy. 2003. *Mount Mitchell and the Black Mountains: An environmental history of the highest peaks in Eastern America*. University of North Carolina Press, Chapel Hill.

Smith, Michael S. 1999. The conflict between "practical utility" and geology: Denison Olmsted, Elisha Mitchell and the 1823 to 1828 Geologic Surveys of North Carolina. *Southeastern Geology* 38: 145–154.

Spencer, Cornelia P. 1866. Diary. (Box 14, folder 159 (volume 3), Cornelia Phillips Spencer Papers #683, Southern Historical Collection, Wilson Special Collections Library, University of North Carolina, Chapel Hill) (p. 138)

Spencer, Cornelia P. 1871. Wanted—a purchaser. *North Carolina Presbyterian* (Fayetteville, NC), new series 4(189): [1], Wednesday, August 16, 1871.

Spencer, Cornelia P. (editor). 1884a. Old times in Chapel Hill. No. 4. [Elisha Mitchell]. *University Monthly* (University of North Carolina, Chapel Hill) 4 (new series): 15–22.

Spencer, Cornelia P. 1884b. The Campus, p. 42 in "Pen and Ink Sketches, Now and Then." [The Author, Chapel Hill]. (Pamphlet in the North Carolina Collection, Wilson Special Collections Library, University of North Carolina, Chapel Hill)

Summerell, Ellen. 1857. Letter of July 15, 1857 to Joseph Summerell. (Box 1, folder 7, Elisha Mitchell Papers #518, Southern Historical Collection, Wilson Special Collections Library, University of North Carolina, Chapel Hill)

Swain, David L. 1823. Letter of January 16, 1823 to Elisha Mitchell. (Box 1, folder 2, Elisha Mitchell Papers #518, Southern Historical Collection, Wilson Special Collections Library, University of North Carolina, Chapel Hill)

Swain, David L. 1842. Letter of October 12, 1842 to Charles Manly. (Box 4, folder 160, University Papers #40005, University Archives, Wilson Special Collections Library, University of North Carolina, Chapel Hill)

Tull, Jethro. 1829. *The Horse-hoeing Husbandry or, a Treatise on the Principles of Tillage and Vegetation*. Published by William Cobbett, London.

Turner, Herbert Snipes. 1971. *The Dreamer, Archibald DeBow Murphey 1777–1832*. McClure Press, Verona, VA. (p. 97)

UNC-A. Business and Finance: Finance Division. Series 1: *Historical Financial Books*, volume 5 (1789–1859), University Archives #40095, Wilson Special Collections Library, University of North Carolina, Chapel Hill. (pp. 361, 379)

UNC Faculty. University of North Carolina, Chapel Hill Faculty. *Minutes of Faculty Meetings*. (University Archives #40106, Wilson Special Collections Library, University of North Carolina, Chapel Hill) (Vol. 3, 1821–1841)

UNC Trustees. 1852. "Trustees of the University of North Carolina, invoice to John B. Tenny, May 1, 1852." (Box 5, folder 181, University of North Carolina Papers #40005, University Archives, Wilson Special Collections Library, University of North Carolina, Chapel Hill)

Vance, Zebulon Vance. 1858. Letter to Miss Margaret Mitchell, May 25, 1858, pp. 37–38 in Frontis W. Johnstown (editor), *The Papers of Zebulon Baird Vance*, vol. 1. State Department of Archives and History, Raleigh, 1963.

Vince, S[amue]l. 1818. *A Treatise on Fluxions*. 5th ed., corrected and enlarged throughout. Printed by J. Smith, Printer to the University, Cambridge, [England].

W., W. D. 1844. University of North Carolina. *Hillsborough Recorder* (Hillsborough, NC) 23(1210): [2–3], Thursday, February 1, 1844. (Reprinted from the *Boston Recorder*)

Watson, Elgiva D. 1984. Elisha Mitchell: A Connecticut Yankee in North Carolina. *Journal of the Elisha Mitchell Scientific Society* 100: 43–49. (pp. 45, 47)

Webb, William P. 1889. Letter of March 25, 1889 to Charles Phillips. (Folder 99, Battle Family Papers #3223-A, Southern Historical Collection, Wilson Special Collections Library, University of North Carolina, Chapel Hill)

Wheeler, Walter H. 1984. Elisha Mitchell as a geologist. *Journal of the Elisha Mitchell Scientific Society* 100: 61–64. (p. 61)

Wright, Perrin. 1992. Measuring Mitchell's mountains. *Journal of the Elisha Mitchell Scientific Society* 108: 1–10.

Yale-College. 1809–1812, 1814, 1816. *Catalogue of the Officers and Students in Yale-College*. Broadsides. November 1809, November 1810, November 1811, November 1812, November 1814, November 1816. (Yale College Catalogues, Yale University Library, Manuscripts and Archives)

Yale-S1. 1813. *Scheme of the Exercises, at the Public Commencement of Yale-College, September 1st, 1813*. O. Steele, Printer, New-Haven, CT. Broadside. (Yale College Orations, Exams, etc., Yale University Library, Manuscripts and Archives)

Yale-S2. 1816. *Scheme of the Exercises, at the Public Commencement of Yale-College, September 11, 1816*. Broadside. (Yale College Orations, Exams, etc., Yale University Library, Manuscripts and Archives)

APPENDIX 3.1.
A BIO-BIBLIOGRAPHY ON ELISHA MITCHELL, INCLUDING SELECTIVE REFERENCES TO HIS EXPLORATIONS OF THE MOUNTAINS IN WESTERN NORTH CAROLINA.[43]

A Stranger. 1857. Lines; [a poem]. *Asheville News* (Asheville, NC), Thursday, July 30, 1857. (*Scrapbook of Charles Phillips*, p. 78)

Anderson W.J. (chairman) and George H. Haigh (secretary). 1857. Death of Prof. Mitchell. *North Carolina Argus*, Fayetteville, Saturday, July 18, 1857. (*Scrapbook of Charles Phillips*, p. 71)

Anderson W.J. (chairman) and Geo. H. Haigh (secretary). 1858. Meeting at Fayetteville. Death of Prof. Mitchell, pp. 30–31 in *A Memoir of the Rev. Elisha Mitchell, D.D., Late Professor of Chemistry, Mineralogy & Geology in the University of North Carolina*. Published by J.M. Henderson, Chapel Hill. (From the *Argus*, July 18, [1857])

Anonymous. [undated]. Elisha Mitchell's thrilling adventure in lofty mountain country recalled by one who assisted in locating body. *News and Observer* (Raleigh). (North Carolina Collection Clipping File through 1975: Biography, vol. 102, pp. 302–303)

Anonymous. [undated]. Last survivor of searchers for Dr. Mitchell dies at 93, James Allen, Yancey County mountaineer, credited with being first to see body of lost explorer, dead in Greensboro. Special to the *Charlotte Observer*. (North Carolina Collection Clipping File, 1976 through 1989: Biography, vol. M (#3), p. 557)

Anonymous. [undated]. The late Professor Mitchell. For the *Presbyterian*. (North Carolina Collection Clipping File through 1975: Biography, vol. 102, pp. 291–292)

Anonymous. [undated]. Prof. Mitchell's monument. (North Carolina Collection Clipping File through 1975: Biography, vol. 102, p. 300)

Anonymous. [undated]. Rev. Dr. Mitchell of Chapel Hill, N.C. (North Carolina Collection Clipping File through 1975: Biography, vol. 102, pp. 289–290)

Anonymous. 1857. Black Mountain again. *News*, Asheville, NC, July 30, Thursday, 1857. (*Scrapbook of Charles Phillips*, p. 84)

Anonymous. 1857. A correction in the physical geography of North Carolina. *Standard* (Raleigh), Saturday, August 29, 1857. (*Scrapbook of Charles Phillips*, p. 89) (Includes information on Elisha Mitchell's measurements of the Black Mountain in North Carolina)

Anonymous. 1857. Death of Prof. Mitchell. [*Asheville Spectator*?], Asheville, NC, ca. July 18, 1857. (*Scrapbook of Charles Phillips*, p. 87)

[43] I have not verified all of the references in this list, including those that appear in the North Carolina Collection Clipping File through 1975: Biography; North Carolina Collection Clipping File, 1976 through 1989: Biography; North Carolina Collection Clipping File, 1990–2009; or in the *Scrapbook of Charles Phillips*. The North Carolina Collection in Wilson Special Collections Library at the University of North Carolina, Chapel Hill holds all of these resources, each of which is fully cited in this list. Two additional resources are the Pack Memorial Library: Newspaper File Collection: Information on Asheville and Buncombe County, compiled through 1998, Asheville-Buncombe Library System, Volume 46 (Mic-Mix) and the G. Hope Summerell Chamberlain Papers at Duke University. Data transcribed from these sources does not represent exact and complete bibliographic citations.

Appendix 3.1. (Continued)

Anonymous. 1857. Death of Prof. Mitchell. Asheville, July 9th, 1857. (*Scrapbook of Charles Phillips*, p. 87)

Anonymous. 1857. Death of Prof. Mitchell. *North Carolinian* (Fayetteville, NC), Saturday, July 18, 1857. (*Scrapbook of Charles Phillips*, p. 71)

Anonymous. 1857. Death of Prof. Mitchell, of North Carolina. *Weekly Picayune* (New Orleans, LA), [1857]. (*Scrapbook of Charles Phillips*, p. 72)

Anonymous. 1857. Distressing dispensation. *Presbyterian* (Philadelphia), Saturday, July, 25, 1857. (*Scrapbook of Charles Phillips*, p. 71)

Anonymous. 1857. Dr. Mitchell of Chapel Hill. *Clinton Independent* (Clinton, NC), Wednesday, July 15, 1857. (*Scrapbook of Charles Phillips*, p. 72)

Anonymous. [1857?]. Dr. Mitchell's library. (North Carolina Collection Clipping File through 1975: Biography, vol. 102, p. 315)

Anonymous. 1857. The highest point of Black Mountain. Gales & Seaton, Wednesday, September 2, 1857. (*Scrapbook of Charles Phillips*, p. 91)

Anonymous. 1857. The late Dr. Mitchell. *North Carolina University Magazine* 7(1): 38–40. (Includes resolutions of the Philanthropic Society and of the Dialectic Society)

Anonymous. 1857. The late Prof. Mitchell. *New York Observer*, September 10, 1857. (*Scrapbook of Charles Phillips*, p. 91)

Anonymous. 1857. The late Prof. Mitchell. *Presbyterian* (Philadelphia), Saturday, August 1, 1857. (*Scrapbook of Charles Phillips*, p. 80)

Anonymous. 1857. The late Professor Mitchell. [*Raleigh Register*?], Friday, July 24, 1857. (*Scrapbook of Charles Phillips*, p. 73)

Anonymous. 1857. Lost in the mountains. *New York Journal of Commerce*, Evening edition, Monday, July 14, 1857. (*Scrapbook of Charles Phillips*, p. 72)

Anonymous. 1857. [Mistakes in article]. *New York Journal of Commerce*, Saturday, July 25, 1857. (*Scrapbook of Charles Phillips*, p. 72)

Anonymous. [1857]. Mitchell's monument. *Asheville News*, July 16, 1857. (North Carolina Collection Clipping File through 1975: Biography, vol. 102, p. 301)

Anonymous. 1857. Painful rumor. *Morning Express* (New York, NY), Tuesday, July 14, 1857. (*Scrapbook of Charles Phillips*, p. 84)

Anonymous. 1857. Proceedings of the citizens of Chapel Hill upon the occasion of the death of the Rev. Dr. Mitchell; The late Professor Mitchell; *Chapel Hill Gazette* (Chapel Hill, NC), Saturday, July 13, 1857. (*Scrapbook of Charles Phillips*, pp. 74–77)

Anonymous. 1857. Professor Mitchell. *New York Morning Express*, Saturday, July 25, 1857. (*Scrapbook of Charles Phillips*, p. 72)

Appendix 3.1. (Continued)

Anonymous. 1857. Professor Mitchell. *Observer* (Fayetteville, NC), Monday, July 20, 1857. *(Scrapbook of Charles Phillips*, p. 79)

Anonymous. 1857. Professor Mitchell. *Standard* (Raleigh), Wednesday, July 22, 1857. (*Scrapbook of Charles Phillips*, pp. 87–88)

Anonymous. 1857. Resolutions of the faculty, [University of North Carolina, Chapel Hill], July 25, 1857. (*Scrapbook of Charles Phillips*, p. 78)

Anonymous. 1857. Resolutions of the students, [University of North Carolina, Chapel Hill], July 22, 1857. (Scrapbook of Charles Phillips, p. 78)

Anonymous. 1857. Rev. Dr. Mitchell of Chapel Hill, N.C. *Central Presbyterian* (Richmond, VA), Saturday, July 18, 1857. (*Scrapbook of Charles Phillips*, p. 72)

Anonymous. 1858. Meeting at Chapel Hill, pp. 28–30 in *A Memoir of the Rev. Elisha Mitchell, D.D., Late Professor of Chemistry, Mineralogy & Geology in the University of North Carolina*. Published by J.M. Henderson, Chapel Hill. (From *Chapel Hill Gazette*, July 13, [1857])

Anonymous. 1858. Removal of Dr. Mitchell's remains. *Chapel Hill Gazette* (Chapel Hill, NC), Saturday, July 17, 1858. (*Scrapbook of Charles Phillips*, p. 103)

Anonymous. 1858. The removal of Dr. Mitchell's remains. [*Weekly*] *Raleigh Register*, Wednesday morning, July 14, 1858, [p. 4]. (By-line reads "By an Eye-Witness.") (*Scrapbook of Charles Phillips*, p. 102)

Anonymous. 1858. Resolution of the students. July 22, 1857. (*Scrapbook of Charles Phillips*, p. 78)

Anonymous. 1858. Resolutions of the faculty, Chapel Hill, July 17, 1857, pp. 36–37 in *A Memoir of the Rev. Elisha Mitchell, D.D., Late Professor of Chemistry, Mineralogy & Geology in the University of North Carolina*. Published by J.M. Henderson, Chapel Hill.

Anonymous. 1858. Resolutions of the students, Chapel Hill, July 22, 1857, pp. 37–38 in *A Memoir of the Rev. Elisha Mitchell, D.D., Late Professor of Chemistry, Mineralogy & Geology in the University of North Carolina*. Published by J. M. Henderson, Chapel Hill.

Anonymous. 1858. Rev. Elisha Mitchell. *North Carolina Journal of Education* 1: 6–11.

Anonymous. 1858. Rev. Elisha Mitchell, D.D. *Hillsborough Recorder* (Hillsborough, NC), Wednesday, January 6, 1858. ("From the *N.C. Journal of Education*.") (*Scrapbook of Charles Phillips*, p. 98)

Anonymous. 1861. Monument to Dr. Mitchell. *North Carolina University Magazine* 10: 507–509. (From *Asheville News*, July 16, 1857)

Anonymous. 1887. A funeral procession up Mount Mitchell. *Youth's Companion*, October 13, 1887, p. 434.

APPENDIX 3.1. (CONTINUED)

Anonymous. 1888. [Note on the address of Dr. W.B. Battle on the placing of a pyramidal shaft at the tomb of Elisha Mitchell]. *University Magazine* (University of North Carolina, Chapel Hill) 21: 35.

Anonymous. 1892. Elisha Mitchell, D.D., pp. 416–417 in *Cyclopedia of Eminent and Representative Men of the Carolinas of the Nineteenth Century*. Brant & Fuller, Madison, WI.

Anonymous. 1897. Mitchell, Elisha, p. 30 in *The National Cyclopaedia of American Biography, Being the History of the United States*, vol. 7. James T. White & Company, New York.

Anonymous. 1908. "Big Tom" Wilson mountaineer, an ideal type. *Charlotte News*, Wednesday, February 12, 1908, p. 2.

Anonymous. 1915. Long lost Mt. Mitchell deed is found after many years. *Asheville Citizen-Times*, November 10, 1915. (North Carolina Collection Clipping File through 1975: Biography, vol. 102, p. 314)

Anonymous. 1922. Elisha Mitchell's thrilling adventure in lofty mountain country recalled by one who assisted in locating body. (Raleigh, Sunday Morning, July 9, 1922) (North Carolina Collection Clipping File through 1975: Biography, vol. 102, pp. 318–319); also *Asheville Citizen*, Tuesday Morning, July 4, 1922 (Pack Memorial Library, Newspaper File Collection: Information on Asheville and Buncombe County, vol. 46)

Anonymous. [1927]. Bronze tablet honors Dr. Elisha Mitchell. (Newspaper clipping in G. Hope Summerell Chamberlain Papers, Duke University; also North Carolina Collection Clipping File, 1976 through 1989: Biography, vol. M (#3), p. 559.

Anonymous. 1927. Dedicate tower on Mt. Mitchell. (Newspaper clipping in G. Hope Summerell Chamberlain Papers, Duke University)

Anonymous. [1927]. Dr. Elisha Mitchell killed 70 years ago. (Newspaper clipping in G. Hope Summerell Chamberlain Papers, Duke University)

Anonymous. [1927]. Elisha Mitchell will be honored. (Newspaper clipping in G. Hope Summerell Chamberlain Papers, Duke University)

Anonymous. [1927]. Marker honoring Mitchell unveiled on mountain top. (Newspaper clipping in G. Hope Summerell Chamberlain Papers, Duke University)

Anonymous. 1927. Mt. Mitchell tower dedicated to memory of peak's explorer. (Special to the *Charlotte Observer*, Mount Mitchell, June 17, [1927]) (Newspaper clipping in G. Hope Summerell Chamberlain Papers, Duke University)

Anonymous. 1927. Tells story of famous burial 70 years back. *Asheville Times*, May 27, 1927. (Pack Memorial Library, Newspaper File Collection: Information on Asheville and Buncombe County, vol. 46)

Anonymous. [1927]. Unveil tablet as tribute to Dr. Mitchell. (Newspaper clipping in G. Hope Summerell Chamberlain Papers, Duke University)

APPENDIX 3.1. (CONTINUED)

Anonymous. 1927. Unveiling of Mitchell tablet recalls explorer's experience. (Newspaper clipping in G. Hope Summerell Chamberlain Papers, Duke University)

Anonymous. 1928. Bronze tablet honors Dr. Elisha Mitchell. *Uplift* 16(38): 11–12. (From the *Asheville Citizen*)

Anonymous. 1928. Elisha Mitchell tablet unveiled on high peak. *Alumni Review* (University of North Carolina) 17(2): 44.

Anonymous. 1929. Man who found Mitchell's body taken by death. James Allen was last survivor of searching party. *Asheville Citizen*, January 4, 1929. (Pack Memorial Library, Newspaper File Collection: Information on Asheville and Buncombe County, vol. 46)

Anonymous. 1932. "Old Mitch." *Chapel Hill Weekly*, Friday, July 22, 1932. (North Carolina Collection Clipping File through 1975: Biography, vol. 102, p. 324)

Anonymous. 1935. Thrilling story of search was told by guide. *Asheville Citizen-Times*, Sunday, June 9, 1935. (North Carolina Collection Clipping File through 1975: Biography, vol. 102, pp. 328–330)

Anonymous. 1949. A biography of Elisha Mitchell written by his granddaughter, Mrs. Hope Chamberlain. (September 16, 1949) (North Carolina Collection Clipping File through 1975: Biography, vol. 102, p. 337)

Anonymous. 1950. UNC to dramatize story of Mitchell's discovery. *Asheville Citizen-Times*, Sunday, January 15, 1950. (North Carolina Collection Clipping File through 1975: Biography, vol. 102, p. 336)

Anonymous. 1957. Another story uncovered on the death of Dr. Elisha Mitchell. *News and Observer* (Raleigh), Thursday Morning, June 27, 1957, [p. 27]. (North Carolina Collection Clipping File through 1975: Biography, vol. 102, p. 352)

Anonymous. 1957. Dr. Elisha Mitchell. *Asheville Citizen* (Asheville, NC), Tuesday, April 16, 1957. (North Carolina Collection Clipping File through 1975: Biography, vol. 102, p. 331)

Anonymous. 1957. Dr. Elisha Mitchell's grave. *Asheville Citizen Times*, June 27, 1957. (Pack Memorial Library, Newspaper File Collection: Information on Asheville and Buncombe County, vol. 46)

Anonymous. 1957. The Elisha Mitchell Centennial. *Asheville Citizen*, Wednesday, June 12, 1957, p. [4]. (North Carolina Collection Clipping File through 1975: Biography, vol. 102, p. 350)

Anonymous. 1957. Ramsey's address on Elisha Mitchell. *Chapel Hill Weekly*, Tuesday, April 20, 1957. (North Carolina Collection Clipping File through 1975: Biography, vol. 102, pp. 353–354)

Anonymous. 1958. Get Mitchell letters. *Durham Morning Herald*, Sunday, May 18, 1958. (North Carolina Collection Clipping File through 1975: Biography, vol. 102, p. 337)

Appendix 3.1. (Continued)

Anonymous. 1960. Mitchell gave name and life to famed peak. [no source given], July 17, 1960. (Pack Memorial Library, Newspaper File Collection: Information on Asheville and Buncombe County, vol. 46)

Anonymous. 1967. Mitchell, Elisha, p. 430 in *Who Was Who in America; Historical volume 1607–1896*. Revised ed. A component volume of *Who's Who in American History*. Marquis Who's Who, Chicago.

Anonymous. 1968. Elisha Mitchell the man—his mountain—his monument. *Historical Foundation News* (Montreat, NC) 24(1): [1], 3, January 1968. (Also in North Carolina Collection Clipping File through 1975: Biography, vol. 102, pp. 360–361)

Anonymous. 1973. Elisha Mitchell, D.D., pp. 416–417 in *Cyclopedia of Eminent and Representative Men of the Carolinas of the Nineteenth Century*. Reprint ed. The Reprint Company, Spartanburg, SC.

Anonymous. 1974. Peak, park named for historic faculty member. *University Gazette* (University of North Carolina) 2(9): 6–7.

Anonymous. 2001. Elisha Mitchell. *Edited Appletons Encyclopedia*. Virtual American Biographies. <http://famousamericans.net/elishamitchell/> (Accessed November 26, 2006)

Anonymous. 2004. The tall legacy of one of Carolina's greats. *Carolina Alumni Review* 93(5): 107.

Anonymous. 2006. Elisha Mitchell. Western North Carolina Heritage. <http://toto.lib.unca.edu/WNC_biography/mitchel_elisha.htm> (Accessed January 20, 2013)

Arthur, Billy. 1997. Claim to the highest peak; the tragic tale of Elisha Mitchell, and his relentless efforts to prove that Mount Mitchell is the highest mountain in the eastern United States, permanently etched his name on the Yancey County alp. *Our State, Down Home in North Carolina* 65(4): 16–19.

Arthur, John Preston. 1914. Mount Mitchell, pp. 297–299 in *Western North Carolina: A history (from 1730 to 1913)*. Edwards & Broughton Printing Company, Raleigh.

Ashe, Samuel A. (editor in chief). 1905. Elisha Mitchell, pp. [384]–391 in *Biographical History of North Carolina from Colonial Times to the Present*, vol. 1. Charles L. Van Noppen, Publisher, Greensboro, NC.

Atkin, Tho[ma]s W. (editor). 1857. The late Professor Mitchell. *Asheville News* (Asheville, NC), Thursday, July 16, 1857. (*Scrapbook of Charles Phillips*, p. 81)

Bailey, Lloyd Richard, Sr. (editor). 1994. *The Heritage of the Toe River Valley*, volume 1. Durham, NC. The search for Professor Mitchell's body [from the *Asheville Spectator*] entry 41 (pp. 44–46); Professor's body found after days of tracking, 'Big Tom' Wilson leads search for Mitchell, entry 42 (pp. 46–48); Mitchell's Peak Park, public letters and papers of Locke Craig, Governor of North Carolina, 1913–1917, entry 43 (pp. 48–49); and Elisha Mitchell, entry 528 (pp. 348–349)

APPENDIX 3.1. (CONTINUED)

Bartholomew, Robert H. 1955. Paging Chapel Hill: Everybody take a look behind Swain Hall. *Charlotte Observer*, July 4, 1955. (Pack Memorial Library, Newspaper File Collection: Information on Asheville and Buncombe County, vol. 46)

Battle, Kemp P. 1907, 1912. *History of the University of North Carolina*. Edwards & Broughton, Raleigh. 2 vols. (Also reprinted by The Reprint Company, Spartanburg, SC in 1974) (Numerous passages concern Elisha Mitchell, including those on the following pages: vol. 1, pp. 249–253, 271, 272, 281, 291, 295, 301–303, 305, 357, 358, 407, 409, 421, 422, 425, 426, 429–431, 441–444, 460, 463, 464, 467, 475, 540, 638, 654, 663, 679–684, 697, 781; vol. 2, pp. 178, 317, 322, 458, 741, 743, 823)

Battle, Kemp P. 1915. Elisha Mitchell, D.D. *Journal of the Elisha Mitchell Scientific Society* 30: 157–164.

Battle, Richard H. 1858. The re-interment of Dr. Mitchell's remains. *Raleigh Register*, July 10, 1858.

Battle, Richard H. 1858. The re-interment of Dr. Mitchell's remains, pp. 46–51 in *A Memoir of the Rev. Elisha Mitchell, D.D., Late Professor of Chemistry, Mineralogy & Geology in the University of North Carolina*. Published by J.M. Henderson, Chapel Hill.

Belk, George W. 1916. *Floreen, or, The Story of Mitchell: A legend of the "Land of the Sky."* Stone Publishing Co., Charlotte, NC.

Blackburn, George T., II, Roger N. Kirkman, and Joseph K.L. Reckford (compilers). 1980. Elisha Mitchell, p. 11 in *The Di & Phi Portrait Index; Catalogue of the portraits of the Dialectic and Philanthropic Literary Societies at the University of North Carolina in Chapel Hill*. Published by order of the Dialectic and Philanthropic Society Foundation, Chapel Hill.

Blackstock, Clarence E. 1912. The Mount Mitchell tragedy. *University of North Carolina Magazine* 29: 19–25.

[Bolton, H.C.] 1891. Sketch of Elisha Mitchell. *Popular Science Monthly* 38: 398–406.

Brawley, James S. 1957. He lost his life giving name to highest peak in eastern U.S. *Sunday Post* (Salisbury, NC), April 14, 1957, [p. 6]. (North Carolina Collection Clipping File through 1975: Biography, vol. 102, p. 356)

Buckley, S.B. (Samuel Botsford). 1859. Mountains of North Carolina and Tennessee. *American Journal of Science and Arts* 27: [286]–293.

Burnett, Fred M. 1960. Elisha Mitchell; The search for Dr. Mitchell's body, pp. 88–106, 182–185 in *This Was My Valley*, [no statement of publisher], Ridgecrest, NC.

Burroughs, Winthrop. 1886. The tragedy of the Black Mountains. *Southern Bivouac* 1: 460–464.

Bursey, Maurice M. 1982. Elisha Mitchell, 1793–1857, pp. 9–14 in *Carolina Chemists: Sketches from Chapel Hill using the records prepared by Horace D. Crockford with recollections by Arthur Roe*. Department of Chemistry, University of North Carolina, Chapel Hill.

APPENDIX 3.1. (CONTINUED)

Buxton, Jarvis (chairman) and John D. Hyman (secretary). 1857. Public meeting [concerning Elisha Mitchell]. *Asheville News* (Asheville, NC), [Thursday, July 16, 1857]. (*Scrapbook of Charles Phillips*, pp. 81–82; North Carolina Collection Clipping File through 1975: Biography, vol. 102, pp. 290–291)

Buxton, Jarvis (chairman) and John D. Hyman (secretary). 1858. Meeting [concerning Elisha Mitchell], pp. 26–28 in *A Memoir of the Rev. Elisha Mitchell, D.D., Late Professor of Chemistry, Mineralogy & Geology in the University of North Carolina*. Published by J.M. Henderson, Chapel Hill. (Reprinted from *Asheville News*)

Chamberlain, G. Hope Summerell. G. Hope Summerell Chamberlain Papers, #RL11429, David M. Rubenstein Rare Book & Manuscript Library, Duke University.

Chamberlain, Hope Summerell. 1926. [Elisha Mitchell], pp. 24–25 in *Old Days in Chapel Hill, Being the Life and Letters of Cornelia Phillips Spencer*. University of North Carolina Press, Chapel Hill.

Chamberlain, Hope Summerell. 1945. "Life Story of Elisha Mitchell, D.D., 1793–1857: Professor in the University of North Carolina from 1818 until his death in 1857." [The Author]. (Typescript in the North Carolina Collection, Wilson Speical Collections Library, University of North Carolina, Chapel Hill)

Chapman, Robert Hett. 1858. A funeral sermon delivered in the Presbyterian Church of Asheville, North Carolina, on the tenth of July, 1857, at the interment of the Rev. Elisha Mitchell, D.D., late professor of chemistry, etc., in the University of the state, pp. [20]–25 in *A Memoir of the Rev. Elisha Mitchell, D.D., Late Professor of Chemistry, Mineralogy & Geology in the University of North Carolina*. Published by J.M. Henderson, Chapel Hill. (Also published as a 12-page pamphlet and printed at the Spectator Office, Asheville)

Clingman, T.L. [1856?]. *Measurements of the Black Mountain; Letter of T.L. Clingman*. G.S. Gideon Print., Washington, D.C.

Clingman, T.L. 1857. The Black Mountain. *Asheville News* (Asheville, NC), Sept. 12, 1857. (*Scrapbook of Charles Phillips*, pp. 92–95)

Clingman, T.L. 1858. Mr. Clingman and Prof. Phillips. *Weekly North Carolina Standard* (Raleigh), [Wednesday, January 20, 1858, p. 4]. (*Scrapbook of Charles Phillips*, pp. 98–99)

Clingman, T.L. 1858. Mr. Clingman and Prof. Phillips. *Raleigh Standard* (Raleigh), [ca. February 1, 1858] (*Scrapbook of Charles Phillips*, p. 101)

Cobb, Collier. 1896. Some beginnings in science. *Appletons' Popular Science Monthly* 49: 763–771.

Colton, H.E. and J.L. Henry (editors). 1857. Appropriate. *Asheville Spectator* (Asheville, NC), Thursday, July 23, 1857. (*Scrapbook of Charles Phillips*, p. 82)

Colton, H.E. and J.L. Henry (editors). 1857. [Reburial of Elisha Mitchell]. *Asheville Spectator* (Asheville, NC), Thursday, August 13, 1857. (*Scrapbook of Charles Phillips*, p. 88)

APPENDIX 3.1. (CONTINUED)

Connor, R.D.W. 1915. Dr. Mitchell—his work in the mountains. *Charlotte Daily Observer*, Sunday, January 24, 1915. (North Carolina Collection Clipping File through 1975: Biography, vol. 102, pp. 316–317)

Constance. 1857. Tribute to the memory of the late Dr. Mitchell, [a poem]. [*Chapel Hill*] *Gazette* (Chapel Hill, NC), July 25, 1857. (*Scrapbook of Charles Phillips*, p. 86)

Craig, Locke. 1915. *Mitchell's Peak and Dr. Mitchell*. Edwards & Broughton Printing Co., Raleigh.

Cummings, A.W. 1857. Supposed death of Rev. Dr. Mitchell. *Knoxville Whig* (Knoxville, TN), Saturday, July 18, 1857. (*Scrapbook of Charles Phillips*, pp. 91–92)

Dellinger, Robert. 2003. *Mitchell's Peak*. Caleb's Press, High Point, NC.

Dexter, Franklin Bowditch. 1912. Elisha Mitchell, pp. 586–589 in *Biographical Sketches of the Graduates of Yale College with Annals of the College History*, vol. 6, Yale University Press, New Haven, CT.

Dimmock, George. 1877. A trip to Mt. Mitchell in North Carolina. *Appalachia; The Journal of the Appalachian Mountain Club* 1: 141–151.

Elisha Mitchell Papers. Southern Historical Collection #518, Wilson Special Collections Library, University of North Carolina, Chapel Hill.

Ellis, Marshall. 2000. A place of extremes; a giant among giants, Mount Mitchell's astounding beauty and even more astounding weather are a wonder to behold. *Our State North Carolina* 68(2): 72–76. (photography by Stephen Schoof)

Emerson, A.J. 1857. The late Dr. Mitchell. *Journal of Commerce* (New York), July 10, 1857. (*Scrapbook of Charles Phillips*, p. 71)

Eulalie. 1857. Lines; [a poem]. *Asheville Spectator* (Asheville, NC), August 8, 1857. (*Scrapbook of Charles Phillips*, p. 88)

Fishburn, C.D. (clerk of the Faculty of Davidson College). 1857. [Preamble and resolutions concerning Rev. Elisha Mitchell, D.D.]. *Chapel Hill Gazette* (Chapel Hill, NC), Saturday, August 1, 1857. (Also included are resolutions of the Philanthropic Society and the Dialectic Society, University of North Carolina.) (*Scrapbook of Charles Phillips*, p. 85)

Fishburn, C.D. (clerk of the Faculty of Davidson College). 1858. Meeting of the faculty of Davidson College, July 18, 1857, pp. 40–41 in *A Memoir of the Rev. Elisha Mitchell, D.D., Late Professor of Chemistry, Mineralogy & Geology in the University of North Carolina*. Published by J.M. Henderson, Chapel Hill.

Gregg, William and Perry Deane Young. 2004. *Mountain of Hope: The story of Elisha Mitchell and the betrayal that led to his death on Mount Mitchell: A play in two acts*. Don Congdon Associates, New York.

Hamilton, J.G. de Roulhac (editor). 1918. [Letter from Elisha Mitchell, Chapel Hill, Jany 23d. 1822], pp. 262–265 in *The Papers of Thomas Ruffin*, vol. 1. Edwards & Broughton Printing Co., Raleigh.

Appendix 3.1. (Continued)

Hamlin, Charles Hunter. 1946. Elisha Mitchell, p. [75]. in *Ninety Bits of North Carolina Biography*. Owen G. Dunn Co., New Bern, NC.

Harris, Wade H. 1935. Famous old guide narrates how searchers found body. (*Charlotte Observer*, [Section 3], Sunday, April 14, 1935, [p. 4]) (North Carolina Collection Clipping File through 1975: Biography, vol. 102, pp. 326–327)

Henderson, Archibald. 1949. *The Campus of the First State University*. University of North Carolina Press, Chapel Hill. (Entries regarding Elisha Mitchell are found on pages 52, 60–62, 69, 83, 84, 87–89, 100, 107, 123–127, 160, 165, 167, 237, 269, and 270.)

Henderson, Archibald. 1957. Elisha Mitchell: Scientist and man. *Journal of the Elisha Mitchell Scientific Society* 73: 204–212.

Henderson, Archibald. 1957. Elisha Mitchell and his mountain. *Greensboro Daily News*, [Sunday], May 26, 1957. (North Carolina Collection Clipping File through 1975: Biography, vol. 102, pp. 347–348)

Henderson, Ida Briggs. 1934. Son of Big Tom Wilson retells story of scientist. *Charlotte Observer*, August 12, 1934. (Pack Memorial Library, Newspaper File Collection: Information on Asheville and Buncombe County, vol. 46)

Henderson, Ida Briggs. 1934. Tells how body of Dr. Mitchell was discovered; Son of 'Big Tom' Wilson retells story of scientist. *Charlotte Observer*, Sunday, August 12, 1934.

Hicklin, J.B. 1932. Participant describes search for Dr. Mitchell. *Charlotte Observer*, Sunday, February 21, 1932. (North Carolina Collection Clipping File through 1975: Biography, vol. 102, pp. 322–323)

Hicklin, W.L. 1930. Search for lost scientist [Elisha Mitchell]. *Atlanta Journal*, April 20, 1930. (North Carolina Collection Clipping File through 1975: Biography, vol. 102, p. 320)

Holmes, J.A. 1885. Mitchell's high peak. *North Carolina Teacher* 3: 109.

Humphries, W.S., Addison Harvey, S.D. Goza and E.S.J. Bell. 1857. [Resolutions], Philanthropic Hall. *Chapel Hill Gazette* (Chapel Hill, NC), Saturday, August 1, 1857. (*Scrapbook of Charles Phillips*, p. 85)

Humphries, W.S., Addison Hervey, W.D. Goza and E.S.J. Bell. 1857. [Resolutions], Philanthropic Hall, July 26, 1857. *North Carolina University Magazine* 1:39.

Humphries, W.S., Addison Hervey, W.D. Goza and E.S.J. Bell (committee). 1858. Meeting of the Philanthropic Society, July 26, 1857, p. 39 in *A Memoir of the Rev. Elisha Mitchell, D.D., Late Professor of Chemistry, Mineralogy & Geology in the University of North Carolina*. Published by J.M. Henderson, Chapel Hill.

Hunter, C.L., W.W. Pharr and Edwin R. Harris (committee). 1858. Meeting of the trustees of Davidson College, July 16, 1857, pp. 39–40 in *A Memoir of the Rev. Elisha Mitchell, D.D., Late Professor of Chemistry, Mineralogy & Geology in the University of North Carolina*. Published by J.M. Henderson, Chapel Hill.

APPENDIX 3.1. (CONTINUED)

Ivey, Mabel. [1927]. Dr. Elisha Mitchell killed 70 years ago. (Lenoir, June 25, [1927]) (Newspaper clipping in G. Hope Summerell Chamberlain Papers, Duke University)

Ivey, Pete. [1957]. Mitchell died 100 years ago. *Asheville Citizen-Times* (Asheville, NC), Sunday, April 14, 1957. (North Carolina Collection Clipping File through 1975: Biography, vol. 102, p. 344)

Ivey, Pete. 1960. Grave of Mitchell's sons, grandsons to be moved. *Asheville Citizen Times*, February 8, 1960. (Pack Memorial Library, Newspaper File Collection: Information on Asheville and Buncombe County, vol. 46)

Jeffrey, Thomas E. 1993. A whole torrent of mean and malevolent abuse: Party politics and the Clingman-Mitchell controversy. *North Carolina Historical Review* 70: [241]–265, [401]–429.

Jennette, B. Culpepper, Jr. [1974]. "Elisha Mitchell: University scientist." [The Author, Chapel Hill, NC]. (Paper written for Physics 37, University of North Carolina, Chapel Hill)

Jones, Thomas Lawrence. 1973. "Elisha Mitchell: The man and his mountain." [The Author, Chapel Hill, NC]. (Honors essay, Department of History, University of North Carolina)

Kephart, Horace. 1913. [Mount Mitchell], p. 57 in *Our Southern Highlanders*. Outing Publishing Company, New York.

King, J.B. 1931. Finding of Mitchell's body by 'Big Tom' Wilson is graphically told. *Asheville Citizen Times*, November 15, 1931. (Pack Memorial Library, Newspaper File Collection: Information on Asheville and Buncombe County, vol. 46)

King, J.B. 1935. Noted hunter led party to pool at falls. *Asheville Citizen*, November 15, 1935. (Pack Memorial Library, Newspaper File Collection: Information on Asheville and Buncombe County, vol. 46)

Knight, Lucian Lamar (editor). 1910. Mitchell, Elisha, pp. 305–306 in *Library of Southern Literature*, vol. 15. Compiled under the direct supervision of southern men of letters. Edwin Anderson Alderman, Joel Chandler Harris (editors-in-chief), Charles William Kent (literary editor). Martin & Hoyt Company, New Orleans.

Lacy, Drury (chairman, committee). 1858. Minute of Synod, pp. 43–44 in *A Memoir of the Rev. Elisha Mitchell, D.D., Late Professor of Chemistry, Mineralogy & Geology in the University of North Carolina*. Published by J.M. Henderson, Chapel Hill.

Lanman, Charles. 1849. *Letters from the Alleghany Mountains*. Geo. Putnam, New-York. (Pp. 192–198, Letter from Elisha Mitchell "To Hon. T.L. Clingman, but originally published in the *New-York Albion*.")

Leinbach, William S. 1949. Was it lawyer Clingman or Dr. Mitchell? *Durham Morning Herald*, Dec. 4, 1949. (North Carolina Collection Clipping File through 1975: Biography, vol. 102, p. 335)

Appendix 3.1. (Continued)

Leinbach, William S. 1965. Dr. Elisha Mitchell and the measuring of Mt. Mitchell. (*Information sheet,* no. 4, North Carolina State Park System) (Based on an article in the *Asheville Citizen-Times*, March 26, 1950)

Lindau, Bill. 1957. Dr. Mitchell's fatal trip related by J.N. Garren. *Asheville Citizen-Times* (Asheville, NC), Sunday, June 9, 1957. (North Carolina Collection Clipping File through 1975: Biography, vol. 102, p. 346)

Lindau, Bill. 1957. Memorial for Mitchell scheduled for June 27. *Asheville Citizen-Times* (Asheville, NC), Sunday, June 16, 1957. (North Carolina Collection Clipping File through 1975: Biography, vol. 102, pp. 349–350)

Lindau, Bill. 1957. Tribute paid Mitchell on centennial of death. *Asheville Citizen*, Friday, June 28, 1957, pp. [1]–2.

M., L.V. 1859. Mount Mitchell—A poem. *North Carolina University Magazine* 8: 121–122.

M., V.O. 1858. Mount Mitchell—June 16, 1858, [a poem], pp. 51–53 in *A Memoir of the Rev. Elisha Mitchell, D.D., Late Professor of Chemistry, Mineralogy & Geology in the University of North Carolina*. Published by J.M. Henderson, Chapel Hill. (From the *University Magazine* [University of North Carolina, Chapel Hill])

MacMillan, Laura (compiler). Elisha Mitchell, 1793–1857, p. 164 in *The North Carolina Portrait Index 1700–1860*. University of North Carolina Press, Chapel Hill.

Mal[l]ett, E. (magistrate police) and P.H. McDade, H.B. Guthrie, J.H. Watson, C.C. Scott (commission) and W.H. McDade (town clerk). 1857. [Resolutions]. *North Carolina University Magazine* 7(1): 40.

Mal[l]ett, E. (magistrate police) and P.H. McDade, H.B. Guthrie, J.H. Watson, C.C. Scott. [1857]. Tribute of respect. (For the *Gazette*, [Chapel Hill, NC]) (*Scrapbook of Charles Phillips*, p. 77)

Mallett, E. (magistrate police) and P.H. McDade, H.B. Guthrie, J.H. Watson, C.C. Scott (commissioners). 1858. Meeting of the Commissioners of Chapel Hill, Chapel Hill, July 13, 1857, pp. 41–42 in *A Memoir of the Rev. Elisha Mitchell, D.D., Late Professor of Chemistry, Mineralogy & Geology in the University of North Carolina*. Published by J.M. Henderson, Chapel Hill.

Manly, Cha[rle]s. [undated]. Professor Mitchell. (North Carolina Collection Clipping File through 1975: Biography, vol. 102, pp. 292–293)

Manly, Cha[rle]s (secretary). 1857. The late Prof. Mitchell—Meeting of the Executive Committee of the Board of Trustees. *Weekly North Carolina Standard* (Raleigh), Wednesday, July 22, 1857, p. [1]

Manly, Cha[rle]s (secretary). 1857. Resolutions of the trustees of the N.C. University. *Chapel Hill Gazette* (Chapel Hill, NC), Saturday, July 25, 1857. (*Scrapbook of Charles Phillips*, pp. 77–78)

APPENDIX 3.1. (CONTINUED)

Manly, Cha[rle]s (secretary). 1858. Resolutions of the trustees of the university, Raleigh, July 4, 1857, pp. [35]–36 in *A Memoir of the Rev. Elisha Mitchell, D.D., Late Professor of Chemistry, Mineralogy & Geology in the University of North Carolina*. Published by J.M. Henderson, Chapel Hill.

Mathes, [Charles] Hodge. [ca. 1950]. *A Saga of the Carolina Hills: Being a true story of the naming of Mt. Mitchell.* Explanation by Rev. M.T. Ray. Geo. O. Ray, [Springfield, MO]

Mauskopf, Seymour H. 1984. Elisha Mitchell and European science: The case of chemistry. *Journal of the Elisha Mitchell Scientific Society* 100: 57–60.

McNab, J.G., J.G. Morehead, Jr. and F.D. Stockton (committee). 1857. [Resolutions], Dialectic Hall, *Chapel Hill Gazette* (Chapel Hill, NC), Saturday, August 1, 1857. (*Scrapbook of Charles Phillips*, p. 85)

McNab, J.G., J.G. Morehead, Jr. and F.D. Stockton (committee). 1857. [Resolutions], Dialectic Hall, July 24, 1857. *North Carolina Magazine* (University of North Carolina) 1: 39–40.

McNab, J.G., J.G. Morehead, Jr. and F.D. Stockton (committee). 1858. Meeting of the Dialectic Society, Dialectic Hall, July 24, 1857, pp. 38–39 in *A Memoir of the Rev. Elisha Mitchell, D.D., Late Professor of Chemistry, Mineralogy & Geology in the University of North Carolina*. Published by J.M. Henderson, Chapel Hill.

McVaugh, Michael. 1984. Elisha Mitchell's library as an index to his scientific interests. *Journal of the Elisha Mitchell Scientific Society* 100: 50–56.

McVaugh, Michael. 1987. Elisha Mitchell's books and the University of North Carolina Library. *Bookmark* 55: [27]–54.

McVaugh, Michael. 1988. Elisha Mitchell's books and the University of North Carolina library. II. *Bookmark* 56: [31]–70.

McVaugh, Rogers, Michael R. McVaugh and Mary Ayers. 1996. *Chapel Hill and Elisha Mitchell, the Botanist*. Botanical Garden Foundation, Chapel Hill, NC.

Mebane, Wm. N. (chairman, committee). 1858. Minute of Presbytery, p. 42 in *A Memoir of the Rev. Elisha Mitchell, D.D., Late Professor of Chemistry, Mineralogy & Geology in the University of North Carolina*. Published by J.M. Henderson, Chapel Hill.

A Memoir of the Rev. Elisha Mitchell, D.D., Late Professor of Chemistry, Mineralogy & Geology in the University of North Carolina. Together with the tributes of respect to his memory, by various public meetings and literary associations, and the addresses delivered at the re-interment of his remains, by Rt. Rev. James H. Otey, D.D., Bishop of Tennessee, and Hon. David L. Swain, LL.D., President of the University. Published by J.M. Henderson, Chapel Hill, 1858. (The memoir was reprinted as an insert in volume 1 (1884) of the *Journal of the Elisha Mitchell Scientific Society*)

APPENDIX 3.1. (CONTINUED)

Merrill, George P. 1906. Elisha Mitchell's work, 1827–1828, p. 285; Mitchell's criticism of Olmsted and Rothe, 1829, p. 286; biographical sketch, p. 706 in *Contributions to the History of American Geology*. (From the *Report of the United States National Museum*, 1904, pp. 189–734)

Merrill, George P. 1924. Mitchell's geological work in North Carolina, pp. 114–115; Mitchell's criticism of Olmsted and Rothe, 1829, pp. 115–116 in *The First One Hundred Years of American Geology*. Yale University Press, New Haven, CT.

Mewborn, Suzanne. 2006. Elisha Mitchell and his mountain. *Tar Heel Junior Historian, North Carolina History for Students* 46(1): 32–33.

Midgette, Nancy. 1999. Mitchell, Elisha (19 Aug. 1793–27 June 1857), pp. 599–600 in John A. Garraty and Mark C. Carnes, general editors. *American National Biography*, vol. 15. Oxford University Press, Oxford, New York.

Mitchell, Elisha. 1845. [Letter from Elisha Mitchell to T.L. Clingman concerning measurements of the elevation of mountains in western North Carolina]. *Highland Messenger* (Asheville, NC), Friday, January 24, 1845. (*Scrapbook of Charles Phillips*, p. 99)

Mitchell, Elisha. [1856]. *Black Mountain, Professor Mitchell's Reply to Mr. Clingman*. [North Carolina]. (Pamphlet in the North Carolina Collection, Wilson Special Collections Library, University of North Carolina, Chapel Hill)

Mitchell, Elisha. 1856. [Letter to the editor, concerning exploration of the Black Mountains, NC]. *Asheville Spectator* (Asheville, NC), Thursday, June 26, 1856. (*Scrapbook of Charles Phillips*, pp. 109–110)

Moore, Ellen J. 1974. Mitchell, Elisha, pp. 420–421 in Charles Coulston Gillispie, editor. *Dictionary of Scientific Biography*, vol. 9. Charles Scribner's Sons, New York.

Morehead, James T. (chairman) and Charles E. Shober (secretary). 1857. Rev. Elisha Mitchell, D.D. *Times* (Greensboro, NC), Thursday, July 23, 1857. (*Scrapbook of Charles Phillips*, p. 73)

Morehead, James T. (chairman) and Charles H. Shober (secretary). 1858. Meeting at Greensboro, pp. 31–33 in *A Memoir of the Rev. Elisha Mitchell, D.D., Late Professor of Chemistry, Mineralogy & Geology in the University of North Carolina*. Published by J.M. Henderson, Chapel Hill. (From the *Times*, August 23, [1857])

Morley, Margaret W. 1913. Mount Mitchell, pp. [302]–314 in *The Carolina Mountains*. Houghton Mifflin Company, Boston.

Moser, Artus. 1922. On Mount Mitchell. *Carolina Magazine*, new series, 39(4): 24–27.

Nevin, Alfred (editor). 1884. Mitchell, Elisha, D.D., pp. 530–531 in *Encyclopaedia of the Presbyterian Church in the United States of America including the northern and southern assemblies*. Presbyterian Encyclopaedia Publishing Co., Philadelphia.

North Carolina Collection Clipping File through 1975: Biography. North Carolina Collection, University of North Carolina at Chapel Hill Library.

APPENDIX 3.1. (CONTINUED)

North Carolina Collection Clipping File, 1976 through 1989: Biography. North Carolina Collection, University of North Carolina at Chapel Hill Library.

North Carolina Collection Clipping File, 1990–2009. North Carolina Collection, University of North Carolina at Chapel Hill Library.

N.Y. Evangelist. 1857. Honor to whom honor is due. *Fayetteville Observer* (Fayetteville, NC), Monday, August 24, 1857. (*Scrapbook of Charles Phillips*, pp. 90–91)

O[lmsted, Denison]. 1857. Obituary—Prof. Elisha Mitchell. *American Journal of Science and Arts*, 2d series, 24: 299.

Otey, Rt. Rev. James H. 1858. A funeral oration at the re-interment of the remains of the Rev. Elisha Mitchell, D.D., late professor of chemistry, mineralogy and geology in the University of North Carolina. On Mount Mitchell, June 15, 1858, pp. [55]–73 in *A Memoir of the Rev. Elisha Mitchell, D.D., Late Professor of Chemistry, Mineralogy & Geology in the University of North Carolina*. Published by J.M. Henderson, Chapel Hill.

P. 1858. Rev. Elisha Mitchell, D.D. *North Carolina Journal of Education* 1: 6–11.

Pack Memorial Library. Newspaper File Collection: Information on Asheville and Buncombe County, compiled through 1990, vol. 46 (Mic-Mix). Pack Memorial Library, Asheville-Buncombe County Library System.

Parker, J.P. 1929. James Allen Picture; North Carolina pioneer and most unique character. *Charlotte Observer*, March 24, 1929. (Pack Memorial Library, Newspaper File Collection: Information on Asheville and Buncombe County, vol. 46)

Parris, John. 1957. An anniversary, tale of discovery and controversy. *Asheville Citizen* (Asheville, NC), Friday, June 28, 1957. (North Carolina Collection Clipping File through 1975: Biography, vol. 102, p. 352)

Parris, John. 1957. How Big Tom Wilson found Mitchell. *Asheville Citizen* (Asheville, NC), Monday, February 11, 1957. (North Carolina Collection Clipping File through 1975: Biography, vol. 102, p. 343)

Paysour, Buck. 1995. Mountain man. *The State, Down Home in North Carolina* 63(6): 33, 35.

Paysour, Conrad. 1960. He climbed. (bibliographical information lacking). (Pack Memorial Library, Newspaper File Collection: Information on Asheville and Buncombe County, vol. 46)

P[hillips], C[harles]. [undated]. The late Prof. Mitchell. For the *Presbyterian*. (North Carolina Collection Clipping File through 1975: Biography, vol. 102, pp. 291–292)

P[hillips], C[harles]. 1857. The Rev. Elisha Mitchell, D.D. *Central Presbyterian* (Richmond, VA), Saturday, August 1, 1857. (*Scrapbook of Charles Phillips*, pp. 79–80; North Carolina Collection Clipping File through 1975: Biography, vol. 102, p. 289)

P[hillips], C[harles]. 1857. Dr. Mitchell, Mr. Clingman and the Black Mountain. *Raleigh Standard* (Raleigh), [ca. December 18, 1857]. (*Scrapbook of Charles Phillips*, pp. 96–97)

Appendix 3.1. (Continued)

P[hillips], C[harles]. [1857–1889]. *Scrapbook, [1857–1889]*. (North Carolina Collection, Wilson Special Collections Library, University of North Carolina, Chapel Hill)

P[hillips], C[harles]. 1858. Dr. Mitchell's investigations among the mountains of Yancey County. *North Carolina University Magazine* 7: [293]–318.

P[hillips], C[harles]. 1858. Memoir, pp. [5]–12 in *A Memoir of the Rev. Elisha Mitchell, D.D., Late Professor of Chemistry, Mineralogy & Geology in the University of North Carolina*. Published by J.M. Henderson, Chapel Hill.

P[hillips], C[harles]. 1858. The Black Mountain again. *Raleigh Standard* (Raleigh), [ca. January 18, 1858]. (*Scrapbook of Charles Phillips*, p. 100)

P[hillips], C[harles]. 1860. Rev. Elisha Mitchell. *North Carolina University Magazine* 9: 569–570. (portrait opposite p. [513]) ("Extract from his *Memoir* by Professor Charles Phillips.")

Phillips, William B. 1888. An address, commemorative of the erection of a monument on Mitchell's high peak in honor of the Rev. Elisha Mitchell, D.D., for thirty-nine years a professor in the University of North Carolina. Delivered before the University Oct. 16th, 1888, by Wm. B. Phillips, class of 1877. *University Magazine* (University of North Carolina, Chapel Hill) 8 (new series): [39]–53.

Phillips, William B. 1888. The erection of the monument to Elisha Mitchell on Mitchell's high peak; abstract of an address delivered before the Elisha Mitchell [Scientific] Society, October 16th, 1888. *Journal of the Elisha Mitchell Scientific Society* 5: [55]–59.

Preston, Wheeler. 1940. Mitchell, Elisha (1793–1857), p. 700 in *American Biographies*. Harper Brothers Publishers, New York and London.

Ramsey, D. Hiden. 1957. Elisha Mitchell and Mount Mitchell. *Journal of the Elisha Mitchell Scientific Society* 73: 199–204.

Rannells, Earl. 1950. Mitchell County, formed in 1861, was named in honor of explorer. *Asheville Citizen-Times*, March 26, 1950. (North Carolina Collection Clipping File through 1975: Biography, vol. 102, p. 315)

Ravenscroft, J.S. and Elisha Mitchell. 1825. *Correspondence Respecting Two Publications Which Appeared in the Raleigh Register in the Month of December, 1824, between Bishop Ravenscroft and Professor Mitchell*. Printed by J. Gales & Son, Printers to the State, Raleigh.

Ravenscroft, J. S. and Elisha Mitchell. 1825. *The Correspondence Published by Professor Mitchell with Bishop Ravenscroft's Answer to His Letter of the 12th of February Last*. J. Gales & Son, Raleigh.

S., D. 1877. A thrilling sketch. *American* (Statesville, NC), November 17, 1877. (*Scrapbook of Charles Phillips*, p. 103)

S., H.M. 1860. Pilgrimage to the grave of Professor Mitchell. *Raleigh Standard*, Wednesday, August 29, 1860. (*Scrapbook of Charles Phillips*, p. 107)

APPENDIX 3.1. (CONTINUED)

S., W.W. 1857. A public loss—Obituary. *Southern Churchman* (Alexandria, VA), July 31, [1857]. (*Scrapbook of Charles Phillips*, p. 89)

Schenck, D. 1897. A thrilling sketch; concerning the search for the body of Rev. Elisha Mitchell who met his death on the mountain that bears his name. (Rich Square, Northampton County, N.C., Thursday, May 27, 1897) (North Carolina Collection Clipping File through 1975: Biography, vol. 102, pp. 310–311)

Schwarzkopf, S.K. 1985. Chapter III. Elisha Mitchell and his early trips to the Black Mountains, pp. 20–34; Chapter V. The Clingman-Mitchell Controversy: claim to the high peak, pp. 49–59 in *A History of Mt. Mitchell and the Black Mountains: Exploration, development, and preservation*. Division of Archives and History, North Carolina Department of Cultural Resources, Division of Parks and Recreation, North Carolina Department of Natural Resources and Community Development, Raleigh.

Sears, Woody. 1957. 2 new stories add to lore about Mt. Mitchell. *Durham Morning Herald* (Durham, NC), Sunday, June 23, 1957. (North Carolina Collection Clipping File through 1975: Biography, vol. 102, p. 351)

Sheppard, Muriel Early. 1932. Big Tom Wilson led party over Mitchell Peak. *Asheville Citizen*, April 17, 1932. (Pack Memorial Library, Newspaper File Collection: Information on Asheville and Buncombe County, vol. 46)

Sigourney, Mrs. L.H. 1859. The mountain burial; [a poem]. *North Carolina University Magazine* 9: 189–190.

Silver, Tim. 1997. Big Tom Wilson. *Wildlife in North Carolina* 61(11): [18]–23. ("When Big Tom Wilson led searchers to the body of Elisha Mitchell in 1857, he cemented his reputation as North Carolina's most famous mountain man.")

Silver, Tim. 2003. *Mount Mitchell and the Black Mountains: An environmental history of the highest peaks in Eastern America*. University of North Carolina Press, Chapel Hill. (Index cites numerous passages concerning Elisha Mitchell, p. 316.)

Smith, C. Alphonso. 1913–1914. Elisha Mitchell, pp. 8–11 in A.C. Smith, Presbyterians in Educational Work in North Carolina since 1813. (Reprint from the *Union Seminary Review*, Richmond, VA, December 1913–1914)

Sondley, F.A. 1928. Removal of body of Dr. Elisha Mitchell, by R.M. Clayton. Notes by F.A. Sondley. (bibliographical information lacking). (Pack Memorial Library, Newspaper File Collection: Information on Asheville and Buncombe County, vol. 46)

Sondley, F.A. 1977. [Mitchell's measurement of the highest peaks of Black Mountain], pp. 530–532; [The Mitchell-Clingman controversy], pp. 556–567 in *A History of Buncombe County, North Carolina*. Reprint of 1930 ed. Reprint Co., Spartanburg, SC. (Includes numerous passages on Elisha Mitchell and Mitchell's Peak)

Spencer, Cornelia P. (editor). 1879. Dr. Mitchell. *Weekly Ledger* (Chapel Hill, NC) 3(7): [2], Saturday, May 31, 1879.

Appendix 3.1. (Continued)

Spencer, Cornelia P. (editor). [1884]. Dr. Mitchell, pp. 13–18 in "Pen and Ink Sketches of the University of North Carolina as It Has Been. Dedicated to the people of the State and to the Alumni of the University." [Chapel Hill, The Author]. (Pamphlet in the North Carolina Collection, Wilson Special Collections Library, University of North Carolina, Chapel Hill)

Spencer, Cornelia P. (editor). 1884. Old Times in Chapel Hill. No. 4. [Elisha Mitchell]. *University Monthly* (University of North Carolina) 4 (new series): 15–22.

Sutton, Ann and Myron Sutton. 1957. The mystery of Elisha Mitchell. *Nature Magazine* 50: 429–432, 424.

Swain, D.L. 1858. A vindication of the propriety of giving the name "Mt. Mitchell," to the highest peak of "Black Mountain:" An address, delivered 16th June, 1858, by Hon. David L. Swain, LL.D., President of the University of N.C., pp. [75]–88 in *A Memoir of the Rev. Elisha Mitchell, D.D., Late Professor of Chemistry, Mineralogy & Geology in the University of North Carolina*. Published by J.M. Henderson, Chapel Hill.

Tripp, Gladys Best. 1938. Lone tombstone marks historic grave. *The Daily Tar Heel*, October 23, 1938. (Pack Memorial Library, Newspaper File Collection: Information on Asheville and Buncombe County, vol. 46)

UNC Faculty. University of North Carolina, Chapel Hill. *Minutes of Faculty Meetings*. (University Archives #40106, Wilson Special Collections Library, University of North Carolina, Chapel Hill) (Vol. 3, 1821–1841)

United States Geological Survey. 1905. *Geologic Atlas of the United States. Mount Mitchell Folio North Carolina – Tennessee. (Mount Mitchell Folio*, no. 124). Description of the Mount Mitchell quadrangle, by Arthur Keith; George W. Stose (editor of geologic maps); S.J. Kobel (chief engraver); Charles D. Walcott (Director, U.S. Geological Survey). Department of the Interior, United States Geological Survey, Washington, DC. (Revised in January 1904, the publication consists of eleven pages of text, two section sheets that cover a columnar section and structure sections, and three maps that cover topography, areal geology, economic geology. The map of topography is the "edition of Mar. 1902, reprinted Feb. 1905.")

Vance, Robert Brank. 1913. [Poem]: Dr. Mitchell's grave. (Friday, August 29, 1913) (North Carolina Collection Clipping File through 1975: Biography, vol. 102, p. 305)

Vance, Z.B. 1857. Death of Professor Mitchell. *Asheville Spectator* (Asheville, NC), Thursday, July 23, 1857. (*Scrapbook of Charles Phillips*, pp. 82–84)

Vance, Z.B. 1858. The search for Professor Mitchell's body, pp. [13]–19 in *A Memoir of the Rev. Elisha Mitchell, D.D., Late Professor of Chemistry, Mineralogy & Geology in the University of North Carolina*. Published by J.M. Henderson, Chapel Hill. (From the *Asheville Spectator*)

Vance, Z.B. [1875?]. Sketches of North Carolina; A brief biography of Professor Mitchell. (North Carolina Collection Clipping File through 1975: Biography, vol. 102, pp. 304–305)

Appendix 3.1. (Continued)

Vance, Z.B., Jas. A. Patton, John A. Dickson, A.S. Merrimon, D. Coleman, and W.M. Shipp. 1857. To the people of North Carolina. *Asheville Spectator* (Asheville, NC), Thursday, August 13, 1857. (*Scrapbook of Charles Phillips*, pp. 88–89)

Vance, Z.B., Jas. A. Patton, John A. Dickson, A.S. Merrimon, D. Coleman and W.M. Shipp (committee). 1858. The re-interment. Proposed monument. To the people of North Carolina, pp. [45]–46 in *A Memoir of the Rev. Elisha Mitchell, D.D., Late Professor of Chemistry, Mineralogy & Geology in the University of North Carolina*. Published by J.M. Henderson, Chapel Hill. (From the *Asheville Spectator*)

W., J. 1857. [Remarks on the death of Rev. Dr. Mitchell]. *New York Journal of Commerce*, Wednesday, July 29, 1857. (*Scrapbook of Charles Phillips*, p. 77)

Warner, Charles Dudley. 2005. *On Horseback: A tour in Virginia, North Carolina, and Tennessee*. Cosmo Classics, New York. (Reprint of the 1885 edition published by Houghton Mifflin and Company) (pp. 93–98)

Watson, Elgiva D. 1984. Elisha Mitchell: A Connecticut Yankee in North Carolina. *Journal of the Elisha Mitchell Scientific Society* 100: 43–49.

Watson, Elgiva D. 1991. Mitchell, Elisha, (19 Aug. 1793–27 June 1857), pp. 281–283 in William S. Powell, editor, *Ditionary of North Carolina Biography*, vol. 4. University of North Carolina Press, Chapel Hill.

Webb, Charles A. 1946. *Mount Mitchell and Dr. Elisha Mitchell*. The Asheville-Citizen Times Co., Asheville, NC.

Webster, W.D. 2002. Elisha Mitchell and the Elisha Mitchell Scientific Society, North Carolina Academy of Science, and *Journal of the Elisha Mitchell Scientific Society*. *Journal of the Elisha Mitchell Scientific Society* 118: 1–11.

Wellman, Manly Wade. 1953. Mountain drama brought Mitchell her name. *News and Observer* [Raleigh, Tuesday], June 30, 1953, [p. 12]. (North Carolina Collection Clipping File through 1975: Biography, vol. 102, pp. 341–342)

Westall, Thomas C. 1857. [Letter of Thomas C. Westall to Z.B. Vance correcting some facts about the death of Elisha Mitchell]. *Asheville Spectator* (Asheville, NC), Thursday, August 6, 1857. (*Scrapbook of Charles Phillips*, p. 86)

Westall, Thomas C. 1889. A chapter of hitherto unwritten history. *Asheville Weekly Citizen*, Thursday, September 26, 1889.

Western North Carolina Heritage. <http://dd1.library.appstate.edu/regional_history/people/Mitchell,2%E. Elisha Mitchell> (Accessed December 27, 2006)

[Wheeler, J.H.] 1883. Rev. Dr. James Phillips, Rev. Dr. Elisha Mitchell, once professors in the University of North Carolina. *University Monthly* (University of North Carolina) 2: 164–166.

[Wheeler, J.H.] 1884. [Elisha Mitchell], pp. 63–64 in *Reminiscences and Memoirs of North Carolina and Eminent North Carolinians*. Columbus Printing Works, Columbus, OH.

Appendix 3.1. (Continued)

Wheeler, Preston. 1940. Mitchell, Elisha (1793–1857), p. 700 in *American Biographies*. Harper & Brothers Publishers, New York.

Wheeler, Walter H. 1984. Elisha Mitchell as a geologist. *Journal of the Elisha Mitchell Scientific Society* 100: 61–64.

Whittle, Gilberta S. [undated]. On the summit of Mt. Mitchell. (North Carolina Collection Clipping File through 1975: Biography, vol. 102, p. 334)

Wilson, Louis Round (editor). 1953. *Selected Papers of Cornelia Phillips Spencer*. Edited with an introduction by Louis Round Wilson. University of North Carolina Press, Chapel Hill. (Entries regarding Elisha Mitchell are found on pages 446, 448, 450, 513, 547, 612, 615, 637, 663, 670, 684, 685, 693, and 695.)

Winston, Pat. 1948. The mountain-climbing professor. *News and Observer*, [Section IV], March 7, 1948, [p. 3]. (North Carolina Collection Clipping File through 1975: Biography, vol. 102, p. 331)

Wright, Perrin. 1992. Measuring Mitchell's mountains. *Journal of the Elisha Mitchell Scientific Society* 108: 1–10.

Wright, Thomas H. (chairman) and John D. Taylor (secretary). 1857. [Resolutions in honor of Prof. Mitchell]. *Wilmington Herald* (Wilmington, NC), August 5, 1857. (*Scrapbook of Charles Phillips*, p. 73)

Wright, Thomas H. (chairman) and John D. Taylor (secretary). 1858. Meeting at Wilmington, pp. 33–34 in *A Memoir of the Rev. Elisha Mitchell, D.D., Late Professor of Chemistry, Mineralogy & Geology in the University of North Carolina*. Published by J.M. Henderson, Chapel Hill. (From the *Wilmington Herald*, August 5, [1857])

Yancey, Noel. 1990. As I recall it; [regarding the Clingman-Mitchell controversy]. *Pilot*, Monday, December 17, 1990. (North Carolina Collection Clipping File, 1990–2009, vol. 34, p. 383)

Youmans, William Jay (editor). 1896. Elisha Mitchell, 1793–1857, pp. 279–289 in *Pioneers of Science in America; Sketches of their lives and scientific work*. Reprinted with additions from the *Popular Science Monthly*. D. Appleton and Company, New York. (portrait opposite p. 279)

Young, Perry. 1999. Research separates Clingman myth, Clingman fact. *Asheville Citizen Times*, Friday, March 12, 1999, p. A6.

Zeigler, Wilbur G. and Ben S. Grosscup. [1883]. [Elisha Mitchell], pp. 230–233 in *The Heart of the Alleghanies or Western North Carolina*. Alfred Williams & Co., Raleigh; William W. Williams, Cleveland, OH.

Appendix 3.2.
Text of Mitchell's ([1840b]) Lecture on Botany from His Booklet, *Natural History* (pp. 2–8).

OF BOTANY.

Object of the Science. Principally to enable us to determine the particular *species* to which a given vegetable belongs.

Certain plants bear a very intimate resemblance to each other, in regard to their aspect, size, duration, time of maturity, the form of their leaves, their action upon the organs of taste, and other sensible characters; the diversities that obtain between different individuals being accidental, and depending on foreign causes. Example—two different stalks of wheat or corn. If a seed from either of these resembling vegetables be covered to a moderate depth with earth, a new plant *of the same kind* will rise from it. The vegetables between which this intimate agreement obtains, are said to belong to the same species, and are supposed to derive their existance from a common organ. Jussieu's definition of a species—"A perennial succession of similar individuals, deriving their origin from successive generation."

In other cases there are observed resemblances, with point of difference; which, though not considerable, are *constant*, and continue when new plants are raised from the seed. These vegetables, nearly related but different, are supposed to be sprung from different original stocks. They are included in the same genus, but referred to different species.

It is the object of the science of Botany to make us acquainted with the Vegetable kingdom, and enable us to determine, by comparing it with its description contained in a Flora, the species to which a particular plant belongs.

Of the Classification and Terminology. The young botanist has first to apply himself to the acquisition of the system of classification and the terminology; the former of which is rendered necessary by the number of the vegetable species, amounting already to more than 50,000, and daily increasing. If a person have a series of volumes containing a description of all these species, or even of the plants of his own neighbourhood amounting to one or two thousand, it must create an intolerable weariness and disgust to be obliged to read carefully through a whole volume, and much more a number of volumes, for the purpose of ascertaining the name of a plant he has just found. *It was to prevent the necessity of this, that a system of classification was originally devised and is now employed.*

The first attempts to classify vegetables were ill directed and unsuccessful. They were arranged under the three heads; of

Appendix 3.2. (Continued)

trees, shrubs, and herbs. Defects of this system. The classes are neither well defined and distinct, nor sufficiently numerous. It is not possible to say, in regard to some vegetables, whether they are trees, shrubs, or herbs. Other schemes of classification, some of them modifications of that just noticed, and others more or less new, were from time to time proposed; but none of them satisfied the Naturalists who were making the vegetable world the object of their study, or enjoyed more than a temporary reputation, until Linnæus published his Fundamenta Botanica in 1735.

Qualities of a good system of classification.

1. It must be founded on diversities, presented by some part or parts, common to all, or nearly all, the objects we would classify.

2. The part or parts on which the classification is founded, must be sufficiently distinct, and not subject to vary.

3. The classes formed must be neither too numerous nor too few.

We cannot, therefore, classify vegetables by their stems or their leaves; because these parts are wanting in some, subject to vary in others, and would not furnish a convenient number of classes.

The only part that is necessarily common to all plants, and fitted in every respect to our purpose, is the *fructification;* meaning by this the whole apparatus employed in producing and perfecting the seed. With this, whatever else it may want, every plant must be provided; since upon its presence the continued existence of that particular species of plant must depend. It had been employed before the time of Linnæus; but this great Naturalist fully established, and laid at the foundation of his system, one important fact, which, though not altogether unknown, had been much neglected by preceding Botanists—*that something analogous to the difference of sex in animals obtains in the vegetable kingdom.* It may seem remarkable, that this truth should have remained so long in obscurity, and mankind have contented themselves with looking on in stupid wonder upon the little apparatus that lies in the bosom of every flower, without troubling themselves to inquire whether it is of any use in the economy of the plant.

It may be stated, as some apology for their ignorance, that most flowers are hermaphrodite; containing both the male and female parts in close proximity. In a few cases, where the sexes are on different trees, the nature of their economy was in some measure understood even by the ancients. Thus we are informed by Herodotus, that in the planting of the Palm groves, in the vicinity of Babylon, care was taken that the greater part of the soil should be occupied by such stocks as were known to produce female flowers and fruit. Here and there, at considerable intervals, they introduced a male tree; which, though yielding nothing itself, rendered those in its neighbourhood productive; and the Babylonians were in the practice of aiding the operations of nature, by collecting branches covered with the male, and attaching them

APPENDIX 3.2. (CONTINUED)

to trees bearing female flowers. Account of a similar experiment with a female palm tree, in the garden of the Royal Academy of Berlin, fecundated by means of a branch brought from Leipsic, ninety miles distant.

The common Lombardy Poplar is another example of this (dioicous) class of vegetables. It is a native of the shores of the Mediterranean, and the tree into which the sisters of Phæton were changed, after having long bewailed his fate on the banks of the Po: and it may be remarked, that this fable of the poets acquires a new interest when we have the figure of this tree before us. There is no other that is so well entitled to stand as the representative of female grace and beauty. It is a male stock that we have in this country. To the same class belong, though less strictly, the persimmon, and the different species of N. American grapes.

In the Maize or Indian corn, the male and female parts are at a distance from each other on the same plant; the former in the top or tassel, and the latter in the silk and ear. If the tassels be removed throughout a whole field of corn just before the period of inflorescence, the production of a crop will be prevented as effectually as by tearing it up by the roots.

Every plant, then, must contain a system of male or female organs, and, a very few cases excepted, it contains both; and if these present (as they do) a considerable diversity in their number, form, and position, they will furnish a foundation for a good system of classification. The number of the stamina varies in different plants from one to more than an hundred. In their length and mode of insertion, there is a similar diversity.

Classes. Linnæus selected *the stamina, or male part of the flower*, to furnish the first grand divisions of his system, distributing the Vegetable world into 24 great classes.

1. The first ten classes comprise such plants as have hermaphrodite flowers, with from one to ten equal or *irregularly* unequal and unconnected stamina. It is from the number of these that they are determined and named. Thus, the class Monandria includes all plants that have only one stamen; the class Diandria, such as have two stamina, &c.

2. The eleventh class includes such as have from 12 to 19 inclusive; the thirteenth, such as have many stamina. Twelve, or half the Linnæan classes, therefore, are determined by the number of the stamina.

3. Eight, the 12th, 14th, 15th, 16th, 17th, 18th, 19th, and 20th, are determined by the comparative length of the stamina, and the manner in which they are inserted upon the receptacle.

4. Plants of the 21st have the male and female parts of fructification on the same plant, but in different flowers; those of the 22d on different plants; those of the 23d combine some of the peculiarities of the last two and of the preceding classes.

5. Besides the vegetables included in these 23 classes, there is

APPENDIX 3.2. (CONTINUED)

a vast number of others, bearing, in common life, the names of Moss, Mushroom, and Sea-weed; whose mode of producing their seed is very obscure, and which were thrown by Linnæus into the single class Cryptogamia.

Orders. Each class is subdivided into orders, founded, through the first 13 classes, on the number of the styles or female part of the flower; in the 14th and 15th, on the fruit; in the 16th, 17th, 18th, 20th, 21st, and 22d, on the number of the stamins; in the 19th, on the nature of the florets.

A Botanist does not, therefore, attempt to determine the name of a plant, unless he finds it in flower. In that case, if he employ the Linnæan method, he turns his attention first of all to the stamina, and determines the class; next to the parts on the diversities of which the orders are founded, and determines the order. He cannot proceed to the investigation of the genus and species till he is acquainted with the terminology of the science.

Terminology is a word derived from the Latin "terminus," meaning, first, a boundary, and secondly, a definition; because, to draw the lines by which a given idea is separated from every other, is to define it. The terminology of a science, therefore, is a collection of all the technical terms belonging to it, with their definitions. These terms relate to all parts of the plant,—the fruit, leaves, stem, and root; and, in the case of Botany, it is necessary that they should be numerous, by reason of the great number of objects, many of them bearing an intimate resemblance to each other, which we have occasion to describe and distinguish by our descriptions. One of the most important services rendered to the science by Linnæus, was that of given copiousness and precision to the language it employs.

Genera. As the whole Vegetable kingdom is separated into class orders, so is each order subdivided into genera, and each genus into species.

The fructification was established by Linnæus as the only foundation of *generic* distinctions. Those plants which agree in this, are included under one genus; those which differ, are separated. It is to be remarked, however, that the agreement which determines the comprehension of different species under the same genus, is not absolute. In making these divisions, the convenience of Botanists is consulted. Where a genus is very large and unwieldy, the separation of a part for the formation of a new one, will be warranted, though the differences in the fructification be such as would be neglected if the species it includes were less numerous. The particular difference on which the separation of a new genus is founded, is called its *essential character.*

Species. Specific differences take their rise from any constant circumstance, not liable to change by culture or other accidents, wherein plants disagree. Linnæus asserts the species to be as many as there were different forms of vegetables called into being at the Creation. Casual differences, produced by culture and other analogous causes, are called varieties.

APPENDIX 3.2. (CONTINUED)

A knowledge of the system of classification and terminology, prepares us for the study of nature by means of a Flora.

A Flora is a dictionary of plants, and is *general* when it contains the names of all known plants and their descriptions: *particular*, when it contains only those of a particular district. Thus, a Flora of Chapel Hill would contain the names of all the plants in our vicinity, arranged according to some method, and with their definitions annexed.

The use of a Flora is exactly the reverse of that of the common dictionary. Both contain all the words they propose to define, accompanied by a definition or description. But the object of the common dictionary is—the word being given to find the definition: with the Flora—a definition or description, or its equivalent, the plant itself, being given, to find the name.

The use of a system of classification is to enable us to refer with readiness to the part of the Flora where a description of a plant, of which we wish to know the name, is to be found. If we employ the Linnæan system, we attend, first of all, to the stamina, and determine the class; and, instead of having a whole volume or more before us, we know the description we are in search of will be found within a limited number of pages. We determine the order, and these undergo a farther reduction—the genus; and we have then to look over but a moderate number of pages, even when the genus is the largest, to find the species or specific description, in connexion with which the name is always given.

If, however, we wish to become accurately acquainted with the plants of a district, we must form an Hortus-Siccus, or Herbarium. This is a collection of plants dried between sheets of bibulous paper. They exhibit the form, and frequently the colour, of the leaves and other parts. Its use is to enable us to compare species which do not flower at the same time of the year, or which we happen to meet with at periods and places remote from each other; and thus obtain an accurate idea of the points of difference when there is a resemblance between them: also to review our studies at our pleasure.

Of Cryptogamous Plants.—The class Cryptogamia, into which Linnæus threw all the vegetables whose fructification is obscure, is larger than any other, and has furnished matter for several distinct sciences. Thus, the Filices or Ferns, the Musci or Mosses, both Hepatic and Frondose; the Fuci, Fungi, and Lichens; have been treated of in separate works. A man of liberal education should be able to distinguish these several orders of the class Cryptogamia from each other, if nothing more.

In arranging these obscure and imperfect vegetable forms, we must adopt principles of classification different from those employed in the first 23 classes; though it is to the fructification that we still look for the distinctive characters of the orders and genera.

The Musci Frondosi, or proper Mosses, elaborate their seed in a cup or capsule, growing generally at the end of a stalk, and

Appendix 3.2. (Continued)

furnished with a lid; and beneath this with a row, and sometimes with two rows, of teeth, which were made by Hedwig the foundation of a new classification of the mosses Hedwig being also an accurate observer, and skilful delineator and describer of the mosses, acquired such a reputation amongst botanists, that he is seldom mentioned but with the epithet "immortal" attached to his name.

Acharius, a Swede, has given a new classification of the Lichens, and a description of many species in a thick quarto volume. Persoon Schweinitz and others have made the Fungi or Mushrooms a particular object of their investigations. The Ferns, Liverworts confervæ, and sea-weeds have had their economy and species amply described and illustrated.

Of the system of Jussieu.—The Linnæan system of classification, being founded on relations of number and magnitude, is more simple than any other, and better adapted to beginners; but determining every thing by the number and position of the stamina, it sometimes separates to a distance from each other plants which, in their habit and most striking characters, are very nearly related.

The whole Vegetable world falls naturally into several great families, the different members of which have an intimate resemblance to each other. Thus, amongst those that are cultivated in our gardens, the Cabbage, Turnip, Radish, Mustard, with some others resembling each other in their manner of growth, inflorescence, flavour of their leaves and fruit, constitute a natural family. The Parsnip, Carrot, Dill, Fennel, Celery, &c. constitute another. The Linnæan system *often* breaks up these natural families, and separates the members to a distance from those plants to which they are most nearly related.

It was to remedy this evil that Jessieu invented his system or method, in which the whole Vegetable kingdom was by him distributed into *one hundred families*, grouped together according to their natural affinities. By the labourers who have succeeded him in the same field, the number of these families has been more than doubled.

The Linnæan system is the most convenient for the beginner; that of Jessieu, by reason of its bringing into view the relationship of plants to each other, will become more of a favourite as our knowledge advances. The former is generally adopted in England, Germany, and the North of Europe; the latter prevails in France, and threatens eventually to supplant its rival in other countries.

It has been objected to Botanists, that they do not exert themselves enough to make their investigations and discoveries of advantage to mankind; that they are more anxious to add another to the thousand of worthless species already known, than to point out a new economical use of one with which they have long been familiar. The reproach does not seem to be merited. They are

Appendix 3.2. (Continued)

8 OF BOTANY.

sufficiently anxious to shew that the objects of their study may be useful to mankind, but an unyielding law of Nature for the most part denies them this satisfaction.

As a part of a liberal education, the study of Botany is to be defended chiefly on the ground that it affords an excellent exercise to the mental powers, generates habits of accurate observation, and furnishes a source of elegant and innocent recreation and amusement.

University of North Carolina.

Chapter 4.

John Kimberly (1817–1882)

Professor of Chemistry Applied to Agriculture and the Arts (1857–1866), absent on leave, 1859
Professor of Agriculture (1875–1876)

Early Life

John Kimberly, born September 1, 1817, in Brooklyn, New York, was the son of David and Elizabeth (Ferris) Kimberly. Little is known about his boyhood years. Young Kimberly enrolled at Yale University in 1833, leaving after the conclusion of his junior year. He completed his undergraduate studies at the University of the City of New York, now the New York University, earning the B.A. degree in 1837 (Anonymous, 1837). Kimberly subsequently removed to Raleigh, North Carolina. There, he studied law (Herrick, 1887), receiving his license to practice in the Superior Court about 1842 ([Coulton], 1842). He never practiced, however.

In the early 1840s, Kimberly was appointed a teacher at Buckhorn Academy[44] in Como (originally known as Buckhorn), Hertford County, North Carolina, where he gained fame as an educator (Moore, 1877). There, he bought the Titus Darden residence and set up a home (Winborne, 1976). After settling in Como, Kimberly married Caroline Amelia Capehart, only daughter of Tristram Capehart, both of nearby Murfreesboro, in 1842. Kimberly and Caroline had four children, two of whom died and two of whom lived—Elizabeth Ferris (Lizzie) and Emma Southall—before Caroline died on July 9, 1848. According to the 1850 census of the "free inhabitants" of Hertford, Ballahack District in Perquimans County, North Carolina, Kimberly was later the principal of Hertford Academy in Murfreesboro (USFC, 1850). On December 8, 1858, Kimberly married Elizabeth (Bettie) Meredith Maney, who was Caroline's cousin and daughter of the Honorable Thomas Maney of Nashville, Tennessee. They had eight children together: Annie Rebecca Maney, Thomas Maney, John, Mary, David, George Maney, Frances Maney (Fannie or Fan), and another child who died (Herrick, 1887; Warner, 1997). There is little published biographical information on Kimberly. Several works provide some insight into his life (Anonymous, 1968; Battle, 1907–1912, 1: 660–661; Bursey, 1982; Bursey and Thomason, 1988; Herrick, 1887; Moore, 1877; Thomason, 1981).

[44] Buckhorn Academy was established in the early 1820s. Although the academy was situated on the grounds of the Buckhorn Baptist Church, it was not church affiliated. It was incorporated by the General Assembly on January 9, 1847. After the academy closed in about the early 1900s, the building was moved to a site on Hill's Ferry Road in Como, where it served as a public school for Black children for about the next half-century (Moore, 1877; Stephenson, 2003, 2005; Whitley, 1988; Winborne, 1976).

Frontispiece Credits: (above) John Kimberly, ca. early 1860s; a photographer's copy, tinted, of Kimberly's portrait donated by his daughter Mary Kimberly in 1948. (John Kimberly Papers #Oversize P-398, Southern Historical Collection, Wilson Special Collections Library, University of North Carolina, Chapel Hill). (below) University of North Carolina campus, ca. 1855. (From *Cyclopaedia of American Literature*, Vol. 2, p. 105, by Evert A. Duyckinck and George L. Duyckinck [1866]).

Harvard Education and Its Influence

From 1853 to 1854, Kimberly was enrolled in the Lawrence Scientific School at Harvard College ([Love], 1901). Founded by botanist Asa Gray and astronomer and mathematician Benjamin Peirce in 1847, the school was established to promote advanced studies in science and literature for graduates and other qualified individuals. A number of innovations in teaching arose from the endeavor, including the instruction of postgraduates in research and the training of teachers, scientific laboratory instruction to undergraduates, the development of museums and herbaria as instructional adjuncts in science, and the introduction of the Bachelor of Science degree at Harvard College on an equal basis with the Bachelor of Arts degree (Love, 1944). Kimberly wished to further his education in subjects offered at the school but not typically available at other colleges (Kimberly, 1853a). When writing to chemistry professor Eben N. Horsford about enrolling there, he asked whether the chemical students had access to the laboratory to analyze soils. He principally desired to pursue instruction in analytical chemistry.

Among Kimberly's mentors and instructors in the Lawrence School were such stellar scientists as Louis Agassiz, Lawrence Professor of Zoology and Geology; Jeffries Wyman, Hersey Professor of Anatomy; and Eben Norton Horsford, Rumford Professor and Lecturer on the Application of the Sciences to the Useful Arts (Fig. 4.1). Kimberly also studied mineralogy under Josiah P. Cooke, Erving Professor of Chemistry and Mineralogy according to North Carolina State Geologist Washington C. Kerr (1854). Swiss-born, Agassiz had received his training in Europe. Joining the new teaching venture at Harvard in 1848, he developed a course of instruction that emphasized individual study and research in geology and zoology, the subject areas of his specialty (Love, 1944). He also took his classes on scientific excursions nearly every other week (HC, 1855). His pedagogical philosophy emphasized that students should study the object itself, thus focusing on nature, not textbooks—a methodology new to American education.

Fig. 4.1. Kimberly's mentors in the Lawrence Scientific School, Harvard University: (left) Louis Agassiz, 1847 (Ernst Mayr Library, Museum of Comparative Zoology Archives, Harvard University, Cambridge, Massachusetts). (center) Eben Norton Horsford (Horsford Family Papers MC5, Institute Archives and Special Collections, Rensselaer Polytechnic Institute). (right) Jeffries Wyman (image [3A], olv347937, Harvard University Archives, Pusey Library).

Kimberly compiled several notebooks while he was a student at Harvard. The course on geology consisted of a series of twenty-seven lectures beginning in early September 1853 and ending in mid-December (Kimberly, 1853b). Instruction included excursions in the field as well as one trip to view a private collection of fossils belonging to John Collins Warren, a noted Boston surgeon who also had an active interest in natural history, including comparative anatomy, geology, and paleontology (Joseph, 1999). Agassiz also enhanced his lectures with specimens, which he placed on exhibit and circulated around the class. The geology course covered a broad spectrum of topics. Agassiz cautioned the students about consulting the literature. He objected to the textbooks of the day because they were essentially abstracts of European works, and they were thus not suitable for the American audience. Other books were either too theoretical or too detailed for the beginning student. He recommended *The Wonders of Geology* by Gideon A. Mantell (1848) and *The Ancient World* by David T. Ansted (1847), as they would awaken the students' interest in geology. Agassiz's lectures covered rocks (their age and alteration), strata, the history of the globe, drift, geological periods, fossil animals and plants, the gradation and succession of animals in the strata of different formations, the succession of vertebrates, and the Carboniferous Period.

Another of Agassiz's courses focused on various subjects in natural history and consisted of twelve lectures, which were delivered Tuesday and Thursday evenings beginning December 6, 1853. The Lowell Institute sponsored the talks (Kimberly, 1853c). Topics included coral reefs, the formation of the peninsula of Florida, the determination of the chronology of modern deposits, materials of the drift, the flora and fauna of the temperate zone, tropical forests, and the races of man.

Jeffries Wyman, another of Kimberly's notable professors, earned his A.B. and M.D. degrees at Harvard in the 1830s and later augmented his education in European colleges, undertaking studies in human anatomy, comparative anatomy, physiology, and zoology in Paris during the summer of 1841. Appointed to the Lawrence School in 1847, he taught anatomy and physiology (Dupree, 1976). Kimberly's course consisted of thirty lectures on comparative physiology that were illustrated by dissections and microscopic observations (HC, 1855).

Finally, Professor of Chemistry Eben Norton Horsford received his early education from the Rensselaer School (now Rensselaer Polytechnic Institute), where he was taught by renowned scientist Amos Eaton. Beginning in 1844, he studied under pioneer chemist Justus von Liebig[45] at the University of Giessen in Germany for two years, only the second American to study in that lab. Horsford's research was pivotal in the transfer of chemical knowledge and expertise from Europe to America (Rezneck, 1972). He was selected as a charter member of the faculty in the Lawrence School. Believing that experimental studies should be balanced with practical applications, he provided day-long instruction in the chemical laboratory and field trips to visit chemical manufacturers, where theory was applied (Love, 1944). Horsford gained fame for developing baking powder. After resigning from Harvard in 1863, he joined the Rumford Chemical Works, where he became a successful chemist (Munday, 1999).

[45] Liebig specialized in research on relative protein content, and he advanced the application of chemistry to industry (Munday, 1999).

Kimberly's ([1854]) laboratory notes from Horsford's chemical class at the Lawrence School survive in manuscript records at UNC. Written in pencil, the notes focus on quantitative analyses that were performed in the first half of 1854 and provide a record of calculations and, in some cases, procedures. Analyses of inorganic substances included silver and gold coins, German silver, gunpowder, natrolite, prehnite, lead glass, and feldspar. Other laboratory experiments covered the ash analysis of Irish potatoes and the analysis of organic material, including oil of turpentine, mineral water, and soil.

The stimulating and innovative environment of scientific inquiry at Harvard inspired Kimberly. He developed an appreciation for the application of chemistry. His future methods of teaching were greatly influenced by his Harvard instructors and included assigning laboratory investigations to undergraduates at UNC, which Professor Hedrick, also a graduate of the Lawrence Scientific School, had as well done.

Teaching in Nashville, Tennessee

About 1855, Kimberly (1855a) settled in Nashville, Tennessee, a city about which he wrote: "No place pleases me more than Nashville—it combines charmingly all the attractions and conveniences both of city and country life." In late summer, he set up a school in a rented house, which he considered to be the "biggest" house in the region (Kimberly, 1856d). After procuring instructional supplies, such as maps and blackboards, he sent out advertisements. Several prominent people wrote supportive letters that he used to establish his credentials. North Carolina governor Thomas Bragg (1855) noted Kimberly's devotion to teaching, his high moral character, and his scholarship. Three of Kimberly's professors from the Lawrence School also vouched for his character and superb abilities. Several days before Kimberly opened the establishment in September, however, his landlord forbade him to commence, because he did not want the dwelling used for school purposes (Kimberly, 1856d). Subsequently, Kimberly decided to build a schoolhouse on a parcel of land on the Franklin Pike. Opening on January 1, 1856, the school building consisted of a large room for his laboratory, which contained his cabinet, apparatus, and chemicals, and two other rooms, one serving as his study, the other for his sleeping quarters. The venture was very promising: he had a "full complement of students," earned nearly $2,000 in the first year (on a $7,000 initial investment), and considered expanding the school by adding a teaching associate the following session. Yet Kimberly became tired of his mode of life as early as March 1856 and desired a change (Kimberly, E., 1856).

Teaching at UNC

Kimberly had learned that Benjamin Sherwood Hedrick might vacate or be dismissed from his professorship of analytical chemistry at UNC. Hedrick, with whom Kimberly (1855a, 1855b, 1856d) had exchanged correspondence in the mid-1850s, was removed from his faculty post on October 11, 1856, because of his views against slavery, which he openly expressed and published in local newspapers (Knapp, 1988). Kimberly (1856a) wrote to UNC professor Elisha Mitchell about the position in the fall of 1856 and to Charles Manly, secretary of the trustees (Kimberly, 1856b). He also sent a letter of application to UNC Professor Charles Phillips, enclosing the three testimonials he had obtained from his Harvard professors who taught in the Lawrence Scientific School (Kimberly, 1856c). Louis Agassiz (1855) remembered Kimberly as being "ardent in his study" and mentioned that Kimberly's lecture notes were the best among all the students he had taught. Jeffries Wyman and Eben N. Horsford (1855) noted Kimberly's experience in teaching, adding that he would "not fail to make himself respected and valued wherever he may reside." These letters most likely assured Kimberly the position, as he was already held in high scholastic esteem by the UNC faculty, which had conferred an honorary Master of Arts degree on him in 1846. The professors (UNC Faculty-T 1855) had praised Kimberly's achievements in teaching, noting that the teachers of the state held him in high esteem and that the students he had prepared for college acknowledged "his ability & faithfulness." Kimberly ([1858]a) was particularly eager to teach at UNC because he wanted to be near his daughters Lizzie and Emma, who were students at St. Mary's School in Raleigh, and he also had a fond attachment to North Carolina, where he had developed connections.

The trustees acted quickly in electing Kimberly as Professor of Chemistry Applied to Agriculture and the Arts (also called Professor of Agricultural Chemistry and Professor of Agriculture and Analytical Chemistry) at a meeting held January 5, 1857 (BOT 6: 202). He was one of several scholars who had been considered for the chair, including J.P. Nelson of Goldsboro, North Carolina. The trustees subsequently set Kimberly's yearly salary at $1,600 (BOT 6: 220) and raised it to $1,700 in 1859 (BOT 6: 246) and to $2,000 in 1864 (BOT 15: 249). On January 6, President Swain (1857a) informed Kimberly of his appointment, noting: "You cannot come on too soon."

Kimberly (1857a) arrived at Chapel Hill on Tuesday evening, February 10, 1857. The village of Chapel Hill and the university were in a thriving state. In 1853, Chapel Hill had undergone appreciable growth and improvements. The main street had been extended, and the land it bisected was divided into building lots. Trees were planted along the sides of the newly formed road, with the prospect of providing shade. A contemporary source enthused, "New houses are springing up in every direction, and the whistlings of the plane, the grating of the saw, the sound of the hammer and the ringing music of the anvil form a pleasing accompaniment to a stroll 'round town'" (Anonymous, 1853). Two years later, another writer (Anonymous, 1855) proclaimed that Chapel Hill was among the prettiest, healthiest, and quietest villages in North Carolina. The broad streets were bordered by stately elms, cedars, and oaks. The buildings were "commodious" and pleasingly situated. The villagers were described as "generally moral, intelligent, enterprising, and liberal." The same writer noted the pure air and said that the village contained no grog-shops and no billiard tables or gaming houses of any kind. Skilled craftsmen abounded, as well as all kinds of mechanics. Professionals were well represented, particularly merchants, apothecaries, and lawyers. In the June 1854 issue of the *North-Carolina University Magazine*, an anonymous (1854) student reporter noted that the village had "eight stores, three tailor shops, several bakeries, a drug-store, two public houses, eight boarding houses, three handsome churches and one in the act of being built, [and] a number of new and very elegant private residences."

The beautiful campus, enhanced by the flourishing village and its services, kindled the admiration and respect of students, faculty, townspeople, and visitors. Its magnificent grove of trees was soothing throughout the seasons. The oaks were particularly noteworthy, providing shade in the summer, swaying in the gentle breezes of the spring, and groaning and creaking during the fury of storms (Anonymous, 1856b). When Kimberly began teaching at UNC, there were six primary buildings on campus. During that year, university officers began to plan for two additional buildings, New East and New West, which were completed in 1861,[46] according to Archibald Henderson (1949) and Rachael Long (1993). Student enrollment had reached an all-time high since the university had opened in 1795, with 435 students in 1857 and 456 a year later. The UNC catalog of 1856–1857 (p. 45) recorded that annual expenses for students ranged from $188 to $249. The breakdown of costs were: tuition, $50; room rent, $10; servant hire, $5; deposit, $4; board for forty weeks, $100 to $140; bed and washing, $16 to $24; wood, $8 to $10; and candles, $5 to $6. In addition to these expenses, each student could expect expenses amounting to $50 to $55 for textbooks during his four years at college.

Upon arriving in Chapel Hill, Kimberly (1857a) was surprised to learn that he had considerable latitude in developing his instructional program, which he termed "carv[ing] out my own work." His enthusiasm became subdued when he entered the laboratory, where confusion reigned. He found "broken and empty bottles, piles of dirt, broken apparatus, everything helter-skelter, upside down, as if there had been a general rebellion and stampede among the chemicals." Although discouraged by the condition of the lab, he began to clean up the room.

Tiring of the cleaning, Kimberly decided to find his class, which was composed entirely of seniors. He found them in recitation and called roll, identifying fifty out of seventy present who would be his students. "Old Dr. Mitchell" interjected: "Plenty to do." At that prompt, Kimberly (1857a) became overwhelmed. He recalled: "I could have bitten his head off at one mouthful." He restrained his feelings, however, and requested that his students assemble in the laboratory the next morning. That evening, Kimberly prepared a lecture for the next day.

Prior to entering the lecture room that day, he had "a sinking of the heart, and such a choking about the throat." He continued:

> The bell rung, my class came flocking in, together with many others to hear the new professor. When all was quiet, one of the professors arose and introduced me. With a nervous effort to compose myself, and a thought about you [*i.e.,* his brother-in-law Henry Maney], I arose. My voice trembled a little at first, but before I had uttered three sentences, I was as composed as if talking with you in your office. There was something inspiriting in the occasion, and if my lecture did me no credit, it did not disgrace me, as a failure would have done.

Later that day, Kimberly met with his class in the laboratory. As there were no chemicals in the lab with which to do experiments and the students had no textbooks, Kimberly discussed the work expected of them in the forthcoming academic session. He was impressed by his class, which he called a "fine body of young men as I ever saw."

[46] Work on the foundations of the buildings began in late October 1858 (Anonymous, 1858). Kemp Battle (1883) reported that the halls were completed in 1859.

In reflecting about his debut at the university, Kimberly (1857a) wrote to his brother-in-law Henry (Hal) Maney:

> I flatter myself I made an impression—Two lectures I have delivered to-day, and with scarcely a note of preparation—This jerking a fellow out of one place, and setting him in another, and then telling him to lecture and talk science, is rather an astounding course of proceeding, particularly when his audience is composed of intelligent well-informed young men, much as I had before me to-day.

Swain's (1857b) early impression of Kimberly as "a modest, timid, industrious well dispersed man" was less flattering, although he saw potential for Kimberly's success with more experience in the classroom and the chemical laboratory.

According to the university catalog of 1856–1857 and those of much of the subsequent decade, his course on agricultural chemistry included the analysis of soils and manures, the assay of ores and minerals, the analysis of mineral water, and the testing of drugs and medicine. The course was in the School for the Application of Science to the Arts.[47] It was an optional course intended for students desiring to become farmers. Students carried out chemical work in the laboratory, which was open to them every day of the week. The experience of undergraduate laboratory instruction was hands-on. The laboratory experience gave students an opportunity to replicate chemical tests given in textbooks. An anonymous (1856a) UNC student praised the practicality of the experiments, because they assisted students in remembering and provided "an ocular demonstration of their truth." Students also attended recitations and lectures. Early in his career at UNC, Kimberly gave one lecture a week, which he prepared "with a great deal of care" (Kimberly, 1858d). After teaching his first session, Kimberly acquired a number of books to support his instruction and to assist his students (Table 4.1). Long after his death, his daughter Mary Kimberly donated forty-three titles (UNC Archives, 1) from Kimberly's private library to the University Library, including a copy of Asa Gray's *Manual of Botany* (1848), as well as several titles cited in Table 4.1. Twenty-three students enrolled in his laboratory during the second session of 1857, with nine men from states beyond North Carolina (Kimberly, 1857–1886) (Table 4.2).

Kimberly used the analytical chemistry lab that was established in 1854 in the basement of Smith Hall (Long, 1993). In planning for the School of the Application of Science to the Arts at UNC, professor Charles Phillips (1854) had enumerated the basic needs for this laboratory: "a furnace with its facilities for sand-baths, evaporating chambers, and distilled waters, balances for considerable and for inconsiderable weights, a supply of retorts, test-tubes and crucibles, together with a stock of chemicals and their necessary vessels." Appropriate furniture was also needed. During Kimberly's tenure at UNC, the laboratory shared the lower story of Smith Hall with the library's "Reading Room" (Anonymous, 1857).

[47] Opening in January 1854, the School for the Application of Science to the Arts prepared students in agricultural chemistry and engineering. Through the practical utility of its instruction, the venture served the state in producing technically trained men and was the "culmination of the development of science in the antebellum University of North Carolina" (Cherry, 1994).

Table 4.1. A List of Book Titles Procured from the Book Dealer H. Ballière (London and Paris)[48] by John Kimberly for Instruction and Research in Agricultural Chemistry, June 26, 1857. (Adapted from a document in the archival papers of John Kimberly. Entries marked with an asterisk (*) denote books that survive in the UNC Library and contain Kimberly's signature.)

Carl F. Plattner and Sheridan Muspratt. 1854. *The Use of the Blowpipe in the Qualitative and Quantitative Examination of Minerals, Ores, Furnace Products, and Other Metallic Combinations*. 3rd edition, revised and further enlarged. J. Churchill, London.	$3.25
G.J. Mulderand, Peter Frederik H. Fromberg, and Finlay Weir Johnston. 1849. *The Chemistry of Vegetable & Animal Physiology*. W. Blackwood and Sons, Edinburgh, London.	$9.00
D.T. Ansted. 1855. *Geology, Mineralogy, and Crystallography; Being a Theoretical, Practical, and Descriptive View of Inorganic Nature*. Houlston and Stoneman, London.	$3.60
Arthur H. Hassall. 1857. *Adulterations Detected; or, Plain Instructions for the Discovery of Frauds in Food and Medicine*. Longmans, Brown, Green, Longmans, and Roberts, London.	$4.50
*J.C. Nesbit. 1856. *On Agricultural Chemistry, and the Nature and Properties of Peruvian Guano*. 8th edition. Longman and Co., London. (Fig. 4.2)	$1.25
W. Marcet. 1856. *On the Composition of Food and How It Is Adulterated, Practical Directions for Its Analysis*. J. Churchill, London.	$2.00
Jas. F.W. Johnston. 1855. *The Chemistry of Common Life*. A new edition revised by Arthur Herbert Church. W. Blackwood and Sons, Edinburgh and London.	$3.00
*G.W. Septimus Piesse. 1856. *The Art of Perfumery, and the Methods of Obtaining Odours of Plants*. Longmans, Brown, Green, and Longmans, London.	$1.25
William Gregory and J. Milton Sanders. 1857. *Handbook of Organic Chemistry; For the Use of Students*. 4th American from the 3rd English edition. A.S. Barnes & Co., New York.	$3.60
*Campbell Morfit. 1856. *A Treatise on Chemistry Applied to the Manufacture of Soap and Candles*. New and improved edition. Parry and McMillan, Philadelphia.	$6.00
Ebenezer Emmons. 1855. *American Geology*. Volume 1. Sprague and Co., Albany.	$3.00
Justus Liebig and William Gregory. 1852. *Animal Chemistry; or, Chemistry in Its Applications to Physiology and Pathology*. Part 1. From the 3rd London edition, revised and greatly enlarged. J. Wiley, New York.	$1.25
Justus Liebig and August W. von Hofmann. 1853. *Hand-book of Organic Analysis; Containing a Detailed Account of the Various Methods Used in Determining the Elementary Composition of Organic Substances*. Walton and Maberly, London.	$1.50
Friedrich Julius Otto. 1857. *A Manual of the Detection of Poisons by Medico-chemical Analysis*. H. Ballière, New York.	$1.75
T. Frederick Hardwich. 1855. *A Manual of Photographic Chemistry, Including the Practice of the Collodion Process*. 2nd edition. J. Churchill, London.	$2.00

[48] Kimberly, John. 1857c. Letter of June 26, 1857 to H. Baillère. (Box 1, folder 9, John Kimberly Papers #398, Southern Historical Collection, Wilson Special Collections Library, University of North Carolina, Chapel Hill).

TABLE 4.1. (CONTINUED)

John Mitchell. 1854. *Manual of Practical Assaying; Intended for the Use of Metallurgists, Captains of Mines, and Assayers in General*. H. Baillière, London.	$5.00
Sheridan Muspratt. [1850s]. *Chemistry, Theoretical, Practical, & Analytical: As applied and relating to the arts and manufactures*. William Mackenzie, Glasgow. Pts. 1, 2, 3.	$7.88

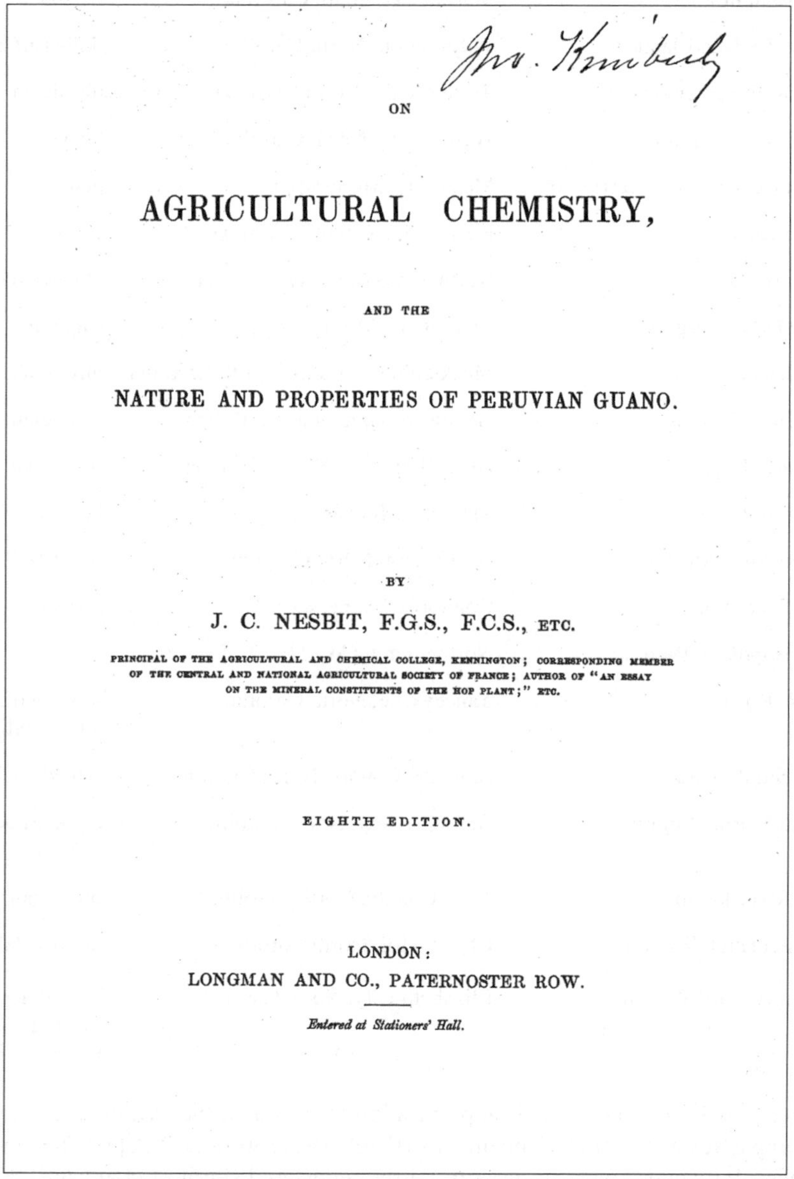

Jno. Kimberly

ON

AGRICULTURAL CHEMISTRY,

AND THE

NATURE AND PROPERTIES OF PERUVIAN GUANO.

BY

J. C. NESBIT, F.G.S., F.C.S., ETC.

PRINCIPAL OF THE AGRICULTURAL AND CHEMICAL COLLEGE, KENNINGTON; CORRESPONDING MEMBER OF THE CENTRAL AND NATIONAL AGRICULTURAL SOCIETY OF FRANCE; AUTHOR OF "AN ESSAY ON THE MINERAL CONSTITUENTS OF THE HOP PLANT;" ETC.

EIGHTH EDITION.

LONDON:
LONGMAN AND CO., PATERNOSTER ROW.

Entered at Stationers' Hall.

Fig. 4.2. Title page from *On Agricultural Chemistry, and the Nature and Properties of Peruvian Guano,* 8th edition, by J.C. Nesbit (1856).

(Courtesy of University Library, University of North Carolina, Chapel Hill).

Table 4.2. A List of Students of the Class of 1857 in John Kimberly's Agricultural Chemistry Course.

Student	Home Address	Profession
Robert John Cannon	Somerville, Tennessee	planter
Thomas Hilliard Christmas	Bolivar County, Mississippi	physician
James Hamilton Evans	Marion CH, South Carolina	---------
Andrew Jackson Flanner	Wilmington, North Carolina	(killed in Civil War)
Daniel McLean Graham	Fayetteville, North Carolina	physician
George Henry Gregory	Washington, North Carolina	lawyer
Pinckney Cotesworth Harrington	Meadville, Mississippi	planter
Eustace Hunt	Pittsylvania County, Virginia	planter
John Calvin Jacobs	Northampton County, Virginia	physician (?)
Louis Meredith Jiggitts	Livingston, Mississippi	surgeon
John Means Lawing	Mecklenburg County, North Carolina	physician
James Early Lindsay	Greensborough, North Carolina	physician
John Early Logan	Greensborough, North Carolina	physician
George M. Lusher	Memphis, Tennessee	lawyer
Norman Allen Morrison	Laurinburg, North Carolina	(died 1859)
Cadwallader Polk	Columbia, Tennessee	farmer
Junius Napoleon Ramsay	Northampton Co., ??	-------
Felix R. Roan	Yanceyville, North Carolina	Register of Deeds, Caswell Co.
Edwin Smith Sanders	Johnston County, North Carolina	(killed in Civil War)
William Moring Sutton	Bertie County, North Carolina	clerk, Superior Court
Henry Roan Thorp	Nash County, North Carolina	physician
Nathaniel Hugh Watson	Chapel Hill, North Carolina	(died 1860)
William Edward Wilson	Elizabeth City, North Carolina	(killed, Roanoke Island, 1862)

The trustees had provided a small appropriation to establish the venture and to purchase suitable apparatus and a stock of chemicals (Kimberly, [1860]e; [1875]a). The laboratory's "appliances, [however,] were necessarily of the rudest and simplest character" (Kimberly, [1875]a). When the trustees authorized the funds to establish the lab, they had envisioned that the facility would become self-sustaining through the assessment of fees, which would subsidize the purchase of new equipment and chemical supplies. Kimberly ([1860]e) observed that numerous students failed to pay their fees, however, and the resulting expenses defaulted to the professor. Hedrick was reimbursed for losses through the deposit fee, but Kimberly was not. He recommended that the trustees should either make an annual appropriation for supplies or collect the laboratory fees along with tuition at the beginning of each term. The latter recommendation was adopted, as confirmed by the UNC catalog of 1859–1860 (p. 41), which noted that each student enrolled in the analytical chemistry laboratory would be assessed a charge of $10 at the start of the term. Students were also required to supply their own glassware, dishes, crucibles, and expensive reagents.

The laboratory was enhanced in February 1858, when Kimberly (1858h) acknowledged the receipt of chemical and philosophical apparatus that the university had purchased from the estate of Elisha Mitchell. The articles included 1 German air pump, 1 Boston air pump, 1 goniometer, 1 Lett electro galvanic apparatus, 2 blowpipes, 1 piece of Iceland spar, 2 agate mortars, 2 analytical balances, 2 torsion balances, 1 pair of reflectors, 1 Daniell's hygrometers, 4 platinum crucibles, and 1 platinum ladle and wire. In addition, he noted receiving a collection of nearly 2,000 mineralogical specimens.

When the building plans for New East were being drafted in 1858, Kimberly suggested that a large laboratory should be part of the building. He thought that the floor plans could be revised by "simply removing the lath & plaster partitions which [would] divide the first floor under the Philanthropic Hall" (Swain, 1858). The resulting area (36 x 36 feet), along with the basement, which was accessible through a trap door, could be made into "one of the best arranged laboratories in the U.S." His suggestions were not adopted.

In his second year of teaching, Kimberly was pleased to learn that his lectures were garnering praise both on campus and in the village. Even Swain sat in on a lecture and praised the presentation at its conclusion. Kimberly (1858d) was nonetheless feeling despondent, however, writing: "I am sick, sick of heart, sick even unto death—disgusted with everything around me, and would not give a snap of my finger for the world-wide reputation of Agassiz, if I am to stay here."

As this statement and other evidence contained in manuscript records held in the Southern Historical Collection at UNC suggest, Kimberly was neither easily satisfied nor often content. He was easily disturbed by everyday incidents, such as the slow delivery of mail, the buzzing of insects, and the glare of sunlight. Each time that he sought a new teaching post, he soon became tired of it, as well as the location in which he lived. He disliked living in Chapel Hill soon after arriving there. Kimberly (1860m) confided in Bettie about his feelings, writing:

> Only think of spending 10 years in Chapel Hill! If I thought that was to be my destiny if life continued, I would soon settle the matter by giving an extra touch to my razor—10 years on Chapel Hill! I mean to leave here, as soon as I can, just as soon as you will let me, and if you say so, I will send my resignation as soon as I can receive your answer. We will go to California.

Kimberly also suffered from rheumatism and was a smoker. In writing to Bettie in July 1860, he noted that he kept to himself, was reserved, and attended to his own business and expected others to do likewise. He was aware that people "regarded [him] as a very mysterious sort of personage, and [was] the object of much wonder and surmise" (Kimberly, 1860f). In a letter of November 10, 1875, to Kemp Battle, Kimberly (1875g) divulged his weakness as being "the blues" (probably depression). About his condition, he wrote: "My colleagues are the cleverest fellows in the world and I owe them much for their efforts to chase away my 'blues'—That's what's the matter—I've got the blues—It frets me to death to think of the demands on my time and my inability to meet them."

Despite—or perhaps because of—his temperament, Kimberly commanded a unique respect among his students. He himself wondered why his students were so well-behaved in his presence. He confirmed this view one afternoon, when he followed his class to a lecture by another professor. He was shocked to learn that a group of young men who were orderly, attentive, and quiet for his lectures could create "so much noise, hubbub and confusion" in his colleague's class. Kimberly (1860o) noted: "Other professors have asked me where the charm is, but I am not curious of every thought upon the subject, except a determination to cut the fellow's throat who disturbed me." Kimberly's good rapport with the students was exemplified by gifts from them. Nineteen members of his analytical chemistry class of 1857–1858 gave him a gold-headed cane to show its "high esteem and affectionate remembrance" of him (Bitting et al., [1858]). His class of 1859–1860 also presented him with a cane in recognizing his "invariably urbane and affable manners" (Committee, [1860]). The students' affection was also apparent in his students' wish to have him as a fraternity advisor. In the spring of 1857, when a group of university students petitioned the Sigma Phi Society to establish a chapter at Chapel Hill, Kimberly offered "his services" (Jenkins and Alston, 1857). (In 1836, he had been elected to New York's Gamma chapter of Sigma Phi [1859].) Although the request failed, another group of students tried to set up a campus chapter and also sought Kimberly's assistance (Singletary et al., 1858). They were also unsuccessful.

Much to Kimberly's chagrin, his teaching load increased after his arrival. When Elisha Mitchell died in June 1857, Kimberly subsequently took over his classes for one session until William James Martin was appointed Professor of Chemistry, Mineralogy, and Geology. Martin, however, did not include botany (or zoology) in his course of instruction (Battle, 1907–1912, 1: 684).[49]

In addition to teaching, Kimberly did research, mostly of an applied nature. With the assistance of students, he determined the quality of pit coal from Deep River, North Carolina; black band iron ore from Chatham County, North Carolina; bituminous slate; and soil (Kimberly, 1857b). He also served the university as an "Officer of the Week."[50] Each

[49] Besides being a chemist, Martin was known as "a diligent student of geology and botany" ([Harrell], 1887). He desired to further his knowledge of the study of plants. In spring 1860 he studied at the Lawrence Scientific School at Harvard University (Harvard, 1901), where he took Asa Gray's lectures on botany and Louis Agassiz's on comparative anatomy (Phillips, J., 1860). According to a letter from Gray (1860), Martin did not complete Gray's class. Gray, however, gave Martin encouragement in studying botany: "Pray keep on with Botany, and consult me on any troubles you meet with. –I mean botanical troubles."

[50] The practice of faculty serving as monitors of the students' behavior in their rooms at night was established in April 1828. Prior to then, tutors were responsible for that duty, but their supervision failed to maintain proper order. Therefore, the trustees directed: "It shall be the duty of the Professors also, in such order and rotation as shall be agreeable to them, to perform similar nightly visitations, and so to arrange them that some one of the Professors shall visit the apartment of every student at the least three times in each week" (BOT 5: 124).

week, a professor would take a turn in carrying out the officer's duties, which included visiting students' rooms, reprimanding the lazy and malicious, and suppressing all noise and disturbances at night. He noted that the officer was "compelled to go nightly to the College, prowl about like a watchman, get up at dawn of day to see that the Chapel and recitation rooms [were] in order, and do all manner of disagreeable work." If Kimberly had known about this "repulsive" duty before being hired, he said he would not have accepted the professorship. He particularly disliked the nightly patrol in search of drunken or riotous students, because he feared getting his "head broke." He claimed that other universities assigned similar responsibilities to tutors (Kimberly, 1860b).

Kimberly also assisted the public with agricultural matters. He received a variety of samples for analysis, such as water samples sent to him by William F. Collins (1857) and limestone from Raleigh resident C.H. Wiley (1858). The latter wanted to know if the limestone had hydraulic qualities. In October 1857, the County Agricultural Society invited Kimberly to deliver an address before their members at its fair. The opportunity gave him a chance to network with leading gentlemen of the region (Phillips, S.F., 1857). Whether Kimberly accepted the invitation to speak is unknown.

Kimberly (1857a) struggled to find a decent place to live. Upon his arrival at Chapel Hill in 1857, he resided at a local hotel, which reeked of filth and grease. He claimed the place had dirty table-cloths, dirty knives and forks, and tumblers, everything was dirty from cellar to garret, including landlord, servants, and eatables. A year later, Kimberly ([1858]a) was still looking for a place to live. In an attempt to get the trustees to assist him toward attaining that goal, he used an offer of a professorship in natural sciences received from Oakland College in Natchez, Mississippi, as a bargaining tool. The offer included living quarters and a better salary than Kimberly earned at UNC. If the trustees would provide him with a house, he told them, "no inducement could tempt me to leave my present position." Despite Kimberly's statement about remaining in Chapel Hill if he had a suitable house, he apparently had told his daughter Emma that he did not like Chapel Hill only a month after moving there (E. Kimberly, 1857). By October 1857, he was complaining to his daughter Lizzie about the "slavish life" (Kimberly, L., 1857). Lizzie suspected a deeper problem existed, and she wrote to her father: "I know that Teaching is not all that troubles you. Tell me Father what it is, am not I old enough for you to confide in me?"

Before the matter was resolved, however, Kimberly left the country. He wished to develop a modern course in agricultural chemistry, but he felt he could only do so by learning the latest methods and techniques from studies in Europe. In January of 1858, he notified the trustees that he desired to take a leave of absence of twelve to eighteen months beginning May 1, 1858 (BOT 15: 222; Kimberly, 1858c). He wanted to visit a number of different institutions of higher learning in the Old World and to study in several laboratories in Germany. He preferred to study with renowned German chemist Justus von Liebig, possibly obtaining a doctoral degree (Kimberly, 1858f). He wanted to study organic chemistry "exclusively" in the laboratory and to attend Liebig's lectures on agricultural chemistry (Kimberly, 1858g). If he could not study there, he desired to go to Göttingen or Giessen.

At their meeting of June 1858, the trustees approved his request for a paid twelve-month leave, stipulating that he give the senior class of 1859 three days of instruction in analytical chemistry in the fall session and the same instruction to the class of 1860 during the last session of its senior year (BOT 6: 226). In the interim, Professor William J. Martin would give lectures on agricultural chemistry (Kimberly, 1858e). Kimberly and his wife subsequently spent most of 1859 in Europe, where he studied at the University of Heidelberg under famed chemist Robert Bunsen.[51]

Kimberly (1860a) left Europe about the end of December 1859, and his wife remained in Germany until May 1860, when he returned there to accompany her home. Kimberly's (1860i) first return trip to America was harrowing, as he recounted in a letter to Bettie:

> We encountered a frightful storm or a succession of storms, against which our gallant vessel had to fight every inch of her way. She was tossed about upon the waves like a cork, now and then shipping a sea which seemed for the time to overwhelm her. Her decks were covered with water constantly. It was only a merciful Providence that rescued us from the perils with which we were surrounded. For twelve days I was so sick I could scarcely raise my head from my pillow.

Upon his return from Europe, Kimberly became interested in living in the house that UNC Professor of Logic and Rhetoric John Thomas Wheat had vacated when he resigned his post in June 1859 (Brawley, 1996). The house was located on the corner of East Franklin Street and Raleigh Street. Nearby was the student infirmary (called "The Retreat") and to its west the Chapel of the Cross, where Wheat was also rector. The house had previously been known as the "Revolutionary Land Warrant" house, so named because it was constructed from the proceeds of land warrants given to the university by the veterans of the Revolutionary War. Back then, the state gave soldiers plots of public land in lieu of pensions or bonuses. When UNC received a number of land warrants, it used the proceeds to build the house (Anonymous, 1923). Among the occupants of the house were John de Berniere Hooper, Eben Alexander, C.W. Bain, James B. Bullitt, and Wallace E. Caldwell (Anonymous, 1923; Anonymous, 1924a). Other occupants included Bishop William Mercer Green, Fisk Brewer, President David L. Swain, and Tom Taylor (former superintendent of campus buildings and grounds) as well as professors James Wills, Henry Van Peters Wilson, and Fordyce M. Hubbard (Anonymous, 1924b). In August of 1860, Kimberly (1860d) still had not seen the entire interior of the house, but he liked the secluded locale and the numerous trees surrounding it. He had some misgivings about the house, particularly its smallness and its appearance of being "inconvenient." Despite these concerns, he had positive feelings about the dwelling's desirable location on Franklin Street, from which it set in about fifty yards.

Kimberly had apparently talked with UNC President David L. Swain about his desire to live in the Wheat house (Fig. 4.3). He reminded Swain to bring up the topic at the next meeting of the trustees (Kimberly, 1860g). Swain and Judge Battle agreed to serve as a committee to formalize a contract, according to the minutes of December 10, 1860. Kimberly had proposed that he be allowed to live in the house rent-free for a term of six years. In return, he would make repairs to the house as needed (BOT 6: 299). According to a document in the

[51] During Kimberly's studies at Heidelberg, Bunsen's investigations focused on photochemical research with Sir Henry Roscoe as well as on the invention of the Bunsen burner (Schacher, 1970).

University Papers (UNC-P), however, Kimberly paid rent on the house from at least January 1, 1862, to July 1, 1866, and he apparently billed the trustees the cost of improvements that he made to the dwelling and yard. The house has since been torn down, and Spencer Hall, a dormitory for women, was built on the site and formally opened in 1925 (Anonymous, 1925).

Fig. 4.3. "The Wheat House," ca. 1892, home of John Kimberly (about 1860 to 1866).
(Album, University of North Carolina; Foster, artist and photographer. Courtesy of North Carolina Collection, Wilson Special Collections Library, University of North Carolina, Chapel Hill).

Kimberly's (1860j) return was very timely, because his class "was in a state of revolt" due to his absence. In addition to developing his lectures, he organized his lab, working from dawn to dusk and "fixing and fussing." By late February, Kimberly's (1860k) laboratory had seventy-two students, and assisting them, as well as preparing for lab setups, kept him occupied. Although he found the class attentive and decent, the large number of students placed a burden on him. In writing to his wife, he noted: "Think of 72 [students] in the laboratory, all asking questions, many idle and noisy, some breaking and smashing, and I trying to plan out work to keep all hands busy." Assisting him were several student volunteers, who did a great amount of work. Based on his European studies, Kimberly enhanced and expanded laboratory instruction at UNC (Bursey and Thomason, 1988).

For this later period of Kimberly's career, it is possible to create a fuller picture of his lecturing style and content than it is for the earlier period. Besides the responsibilities of preparing and directing laboratory sessions, Kimberly (1860l) gave recitations and lectures. The former did not require much preparation, he admitted, but "a lecture requires days of reading and thinking." He explained how he prepared his lectures. Each morning after breakfast, he would assemble his books, read them, and take notes until evening. Subsequently, he would think about the subject and organize the relevant facts in his mind. He considered this "the hardest work." He set his academic standards at a very high level. Kimberly (1860m) wrote to Bettie expressing his feelings about lecturing:

> I would rather be whipped any time than get up to lecture, for I never satisfy myself, never come up to the standard I think I should, and I always imagine that some listener is saying to himself, "what a fool that fellow is." It is an easy matter to lecture, where there are experiments to perform, for they [*i.e.*, there] are so many breaks and fresh starting points, but the discussion of scientific theories for an hour at a time requires a man to have all his thoughts about him, and these arranged in a manner so methodical as to be available at a moment.

Housed in Kimberly's archival records at UNC are two incomplete sets of his lecture notes from fourteen classes taught in the latter part of the fall term of 1860, when he presented nearly thirty lectures, and the beginning of the following fall term (Kimberly, 1860h; 1857–1886). Kimberly (1860n) regretted that he did not prepare extensive lecture notes but said there was never sufficient time to complete them. His notes were principally outlines of topics to be discussed, but they offer a glimpse into the nature of his instruction that is particularly noteworthy, as the lectures were given after his studies in Europe, where he gained new insights not only into agricultural chemistry but also into how to teach it.

Kimberly (1857–1886) introduced his class on agricultural chemistry by explaining the subject's practical benefits, pointing out that it taught about matters from the germination of seeds to growing food. He posed the question: "What can chemistry do for agriculture?" In defining agricultural chemistry, he noted that it included physics, geology, vegetable and animal physiology, and meteorology. He concluded the first lecture with a discussion of the history of agriculture, which topic extended through his second lecture. His subsequent six lectures were about plants, including their organic constituents (two classes), inorganic constituents, sources of carbon, sources of oxygen, hydrogen, and nitrogen, and physiology, in which he highlighted the functions of the root, stem, and leaves.

The concluding lectures of Kimberly's class encompassed talks on different sources of soil enrichment. He gave at least two classes about lime, in which he discussed its occurrence, its forms used in agriculture, and its effect on the land. Next, he talked about common salts, noting that the eastern countries used them. He gave one lecture on bones and how they were processed for use as a soil amendment. His presentations on animal waste products focused on guano and farmyard manures. About guano, he provided analyses on the composition of various kinds, including Peruvian, Cuban, and Patagonian. Kimberly (1860m) also lectured on the action of the atmosphere on soil.

As the dark clouds of the Civil War descended over Chapel Hill in the spring of 1861, great numbers of students abandoned the university to fight on behalf of the Confederacy. So many took up arms that the remaining students petitioned university officials to close the campus. That effort was unsuccessful, and the campus remained open throughout the war. Of the fourteen members of the faculty, six went off to war, and eight remained (Battle, 1917), including Kimberly.

By August 1861, Martin had formed a company to fight for the cause of the Confederacy. Kimberly (1861a) realized that Martin's teaching duties would fall on him. His only stipulation was that Martin should continue to receive his salary while on leave. Kimberly (Fig. 4.4) remained deeply indebted to Martin, who had substituted for him while he took an annual leave of absence in Europe in 1859. He was therefore willing to accept Martin's teaching responsibilities even though he considered his present duties "onerous" and though he would not receive additional pay. After he accepted the additional classes, his days at the university became endless. He arrived on campus early in the morning and left at night. Despite the hard times of the university, including a low enrollment and scarcity of money, Kimberly was displeased that the university sought the "benefit of [his] services without paying for them." He confided in his wife, "I become perfectly furious whenever I think of the contemptible littleness, which characterizes every step in the management of this institution."

Fig. 4.4. John Kimberly, ca. late 1850s.

(From *A History of the University of North Carolina* by Kemp P. Battle [1907–1912], vol. 1, plate opposite p. 550).

Kimberly (1860c) was particularly critical of President Swain, whom he called "Governor." In writing to Bettie, he vented his opinions about the university leader and also pointed out several needed changes:

> Our Gov. goes in for saving money, and will grind the faces of the poor professors as much as they will allow him. He signalizes his administration by piles of brick and mortar, working his professors to death on small salaries, and depriving the college of educational facilities. We want books, apparatus, fixtures, higher salaries, exemption from menial duties, which we will never get so long as they cost money, and he holds the reins. He is nearly as deaf as a post, and still pretends to hear his Classes, while they take the opportunity to kick up all sorts of rows.

UNC professor of mathematics and historian Archibald Henderson (1949) summarized the state of affairs of the university and Chapel Hill during the war in his chapter titled "The Valley of the Shadow: 1861–1865." With each successive year of the conflict, student enrollment dwindled. Food and medical supplies became depleted, and the university had a hard time paying the faculty. In his correspondence, Kimberly (1861b) mentioned the growing scarcity of food. In the fall of 1861, he wrote that flour and meal were cheap, but butter was scarce. Chickens were unavailable, but beef could be procured. The decline of the university brought with it the decline of the village, which depended on the vitality of the college (Spencer, [1866]).

Kimberly supported the Confederacy and its cause. He had at least two Black servants who worked in his home: Milly, a washerwoman, and Spencer, a cook. Kimberly (1861a) appreciated their good service, but his attitude toward them was typical of a Southern gentleman of the period. He noted that Spencer was "a good cook and seem[ed] willing to do all he [could], but he ha[d] been so long his own master, he require[d] watching—with some pains and a few good whippings he would make a valuable servant." Historians Maurice Bursey and Sallie Thomason (1988) noted that Kimberly's contribution to aid the war effort was "singular in North Carolina." In addition to teaching at UNC during the war, he founded a business that produced nitric and sulfuric acid to supply the Confederate States Army and various companies in the state. Nitric acid was necessary to make gunpowder. Chemical manufacturers throughout the Confederacy sought his chemical expertise.

With the defeat of the South, Cornelia Spencer (1853–1906) recorded her observations and feelings in one of her diaries. She noted that the entire framework of the Southern social system had collapsed. The defeat seemed very sudden. The angst and humiliation that flowed from her pen expressed the sentiments of those who had supported the secessionist cause. The following passages provide an example of those feelings,[52] as well as a glimpse of the state of affairs in Chapel Hill when the South fell to Union forces:

> Apr. 17, 1865. To-day—being Easter Monday the Federal troops under Gen. Atkins—part of Kilpatrick's command entered C[hapel] H[ill] & took quiet possession![53] A memorable day! Every house has a guard! What a most agitating week the past has been. Johnson's army in full retreat. What sorrow—what anguish for us—Lee defeated & surrendered—past all hope—all praying for–Sherman in hot pursuit now of the last hope of the Confederacy—Johnson.
>
> [April] 30 [1865]. It is all over! Johnson has surrendered—Peace is come—but what a peace! . . . The South prostrate—exhausted—helpless! It is all over indeed. . . . The army is still here. The country is being devastated by them. Our beautiful groves are full of them & their horses—our streets—our houses. All work is suspended—we seem

[52] Twenty-one years after entering these comments in her diary, Spencer (1853–1906) considered that she was more prone to writing down "gloomy thoughts, feelings, experiences" than brighter ones. The Civil War intensified those gloomy feelings.

[53] On April 17, 1865, Kimberly (1865) also wrote about Kilpatrick's troops, which "entered with music and flying colors." He wrote: "A special guard was sent me immediately, and our citizens, as far as I know, have been treated with the greatest kindness and consideration." He also pointed out that the Hoke and Chatham divisions with Wheeler's cavalry passed through the Chapel Hill area. Kimberly claimed that Wheeler's men stole "Fetter's and Sugg's mules and everything on four legs" as well as his corn, which had been brought to the mill. He and Judge Battle stood guard all night, passing back and forth between their houses and occasionally walking through town, which was desolate. Kimberly also checked his quarters at the college and discovered that his drawers had been rifled but nothing of value taken.

> to be aghast & stunned. The ruin has come so suddenly. Every now & then squads of Lee's men come walking sadly & slowly along—paroled—going home—Home! Many of them, have no home left to go to! Houses are burned—families dispersed.
>
> May 7th. The Federal forces left last Wed[nesday], 3d. A guard of 35 is left here for the U. . . . The same quiet & stillness reigns over our groves—.The stillness of a subjugated & ruined people. . . . Our Confederacy is no more. It seems a year since the Yankee army entered the place. The negroes are all leaving or have left their homes. They seem to show no feeling or attachment for their owners—those who have raised & fed & clothed them. . . . And no one has any money—very few indeed have corn or anything to eat.

In reminiscing about the departure of the government troops, Spencer (1854–1868) recalled the bleak scene in Chapel Hill:

> I walked this evening, alone, all over the hill back of our house where a [regiment] was lately encamped. Not a blade of grass–a shrub or leaf of any kind is to be found there. The ground is stamped as smooth & bare as a floor. A profound depression has seized upon me. I see before us only humiliation, privation & a life of continued toil. This Southern land is ruined for this & the coming generation. I feel the overthrow of the Confederacy as keenly as if I had embarked in secession with all my heart at the first, which was not so.

One incident of particular note occurred to Kimberly during the war: In the summer of 1863, he narrowly escaped serious injury in a laboratory accident. According to his wife, Bettie Kimberly (1863), he had gone into the lab to arrange and prepare his class for the new session. While he worked, a vessel containing one hundred pounds of sulphuric acid burst and spattered all over him. He immediately removed some of his clothes and ran to the well. A servant had just pulled up a bucket of water and washed off Kimberly, which saved him from severe harm. Although his skin was badly burned and his clothing was riddled with holes, his eyes were not affected.

As the war drew to a close, Kimberly resigned; there is some question as to the exact date. According to an account of university disbursements for November 20, 1864 to November 20, 1865, Kimberly was paid for teaching during that period of time (Manly, 1865). Records of disbursement for the first session of 1866 do not include Kimberly's name, however (Fetter, 1866). Writing from Chapel Hill in mid-March 1866, professor Charles Phillips (1866) noted: "Prof. Kimberly is here, but is not an acting Prof. at present." It is reasonable to assume that he ceased teaching at UNC at the end of 1865 or in early 1866.

In the summer of 1866, Kimberly (1866a) traveled to the western part of the state, where he looked for land to farm. One of his first stops was in Charlotte, followed by stops at Lincolnton and White Sulphur Springs in Catawba County. He complained about the land, stating that he would not give ten cents per acre, because "the soil like the people seem[ed] poverty stricken." He looked forward to scouting the region between Morganton and Asheville, which was reportedly beautiful and fertile. Upon reaching Asheville and the vicinity, Kimberly (1866b) became excited about the area, writing: "The more I see of this country, the more I am impressed with its beauty and better pleased. It is not the garden spot of N.C., it is the garden spot of the world." His spirits were elevated with visions of building a house there and establishing a farm; he wrote that he was "intoxicated and excited" by the prospect of purchasing land in the area.

According to the register of deeds in Asheville, North Carolina, Kimberly bought a six-hundred-acre tract of land near Asheville that had formerly belonged to lawyer and soldier John W. Woodfin (LGB 35: 255). David L. Swain loaned Kimberly $20,000 to purchase the property. There, he cultivated the land called Glen Creek Farm and raised livestock. The farm now constitutes a part of North Asheville encompassing an area from Coleman Avenue north to Beaverdam, east to the summit of the mountain, and west to Merrimon Avenue (Thomason, 1981).

Kimberly and his family left Chapel Hill for Asheville on October 16, 1866. Several days later, Cornelia Spencer (1853–[1906]) reflected on her former neighbors:

> I passed thro' Mr. Kimberly's yard–deserted & forlorn. I knew very little of the family, yet there is something touching in looking at a place from which late neighbors have departed forever. "Their place shall know them no more." A pile of broken bottles, retorts, jars & crucibles were all that is left of our late Prof. of Agric. Chem. The place is very much dilapidated.

Spencer was not the only member of the community Kimberly kept at arm's length. Professor Charles Phillips (1876c) wrote years later of Kimberly's reserved nature, "His life is a mystery to us—so lonely."

In August 1867, Kimberly's teaching post still remained vacant. At that time, Solomon Pool (1867), who then served as U.S. Deputy Appraiser in North Carolina but had ties with influential friends of the university, wrote to Benjamin Hedrick to offer him a professorship in chemistry. The political climate had drastically changed since Hedrick had been discharged from the UNC faculty. These changes are discussed in the chapter on George Dixon, including the closure of the university. Pool considered Hedrick to be the only native North Carolinian "who [was] both loyal & qualified in literary point of view, to whom my mind reverts." Hedrick declined the offer, however.

Teaching at the "New" University

In 1874, Kemp P. Battle became a trustee of the university, having previously served in that capacity from 1862 to 1868. He had gained fame as a successful lawyer in Raleigh, and he was instrumental in the movement to revive the university (Winston, 1895). Battle also worked to procure funding for the reopening of UNC. During its closure,[54] the college buildings fell into disrepair, and the campus grounds were overgrown. The "instruments belonging to the laboratory [were] very much scattered and injured, they having been exposed to the public" (BOT 7: 209). By early March 1875, work was well underway in repairing the buildings and clearing the grounds. A month later, a local newspaper reporter noted the repairs and observed that the grounds were being cleaned up, shrubs and trees trimmed, and leaves and debris removed (Anonymous, 1875). The work was also eagerly surveyed by Cornelia Spencer (1875a), who wrote: "The College Buildings look[ed] splendidly in all the glory of fresh paint and whitewash. A small army of workmen [was] busy on them and on the houses of the Faculty. The Hotel [was] nearly complete."

[54] The closure of the university is described in the chapter on George Dixon.

The Board of Trustees met on May 4 and 5, 1875, to plan for the reorganization of the university (Battle, 1875). Among the resolutions adopted were the following: that the university would open on the first Monday of September 1875; that tuition would be $60 per term and room rent, servant hire, and deposit money $10; that six colleges would be established: Agriculture, Engineering and the Mechanic Arts, Natural Science, Literature, Mathematics, and Philosophy; that the colleges would be subdivided into schools, implemented as funding would permit. In the College of Natural Science, three schools were envisioned: Chemistry, Zoology and Botany, and Geology and Mineralogy. In the College of Agriculture, three schools were planned: Scientific Agriculture, Practical Agriculture, and Horticulture. The board also planned to establish a farm and garden in connection with the Colleges of Agriculture, and of Engineering and the Mechanic Arts.

The academic structure was modeled in part after that of the University of Virginia. Departments (called colleges) were developed, including an agricultural college. Four courses of study led to the following diplomas: A.B. (Bachelor of Arts), A.M. (Master of Arts), B.S. (Bachelor of Science), and B.Agr. (Bachelor of Agriculture) (Phillips, 1875a). The trustees, particularly several prominent members (Paul C. Cameron, Rev. Dr. Neill McKay, and Col. William L. Saunders), were very supportive of the establishment of the agricultural program. Battle thought "that every attention [should] be paid to the future farmers, gardeners, fruiterers, stock raisers, wool growers & cheese factor[ies]." The college's farm might even supply the faculty with their vegetables and mutton (Phillips, 1875b).

The agricultural department aimed to comply with the "Land Scrip." The Morrill Land-Grant Act, which the U.S. Congress passed on July 2, 1862, provided for the establishment of colleges of agriculture and mechanic arts. Each state was granted 30,000 acres of public land for each senator and representative as of 1860. Based on that formula, North Carolina was entitled to 270,000 acres of land. The property was to be sold and the proceeds invested in state or federal bonds. The income accrued from the investments, called the Land-Scrip Fund, would be used to support an agricultural college.

In February 1866, the North Carolina General Assembly adopted a joint resolution to accept the 270,000 acres (NC Laws, 1867). The stipulations of the act included: (1) that the transfer of the scrip to the trustees of the university was contingent on the proviso that the university provide courses in agriculture and mechanic arts without excluding other scientific and classical studies, including military tactics; (2) that the trustees were authorized to dispose the land scrip and use the proceeds to establish two academic chairs to fulfill the federal act; (3) that each county was allowed to send one indigent student to UNC, free of tuition and room rent; and (4) that students could be admitted into classes of agriculture and mechanic arts without the standard training, normally required for admission. Young (1934) and Battle (1907–1912, 1: 756–758) provide background on the land scrip at UNC.

Kimberly had shown an interest in procuring a professorship at the reorganized university. On May 27, 1875, he formally applied for the chair of agriculture and applied chemistry (Kimberly, 1875f). In June, the trustees elected the members of the faculty by ballot. Kimberly and William A. Allen were the contenders for the professorship in agriculture (also noted as professor of the College of Agriculture). On June 16, the trustees elected Kimberly. The annual salary of each faculty member was set at $1,500 and was to be paid semi-annually. Faculty who were heads of households were also furnished with residences or "commutation" in addition to their salaries (BOT 7: 225, 227). In writing to Kimberly, Charles Phillips (1875a) said that the "great work before us" was "a work of patience–of energy and of tact." He felt confident that the faculty would have the "hearty cooperation of the Trustees, & but little interference from them." Although the operating budget would likely be sparse, he thought it could be bolstered as the institution built up public confidence. Phillips told Kimberly that his instruction would be focused on agriculture, but its details would be developed later. He counted on Kimberly to use his prior experience and expertise in developing the course of studies. Students applying to the College of Agriculture were required to have fulfilled the prerequisites in English, geography, and arithmetic.

Kimberly, who was occupied with harvesting crops on his Asheville farm, quickly replied to Phillips, reminding him how eager he was to develop the agricultural curriculum. He also expressed his wish that the department not be a "mere nominal thing, an appendix to secure the Congressional appropriation." He intended to work like a Trojan and struggle like a tiger, he said, in developing the department (Kimberly, 1875b).

The preparations for reopening the university proceeded well, and classes started on September 6. Ceremonies marking the occasion were held on September 15. Chapel Hill celebrated a holiday, and businesses closed for the event. Writing under the alias of a "Friend," Cornelia Spencer (1875b) declared happily: "The old college ha[d] donned her new dress, a delicate fawn relieved with brown." In receiving a Doctor of Laws degree at the University Day ceremonies in 1910, educational administrator Daniel Harvey Hill (1910) reminisced about the momentous day in 1875. He recalled the celebratory atmosphere throughout the state and particularly the fervor of Chapel Hillians, as well as the trustees and alumni, who "were simply delirious with joy." He continued: "Flags, ribbons and garlands of flowers hung from every gate, door and window." The occasion was cause for the people of Chapel Hill to smile again (Phillips, C., 1875a).

According to the UNC catalog of 1875–1876 (p. 21), the plan (Fig. 4.5) of instruction for the College of Agriculture encompassed an array of subjects important to agricultural studies. Under the plan, students could earn a Bachelor of Agriculture degree in three years. Kimberly publicized the aims and objectives of the agricultural program in several ways. He issued a circular, the text of which was published in the *Raleigh Daily News*. The article is the only known publication by Kimberly (1875c). He also lectured on the aims of the program (to advance agriculture by scientific methods) at a meeting of the North Carolina Agricultural Society held October 13, 1875 (Kimberly, 1875e). He believed, he said, that men would be "learned in farming" just as they were trained in medicine and law. He desired "to go before our brother farmers and talk with them as I would talk with them here in my room."

Kimberly (1875c) noted that the program's method of instruction would focus on lectures presented by professors and experienced farmers who had attained specialties in their field. Of particular note was Kimberly's intention to bring under cultivation nearly 800 acres of land owned by the university. A large section of the land would be used for pasture and for cultivating different species of grasses, another for growing staple crops, and the rest for a garden and cultivation of fruit trees. The farm would serve as a practical laboratory where students could apply their theoretical knowledge. Kimberly also desired to conduct experiments on the chemical and mechanical treatment of the soil, "care" of stock, and the growth of vegetables. The results, he envisioned, would benefit the state's farmers.

Toward his goal of developing the agricultural program, Kimberly ([1875]d) requested a budget of $1,000 from the trustees in about December 1875. He desired to purchase a team, a wagon, agricultural tools, a plow, and wagon gear to clear out and fence twenty acres of land to be cultivated. The acreage would be used as a fruit orchard and would be enclosed by an osage orange hedge, which would protect the trees and their harvest from outside depredation. He also proposed that the Dialectic and Philanthropic Societies should control the enterprise. Besides bringing profits, the venture would serve in the instruction of budding, grafting, and culturing fruit. Needy students could provide the labor to help defray their educational expenses. Kimberly (1875c) hoped the students would gain an appreciation for the necessity of raising their own produce, promoting a diversity of crops, growing grasses for the purpose of improving soils, improving the breeds of domestic livestock, and promoting the cultivation of fruit. The subsequent spring, he proposed increasing the acreage planted in grass to create a pasture.

In the same budget request, Kimberly ([1875]d) expressed his desire to upgrade the practical chemistry laboratory, which was "entirely unfitted" for current chemical instruction and investigations. He believed that an improved lab would provide the program with a "respectable rank" in collegial circles but that to continue with the present appliances would "make us the laughing stock of every college in the land." In December 1875, Kimberly recommended moving the lab to the lower floor of the New East Building, where the rooms could be renovated for only a modest cost. There, he wanted a "work-room, class-room, weighing-room, furnace-room, and chemical room." He estimated that the cost of moving his lab would be about $1,500 (BOT 7: 249). Kimberly ([1875]d) also requested $1,300 for the following articles: desks for students, $150; a water tank with iron pipes, faucets, and basins, $150; a chemical furnace with distilling apparatus, air-baths, etc., $400; a portable furnace, $50; a sand-bath, drying chambers, hood, etc., $50; and chemical ware, $500. The trustees discussed Kimberly's request for expenditures but postponed any resolution on it.

Kimberly's attempt to modernize the agricultural chemistry laboratory was unsuccessful. At their June 1, 1876, meeting, the trustees authorized a scant $200 for the lab (BOT 16: 9). In an effort to provide additional funds for the lab, a committee consisting of Bartholomew Figures Moore, Seaton Gales, and Kemp Battle was constituted to raise contributions to purchase apparatus. By at least early March 1876, according to professor Charles Phillips (1876c), Kimberly had become dissatisfied with the meager appropriations made to his department. In late May 1876, he (Kimberly, [1876]b) openly expressed his concern about advancing instruction in practical agriculture at the university. He wrote: "The popular voice demands it, and the cry will not be hushed until some demonstration is made in that direction." The low level of support for Kimberly's work caught the attention of people and organizations within the state. There was a growing concern that the university was not meeting its commitment of the Land Scrip. When the State Grange criticized the university's use of the Land Grant funds, Kemp Battle defended the program at UNC (Lockmiller, 1939).

University of North Carolina.

The exercises of this Institution will be renewed on the 1st Monday in September next. The term will continue, with two weeks vacation at Christmas, until the 2nd Thursday in June, 1876.

The charge for tuition is $60 for the term, with $10 for room rent. One half of the above amount ($35) payable in advance, the other half on or before the opening of the session in January. Provision will be made for gratuitous instruction of meritorious students, who are unable to pay.

The buildings have been repainted and repaired, inside and out, and rooms will be ready by the time above mentioned, for several hundred students. The Campus is being placed in order, and in a few weeks the College property will be not inferior in beauty and fitness for educational uses to any in the Union.

Either of the following courses may be pursued by the student, at his own election, with the consent of his parent or guardian.

1. The *Agricultural course* of three years, in which by proper diligence, the student can obtain the degree of Bachelor of Agriculture. B. Ag'r.

The requisites for admission into this course are a competent knowledge of Arithmetic, the English Language and Geography.

Instructions in the course will be given in Theoretical and Practical Agriculture, English Language and Literature, Botany, Zoology, Chemistry, Mineralogy and Geology, Mathematics, Engineering, Political Economy, Constitutional Law, &c.

2. The *Science course*, leading in three years by proper diligence to the degree of "Bachelor of Science." B. S.

In this course instructions will be given in English Language and Literature, French and German, History, Mathematics, Chemistry, Logic and Rhetoric, Zoology, Natural Philosophy, Mineralogy and Geology, Mental Science, &c.

The requisites for admission into this course are a competent knowledge of the elements of the English Language and of Geography, of Arithmetic and Algebra, through Equations of the 2nd degree.

3. The *course of Arts*, corresponding to the usual College curriculum, leading after a four years' course to the degree of "Bachelor of Arts." A. B.

In this course instruction will be imparted in Latin, Greek, French and German, Mathematics, Natural Philosophy, History, Logic and Rhetoric, Astronomy, Mineralogy and Geology, Chemistry, Mental and Moral Science, International Law, Political Economy, English Literature, &c.

The requisites for admission into this course are the same as in the Science course and in addition, Latin Grammar, Prosody and Composition, 4 Books of Cæsar, 5 Books of Virgil's Aeneid, Ovid, Sallust or Cicero's Orations, according to the choice of the applicant, 4 Books of Xenophon's Anabasis (or Memorabilia,) and 2 Books of Homer's Iliad, or the equivalent.

The Bible will be taught in all the foregoing courses.

Besides the foregoing a student may pursue what may be called an "Optional course," in which he may avail himself of the instructions of such Professors, as he may elect, with the concurrence of his parent or guardian and of the Professor of the departments chosen. "Certificates of Proficiency" in the branches pursued may be obtained, but no degree.

5. An advanced course will be hereafter prescribed in which a graduate of Arts may obtain, upon a satisfactory examination, the degree of A. M. This degree will not be conferred as an honorary distinction as heretofore.

Instruction in Military Tactics will be given in all the courses.

Attendance on religious services on Sunday morning will be required, at such Church as the student shall choose, notes being made of the attendance, and also of attendance on the daily prayers, and the absences reported to the parents or guardians.

Students will be required to observe the utmost decorum and courtesy towards each other at all times and the practice of "*hazing*" will be rigorously suppressed.

No secret societies or clubs, other than the two well known Literary Societies, (Dialectic and Philanthropic), will be allowed without the express permission of the Trustees of the University.

The Faculty will remove from the University any student who is negligent of his duties, or exerts an evil influence on his fellows.

Students should be sixteen years of age before entering the Institution.

Fig. 4.5. Circular on the reopening of the University of North Carolina at Chapel Hill, 1875.

(From *Scrapbook* compiled by Charles Phillips, [1857–1889] sheet inserted between pages 55 and 56; courtesy of North Carolina Collection, Wilson Special Collections Library, University of North Carolina, Chapel Hill).

Good board can be had at Chapel Hill, for from $12 to $15 per month, and this can be cheapened, by the formation of messes. The following is the list of Professors already appointed. Probably a President will be elected before the session will open. Other Professors will be added when needed.

THE FACULTY:

Rev. Charles Phillips, D. D., Professor of the College of Mathematics.

Mr. J. DeB. Hooper, Professor, } College of Literature.
Mr. Geo. T. Winston, Adjunct Professor, }

Mr. John Kimberly, Professor of the College of Agriculture.

Rev. A. W. Mangum, Professor of the College of Philosophy.

Rev. A. F. Redd, Professor of the College of Natural Science.

Mr. Ralph H. Graves, Jr., Professor of the College of Engineering and the Mechanic Arts.

Andrew Mickle, Chapel Hill, Bursar.

List of Trustees of the University of North Carolina:

His Excellency CURTIS H. BROGDEN, Governor and Chairman *ex-officio*.

D M CARTER, Esq., Beaufort, since removed to Wake.
Hon MILLS L EURE, Gates.
W F MARTIN, Esq., Pasquotank.
W B SHAW, Esq. Currituck.
Hon LOUIS HILLIARD, Pitt.
JOHN A MOORE Esq, Washington.
Rev C B HASSEL, Martin.
DUNCAN C WINSTON, Esq., Bertie.
WM H JOHNSTON, Esq., Edgecombe.
JOHN E DUGGER, Esq., Warren,
W T FAIRCLOTH, Esq., Wayne.
Hon GEO. GREEN, Craven.
WM H DAY, Esq., Halifax.
Prof. J DE B HOOPER, Wilson.
ROBERT B PEEBLES, Esq., Northampton.
Rev B S HARDY, Greene.
FORNEY GEORGE, Esq., Columbus.
W L SAUNDERS, Esq., New Hanover.
R W NIXON, Esq, Onslow.
B F GRADY, Esq., Sampson.
Rev. Dr. NEILL MCKAY, Harnett.
Dr JOHN MCIVER, Moore.
Hon T D MCDOWELL, Bladen.
R W MILLARD, Esq., Duplin,
J H THORPE, Esq., Nash.
J S AMIS, Esq., Granville
PAUL C CAMERON, Esq., Orange.
Hon B F MOORE, Wake, in place of C B SANDERS, Esq., deceased
Hon JOHN MANNING, Chatham.
KEMP P BATTLE, Esq., Wake.
Hon JOSEPH J DAVIS, Franklin.
Hon WM A GRAHAM, Orange.
H CLAY THOMAS, Esq., Davidson.
RUFUS FRAZER, Esq., Randolph.
M S ROBBINS, Esq., Randolph.
J A GILMER, Esq., Guilford,
Hon JOHN KERR, Caswell,
Dr J L DUSENBURY, Davidson.
Hon DAVID S REID, Rockingham.
JAMES A GRAHAM, Esq., Alamance.
ARCHIBALD PURCELL, Esq., Robeson,
JAMES C MARSHALL, Esq., Anson.
WALTER L STEELE, Esq., Richmond.
Hon S H WALKUP, Union.
Hon Z B VANCE, Mecklenburg.
JOHN F HOKE, Esq., Lincoln.
PAUL B MEANS, Esq., Cabarrus.
W J EWING, Esq., Montgomery.
J F GRAVES, Esq., Surry.
JOSEPH WILLIAMS, Esq., Yadkin.
W KELLY GIBBS, Esq., Davie.
Rev GEORGE B WETMORE, Rowan.
R L PATTERSON, Esq., Forsythe.
E HAYNE DAVIS, Esq., Iredell.
Dr. C L COOK, Wilkes.
G H HAMILTON, Esq., Ashe.
D COLEMAN, Esq., Buncombe.
Rev C D SMITH, Macon,
S MCD TATE, Esq., Burke.
W L TWITTY, Esq., Rutherford.
J D HYMAN, Esq., Henderson.
E W JONES, Esq., Caldwell,
W C BOWMAN, Esq., Mitchell.
J O HICKS, Esq, Clay.

Executive Committee of Board of Trustees for 1875:

HON. WILLIAM A. GRAHAM, Chairman.

COL. W. L. SAUNDERS,
PAUL C. CAMERON, ESQ.,
KEMP P. BATTLE, ESQ.,
REV. DR. NEILL MCKAY,
D. M. CARTER, ESQ.,
JOHN A. GILMER, ESQ.,

KEMP P BATTLE, Secretary and Treasurer,
Raleigh, N. C.

Daily News Job Presses, Raleigh, N. C.

Fig. 4.5 (Continued from previous page)

In reporting about his first year of instruction since the reopening of the campus, Kimberly ([1876]b) provided insight into his agricultural and chemical teaching. He had ten students (Charles Bond, Charles Coleman Covington, Robert Henry Davis, Neill McKay, Jr., Prentys McKay, John Walker Mallett, Rufus Grant Merritt, Henry Barber Nixon, James Whedbee Nixon, and John Henry Sherrod) who pursued the agricultural course of studies that ended on May 30, 1876. Instruction commenced with a course on physical geography, with recitations three times a week until the assigned textbook, Maury's *Physical Geography* (1873), had been completely reviewed. The agricultural class also joined students of the Scientific Course taught by other professors (and possibly Kimberly). They attended recitations five times a week in chemistry, human physiology, and vegetable physiology. The required textbooks included Dalton's *Physiology* (1875) and Johnson's *How Crops Grow* (1868). In chemistry, instruction was chiefly by lectures accompanied by experiments. Kimberly considered chemistry "the groundwork for thorough and efficient instruction." He believed that the success of the agricultural college depended on chemical science with facilities to support practical and theoretical instruction. By the spring of 1876 he had twenty-five students. The spring course probably focused heavily on botanical topics, because the students' grades are recorded under the heading of "Botany" in the university grade book for that session (UNC Archives, 2; Appendix 1).

It is noteworthy that the university used a new system of grading in the spring session of 1876. The UNC faculty officially adopted the new grading system in July 1876. Marking was established on a scale of 100 (highest or best) to 0 (lowest or worst), with the possibility of fractions. In order to earn a certificate, degree, or diploma, students had to earn a whole mark of 70 or higher. Those with an average of 95 or higher would earn first distinction, those with 90 to 94 second distinction, and those with 85 to 89 third distinction (UNC Faculty 6: [398]).

Kimberly's instruction in the first year did not meet the expectations of the university leaders. Rather suddenly, he resigned his position. In a one-sentence letter written on June 1, 1876, Kimberly (1876a) proclaimed: "I herein tender my resignation of the Professorship of Agriculture in the University of the State." When writing about his "unpleasant university duties," Battle (1945) provided insight into Kimberly's resignation. The Visiting Committee, of which Battle was chair in the spring of 1876, noted in its report that a professor of agriculture could be hired from a Land Grant college at half the salary that Kimberly earned. Kimberly's insufficient background in botany, physiology, and related subjects encouraged the committee to look more closely at his post, and they ultimately recommended that he be replaced. Next, the trustees met in May, at which time they concurred with the committee's view. Battle proposed that a special committee be formed to interview Kimberly. He subsequently explained the situation to Kimberly, who "wrote his resignation without a word of either protest or acquiescence." On the same day that Kimberly wrote his letter of resignation, the trustees formally accepted it (BOT 7: 262), and on July 21, the executive committee approved that his salary be paid through the end of the year (BOT 16: 24).

Cornelia Spencer (1876a) and other Chapel Hillians were surprised to learn of Kimberly's resignation. Ever since he returned to Chapel Hill, however, his health had been poor. The trustees were disappointed that Kimberly devoted too much attention to chemistry and not enough to fieldwork. In writing to Cornelia Spencer, Kemp Battle (1876) revealed his thoughts on the episode: It was generally thought that the professor of agriculture should be "many-sided," "versed in many sciences," and "an expert in organizing and conducting agric[ultural] Colleges," as well as "quick and versatile in mind and body." Kimberly did not fit this model. Battle deemed him to be ill-tempered, in poor health, and lacking adequate training as an agricultural professor, but he regretted that Kimberly felt "wounded." Professor Charles Phillips (1876a) wrote to Kemp Battle that Kimberly "went off . . . broken hearted—because of the insult—being for the first time regarded as incompetent *in presenti et in futuro*, and that contrary to my advice & that of others he has told all about that he was 'forced' to resign." Although Phillips preferred to replace Kimberly with a North Carolinian, he did not have a name to recommend. Professor Adolphus W. Mangum (1876) was also concerned that unpleasant publicity that might be created by the incident. In the case of negative exposure, he suggested to Battle that they be prepared with a statement, such as that Kimberly "& the Trustees had different ideas to the management of his work." This reflected the truth, he felt, and should not hurt Kimberly. By the end of June, Phillips (1876b) had received a long letter from Kimberly, who seemed to have mellowed and "manifest[ed] good temper—a Christian spirit."

Kimberly returned to his farm in Asheville. He had considered reviving the Female Institute at Asheville,[55] but this never happened. By August 1876, he seemed cheerful, according to Cornelia Spencer (1876b). In early January 1877, Kimberly (1877) wrote to the Honorable William H. Battle (father of Kemp Battle) and mentioned that he was teaching his children at home as time permitted.

Natural History Collections

During Kimberly's tenure at UNC, some efforts were made to build up and organize the collections of natural history specimens, chiefly geological. About early 1861, North Carolina State Geologist Ebenezer Emmons deposited at UNC "a valuable collection of Geological Specimens, which contribute[d] interesting truths to Science, and illustrate[d] both directly and indirectly, the physical peculiarities of North Carolina" (UNC Faculty 6: 243–244). The specimens were added to the cabinet of natural history. The faculty was deeply grateful for the donation, saying that it would be a useful tool in instruction.

In 1875, Kimberly (1875g) noted that "Dr. P"[hillips] had found a cabinet of specimens and exhibit cases given to the university by the state. (These may have been the ones given to the university in 1861.) They had been considered lost. He also mentioned the need to relabel and rearrange specimens in the Vienna cabinet. Kimberly proposed finding a suitable room for developing the collections into a museum. He even considered establishing a State Geological Museum should the North Carolina Legislature donate state geologist Washington C. Kerr's collection to the university.

[55] Kimberly most likely was referring to the Asheville Female College, which was noted for instructing "young ladies in the languages and the arts so that they may be able to perform more fully their duties in circles of refinement and culture" (Wright, 2006). Established in the early 1840s, the college operated under several names. It closed during the Civil War and reopened in 1866, only to experience financial difficulties. The institution became the Asheville Female College in 1869. In 1901, the school, which was then called the Asheville College for Girls & Young Women, closed. The city of Asheville bought the college building in 1907.

Kimberly (1876b) believed that the museum collections could play an educational role in agriculture and the sciences, as well as attract visitors to the campus. Toward the improvement of the collections, he recommended additional shelving and cases for the geological and mineralogical specimens, which were still not arranged. He further considered the development of an agricultural museum to aid instruction in that discipline. He considered its development possible through contributions from manufacturers of agricultural tools and commercial fertilizers and from the state's farmers with the donation of plants, and "whatever is curious and remarkable in vegetable growth." Kimberly's resignation in mid-1876 deferred the establishment of a museum room and the organization of natural history specimens to the future professor of natural history William Henry Smith.

Legacy

John Kimberly died on March 6, 1882, at age sixty-four, at his home in Asheville. According to a newspaper obituary, he had retired the previous night "in his usual health but was subject[ed] to serious affection of the throat, which was probably the cause of death" (Anonymous, [1882]). He was preceded in death by his second wife, Bettie, who died on March 24, 1876, of pneumonia.[56] They were originally buried in the Trinity Episcopal graveyard in Asheville but were later (August 5, 1899) reinterred at the Riverside Cemetery, also in Asheville (Fig. 4.6).

John Kimberly was a noted educator. His teaching career began in private academies in Perquimans and Hertford Counties, North Carolina, and later developed in Nashville, Tennessee. He furthered his scientific education at the Lawrence Scientific School, Harvard University, where he studied under such scientific luminaries as Louis Agassiz and Eben N. Horsford. He subsequently taught at the University of North Carolina, serving first as professor of agricultural chemistry. His pedagogical methodology included lectures, recitations, and laboratory work for students. A number of his lectures were on botany, principally physiological aspects. In 1859, he took a leave of absence to study current theory and laboratory methods in agricultural chemistry in Germany. About 1866, Kimberly resigned his professorship, but he returned to UNC in 1875 to teach agriculture. His agricultural instruction was based chiefly on lectures and emphasized the theoretical and practical aspects of the subject. In teaching students the principles of agriculture, he considered the role of chemistry to be essential. He tried to modernize the chemical laboratory, but his efforts were unsuccessful. He served the public by analyzing water and minerals, but he did not extensively engage in basic scientific investigations, and he did not publish much. He established a company that produced sulfuric, nitric, and other chemicals, which were used to aid the Confederacy. Due to his poor health, his ill-temper, and his deficient background in botany and related subjects, he was encouraged to resign in 1876. He retired to Asheville, North Carolina, where he pursued a livelihood as a farmer. Kemp P. Battle (1907–1912, vol. 1) considered John Kimberly to be "a man of superior talent" (p. 660) and "of distinguished manners [. . .] accomplished in the department of Chemistry applied to the Arts" (p. 785).

[56] Mr. and Mrs. Kimberly were not in accord ever since he returned to UNC in 1875. Cornelia Spencer (1876a) claimed: "They differed seriously. He never saw her after he came to C.H. last summer! Did not go to his family in the winter vacation—and when she was ill." She had directed that Professor Kimberly should be notified of her death only after her burial.

Fig. 4.6. Gravestone of John and Bettie Kimberly, with inset of the dates for John Kimberly, Riverside Cemetery (section F, lot 41, grave 1), Asheville, North Carolina, June 2005.

(Courtesy of David Olson).

References Cited

(References that conclude with page numbers within parentheses indicate sources from where quotes and pertinent information were obtained.)

Agassiz, Louis. 1855. Letter of August 25, 1855 to Dr. I.T. Young. (Box 5, folder 199, University Papers #40005, University Archives, Wilson Special Collections Library, University of North Carolina, Chapel Hill)

Anonymous. 1837. [Program]: *Commencement of the University of the City of New-York, Thursday, 20th July, 1837*. [University of the City of New-York]. (University of New York Archives)

Anonymous. 1853. [Our village], p. 46 in Editorial table. *North-Carolina University Magazine* 2: 44–48.

Anonymous. 1854. The best opening for a newspaper and printing-establishment in North Carolina, pp. 229–230 in Editorial table. *North-Carolina University Magazine* 3: 229–240.

Anonymous. 1855. Wood and women, pp. 430–431 in Editorial table. *North-Carolina University Magazine* 4: 421–432.

Anonymous. 1856a. Chemistry at the university, pp. 139–140 in Editorial table. *North-Carolina University Magazine* 5: 135–144.

Anonymous. 1856b. The campus, pp. 378–379 in Editorial table. *North-Carolina University Magazine* 5: 376–[384].

Anonymous. 1857. Commencement week. *Chapel Hill Literary Gazette* (Chapel Hill, NC) 1(8): [2], Saturday, June 6, 1857.

Anonymous. 1858. The University. *Hillsborough Recorder* (Hillsborough, NC) 37(1962): [3], Wednesday, October 20, 1858.

Anonymous. 1875. Chapel Hill. *Hillsborough Recorder* (Hillsborough, NC), new series 3(26): [3], April 28, 1875.

Anonymous. [1882]. [Obituary on John Kimberly], p. 21 in Kemp P. Battle, *Scrapbook of North Carolina History and Biography*, 1880–1894, vol. 1. (Newspaper clipping without complete bibliographical data, North Carolina Collection, Wilson Special Collections Library, University of North Carolina, Chapel Hill)

Anonymous. 1923. Old landmark to go. Women's building in lot next to Episcopal Church. *Chapel Hill Weekly* (Chapel Hill, NC) 1(9): 3, Thursday, April 26, 1923.

Anonymous. 1924a. Women stay put. Trustees hold to first decision on site of building. *Chapel Hill Weekly* (Chapel Hill, NC) 2(19): 1, Thursday, July 10, 1924.

Anonymous. 1924b. Eben Alexander home place to be resurrected. *The Tar Heel* (University of North Carolina, Chapel Hill) 33(11): 1, Saturday, October 25, 1924.

Anonymous. 1925. Formal opening of woman's dormitory. *Chapel Hill Weekly* (Chapel Hill, NC) 3(41): 3, Friday, December 11, 1925.

Anonymous. 1968. Kimberly, John, unpaged entry in Rossiter Johnson (editor in chief) and John Howard Brown (managing editor), *The Twentieth Century Biographical Dictionary of Notable Americans*, vol. 6. Gale Research Co., Detroit.

Ansted, David T. 1847. *The Ancient World, or, Picturesque Sketches of Creation*. Lea & Blanchard, Philadelphia.

Battle, Kemp P. 1875. [Broadside]. *Proceedings of the Meeting of the Trustees of the University, Held in Raleigh, May 4th and 5th, 1875*. [no statement of publisher or place]

Battle, Kemp P. 1876. Letter of June 15, 1876 to Mrs. [Cornelia P.] Spencer in "Letters and Writings of Cornelia Phillips Spencer, April 1876–August 1877," transcript vol. 9. (Cornelia Phillips Spencer Papers #683, Southern Historical Collection, Wilson Special Collections Library, University of North Carolina, Chapel Hill)

Battle, Kemp P. 1883. *An Address on the History of the Buildings of the University of North Carolina. Delivered on University Day, 1883 in Gerrard Hall*. Thomas, Reece & Co., Printers, Greensboro, NC. (p. 14)

Battle, Kemp P. 1907–1912. *History of the University of North Carolina*. Printed by Edwards & Broughton, Raleigh. 2 vols. (Vol. 1, pp. 660, 684, 785; portrait of John Kimberly on plate opposite p. 550)

Battle, Kemp P. 1917. War clouds over the campus, '61, '98, '17. *University Magazine* (University of North Carolina, Chapel Hill), new series 35: 104–111.

Battle, Kemp P. 1945. *Memories of an Old-time Tar Heel*. Edited by his son William James Battle. University of North Carolina Press, Chapel Hill. (p. 255)

Bitting, [John H.], [Stephen W.], Isler, F[rancis M.] Johnson, and others. [1858]. Letter [to John Kimberly], undated. (Box 1, folder 12, John Kimberly Papers #398, Southern Historical Collection, Wilson Special Collections Library, University of North Carolina, Chapel Hill)

BOT. Board of Trustees, University of North Carolina, Chapel Hill. *Minutes of the Board of Trustees*. (University Archives #40001, Wilson Special Collections Library, University of North Carolina, Chapel Hill) (Vol. 5, December 1822–December 1840; vol. 6, December 1841–June 1868; vol. 7, July 1868–March 1883; vol. 15, 1835–1873; vol. 16, 1875–1881)

Bragg, Thomas. 1855. Letter of August 21, 1855 [to whom it may concern]. (Box 1, folder 6, John Kimberly Papers #398, Southern Historical Collection, Wilson Special Collections Library, University of North Carolina, Chapel Hill)

Brawley, James S. 1996. Wheat, John Thomas, pp. 164–165 in William S. Powell (editor), *Dictionary of North Carolina Biography*, vol. 6. University of North Carolina Press, Chapel Hill.

Bursey, Maurice M. 1982. John Kimberly, pp. 17–20 in Maurice M. Bursey, *Carolina Chemists, Sketches from Chapel Hill using the records prepared by Horace D. Crockford, with recollections by Arthur Roe*. Department of Chemistry, University of North Carolina, Chapel Hill. (portrait, p. 18)

Bursey, Maurice M. and Sallie Carter Thomason. 1988. Kimberly, John (1 Sep. 1817–6 Mar. 1882), pp. 361–362 in William S. Powell (editor), *Dictionary of North Carolina Biography*, vol. 3. University of North Carolina Press, Chapel Hill.

Cherry, Thomas Kevin B. 1994. Bringing science to the South: The School for the Application of Science to the Arts of the University of North Carolina. *History of Higher Education Annual* 14: 73–99. (pp. 91–92)

Collins, William F. 1857. Letter of March 24, 1857 to John Kimberly. (Box 1, folder 8, John Kimberly Papers #398, Southern Historical Collection, Wilson Special Collections Library, University of North Carolina, Chapel Hill)

Committee (Sydenham B. Alexander, Norfleet Smith, and Stephen S.K. Watkins). [1860]. Letter of April 26, [1860] to John Kimberly. (Box 1, folder 6, John Kimberly Papers #398, Southern Historical Collection, Wilson Special Collections Library, University of North Carolina, Chapel Hill)

[Coulton?],[57] A. 1842. Letter of July 4, 1842 to Mrs. [Rebecca] Maney. (Box 1, folder 3, John Kimberly Papers #398, Southern Historical Collection, Wilson Special Collections Library, University of North Carolina, Chapel Hill)

Dalton, John C. 1875. *A Treatise on Human Physiology: Designed for the use of students and practitioners of medicine*. 6th ed., revised and enlarged. H.C. Lea, Philadelphia.

Dupree, A. Hunter. 1976. Wyman, Jeffries, pp. 532–534 in Charles Coulston Gillispie (editor in chief), *Dictionary of Scientific Biography*, vol. 14. Charles Scribner's Sons, New York.

Duyckinck, Evert A. and George L. Duyckinck. 1866. *Cyclopaedia of American Literature; Embracing personal and critical notices of authors, and selections from their writings. From the earliest period to the present day.* With portraits, autographs, and other illustrations. Charles Scribner, New York.

Fetter, M. 1866. Letter of May 31, 1866 to Chas. Manly. (Box 6, folder 223, University Papers #40005, University Archives, Wilson Special Collections Library, University of North Carolina, Chapel Hill)

Gray, Asa. 1848. *A Manual of the Botany of the Northern United States, from New England to Wisconsin and South to Ohio and Pennsylvania Inclusive, (the Mosses and Liverworts by Wm. S. Sullivant) Arranged According to the Natural System*. J. Munroe, Boston.

Gray, Asa. 1860. Letter of July 1860 to William J. Martin. (Col. William Joseph Martin Collection #DC026, Davidson College)

[Harrell, Eugene G.], editor. 1887. Colonel Wm. J. Martin, M.A., LL.D. North *Carolina Teacher* 5: 182–185.

Harvard University. 1901. *List of Students of the Lawrence Scientific School 1847–1900*. Published by the University, Cambridge, MA.

HC. Harvard College. 1855. *Twenty-ninth Annual Report of the President of Harvard College to the Overseers, Exhibiting the State of the Institution for the Academical Year 1853–1854.* Metcalf and Company, Printers to the University, Cambridge, MA. (p. 19; Appendix 18.4, Lawrence Scientific School)

57 Surname is illegible, but it appears to read Coulton.

Henderson, Archibald. 1949. *The Campus of the First State University*. (*The University of North Carolina Sesquicentennial Publications*). University of North Carolina Press, Chapel Hill. (pp. 152, 181–188)

Herrick, Henry (Secretary). 1887. John Kimberly, pp. 168–169 in *Record of the Class of 1837 in Yale University, with a Notice of Their Meetings in 1877, 1882, 1887*. 7th ed. Privately Printed—Not Published. Weaver, Shipman and Company, Printers, Springfield, MA.

Hill, Daniel Harvey. 1910. President Hill's address, pp. 16–18 in University Day at Chapel Hill. (*The University of North Carolina Record,* no. 85). The University Press, Chapel Hill.

Jenkins, J.M. and Phil G. Alston. 1857. Letter of May 16, 1857 to Sigma Phi Society. (Box 1, folder 9, John Kimberly Papers #398, Southern Historical Collection, Wilson Special Collections Library, University of North Carolina, Chapel Hill)

Johnson, Samuel W. 1868. *How Crops Grow. A treatise of the chemical composition, structure, and life of the plant, for all students of agriculture*. Orange Judd & Company, New York.

Joseph, D. George. 1999. Warren, John Collins, pp. 724–725 in John A. Garraty and Mark C. Carnes (general editors), *American National Biography*, vol. 22. Oxford University Press, New York, Oxford.

Kerr, Washington C. 1854. Letter of April 7, 1854 to Benjamin S. Hedrick. (Box 1, Benjamin S. Hedrick Papers, Rare Book, Manuscript, and Special Collections Library, Duke University)

Kimberly, Bettie. 1863. Letter of July 19, 1863 to Sis [*i.e.*, Bettie's sister Annie Maney Sehon]. (Box 3, folder 42, John Kimberly Papers #398, Southern Historical Collection, Wilson Special Collections Library, University of North Carolina, Chapel Hill)

Kimberly, Emma. 1856. Letter of March 7, 1856 to Father [*i.e.*, John Kimberly]. (Box 1, folder 6, John Kimberly Papers #398, Southern Historical Collection, Wilson Special Collections Library, University of North Carolina, Chapel Hill)

Kimberly, Emma. 1857. Letter of March 11, 1857 to Father [*i.e.*, John Kimberly]. (Box 1, folder 8, John Kimberly Papers #398, Southern Historical Collection, Wilson Special Collections Library, University of North Carolina, Chapel Hill)

Kimberly, John. 1853a. Letter of February 21, 1853 to Eben Norton Horsford. (Eben Norton Horsford Papers, Rensselaer Institute Archives and Special Collections)

Kimberly, John. 1853b. ["Notes from Lectures on Geology presented by Louis Agassiz, 1853"]. (Box 6, folders 66 and 67, volumes 1-B and 2 [first part], John Kimberly Papers #398, Southern Historical Collection, Wilson Special Collections Library, University of North Carolina, Chapel Hill)

Kimberly, John. 1853c. *Lowell Institute, Twelve Lectures on Various Subjects of Natural History by Professor Agassiz*. (Box 6, folder 67, volume 2 [second part], John Kimberly Papers #398, Southern Historical Collection, Wilson Special Collections Library, University of North Carolina, Chapel Hill)

Kimberly, John. [1854]. ["Notebook compiled by] John Kimberly, Chemical Laboratory, L. S. School, Cambridge, Mass." (Box 7, folder 77, volume 11, John Kimberly Papers #398, Southern Historical Collection, Wilson Special Collections Library, University of North Carolina, Chapel Hill)

Kimberly, John. 1855a. Letter of April 15, 1855 to Benjamin S. Hedrick. (Box 1, folder 8, Benjamin Sherwood Hedrick Papers #325, Southern Historical Collection, Wilson Special Collections Library, University of North Carolina, Chapel Hill)

Kimberly, John. 1855b. Letter of July 7, 1855 to Benjamin S. Hedrick. (Box 1, folder 8, Benjamin Sherwood Hedrick Papers #325, Southern Historical Collection, Wilson Special Collections Library, University of North Carolina, Chapel Hill)

Kimberly, John. 1856a. Letter of November 15, 1856 to Elisha Mitchell. (Box 5, folder 198, University Papers #40005, University Archives, Wilson Special Collections Library, University of North Carolina, Chapel Hill)

Kimberly, John. 1856b. Letter of December 1, 1856 to Charles Manly. (Box 5, folder 199, University Papers #40005, University Archives, Wilson Special Collections Library, University of North Carolina, Chapel Hill)

Kimberly, John. 1856c. Letter of December 1, 1856 to Charles Phillips. (Box 5, folder 199, University Papers #40005, University Archives, Wilson Special Collections Library, University of North Carolina, Chapel Hill)

Kimberly, John. 1856d. Letter of March 9, 1856 to Benjamin S. Hedrick. (Box 2, folder 10, Benjamin Sherwood Hedrick Papers #325, Southern Historical Collection, Wilson Special Collections Library, University of North Carolina, Chapel Hill)

Kimberly, John. 1857a. Letter of February 12, 1857 to Hal [*i.e.*, Mr. Henry Maney]. (Box 1, folder 7, John Kimberly Papers #398, Southern Historical Collection, Wilson Special Collections Library, University of North Carolina, Chapel Hill)

Kimberly, John. 1857b. Private researches. (Box 6, folder 68, volume 3, John Kimberly Papers #398, Southern Historical Collection, Wilson Special Collections Library, University of North Carolina, Chapel Hill)

Kimberly, John. 1857c. Letter of June 26, 1857 to H. Ballière. (Box 1, folder 9, John Kimberly Papers #398, Southern Historical Collection, Wilson Special Collections Library, University of North Carolina, Chapel Hill)

Kimberly, John. 1857–1886. [Notebook containing miscellaneous writings presented in longhand]. (Box 6, folder 68, volume 3, John Kimberly Papers #398, Southern Historical Collection, Wilson Special Collections Library, University of North Carolina, Chapel Hill)

Kimberly, John. [1858]a. Letter of "Friday night" [about January 8, 1858] to D.L. Swain. (Box 2, folder 15, John Kimberly Papers #398, Southern Historical Collection, Wilson Special Collections Library, University of North Carolina, Chapel Hill)

Kimberly, John. [1858]b. Letter of April 8, 1858 to Charles Manly. (Box 2, folder 14, John Kimberly Papers #398, Southern Historical Collection, Wilson Special Collections Library, University of North Carolina, Chapel Hill)

Kimberly, John. 1858c. Letter of January 9, 1858 to the Board of Trustees of the University of North Carolina. (Box 5, folder 206, University Papers #40005, University Archives, Wilson Special Collections Library, University of North Carolina, Chapel Hill)

Kimberly, John. 1858d. Letter of February 17, 1858 to Hal [Kimberly's brother-in-law Henry Maney]. (Box 1, folder 12, John Kimberly Papers #398, Southern Historical Collection, Wilson Special Collections Library, University of North Carolina, Chapel Hill)

Kimberly, John. 1858e. Letter of May 3, 1858 to the Board of Trustees of the University of North Carolina. (Box 2, folder 15, John Kimberly Papers #398, Southern Historical Collection, Wilson Special Collections Library, University of North Carolina, Chapel Hill)

Kimberly, John. 1858f. Letter of September 9, 1858 to Eben Norton Horsford. (Eben Norton Horsford Papers, Rensselaer Institute Archives and Special Collections)

Kimberly, John. 1858g. Letter of September 25, 1858 to Eben Norton Horsford. (Eben Norton Horsford Papers, Rensselaer Institute Archives and Special Collections)

Kimberly, John. 1858h. [Receipt of articles of chemical and philosophical apparatus, February 1, 1858]. (Box 5, folder 207, University Papers #40005, University Archives, Wilson Special Collections Library, University of North Carolina, Chapel Hill)

Kimberly, John. 1860a. Letter of February 4, 1860 to Annie [Kimberly's sister]. (Box 2, folder 27, John Kimberly Papers #398, Southern Historical Collection, Wilson Special Collections Library, University of North Carolina, Chapel Hill)

Kimberly, John. 1860b. Letter of February 26, 1860 to Bettie [Kimberly]. (Box 5, folder 57, John Kimberly Papers #398, Southern Historical Collection, Wilson Special Collections Library, University of North Carolina, Chapel Hill; transcription by Seth L. Warner, August 22, 1995)

Kimberly, John. 1860c. Letter of March 11, 1860 to Bettie [Kimberly]. (Box 5, folder 57, John Kimberly Papers #398, Southern Historical Collection, Wilson Special Collections Library, University of North Carolina, Chapel Hill; transcription by Seth L. Warner, September 23, 1995)

Kimberly, John. 1860d. Letter of August 18, 1860 to Bettie [Kimberly]. (Box 5, folder 58, John Kimberly Papers #398, Southern Historical Collection, Wilson Special Collections Library, University of North Carolina, Chapel Hill; transcription by Seth L. Warner, August 12, 1995)

Kimberly, John. [1860]e. The laboratory supplies for the Department of Analytical Chemistry. (undated items). (Box 2, folder 19: John Kimberly Papers #398, Southern Historical Collection, Wilson Special Collections Library, University of North Carolina, Chapel Hill)

Kimberly, John. 1860f. Letter of July 29, 1860 to Bettie [Kimberly]. (Box 5, folder 58, John Kimberly Papers #398, Southern Historical Collection, Wilson Special Collections Library, University of North Carolina, Chapel Hill; transcription by Seth L. Warner, August 2, 1995)

Kimberly, John. 1860g. Letter of November 30, 1860 to D.L. Swain. (Box 6, folder 217, University Papers #40005, University Archives, Wilson Special Collections Library, University of North Carolina, Chapel Hill)

Kimberly, John. 1860h. ["Lecture Notes on Agricultural Chemistry"]. (Box 6, folder 67, volume 2, John Kimberly Papers #398, Southern Historical Collection, Wilson Special Collections Library, University of North Carolina, Chapel Hill). (The lecture notes are inserted at the end of the volume.)

Kimberly, John. 1860i. Letter of January 17, 1860 to Bettie [Kimberly]. (Box 5, folder 57, John Kimberly Papers #398, Southern Historical Collection, Wilson Special Collections Library, University of North Carolina, Chapel Hill; transcription by Seth L. Warner, September 10, 1995)

Kimberly, John. 1860j. Letter of January 29, 1860 to [Bettie Kimberly]. (Box 5, folder 57, John Kimberly Papers #398, Southern Historical Collection, Wilson Special Collections Library, University of North Carolina, Chapel Hill; transcription by Seth L. Warner, September 15, 1995)

Kimberly, John. 1860k. Letter of February 22, 1860 to Bettie [Kimberly]. (Box 5, folder 57, John Kimberly Papers #398, Southern Historical Collection, Wilson Special Collections Library, University of North Carolina, Chapel Hill; transcription by Seth L. Warner, September 12, 1995)

Kimberly, John. 1860l. Letter of August 12, 1860 to Bettie [Kimberly]. (Box 5, folder 58, John Kimberly Papers #398, Southern Historical Collection, Wilson Special Collections Library, University of North Carolina, Chapel Hill; transcription by Seth L. Warner, October 17, 1995)

Kimberly, John. 1860m. Letter of August 22, 1860 to Bettie [Kimberly]. (Box 5, folder 58, John Kimberly Papers #398, Southern Historical Collection, Wilson Special Collections Library, University of North Carolina, Chapel Hill; transcription by Seth L. Warner, October 17, 1995)

Kimberly, John. 1860n. Letter of August 23, 1860 to Bettie [Kimberly]. (Box 5, folder 58, John Kimberly Papers #398, Southern Historical Collection, Wilson Special Collections Library, University of North Carolina, Chapel Hill; transcription by Seth L. Warner, October 17, 1995)

Kimberly, John. 1860o. Letter of August 9, 1860 to Bettie [Kimberly]. (Box 5, folder 58, John Kimberly Papers #398, Southern Historical Collection, Wilson Special Collections Library, University of North Carolina, Chapel Hill; transcription by Seth L. Warner, August 14, 1995)

Kimberly, John. 1861a. Letter of August 10, 1861 to Bettie [Kimberly]. (Box 3, folder 34, John Kimberly Papers, Southern Historical Collection #398, Wilson Special Collections Library, University of North Carolina, Chapel Hill)

Kimberly, John. 1861b. Letter of September 29, 1861 to Bettie [Kimberly]. (Box 3, folder 35, John Kimberly Papers #398, Southern Historical Collection, Wilson Special Collections Library, University of North Carolina, Chapel Hill)

Kimberly, John. 1865. Letter of April 17, 1865 to Bettie [Kimberly]. (Box 4, folder 47, John Kimberly Papers #398, Southern Historical Collection, Wilson Special Collections Library, University of North Carolina, Chapel Hill)

Kimberly, John. 1866a. Letter of July 13, 1866 to Bettie [Kimberly]. (Box 4, folder 48, John Kimberly Papers #398, Southern Historical Collection, Wilson Special Collections Library, University of North Carolina, Chapel Hill)

Kimberly, John. 1866b. Letter of July 28, 1866 to Bettie [Kimberly]. (Box 4, folder 48, John Kimberly Papers #398, Southern Historical Collection, Wilson Special Collections Library, University of North Carolina, Chapel Hill)

Kimberly, John. [1875]a. Letter to the Trustees of the University of North Carolina. (Box 48, folder 1728, University Papers #40005, University Archives, Wilson Special Collections Library, University of North Carolina, Chapel Hill). (Although the letter is undated, it is probably from December 1875.)

Kimberly, John. 1875b. Letter of July 12, 1875 to Charles Phillips. (Box 5, folder 78, Battle Family Papers #3223-A, Southern Historical Collection, Wilson Special Collections Library, University of North Carolina, Chapel Hill)

Kimberly, John. 1875c. An interesting paper concerning the University of North Carolina. *Daily News* (Raleigh) 8(21): [1], Friday, September 24, 1875.

Kimberly, John. [1875]d. [Request] to the Trustees of the University of N. Carolina. (undated, probably about summer 1875) (Box 48, folder 1728, University Papers #40005, Wilson Special Collections Library, University of North Carolina, Chapel Hill)

Kimberly, John. 1875e. State Agricultural Society; the meeting last evening [*i.e.*, October 13, 1875]. *Daily Sentinel* (Raleigh) 21(60): [2], Thursday, October 14, 1875.

Kimberly, John. 1875f. Letter of May 27, 1875 to Honorable Board of Trustees of the University of North Carolina. (Box 7, folder 262, University Papers #40005, University Archives, Wilson Special Collections Library, University of North Carolina, Chapel Hill)

Kimberly, John. 1875g. Letter of November 10, 1875 to Kemp P. Battle. (Box 7, folder 273, University Papers #40005, University Archives, Wilson Special Collections Library, University of North Carolina, Chapel Hill)

Kimberly, John. 1876a. Letter of resignation to the Board of Trustees of the University of North Carolina, June 1, 1876. (Box 7, folder 279, University Papers #40005, University Archives, Wilson Special Collections Library, University of North Carolina, Chapel Hill)

Kimberly, John. [1876]b. [A Report on the] Agricultural College [to] Prof. J[ohn] De B[erniere] Hooper, Chairman pro tem of the Faculty of the University of N. Carolina, [May 30, 1876]. (Box 7, folder 278, University Papers #40005, University Archives, Wilson Special Collections Library, University of North Carolina, Chapel Hill)

Kimberly, John. 1877. Letter of August 8, 1877 to Wm. H. Battle. (Box 6, folder 83, Battle Family Papers #3223-A, Southern Historical Collection, Wilson Special Collections Library, University of North Carolina, Chapel Hill)

Kimberly, Lizzie. 1857. Letter of October 17, 1857 to Father [*i.e.*, John Kimberly]. (Box 1, folder 10, John Kimberly Papers #398, Southern Historical Collection, Wilson Special Collections Library, University of North Carolina, Chapel Hill)

Knapp, Sharon E. 1988. Hedrick, Benjamin Sherwood, p. 95 in William S. Powell (editor), *Dictionary of North Carolina Biography*, vol. 3. University of North Carolina Press, Chapel Hill.

LGB. *Land Grantee Book* 35: 255. (Register of Deeds, Asheville, Buncombe County, North Carolina)

Lockmiller, David Alexander. 1939. The establishment of the North Carolina College of Agriculture and Mechanic Arts. *North Carolina Historical Review* 16: 273–295.

Long, Rachael. 1993. *Building Notes, University of North Carolina at Chapel Hill*. Updated and revised. [Facilities Planning and Design, University of North Carolina, Chapel Hill]. (pp. 413, 506)

[Love, James Lee]. 1901. *Harvard University, List of Students of the Lawrence Scientific School 1847–1900*. Published by the University, Cambridge, [MA].

[Love, James Lee]. 1944. *The Lawrence Scientific School in Harvard University 1847–1906*. [The Author], Burlington, NC.

Mangum, Adolphus W. 1876. Letter of June 15, 1876 to Mr. [Kemp P.] Battle. (Box 5, folder 80, Battle Family Papers #3223-A, Southern Historical Collection, Wilson Special Collections Library, University of North Carolina, Chapel Hill)

Manly, Chas. 1865. "Annual ac[count]-current of Chas. Manly, Treasr., 1865." (Box 6, folder 222, University Papers #40005, University Archives, Wilson Special Collections Library, University of North Carolina, Chapel Hill)

Mantell, Gideon Algernon. 1848. *The Wonders of Geology; or A familiar exposition of geological phenomena*. 6th ed. H.G. Bohn, London.

Maury, Matthew F. 1873. *Physical Geography*. University Publishing Co., New York.

Moore, J.W. 1877. Historical sketches of Hertford County. Chapters XXXII and XL. *Murfreesboro Enquirer* (Murfreesboro, NC) 2(29): [2], Thursday, May 17, 1877; 2(40): [2], Thursday, August 2, 1877. (A reprint of Moore's sketches was published under the title *Historical Sketches of Hertford County by Liberty Shield Press*, Winston, North Carolina, in 1998.)

Munday, Pat. 1999. Horsford, Eben Norton, pp. 227–228 in John A. Garraty and Mark C. Carnes (general editors), *American National Biography*, vol. 11. Oxford University Press, New York, Oxford.

NC Laws, 1867. Chapter II. An act to transfer the Land Scrip given by the United States to the State of North Carolina, pp. 5–6 in *Public Laws of the State of North Carolina, Passed by the General Assembly at the Sessions of 1866–'67*. Wm. E. Pell, State Printer, Raleigh.

Nesbit, J.C. 1856. *On Agricultural Chemistry, and the Nature and Properties of Peruvian Guano*. Longman and Co., London.

Phillips, Charles. 1854. [Letter of January 2, 1854 to President David L. Swain], pp. 90–95 in Anonymous, The School for the Application of Science to the Arts. *North-Carolina University Magazine* 3: 89–95.

P[hillips], C[harles] (compiler). [1857–1889]. *Scrapbook, [1857–1889]*. (North Carolina Collection, Wilson Special Collections Library, University of North Carolina, Chapel Hill)

Phillips, Charles. 1866. Letter of March 12, 1866 to [Joseph Bruckner Killebrew]. (Box 1, folder 2, Cornelia Phillips Spencer Papers #683, Southern Historical Collection, Wilson Special Collections Library, University of North Carolina, Chapel Hill)

Phillips, Charles. 1875a. Letter of July 8, 1875 to John Kimberly. (Box 4, folder 55, John Kimberly Papers #398, Southern Historical Collection, Wilson Special Collections Library, University of North Carolina, Chapel Hill)

Phillips, Charles. 1875b. Letter of July 15, 1875 to John Kimberly. (Box 4, folder 55, John Kimberly Papers #398, Southern Historical Collection, Wilson Special Collections Library, University of North Carolina, Chapel Hill)

Phillips, Charles. 1876a. Letter of June 12, 1876 to Kemp P. Battle. (Box 7, folder 280, University Papers #40005, University Archives, Wilson Special Collections Library, University of North Carolina, Chapel Hill)

Phillips, Charles. 1876b. Letter of June 30, 1876 to Kemp P. Battle. (Box 7, folder 280, University Papers #40005, University Archives, Wilson Special Collections Library, University of North Carolina, Chapel Hill)

Phillips, Charles. 1876c. Letter of March 3, 1876 to K[emp Battle]. (Box 7, folder 276, University Papers #40005, University Archives, Wilson Special Collections Library, University of North Carolina, Chapel Hill)

Phillips, J[ames]. 1860. Letter of May 28, 1860 to Tinnie in "Letters and Writings of Cornelia Phillips Spencer, December 1839–July 1866," transcript vol. 1. (Cornelia Phillips Spencer Papers #683, Southern Historical Collection, Wilson Special Collections Library, University of North Carolina, Chapel Hill)

Phillips, S[amuel] F[ield]. 1857. Letter of October 2, 1857 to Prof. [John] Kimberly. (Box 1, folder 10, John Kimberly Papers #398, Southern Historical Collection, Wilson Special Collections Library, University of North Carolina, Chapel Hill)

Pool, Solomon. 1867. Letter of August 20, 1867 to B.S. Hedrick. (Box 4, folder 62, Battle Family Papers #3223-A, Southern Historical Collection, Wilson Special Collections Library, University of North Carolina, Chapel Hill)

Rezneck, Samuel. 1972. Horsford, Eben Norton, pp. 517–518 in Charles Coulston Gillispie (editor in chief), *Dictionary of Scientific Biography*, vol. 6. Charles Scribner's Sons, New York.

Schacher, Susan G. 1970. Bunsen, Robert Wilhelm Eberhard, pp. 586–590 in Charles Coulston Gillispie (editor in chief), *Dictionary of Scientific Biography*, vol. 2. Charles Scribner's Sons, New York.

Sigma Phi. 1859. *Catalogue of the Sigma Phi*. T.R. Marvin and Son, Boston. (p. [45])

Singletary, Richard W., Benjamin G. Smith, R.W. Cole, and others. 1858. Letter of February 1, 1858 to John Kimberly. (Box 1, folder 12, John Kimberly Papers #398, Southern Historical Collection, Wilson Special Collections Library, University of North Carolina, Chapel Hill)

Spencer, Cornelia Phillips. 1853–1906. *Journal of Cornelia Phillips Spencer, 1853–1906*. (Box 14, folder 158, volume 2, Cornelia Phillips Spencer Papers #683, Southern Historical Collection, Wilson Special Collections Library, University of North Carolina, Chapel Hill) (pp. 137–139, 147)

Spencer, Cornelia Phillips. 1854–1868. *Journal of Cornelia Phillips Spencer, 1854–1868*. (Box 14, folder 159, volume 3, Cornelia Phillips Spencer Papers #683, Southern Historical Collection, Wilson Special Collections Library, University of North Carolina, Chapel Hill) (p. 47).

Spencer, Cornelia Phillips. [1866]. Letter of [ca. 1866] to [Ella Summerell] in "Letters and Writings of Cornelia Phillips Spencer, August 1866–May 1869," transcript vol. 2. (Cornelia Phillips Spencer Papers #683, Southern Historical Collection, Wilson Special Collections Library, University of North Carolina, Chapel Hill)

Spencer, Cornelia Phillips. 1875a. The University, pp. 683–684 in Louis Round Wilson (editor), *Selected Papers of Cornelia Phillips Spencer*. University of North Carolina Press, Chapel Hill.

[Spencer, Cornelia Phillips]. 1875b. University of North Carolina. *Daily News* (Raleigh) 8(16): [1], Saturday, September 18, 1875.

Spencer, Cornelia Phillips. 1876a. Letter of June 5, 1876 to Ella in "Letters and Writings of Cornelia Phillips Spencer, April 1876–August 1877," transcript vol. 9. (Cornelia Phillips Spencer Papers #683, Southern Historical Collection, Wilson Special Collections Library, University of North Carolina, Chapel Hill)

Spencer, Cornelia Phillips. 1876b. Letter of August 11, 1876 to Mrs. [David L.] Swain. (Cornelia Phillips Spencer Personal Collection #24, State Archives of North Carolina)

Stephenson, Frank. 2003. *Hertford County*. (*Images of America*). Arcadia Publishing, Charlestown, SC. (pp. 40, 44, 45)

Stephenson, Frank. 2005. Electronic mail to William R. Burk, September 13, 2005.

Swain, D[avid] L. 1857a. Letter of January 6, 1857 to John Kimberly. (Box 1, folder 7, John Kimberly Papers #398, Southern Historical Collection, Wilson Special Collections Library, University of North Carolina, Chapel Hill)

Swain, D[avid] L. 1857b. Letter of August 1, 1857 to North Carolina Governor Manly. (Box 5, folder 203, University Papers #40005, University Archives, Wilson Special Collections Library, University of North Carolina, Chapel Hill)

Swain, D[avid] L. 1858. Letter of January 27, 1858 to [William A. Graham]. (Box 10, folder 177, William A. Graham Papers #285, Southern Historical Collection, Wilson Special Collections Library, University of North Carolina, Chapel Hill)

Thomason, Sallie Carter. 1981. Professor John Kimberly & family, pp. 250–251 in Doris Cline Ward (editor); Charles D. Biddix (associate editor), *The Heritage of Old Buncombe County*, Volume I—1981, Old Buncombe County Genealogical Society, Asheville, NC.

UNC Archives, 1. *Library Accession Register*. (Academic Affairs Library, Acquisitions Department, Subgroup 3: Accession Registers, Series 1: Main Collection, volume 60, accession entries 417,954 to 417,982; volume 61, accession entries 418, 322–418,334 and 418, 451, University Archives #40057, Wilson Special Collections Library, University of North Carolina, Chapel Hill)

UNC Archives, 2. *University Affairs, Student Records & Faculty Reports,* Subgroup 1: Academic Records, vol. 4, 1875–1885. (University Archives #40131, Wilson Special Collections Library, University of North Carolina, Chapel Hill)

UNC Faculty. University of North Carolina Faculty. *Minutes of Faculty Meetings*. (University Archives #40106, Wilson Special Collections Library, University of North Carolina, Chapel Hill) (Vol. 6, 1856–1885)

UNC Faculty-T. 1855. [Testimonial on John Kimberly, September 10, 1855]. (Box 1, folder 6, John Kimberly Papers #398, Southern Historical Collection, Wilson Special Collections Library, University of North Carolina, Chapel Hill)

UNC-P. Manuscript, [ca. July 1, 1866]. (Box, 6, folder 223, University Papers #40005, University Archives, Wilson Special Collections Library, University of North Carolina, Chapel Hill)

USFC. *United States Federal Census*. 1850. Perquimans County, North Carolina. (p. 403)

Warner, Seth L. (compiler). 1997. The Family of Daniel Southall: Four generations. (Box 5, folder 57, John Kimberly Papers #398, Southern Historical Collection, Wilson Special Collections Library, University of North Carolina, Chapel Hill)

Whitley, R.G. 1988. Buckhorn Academy. Como, pp. 53–54 in *Hertford County Reflections, the Way It Was*. Hertford County Committee for America's Four Hundredth Anniversary, [no statement of publisher or place]. (also pp. 51–52)

Wiley, C.H. 1858. Letter of January 21, 1858 to John Kimberly. (Box 1, folder 12, John Kimberly Papers #398, Southern Historical Collection, Wilson Special Collections Library. University of North Carolina, Chapel Hill)

Winborne, Benjamin B. 1976. *The Colonial and State History of Hertford County, North Carolina. With an added foreword, index, table of contents, and list of illustrations*. Genealogical Publishing Co., Inc., Baltimore. (pp. 117, 119–122)

Winston, Geo. T. 1895. I. The revival of the university in 1875, pp. [48]–49 in Benefactions by the alumni. *Alumni Quarterly* (University of North Carolina, Chapel Hill) 1(2): [48]–57.

Wright, Ann S. 2006. The Asheville Female College, p. 70 in William S. Powell (editor); Jay Mazzocchi (associate editor), *The Encyclopedia of North Carolina*. Published in association with the University of North Carolina at Chapel Hill Library [and] the University of North Carolina Press, Chapel Hill.

Wyman, Jeffries and E.N. Horsford. 1855. Letter of September 10, 1855 to unidentified person of Nashville, Tennessee. (Box 5, folder 199, University Papers #40005, University Archives, Wilson Special Collections Library, University of North Carolina, Chapel Hill)

Young, Wade Phillips. 1934. A History of Agricultural Education in North Carolina. A thesis submitted to the faculty of the University of North Carolina in partial fulfillment of the requirements for the degree of Doctor of Philosophy in the Department of Rural Social-Economics. Chapel Hill. (pp. 88–98)

George Dixon

Chapter 5.

George Dixon (1812–1904)

Professor of Agriculture (1869/1870)

Early Years in England

George Dixon, born February 10, 1812, in Staindrop, near Darlington, County Durham, England, was the son of Ralph and Ann (Booth) Dixon. His father served in the Napoleonic wars and later became a sergeant. He was discharged in 1810 following a yearlong convalescence from a wound received in Spain. Removing to Staindrop, Ralph Dixon joined the Society of Friends and became a flour miller (Thistlethwaite, 1936).

Young Dixon attended the day schools of Staindrop. At age seventeen, he contracted typhus. Believing that physical activity would benefit Dixon's health, his parents arranged a three-year apprenticeship for him at a nearby farm. There, he avidly learned about diverse aspects of agricultural practices: "ploughing, sowing, reaping, mowing, hedging, ditching, feeding the stock, milking the cows, tending the sheep and lambs, and shearing them of their fleece" ([Dixon], 1891). After concluding the apprenticeship, Dixon returned home, working days as a farm laborer and studying nights. He decided to become a teacher. In preparing himself in the Lancasterian system of education,[58] he gained appropriate skill under the mentorship of William Shotton at the Skinnergate School in Darlington. Dixon then taught at the Bishop Auckland School, but he soon returned to Skinnergate School, where he continued to teach. Subsequently, he became master of the Old Shildon School. At this time, he married Alice Swinburn on February 11, 1834, with whom he had six children: John (III), Ralph (II), Catherine, Alice, Sarah Ann, and George (III) (Thistlethwaite, 1936). Returning to the school at Bishop Auckland, Dixon taught there for the next five years.

When the North of England Agricultural School (later Great Ayton School) in North Yorkshire was established by the Quakers in 1841, Dixon and his wife were appointed as master and mistress, respectively. The students' experiences included both labor and learning. Dixon emphasized natural history, particularly botany. Toward developing a method of teaching botany at Ayton, he wrote *The Hand Book to the Herbarium* (Dixon, 1845) (Fig. 5.1), now a very scarce volume, copies of which are held at the British Library and the National Library of Scotland. In 1888, Dixon (1888) published a second edition of the textbook but changed the title to *The Class-book of Botany, and Hand-book to the Herbarium*. The

[58] The Lancasterian (also spelled Lancastrian) system was a method of elementary education developed in nineteenth-century England to provide schooling that was affordable to the poor and underprivileged. In one classroom, students were grouped into units of ten individuals. Each group was headed by a monitor who was an older and advanced student directly instructed by the schoolmaster.

Frontispiece Credits: (above) George Dixon, ca. 1880. (below) Early scene of Hampton Institute, showing Academic Hall as the large building in the center, ca. 1870. (Portrait from picture collection of early faculty; campus scene from photographic files; courtesy of Hampton University Archives, Hampton University, Hampton, Virginia).

following description is based on the second edition, which is itself a very scarce book, and which contains the introduction and preface from the first edition. Dixon thought students of botany should have a volume that listed the common and scientific names of the British plants, arranged in taxonomic order. He explained that as he knew of no such book, he set out to write one. Dixon aimed to provide an affordable text for students and for others, including "the labouring poor," whom he hoped might be induced to take up botany as a hobby. He used the artificial, or sexual, system of Linnaeus, rather than the natural system, in classifying plants. Nonetheless, he also introduced readers to the natural method, which he thought was useful for arranging a herbarium.

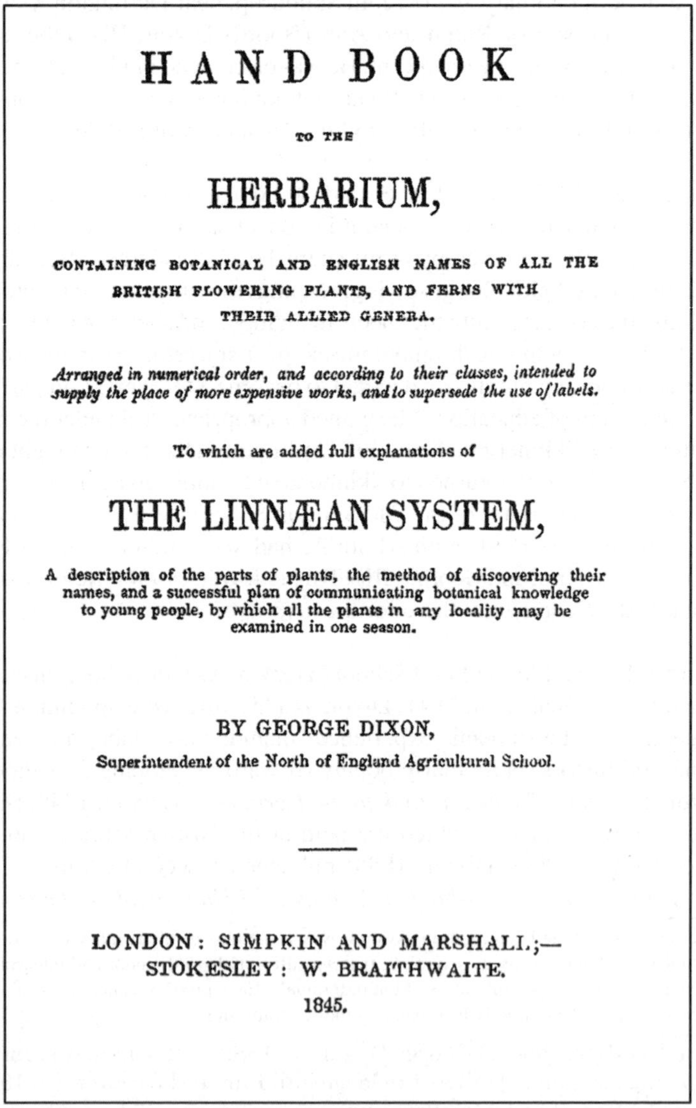

HAND BOOK

TO THE

HERBARIUM,

CONTAINING BOTANICAL AND ENGLISH NAMES OF ALL THE BRITISH FLOWERING PLANTS, AND FERNS WITH THEIR ALLIED GENERA.

Arranged in numerical order, and according to their classes, intended to supply the place of more expensive works, and to supersede the use of labels.

To which are added full explanations of

THE LINNÆAN SYSTEM,

A description of the parts of plants, the method of discovering their names, and a successful plan of communicating botanical knowledge to young people, by which all the plants in any locality may be examined in one season.

BY GEORGE DIXON,

Superintendent of the North of England Agricultural School.

LONDON: SIMPKIN AND MARSHALL;—
STOKESLEY: W. BRAITHWAITE.
1845.

Fig. 5.1. Title page from *Hand Book to the Herbarium* ... by George Dixon (1845).

(Courtesy of the National Library of Scotland).

In their natural history studies, Dixon's pupils collected, identified, and pressed plants for a herbarium. This method of learning "direct[ed] them to Nature herself, instead of depending on books" (Anonymous, 1891). Students were inspired in their collecting by an award of five shillings for finding a plant new to the area. In later years, students were also instructed to collect mosses, liverworts, lichens, fungi, and seaweeds.

The concluding six pages of the second edition advertise products that Dixon offered through the Cleveland Natural History Supply Depot at Great Ayton, which he established and managed. His aim was "to supply Working Men, Young Students, Natural History Classes, and Home Students, all the books and requisites necessary for the successful study of Botany, Entomology, and Conchology, at the *least possible cost*" (Dixon, 1888). He offered books, catalogs, collections of specimens, apparatus, and other items. Among the apparatus for sale was a plant press that Dixon had designed and patented. Measuring sixteen inches by eleven-and-one-half inches, the press did not need screws or weights and conveniently fit endwise on shelves. He also sold portfolios, which he had patented.

When Dixon reduced his teaching responsibilities (but did not give up his superintendency) at the Ayton School in 1857, he embarked on various business projects. Among these were the manufacture of bricks and draining tiles, the rental of farmland, the quarrying of stone, and the operation of a steam sawmill (Watson, 1941).

In England, Dixon avidly pursued avocations in natural history, particularly botany and conchology (the study of mollusks); he collected lichens and land mollusks. His earliest journal publication was on the lichen *Lecanora rubra*. Mark Seaward (1987, 2004b), Professor of Environmental Biology at the University of Bradford in England, noted that Dixon influenced the botanical pursuits of William Mudd, "Yorkshire's most gifted 19th century lichenologist." Seaward (2004a) discovered that Dixon's specimens of lichens are housed in the Manchester Herbarium, University of Manchester, England.

Conchology was another area of natural history of which Dixon was a devotee. In 1858, Dixon and James William Watson (1858) published *A Descriptive Manual of British Land and Fresh-water Shells, Containing Descriptions of All the Species*. According to Mark Seaward (2004a), Dixon wrote the text and Watson illustrated it. The authors noted that pupils at the Ayton School found the study pleasing, and they encouraged principals at other schools to introduce their classes to the study. They wrote the volume in order to provide a single, affordable, descriptive guide to the terrestrial and freshwater mollusks of Britain and to popularize their study.

After his wife died on March 25, 1865, Dixon resigned his post at Ayton on February 5, 1866. He decided to part with his past by sailing to America, where he desired to aid the Southern freedmen. Even prior to his wife's death, Dixon had considered taking a trip to the United States to visit relatives. He had already been assisting the freedmen by raising funds and collecting clothing for them in England. He had further contemplated the possibility of establishing schools in America by offering education and work, a system he had successfully used at Ayton (Watson, 1941). The English Friends' Freedmen's Aid Society sponsored his trip ([Marshall], 1885).

On February 7, 1866, three days before his departure, Dixon was honored at Darlington, England, by nearly forty of his former pupils and a number of members of the Great Ayton Committee, including Isaac Sharp and John Pease, who, like Dixon, had been associated with the school since its establishment. Dixon was presented with a purse containing twenty guineas and testimonials that praised the "zeal and lively interest" he extended to his students (Anonymous, 1866a; Anonymous, 1866b). Dixon, who was accompanied by his daughter Sarah Ann Dixon, arrived in New York City on February 23, 1866, on the H. M. S. *Cuba* (Ancestry, 2003).

Assisting the Freedmen

Toward the end of 1863, the Friends of Philadelphia had distributed a circular (Fig. 5.2) that announced their plans to assist the Southern freedmen and to raise funds for this endeavor. Their goal was to improve the freedmen's physical condition and to provide them with an education. These efforts led to the establishment of the Friends' Association of Philadelphia and Its Vicinity for the Relief of Colored Freedmen (EBM I: no. 1), later called the Friends' Freedmen's Association of Philadelphia.

Noteworthy among the early organizers was Yardley Warner of Germantown, Pennsylvania. He raised funds, set up schools in the southern states, and served as the temporary superintendent of the freedmen's schools in North Carolina ([Warner, Y.], 1865b). In developing schools, he traveled extensively. During his journeys, he encountered resistance from white people, particularly in the early years. He withstood being fired upon with shotguns and assailed by stones. About 1865, at an area in Greensboro, North Carolina, later known as Warnersville, Warner began a venture to build houses for the freedmen, develop gardens, and establish a schoolhouse (Warner, S.A., 1957). While organizing these endeavors, he temporarily lived and taught there.

In late August 1865, Warner made a report to the association's instruction committee in which he recommended locations for establishing a number of schools (EBM I: no. 2). He was very optimistic about the undertaking in North Carolina. In a letter of November 16, 1865, he noted the prospects of developing several schools, "one at Goldsboro [with] 300 [estimated] scholars, Hillsboro 200, Chapel Hill 75 or 100, Durham 25 or 30, Haw River Station one in view" ([Warner, Y.], 1865b). By early 1866, the association had enrolled nearly 1,900 pupils in North Carolina (EBM I: no. 3). Sites for schools were selected along the North Carolina Railroad "from New-Bern to Charlotte, and adjacent to that from Greensboro to Danville, [Virginia], at distances from those roads, varying from twenty-five miles down" ([Warner, Y.], 1865a) (Fig. 5.3). The proximity of the schools to railroad lines made them "readily accessible" (Heckman, 1946). The ownership of the schoolhouses varied: some were purchased by the association from private owners or from the federal government, such as former barracks; some were owned by the teachers or by the freedmen; some were rented through the United States Freedmen's Bureau; and some were offered rent-free by individuals. The schools in Danville, Charlotte, Greensboro, and Salisbury had mission houses connected with them; these provided rooming for teachers and lodging for visitors. The schools typically opened about September and continued to operate until early June, when the hot weather began. Teachers usually took a two-month vacation during the summer.

AN APPEAL

To Friends and others on behalf of the Destitute Freedmen of the South.

DEAR FRIENDS :—

In the midst of a war which is desolating a large portion of our land, we have been permitted to dwell in our homes in peace. A more than common prosperity has attended the labors of the past year. While thankfully acknowledging these blessings at our Heavenly Father's hand, is it not well for us to inquire if we are fulfilling the duties which these solemn days present to us? God, in His providence, is leading up from the house of bondage a people whose condition demands our tenderest sympathy and regard. Singly and in families, by tens and hundreds, they have come within the lines of our armies, until they now number nearly two hundred thousand souls. Crushed and dispirited by long years of oppression, they know not how to supply their simplest wants, and without instruction and encouragement their case will be pitiable indeed. The recital of the privations endured by some of them would move the stoutest heart. Shall we not give of our abundance to relieve the wants of these suffering ones?

As unwilling participants in the sin of Slavery, let us not forget that we owe them a debt incurred by using the products of their unrequited toil. As a religious body, our faith impels us to listen to their cry. As a Society, we are traditionally reminded that our fathers first gave this race the boon of freedom, and taught them to look to us for succor. Let them not look in vain.

An Association of Friends has been formed having their relief in view. It is our purpose to select some field of labor not already occupied by any similar body, and, after supplying their physical wants, to do something towards educating them for the responsibilities of their new life. In the meantime we shall use the means entrusted to us for cases of urgent need wherever found. Donations in money, clothing, and materials will be received by either of the undersigned members of the Executive Committee.

SAMUEL R. SHIPLEY, President, 112 Chestnut Street.

JOHN B. GARRETT, Secretary, 400 Chestnut Street.

RICHARD CADBURY, Treasurer, 117 Chestnut Street.

MARMADUKE C. COPE, 1312 Filbert Street.
ANTHONY M. KIMBER, 26 South Delaware Avenue.
PHILIP C. GARRETT, 400 Chestnut Street.
THOMAS SCATTERGOOD, 413 Spruce Street,
BENJAMIN COATES, 127 Market Street.
WILLIAM EVANS, JR., 252 South Front Street.
JAMES WHITALL, 410 Race Street.
JOHN S. HILLES, 27 North Juniper Street.
ELLISTON P. MORRIS, Germantown.
J. WISTAR EVANS, 817 Arch Street.
JOEL CADBURY, JR., 403 Arch Street.
GEORGE VAUX, 1715 Arch Street.
HENRY HAINES, 1202 Girard Avenue.
CHARLES RHOADS, 513 Pine Street.
DR. JAMES E. RHOADS, Germantown.
EDWARD BETTLE, Haddonfield, N. J.
GEORGE S. GARRETT, Upper Darby, Delaware Co., Pa.
ASHTON RICHARDSON, Wilmington, Delaware.
RICHARD MOTT, Burlington, N. J.
WILLIAM M. CANBY, Chadd's Ford, Pa.

Friends generally, of both sexes, are invited to attend a meeting of the Association to be held in the Committee Room of the Arch Street Meeting House, on fifth day evening, the 12th inst., at 8 o'clock, when the condition and necessities of the Southern Freedmen, and the objects of this Association, will be more fully explained.

Philadelphia, Eleventh Month, 1863.

Fig. 5.2. Circular: An Appeal to friends and others on behalf of the destitute freedmen of the South (November 1863).

(From Executive Board Minutes, Friends' Association of Philadelphia and Its Vicinity for the Relief of Colored Freedmen, 1863–1866, p. 10; courtesy of the Friends Historical Library, Swarthmore College, Swarthmore, Pennsylvania).

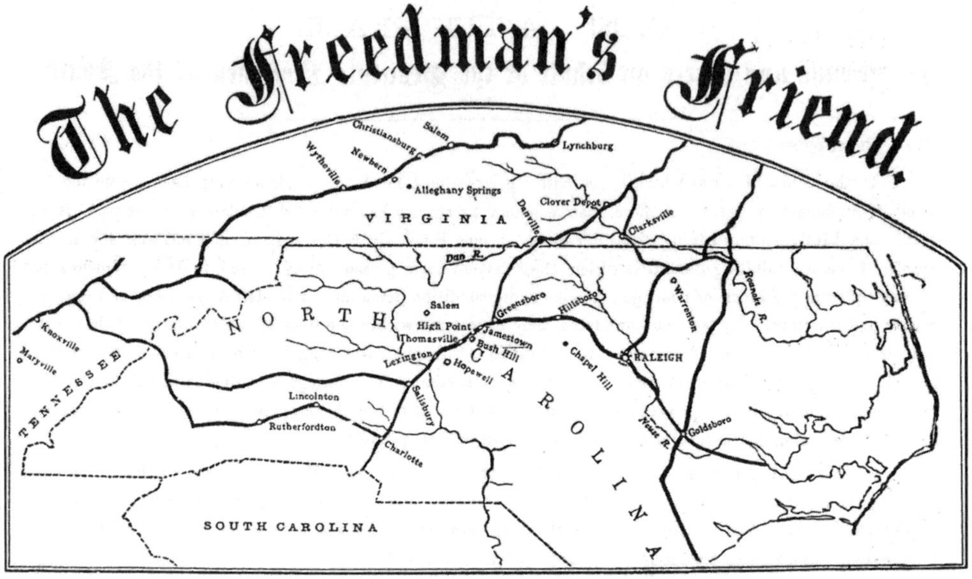

Fig. 5.3. Map showing locations of schools established in North Carolina and Virginia along the North Carolina Railroad (bold lines) by the Friends' Freedmen's Association of Philadelphia.

(From masthead of *Freedman's Friend*, no. 33, 1872).

Soon after his arrival in America, Dixon met with members of the association's instruction committee in Philadelphia. He presented his credentials from the Society of Friends and others from his homeland, which proved satisfactory. At the meeting, Dixon expressed his ideas for promoting industrial schools as well as literary education among the freedmen (ICM I: no. 1), and he offered his services "without remuneration" (EBM I: no. 4). (Apparently, Dixon received a nominal income from property in Ayton, which he leased.) He then proceeded to Greensboro, where he consulted with Thomas Conard about his future undertakings. He wasted little time in planting and landscaping around the mission house, in removing tree roots, and in planting flowers and fruit trees. He had hoped that the improvements would serve as a model for the freedmen. In writing about the freedmen's eagerness to grow their own crops, Dixon (1866a) referred to the crops as "THEIR FIRST CROP IN FREEDOM." His daughter Sarah Ann was also involved with aiding the freedmen, serving as a teacher at the Greensboro school and later as its principal.

In late March, Dixon made a reconnaissance of the schools at Salisbury and Charlotte, North Carolina, as well as that in Danville, Virginia. Although he was impressed by the industry of the teachers and the application of the students, he was concerned about the dilapidated buildings (Dixon, 1866a). Dixon also encountered threats from those who opposed the Friends' efforts. Eunice Congdon, who was then the school mistress in Danville, recounted comments made by the "rebs." They uttered such unsettling admonitions as their intentions to "fire the building, to put a stop to the __ nigger teaching" (Congdon, 1866). One night in June 1866, a rebel snuck into Congdon's bedroom through a window. Realizing what had happened, a servant summoned Dixon, whose approach scared away the intruder. It was surmised that the reb "intended to murder, plunder the place, and then set fire to the buildings" (Dixon, 1866b). Afterward, the freedmen nightly kept watch outside the buildings from 11 P.M. until dawn. Dixon remained at Danville until late June, when the intensity of threats against Congdon and the school subsided. He then returned to Greensboro. On November 14, 1866, Dixon and Eunice Congdon were married.

At the August 7, 1866, meeting of the association's executive board, Dixon was appointed superintendent of the schools in North Carolina and in Danville, Virginia, the group of schools that was collectively designated the North Carolina circuit. He relieved "F. A. Fiske of that capacity" (EBM I: no. 5; ICM I: no. 3). In addition to establishing schools, Dixon was instrumental in maintaining and improving existing ones. In February 1868, the executive board of the association authorized him to build fences around all of its schoolhouses that were not then enclosed (ICM II: no. 1). His responsibilities were numerous and varied. He made recommendations for hiring teachers, confirming their qualifications and advising on salaries. He obtained and distributed goods (seeds, clothing, fabric, shoes, stockings, and blankets) and educational supplies (books, tracts, Bibles, Testaments, slates, globes, furniture, and mathematical tables). (Although students were supplied with religious literature, it was supposedly "read without comment" [Hood, 1869]). Dixon periodically visited schools in his circuit to assess the needs of students, teachers, and buildings. He would frequently write to and occasionally visit the association's instruction committee in Philadelphia, updating it on the conditions at the schools. A number of his letters were reprinted or extracted in the Friends' publications, including the *Freedman's Friend* and the *Friends' Review*. Over two hundred letters written to Dixon by association managers are preserved as letterpress copies at the Friends Historical Library at Swarthmore College.

The dedicated work of Dixon and other fieldworkers of the Freedmen's Association of Philadelphia in developing educational opportunities for the freedmen in the North Carolina district was apparent. By the end of 1866, there were thirteen schools in the district at Chapel Hill, Charlotte, Danville (Virginia), Greensboro, Hillsborough, Jamestown (Guilford County), Lexington, Lincolnton, Mebanesville, Mt. Vernon, Raleigh, Salisbury, and Thomasville (SS, no. 1). In 1867, sixteen schools enrolled over 1,860 day students and 729 evening students (Table 5.1). In 1869, there were twenty-five schools with 2,475 pupils (Hood, 1869). A year later, there were thirty schools, probably the peak number. The added locations in North Carolina included Boon Hill, Brookside, Cameron, Cedar Grove, Clarksburg, Fifth Creek, High Point, Hopewell, Huntsville, Mocksville, Morrisiana, New Garden, New Zion, Pittsburg, Rutherford[ton], Salem, Warrenton, Yadkin Grove, and Zion (EBM II: no. 1). In early 1870, dwindling revenue and resources necessitated the curtailment of expenditures. The instruction committee therefore decided to eliminate a number of the smaller and less important schools in North Carolina and Virginia. The North Carolina schoolhouses that remained in operation were those in Charlotte, Goldsboro, Greensboro, Hillsborough, Lincolnton, Salisbury, Thomasville, and Warrenton. Alfred H. Jones of China, Maine, who later replaced Dixon as superintendent of the North Carolina district, reviewed the remaining schools and decided which would be continued and which abandoned (ICM II: no. 2). According to a note in the November 1870 issue of the *Freedman's Friend*, the other schools that escaped closure were at Chapel Hill, High Point, Hopewell, Jamestown, Lexington, Raleigh, Rowan County (three schools), and Rutherfordton (Anonymous, 1870a).

TABLE 5.1. LIST SHOWING ATTENDANCE AT FRIENDS' FREEDMEN'S SCHOOLS IN THE NORTH CAROLINA DISTRICT FOR 1867.
(ADAPTED FROM *THE FREEDMAN'S FRIEND* 1(16): [22], 1867)

Location	Day Students	Evening Students	Teachers
Chapel Hill	105	95	2
Charlotte	189	60	3
Danville (Virginia)	394	203	4
Goldsboro	143	84	3
Greensborough	181	60	3
Hillsborough	187	138	2
Hopewell	112		1
Jamestown	63		1
Lexington	46		1
Lincolnton	108		2
Mebanesville	114		2
Raleigh	52		2
Salisbury	113	25	2
Thomasville	53	64	2
Total	1,860	729	30

In August 1868, the Dixons notified the association of their intention to resign their posts, relinquishing their duties in October. They wanted to travel to Jamaica to help the local people.[59] Following the appointment of Alfred H. Jones as the new superintendent of the North Carolina district (EBM II: no. 2), the Dixons set out for the West Indies in December.

The Dixons, including Sarah Ann, intended to serve in Jamaica much as they had in North Carolina. Their goal was to improve the schools there and to introduce teaching methods that they had successfully used in America. George was interested in improving the gardening and farming practices of the people. They also served as missionaries through meetings with the Jamaicans "to excite interest on the subjects of education, temperance, &c., and generally for the promotion of morality and religion" (Anonymous, 1868). En route, they were detained in Cuba; however, they finally arrived in Jamaica, probably in February 1869 (Anonymous, 1869a). At Kingston, they loaded their vehicle to capacity with tracts and books and began their visits to a number of schools and locations. Dixon (1869a) pointed out the need for books, maps, and atlases, as well as other items.

Prior to his departure for Jamaica, Dixon had purchased property,[60] which he named Oakland Lodge, near the meeting house in New Garden, Guilford County, North Carolina, and next

[59] Ever since a supposed "rebellion of colored people" was put down cruelly and illegally at Morant Bay, Jamaica, the consciences of citizens in Great Britain had been acutely aroused on behalf of the Black people of Jamaica (Anonymous, 1865).

[60] According to records at the Register of Deeds of Guildford County, North Carolina (DBG), Dixon purchased a number of tracts of land, ranging from 25 to 221 acres between 1868 and 1875.

to the boarding school (Dixon, E., 1868). The property contained nearly 221 acres of land (DB-G, no. 1). From New Garden, George Dixon sent a letter dated October 12, 1869 to the Freedmen's Association. The letter included a brief passage about the Dixons' next undertaking, the operation of a small day school at New Garden. Eunice also held an evening school in their house (Dixon, 1869b).

George and Eunice Dixon probably left New Garden at the end of December 1869 or early January 1870 to live in Chapel Hill, where Dixon accepted a faculty appointment at UNC. Soon after their arrival there, the freedmen's school had an opening for a teacher. Eunice wished to teach at the school and was offered the post.

The Freedmen's School in Chapel Hill

Yardley Warner's recommendation of late 1865 for the development of a freedmen's school in Chapel Hill became a reality. Through a public sale held in Orange County, North Carolina, on January 2, 1866, the Freedmen's Association obtained an acre of land[61] that had belonged to the late John M. Craig in Chapel Hill. The successful bid ($50.00) had been placed in trust by Benjamin Craig and Green Cordal on behalf of Elliston P. Morris, Anthony M. Kimber, and Richard Cadbury of the Philadelphia Association "for the only proper use of the Association of Friends of Philadelphia and Its Vicinity for the Relief of Colored Freedmen forever" (DB-H, no. 1). The transaction was consummated on August 3, 1867, in the presence of Charles Phillips. The site of the property was then at the northwest end of Franklin Street, described in the deed as follows:

> Beginning at a stone opposite the North West end Franklin Street in Chapel Hill and running thence with a Street (unnamed) North Twenty Two and a half degrees west (N 22 ½° W) Thirty four and One third yards (34 1/3 yds.) to a stone thence South Sixty Seven and a half degrees (S 67½° W) West One hundred and Forty One yards (141 yds.) to a stone. Thence South Twenty two and one half degrees East (S 22½° E) Thirty four and One third yards (34 1/3 yds.) to a stone Thence North Sixty-Seven and one half degrees East (N 67½° E) One hundred and Forty one yards (141 yds.) to the beginning.

As early as March 1866, the Friends had been trying to begin a school in Chapel Hill (Phillips, 1866). In October 21, 1866, Dixon (1866c) noted that the Chapel Hill school had not yet been put into operation. When the school opened its doors, probably in November, Mary S. Bowers and her husband, W.G. Bowers, served as the supervisors (SS, no. 2). Alfred B. Seckenby taught at the school until November 1867. The day school enrolled fifty-eight pupils and the evening school thirty-two.

In writing to Dixon in November 1867, Mary Bowers (1867) described the state of affairs at the school and in the town. Her comments elucidate the condition of the freedmen in Chapel Hill and vicinity at that time. In general, she thought they were faring fairly well. Describing the freedmen as very industrious and hardworking, she wrote that they all ultimately strove to acquire a little plot of land. She also wrote about one man who remained with his master during the year after the surrender. As payment for feeding the man's four children, the

[61] Cornelia Spencer (1878) referred to the school site as being "on land of the old Craig farm." Behind the school was a path that students took to "Hunter's Spring," noted for its water. Before the Civil War, the area was known for its wildflowers and captivating flora.

former master took all his pay. At the year's end, the man left his master, devoted himself to hard work, and ultimately harvested twenty bushels of wheat and twenty barrels of corn and raised six hogs ready for slaughter. He found a house, which he bought and paid for in full. Finally, by contrast, Bowers related a story about a plantation situated four miles away from the school. The owner asked the workers to remain on his premises and to build their own houses. He even helped them build a schoolhouse.

Bowers[62] resigned her position in December 1867 (FFA, no. 1). Around February 1868, Caroline B. Loveland filled the vacancy (FFA, no. 2), but she resigned in September. After Loveland's resignation, Mary B. Bowman and her sister Sarah Bowman were offered the teaching posts (ICM II: no. 3), but they declined. Instead, Fanny C. Colver accepted the position (FFA, no. 3). In a visit to the schoolhouse, Superintendent Alfred H. Jones (1869) was impressed by her faithfulness in teaching. She was, he said, "much esteemed by the Blacks and respectfully treated by the whites." By November 1869, the post was once again vacant (Anonymous, 1869b). Soon after, William H. Williams became schoolmaster (ICM II: no. 4), but his performance was unsatisfactory, and he was transferred to a school in Wytheville, Virginia (FFA, no. 4). UNC Professor of Greek Fisk P. Brewer (1870) confirmed the difficulties at the school by noting that the "colored people seem[ed] very much neglected" there. Superintendent Jones temporarily looked after the school. After expressing a desire to head the school in early 1870, Eunice Dixon became the schoolmistress (FFA, no.5). When the Dixons decided to leave Chapel Hill around June 1870 (discussed later in the chapter), George recommended that Wilson Caldwell take charge until September 1, when Eliza Thompson would become the teacher (ICM II: no. 5).

In the summer of 1870 or thereabouts, the state "adopted" the Chapel Hill school and assumed most of its financial needs (EBM II: no. 3) (Fig. 5.4). As a public school, however, its classes were limited to only four or five months of the year, and the teachers were poorly paid (EBM II: no. 4). The Freedmen's Association provided limited assistance by supplying textbooks (ICM II: no. 6), but it withdrew its support in 1874. Although Alfred H. Jones recommended in 1873 that the association take charge of the Chapel Hill school again, the executive board decided against the idea. The board preferred to focus on managing schools already in operation rather than extend its work (ICM II: no. 7).

In December 1880, the executive board of the Freedmen's Association resolved to sell the property to Colored School District No. 42 of Orange County, North Carolina. George Tayloe Winston, W.I. Newton, and Abner Benton Roberson represented the district in the sale. The deed of sale, dated March 22, 1881, included a stipulation that "the said property [was] to be always used for a public school from which the Colored people [should] not be excluded" (DB-H, no. 2). The Board of Education established grammar and high school facilities for the Blacks of Chapel Hill and Carrboro. During the early 1940s, the property became idle.

[62] A decade later, Mrs. Mary Bowers operated a boardinghouse in Chapel Hill. It has been mentioned elsewhere that Cornelia Spencer noted that Bowers's establishment, formerly the Weavers' house located next to the Baptist church, was a dilapidated place. Because Bowers kept a jug of whiskey near the door, the university faculty would not allow students to frequent her place.

ORDER OF EXERCISES FOR
THE CELEBRATION OF THE

FOURTH OF JULY,

BY THE

COLORED CITIZENS

OF CHAPEL HILL N. C.

The Procession will form at 9 O'clock A. M. in the Grove, in front of Mr Charles Mallett's, and proceed thence to the Freedmen's School House.

The Exercises will be opened with Prayer by
EDMUND COLE.

Reading of the Declaration of Independence by
JAMES SHAW.

Oration by
REV. G. W. PUREFOY.

Addresses by
Messrs. J. WATSON, M. J. W. M'CAULEY & Others.

After which the School House will be presented, through Jordan Swain, to the School Committee.

We cordially extend an invitation to our white friends to be present.

Jordan Swain, Chairman.
James Shaw, Secretary.

FRED LANE, Chief Marshal,

WILSON CALDWELL
BENJ. J. CRAIG
GEORGE TRICE
JULIUS SMITH
MOSES SPRUILL
} Ass't. Marshals.

JORDAN EMERSON
JOHN CALDWELL
COLUMBUS OLDHAM
AMOS ANDERSON
} Com. of Arrangements

POMPEY BURGWIN. Flag Bearer.

Fig. 5.4. Circular: Order of exercises for the celebration of the fourth of July [ca. 1870], by the colored citizens of Chapel Hill, N.C. Following the event, the Freedmen's School House was presented to the School Committee.

(From Box 4, folder 52, John Kimberly Papers #398, Southern Historical Collection, Wilson Special Collections Library, University of North Carolina, Chapel Hill).

In the intervening years, a road connecting Chapel Hill and Carrboro bisected the property. The construction and pavement of the road resulted in an assessment of $1,700.00. The board subsequently took advantage of two offers to buy the land. The trustees of the St. Paul A.M.E. Church of Carrboro bought the property northeast of the road, and the Service Insurance and Realty Company of Chapel Hill purchased the remaining land southwest of the road. From the proceeds, the Board of Education paid the assessment fee and used the remaining money for a "Community Center for Colored People, for a Colored Nursery" (DB-H, no. 3). This arrangement fulfilled the original intent of the Freedmen's Association that the property should benefit African Americans.

Additional background on the Freedmen's School in Chapel Hill has been published. Two noteworthy sources are a newspaper article by historians Robert J. Gwyn and David J. Brunn (1979) and a website by Steve Rankin (2014). The website includes several photographs of the school and its students.

In addition to serving the freedmen under the sponsorship of the Friends' Freedmen's Association of Philadelphia, Dixon assisted the Bureau of Refugees, Freedmen and Abandoned Lands in North Carolina, also known as the Freedmen's Bureau. In the bureau, he became the Assistant Superintendent of Education in North Carolina about early February 1867. Although the post did not provide monetary compensation, it covered travel expenses in connection with service to the freedmen in the state. The bureau also turned over a number of buildings in Greensboro, North Carolina, to Dixon in assisting the freedmen (BOR).

Teaching at UNC

In August 1868, over a year before Dixon's appointment at UNC, the Chapel Hill scene was markedly peaceful. The main street was shaded by pendulant branches of stately elm trees; wagons were stationed at random locations from which vendors sold hard cider and watermelons. According to one account, "Old black mammies in red bandannas sat on the curb and sold ginger cakes nearly a foot long for five cents apiece" (Long, 1939). The town's white boys proudly wore Horatio Seymour and Francis Blair political buttons, whereas the Black boys displayed those of Ulysses S. Grant and Schuyler Colfax. There were eight to ten thriving stores, four churches (in addition to the university's chapel), six schools, and about 1,000 residents (Spencer, 1869a).

This peaceful scene belied the state of affairs of the university and its bleak political circumstances. When the Republicans, including William Woods Holden as a candidate for governor, were elected to state office in the spring of 1868, a difficult period for the university commenced. The newly elected carpetbag government in Raleigh replaced the nonpolitical Board of Trustees with a political board, which was composed entirely of Republicans, all with a loyalist outlook (Connor, 1931). The rationale for this action was to transfer the management of the university from the aristocrats who favored the old order to the reconstructionists "who championed the cause of the common people" (Henderson, 1949). Among the early actions of the new government was the closure of the university, which was subsequently overseen by a company of Black troops (Spencer, 1869a). In order to install a partisan faculty, university authorities were summarily dismissed that summer.

About this time, Dixon read Holden's inaugural address in the *Standard*, a Raleigh newspaper. He requested extra copies of the newspaper from Holden to share with his contacts among the Society of Friends in Europe. Dixon desired to promote an interest in North Carolina to encourage emigration to the state. He, himself, intended to settle in North Carolina, which he called his "adopted state" (Dixon, 1868). He further broached the topic of obtaining a position.

By December, Spencer (1868) described Chapel Hill as the "Deserted Village of the South." About twenty respected families planned to move away, and local businesses were stagnating. There were ten stores, but two years later only three were still in business (Jones et al., 1871). The economic situation was so austere that by March 1870 neither the faculty nor the university could get credit in the remaining local stores (Spencer, 1870b). The condition of the environment was also bleak. According to former student Eugene Lindsay Morehead (1869–1871), the campus trees had been greatly plundered for fuel. He wrote in his diary for July 26, 1869: "They have cut down many of the magnificent oaks of the place—turned them into fire wood."

Plans evolved to reorganize the academic institution with the appointment of loyalist faculty. On January 1, 1869, the Executive Committee (BOT 15: 305-B) chose Rev. Solomon Pool as president of the university, paying him an annual salary of $2,000.00 and offering a yearly allowance of $120.00 for house rent. The Executive Committee next recommended establishing chairs or departments of instruction in the following areas: (1) Latin Language and Literature; (2) Greek Language and Literature; (3) Modern Languages; (4) Logic, Rhetoric, Political Economy and History (Ancient and Modern); (5) Mathematics; (6) Chemistry, Mineralogy and Geology; (7) Natural and Mental Philosophy; (8) Applied Sciences; and (9) Law. Upon taking a vote, committee members appointed David Settle Patrick, Fisk Parsons Brewer, James A. Martling, and Alexander McIver to chair departments. Other professors would be appointed later. They set the newly appointed professors' annual salaries at $1,500, including house rent (BOT 7: 53). Although classes opened on March 3, Cornelia Spencer (1869c) wrote with dismay that no students were observed, nor could any "be detected even with the aid of a magnifying glass." Attracting students to the campus had become difficult. Parents lacked the monetary means to send their sons to college, and many, especially those opposed to the new state government headed by William Woods Holden, had lost trust in the university. The 1869/1870 academic year, in which nearly fifty-five students enrolled, commenced on August 18, 1869 (Battle, 1907–1912, 2: 24). Typical expenses for each session were estimated from $86.25 to $97.50 (Fig. 5.5). In addition, there was the cost of textbooks.

On November 16, 1869, Governor Holden (1870) reported to the General Assembly that the trustees of the university would "organize the colored department" as soon as possible. He considered its formation to be a matter of justice and necessity. The department was to have "a Principal with as many assistants as may be necessary; Model Farm, Work-shops, &c." (Ashley, 1869). "Agriculture, Theoretical and Practical" was one of five proposed units within the "Colored Department." Most likely, the prospect of teaching the freedmen enticed Dixon to seek the professorship of agriculture at UNC. From earlier correspondence with Holden, Dixon's desire for a position was known to him. In reporting about his appointment at Chapel Hill, the editors of the Quaker journal *Friend* (London) prematurely noted, "The freed people are now admitted there on equal terms with the whites" (Anonymous, 1870b). Measures to set up a college for the Black students to be connected with the university were never put into effect under Pool's tenure as president, however (Henderson, 1949).

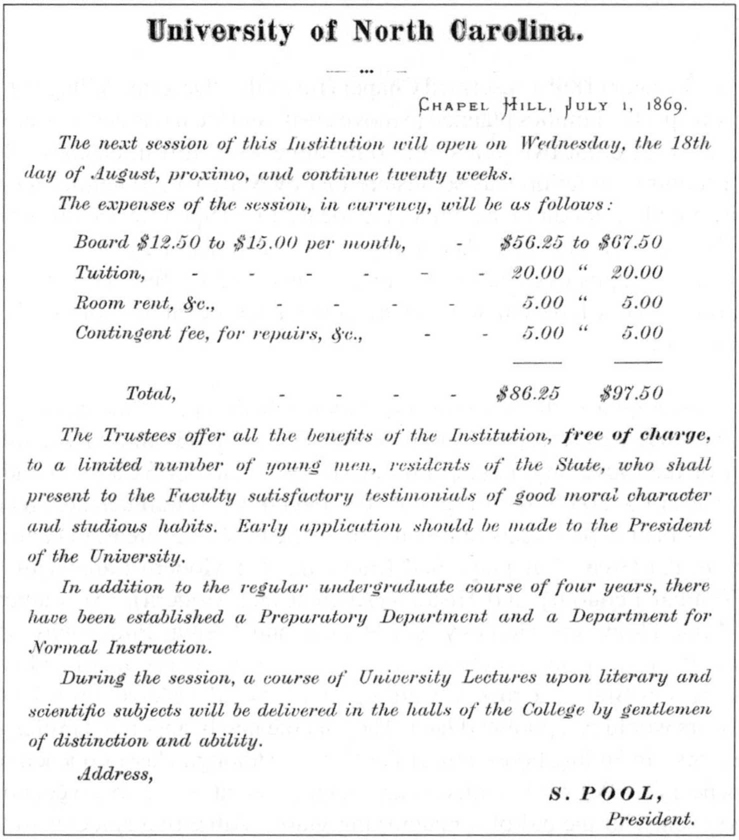

University of North Carolina.

Chapel Hill, July 1, 1869.

The next session of this Institution will open on Wednesday, the 18th day of August, proximo, and continue twenty weeks.

The expenses of the session, in currency, will be as follows:

Board $12.50 to $15.00 per month,	*$56.25 to*	*$67.50*
Tuition,	*20.00 "*	*20.00*
Room rent, &c.,	*5.00 "*	*5.00*
Contingent fee, for repairs, &c.,	*5.00 "*	*5.00*
Total,	*$86.25*	*$97.50*

The Trustees offer all the benefits of the Institution, ***free of charge,*** *to a limited number of young men, residents of the State, who shall present to the Faculty satisfactory testimonials of good moral character and studious habits. Early application should be made to the President of the University.*

In addition to the regular undergraduate course of four years, there have been established a Preparatory Department and a Department for Normal Instruction.

During the session, a course of University Lectures upon literary and scientific subjects will be delivered in the halls of the College by gentlemen of distinction and ability.

Address,

S. POOL,
President.

Fig. 5.5. Circular: University of North Carolina, Chapel Hill, July 1, 1869.
(From Box 1, folder 17, William J. Clarke Papers #153, Southern Historical Collection, Wilson Special Collections Library, University of North Carolina, Chapel Hill).

At their meeting of December 1, 1869, the trustees recommended that the creation of the Department of Agriculture and Mechanic Arts[63] be referred to the Executive Committee for approval (BOT 7: 105). On the motion of Henry J. Menninger, George Dixon was appointed professor of agriculture by the UNC trustees at its meeting of December 9, 1869 (BOT 15: 320). His election was in compliance with the university's obligation to fulfill the stipulations of the Land Grant Act of 1862.

Several days after Dixon had been named to the professorship in agriculture, the faculty discussed the appointment and recommended that President Pool write a letter of

[63] The university had prematurely advertised that a special course in "Agriculture and the Mechanic Arts" would commence on September 6, 1869 (Anonymous, 1869c).

congratulations and welcome to Dixon (UNC Faculty 6: 352). Having superintended the freedmen's school in Chapel Hill from 1866 to 1868, Dixon was already known by a number of townspeople. Cornelia P. Spencer wrote to her sister-in-law Laura Battle Phillips about the newly appointed faculty member on December 18, 1869. Apparently rumors had spread that a "negro" had filled the position. Spencer continued the suspense, writing: "Who do you & C.[harles] think the new Prof. is? How amused you will be. No less than your friend—Friend Dickson [*sic*] the English Quaker—negro school teacher & agent! I hope you have not heard it before, so I may enjoy the fun of seeing your amaze. You must make your own comments. I have nothing to add" (Spencer, 1869b). Spencer (1870a) later reflected on Dixon, calling him "as green a specimen of a fresh-caught John Bull" as she had ever met. She considered him "more ignorant of the people he had come among than semi-educated Englishmen usually are."

UNC Professor of Mathematics Charles Phillips (Spencer's brother) had more favorable views of Dixon than Spencer did. He had great respect for the Friends' assistance of the freedmen. In writing to Kemp Battle about some Quaker literature that Dixon had given him, Phillips said he looked at it and cried over it. Revealing his feelings, he wrote: "Bless God that there are some good people left—who sympathize with & hold down a hand to those in a lower level. The spirit breathing through these pages is admirable" (Phillips, 1867).

The social and political views of Spencer and Phillips influenced their divergent opinions of Dixon. Spencer held staunch southern sympathies and was wary of northern intervention in the state. After reading the book *The Story of the Great March*,[64] she felt as though she had enveloped herself in a blister. Spencer (1854–1868) further wrote in her diary: "How I do hate the whole Yankee nation as I lay such a book down—& feel the bitterness of death roll over me afresh as I contemplate the southern failure." In 1868, she was displeased with the state's Republican government and those who reorganized the university. She blamed Holden's "Negro guard" for pillaging the society libraries at UNC. Moreover, Spencer held the highest disdain for the loss of the university's honor, love, obedience, and troops of friendship. Phillips and his wife Laura, by contrast, had been strong advocates of aiding the African Americans. Of particular note, they established a "Negro Sunday school" at the Chapel Hill Presbyterian Church (Martin, 1994). Margaret Mitchell (Elisha Mitchell's daughter) as well as some of the university students assisted with the teaching (Phillips, 1866). In supporting this cause, Phillips secured books from New York, including a second shipment of volumes in 1866 (Spencer, 1866).

When Dixon's appointment to the faculty was approved, the College of Agriculture and the Mechanic Arts was still being developed. Dixon was charged to instruct students in agriculture, chemistry, and natural history (chiefly botany) (Fig. 5.6). The required text in botany was Eaton and Wright's (1840) *Botany*. Classroom instruction probably included the use of microscopes (UNC Faculty 6: 366). Although little else is known about his botanical lessons at UNC, it is possible that he required his students to develop an herbarium, an exercise he had assigned to students at Ayton.

[64] In the book, author George Ward Nichols (1865), who was a brevet major and aid-de-camp to General Sherman, included a description of the burning of Atlanta and the subsequent march into the Carolinas, where the Union troops captured Fayetteville, Goldsborough, and Raleigh, North Carolina.

CATALOGUE

OF THE

TRUSTEES, FACULTY AND STUDENTS

OF THE

University of North Carolina,

1869–'70.

RALEIGH:

"STANDARD" STEAM BOOK AND JOB PRINT.

1870.

IV.—COLLEGE OF AGRICULTURE AND THE MECHANIC ARTS.

This College is not yet fully organized.

The Professor of Agriculture will give instruction in his department, and also in Chemistry and Natural History. Draper's Chemistry, Emmons' Geology, and Eaton's Botany will be used as text books.

No previous literary attainments are required for the study of Agriculture. The text-books used will be Johnston's series, and Caldwell's Analytical Chemistry.

Fig. 5.6. *Catalogue of the Trustees, Faculty and Students of the University of North Carolina, 1869–'70* (1870). (above) Title page; (below) extract from page 13.

(Courtesy of North Carolina Collection, Wilson Special Collections Library, University of North Carolina, Chapel Hill).

In agriculture, the recommended texts were Johnston's series of agricultural books and George Chapman Caldwell's (1869) agricultural chemical analysis. James Finlay Weir Johnston was known for successfully culling current scientific discoveries in agriculture and presenting them to readers for practical application in the field (G[oodwin], 1967–1968). Among the books probably used by Dixon were Johnston's *Lectures on the Applications of Chemistry and Geology to Agriculture* ([1850]) and *Notes on North America* (1851). The former volume has four parts, each preceding part preparing the student for the subsequent one. The following topics were covered: Part 1, organic substances and parts of plants, and the process by which plants assimilate the substances; Part 2, inorganic substances of plants, the nature of soils, and relationship of geology to agriculture; Part 3, the improvement of soil by mechanical and chemical means; and Part 4, the vegetable products and their use in feeding animals.

At a faculty meeting of August 3, 1869, it was decided that the professors would present a series of special lectures during the 1869/1870 academic year on various subjects, including "Mental and Moral Philosophy, Astronomy, Physiology, Agricultural Chemistry, [and] Botany" (UNC Faculty 6: 341). As part of these lectures, Dixon gave a talk on botany before breakfast on February 18, 1870 (UNC Faculty 6: 359). In addition to his classroom instruction duties, Dixon was assigned Bible readings in the university chapel a couple of times each week (UNC Faculty 6: 367). Although it was not part of his university responsibilities, Dixon also taught Sunday school to a small group of freedmen at a local church (Brewer, 1870).

In carrying out his teaching responsibilities in agriculture, Dixon desired a model farm[65] and workshop. At the trustees' meeting of January 26, 1870, President Pool presented a proposal for establishing such a farm on campus (BOT 15: 322). The trustees asked Professor Dixon to present the idea before the General Assembly and meanwhile postponed a discussion of the possible location of the farm. In preparing for the creation of the farm, Dixon and professor McIver had surveyed nearly one hundred acres of land "back of the College buildings" and had begun contracting to have the trees removed (Spencer, 1870a). Apparently some trees south of South Building had already been felled (Henderson, 1949). A committee charged by the House of Representatives to report on the condition of the university observed that nearly twelve acres of land had been clearcut for the farm. They were displeased with what they saw, noting about the model farm, "A misnomer more glaringly ludicrous could not possibly have been devised" (Jones et al., 1871). Nonetheless, they softened their critical assessment by commending Dixon for his farsighted wisdom and the breadth of his agricultural knowledge. Dixon's efforts to develop a model farm were unsuccessful, both because the Executive Committee "did not see their way to raise the money" (Dixon, 1870) and because the trustees had been influenced by the negative report made by the visiting committee. Another factor, no doubt, was some critical commentary made in an editorial letter by Cornelia Spencer (1870a).

On April 14, 1870, Dixon (1870) wrote a letter to Robert Walter Lassiter, secretary and treasurer of the Board of Trustees, in which he expressed his intention to visit relatives in England during the summer break. He wished to bring back to Chapel Hill "chemical tables, charts &c." to use in his lectures at UNC. He also desired to promote the emigration of capitalists to the state. Dixon offered to take a university microscope, which was broken, to

[65] A model farm was a demonstration farm for instruction in modern methods of farming.

England to have it repaired (UNC Faculty 6: 366). George and Eunice Dixon probably left North Carolina at the time of the commencement (June 8 and 9).

Whether Dixon ever returned to his teaching post at the university has been a topic of controversy. Battle (1907–1912, 2: 11) noted that Dixon reneged on his promise to return to his professorship.[66] Although he did not provide references, UNC chemistry professor Maurice Bursey (1982) correctly set the record straight by stating that Dixon returned to UNC and taught in the fall of 1870. Several sources substantiate Bursey's claim. Dixon attended faculty meetings at least in late September and throughout October (UNC Faculty 6: 369, [370]). Another supportive clue is an entry in the treasurer's report of 1870, noting that part of Dixon's salary ($175.00) was paid on September 15 (BOT 7: 120).

Toward the end of 1870, the financial outlook of the state was bleak, and monetary support for the campus was virtually nonexistent. Fueling the demise of the university was the lack of support from the alumni and friends of the university, who were displeased over the political control of the institution. Recognizing that the university had "sunk to the level of an old-field school," the UNC Board of Trustees stopped paying faculty salaries (Henderson, 1949). In February 1871, the university closed its doors. When the university was nearing closure, Dixon returned to Oakland Lodge. As his name is lacking from the treasurer's report that includes disbursements for December 1870, it is likely that Dixon left Chapel Hill before then (Lassiter, 1872). In 1868, Dixon (1877) had bought property at Raleigh Cross-roads (also cited as Raleigh X Roads) about two and a half miles from New Garden (Dixon, 1871; DB-G, no. 2). There, he assisted with the creation of a new colony of freedmen. He divided the land into six farms for the freedmen and built new houses on three of them, the other three farms already having houses on them. He and his wife founded a school for the freedmen, who set up a meeting house. According to Dixon (1877), the community became "a very thriving little colony."

Departure from North Carolina and Later Years at Hampton Institute

George and Eunice Dixon probably left North Carolina at the end of 1871 (FFA, no. 6). Again en route to England, they visited Hampton, Virginia, where the Hampton Normal and Agricultural Institute had been established in 1868 under the guidance of Samuel Chapman Armstrong, who became its first principal. The school was almost exclusively for Black men and women. Eunice's brother, General Congdon, had arranged for her and her husband to meet with Armstrong, who had also been a general. The Dixons wished to see the progress of the fledgling institute, which was founded for the primary purpose of training African American schoolteachers. Now known as Hampton University, the institute had an educational mission to train teachers and also to provide work responsibilities, including farming. After they returned to England, the Dixons solicited their contacts among English and Irish Friends for funds to support students at Hampton (Anonymous, 1873). They raised over seven thousand dollars, which would later pay for over sixty of their former students to attend Hampton Institute (Anonymous, 1884). Early in the spring of 1873, they returned to America and

[66] Dixon's contributions to the university appear to have been misunderstood by other historians as well. Phillips Russell (1949) claimed that Dixon was "in Chapel Hill only a few months."

settled in Hampton, where they were associated with the Normal and Agricultural Institute. There, they oversaw "the students who were being educated through [the] funds" that they had raised (Anonymous, 1884). They lived in quarters in Hampton Institute's Butler House until 1877, when Dixon bought a lot (#3 on Hampton River, Ferry Point) (ECC, no. 2). There, he built Raby Lodge, which served as the Dixons' residence until they left Hampton.

George Dixon had fewer educational responsibilities at the institute than Eunice did. He was a lecturer in agriculture from 1873 to 1875, giving "lessons and lectures in Botany, Geology, and Agricultural Chemistry, &c.," for which service he received no payment ([Dixon], 1875a). Dixon's (1874b) "apprenticeship with one of the best farmers in the County of Durham, England," provided him with practical agricultural experience. He also kept current with the advancements in theoretical farming through reading. While Dixon resided in North Carolina, the Society of Friends supplied him with relevant literature, such as *Gardener's Monthly* [*and Horticultural Advertiser*] (ICM I: no. 2). His background served him well as a lecturer in agriculture at Hampton Institute (as it had in North Carolina).

While living in Hampton, Dixon wrote articles on agricultural and botanical topics in the school newspaper, the *Southern Workman*. According to historian Linda Bart Tucker (1997), Dixon's papers were the only original scientific ones to be published by anyone at the institute during its first twenty-five years of existence. Several of the columns were about soils and their drainage as well as plowing. In his publications, he drew heavily upon his knowledge of British farming practices to provide examples for southern farmers. Several of his articles educated the farmer about plant structure and physiology, including leaf structure, plant food, and "unorganized" substances. The series of publications by Dixon was intended to provide southern farmers with "hints and facts" about improving their farmlands and their harvests (Anonymous, 1874).

Dixon also took a keen interest in local business ventures. After the Civil War ended, northerners had been drawn to Hampton by economic opportunities there, particularly in the oyster industry (Blair, 1975). By the time Reconstruction ended, fishing and oyster harvesting had become the most important industries for Hampton (Engs, 1979). Dixon was clearly aware of the lucrative prospects of the seafood industry. Fulfilling his entrepreneurial inclination while simultaneously assisting freedmen, he devoted significant time and money to establishing the Hampton Bar Oyster Company, Produce Merchants, of which he was the manager in about 1874 (Fig. 5.7). His store was probably the one referred to in the records of the Elizabeth City County Courthouse (ECC, no. 3). It was situated on land owned by James Sands Darling (who, with his son Frank W. Darling, became successful oyster dealers in Hampton) and was opposite the building known as Darling's Old Mill. The venture, which was a produce and commission store, was started as a way to provide the freedmen with a more lucrative means of making a living. Dixon had observed that the poor men and older boys who lived in the vicinity of Fortress Monroe and Hampton would paddle out in little canoes and harvest oysters with long-handled tongs. Although the oysters were of a high quality, known as the "Hampton Bar" oysters, the harvesters were given insufficient compensation from "a company, who form[ed] a 'ring,' and just g[a]ve them what they please[d], paying them at their stores and liquor-shops" ([Dixon], 1875a). Through his company, Dixon paid nearly twice as much for the oysters in cash. In 1875, Dixon (1875b) mentioned that the cooperative was successful and that he would continue his efforts on behalf of the freedmen for another year. Tucker (1997) noted that the company also offered an outlet for "farmers to market their products." In addition, Dixon established a temperance coffee house for the local freedmen (Anonymous, 1905). Throughout his life, Dixon was a proponent of temperance and a reformer (B., 1904). At twenty-one years of age, he established a temperance society in Staindrop (Anonymous, 1904b).

Hampton Bar Oyster Company,
PRODUCE MERCHANTS,
HAMPTON, VA.
GEORGE DIXON, Manager.

Fig. 5.7. Stationery heading. Letter head of stationery that George Dixon used for the Hampton Bar Oyster Company. (From correspondence file of George Dixon).

(Courtesy of Hampton University Archives, Hampton University, Hampton, Virginia).

In expanding the business, Dixon considered sending fresh oysters to cities in the United States. Steamers from Baltimore, Philadelphia, New York, and Boston could provide inexpensive and quick transportation. From these points, oysters could be brought to inland towns of the middle, northern, and western states. He also considered canning oysters for shipment to England. Having observed the practices of the largest oyster companies in the region of Hampton, Dixon believed he could develop a cleaner and more hygienic processing environment than the other companies. He proposed to "send [oysters] *in the shell* in barrels containing two and a half bushels, or in bushel boxes to private families; and *opened*, in tubs or cans, from one gallon upwards" (Dixon, 1874a). In this effort, Dixon leased a warehouse near the county wharf in 1876 (ECC, no. 1). He considered erecting sheds and drawing water from a well on the land through pipes placed underground. Whether the enlarged business venture succeeded is unknown; no further information about the undertaking has been found in the published literature.

Eunice Dixon, meanwhile, held several positions at the Hampton Institute: she was in charge of Butler School (1873–1875), in charge of the preparatory department (1875–1877), head of primary classes (1877–1879), and graduate correspondent (1880–1882) (HU-A, no. 1). Among the students whom the Dixons taught and befriended was Booker T. Washington, who became a noted educator.

Although there is no documented evidence of Dixon's botanizing in North Carolina, he assiduously collected plants growing in the region of Hampton soon after his arrival in

Virginia. Within two years, he had collected and identified nearly five hundred species of plants while taking "rambles on foot in the country" (Dixon, 1874c). He dried and prepared specimens, which he had intended to arrange into a "*hortus siccus*" (a dry garden, *i.e.*, an herbarium) and to deposit them in the museum at Hampton Institute. The specimens did not survive; they were probably destroyed in 1879, when the Academic Hall, which housed the museum, was consumed by fire. Nonetheless, lists of the species of plants he collected were published as a series of seven articles in the 1874 volume of the *Southern Workman*. The plants (Appendix 5.1) are recorded by the month in which they bloom; scientific and common names are used. Dixon had hoped to publish a pamphlet that would list the plants as a guide for botanists and naturalists. If the plant list was published, no copy of it has been found.

About March 1880, the Dixons returned to England for a rest and an opportunity to visit relatives. They also desired to raise money to aid in rebuilding Academic Hall (Dixon, 1880a). Serving as the central building on the campus, the hall had contained the school and recitation rooms, administrators' offices, an assembly room, the laboratory, the museum, the library, and the dormitories for the preparatory students (Armstrong, 1880). The Dixons returned to Hampton at the end of the year (Dixon, 1880b). After the passage of four years, they decided to retire to England. Before leaving Virginia, Dixon sold his residential and business property in Hampton (ECC, no. 3, 5).

Later Years in England

George and Eunice Dixon returned to "the Nutshell" (Fig. 5.8), their home in England, toward the end of 1884. In retirement, Dixon devoted time and attention to the boys and girls at the Ayton School and also to his grandchildren (Anonymous, 1904a). His passion for botany did not diminish with age; he took botanical excursions up to twelve miles long at the age of eighty. On May 16, 1904, George Dixon died at age ninety-two. He was buried three days later in the Friends Burial Ground, Great Ayton (Fig. 5.9).

Nearly a year after Dixon's death, his wife Eunice expressed a desire to return to Hampton Institute. Her wishes came true in June 1905, when she sailed to America (Anonymous, 1905). She spent the last two years of her life at Hampton Institute, where she died on February 27, 1907, at age eighty-six (Anonymous, 1907). She was buried in the institute's cemetery, her plot being marked with an impressive black granite tombstone (Anonymous, 1911).

Legacy

George Dixon was a man of seemingly limitless energy. His Quaker desire to serve others, particularly the poor and underprivileged, guided him throughout his life. At Ayton School in England, his outreach included a commitment to teaching and inspiring youth in the study of natural history, especially in botany. He also had an acumen for business ventures both in England and in the United States. He lived in the United States nearly seventeen years, during which time he served the freedmen first by superintending schools, primarily in North Carolina, and subsequently by teaching in Virginia at Hampton Institute and by providing financial assistance in the region of Hampton. He was an avid collector of plants. His list of plants found in the region of Hampton is one of the earliest efforts of a botanist to complete a floristic study in Virginia.

Fig. 5.8. The Nutshell, Dixon's home at Great Ayton, Middlesbrough, North Yorkshire, England, March 2005.

(Courtesy of Ben J.G. Avison).

Fig. 5.9. Gravestone of George Dixon, Cemetery of the Friends Meeting House, Great Ayton, Middlesbrough, North Yorkshire, England, March 2005.

(Courtesy of Ben J.G. Avison).

In between his superintendence in North Carolina and his work at Hampton, he taught agriculture, botany, and chemistry at the University of North Carolina. In botany, his instruction probably made use of microscopes and most likely required students to make herbaria. After returning to England in 1884, he continued to assist in educating children.

Dixon staunchly adhered to his goal of helping the less fortunate. In England, he reached out to the working class. In America, his pedagogic work was challenged by the financial adversity of post-Civil War Reconstruction and by some southern prejudices against aiding the Black freedmen. Nonetheless, he succeeded in his efforts to educate youth and others, particularly in botany and agriculture.

References Cited

(References that conclude with page numbers within parentheses indicate sources from where quotes and pertinent information were obtained.)

Ancestry. 2003. Ancestry.com. New York Passenger Lists, 1851–1891 [database online]. MyFamily.com, Inc., Provo, Utah. ("Original data: *New York Passenger lists of vessels arriving at New York, New York, 1820–1897*. Micropublication M237, Rolls #95-580. National Archives, Washington, DC.")

Anonymous. 1865. Jamaica. *Friends' Review* 19(18): 280–281.

Anonymous. 1866a. Arrival of Dixon. *Friends' Review* 19(31): 488–489.

Anonymous. 1866b. Testimonial to George Dixon. *Friend* (London) 6: 52.

Anonymous. 1868. [Introduction to] Letter from George Dixon on the position of the freedmen in the Southern states of America. *Friend* (London) 8: 250.

Anonymous. 1869a. G. Dixon in Cuba. *Friend* (London) 9: 77–78.

Anonymous. 1869b. Report of the Committee on Instruction. *Freedman's Friend* 1(21): [1]–2.

Anonymous. 1869c. Agricultural course at the N.C. University. *Hillsborough Recorder* (Hillsborough, NC) 49(30): [2], Wednesday, August 18, 1869.

Anonymous. 1870a. [Freedmen's schools]. *Freedman's Friend* 1(24): 4.

Anonymous. 1870b. [Note on George Dixon]. *Friend* (London) 10: 93.

Anonymous. 1873. Hampton Agricultural College Fund for training coloured teachers. *Friend* (London) 13: 37–38.

Anonymous. 1874. Agriculture in the South. *Southern Workman* 3(9): 66.

Anonymous. 1884. Farewell to friends. *Southern Workman* 13: 112.

Anonymous. 1891. *History of Great Ayton School*. Published for the Jubilee Committee, June 25th, 1891. Jordison & Son, Middlesbrough.

Anonymous. 1904a. George Dixon, pp. 43–51 in *The Annual Monitor for 1905, or Obituary of the Members of the Society of Friends in Great Britain and Ireland for the Year 1904*. New Series, no. 63. Headley Bros., London.

Anonymous. 1904b. [Obituary]. Mr. George Dixon. *Leeds Mercury, Weekly Supplement*, no. 20,634, p. 6, Saturday, May 21, 1904.

Anonymous. 1905. [Notes on Eunice Congdon Dixon]. *Southern Workman* 34: 456–458.

Anonymous. 1907. [Death of Mrs. Eunice Congdon Dixon]. *Southern Workman* 36: 247–248.

Anonymous. 1911. In the school cemetery. *Southern Workman* 40: 386.

Armstrong, S.C. 1880. [Text of circular about Academic Hall, from S.C. Armstrong, January 1880]. *Friend* (London) 20: 83.

Ashley, S.S. 1869. *Annual Report of the Trustees of the University of North Carolina 1869*. [no statement of publisher or place]. (North Carolina Collection, Wilson Special Collections Library, University of North Carolina, Chapel Hill). (p. 29)

B., C. 1904. A veteran reformer. 90 years old. Reprint "from the *South Durham and Cleveland Mercury*, 'February 14th, 1902.'"

Battle, Kemp P. 1907–1912. *History of the University of North Carolina*. Printed by Edwards & Broughton, Raleigh. 2 vols. (Vol. 2, pp. 11, 24)

Blair, Gladys A. 1975. *Northerners in the Reconstruction of Hampton, Virginia, 1865–1870.* A thesis presented in partial fulfillment of the requirements for the degree of Master of Arts, Department of History, Old Dominion University. (p. 50)

BOR. Bureau of Refugees. *Records of the Bureau of Refugees, Freedmen, and Abandoned Lands.* U.S. National Archives & Records Administration, Record Group 105. (NARS Microfilm Publication #M843.1; copy at the North Carolina Office of Archives & History–Raleigh, Positive Microfilm F 15.1). (Entries on Dixon include the following: Letters Sent, vol. 1 [July 4, 1865–July 13, 1867], entries 299, 369, 423 on Reel 1 and vol. 3 [August 1–December 31, 1868], entries 182, 214 on reel 2; Endorsements Sent, vol. 1 [July 11, 1865–September 20, 1866], entry 227 on reel 3 and vol. 2 [September 27, 1866–July 24, 1868], entries 77, 84, 92 on reel 4)

BOT. Board of Trustees, University of North Carolina, Chapel Hill. *Minutes of the Board of Trustees.* (University Archives #40001, Wilson Special Collections Library, University of North Carolina, Chapel Hill) (Vol. 7, July 1868–March 1883; vol. 15, 1835–1873)

Bowers, Mary. 1867. [Letter to] kind and esteemed friend Dixon. *Freedman's Friend* 1(16): [5–6].

Brewer, Fisk P. 1870. Letter of March 26, 1870 to Rev. E.P. Smith. (American Missionary Association Papers #M-3743, Reel #2: Fisk P. Brewer Letters, 1851–1877, Southern Historical Collection, Wilson Special Collections Library, University of North Carolina, Chapel Hill)

Bursey, Maurice M. 1982. George Dixon, 1812–1904, pp. 24–27 in *Carolina Chemists: Sketches from Chapel Hill, using the records prepared by Horace D. Crockford. With Recollections by Arthur Roe.* The Department of Chemistry, University of North Carolina, Chapel Hill.

Caldwell, G.C. (editor). 1869. *Agricultural Qualitative and Quantitative Chemical Analysis.* Orange Judd and Co., New York.

Congdon, Eunice. 1866. Letter of June 7, 1866, p. 108 in Letters from our teachers and others. *Freedman's Friend* 1(15): 108–109.

Connor, R.D.W. 1931. The University of North Carolina; An historical sketch. *Yackety Yack* 41: 3–8.

DB-G. Deed Book. (Register of Deeds, Greensboro, Guilford County, North Carolina). No. 1. Deed Book 39, pp. 448–450.—no. 2. Deed Book 44, pp. 102–103. (From 1868 to 1875, the Dixons purchased a number of parcels of property in Guildford County, North Carolina, which they later sold between 1870 and 1875)

DB-H. Deed Book. (Register of Deeds, Hillsborough, Orange County, North Carolina). No. 1. Deed Book 37, pp. 453–454.—No. 2. Deed Book 47, pp. 249–251.—No. 3. Deed Book 123, pp. 94, 94-A and 163–164.

Dixon, Eunice. 1868. [Letter from] New Garden, N.C., Oakland, 11th Mo. 13, 1868, p. 5 in Correspondence. *Freedman's Friend* 1(18): 3–6.

Dixon, George. 1845. *Hand Book to the Herbarium Containing Botanical and English Names of All the British Flowering Plants, and Ferns with Their Allied Genera.* Simpkin and Marshall, London; W. Braithwaite, Stokesley. *Report of Lectures on the Natural History of Plants Yielding Food; with incidental remarks on the functions and disorders of the digestive organs. Delivered by Edwin Lankester.* Published by John Churchill, Soho. 1845. (copy at National Library of Scotland)

Dixon, George. 1866a. Letters from George Dixon. *Freedman's Friend* 1(14): 103–105. (pp. 103–104)

Dixon, George. 1866b. Letters from George Dixon. *Freedman's Friend* 1(15): 112–114. (pp. 113–114, letter of June 13, 1866)

Dixon, George. 1866c. Letter of October 21, 1866 to Edward Pease. Typescript of original by Margaret Alderson, archivist at Great Ayton School. (Copy in author's files and at County Records Office, County Hall, North Yorkshire County Council, Northallerton, North Yorkshire, England)

Dixon, George. 1868. Letter of July 12, 1868 to Governor William Woods Holden, pp. 331–332 in Horace W. Raper (editor) and Thornton W. Mitchell (assistant editor), The Papers of William Woods Holden, vol. 1, 1841–1868. Division of Archives and History, North Carolina Department of Cultural Resources, Raleigh, 2000.

Dixon, George. 1869a. [Extract from letter sent by George Dixon, dated April 26, 1869, from Jamaica], pp. 183–185 in Robert Alsop, George Dixon and the schools in Jamaica. *Friend* (London) 9: 183–185.

Dixon, George. 1869b. [Letter], Oakland Lodge, 11th month 13, 1868, p. 5–6 in Miscellaneous letters. *Freedman's Friend* 1(21): 5–6.

Dixon, George. 1870. Letter of April 14, 1870 to R.W. Lassiter. (Box 6, folder 247, University of North Carolina Papers #40005, University Archives, Wilson Special Collections Library, University of North Carolina, Chapel Hill)

Dixon, George. 1871. Friends' Freedmen's Association. *Friend* (London) 11: 222. (Letter from Dixon, dated February 20, 1871)

Dixon, George. 1874a. Oysters. *Friends' Review* 28: 188. (Letter from Dixon, dated October 23, 1874, Hampton, Virginia)

Dixon, George. 1874b. Hampton. To the editors of the *Southern Workman*. *Southern Workman* 3: 56.

Dixon, George. 1874c. The Flora of Hampton. *The Southern Workman* 3(5): 39, 3(6): 43, 3(7): 56, 3(8): 64, 3(9): 67, 3(10): 79, 3(11): 83. (George Dixon collected and identified nearly 500 species of plants growing in the region of Hampton, Virginia. A list of the species was serially published in seven issues of volume 3 of *The Southern Workman*, according to the time of blooming. *The Southern Workman*, published by the Hampton Institute (now Hampton University) Press, was issued monthly. Dixon's flora has escaped the attention of many botanists most likely because of the relatively limited circulation of the journal.)

[Dixon, George]. 1875a. [Letter of December 6, 1874 to George Sturge], p. 23 in *The American Freedmen,* Hampton College. *Friend* (London) 15: 23.

Dixon, George. 1875b. [Letter] to the editor of *Friends' Review*; Hampton oysters. *Friends' Review* 29: 107.

Dixon, George. 1877. Letter of October 19, 1877 to General [Samuel C.] Armstrong. (General Correspondence, Samuel C. Armstrong, Hampton University Archives, Hampton University, Hampton, Virginia)

Dixon, George. 1880a. Letter of March 9, 1880 from Great Ayton, Northallerton Normal. *Friend* (London) 20: 83.

Dixon, George. 1880b. The Hampton Normal and Agricultural Institute. *Friend* (London) 20: 269. (Letter from George and Eunice Dixon, dated September 17, 1880, Agricultural School, Great Ayton, via Northallerton, Yorkshire, England)

Dixon, George. 1888. *The Class-book of Botany, and Hand-book to the Herbarium.* 2nd ed. S. Harris & Co., London; Harrison Penney, Darlington. (The last series of pages consists of "advertisement" for the Cleveland Natural History Supply Depot.)

[Dixon, George]. 1891. Chapter II, Early life of George Dixon, pp. [4]–10 in [Anonymous], *History of Great Ayton School*. Published for the Jubilee Committee, June 25th, 1891. Printed by Jordison & Company, Middlesbrough and London. (portrait opposite p. [4]; quotes, pp. [4], 70)

Dixon, George and [James William] Watson. 1858. *A Descriptive Manual of British Land and Fresh-water Shells, Containing Descriptions and Figures of All the Species.* Harrison Penney, Printer, Darlington. ("Printed for Dixon and Watson, Ayton School, near Stokesley.") (Cover title reads: "Dickson [*sic*] & Watson's *Descriptive Manual of British Land and Fresh Water Shells with Plates of all the Species, and Genera of Animals*.")

Eaton, Amos and John Wright. 1840. *North American Botany; Comprising the native and common cultivated plants, north of Mexico. Genera arranged according to the artificial and natural methods.* 8th ed. E. Gates, Troy, NY.

EBM. *Executive Board Minutes*, Friends' Association of Philadelphia and Its Vicinity for the Relief of Colored Freedmen. (Friends' Freedmen's Association of Philadelphia, Records 1863–1982, #RG 4/024, series 2, Friends Historical Library, Swarthmore College, Swarthmore, Pennsylvania). I. Minutes for 1863–1866. no. 1. "An Appeal to Friends and Others on Behalf of the Destitute Freedmen of the South," Philadelphia, November 1863. (p. 10).—no. 2. September 5, 1865 (p. 137).—no. 3. January 2, 1866 (pp. 152–153).—no. 4. April 3, 1866 (p. 159).—no. 5. August 7, 1866 (p. 177).–II. Minutes for 1866–1872. no. 1. Report (p. 150).—no. 2. November 10, 1868 (p. 96).—no. 3. October 11, 1870 (p. 166).—no. 4. Report (p. 186)

ECC. Deed Book, Elizabeth City County, Deed Books, Circuit Court, Hampton Courthouse, Hampton, Virginia. No. 1. Book 4, pp. 402–404.—no. 2. Book 5, p. 178.—no. 3. Book 12, pp. 267–269.—no. 4. Book 8, pp. 25–26.—no. 5. Book 12, pp. 212–214.

Engs, Robert Francis. 1979. *Freedom's First Generation; Black Hampton, Virginia, 1861–1890.* University of Pennsylvania Press, [Philadelphia]. (p. 166)

FFA. Friends' Association of Philadelphia and Its Vicinity for the Relief of Colored Freedmen. Correspondence. (Friends' Freedmen's Association of Philadelphia, Records 1863–1982, #RG 4/024, series 7, Friends Historical Library, Swarthmore College, Swarthmore, Pennsylvania). (Letterpress letters in bound volumes). No. 1. Letter dated December 12, 1867, from M.E. Shearman (Actuary), to George Dixon, (box 7, volume 1867–1868, p. 126).—no. 2. Letter dated February 12, 1868, from M.E. Shearman (Actuary) to George Dixon, (box 7, volume 1867–1868, p. 225).—no. 3. Letter dated September 30, 1868, from M.E. Shearman (Actuary) to George Dixon, (box 7, volume 1867–1868, p. 224).—no. 4. Letter dated January 17, 1870, unsigned (probably from C.C. Howard) to George Dixon, (box 8, volume 1869–1870, p. 150).—no. 5. Letter dated February [8], 1870, from M.E. Shearman (Actuary) to George Dixon, (box 8, volume 1869–1870, p. 217).—no. 6. Letter dated [September] 1871, from M.E. Shearman (Actuary) to George Dixon, (box 8, volume 1871–1872, p. 78)

G[oodwin], G[ordon]. 1967–1968. Johnston, James Finlay Weir (1796–1855), pp. 953–954 in Sir Leslie Stephen and Sir Sidney Lee (editors), *The Dictionary of National Biography*, vol. 10 (Howard—Kenneth). Oxford University Press.

Gwyn, Robert J. and David J. Brunn. 1979. The Old Quaker School. *Chapel Hill Newspaper* 57(184): Section A: 5A, Sunday, August 5, 1979.

Heckman, Oliver S. 1946. Pennsylvania Quakers in Southern Reconstruction. *Pennsylvania History* 12: 248–264. (p. 257)

Henderson, Archibald. 1949. *The Campus of the First State University*. (*University of North Carolina Sesquicentennial Publications*). University of North Carolina Press, Chapel Hill. (pp. 191, 193–195)

Holden, W.W. 1870. Governor's message. Jo. W. Holden, State Printer and Binder, Raleigh. (Document No. 1, *Executive and Legislative Documents Laid Before the General Assembly of North Carolina, Session 1869–'70*). (At head of title "Ordered to be Printed, M.S. Littlefield, State Printer and Binder). (p. 6)

Hood, J.W. 1869. Report of Rev. J.W. Hood, pp. 16–27 in *Report of the Superintendent of Public Instruction of North Carolina, for the Year 1869*. M.S. Littlefield, State Printer and Binder, Raleigh. (pp. 21–23)

HU-A. Hampton University Archives, Hampton University, Hampton, Virginia. No. 1. Faculty card on Mrs. George (Eunice C.) Dixon.

ICM. *Instruction Committee Minutes*. Friends' Association of Philadelphia and Its Vicinity for the Relief of the Colored Freedmen. (Friends' Freedmen's Association of Philadelphia, Records 1863–1982, #RG 4/024, series 3, Friends Historical Library, Swarthmore College, Swarthmore, Pennsylvania). I. Minute Book 1864–1867. no. 1. March 13, 1866.—no. 2. April 3, 1866.—no. 3. August 7, 1866. II. Minute Book 1868–1875. no. 1. February 3, 1868.—no. 2. August 15, 1870.—no. 3. September 7, 1868.—no. 4. November 22, 1869.—no. 5. June 13, 1870.—no. 6. January 13, 1873.—no. 7. November 10, 1873.

Johnston, James F.W. [1850]. *Lectures on the Applications of Chemistry and Geology to Agriculture*. New edition, with an appendix. C.M. Saxton, Agricultural Book Publisher, New York.

Johnston, James F.W. 1851. *Notes on North America; Agricultural, economical, and social*. Charles C. Little and James Brown, Boston; William Blackwood and Sons, Edinburgh and London.

Jones, A.H. 1869. Letter of March 11, 1869 from Hillsboro, North Carolina, p. 4 in Correspondence, from letters of A.H. Jones, Superintendent. *Freedman's Friend* 1(20): 4–8.

Jones, Edmund, Thomas A. McNeill and E.B. Lyon. 1871. *Report of the Special House Committee to Investigate the Affairs of the University*. James H. Moore, State Printer and Binder, Raleigh. 6 pp. (Document no. 32, *Executive and Legislative Documents Laid before the General Assembly of North Carolina, Session 1870–'71*). (p. 3)

Lassiter, R.W. 1872. *Report of the Treasurer of the University of North Carolina*. Theodore N. Ramsay, State Printer and Binder, Raleigh. (Document no. 38, *Executive and Legislative Documents Laid before the General Assembly of North Carolina, Session 1871–'72*). (p. 6)

Long, Augustus White. 1939. *Son of Carolina*. Duke University Press, Durham, NC. (p. 30)

[Marshall, J.F.B.]. 1885. Hampton Institute, Virginia. *Friend* (London), new series 25: 66.

Martin, Katherine F. 1994. Phillips, Charles, pp. 89–90 in William S. Powell (editor), *Dictionary of North Carolina Biography,* vol. 5. University of North Carolina Press, Chapel Hill.

Morehead, Eugene Lindsay. 1869–1871. [Diary of Eugene Lindsay Morehead, March 4, 1869–March 31, 1871]. (Southern Historical Collection #992-z, Wilson Special Collections Library, University of North Carolina, Chapel Hill) (p. 25)

Nichols, George Ward. 1865. *The Story of the Great March; From the diary of a staff officer.* Harper and Brothers, New York.

Phillips, Charles. 1866. Letter of March 12, 1866 to [J.B. Killibrew]. (Box 1, folder 2, Cornelia Phillips Spencer Papers #683, Southern Historical Collection, Wilson Special Collections Library, University of North Carolina, Chapel Hill)

Phillips, Charles. 1867. Letter of February 14, 1867 to Kemp P. Battle. (Box 4, folder 60, Battle Family Papers #3223-A, Southern Historical Collection, Wilson Special Collections Library, University of North Carolina, Chapel Hill)

[Rankin, Steve]. 2014. The Quaker Freedmen's school in Chapel Hill. <http://freepages.history.rootsweb.ancestry.com~orangecountync/places> (Accessed November 26, 2014)

Russell, Phillips. 1949. *The Woman Who Rang the Bell; The story of Cornelia Phillips Spencer*. University of North Carolina Press, Chapel Hill. (p. 114)

Seaward, Mark R.D. 1987. 300 years of Yorkshire lichenology. *Naturalist* 112: 37–52. (George Dixon [1812–1904], pp. 41–42; portrait)

Seaward, Mark R.D. 2004a. Letter of May 30, 2004 to William R. Burk. (author's files)

Seaward, Mark R.D. 2004b. Mudd, William (1829–79), pp. 1433–1434 in Bernard Lightman (general editor), *The Dictionary of Nineteenth-century British Scientists*, vol. 3. University of Chicago Press, Chicago.

Spencer, Cornelia Phillips. 1854–1868. *Journal of Cornelia Phillips Spencer, 1854–1868*. (Box 14, folder 159: volume 3, Cornelia Phillips Spencer Papers #683, Southern Historical Collection, Wilson Special Collections Library, University of North Carolina, Chapel Hill) (p. 106).

Spencer, Cornelia P. 1866. Letter of March 10, 1866 to Eliza [Mitchell]. (Cornelia Phillips Spencer Personal Collection #7, book I, p. 6, State Archives, Raleigh, NC)

Spencer, Cornelia P. 1868. Under the poplar. (From *The Sentinel* [Raleigh], December 1, 1868) in "Letters and Writings of Cornelia Phillips Spencer, August, 1866–May, 1869," transcript volume 2. (Cornelia Phillips Spencer Papers #683, Southern Historical Collection, Wilson Special Collections Library, University of North Carolina, Chapel Hill)

Spencer, Cornelia P. 1869a. Pen and ink sketches of the University of North Carolina, as it has been. Dedicated to the people of the state, and to the alumni of the university: no. 15. *Semi-Weekly Sentinel* (Raleigh) 3(106): [1], Wednesday, July 7, 1869.

Spencer, Cornelia P. 1869b. Letter of December 18, 1869 to Laura [Phillips]. (Box 1, folder 4, Cornelia Phillips Spencer Papers #683, Southern Historical Collection, Wilson Special Collections Library, University of North Carolina, Chapel Hill)

Spencer, Cornelia P. 1869c. Letter of March 6, 1869 to Mrs. David L. Swain. (Cornelia Phillips Spencer Personal Collection #7, book I, pp. 30–31, State Archives of North Carolina)

Spencer, Cornelia P. 1870a. [Letter to the editor of *The Sentinel*, February 1, 1870]. *Daily Sentinel* (Raleigh, NC) 5(141): [3], Thursday, February 3, 1870.

Spencer, Cornelia P. 1870b. Letter of March 27, 1870 to Mrs. Swain. (Cornelia Phillips Spencer Personal Collection #7, Book I, pp. 50–51, State Archives of North Carolina)

Spencer, Cornelia P. 1878. Our springs. *University Magazine* (University of North Carolina, Chapel Hill) 1: 20–21.

SS. *Statistics of Schools 1866–1868*. Friends' Association of Philadelphia and Its Vicinity for the Relief of Colored Freedmen. (Friends' Freedmen's Association of Philadelphia, Records 1863–1982, #RG 4/024, series 5, Friends Historical Library, Swarthmore College, Swarthmore, Pennsylvania). No. 1. North Carolina Circuit. Term 1866–1867 (pp. 2–3).—no. 2. Chapel Hill, N.C. (p. 3)

Thistlethwaite, Bernard. 1936. *The Bax Family; An account of the early Quaker Baxes of Capel and Ockley in the county of Surrey, with particulars of some of their descendants and notes relating to a number of allied families*. With an introduction by Alfred Neave Brayshaw. Headley Brothers, London.

Tucker, Linda Bart. 1997. *Science at Hampton Normal and Agricultural Institute, 1868–1893*. A dissertation submitted to the Johns Hopkins University in conformity with the requirements for the degree of Doctor of Philosophy. (Facsimile by UMI Dissertation Services, A Bell & Howell Company, Ann Arbor, MI, 2001) (pp. 87, 130)

UNC Faculty. University of North Carolina, Chapel Hill Faculty. *Minutes of Faculty Meetings*. (University Archives #40106, Wilson Special Collections Library, University of North Carolina, Chapel Hill) (Vol. 6, 1856–1885)

Warner, Stafford Allen. 1957. *Yardley Warner, the Freedman's Friend; His life and times. With his journal and letters reproduced in an appendix*. With an introduction by Janet Whitney. The Wessex Press, Didcot, [England]. (p. 64)

[Warner, Yardley]. 1865a. Report of Yardley Warner's visit to N. Carolina, &c. *Freedman's Friend* 1(11): 77–79. (p. 77)

[Warner, Yardley]. 1865b. Report of Yardley Warner's visit to N. Carolina, &c. [continued from last number]. *Freedman's Friend* 1(12): 88–89. (p. 88)

Watson, G. Alston. 1941. Chapter V., George Dixon: 1841–1865, pp. 39–52 in G. Alston Watson, *Ayton School, the Centenary History 1841–1941*. Printed by Headley Brothers, Kingsway. (portrait, opposite p. 39; pp. 44–45)

APPENDIX 5.1.

THE FLORA OF HAMPTON; A LIST OF PLANTS COLLECTED BY GEORGE DIXON IN THE VICINITY OF HAMPTON, VIRGINIA, IN 1874, ARRANGED ACCORDING TO TIME OF BLOOMING.

(PUBLISHED IN THE *SOUTHERN WORKMAN*, VOLUME 3, NOS. 5–11. 1874. EXCEPT FOR MINOR CHANGES IN OBVIOUS SPELLING ERRORS, THE TRANSCRIPTION FOLLOWS THAT OF THE PUBLISHED VERSION WITH COMMON NAMES GIVEN IN ITALICS).

(p. 39).

JANUARY.

Draba verna—*whitlow grass*. Lamium amplexicaule—*hen-bit*. Capsella bursa-pastoris—*shepherd's purse*.

FEBRUARY.

Senebiera didyma—*carpet cress*. Stellaria media—*chick weed*. Taraxacum dens-leonis—*dandelion*. Poa annua—*early pasture grass*.

MARCH.

Oxalis stricta—*yellow wood-sorrel*. Houstonia caerulea—*innocence*. Acer rubrum—*red maple*. Asarum virginicum—*wild ginger*. Epigaea repens—*trailing arbutus*. Vinca minor—*smaller periwinkle*. Viola cucullata—*hooded violet*. Amelanchier canadensis—*shad bush*. Ranunculus bulbosus—*bulbous buttercup*. Antennaria plantaginifolia—*mouse-ear everlasting*. Prunus americana—*wild plum*. Muscari moschatum—*grape hyacinth*. Fragaria virginiana—*wild strawberry*. Arabis thaliana—*rock cress*. Vicia sativa—*partridge pea*. Fumaria officinalis—*fumitory*. Cardamine hirsuta—*hairy cardamine*. Cerastium viscosum—*sticky chickweed*. Barbarea vulgaris—*winter cress*. Salix babylonica—*weeping willow*. Viola tricolor—*pansy*.

APRIL.

Zephyranthus atamasco—*amaryllis*. Veronica arvensis—*field speedwell*. Ranunculus aquatilis—*water crows-foot*. Ranunculus sceleratus—*celery crows foot*. Nepeta glechoma—*ground ivy*. Luzula campestris—*wood rush*. Pinus taeda—*old field pine*. Benzoin odoriferum—*spicewood*. Podophyllum peltatum—*mandrake*. Myrica cerifera—*bayberry*. Vaccinium corymbosum—*blue bilberry*. Veronica serpyllifolia—*thyme leaved speedwell*. Allium striatum—*streaked garlic*. Gnaphalium purpureum—*purple everlasting*. Gelsemium sempervirens—*yellow jessamine*. Cerastium vulgatum—*mouse-ear chickweed*. Lithospermum arvense—*gromwell*. Rumex acetosella—*sheep sorrel*. Sonchus asper—*sow thistle*. Potentilla norvegica—*cinquefoil*. Ranunculus abortivus—*spearwort*. Populus tremuloides—*aspen poplar*. Viola blanda—*smooth sweet violet*. Viola palmata—*hand-leaved violet*. Claytonia virginica—*spring beauty*. Sassafras officinale—*sassafras*. Cornus florida—*dogwood*. Pyrus arbutifolia—*choke berry*. Senecio tomentosus—*woolly groundsel*. Listera convallarioides—*purple tway-blade*. Iris verna—*vernal iris*. Narcissus poeticus—*poet's narcissus*. Pyrus coronaria—*sweet-scented crab tree*. Cercis canadensis—*Judas tree*. Ranunculus acris—*buttercup*. Valerianella olitoria—*lamb lettuce*. Plantago virginica—*lesser plantain*. Rubus trivialis—*southern dewberry*. Nyssa uniflora—*swamp tulepo*. Trifolium repens—*white clover*. Anthoxanthum odoratum—*sweet vernal grass*. Krigia caroliniana—*dwarf dandelion*. Leucanthemum vulgare—*ox-eye daisy*. Plantago lanceolata—*ribwort*.

Appendix 5.1. (Continued)

(p. 43).

APRIL–CONTINUED.

Poa pratensis—*spear grass*. Potentilla canadensis—*Canadian cinquefoil*. Quercus virens—*live oak*. Trifolium procumbens—*white clover*. Asplenium ebeneum—*ebony spleenwort*. Arisaema triphyllum—*Jack-in-the-pulpit*. Ranunculus pusillus—*slender crowfoot*.

MAY.

Azalea nudiflora—*early honeysuckle [i.e., azalea]*. Vicia hirsuta—*hairy vetch*. Pteris aquilina—*brake*. Osmunda cinnamomea—*cinnamon fern*. Senecio aureus—*golden groundsel*. Salvia lyrata—*wild sage*. Salix nigra—*black willow*. Sagina elliottii—*pearlwort*. Linaria canadensis—*blue toadflax*. Dactylis glomerata—*orchard grass*. Ranunculus parviflorus—*small-flowered ranunculus*. Malva rotundifolia—*round-leaved mallow*. Cirsium lanceolatum—*common thistle*. Sisymbrium officinale—*hedge mustard*. Lepidium virginicum—*wild pepper grass*. Polygonatum biflorum—*two-flowered Solomon's seal*. Onoclea sensibilis—*sensitive fern*. Cypripedium acaule—*low ladies' slipper*. Lonicera sempervirens—*perennial-leaved woodbine*. Trifolium pratense—*red clover*. Galium trifidum—*ladies' bed-straw*. Medeola virginica—*Indian cucumber root*. Pyrus americana—*mountain ash*. Oxalis violacea—*violet wood sorrel*. Carex straminea—*straw sedge*. Geranium carolinianum—*Carolina crane's bill*. Cardamine hirsuta—*common bitter cress*. Nyssa multiflora—*many-flowered tupelo*. Juglans nigra—*black walnut*. Morus alba—*white mulberry*. Vaccinium stamineum—*deer berry*. Kalmia latifolia—*broadleaved laurel*. Hieracium venosum—*vein-leaf hawkweed, bloodwort*. Asparagus officinalis—*asparagus*. Myosotis stricta—*forget-me-not*. Sisyrinchium bermudiana—*blue-eyed grass*. Gratiola virginica—*Virginia water hyssop*. Rubus villosus—*high blackberry*. Cerasus serotina—*black or wild cherry*. Alopecurus geniculatus—*floating foxtail grass*. Andromeda racemosa—*racemed andromeda*. Vaccinium corymbosum—*high blueberry*. Larix americana—*tamarack, American larch*. Silene antirrhina—*snapdragon catchfly*. Rhus toxicodendron—*poison oak*. Trifolium procumbens—*yellow clover*. Smilax glauca—*green brier*. Silene pensylvanica—*pink catch-fly*. Helianthemum canadense—*cistus, rock rose*. Oenothera sinuata—*evening primrose*. Quercus cinerea—*upland willow-oak*. Hordeum pusillum—*small wild barley*. Opuntia vulgaris—*prickly pear, Indian fig*. Gleditschia triacanthos—*honey locust*. Philadelphus inodorus—*false syringa*. Tradescantia virginica—*spiderwort*. Anagallis arvensis—*scarlet pimpernel*. Viburnum opulus—*guelder rose*. Daucus carota—*wild carrot*. Trisetum palustre—*three-awned bog grass*. Rumex crispus—*yellow dock*. Danthonia spicata—*Danthoin's oat grass*. Rumex verticillatus—*water dock*. Euonymus americanus—*American burning bush*. Agrostis scabra—*hair grass*. Bromus secalinus—*cheat or chess*. Ilex opaca—*holly tree*. Smilacina racemosa—*clustered Solomon's seal*. Achillea millefolium—*yarrow milfoil*. Osmunda regalis—*flowering fern*. Aspidium acrostichoides—*shield fern*. Centaurea nigra—*blue bottle, bachelor's button*. Avena praecox—*wild oat*. Elymus virginicus—*wild rye*. Filago germanica—*cudweed*.

(p. 56).

MAY–CONTINUED.

Agrostemma githago—*corn cockle*. Specularia perfoliata—*clasping bell-flower*. Veronica americana—*brooklime*. Polygonum aviculare—*bird's knot grass*. Lolium temulentum—*poison darnel*. Juncus tenuis—*slender rush*. Juncus tenuis, variety dichotomus—*two-stemmed slender rush*. Juncus acuminatus—*pointed rush*. Juncus effusus—*soft bulrush*. Carex straminea—*straw sedge*. Carex hystericina.

APPENDIX 5.1. (CONTINUED)

JUNE.

Carex vulpinoidea—*fox sedge*. Carex debilis—*slender sedge*. Carex miliacea—*many-flowered sedge*. Eleocharis palustris—*marsh club rush*. Erigeron annuus—*flea bane, white weed*. Salvia officinalis—*wild sage*. Ornithogalum croceum—*yellow star of Bethlehem*. Sambucus canadensis—*black elder*. Lonicera periclymenum—*woodbine*. Samolus valerandi—*water pimpernel*. Trifolium arvense—*hare's foot trefoil*. Echium vulgare—*viper's buglos, blue thistle*. Arnica nudicaulis—*naked-stemmed arnica*. Aster subulatus—*slender sea aster*. Maruta cotula—*May weed*. Rumex maritimus—*golden dock*. Datura stramonium—*thorn apple, jimson weed*. Lolium perenne—*ray darnel*. Holcus lanatus—*soft grass*. Elymus europaeus—*lyme grass, wild rye*. Bromus ciliatus—*brome grass*. Verbascum blattaria—*moth mullein*. Hypericum perforatum—*perforated St. John's wort*. Allium canadense—*meadow garlic*. Phleum pratense—*timothy, or herd's grass*. Bromus unioloides—*rescue grass*. Verbena spuria—*vervain*. Centaurea nigra—*black knap weed*. Typha angustifolia—*cat-tail, reed mace*. Scutellaria integrifolia—*entire-leaved skull cap*. Erigeron divaricatus—*branched flea bane*. Myriophyllum scabratum—*water milfoil*. Rosa lucida—*shining, or wild rose*. Solanum carolinense—*horse nettle*. Pastinaca sativa—*parsnip*. Saururus cernuus—*lizard-tail*. Azalea viscosa—*white azalea*. Spergula arvensis—*corn spurry*. Linaria vulgaris—*toad flax*. Setaria glauca—*bottle grass, or foxtail panic*. Cirsium horridulum—*horrid thistle*. Phytolacca decandra—*poke weed, pigeon berry*. Agrostis vulgaris—*red top, herd's grass*. Panicum latifolium—*broad-leaved panic grass*. Verbascum thapsus—*mullein*. Panicum gibbum—*gibbus-panic grass*. Leontodon autumnalis—*autumnal hawkbit*. Glyceria nervata—*nerved manna grass*. Asclepias tuberosa—*butterfly weed*.

(p. 64).

JUNE–CONTINUED.

Briza media—*quaking grass*. Baptisia tinctoria—*wild indigo*. Prunella vulgaris—*self-heal, heal-all*. Liriodendron tulipifera—*white wood, tulip tree*. Magnolia glauca—*white bay, beaver tree*. Liquidambar styraciflua—*sweet gum*. Linum virginianum—*yellow flax*. Andromeda arborea—*sorrel tree*. Medicago sativa—*lucerne, medick*. Melilotus albus—*white melilote clover*.

JULY.

Cirsium lanceolatum—*common thistle*. Polygonum persicaria—*smart weed*. Delphinium consolida—*larkspur*. Forsteronia difformis—*Forsteronia*. Silene armeria—*catch fly*. Sericocarpus conyzoides—*white-tipped aster*. Spiranthes gracilis—*slender ladies' tresses*. Lobelia boykinii—*slender lobelia*. Lysimachia quadrifolia—*four-leaved loosestrife*. Sedum pulchellum—*white stonecrop*. Vitis vulpina—*fox grape*. Vitis aestivalis—*summer grape*. Tecoma radicans—*trumpet flower*. Castanea vesca—*chestnut*. Plantago major—*larger plantain*. Pogonia ophioglossoides—*snakemouth arethusa*. Rubus occidentalis—*black raspberry, thimbleberry*. Scutellaria pilosa—*hairy skull-cap*. Asclepias phytolaccoides—*poke-leaved silkweed*. Stylosanthes elatior—*pencil flower*. Zanthoxylum americanum—*prickly ash*. Polypogon monspeliensis—*polypog grass*. Salsola kali—*saltwort*. Hydrocotyle interrupta—*pennywort, navelwort*. Geum album—*white avens*. Typha latifolia—*broad-leaved cat-tail*. Phryma leptostachya—*lop seed*. Centaurea calcitrapa—*star thistle*. Rhexia virginica—*deer grass*. Diospyros Virginiana—*persimmon, seed plum*. Castanea pumila—*chinquapin*. Quercus phellos—*willow oak*. Hypericum mutilum—*St. John's wort*. Ilysanthes refractra—*ilysanthes*. Diodia virginiana—*Virginian diodia*. Cephalanthus occidentalis—*button bush*. Sanicula marilandica—*Maryland sanicle*. Sericocarpus tortifolius—*twisted-leaved aster*.

APPENDIX 5.1. (CONTINUED)

Callicarpa americana—*French mulberry*. Allium mutabile—*garlic, wild onion*. Borrichia frutescens—*sea oxeye*. Discopleura capillacea—*bishop weed*. Martynia proboscidea—*unicorn plant*. Elymus striatus—*lyme grass*. Centrosema virginianum—*sea pea*. Teucrium canadense—*germander*. Cnidoscolos stimulosus—*spurge nettle, tread softly*. Clethra alnifolia—*sweet pepper bush*. Scabiosa atropurpurea—*purple scabish*. Lactuca elongata—*trumpet milkweed*. Rhus quercifolia—*oak-leaved sumac*. Chimaphila maculata—*prince's pine, bitter wintergreen*. Sanicula canadensis—*sanicle*. Atropa bella-donna—*deadly nightshade*. Mentha piperita—*peppermint*. Ludwigia alternifolia—*seed box*. Passiflora caerulea—*blue passion flower*.

(p. 67).

JULY–CONTINUED.

Sabbatia corymbosa—*American centaury*. Paspalum laeve, variety undulosum—*crab grass*. Paspalum digitaria—*finger-grass*. Sagittaria variabilis—*arrow-head*. Polygala nuttallii—*Nuttall's milk-wort*. Pteris aquilina, variety caudata—*Southern brake*. Woodwardia angustifolia—*narrow-leaved woodwardia*. Pycnanthemum aristatum—*spiny horse-mint*.

AUGUST.

Ludwigia palustris—*water purslane*. Cuscuta gronovii—*dodder, love vine*. Boehmeria cylindrica—*false nettle*. Linum diffusum—*wild yellow flax*. Mentha viridis—*spearmint*. Conium maculatum—*poison hemlock*. Hypericum corymbosum—*Saint John's-wort*. Mentha arvensis—*corn mint, field mint*. Linum usitatissimum—*flax*. Hypericum ellipticum—*orange-colored Saint John's-wort*. Desmodium nudiflorum—*naked bush trefoil*. Galactia mollis—soft galactia. Apocynum androsaemifolium, variety incanum—*dog's bane*. Nepeta cataria—*catnep*. Lappa major—*burdock*. Humulus lupulus—*hop*. Tanacetum vulgare—*tansy*. Saponaria officinalis—*bouncing bet*. Lycium barbarum—*matrimony vine*. Verbena urticifolia—*nettle leaved vervain*. Polypremum procumbens—*branched polypremum*. Agrostis scabra, variety perennans—*perennial bent grass*. Mollugo verticillata—*carpet-weed*. Ascyrum stans—*Saint Peter's-wort*. Pluchea bifrons—*marsh flea-bane*. Milkania scandens—*climbing boneset*. Salicornia herbacea—*saltwort, sea samphire*. Euphorbia maculata—*spotted-leaved spurge*. Phaseolus helvolus—*slender wild bean vine*. Ipomoea lacunosa—*purple bind-weed*. Ipomoea pandurata—*wild potato, man-of-the-earth*. Melothria pendula—*creeping cucumber*. Hypericum sarothra—*slender Saint John's-wort*. Apios tuberosa—*ground nut*. Cacalia atriplicifolia—*orache caraway, tassel flower*. Cassia chamaecrista—*sensitive pea*. Mimulus alatus—*winged monkey flower*. Lycopus europaeus—*water hoarhound*. Amaranthus retroflexus—*pig-weed, rough amaranth*. Pyrrhopappus carolinianus—*false dandelion*. Abutilon avicenna—*Indian mallow*. Asclepias verticillata—*whorled asclepias, dwarf milk-weed*. Tillandsia usneoides—*black moss, Spanish moss*. Galactia glabella—*milky-way plant*. Agrimonia eupatoria—*agrimony*. Elodes virginica—*marsh Saint John's-wort*. Oenothera rhombipetala—*evening primrose*. Eupatorium sessilifolium—*upland boneset*. Penthorum sedoides—*Virginia stonecrop*. Eupatorium maculatum—*purple boneset*. Elephantopus tomentosus—*elephant's foot*. Gnaphalium polycephalum—*sweet-scented life-everlasting*. Lobelia inflata—*Indian tobacco*. Lobelia puberula—*hairy lobelia*. Chenopodium anthelminticum—*worm-seed*.

(p. 79)

SEPTEMBER.

Ipomoea commutata—*false bindweed*. Eupatorium hyssopifolium—*hyssop-leaved boneset*. Vernonia noveboracencis—*ironweed*. Solidago canadensis, variety procera—*Canadian golden-rod*. Xanthium strumarium—*clotweed, clot-bur*. Ambrosia artemisifolia—*bitterweed, hogweed*.

APPENDIX 5.1. (CONTINUED)

Cirsium lanceolatum—*common thistle*. Brizopyrum spicatum—*spike grass*. Erianthus brevibarbis—*woolly beard grass*. Cucumis melo—*musk melon*. Diodia virginiana—*Virginian diodia*. Polygonum acre—*water pepper, biting knotweed*. Polygonum pennsylvanicum—*knee knotweed*. Eupatorium serotinum—*boneset*. Kosteletzkya virginica—*sweat weed*. Quamoclit coccinea—*Jasmine bindweed*. Pluchea camphorata—*ploughman's-wort*. Conoclinium coelestinum—*mist flower*. Polymnia uvedalia—*leaf cup*. Baccharis halimifolia—*sea groundsel tree*. Andropogon scoparius—*broom grass*. Amaranthus hybridus—*purple pigweed*. Portulaca oleracea—*purslane*. Eupatorium hyssopifolium, variety linearifolium—*narrow-leaved boneset*. Xyris fimbriata—*yellow-eyed grass*. Nabalus albus—*lion's foot, white lettuce*. Eupatorium purpureum—*purple boneset*. Eragrostis [hirsuta?]—*hairy meadow grass*. Eragrostis purshii—*slender pasture grass*. Antennaria margaritacea—*everlasting*. Helianthus rigidus—*rough-leaved sunflower*. Ascyrum stans—*St. Peter's-wort*. Aster flexuosus—*salt marsh aster*. Phaseolus diversifolius—*wild kidney bean*. Gerardia purpurea—*purple Gerardia*. Spartina alterniflora—*soft marsh grass*. Diodia teres—*buttonweed*. Euphorbia polygonifolia—*knotgrass spurge*. Cardiospermum halicacabum—*heart seed, balloon vine*. Bidens bipinnata—*Spanish needles*. Lobelia syphilitica—*blue cardinal flower*. Bidens chrysanthemoides—*bur marigold, daisy beggar ticks*. Ludwigia linifolia—*narrow-leaved seed box*. Chrysopsis mariana—*golden aster*. Eupatorium aromaticum—*sweet-scented boneset*. Lespedeza sessiflora—*bush clover*. Desmodium canescens—*tick trefoil*. Rhus [copalinum?]—*sumac*. Lobelia puberula—*puberulent lobelia*. Ludwigia alata—*winged seed box*. Atriplex hastata—*orach*. Galium hispidulum—*rough bedstraw*. Aster patens—*spreading aster*. Tripsacum dactyloides—*sesame grass*. Verbesina siegesbeckia—*crown beard*. Uniola paniculata—*sea-side oat*. Helianthus tuberosus—*Jersusalem artichoke*. Solidago tenuifolia—*narrow-leaved golden-rod*. Salicornia mucronata—*dwarf saltwort*. Monarda punctata—*horse-mint*. Lespedeza violacea—*violet bush clover*. Commelina virginica—*Virginian day flower*. Eclipta erecta—*erect eclipta*. Bidens frondosa—*beggar's tick*.

(p. 83).

SEPTEMBER—CONTINUED.

Cunila mariana—*ditany*. Impatiens fulva—*orange-colored touch-me-not*. Coreopsis auriculata—*broad-eared tick seed*. Eupatorium fistulosum—*trumpet weed*. Hibiscus moscheutos, variety flavescens—*yellow marsh hibiscus*. Hypericum mutilum—*dwarf St. John's-wort*.

OCTOBER.

Woodsia obtusa—*rock polypod*. Fagopyrum esculentum—*buckwheat*. Campanula divaricata—*bell-flower*. Lisianthus coeruleus—*false pennyroyal*. Dianthus barbatus—*burnet-pink*. Tragia urticifolia—*nettle-leaved tragia*. Verbesina siegesbeckia—*crown beard*. Gaura biennis—*biennial gaura*. Oenothera glauca—*evening primrose*. Gerardia tenuifolia—*slender leaved Gerardia*. Collinsonia punctata—*spotted collinsonia*. Nabalus altissimus, variety ovatus—*oval leaved drop flower*. Gentiana quinqueflora—*five-flowered gentian*. Liatris squarrosa—*blazing star*. Lespedeza repens—*creeping bush clover*. Calamintha repens—*calaminth*. Symphoricarpos vulgaris—*purple wolfberry*. Helenium autumnale—*autumnal sneezewort*. Helenium parviflorum—*small flowered sneezewort*. Heliopsis laevis—*ox-eye*. Gentiana andrewsii—*Andrew's gentian*. Chrysopsis graminifolia—*grass leaved chrysopsis*. Eupatorium perfoliatum—*thoroughwort, boneset*. Chelone glabra—*turtlehead, snakehead*. Mirabilis jalapa—*four o'clock, marvelworts*. Eupatorium foeniculaceum—*dog fennel*. Gentiana saponaria—*soap gentian*. Oplismenus crus-galli—*barn-yard grass*. Monotropa uniflora—*Indian pipe*. Solidago bicolor—*white golden rod*. Aster corymbosus—*corymbed aster*. Polygonum sagittatum—*scratch grass*.

W. H. Smith.

Chapter 6.

William Henry Smith (1846–1925)

Professor of Natural History (1876–1877)

Early Years and Education

Published biographical information on William Henry Smith is meager (Anonymous, 1903; Anonymous, 1925; Jenks, 1912). The most comprehensive account of his life is a handwritten biographical essay (probably by Smith himself) housed in the Bentley Historical Library at the University of Michigan (BHL). Born May 9, 1846, on a farm near Three Rivers, Michigan, William Henry Smith was the son of Benjamin and Mary Ann (Tomlinson) Smith. He attended the district school of his locality and continued his education at Three Rivers public schools. By his thirteenth birthday, he had lost both of his parents within a period of three years. Smith then worked variously on a farm, in a brickyard, and in the woods of northern Michigan as a lumberman. On February 10, 1864, he enlisted as a private in Company K of the 26th Michigan infantry, transferring to Company B on April 18, 1864 (Turner, 1905). He took part in Civil War campaigns from the Battle of the Wilderness to Appomattox, Virginia. When discharged on June 4, 1865, at Alexandria, Virginia, he stood five feet, six inches tall, was of light complexion, and had light hair (BHC). He returned to Three Rivers, where he resumed his high school education. There, superintendent W.A. Simpkins inspired and guided Smith in his professional pursuits.

In the fall of 1866, Smith entered the preparatory department of Kalamazoo College, a small school that the Northwest Territorial Legislature had chartered in 1834. Situated in scenic Kalamazoo, Michigan, on a plateau covered with native oaks, the institution in the mid-1860s was composed of two separate colleges, one for men and another for women. Of the two courses of instruction offered (classical and scientific), Smith chose the former. In addition to Greek and Latin, he was required to take several courses in mathematics as well as botany, physics, chemistry, and astronomy (KC). Smith's academic environment included a cabinet of natural history specimens to which new accessions were being received. He was active in college associations. He served as librarian of the Sherwood Rhetorical Society, which held regular meetings for discussions and rhetorical exercises. He also was a member of the Soldiers' Memorial Society, consisting of Kalamazoo College men who had served in the United States Army or Navy. At his graduation ceremonies in June 1871, Smith gave an oration on astronomer John Kepler (Anonymous, 1871) and received the B.A. degree.

About the fall of 1871, Smith accepted the professorship in Greek and German at Central University of Iowa (now called Central College) and remained there probably through

---

Frontispiece Credits: (above) William Henry Smith, ca.1871. (Courtesy of Kalamazoo College Archives). (below) Mason Hall, University of Michigan, 1871; in 1856, the North Wing housed the university library and museums. (#BL005277 Bentley Image Bank, Bentley Historical Library, University of Michigan, Ann Arbor).

the spring (or summer) of 1872. He then undertook graduate work at the University of Michigan, earning the M.A. in science in 1873. His classes included two geology courses, moral philosophy, international law, and zoology (UM, [1873]). Smith joined the staff of the University of Michigan Museum (Frontispiece of chapter 6, lower image) about the beginning of 1873. His title was variously given as Assistant or Custodian of the Museum. His work there involved mostly zoological specimens, although he collected and curated some geological and mineralogical specimens (Smith, 1874). He was assigned to work six hours each day but actually averaged eight hours per day. Mark Walrod Harrington (1874)—then an assistant professor of geology, zoology, and botany and later a well-known meteorologist and astronomer—noted that Smith's "industry, efficiency, and trustworthiness can not be too highly commended." Harrington (Fig. 6.1) became a valued mentor to Smith. Among his duties, Smith (1874) worked on a variety of tasks in the Zoological Department, including identifying spiders, shells, and birds' eggs; rearranging the Zoological Hall; assisting Professor Harrington in straightening out the Zoological Working Room and in working on the classroom; and reviewing a course of study with five seniors by studying and dissecting specimens. In the Geological Department, he labeled numerous mineralogical, lithological, and paleontological specimens, cleared up a room on the fourth floor by transferring several thousand lithological specimens to the attic, helped Professor Harrington by moving duplicate fossils to the Zoological Working Room, into which he also transferred Joseph B. Steere's collection of fossils, and assisted Eugene Woldemar Hilgard (professor of mineralogy, geology, zoology, and botany) with preparations for a class and examinations. Smith was also a donor of specimens, not only of amphibians, but also of helices, bones of a horse and ox, snakes' eggs, and rocks. In the summer of 1874, he and colleague John Foster Eastwood relabeled and arranged specimens in the museum (Pettee, 1875). Later that summer, Smith became a teacher of the sciences in the high school of Omaha, Nebraska. There, he continued collecting specimens of fossils, some of which he later donated to the University of Michigan Museum (Pettee, 1876).

In 1875, Smith returned to Ann Arbor, where he studied under the stimulating tutelage of Mark W. Harrington, William Henry Pettee (professor of mining engineering), and Albert Benjamin Prescott (professor of organic and applied chemistry). He also associated with Joseph Beale Steere (zoologist, paleontologist, and pioneer student of animal geography). Smith's doctoral research on the tailed amphibians is among the earliest significant herpetological investigations in Michigan. His studies culminated in 1876 with the completion of his dissertation[67] "The Zoology of Anoura and Caecilia" (Smith, 1876e) and the conferral of the Ph.D. Smith (1877k) dedicated the book, published in 1877, under the title *The Tailed Amphibians, Including the Caecilians* (Fig. 6.2) to his respected professor, Mark Harrington. He wrote: "To Prof. M.W. Harrington, my esteemed teacher and friend, to whose care, zeal and powers of observation I owe much of what I may be in life, in grateful recognition of the many favors received, this little volume is respectfully inscribed." His work included "all known species of caecilians [and presented] one new form named *Caecilia gelatinosa*, [which was] not identifiable" (Taylor, 1968). Smith and classmate Victor Clarence Vaughan were the first students at Michigan to receive the Doctor of Philosophy degree (Sensemann, 1923). With his doctorate in hand, Smith sought a professorship.

[67] I was unable to find a copy of Smith's original dissertation of 1876 at the University of Michigan—Ann Arbor campus.

Fig. 6.1. Mark Walrod Harrington, 1876. William H. Smith's mentor and friend. Caption reads: "Instructor-Zoology and Botany."

(From *Class Album 1876*. Courtesy of the Bentley Historical Library, University of Michigan, Ann Arbor).

An impressive and influential aspect of Smith's academic environment at Ann Arbor was clearly his association with the University Museum. The enterprise provided him with skills in curating zoological and mineralogical specimens and broadened his knowledge of the systematics of animals and the determination of minerals. Smith's academic mentor and friend Mark Harrington took special interest in curating the botanical collection. It is possible to speculate that through Smith's curatorial position and his association with Harrington, he gained a general knowledge about collecting and preserving plants. The 1872–1873 UM catalog, also called calendar (UM-C, no. 1), describes the natural history collections in the museum. The botanical collection consisted of 35,000 specimens, including a collection of 9,000 specimens from Michigan, the Douglass Houghton Herbarium of 1,800 species, the Sager Herbarium of 1,200 species and 5,000 specimens, the Dr. Ames Herbarium of 7,000 specimens and 10,500 duplicates, a collection of wood samples, and a collection of seeds of native and cultivated plants. By 1875, the botanical collection held nearly 70,000 specimens (UM-C, no. 2). Smith would later teach botany at UNC and would require his students to make a herbarium of local plants, and he later noted his experience in preserving local plants (Smith, 1876a).

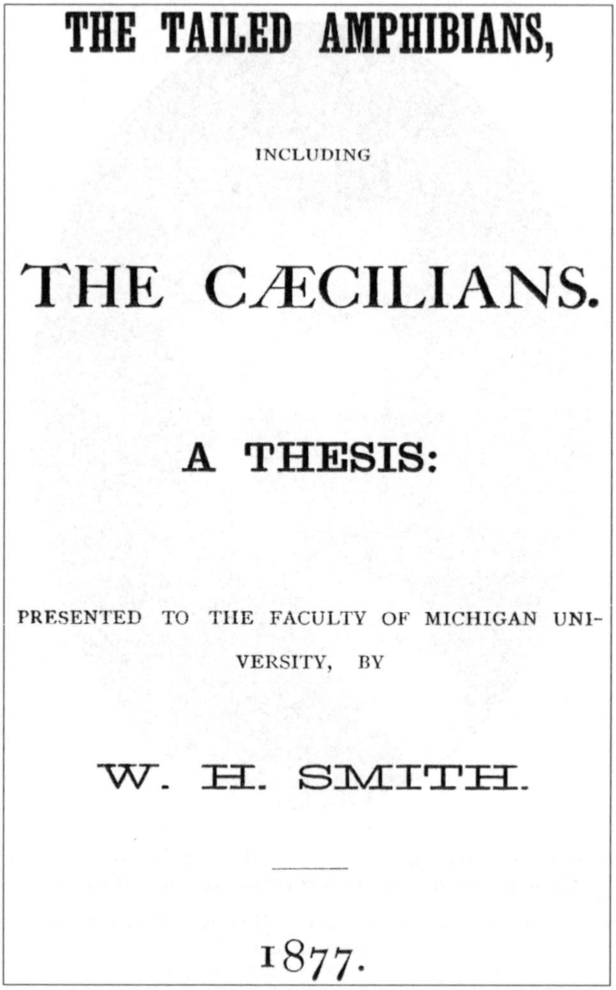

THE TAILED AMPHIBIANS,

INCLUDING

THE CÆCILIANS.

A THESIS:

PRESENTED TO THE FACULTY OF MICHIGAN UNI-
VERSITY, BY

W. H. SMITH.

1877.

Fig. 6.2. Title page of *The Tailed Amphibians, including the Caecilians. A Thesis.* By William H. Smith (1877). (Published version of the author's dissertation, originally titled *The Zoology of Anoura and Caecilia* [1876]).

(Courtesy of North Carolina Collection, Wilson Special Collections Library, University of North Carolina, Chapel Hill).

Another scholarly influence that incited and fostered Smith's interests in science was the Ann Arbor Scientific Association, which was established in April 1875. The organization provided a forum for public lectures on scientific and medical topics as well as an opportunity for networking among local scholars and citizens. The presentations covered a variety of subjects, including elevated land features in the Arctic region, freshwater shells of Ann Arbor and vicinity, aromatic organic compounds and their significance in plants, the nature of life, red snow, physical therapy of hearing, Iceland, the connection between organic germs and disease, and the history of the theory of spontaneous generation. During the first year of the association, members showed an interest in documenting the local flora and fauna. Three committees, each charged to prepare a list of organisms within a four-mile radius of

Ann Arbor, focused on (1) plants, (2) freshwater and terrestrial shells, and (3) vertebrates, excluding fishes. Smith was elected a member at the association's meeting of February 5, 1876 (AASA).

Teaching at UNC

In 1876, a professorship in natural history was established at UNC to support studies in agriculture and the mechanic arts (Battle, 1907–1912, 2: 122). In that summer, President Battle and State Geologist and Professor of Geology Washington Caruthers Kerr visited several institutions to gain a perspective on the university's fulfillment of the provisions of the Land Scrip regarding agricultural and mechanical studies. The minutes of the July 21, 1876, meeting of the UNC Board of Trustees (BOT 16: 25–26) note that "the President is authorized to visit the Virginia Agricultural and Mechanical College at Blacksburg with the view to inform himself of the organization of that institution, and be instructed to make special inquiries for one qualified to impart instruction particularly in Natural History and be instructed to report the names of such persons as may be so qualified to the Committee." Battle's efforts to provide an education for farmers were developing but still lacked a defining direction. One faculty member, Professor Charles Phillips (1876), wondered whether the focus of the program would be preparatory or professional.

The university wasted little time in filling the new professorship. Among the men being considered for the position were Eugene H. Beck of Georgia and William H. Smith of Michigan. Two other names (Charles Youdan Lacy and William Rane Lazenby) had come to the attention of UNC Professor George T. Winston from a letter sent by Cornell University President Andrew Dickson White (1876). In responding to a letter from Winston, Lazenby (1876) mentioned that although he had a good position at Cornell and fairly good prospects for promotion, he did not like the cold weather in Ithaca, which he feared might cause him to get rheumatism. At a special meeting of the faculty convened on July 17, 1876, it was resolved that President Battle should recommend William Henry Smith for the professorship (UNC Faculty 6: 399). At the August 14, 1876 meeting of the trustees, President Battle presented testimonials on behalf of Smith, who was highly recommended by his Ann Arbor professors. Mark W. Harrington stated that Smith was "a diligent, conscientious & enthusiastic student & a Christian gentleman," and Eugene W. Hilgard spoke highly of Smith's "scientific attainments & gentlemanly qualities" (Smith, 1876b). An anonymous (1876b) reporter for the Raleigh *Daily News* noted that Prof. Smith was well qualified for the position and "young, enthusiastic, and a hard worker." He had developed a good reputation in the field of natural history. On August 14, 1876, the trustees duly elected Smith as Professor of Natural History in the Department of Agriculture with an annual stipend of $1,200 (BOT 16: 28).

In writing to President Battle to apply for the position, Smith (1876b) had noted that his background included farm experience and reading several works such as Mechi's *How to Farm Profitably* (1864), Pendleton's *Scientific Agriculture* (1875), and Johnson's *How Crops Grow* (1868). He concluded, however, by mentioning that his "work in the main ha[d] been pure natural history." Although he had practical experience working in the University of Michigan Museum, identifying and labeling animal and mineralogical specimens, Smith's experience in botany seems to have been weak. Smith (1876c) noted in a subsequent letter that he was encouraged in accepting the faculty position by the kind letters he had received from Battle and UNC professors Charles Phillips (professor of mathematics) and Alexander F. Redd (professor of physics and chemistry). He began his journey to Chapel Hill on August 23, 1876, and anticipated arriving on Saturday, September 2, two days before he began teaching. There, he joined a faculty of eleven other men. Soon after his arrival, word was spreading in the village that he was "great upon bugs, butterflies & caterpillars of all sorts" (Spencer, 1876a).

When Smith arrived in Chapel Hill, the UNC campus had changed little physically since the university reopened in 1875. Cornelia Spencer (1876b) thought that the village still had not recovered socially from the "break-up" of the university. She discerned that of the new families who moved into Chapel Hill, some had not assimilated into its social fabric and they seemed less neighborly.

The first half of 1877 heralded two noteworthy events that had a significant impact on the university. First, Zebulon B. Vance took office as governor in January 1877, indicating that rule of the state by carpetbaggers had ended. Vance's efforts, as well as those of Kemp P. Battle and North Carolina lieutenant governor Thomas J. Jarvis, were aimed at improving the state by promoting scientific agriculture and basic agricultural education, expanding the university, and assessing the state's natural resources (Young, 1934).

The other event concerned the advancement of agricultural research at the university in early 1877, which owed much to the efforts of President Battle. The North Carolina Department of Agriculture established the Agricultural Experiment and Fertilizer Control Station in early 1877. The station was directed to analyze fertilizers and their products and to conduct experiments on the nutrition and growth of plants. It was the second station of its kind in the country, the first being in Connecticut. Several sources provide historical background on the station and its accomplishments (Battle, 1907–1912, 2: 136–139; Dabney, 1882; Ledoux, 1879, 1880; True, 1937; Young, 1934). To place the role of the station in perspective with the advancement of agricultural research at UNC at this time, a summary of the venture is hereby given.

Albert Reid Ledoux, who was appointed the first director, began his duties on April 19, 1877. He had graduated from the Columbia School of Mines in 1873 and subsequently studied chemistry in Europe, where he received the Ph.D. from the University of Göttingen in 1875. In the basement of Smith Hall, which housed the library on its main floor, the station at first shared research facilities with the Laboratory of Chemistry but later occupied four rooms (a commodious lab room, a balance room, an assay room, and a darkroom for using the polariscope,[68] which had been purchased in Europe).

During three years of work, Ledoux and his assistant chemists performed 880 analyses. In addition to making chemical determinations, the station reached out to the farmers through publications, which included a series of bulletins on the "Analysis and Valuations of Fertilizers," sugar beets, and composting, as well as annual reports. Articles in local

[68] The polariscope, which measured the amount of sugar contained in beets and other plants, was imported expressly for the station (Anonymous, 1878b).

newspapers conveyed the results of Ledoux's analyses and the text of circulars that he issued. Although his responsibilities did not include teaching, Ledoux gave special lectures to inform students and the public about the work of the station. For example, on March 29, 1879, he delivered a talk before UNC students on "What we are doing at the Experiment Station." A newspaper reporter (P., 1879) for the *Charlotte Democrat* summarized the lecture in an article to provide practical information for farmers. By 1880, Ledoux and his assistants offered an array of services (Fig. 6.3).

The overwhelming success of the agricultural research station was evident from the opinions and commentary of the state's citizens and the press. Comments included those of Cornelia P. Spencer, who praised Ledoux—a friend of hers—noting that he was "a great acquisition to the working force of the State" (Love, 1921; Holt and Cheek, 1880). In a letter to the editor of the *Observer* dated July 9, 1877, John Smith (1877) of Fayetteville, North Carolina, claimed that the chemical analyses of fertilizers had "already saved the farmers of North Carolina *a million of dollars*." William B. Phillips (1880), a chemical assistant at the station who was the son of UNC professor Charles Phillips and Laura Battle Phillips, wrote that "Dr. L. has ma[d]e himself very popular here in N.C. & has given quite an impetus to scientific farming."

Ledoux tendered his resignation to the North Carolina Board of Agriculture to take effect April 17, 1880. The board was so reluctant to let him leave that it offered him $1,000 more in salary, which would have made his salary the second highest in the state (Phillips, W.B., 1880). When Ledoux departed, Charles William Dabney, Jr. then took charge. Like Ledoux, he had also received a doctorate in chemistry from the University of Göttingen. One newspaper reporter (Anonymous, 1880) considered Dabney "the man for the place, active, zealous, an accomplished Chemist, a good man of business, courteous and affable." Dabney continued the experimental lab work that his predecessor had established.

Soon after Dabney assumed the directorship, the North Carolina Board of Agriculture requested a building for its work, including accommodations for the experimental station. When the legislature approved the request in March 1881, the station was moved to Raleigh and resumed full operation by the end of that year. (According to UNC student Stephen B. Weeks (1886), "the laboratory under Smith Hall ha[d] been deserted" by the fall of 1886 at latest. The gas pipes connecting that lab and the physics building had even been removed.) In 1885, the Board of Agriculture approved a resolution to develop an experimental farm[69] in connection with the station (McGehee, 1887). George Dixon's efforts to promote agricultural studies with an experimental farm were at last realized, although not on the Chapel Hill campus. When the North Carolina College of Agriculture and Mechanic Arts (now North Carolina State University) opened its doors in 1889, the station was relocated there.

Whether there was any substantive interaction between the professor of natural history and the station and its director is conjectural, as no historical documentation has been found to support that possibility. Nonetheless, Smith taught courses in natural history with particular emphasis on agriculture. Just as the station's work benefited the state's farmers so Smith's teaching enhanced agriculture.

[69] Ledoux had proposed the establishment of an experimental farm at UNC during his earlier years at UNC (Young, 1934).

THE NORTH CAROLINA

State Experiment Station.

THE ADVANTAGES IT OFFERS TO

FARMERS AND OTHERS.

1. Free analyses of all Chemicals used in Composting, provided the sender secures the samples according to printed instructions furnished by the Station, and prepays the express charges.

2. Free analyses of all Fertilizers legally on sale in the State, provided the samples are secured by an authorized agent of the Department of Agriculture or after correspondence with the Director.

3. Free analyses of all Well Waters, articles of Food and Drugs, Disinfectants, &c., when the analyses are desired for Sanitary puposes, the permission of the Secretary of the State Board of Health is secured, and the express charges on the sample are prepaid.

4. Free analyses and examination in all cases of probable Poisoning when the suspected article is sent according to printed instruction, and by order of Coroners and County Superintendents of Health.

5. Free analyses of Marls when desired by farmers.

6. Free analyses of Mineral Waters when sent with the permission of tbe State Geologist. and the express charges are prepaid.

7. Free identification and qualitative analyses of all Minerals and Ores. For a complete or quantitative analysis the permission of the State Geologist must be obtained.

8. Free tests of the germinating or sprouting capacity of all Seeds of grains, grasses, flowers or vegetables, with a report as to their purity, and identification of any weed or other foreign seeds present. Seed samples are best sent by mail.

9. Free identification of useful or injurious Grasses.

10. Free identification of Insects injurious to Vegetation, and reports upon the best means of destroying them.

11. Correspondence will be promptly attended to upon all subjects directly or indirectly related to Agriculture.

12. Printed reports will be mailed free of charge upon request upon the following subjects: Directions for composting; formulas for different soils and crops; analyses of chemicals and fertilizers legally on sale in North Carolina; directions for utilizing bones; for making vinegar; for growing sugar-beet; upon drinking waters; the value of pine straw; the history, use and value of the cow-pea; the composition of marls and soils; the extraction of sassafras oil; the history and use of artificial manures, &c., &c.

The Experiment Station having been fully equipped by order of the Board of Agriculture, for the prosecution of the various branches of work above mentioned, our farmers are urgently requested to avail themselves of these advantages. Address all letters to

The N. C. Agricultural Experiment Station,

CHAPEL HILL, N. C.

Fig. 6.3. Outside back cover of *What Is Said about the North Carolina Agricultural Experiment Station*. [By Thomas M. Holt and Wm. H. Cheek] (1880).

(Courtesy of North Carolina Collection, Wilson Special Collections Library, University of North Carolina, Chapel Hill).

The UNC catalog of 1876–1877 (pp. 33–35) outlined the new organization of instruction. The College of Natural History was composed of four schools: (1) Botany, (2) Zoology, (3) Veterinary Science, and (4) Mineralogy and Geology. It provided courses "with special reference to the agricultural interests" of North Carolina. Its intentions also were to prepare students for further scientific education or for entrance into medical school. Other objectives of the college included "the use of plants in medicine and agriculture." Instruction consisted of textbooks, lectures, and laboratory work. The catalog describes the curriculum of the School of Botany, which was Smith's responsibility.

> The work in this science, obligatory upon all the students for the first session, is as follows: Morphology and Histology of plants, including structure and arrangement of their parts; Physiology or the laws of plant life, growth and reproduction; the principles of classification, and the use of the prominent species in each order. In addition to this, each student is required to collect, identify, preserve, and familiarize himself with the habits and uses of at least seventy-five species of plants.
>
> The work of the second session will be a continuation of Structural, Physiological and Systematic Botany. In this session special attention will be given to the analysis of the grasses and cryptogams and also to the various applications of the science.
>
> Text-books and works of reference: Gray's *Lessons and Manual* [1871a], Gray's *Structural and Systematic Botany* [1871b], Harrington's *Plant Analysis* [1876], Johnson's *How Crops Grow* [1868], Johnson's *How Crops Feed* [1870], Darlington's *American Weeds and Useful Plants* [1865], Sachs' *Text-book of Botany* [1875], Sachs' *Experimental-Physiologie der Pflanzen* [1865], Cooke's *Hand-book of British Fungi* [1871], Chapman's *Flora of the Southern States* [1860], Porcher's *Resources of the Southern Fields and Forests* [1863].[70]

In scientific instruction, Smith thought that the students should thoroughly examine specimens themselves. He believed that "book knowledge, however good, when compared to an examination of the object is like the dry bones in the prophet's vision" (Smith, 1877k). On September 4, 1876, Smith's first day on the job, a special faculty meeting was convened to consider courses of study that would shape a busy teaching load for him. It was resolved that during the second session of the first year of the classical course, the following were required: three recitations of Greek per week, four of mathematics, four of Latin, three of botany, and one of rhetoric. Zoology would be taught five times per week during the first year of the scientific course and botany three times per week during the second session of the first year and five times per week during the first session of the second year (UNC Faculty 6: [404]). On several occasions, students were given permission to drop or substitute courses for botany. In meeting the needs of the agriculturist, the university's courses in chemistry and botany touched on the theoretical and practical. Instruction in chemistry included studies on the composition and analysis of soils and manures as well as laboratory work; botany included studies on plants grown for the use of man, their mode of growth, and their diseases (Battle, 1876b).

The financial austerity of the times was evident. In preparing to give his recitations in a room in South Building beginning in January 1877, Smith (1877a) reminded President Battle that the room needed a stove "of pretty large size & about 26 ft. of pipe." Instead of spending

[70] Citations for these textbooks are included in the References Cited section of this chapter.

funds for the purchase of new piping, Smith proposed salvaging unused sheet iron pipe as long as it would fit the stove acquired.

The University Museum

Smith played an important role in consolidating the University Museum collections and overseeing the renovation of the third-floor room in Old East that was formerly used by the Philanthropic Society as a library (Anonymous, [1876]d) and would now serve as the quarters for the museum. Old East at this time also housed dormitories and lecture rooms (Southgate, 1877). In November 1876, Battle reported that the "handsome [and well-lighted] hall ha[d] been set apart for the Museum and [they] earnestly request[ed] the friends of the institution . . . in filling it with samples showing the wonderful resources of the State. [They] wanted animals, plants, minerals, ores, soils, and whatever else, natural or artificial, which [would] illustrate these resources and be useful for teaching our students" (Anonymous, 1876a; Anonymous, [1876]c).

In arranging the collections, Smith took specimens from the drawers of the Vienna cabinet of nearly 2,000 minerals to display on the shelves, as well as specimens of minerals and ores that had been collected by Dr. Ebenezer Emmons and transferred to the university during the Civil War (Anonymous, [1876]d). As early as mid-December 1876, professor George T. Winston (1876) favorably commented on Smith's efforts to develop the museum, stating: "Smith is making the Museum 'a thing of beauty.'" Several townspeople (Miss June, Miss Lucy, and Alex and [Jim?] Phillips) assisted Smith in organizing the collections. During the Christmas vacation of 1876, Smith (1876d) intended to label the mineralogical specimens of the Vienna collection, for which he requested a supply of printed labels. Smith (1877b) notified Battle in 1877 that work on renovating the room would begin at the earliest practicable time. By mid-March 1877, Kinsey Bowles had begun plastering and whitewashing the room (Smith, 1877c). Prof. Smith had also succeeded in organizing and labeling the Vienna cabinet.

The museum housed a collection of 1,000 kinds of seeds, which W.C. Kerr had acquired from exhibitors at the Centennial Exposition[71] held in Philadelphia in 1876 (UNC-C; Kerr, 1876). There were "120 varieties of grain and seed from Egypt: wheat, rye, oats, and other seeds; about 100 different kinds from Australia. Also various sorts from Liberia, Algier, Cape of Good Hope, Spain, Norway, Sweden, France, Italy, Brazil. From the last mentioned country are, besides tropical products peculiar to that country, products of palm trees, silk, &c. Then, likewise, the famous Black Sea wheat and the macaroni wheat of Italy, cotton from Italy and Egypt and Russia, seeds and medical plants and fabrics from China" (Anonymous, [1876]d). (Kerr suggested that the seeds should be distributed throughout the state for experimentation.) Each botany student was required to grow plants from these foreign seeds, studying their germination, growth, ripening, and other stages of development. They also analyzed (*i.e.*, identified) forty plants and contributed four specimens of wood to the museum (Anonymous, [1876]d).

[71] The Centennial Exhibition, held in Philadelphia from May 10 to November 10, 1876, was a celebration of America's one hundredth anniversary of its nationhood. Exhibitions of the United States as well as those of fifty other countries featured specimens of their products in the following departments: (1) mining and metallurgy, (2) manufactures, (3) education and science, (4) art, (5) machinery, (6) agriculture, and (7) horticulture. The successful venture attracted nearly ten million people (Ingram, 1876).

In developing the University Museum, Battle, Kerr, and Smith made inquiries of scientists, including those at the Smithsonian Institution. In October 1876, Battle had written to Joseph Henry, Secretary of the Smithsonian, about acquiring duplicates[72] from the centennial exposition of 1876 (Battle, 1876a). Henry (1876) replied that the specimens would be sent when Congress allocated the funds necessary for that purpose. However, it seems unlikely that UNC ever received the specimens. According to Spencer F. Baird ([1877?]), assistant secretary of the Smithsonian, Congress did not appropriate the funds for assembling sets of specimens and distributing them. Kerr, who also served as a Centennial judge (Chamberlain, 1926), took an active role in acquiring other specimens for the museum. He had secured some mineral specimens while attending the exhibition. Among the items acquired were the following: "antimony, nickel, copper and tin ores, &c. from Australia; Hone stone, lead, iron, quartz, crystals &c., from Arkansas; lead ore, &c. from Wisconsin and Iowa; coal, iron ore, phosphate of lime, &c., from Canada; iron ore, cannel coal, &c. from Kentucky; copper and iron ore, coal, kaoline, &c., from Tennessee; iron and ores, &c., marble, &c., from Alabama and Michigan; iron, manganese, &c., ore from Pennsylvania" (Anonymous, [1876]d). Despite his success in obtaining specimens from the exposition, Kerr was disappointed that he had not obtained a greater number of samples. Smith also maintained contacts with the Smithsonian Institution. He sought assistance from Joseph Henry in determining the identity of a leaf with a variety of nut galls growing on it (Henry, 1877).

President Battle pointed out that people should not be reluctant to send too many specimens, because surplus objects could be used in exchanges with other museums (Anonymous, 1876a). North Carolina, he noted, had abundant natural resources but needed to develop them and educate the public about them. Concerning the latter, a newspaper reporter wrote: "We must educate our people as well from the great book of nature as from the text books of schools and colleges, and infuse into them the new blood of push and energy" (Anonymous, [1876]c). The museum, therefore, would serve for instructing not only UNC students but also the people of the state when they visited the campus. The faculty was proud of the museum, which was garnering attention as being attractive and instructive. It recommended that President Battle be authorized to hire a student for $5.00 to manage the museum and entertain visitors during the June 1877 commencement (UNC Faculty 6: 423).

As an incentive to encourage citizens to donate specimens of birds, insects, plants, reptiles, soils, ores, and other objects, Smith (1876a) wrote a fifteen-page "Circular to the Friends of the University of North Carolina in Regard to Specimens of Natural History" (Fig. 6.4), which was issued on October 30, 1876. He provided directions for skinning birds, and collecting and preserving plants, animals, and minerals. Regarding the preservation of plants, particularly those used as food by insects, he instructed that they "should be prepared by placing them between any porous paper of several layers in thickness enclosed either by two wire gauzes hung in the wind or by a board above and below with a weight." He continued:

> In my own work I use common newspaper, first placing upon the floor three or four papers then a plant, then more papers, then another plant, and thus alternating. Finally a board placed over the top with a stone or some books upon it completes the apparatus. The paper needs to be changed for ordinary plants once a day for about a week after

[72] The Smithsonian Institution received sixty freight car loads of materials, including natural history specimens, when the Philadelphia Centennial Exposition of 1876 closed (Rothenberg, 2002).

> which they can be packed between layers of paper and forwarded without farther attention. The wire gauze hung in the wind is better as it obviates the necessity for so frequent a change of papers, in this case if the paper used is very porous, as blotting paper, change may be altogether dispensed with. In case the plant is too large to be pressed entire it may be broken into segments and these preserved together, or if its size will not admit of that then press the flower, leaf, and part of its root. Fruits or products of vegetables rarely require any preparation, but in the case of small seeds they can be placed in vials.

Smith also reminded the reader that each specimen should be labeled, including common name (if known), locality, and date of collection.

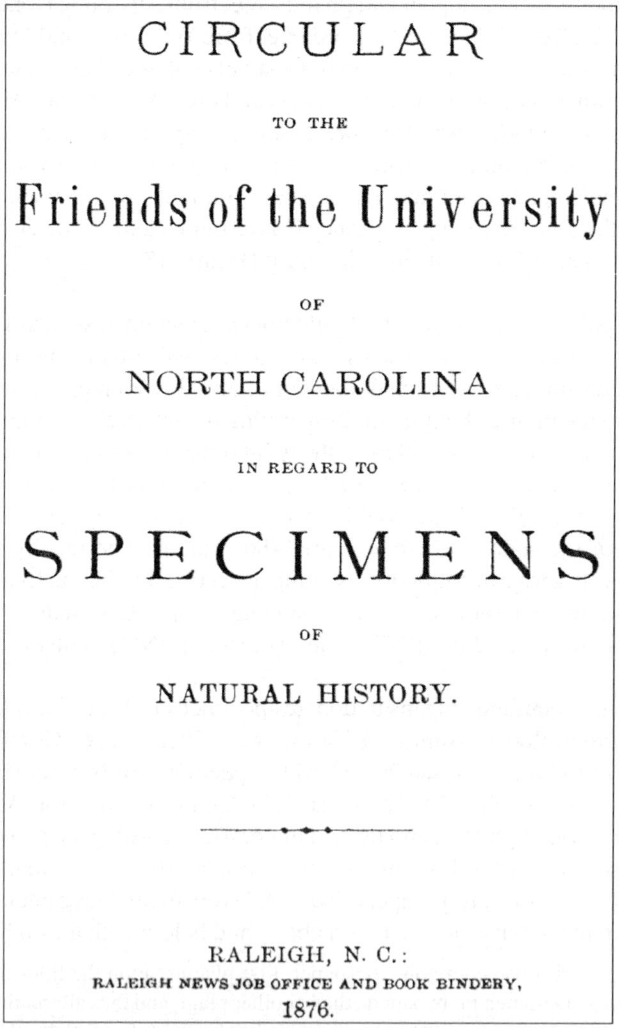

CIRCULAR

TO THE

Friends of the University

OF

NORTH CAROLINA

IN REGARD TO

SPECIMENS

OF

NATURAL HISTORY.

RALEIGH, N. C.:

RALEIGH NEWS JOB OFFICE AND BOOK BINDERY,

1876.

Fig. 6.4. Title page of *Circular to the Friends of the University of North Carolina in Regard to Specimens of Natural History*. [By William H. Smith] (1876).

(Courtesy of the North Carolina Collection, Wilson Special Collections Library, University of North Carolina, Chapel Hill).

A month after the *Circular* was published, notices on its issuance appeared in area newspapers such as the *Raleigh Sentinel*, the *Raleigh News*, and the *Hillsborough Recorder* (Fig. 6.5). Readers could obtain a copy of the publication from Smith. Mr. DeLorme McLeod (1876), a surgeon and physician in Raleigh, North Carolina, requested a copy. In fact, he stated that if the university anticipated the employment of a collector, he would be very interested in filling that role.

UNIVERSITY OF NORTH CAROLINA.

CHAPEL HILL, N· C; Nov· 29, 1876.

The following directions for skinning birds, collecting and preserving insects, plants, reptiles, minerals, &c; have been prepared by W. H. Smith, Professor of Natural History in the University, and are printed for the use of those who are willing to aid in equipping the Museum of the University. A handsome hall has been set apart for the Museum, and we earnestly request the friends of the Institution to co-opperate with us in filling it with samples showing the wonderful resources of the State. We want animals, plants, minerals, ores, soils, and whatever else, natural or artificial, which will illustrate those resources and be useful for teaching our students.

Our friends should not be afraid or sendidg us too many things. Surplus specimens can be exchanged with other museums. We have already received offers to furnish us with the products of the West, South America and the Old World in exchange for those of North Carolina, and we feel sure that at no distant day we will have a Museum which the people of the State will visit with pride and from which they will derive substantial benefit.

KEMP P. BATTLE, *President.*

The pamphlet can be had on application to Prof. W· H. Emitt Chapel Hill, N. C.

Fig. 6.5. Notice on the *Circular to the Friends of the University of North Carolina...* by W.H. Smith (1876).

(*Hillsborough Recorder* [Hillsborough, NC], new series 5(1): [3], Wednesday, November 29, 1876).

Smith's efforts to acquire natural history specimens were very successful, as confirmed by a list of donations enumerated in the UNC catalog of 1876–1877 (pp. 44–45). In addition to the seed collection, several botanical accessions were noted: silicified wood from the bottom of the Tar River from Dr. Newsom Jones Pittman, Tarboro; a night-blooming cereus from James Haywood Southgate; and wood samples from Isaac Montrose Taylor, Edward Benson Engelhard, Bennett Franklin Hester, David Bell, and Kemp Plummer Battle, Jr. Among specimens donated in 1877 were fifty-three samples of rocks and minerals collected in the Ann Arbor, Michigan area and presented by Prof. Smith.

Leaving UNC

By March 1877, Smith was confronted with a personal problem. There is no reference to the incident in such primary historical sources as the meeting minutes of the UNC Faculty or the Board of Trustees. The earliest documentation alluding to the problem appears in a letter that Adolphus W. Mangum (1877), Professor of Moral Philosophy, History, and English Literature, wrote to Battle on March 27, 1877. Mangum stated:

> I have not seen Prof. S. today, since you left. I learn he is much incensed against me. He does me great injustice. Some one has given him false information as to my connection with his case. I sincerely pity him; & if he does not give me greater cause than I yet have, I shall be kindly quiet. I have acted the best I could, for all concerned; & as I was not at liberty to give my informant's name, & further, had only such information as engendered suspicion, I hold that I did not do wrong in delaying to inform him of what I had heard. I feel gloomy over the case, but do not think it will do us material damage before the public. . . . I have heard today that Prof. S. intends to leave tomorrow.

In a letter to President Battle written in Raleigh, North Carolina, and dated March 30, 1877, Smith (1877d) wrote: "Believing that my farther connection with the University would not be conducive to the best interests of the institution, I hereby through you, tender the trustees my resignation of the Professorship of Natural History with which you have honored me." The UNC Trustees, however, had accepted his resignation at their meeting of March 29, 1877 (BOT 16: 41). Nearly three weeks later, Smith was in Detroit, Michigan, from where he wrote to Battle. His letter alluded to the problem that had contributed to his resignation. Smith (1877e) confided in Battle, saying, "After careful thinking over this matter I am satisfied that this attack on me would not have been made had I been a southern man. However, I think there were other things combined with that. I wish that you without stirring the matter up could ascertain who the man was to whom [Professor] Redd alluded & what he claimed to know."

The best clue to the nature of the incident that led to Smith's resignation is found in President Battle's book of memories (Battle, W.J., 1945). In a chapter titled "Unpleasant University Duties," he briefly discussed several experiences that tested his administrative prowess. He wrote: "A professor from Michigan was reported to have been too intimate with a . . . negro cook." When Battle confronted the professor with this story, the professor did not deny it. In protecting the professor's identity, Battle refers to him as "Mr. Blank" but alludes to his alma mater as being in Ann Arbor (*i.e.*, the University of Michigan). Of the faculty who taught during Battle's presidency, the only known professor who had received a degree from the University of Michigan was William H. Smith.

Smith's abrupt departure from Chapel Hill in 1877 raised questions then and throughout the first half of the twentieth century. In a letter to President Battle, lawyer, politician, and businessman Walter L. Steele (1877) of Rockingham, North Carolina, mentioned having read about Smith's resignation in "a mere local newspaper." He wondered what had happened to Smith. Sixty-three years after Smith's departure from Chapel Hill, UNC Professor of Zoology Robert Ervin Coker (1948) was likewise curious about Smith's fate, noting there was "some mystery regarding his brief career in the University."

From the evidence that Kemp Battle provided and statements made by Smith, it seems likely that Smith abandoned his professorship because he had (or was accused of having) an inappropriate interracial relationship with a Black cook. This appears plausible as Smith (1877e) himself stated that the accusation against him would have been different had he been a southerner. Nonetheless, another possible reason for Smith's departure may have been an accusation of his having an inappropriate cross-class, premarital sexual affair with the cook.

Adjustments were made for the classes taught by Smith (1877e), who recommended that Assistant Professor of Mathematics and Chemistry Carey Dempsey Grandy take over his botany class and have the students plant seeds. Smith (1877h) also entrusted his books and belongings to the care of Grandy, who was to ship them to Michigan; however, by September 5, 1877, Smith still had not received his possessions. In May 1877, the faculty agreed to approve the work of Smith's students as Prof. Smith was absent and no examinations were held, and to commend them at commencement for their work (UNC Faculty 6: 423). A newspaper reporter from the *Raleigh Observer* noted that Smith's students of botany and zoology "were not subjected to the ordeal of a final examination, but special work was assigned them, which they performed . . . [They] were especially commended for their zeal and success in collecting specimens illustrative of their studies for the Museum of Natural History" (Anonymous, [1877]). According to the university grade books (Appendix 1), Smith's botany class was composed of twenty-three students, including President Battle's son Kemp P. Battle, Jr.

After Smith's departure, President Battle desired to see that work organizing the University Museum was completed. Toward that goal, Cornelia Spencer reluctantly accepted his request to finish arranging the various collections. She brought her daughter June as well as her niece Lucy Phillips with her (Phillips, L., 1877). The work was completed in time for the June commencement.

In responding to a letter from Battle, Smith (1877f) wrote on May 15, 1877, that he enclosed his [official] resignation but that he might "yet arrange to return to C[hapel] H[ill]." The incident that led to his resignation still bothered him greatly. He told Battle, "If the persons or person seeking my detriment continue[d] the present course of procedure, it is bound to ruin me & I might as well meet it as to go on endeavoring to get over the matter quietly. Personally I do not feel afraid of the facts & would never have left C. H. had it not been for the urging of others."

In July 1877, Smith (1877g) requested the balance of his salary. In notifying Battle on September 5, 1877, that he still had not received his pay, Smith (1877h) alluded to a possible reason for the delay: "I did not suppose that the question of legal claim would ever be raised.

I supposed that there was too much honor among the southern people to neglect paying a professor in one of their state universities. I am sure the thing would not happen in any northern institution." In early October, Smith (1877i) acknowledged receipt of his salary payment. In his last known letter to Battle, Smith (1877j) was relieved and "glad to have . . . arrangements with the southern people closed up." He did not, however, hold a grudge against the university. He had gratefully remembered the university by donating a copy of his book on the tailed amphibians. The volume is inscribed: "Presented to the Dialectic Society by W. H. Smith, L.L.D. Professor of Natural History in the University. 'In grateful memory,' May 30th, 1877." Whatever personal issues tarnished Smith's reputation, the editor of the *University Magazine* (Chapel Hill) provided some praise in the April 1878 issue: "Dr. Smith was an excellent professor, and it is much to be regretted that he could not be persuaded to remain with us" (Anonymous, 1878a).

Later Life and Legacy

After reestablishing himself in Ann Arbor, Smith entered the medical school at his alma mater. A year later, he accepted a professorship of zoology and geology at Vassar College, serving from February to June 1878. Resuming his studies at Michigan, he earned the M.D. in March 1879. Beginning prior to completing his medical degree and extending into 1880, he conducted research for the Ohio Geological Survey under the direction of paleontologist and geologist John S. Newberry. Smith (1882) published a report on the reptiles and amphibians of that state. Smith (1879) also published a "Catalogue of the reptilia and amphibia of Michigan," based in great part on specimens in the University of Michigan Museum and those collected by Smith and others.

In 1880, Dr. Smith started his medical practice (Fig. 6.6) in St. Clair, Michigan, to which he devoted the remaining years of his life. In 1900, he was licensed to practice medicine in Michigan (Hafner, 1993). From 1913 to 1919, he served as the District Medical Inspector of St. Clair. Smith published on diverse topics: gold standards (1896), poetry (1894), and water pollution (Pugh, 1975). He also provided words to patriotic songs (Smith, 1887; 1888). In writing the words to forty-nine songs set to familiar tunes, he aimed to provide "G.A.R. [Grand Army of the Republic] Posts and kindred organizations with songs appropriate to all occasions whether Memorial Day, the camp-fire, or re-unions" (Smith, 1887). Between 1879 and 1921, Dr. Smith published nearly thirty articles, which were predominantly on public health (Table 6.1). According to a note in the *College Index* published by Kalamazoo College, Smith also wrote a number of articles for "Science Notes," a featured column of the *Christian Herald* (Anonymous, 1885).

In addition to publishing, Smith (Fig. 6.7) held memberships in professional organizations. He was elected to the National Association of Education in 1874 and the American Association for the Advancement of Science in 1875. He also belonged to the American Medical Association. He was a member of two fraternal societies, the Knights of Pythias and the Grand Army of the Republic. At age seventy-nine, Smith died of a stroke on November 2, 1925, at the home of Mr. and Mrs. Solomon P. Gilbert, where he had been living (Anonymous, 1925). He is buried in Hillside Cemetery, St. Clair, Michigan (Fig. 6.8). Remaining single his entire life, he had no immediate family at the time of his death.

Table 6.1. A List of Titles of Articles on Medicine and Public Health Written by William Henry Smith, Arranged Chronologically under Journal Titles.

I. *Physician and Surgeon.*

1879. Practical mode of studying the heart. 1: 385–395.

1881. Case of dysentery. 3: 481–484.

1882. Doses of certain preparations that are not given in Stewart's dose-book. 4: 4–18.

1882. Notes of cases in private practice. 4: [193]–196.

II. *Public Health.* The Michigan State Board of Health, new series.

1913. The work of the health officer. 1: 8–12.

1914. Our legacy from savagery. 2: [497]–500.

1915. A public health sentiment—How best attended. 3: 11–15.

1915. School sanitation. 3: 70–76.

1915. Spring cleaning. 3: 170–176.

1915. Summer resort sanitation. 3: 224–229.

1915. The care of the baby. 3: [263]–270.

1915. The sanitary treatment of diphtheria. 3: 501–506.

1915. Certain facts about cancer. 3: 555–557.

1915. The responsibility of parenthood. 3: 624–630.

1916. Winter and contagious diseases. 4: [581]–586.

1916. The rights of children. 4: 151–155.

1916. Preparedness for life. 4: [257]–261.

1916. Summer resorts and hygiene. 4: [315]–321.

1916. A meritorious war. 4: 435–441.

1917. Medical triumphs over diphtheria. 5: 538–543.

1918. Our people in school. 6: 152–158.

1918. Measles and whooping cough. 6: 428–432.

1919. The duty of the state. 7: 54–59.

1919. Swat the disease-spreading flies. 7: [199]–204.

1919. Smallpox or causes for thankfulness. 7: [489]–496.

1920. Our aristocracy of health. 8: 207–213.

1920. The value of bathing. 8: 306–310.

1921. Prophets of health. 9: [401]–404.

III. *Journal of the Michigan State Medical Society.*

1918. The future of medicine. 17: 52–56.

Fig. 6.6. Façade of building housing Smith's medical office and for a number of years his residence, 111 Jay Street, St. Clair, Michigan, 1910. (Entrance door [lower arrow] to Smith's quarters on the second floor [upper arrows]).

(Courtesy of Ronald H. Brenner, North Street, Michigan).

William Henry Smith had a difficult early life, working his way through the last years of high school and throughout college, graduate school, and medical training. He was the first UNC professor to hold the Ph.D. at the time of being hired and presumably the first in North Carolina to instruct in the art of taxidermy (through his circular regarding preparing specimens for the University Museum). Having gained practical curatorial skills working in the University of Michigan Museum, he was eminently qualified for his efforts to consolidate and organize the natural history collections of the museum at UNC. His botanical instruction included plant collecting and laboratory work in which students observed the development of plants and identified them. Students were required to donate their collections of local plants to the UNC museum. In his later years, Smith served for over forty years as a physician in St. Clair, Michigan, where he gained fame for taking a stand to improve the city's condition and served as health officer (Pugh, 1975). Although his tenure at UNC lasted only seven months, William Henry Smith was deemed "a teacher of decided merit, quite an accomplished expert in his department" (Battle, 1907–1912, 2: 122).

Fig. 6.7. William Henry Smith, ca. 1910.

(From GAR William Sanborn Post, composite photo, courtesy of the Port Huron Museum, Port Huron, Michigan).

Fig. 6.8. Gravestone of William Henry Smith, Hillside Cemetery, St. Clair, Michigan, block #61, lot #1, grave #7.

(Courtesy of Ronald H. Brenner, North Street, Michigan).

References Cited

(References that conclude with page numbers within parentheses indicate sources from where quotes and pertinent information were obtained.)

AASA. Ann Arbor Scientific Association. *The Constitution and By-laws of the Ann Arbor Scientific Association with the Proceedings for the Year Ending May 1, 1876*. Courier Steam Printing House, Ann Arbor. ("List of Members," pp. [8]–9)

Anonymous. 1871. Kalamazoo College—closing exercises. *Kalamazoo Gazette* (Kalamazoo, MI) 38(12): 4, Friday, June 23, 1871.

Anonymous. 1876a. A Museum for Chapel Hill. Circular letter from the President of the University. *Raleigh News* (Raleigh) 10(71): [4], Wednesday, November 22, 1876.

Anonymous. 1876b. Professor of Natural History in the University. *Daily News* (Raleigh) 9(155): [1], Thursday, August 31, 1876.

Anonymous. [1876]c. University Museum of Natural History, p. 24 in Kemp P. Battle, *Scrapbook of Material Relating to the University of North Carolina, 1869–1893*, vol. 2. (Newspaper clipping probably from the *Raleigh Sentinel* (Raleigh), November 22, 1876). (North Carolina Collection, Wilson Special Collections Library, University of North Carolina, Chapel Hill)

Anonymous. [1876]d. Museum of the University, p. 31 in Kemp P. Battle, *Scrapbook of Material Relating to the University of North Carolina, 1869–1893*, vol. 2. (Newspaper clipping without complete bibliographical data). (North Carolina Collection, Wilson Special Collections Library, University of North Carolina, Chapel Hill)

Anonymous. [1877]. Chapel Hill commencement. *Observer* (Raleigh) 1(29): [2], Tuesday, June 12, 1877.

Anonymous. 1878a. Wm. H. Smith, Ph.D., of Michigan. *University Magazine* (University of North Carolina, Chapel Hill) 1(2): 70.

Anonymous. 1878b. Commencement exercises at the University of North Carolina, Chapel Hill, June 5, 1878. *Observer* (Raleigh) 3(114): [3], Thursday, June 6, 1878.

Anonymous. 1880. The N.C. University, &c. Chapel Hill, N.C., Nov. 23, 1880. *Charlotte Democrat* (Charlotte, NC) 29(1461): [3], Friday, December 3, 1880.

Anonymous. 1885. W.H. Smith. *College Index* (Kalamazoo College) 8(2): 27, November 1885.

Anonymous. 1903. William Henry Smith, p. 14 in *Historical Catalogue of the Students of Kalamazoo College and of Kalamazoo Theological Seminary, 1851–1902*. Ihling Bros. & Everard, Publishers, Kalamazoo, MI.

Anonymous. 1925. Death ends career of St. Clair doctor. *Port Huron Times-Herald* (Port Huron, MI). Monday, November 2, 1925, p. 14.

[Baird, Spencer F.] [1877?]. Letter to Kemp P. Battle, undated (probably 1877). (Record Unit 53, Assistant Secretary [Spencer F. Baird], 1850–1878, Outgoing correspondence, vol. 73, p. 324, Smithsonian Institution Archives)

Battle, Kemp P. 1876a. Letter of October 30, 1876 to Joseph Henry. (Record Unit 28, Office of the Secretary [Spencer F. Baird], 1879–1882, Incoming correspondence, #2693, Smithsonian Institution, Washington, DC)

Battle, Kemp P. 1876b. University of North Carolina. University's position of the University Agricultural Department, p. 25 in Kemp P. Battle, *Scrapbook of Material Relating to the University of North Carolina, 1869–1893*, vol. 2. (Newspaper clipping published about November 1, 1876, without complete bibliographical data). (North Carolina Collection, Wilson Special Collections Library, University of North Carolina, Chapel Hill)

Battle, Kemp P. 1907–1912. *History of the University of North Carolina*. Printed by Edwards & Broughton, Raleigh. (Vol. 2, pp. 122, 136–139)

Battle, William James (editor). 1945. *Memories of an Old-Time Tar Heel by Kemp Plummer Battle, President of The University of North Carolina 1876–1891*. University of North Carolina Press, Chapel Hill.

BHC. Burton Historical Collection, Detroit Public Library. (Records of Service of Michigan Volunteers in the Civil War 1861–1865)

BHL. Bentley Historical Library, The University of Michigan, Ann Arbor. Necrology file on William H. Smith, collected by the Office of Alumni Records. (Includes an autobiographical essay, 4 pp.)

BOT. Board of Trustees, University of North Carolina, Chapel Hill. *Minutes of the Board of Trustees*. (University Archives #40001, Wilson Special Collections Library, University of North Carolina, Chapel Hill) (Vol. 16, 1875–1881)

Chamberlain, Hope Summerell. 1926. *Old Days in Chapel Hill; Being the life and letters of Cornelia Phillips Spencer*. University of North Carolina Press, Chapel Hill. (p. 227)

Chapman, A.W. 1860. *Flora of the Southern United States: Containing abridged descriptions of the flowering plants and ferns of Tennessee, North and South Carolina, Georgia, Alabama, Mississippi, and Florida: Arranged according to the natural system. The Ferns by Prof. Daniel C. Eaton*. 1st ed. Ivison, Blakeman, Taylor & Co., New York.

Coker, Robert Ervin. 1948. The Department of Zoology, University of North Carolina. *Bios* 18: 211–224.

Cooke, M.C. 1871. *Handbook of British Fungi, with Full Descriptions of All the Species, and Illustrations of the Genera*. Macmillan and Co., London, New York.

Dabney, Charles W., Jr. 1882. [Historical information], pp. 6–27 in *Annual Report of the North Carolina Agricultural Experiment Station, for 1881*. Printed by Order of the Board of Agriculture. Ashe & Gatling, State Printers and Binders, Presses of Edwards, Broughton & Co., Raleigh.

Darlington, William. 1865. *American Weeds and Useful Plants; Being a second and illustrated edition of agricultural botany: An enumeration and description of useful plants and weeds, which merit the notice, or require the attention of American agriculturists*. Revised with additions by George Thurber. Orange Judd and Company, New York.

Gray, Asa. 1871a. *Gray's Lessons in Botany and Vegetable Physiology: To which is added a copious glossary, or dictionary of botanical terms; Manual of the Botany of the Northern United States*. Ivison, Blakeman, Taylor, New York.

Gray, Asa. 1871b. *Introduction to Structural and Systematic Botany, and Vegetable Physiology,* Being a 5th and Revised Edition of *The Botanical Text-book*. Ivison, Blakeman, Taylor, New York.

Hafner, Arthur W. (editor). 1993. *Directory of Deceased American Physicians 1804–1929*, vol. 2. American Medical Association, Chicago.

Harrington, Mark W. 1874. Report of Prof. M.W. Harrington, pp. 11–13 in *Report of Prof. E.W. Hilgard on the Department of Geology, Zoology and Botany for 1873–4, University of Michigan*. Dr. Chase's Steam Printing House, Ann Arbor. (p. 12) (Box FiMu C450b, Folder 1873–4, Report of Operations in the Museum of the University of Michigan, Bentley Historical Library, University of Michigan, Ann Arbor)

Harrington, Mark W. 1876. *The Analysis of Plants Intended for Schools and Colleges and for the Independent Botanical Student*. Sheehan, Ann Arbor, MI.

Henry, Joseph. 1876. Letter of November 9, 1876 to Kemp P. Battle. (Box 7, folder 287, University Papers #40005, University Archives, Wilson Special Collections Library, University of North Carolina, Chapel Hill)

Henry, Joseph. 1877. Letter of November 9, 1876 to Prof. W. H. Smith. (Record Unit 33, Smithsonian Institution Archives, Office of the Secretary, Correspondence, 1865–1891, vol. 53, p. 315)

Holt, Thos. M. and Wm. H. Cheek. 1880. *What Is Said about the North Carolina Agricultural Experiment Station. Opinions of the Press and the People*. Published by Order of the Board of Agriculture. M. Hale, and Edwards, Broughton & Co., Raleigh.

Ingram, J.S. 1876. *The Centennial Exhibition, Described and Illustrated*. Hubbard Bros., Philadelphia.

Jenks, William Lee. 1912. William H. Smith, M.D., pp. 789–790 in *St. Clair County, Michigan: Its history and people*, vol. 2. Lewis Publishing Company, Chicago.

Johnson, Samuel W. 1868. *How Crops Grow. A Treatise of the Chemical Composition, Structure, and Life of the Plant, for All Students of Agriculture*. Orange Judd and Company, New York.

Johnson, Samuel W. 1870. *How Crops Feed: A treatise on the atmosphere and the soil as related to the nutrition of agricultural plants*. Orange Judd and Company, New York.

KC. 1868. Kalamazoo College. *Catalogue of the Officers and Students of Kalamazoo College, Kalamazoo, Michigan. September, 1866, to December, 1867*. Clark & Cadman's Power-Presses, Kalamazoo. (pp. 16–17)

Kerr, W.C. 1876. Letter of November 20, 1876 to Kemp P. Battle. (Box 7, folder 287, University Papers #40005, University Archives, Wilson Special Collections Library, University of North Carolina, Chapel Hill)

Lazenby, William R. 1876. Letter of July 28, 1876 to George T. Winston. (Box 7, folder 282, University Papers #40005, University Archives, Wilson Special Collections Library, University of North Carolina, Chapel Hill)

Ledoux, Albert R. 1879. History of the station, pp. 16–22 in *Annual Report of the North Carolina Agricultural Experiment Station, for 1879*. The Observer, State Printer and Binder, Raleigh, NC.

Ledoux, Albert R. 1880. History of the station, continued from the *Annual Report of 1879*, pp. 16–21 in *Third Annual Report of the North Carolina Agricultural Experiment Station, for the Year Ending April 15th, 1880*. Hale and Edwards, Broughton & Co., State Printers and Binders, Raleigh, NC.

Love, James Lee. 1921. [Diaries, Volume 6]. (Box 6, folder 70, James Lee Love Papers #4139, Southern Historical Collection, Wilson Special Collections Library, University of North Carolina, Chapel Hill). (p. 23)

Mangum, Adolphus W. 1877. Letter of March 27, 1877 to Kemp P. Battle. (Box 6, folder 83, Battle Family Papers #3223-A, Southern Historical Collection, Wilson Special Collections Library, University of North Carolina, Chapel Hill)

McGehee, Montford. 1887. Report of the Board of Agriculture, pp. [3]–16 in *Report of the Board of Agriculture to the General Assembly of North Carolina, Session 1887*. P.M. Hale, State Printer and Binder, Presses of E.M. Uzzell, Raleigh. (pp. 11–12)

McLeod, DeLorme. 1876. Letter of November 21, 1876 to William H. Smith. (Box 7, folder 287, University Papers #40005, University Archives, Wilson Special Collections Library, University of North Carolina, Chapel Hill)

Mechi, John Joseph. 1864. *How to Farm Profitably; Or, the sayings & doings of Mr. Alderman Mechi*. New ed. Routledge, Warne & Routledge, London, New York.

P., A.L. 1879. The Experimental Agricultural Station. Correspondence of the *Charlotte Democrat*. Chapel Hill, N.C., March 31, '79. *Charlotte Democrat* (Charlotte, NC) 27(1375): [2], Friday, April 4, 1879.

Pendleton, Edmund Monroe. 1875. *Text-book of Scientific Agriculture*. A.S. Barnes, New York.

Pettee, W.H. 1875. [Report] to the President of the University, pp. 474–475 in *University of Michigan, Proceedings of the Board of Regents* (1870–1876). <http://www.hti.umich.edu/u/umregproc/>

Pettee, W.H. 1876. [Report] to the President of the University, Oct[ober] 1876. (University of Michigan Museum, Box 3, folder 1863–1911, the Bentley Historical Library, University of Michigan, Ann Arbor)

Phillips, Charles. 1876. Letter of June 20, 1876 to Kemp P. Battle. (Box 5, folder 80, Battle Family Papers #3223-A, Southern Historical Collection, Wilson Special Collections Library, University of North Carolina, Chapel Hill)

Phillips, Laura C. 1877. Letter of May 3, 1877 to Mrs. David L. Swain. (Box 1, folder 8, Cornelia P. Spencer Papers #683, Southern Historical Collection, Wilson Special Collections Library, University of North Carolina, Chapel Hill)

Phillips, W[illia]m B. 1880. Letter of January 25, 1880 to Lucy [P. Phillips]. (Box 1, folder 5, Charles Phillips Papers #2462, Southern Historical Collection, Wilson Special Collections Library, University of North Carolina, Chapel Hill)

Porcher, Francis Peyre. 1863. *Resources of the Southern Fields and Forests, Medical, Economical, and Agricultural: Being also a medical botany of the Confederate States; With practical information on the useful properties of the trees, plants, and shrubs*. Prepared and published by order of the Surgeon-General, Richmond, Va. West and Johnston, Richmond, VA.

Pugh, Mary Jo. 1975. Michigan Ph.D. almost a century old. *Rackham Reports from the Graduate School of the University of Michigan* 2(1): [4–6], 1975.

Rothenberg, Marc. 2002. Joseph Henry Papers, Smithsonian Institution. (Electronic mail to William R. Burk, January 2, 2002)

Sachs, Julius. 1865. *Handbuch der Experimental-Physiologie der Pflanzen. Untersuchungen über die Allgemeinen Lebensbedingungen der Pflanzen und die Functionen Ihrer Organe*. W. Engelmann, Leipzig.

Sachs, Julius. 1875. *Text-book of Botany. Morphological and Physiological*. Translated and annotated by Alfred W. Bennett, assisted by W.T. Thiselton Dyer. At the Clarendon Press, Oxford, [England].

Sensemann, Harley L. (editor). 1923. *Catalogue of Graduates, Non-graduates, Officers, and Members of the Faculties 1837–1921*. Published by the University, Ann Arbor, MI. (p. 285)

Smith, John. 1877. Letter from Fayetteville. About agricultural matters. Fayetteville, July 12, 1877. *Observer* (Raleigh) 2(50): [2], Thursday, July 12, 1877.

Smith, William H. 1874. Report of Work Done by Mr. W.H. Smith, Custodian of Museum [1873–1874], pp. 361–363 in *University of Michigan, Proceedings of the Board of Regents (1870–1876).* <http://www.hti.umich.edu/u/umregproc/>

Smith, William H. 1876a. *Circular to the Friends of the University of North Carolina in Regard to Specimens of Natural History*. Raleigh News Job Office and Book Bindery, Raleigh. (p. 10)

Smith, William H. 1876b. Letter of July 4, 1876 to Kemp P. Battle. (Box 7, folder 282, University Papers #40005, University Archives, Wilson Special Collections Library, University of North Carolina, Chapel Hill). (Letter contains annotations of extracts of recommendations from several people on behalf of Smith's application for the UNC professorship in natural history.)

Smith, William H. 1876c. Letter of August 23, 1876 to Kemp P. Battle. (Box 7, folder 284, University Papers #40005, University Archives, Wilson Special Collections Library, University of North Carolina, Chapel Hill)

Smith, William H. 1876d. Letter of November 27, 1876 to Kemp P. Battle. (Box 5, folder 81, Battle Family Papers #3223-A, Southern Historical Collection, Wilson Special Collections Library, University of North Carolina, Chapel Hill)

Smith, William H. 1876e. *The Zoology of Anoura and Caecilia*. Ph.D. dissertation. University of Michigan, Ann Arbor. (I was unable to find a copy of Smith's original dissertation of 1876 at the University of Michigan—Ann Arbor campus.)

Smith, William H. 1877a. Letter of January 9, 1877 to Kemp P. Battle. (Box 8, folder 290, University Papers #40005, University Archives, Wilson Special Collections Library, University of North Carolina, Chapel Hill)

Smith, William H. 1877b. Letter of January 17, 1877 to Kemp P. Battle. (Box 8, folder 290, University Papers #40005, University Archives, Wilson Special Collections Library, University of North Carolina, Chapel Hill)

Smith, William H. 1877c. Letter of March 10, 1877 to Kemp P. Battle. (Box 8, folder 292, University Papers #40005, University Archives, Wilson Special Collections Library, University of North Carolina, Chapel Hill)

Smith, William H. 1877d. Letter of March 30, 1877 to Kemp P. Battle. (Box 8, folder 292, University Papers #40005, University Archives, Wilson Special Collections Library, University of North Carolina, Chapel Hill)

Smith, William H. 1877e. Letter of April 20, 1877 to Kemp P. Battle. (Box 8, folder 294, University Papers #40005, University Archives, Wilson Special Collections Library, University of North Carolina, Chapel Hill)

Smith, William H. 1877f. Letter of May 15, 1877 to Kemp P. Battle. (Box 8, folder 295, University Papers #40005, University Archives, Wilson Special Collections Library, University of North Carolina, Chapel Hill)

Smith, William H. 1877g. Letter of July 16, 1877 to Kemp P. Battle. (Box 8, folder 301, University Papers #40005, University Archives, Wilson Special Collections Library, University of North Carolina, Chapel Hill)

Smith, William H. 1877h. Letter of September 5, 1877 to Kemp P. Battle. (Box 8, folder 306, University Papers #40005, University Archives, Wilson Special Collections Library, University of North Carolina, Chapel Hill)

Smith, William H. 1877i. Letter of October 4, 1877 to Kemp P. Battle. (Box 8, folder 309, University Papers #40005, University Archives, Wilson Special Collections Library, University of North Carolina, Chapel Hill)

Smith, William H. 1877j. Letter of November 30, 1877 to Kemp P. Battle. (Box 8, folder 312, University Papers #40005, University Archives, Wilson Special Collections Library, University of North Carolina, Chapel Hill)

Smith, William H. 1877k. *The Tailed Amphibians, Including the Caecilians. A Thesis: Presented to the faculty of Michigan University,* by W. H. Smith. Herald Publishing House, Detroit, MI. (The cover title reads: "Urodela and Caecilia."). (p. [5])

Smith, William H. 1879. *Catalogue of the Reptilia and Amphibia of Michigan. Supplement to Science News*, [Salem, MA].

Smith, William H. 1882. Report on the reptiles and amphibians of Ohio, pp. [629]–734 in *Report of the Geological Survey of Ohio*, vol. 4 (Zoology and Botany, Part I. Zoology). Nevins & Myers, State Printers, Columbus, OH.

Smith, William H. 1887. *Complete G.A.R. Song Book*. [words only]. Published by Capt. Robt. W. Haskin, Herald Publishing House, Detroit, MI. (p. 3). (Microfiche copy from Library of Virginia, Richmond, Virginia, filmed by NYT Microfilming Corporation of America, from the original in the State Historical Society of Wisconsin)

Smith, William H. 1888. *Campaign Songs for 1888*. W.H. Smith, St. Clair, MI. (words only). (not seen)

Smith, William H. 1894. *The Frailties of Humanity and Other Poems*. The Irving Company, New York.

Smith, William H. 1896. *The Effects of the Gold Standard, or Bimetallists' Catechism*. (*American Politics*, no. 6). Charles H. Kerr & Company, Chicago.

Southgate, James. 1877. Letter of February 24, 1877 to Kemp P. Battle. (Box 8, folder 291, University Papers #40005, University Archives, Wilson Special Collections Library, University of North Carolina, Chapel Hill)

Spencer, Cornelia P. 1876a. Letter of September 30, 1876 to Mrs. Swain. (Cornelia P. Spencer, Personal Collection #74, State Archives of North Carolina)

Spencer, Cornelia P. 1876b. Letter of December 8, 1876 to Ellen. (Box 1, folder 7, Cornelia Phillips Spencer Papers #683, Southern Historical Collection, Wilson Special Collections Library, University of North Carolina, Chapel Hill)

Steele, Walter L. 1877. Letter of April 6, 1877 to Kemp P. Battle. (Box 6, folder 83, Battle Family Papers #3223-A, Southern Historical Collection, Wilson Special Collections Library, University of North Carolina, Chapel Hill)

Taylor, Edward Harrison. 1968. *The Caecilians of the World; A taxonomic review*. University of Kansas Press, Lawrence. (p. 835)

True, Alfred Charles. 1937. North Carolina Experiment Station, pp. 89–91 in *A History of Agricultural Experimentation and Research in the United States 1607–1925, Including a History of the United States Department of Agriculture*. (*United States Department of Agriculture Miscellaneous Publication*, no. 251). Government Printing Office, Washington, D.C.

Turner, George H. (compiler). 1905. *Record of Service of Michigan Volunteers in the Civil War, 1861–1865*. [*Twenty-Sixth Infantry*]. Michigan Adjutant-General's Department, Ihling Bros. & Everard, [Lansing, Michigan]. (Copy seen at the Bentley Historical Library, University of Michigan, Ann Arbor)

UM. University of Michigan, Transcripts Office. [1873]. Transcript of the courses taken by William H. Smith. (A copy of Smith's transcript is on file in the Transcripts Office, University of Michigan, Ann Arbor. Smith was designated as a "Resident Graduate" in the "Scientific" curriculum.)

UM-C. *Calendar of the University of Michigan*. Published by the University, Ann Arbor. no. 1. 1872–3. ("Botanical Collection," pp. 13–14).– no. 2. 1875–6. ("Botanical Collection," pp. 15–16)

UNC-C. 1877. *Catalogue of the Trustees, Faculty and Students of the University of North Carolina 1876-'77*. University of North Carolina, Chapel Hill. (p. 44)

UNC Faculty. University of North Carolina at Chapel Hill Faculty. *Minutes of Faculty Meetings*. (University Archives #SV-40106, Wilson Special Collections Library, University of North Carolina, Chapel Hill) (Vol. 6, 1856–1885)

Weeks, Stephen B. 1886. [The laboratory under Smith Hall], p. 28 in College record. *University Magazine* (University of North Carolina, Chapel Hill), new series 6: 27–32.

White, Andrew D. 1876. Letter of July 18, 1876 to Professor George T. Winston. (Box 7, folder 282, University Papers #40005, University Archives, Wilson Special Collections Library, University of North Carolina, Chapel Hill)

Winston, George T. 1876. Letter of December 15, 1876 to [Kemp P. Battle]. (Box 5, folder 82, Battle Family Papers #3223-A, Southern Historical Collection, Wilson Special Collections Library, University of North Carolina, Chapel Hill)

Young, Wade Phillips. 1934. Chapter VIII. The Experiment Station, pp. 142–188 in *A History of Agricultural Education in North Carolina*. A thesis submitted to the faculty of the University of North Carolina in partial fulfillment of the requirements for the degree of Doctor of Philosophy in the Department of Rural Social-Economics. Chapel Hill.

Chapter 7.

Frederic William Simonds (1853–1941)

Professor of Geology, Zoology, and Botany (1877/1878–1880/1881)
Professor of Botany and Physiology in the Medical School (1879/1880–1880/1881)
Professor of Botany in the Pharmacy School (1880/1881)

Early Years and College Education

Frederic William Simonds was born on July 3, 1853, in Charlestown, Massachusetts (now part of Boston), to Andrew Wait and Ellen Rachel (Cox) Simonds. He attended the old Bunker Hill School of his birth town. In 1864, the Simonds family moved to Richmond, Indiana, where young Simonds completed his secondary schooling in 1871. He was valedictorian of his class. He then became an assistant to the city engineer. When his supervisor was appointed chief engineer of the Cincinnati, Richmond, and Fort Wayne Railroad, Simonds joined him as his assistant on the survey of location. As the reputation of the newly founded Cornell University in Ithaca, New York, reached the West, he considered extending his education, mostly due to his mother's urging (Simonds, [1908]a). He left the survey to enter Cornell in September 1871.

At Cornell, Simonds was struck with the scholastic enthusiasm of an active group of scientists. As an undergraduate major in natural history, Simonds enrolled in numerous scientific courses in the following disciplines: botany, chemistry, comparative anatomy, entomology, geology, and physics (CU-RS). He was a pupil of Cornell Professor of Zoology Burt Green Wilder (Anonymous, 1877b). Paleontologist and geologist Charles Fred Hartt influenced young Simonds's interests in geology. Wilder and Hartt had been students of the famed naturalist Louis Agassiz. Although geology became a primary focus in his professional career, Simonds also took an avid interest in zoology and botany while he was a college student.

Cornell offered a stimulating educational and research environment in botany, which no doubt influenced Simonds. Nearly a year after he arrived at Cornell, the botany department moved into new quarters in the west wing of Sibley College (Fitzpatrick, 1926). The lecture room of botany was on the second floor and the botanical museum on the third (CU-R). A small herbarium, plant models, and charts augmented his botanical studies. A botanical laboratory provided an additional dimension to his educational experience. His botanical mentors were Albert N. Prentiss and William R. Dudley, the latter whom he befriended. Although Simonds

Frontispiece Credits: (above) Frederic William Simonds, 1875. (Cobb Family Papers, 1792–1975 #P-4008/115, Southern Historical Collection, Wilson Special Collections Library, University of North Carolina, Chapel Hill). (below) UNC campus, ca. 1877; impression by W.T. Blackwell & Co., Proprietors, Durham Job Printing Establishment, Durham, North Carolina. (Box 8, folder 300, University Papers #40005, University Archives, Wilson Special Collections Library, University of North Carolina, Chapel Hill).

is not listed as having been a pupil of Dudley's (DMV, 1913), he interacted with Dudley, who was his classmate and subsequently an instructor in botany. They also were members of the Natural History Society. After leaving Cornell, Simonds remained in contact with Dudley.

In each of his first three years at Cornell, Simonds took a course in botany. For the spring trimester course in botany, Prentiss required students to attend lectures, compile notebooks, examine plants, and use the textbook "Gray's *Manual and Lessons in Botany*, fifth edition, bound in one volume." Based on this text, the lectures consisted of explanations of botanical tenets and were accompanied by illustrations and plant specimens as visual aids. Students were also required to compile a notebook founded on the subject content of the lectures and accompanying illustrations. Theory was reinforced through the examination and identification of plant specimens in the laboratory (Prentiss, 1871). Simonds would later use a similar teaching methodology when he taught natural history at UNC.

Another factor influencing Simonds's interest in science was the Natural History Society. Joining the society in 1872, Simonds became an active member, serving as curator of zoology (1872–1873), recording secretary (1873–1874), and president (1874–1875). According to notices in the *Cornell Era* (a student-run weekly journal of commentary) and entries in the society's proceedings (NHSR), Simonds gave a number of presentations on such topics as a two-legged horse, its anatomy and habits; erosion and denudation caused by freshwater and volcanoes. As a young, inquisitive member of the society, Simonds gained experience in leadership, scientific knowledge, and opportunities to exchange ideas with other enthusiastic and highly motivated scholars.

A further invigorating aspect of Simonds's scholarly environment was his association with classmates who were dedicated to scientific research and inquiry. Among these individuals were several who became notable plant scientists: William Russel Dudley (also an instructor of botany during Simonds's junior and senior years), William Ashbrook Kellerman, William Arnon Henry, Charles Youdan Lacy, William Rane Lazenby, Charles Frederick Millspaugh, and William Trelease. Two other colleagues, John Casper Branner and Herbert Edson Copeland, also demonstrated botanical interests. Another classmate, David Starr Jordan, who showed early interests in botany at Cornell, became a renowned educator and scientist, particularly as an ichthyologist.

While at Cornell, Simonds compiled a scrapbook (CU), which notes that he was a member of the guard of honor when the body of Ezra Cornell lay in state in Library Hall. A black mourning badge worn on December 12, 1874, by Simonds, First Lieutenant of Company E, Cornell University Cadet Corps, is pasted in the scrapbook. The book is currently housed in the special collections of the Carl A. Kroch Library, Rare and Manuscript Collections at Cornell University. Excelling in his scientific studies, Simonds ([1908]a) received the President's Prize in Zoology (1872) and the President's Prize in Geology (1875). He earned a B.S. degree in natural history (1875) and an M.S. degree in geology (1876). At the seventh annual commencement exercise at Cornell, held on June 17, 1875, he presented an essay on natural history, entitled "The Geology of Ithaca and the Vicinity" (Cornell University, 1875). The essay also constituted his bachelor's thesis (Simonds, 1875) and was later the subject of his master's research (Simonds, 1877b).

Subsequent to his schooling at Cornell, Simonds spent a season at Woods Hole, Massachusetts, where he assisted paleontologist and marine biologist Alpheus Hyatt, who was himself a visiting researcher from the Museum of the Boston Society of Natural History. According to an anonymous (1941a) writer, Simonds's "experience there was lifelong in its effect." From 1875 to 1877, he served as an instructor in geology and paleontology at Cornell. During this time, he began studies toward a doctorate in geology.

The University of North Carolina at Chapel Hill

In June 1877, UNC President Kemp P. Battle presented to the Executive Committee of the Board of Trustees applications and letters of reference for scholars seeking the natural history position vacated by William Henry Smith. Among the pool of applicants were Alfred A. Bennett, C.C. Frederick, Frederic W. Simonds, and Lucien Marcus Underwood (who became a noted botanist). Victor C. Vaughan (1877), William H. Smith's classmate at the University of Michigan, had also inquired about the position in late April 1877. Simonds's testimonials included supportive comments from William Channing Russel (1877), acting president of Cornell, who stated that Simonds was "a young man of a great deal of will and determination and principle." He also added that Simonds was "a regular attendant at church from which [he] conclude[d] that he [was] interested in moral and religious growth." Simonds's academic credentials so impressed the UNC Board of Trustees that he "was unanimously elected to fill said Professorship to take effect from the 15th August 1877 at a salary of $1,250 per annum" (BOT 16: 54).

Before arriving in Chapel Hill, Simonds married Norma Anna Wood of Syracuse, New York (niece of Mrs. Ezra Cornell) on August 21, 1877. They had three children, Fred Hartt, Anna Wood, and Andrew Warren. Apparently, his son Fred Hartt was named in honor of Cornell professor of geology Fred Hartt, Simonds's mentor in geology.

When Simonds arrived in Chapel Hill, the University campus was composed of eight primary buildings (Frontispiece of chapter 7, lower image) with twelve professors (including Simonds) instructing a student body of 169 men. In September 1877, a newspaper reporter (S[lade], 1877) described the village of Chapel Hill as having a scattering of houses with an occasional store to provide students with their needs. "In whatever direction the eye may wander, glimpses of hills and cornfields are taken in, and the hum of insect nature, both night and day, continually charm the ear. Behind the College buildings are the base ball grounds and the gymnasium, where the members of the Athletic Association, which [had] been organized among the students, practice[d] regularly during the late hours of the afternoon." UNC freshman Henry Elias Faison (1877) gave additional insight about the Chapel Hill landscape, writing, "Our vast fields of cotton give place to broad hillsides of wheat and our valuable timber land gives way to hills of oak & hickory, studded here and there with a Rosemerry pine."

By 1879, the village had 700 inhabitants and contained eight stores, three drugstores, eight or ten mechanics' shops, one hotel, numerous private boardinghouses, six churches, and four or five schools for the children of the town and its vicinity (Heitman, 1879). The UNC catalog of 1877–1878 (p. 21) informed students that typical expenses for each of two terms (the first beginning July 13, 1877, the second on January 3, 1878) were $30 for tuition and

$5 for room rent. Other expenses, including washing, board, wood, lights, and books, ranged from $58.50 to $88.50 per semester.

Slade (1877) informed readers that the practice of hazing had nearly died out by 1877. "The unsophisticated 'nervies,' who have heretofore trembled at the slightest intimation of blacking from one of the old boys, now when lodging within the College buildings, sleep during their first few nights in quiet and contentment." Another observer also noted that hazing was no longer inflicted on the "Fresh" (Bill, 1878). But in a letter to his sister Belle Alderman, UNC student William C. Alderman (1878) mentioned that although blacking had been abolished, hazing still existed. Henry E. Faison (1877) reported that a band of "old boys" had intended to black the "Fresh"; however, the ringleaders were too intoxicated to carry out the act.

During 1877, several noteworthy events occurred. The Normal School was established that year for the training of teachers in the summers, the first of its type in the country. Women were first admitted to UNC at this school, which is discussed in chapter 10. University Day was instituted on October 12, 1877 (Fig. 7.1) to commemorate the laying of the cornerstone of Old East. It is a tradition still followed to the present. At the 1879 celebration, Simonds gave a talk. When the Board of Trustees approved a Medical School on February 12, 1879 (BOT 16: 61) and a Pharmacy School on September 23, 1880 (BOT 16: 81), Simonds became a faculty member in these newly created schools.

Another development at UNC was the opening of the graduate program, which had been initially considered in 1875. The stipulations for earning the master's degree were noted in the UNC catalog of 1876–1877 (p. 22): "The Master's Degree will be open to any Post-Graduate who has taken the Bachelor's Degree in the same Courses in the University or in any institution whose requirements are equal to our own, provided he shall pass one year in post-graduate studies at the University, shall pass his examinations and present a satisfactory thesis on some subject within the department of study especially pursued." In 1877, degrees for the master of science, master of arts, and doctor of philosophy were first announced (Knight and Adams, 1946); the latter two degrees were first conferred in 1883. Not until a Department of Botany was established in 1908 were the first graduate degrees in botany awarded: the master of arts in 1912 and the first doctoral degree in 1923.

President Battle (1878a) reported on the achievements in advancing scientific instruction and research on campus. In 1877, a room was fitted up as a natural history laboratory to promote practical instruction. "Branches relating to agriculture and mechanic arts" had been particularly enhanced by practical teaching. Other scientific departments were strengthened. Person Hall, which had burned in spring 1877, was rebuilt. It was furnished with gas, a recitation room in physics, and a lecture room in chemistry. Battle noted the hall was "handsome and commodious." The chemical laboratory was improved by additions to the working room and appliances and by the construction of a gas works (not only for chemistry, but also for physics) and the installation of a force pump and water pipes.

A year after Simonds's arrival at UNC, Battle (1878b) prepared a broadside for distribution to university supporters in an attempt to raise funds for the institution. In highlighting the developments of the campus, he reported that "laboratory practice is required in the

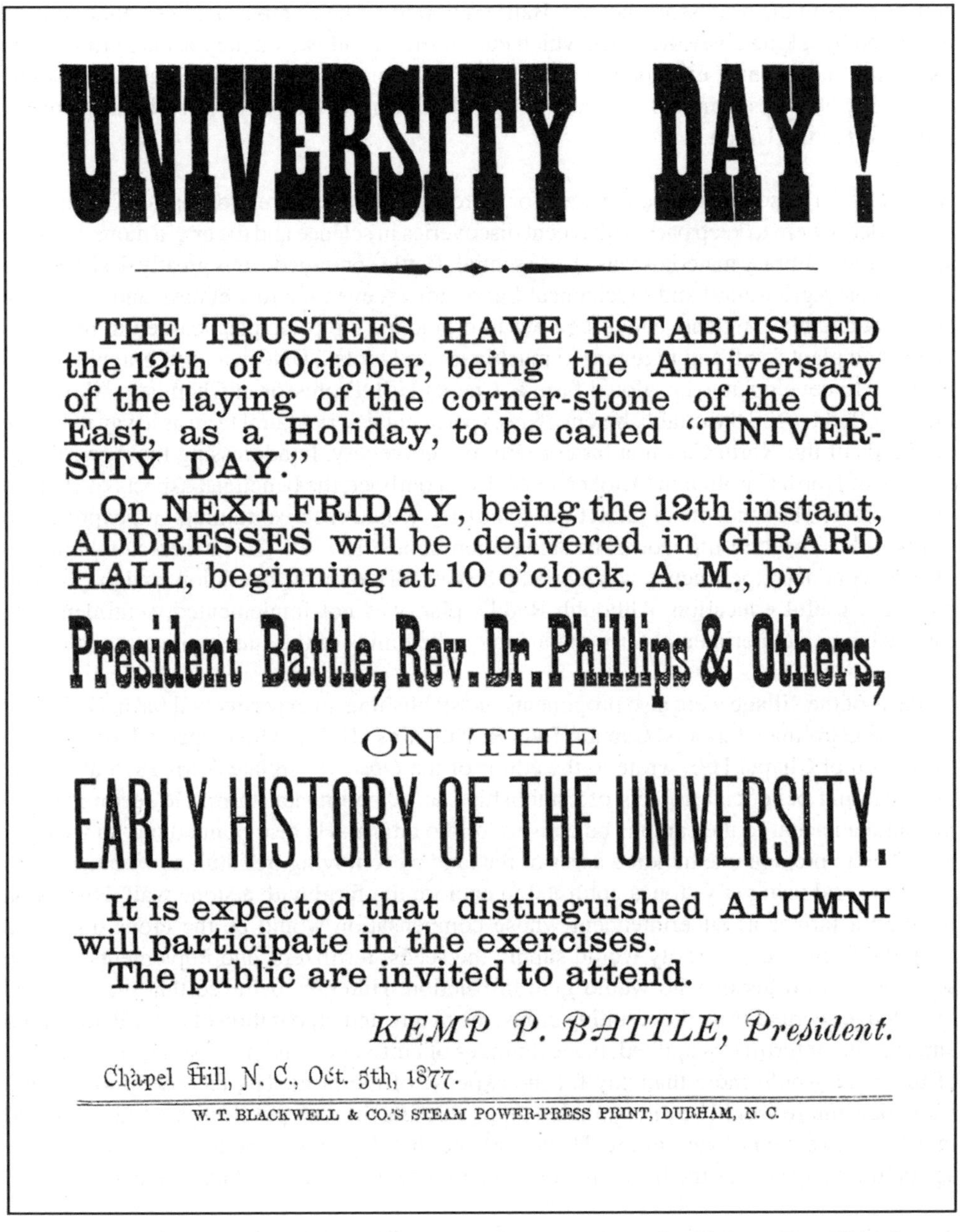

UNIVERSITY DAY!

THE TRUSTEES HAVE ESTABLISHED the 12th of October, being the Anniversary of the laying of the corner-stone of the Old East, as a Holiday, to be called "UNIVERSITY DAY."

On NEXT FRIDAY, being the 12th instant, ADDRESSES will be delivered in GIRARD HALL, beginning at 10 o'clock, A. M., by

President Battle, Rev. Dr. Phillips & Others,

ON THE

EARLY HISTORY OF THE UNIVERSITY.

It is expectod that distinguished ALUMNI will participate in the exercises.

The public are invited to attend.

KEMP P. BATTLE, President.

Chapel Hill, N. C., Oct. 5th, 1877.

W. T. BLACKWELL & CO.'S STEAM POWER-PRESS PRINT, DURHAM, N. C.

Fig. 7.1. Circular: University Day! October 12, 1877, at the University of North Carolina, Chapel Hill.

(Box 8, folder 309, University Papers #40005, University Archives, Wilson Special Collections Library, University of North Carolina, Chapel Hill).

departments of Chemistry and Zoology, and Botany, Mineralogy and Geology." Physical plant improvements included recoloring and repairing the buildings, rearranging the books in the library, fitting up recitation rooms with good blackboards, and repainting and adding new porches to the professors' houses. Battle reported, "The Campus has been enlarged and improved by taking away the fence which cut off the 'meadow,' on the East next to Raleigh road, and 'the Avenue' extended to that road. The Campus is now wonderfully handsome, containing 50 acres, perhaps the finest grove of ancient forest trees owned by any literary institution in the Union."

Battle (1878a) also mentioned the need for better telescopes and microscopes. If the faculty and students were to keep pace with recent discoveries in science and the arts, a more vigorous acquisition of library materials was also essential. Battle continued, "It is greatly desirable too to establish Agricultural and Mechanical Laboratories, even if a model farm and workshop be not attainable. Students could then put into practice the principles learned in regard to plants and plant food, and in regard to machinery and tools." UNC was still grappling with the need to develop an educational farm at Chapel Hill. Professor of Chemistry Alexander Fletcher Redd urged the establishment of an experimental agricultural farm as a way to bridge the needs of the North Carolina farmer with the university. In addressing the Agricultural Society of North Carolina in October 1877, Redd outlined the beneficial aspects of such an undertaking. The students of analytical chemistry would receive training and exposure to "a high ideal for their life work," he said (Anonymous, 1877d). The farmers would have a place to conduct experiments, which would be subsidized by the operation, and they would receive a useful education. Although Redd's plan was not implemented, sentiments for establishing an experimental farm remained on the minds of the state's citizens.

Citizens of the village were also proponents of establishing an experimental farm. On behalf of North Carolina's farmers, Captain John R. Hutchins (1878), who owned a large farm in the region of Chapel Hill, wrote to the editor of the *Observer* in late April 1878 about the practical and beneficial aspects of establishing an experimental farm. He recommended that land adjacent to the campus be converted into a farm. He also pointed out how easy it would be to prepare one hundred acres of the land by removing tree stumps with the aid of dynamite and using any stones upturned to enclose the farm with a stone wall. His vision featured a farmer as superintendent whose compensation would be the crops harvested from the farm; the university would supply the seeds, fertilizers, and implements and the services of a professor who would give instruction. Hutchins believed that the recorded knowledge accrued from the resulting crops, the nature and preparation of the soil, the brand and amount of fertilizers applied, the techniques of cultivation and harvesting, and the yield of the crops would more than pay for the expense of the farm. The recorded information from the venture would be disseminated to the farmers through publications and would aid in the improvement of agriculture. He considered that the "Experiment Farm" would serve agricultural students as the hospital and dissecting room served medical students.

Although an experimental farm was never attained at Chapel Hill, the North Carolina Board of Agriculture successfully passed a resolution in December 1885 to establish a farm in connection with the Experiment Station in Raleigh. Its early work included tests on varieties of cotton, the results of manuring cotton, and the distance of planting on maturation (Young, W., 1934).

Teaching at UNC

In accepting the professorship, Simonds (1877a) informed President Battle that he expected to arrive in Chapel Hill toward the end of August 1877. He also inquired about funding for acquiring supplies such as alcohol and bottles for preserving specimens. He believed that "by the use of specimens the value of instruction may be increased tenfold. That which can be seen creates often times a greater impression than that which is heard or read only." He continued by noting that he desired to establish laboratories in natural history. He added that there would be costs, including those for the purchase of books and travel. He anticipated that his courses in entomology and vegetable physiology would be of a "practical use to the farmers." Soon after Simonds and his wife arrived in Chapel Hill, Cornelia P. Spencer (1877a) wrote to her friend Mrs. David L. Swain, saying, "They are quite young people—& very Yankee looking!"[73] After attending some of his classes, Spencer (1877b) soon realized the extent of Simonds's professional talents, writing, "in him the University has secured a prize." She also noted that Simonds was an expert in natural history, "fresh from the fountain head of this" discipline (Spencer, [1877]c). She praised his enthusiasm and eagerness in instructing students.

Within the College of Natural History, Simonds was the sole instructor in the School of Geology, the School of Zoology, and the School of Botany. His academic quarters were in the old Phi Library in the north end of Old East (Battle, 1907–1912, 2: 639). According to a crude diagram sketched on a page of a volume containing UNC student records, the museum in 1879 was still housed on the third floor of Old East and Simonds's station was on the second floor, directly below the museum (UNC Student Records, vol. 11). Concerning his method of teaching, Cornelia Spencer ([1877]c), wrote that "he employ[ed] the lecture system entirely in the departments of zoology, geology, and botany; discard[ed] the old-time method of teaching by text-book routine, present[ed] the best thoughts and latest discoveries of the best authorities." His students gained hands-on experience in the laboratory by analyzing plants or dissecting animals. They also collected specimens in the field. Simonds was a master artist at the blackboard, providing illustrations of plants and animals as aids in instructing students. Students were required to take notes from his lectures and to copy his illustrations.

Simonds seems to have made a good impression on those he met and instructed early in his career. A reporter from the *Hillsborough Recorder* wrote: "The new Professor gives universal satisfaction. He is six feet, two, very handsome, can handle spiders, snakes, &c., with impunity ... He solicits specimen birds, bugs, and curiosities for the Museum. A live Alligator was received a few days ago" (Anonymous, 1877a). Another reporter praised Simonds highly following a lecture on blood and the blood vessels that he gave at the Mebaneville (North Carolina) Academy (M., 1879). "His style and manner are attractive, his knowledge extensive and thorough. ... The young professor made a fine impression for himself and for the University. In re-establishing the University, the trustees have shown nowhere more wisdom and given nowhere better omens of success than in the infusion of

[73] In writing to her daughter June, Spencer (1876) commented that she "never did like the Germans anymore than I like the Yankees."

vigorous young blood into the faculty, thus combining the old and the new and getting the benefit of both."

Simonds's breadth of knowledge, the enthusiastic execution of his delivery, and his artistic drawings gained attention both at the university and in the region. In his history of UNC, Kemp Battle (1907–1912, 2: 105) recounted supportive commentary about Simonds's presentations. Simonds delivered a lecture at the State Fair before the State Agricultural Society in Raleigh, where he made a deep impression on the audience. One of the listeners, David Kerr, uncle of North Carolina State Geologist Washington C. Kerr, gave Simonds high praise: "The best thing I saw at the Fair was that Chapel Hill Professor at the blackboard, drawing the figures in his lecture with both hands."

Contemporary newspaper articles described Simonds's teaching as enthusiastic and augmented by his superb illustrations. He was very accomplished in freehand drawing and was noted, as Kerr suggests, for drawing on the chalkboard using both hands simultaneously. In 1877, one reporter noted that "in the Lecture Hall the black boards were covered with large drawings in colored crayon. ... These drawings, and the lectures accompanying, students are required to copy in their note-books and to become thoroughly familiar with by the next recitation. Prof. Simonds does not depend upon mere text-book teaching" (Anonymous, 1877e). The drawings were a distinctive feature of his instruction, which Simonds called "the condensation of thought into diagrams" (Spencer, [1877]c).

The lecture notebooks of students Edwin Anderson Alderman, Ernest Haywood, and William George Randall survive in the Southern Historical Collection at UNC. They provide details of the lectures delivered and illustrated by Simonds. Also surviving are Simonds's (1878b, 1879b) syllabi for his courses in Physical Geography and for Geology and General Zoology, as well as a three-page synoptic outline for the latter course (Simonds, 1878c) and a "Schedule of Lectures on Physiology & Hygiene" (Simonds, 1880c). As editor of the *North Carolina Medical Journal*, physician and botanist Thomas Fanning Wood (1879) reproduced in the *Journal* Simonds's "Syllabus of a Course of Twenty-five Lectures on Human Physiology and Hygiene." A ten-page booklet of Simonds's (1880b) General Geology course outline is held in the UNC North Carolina Collection. If a syllabus had been prepared for general botany, it does not survive in the UNC archives.

The UNC catalog of 1877–1878 (p. 40) described Simonds's responsibilities in teaching botany, which consisted of two courses: Physiological Botany and Vegetable Physiology. The former course emphasized a description of the characteristics of plants as well as their identification. I have not found documentation to indicate that any students enrolled in the advanced botany course on vegetable physiology.

According to the UNC catalog of 1877–1878 (p. 40), the following are descriptions of courses in the School of Botany.

3. SCHOOL OF BOTANY. PROF. SIMONDS.

In this School are included the following:

I. *A course of twenty Lectures upon Physiological Botany, delivered during the last half of the second session.*

Students taking this course are required to analyze and preserve at least fifty species of wild plants, twenty-five of which must be described according to Harrington's *Plant Analysis* [1876]. These are to be the property of the University Museum.

Books of Reference: Gray's *Lessons* [1857] and *Manual of Botany* [1867]; Gray's *Structural and Systematic Botany* [1876]; Oliver's *Elementary Botany* [1869]; Sach's *Text-Book of Botany* [1875]; Johnson's *How Crops Grow* [1880]; Johnson's *How Crops Feed* [1879]; Porcher's *Resources of Southern Fields and Forests* [1863].[74]

II. An advanced course of instruction in Vegetable Physiology, together with original investigations upon the growth of Plants.

This course can be pursued only by those who have completed the preceding.

In April 1878, a reporter (Anonymous, 1878d) of the Chapel Hill *Weekly Ledger* wrote: "The department of Botany is undoubtedly the most attractive feature in the instruction at present." The journalist was impressed that Simonds's presentations were not only theoretical but practical. He delivered two lectures each week and illustrated every detail by effective and clearly executed diagrams. This method of using illustrations reinforced the principles on the student's mind through an association of ideas. The practical aspect of his course consisted of the analysis and description of at least fifty plants that grew in the vicinity of Chapel Hill, which exercise is discussed later in this chapter.

In the academic year 1879/1880, botanical instruction consisted of ten lectures given to sophomores and included laboratory work, which involved plant analysis as discussed above and outlined in the university catalog course description. Simonds's predecessor William H. Smith implemented laboratory work in his botanical instruction but did not remain in his position long enough to fully develop that aspect of his teaching. Apparently, botanical instruction was reduced from twenty lectures in 1877–1878, to fifteen lectures in 1878–1879 and then to ten lectures the following academic year.

Among Ernest Haywood's class notes are notebooks from Simonds's classes in chemistry, geology, zoology, and botany. The botany notebook consists of the detailed texts of nine lectures accompanied by diagrams and illustrations, many drawn in color (Haywood, [1880]). According to a large ledger book in the UNC Archives (UNC Student Records 4: 34), Haywood enrolled in botany in the spring of 1880. Except for the first two lectures, the notes for each lecture begin with a meticulous table (Fig. 7.2), most likely derived from drawings that Simonds made on the blackboard, followed by descriptive text that refers to the table. On page three of his notebook, Haywood recorded how Simonds introduced students to plants:

> Plants are universally distributed. Wherever we go we find some forms; at the equator & in the Torrid regions, in great abundance & luxuriance, diminishing in size & abundance as we go N. until very far N. we find only mosses, lichens &c. They are also found in greater luxuriance at the foot of mts. & diminish towards the summit. They exist not only in well-watered regions but also in waste places & deserts. They grow also in H_2O as the H_2O lily & some are microscopic. All plants may be classed into two great groups:

[74] Citations for these textbooks are included in the References Cited section of this chapter.

Plants	Phenogams	-	Flowering
	Cryptogams	-	Flowerless

The general contents of each lecture were as follows:

I. Flowers, including four types (perfect, complete, regular, and symmetrical).

II. Floral envelopes and flower types (imperfect, incomplete, irregular, unsymmetrical).

III. Stamens (cohesion, adhesion, length, parts); Anthers (attachment and dehiscence); Pistils (parts, kinds: simple and compound); Flowers according to stamens and pistils.

IV. Fruits (simple and multiple), their dehiscence; Ovules and seeds (arrangement, attachment, types); Seeds (coats, nucleus, markings, expansions, and albumen of nucleus); Plants (monocotyledonous, dicotyledonous, polycotyledonous); Exogens.

V. Leaf (parts, venation: net and parallel); Compound Leaves; Phyllotaxy; Vernation (arrangement of leaves in the bud).

VI. Flowers (terminal, axillary): attachment; Inflorescence (indeterminate and determinate); Aestivation; Seeds (parts, constituency, germination, cotyledons); Motion in plants.

VII. Comparison of plants (size, duration, texture, form); Appendages of trunk (branches and buds, thorns, tendrils, aerial roots); Roots (primary and fleshy); Underground stems.

VIII. Plant organs (roots, stems, leaves); Stem; Leaf; Tissue; Exogens.

IX. Leaves (structure); Sap; Products of assimilation.

These lectures introduced Simonds's students to the principal structures of plants; he prepared them in using the required textbook, *The Analysis of Plants*, by Mark Walrod Harrington (1876), Assistant Professor of Botany at the University of Michigan. In the introduction to the book, Harrington expressed his belief that botany mentally challenged the students as effectively as the disciplines of classics and mathematics did. In fact, he thought botany provided superior training "to the powers of observation and critical judgment." The text consisted of a three-page "classified vocabulary" and a set of blank forms in which students would fill in the character states of plants (Fig. 7.3). The terms used in plant identification were listed under the following plant structures: roots, stems, leaves, inflorescences, flowers, calyx, corolla, stamens, pistils, fruit, and seeds. In his lectures, Simonds described and illustrated the character states of these plant structures. After learning the required terminology, students completed the forms, which were adapted to Asa Gray's botanical textbooks (as well as other texts of the day), wherein students would determine the species of plants being identified.

William G. Randall (1880) enrolled in Simonds's general zoology class in the fall of 1880. Randall himself was a skilled artist, later noted for his portraiture. His class notes contain excellent reproductions of Simonds's drawings (Fig. 7.4). Classmate James Lee Love (1921), who became professor of mathematics at UNC, provides remarkable insight on Randall, whom he considered his most interesting classmate and the most lovable and best-liked man in the class. While at Carolina, Randall was known for his crayon portraits. Augustus White Long (1939) also provides personal reminiscences about Randall.

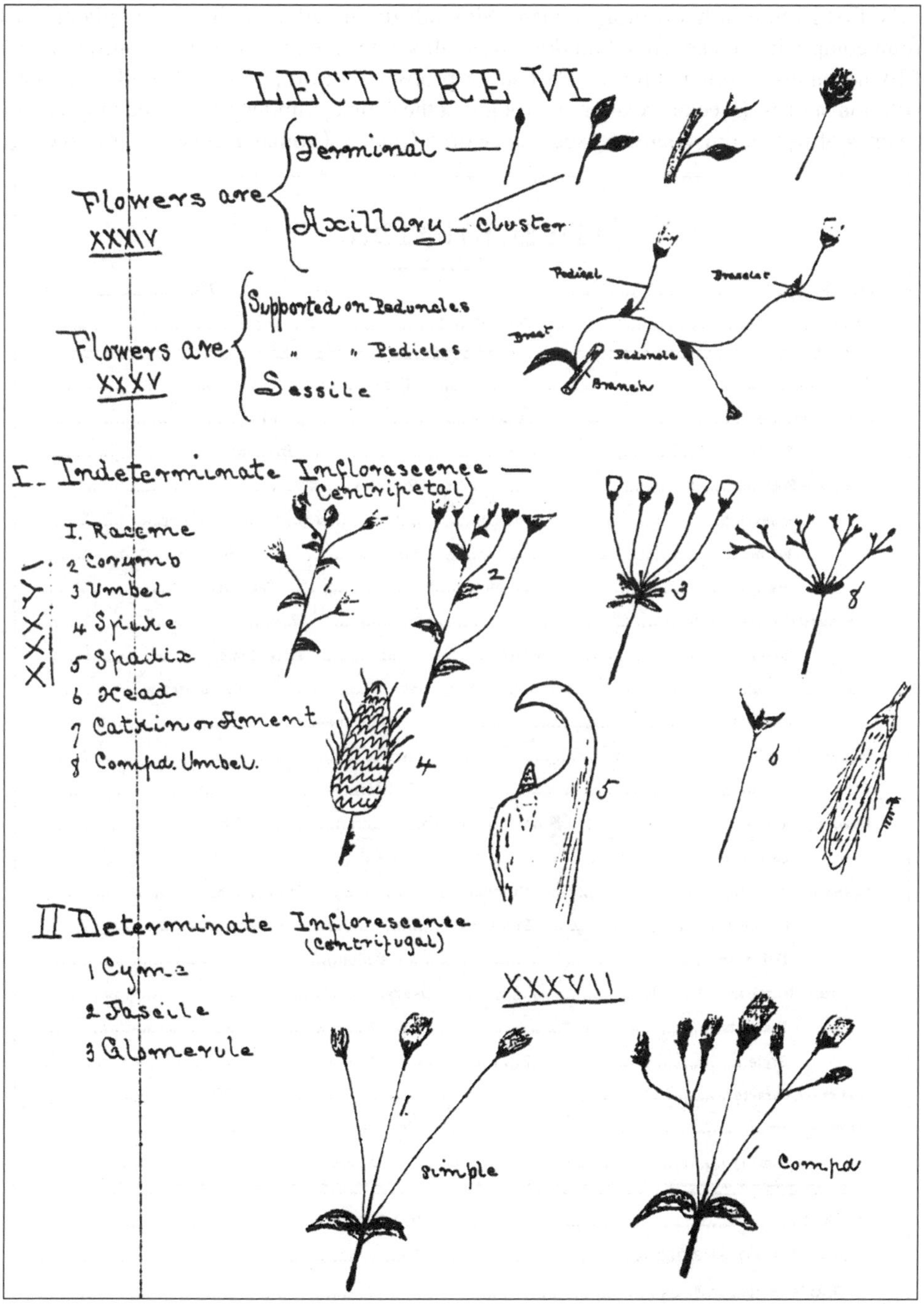

Fig. 7.2. Drawings of types of inflorescences by UNC student Ernest Haywood.

(From Notebook of lectures on botany by Frederic W. Simonds, compiled by Ernest Haywood, spring 1880, p. 31; Collection of Haywood Family #1290, Southern Historical Collection, Wilson Special Collections Library, University of North Carolina, Chapel Hill).

In the first lecture of his zoology course, Simonds discussed the animals and plants, each group comprising a respective kingdom. Randall's notes mention a possible third kingdom of living things, Protisticum (*i.e.,* Protista or protist), first proposed in 1866 by the noted evolutionist Ernst Haeckel. Scattered throughout the course, Simonds presented biographical sketches of influential scientists, such as George Cuvier, Thomas Henry Huxley, and Carl

DESCRIPTION.

DATE.................... **NO**....................

SIZE.................... **HABITAT**....................

Root.—Character.................... Form....................

Stem.—Duration.................... Character.................... Shape....................

Surface.................... Habit.................... Branches....................

Leaves.—Petiole.................... Stipules....................

Venation.................... Position.................... Color....................

Form.................... Apex.................... Base.................... Surface....................

Margin.................... Compound.................... Tendril....................

Inflorescence.—Character.................... Kind....................

Bracts.................... Scales.................... Receptacle....................

Flower.—Peduncle or Pedicel.................... Regularity.................... Symmetry....................

Completeness.................... Perfectness....................

Calyx.—Number of Sepals.................... Form....................

Duration.................... Relations....................

Corolla.—Color.................... Number of Petals.................... Shape....................

Estivation....................

Stamens.—Insertion.................... Number.................... Proportions....................

Connection.................... Filament.................... Anther....................

Dehiscence.................... Pollen....................

Pistils.—Number of Carpels.................... Ovary....................

Placenta.................... Ovule....................

Style.................... Stigma....................

Fruit.—Character.................... Dehiscence.................... Kind....................

Seed.—Form.................... Surface....................

Embryo.................... Albumen....................

SERIES.................... **CLASS**....................

GENUS AND SPECIES....................

COMMON NAME....................

REMARKS....................

Fig. 7.3. Blank form used by students to analyze plants.

(From *The Analysis of Plants*, by Mark W. Harrington [1876]).

Linnaeus. Although the lectures were structured on a classification basis, starting with the lowest forms and ending with the highest forms of animals, the concluding lecture elucidated the geographical distribution of animals. Following the zoological course was that of physiology, described in a newspaper article as "taught in the same clear and practical style. Animals are dissected. . . . After physiology comes botany, in studying which the students are required to work with the Professor and under his directions in the analysis of plants—no theoretical teaching, which is not supplemented by the practical" (Anonymous, [1879]a). Simonds also made use of the microscope in his instruction.

Although not as highly illustrated as the notebooks compiled by Randall and Haywood, Alderman's (1880) notes from Simonds's course on physiology and hygiene, which consisted of twenty-five lectures, included some color illustrations. Those of the human heart and a longitudinal section of a tooth are noteworthy. A schedule of the lectures for this course, given to sophomores in the spring of 1880, is pasted on the inside front cover of the notebook.

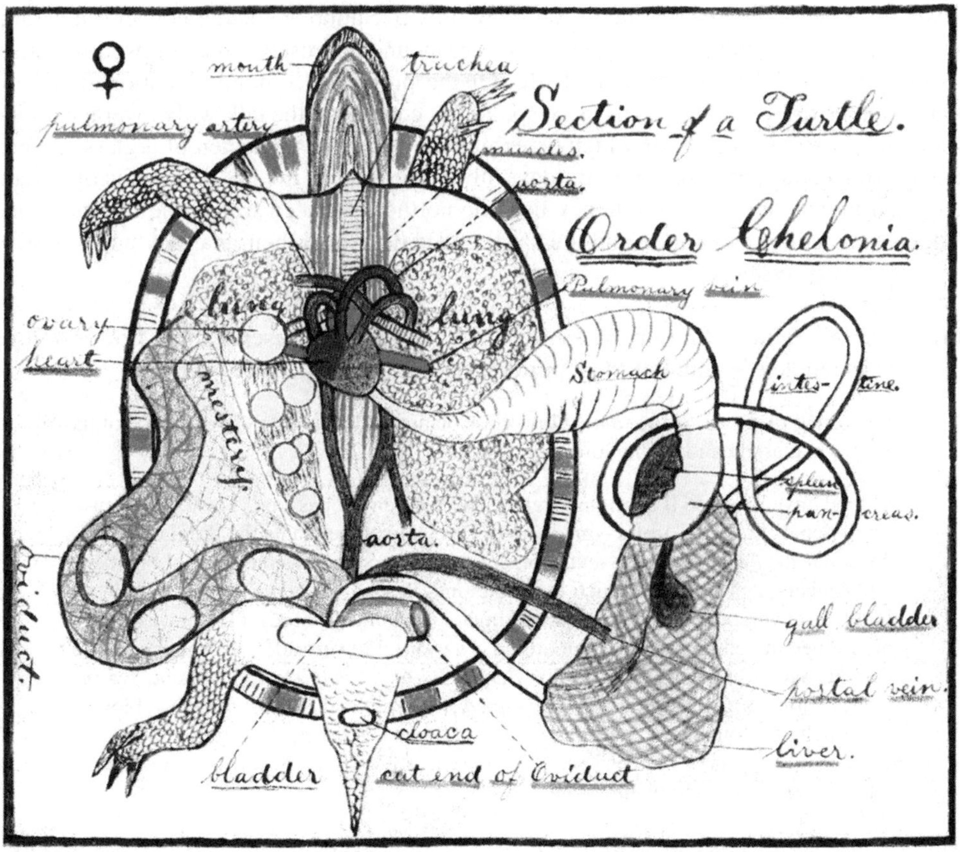

Fig. 7.4. Illustration of a dissected turtle drawn by UNC student William G. Randall.

(Although a zoological topic, the drawing shows the detail of Randall's illustrations. From Lectures on Zoology by Prof. Simonds; [notebook compiled by William G. Randall, 1880], p. 202; Southern Historical Collection #3580-z, Wilson Special Collections Library, University of North Carolina, Chapel Hill).

Instruction in geology and mineralogy consisted of three courses: general geology (thirty lectures), resources of North Carolina (five lectures), and laboratory instruction in determinative mineralogy and blow-pipe analysis. In his lecture on historical geology, Simonds noted his views on Charles Darwin's theory of evolution, which had clearly influenced his perspective in science. According to UNC student Henry Faison (1879), Simonds's views were inclined toward the "evolutionist theory." Simonds lectured how previous forms of plants and animals had been preserved as fossils. Faison was captivated by these new theories, telling his father, "It is very curious to know that the animal and plant life of forgotten eras are preserved perfectly in the rocks as fossils. Tis a beautiful study; and the establishment of the science stands as an everlasting monument to the progress of man."

While the university was expanding its educational mission, Simonds assumed additional teaching responsibilities. In the Normal School, he gave five lectures on natural history in 1879 and two on zoology and one on "Curious Flowers" in 1880. An anonymous newspaper reporter (Anonymous, 1879b) described the content of Simonds's natural history lectures given in 1879. In his first lecture of Saturday, July 5, Simonds noted a new era in natural history had opened thirty years ago with the compound microscope that allowed scientists to study the "low" forms of animal life. Of the protozoa, Simonds mentioned several kinds, and he drew the organisms as he spoke. He concluded the talk with coverage on the foraminifera and sponges, for which he exhibited a stunning specimen of a glass sponge. His second lecture, presented on Tuesday, July 8, focused on the life and habits of several common insects, including the honey bee, the northern tobacco worm, the mosquito, the Rocky Mountain grasshopper, and the ant lion. He described their anatomy and digestive and circulatory systems.

Presenting his third lecture on Wednesday, July 2, Simonds described the flower and its parts. The complete account of the reporter's synopsis of the lecture follows.

> Flowering plants or *Phenogams*. Flowerless plants or *Cryptogams*. Importance of the flower in classification. The flower consists of
>
> I Floral Envelopes and II Essential Organs. The floral envelopes are 1. The Calyx—made up of Sepals. 2. The Corolla—made up of Petals. In some plants the floral envelopes constitute the *Perianth*. The essential organs are 1. The Stamens, consisting of Filaments and Anthers, 2. The Pistil, which is divided into Stigma, Style and Ovary. Within the latter are developed the ovules which ultimately become the seeds. Flowers, as leaves, are formed in buds. The time of blooming is various. Flowers differ very much in size. The standard flower is that of the Flax which is *Perfect*, *Complete*, *Regular* and *Sym[m]etrical*. Flowers may be *Imperfect*, *Incomplete*, *Irregular* and *Unsym[m]etrical*. Flowers are usually built upon a certain plan—that of "fives" is the most common. Many are built upon the plan of "threes" and "fours." The office of the flower is to provide for the reproduction of the plant.

The two concluding talks on natural history were on animals and concerned respiration and circulation (blood, blood vessels, and the heart), presented on Thursday, July 10, and Friday, July 11, respectively (Anonymous, 1879d). Simonds explained the differences in the hearts of invertebrates and vertebrates, including fish, reptile, and mammal.

When the School of Medicine was established in 1879, Simonds concurrently became a "Professor of Botany and Physiology," teaching the first-year medical students. The school primarily prepared students to attend lectures at medical colleges at that time. Students took zoology, physiology, and botany as auxiliary courses and paid a fee of $15.00 for the academic year (BOT 16: 61). In the Pharmacy School, Simonds was "Professor of Botany." In the spring session, he gave three lectures on structural and physiological botany each week. When the school was set up, its prospectus emphasized that in botany courses students would especially focus on "the analysis of plants and the making of herbaria" (BOT 16: 82). For identifying plants, students were to consult *Gray's Manual* (1867) and *Lessons* (1879) or Bessey's *Botany* (1880) as a textbook. Bessey's textbook was very significant, because it introduced American students of botany to the current scientific work of a number of European botanists. The book was based in great part on the *Lehrbuch der Botanik* by the eminent German botanist Julius Sachs (1868). The information presented by Sachs was founded on his original investigations in morphology and anatomy as well as physiology. Bessey's book fostered a new generation of American thought in botany.

Simonds's medical and pharmacy students also had access to the university's libraries and museum collections. The annual fee for Materia Medica and Pharmacy was $30.00 (Botany, $5; Chemistry, $25; and Diplomat, $2). Beginning with the fall term of 1880, Simonds added a class in freehand drawing and one in mineralogy (BOT 7: 312). On a temporary basis, Simonds worked in 1880 for the U.S. Tenth Census, collecting statistics on the mica industry of North Carolina. Concurrently, he continued his doctoral studies, which culminated in a Ph.D. in geology awarded in 1879 by Syracuse University, where he had been allowed a non-residency status but was required to take stringent examinations. No extant copy of his dissertation can be found.

A glimpse into the extent of Simonds's teaching responsibilities is found in the academic report of his instruction in 1879–1880, which is recorded in the minutes of the UNC Board of Trustees (BOT 7: 350–352). Of the seventy-nine students that Simonds instructed, seventeen took geology, fifteen mineralogy, forty-seven zoology, fifty-two physiology, and forty-nine botany. Another source of information on Simonds's botanical instruction is the student grade ledgers housed in the UNC Archives. From 1878 to 1880, he taught botany each spring semester, instructing 122 students (29 in 1878, 56 in 1879, and 47 in 1880) (Appendix 1). Among these young men who became notable citizens were: Edwin Anderson Alderman (who served as UNC president from 1896 to 1900), Charles Duncan McIver (who was the founder and first president of State Normal and Industrial School for Girls, now UNC at Greenboro), and two of President Battle's sons, Kemp Plummer Battle, Jr. (who became a noted physician and medical teacher), and Herbert Bemerton Battle (who served the North Carolina Agricultural Experiment Station, first as assistant chemist and later as state chemist and director).

Botanical and Entomological Endeavors

According to the UNC catalog of 1877–1878 (pp. 20–21), the university herbarium contained not only specimens of the local flora but also "a fine set of native woods, carefully prepared and labeled; and a large collection of seeds, grains and fibres from many countries, secured for the Museum by the State Geologist, some of which [would] be employed in the experiments of the students in Botany. By an arrangement with the State Board of Agriculture, the University

[was] to receive duplicates of its collections, so that ample opportunity [would] be given for familiar acquaintance with the resources of the State." A Raleigh newspaper reporter (Anonymous, 1878c) also commented on the "well-represented" flora of the herbarium. The dried plants (and insect collections) were part of the museum, then occupying a large room ("Library Hall," formerly used by the Philanthropic Society) on the third floor of Old East. In a broadside, President Battle (1878b) stated: "The Museum is of great assistance in instruction, particularly in Mineralogy and other branches of Natural History." In writing to Harvard University botanist Asa Gray in early September 1879, Simonds (1879c) mentioned that the press of academic duties prevented him from developing the herbarium. Nonetheless, the herbarium was increased by collections of plant specimens from students in Simonds's botany class and from acquisitions that he procured from the U.S. Commissioner of Agriculture, such as a box of "botanical specimens" (LeDuc, 1878) and a collection of grasses (Simonds, 1880d, 1880e). Other donations to the herbarium, which are noted in the UNC catalog of 1877–1878 (p. 51), included an unusual form of wiregrass from President Battle and native woods from A.T. McIver.

In teaching his first botany class (in the spring of 1878), Simonds (1879c) apparently inspired his students to conduct botanical excursions into the region around Chapel Hill. Several students reported the occurrence of a beautiful rhododendron not far from campus. Somewhat incredulous that such a shrub was found, Simonds (1879a) himself visited the area, which was on north-facing rocky slopes above Morgan Creek, nearly three miles from the village. In describing the find to botanist Asa Gray, Simonds (1879c) wrote: "The steep bank of Morgan's Creek was burried [*sic*] in a mass of evergreen leaves among which were clusters of flowers varying from light to deep purple. The plant was a *Rhododendron*. I examined the flowers and sent specimens to my friend Professor Dudley[75] of Cornell University." Simonds and Dudley thought the species to be *Rhododendron catawbiense* Michaux. Simonds sent Asa Gray specimens from two collections, one made on April 30, 1879 (Fig. 7.5) and another on April 26, 1880. Both specimens are preserved in the Gray Herbarium at Harvard University. Enthused by the discovery, Simonds (1879a) published an article about the shrub.

Although Elisha Mitchell had collected this species on May 20, 1820, in the same vicinity, his find is known only from an entry in his private journal (McVaugh et al., 1996). The area of collection was popularly known as Laurel Hill. Even up to 1896, students frequented the steep slope on the southern side of Morgan's Creek on local excursions and gathered the shrub's blossoms in a reckless manner and in large quantities (Anonymous, 1895; Anonymous, 1896). Apparently concern was growing to preserve the stand of specimens, because the plant was rare in this region of the state. The species is different from the usual rhododendron (*Rhododendron maximum*) of the North Carolina mountains in having broader leaves and deeper pink flowers. In the mountains of western North Carolina, *R. catawbiense* (called rosebay) blooms before other species of rhododendrons (Hunt, 1873). It is also found close to Occoneechee Mountain near Hillsborough, North Carolina. In 1919, botanist William C. Coker (1919) reviewed the distribution of the species in North Carolina, including Simonds's record, and described the shrubs from Chapel Hill and several other local populations as a new form, *R. catawbiense insularis*.

[75] Several specimens, most likely those sent by Simonds to Dudley, survive in Dudley's herbarium, now housed in the California Academy of Sciences Herbarium.

Fig. 7.5. Specimen of *Rhododendron catawbiense* Michaux collected in Chapel Hill, North Carolina, April 30, 1879, by Frederic W. Simonds.

(Courtesy of the Gray Herbarium [GH], Harvard University Herbaria).

Simonds's botanical endeavors included excursions to the mountains of North Carolina, such as his observation of the flora of Roan Mountain in 1880 (Chickering, 1880). His botanical publications include the article on *Rhododendron* (1879a) already discussed as well as several book chapters: "Vegetation" of Texas (1905), the "Animals and Plants" of Texas (1908b), and "The Distribution of Useful Plants" (1908b). The remainder of his publications, which number slightly over forty, chiefly cover geological topics.

Besides sharing botanical specimens with correspondents, Simonds was also interested in zoological specimens. Spencer Fullerton Baird (1880a, 1880b), Assistant Secretary of the United States National Museum, wrote to Simonds on at least two occasions. He requested natural history objects (including birds' eggs, particularly of hawks) and thanked Simonds for sending skulls and skeletons.

Simonds also devoted considerable time to entomological studies, including the acquisition of literature. He requested and procured relevant literature on insects such as Comstock's (1879) "Report upon Cotton Insects" (Simonds, 1880d, 1880e). Simonds's teaching at UNC included field collections; the UNC catalog of 1877–1878 (p. 51) noted that his class deposited a cabinet of insect specimens from Chapel Hill and environs in the university museum. The specimens, which were arranged in systematic order by species, consisted of "splendid collections of butterflies, moths, beetles, and other insects" (Anonymous, [1878] a), as well as grasshoppers, hornets, wasps, and spiders (Anonymous, 1877e). Simonds and his students in zoology also collected and deposited in the museum other animals, such as bats, snakes, frogs, mice, and moles. Specimens were either "impaled on cork or comfortably coiled up in wide-mouthed vials" (Anonymous, 1877e).

Simonds desired to educate farmers about insect pests of crops. His efforts were among the earliest such work in the state. In the fall of 1877, Simonds (Anonymous, 1877c) gave an address before the Agricultural Society of North Carolina, which was held in conjunction with the State Fair in Raleigh. In the evening of October 17, he delivered a lecture on the "Science of Entomology." Simonds encouraged farmers to become knowledgeable about insects harmful to crops and to learn ways known to control them. In an article, which is most likely his first publication after arriving in Chapel Hill, Simonds (1878a) reported that under his supervision twenty cases of insects from the Chapel Hill area had been collected. In the paper, he recommended that the state's citizens send him insect specimens from North Carolina, particularly "those injurious to vegetation." He further recommended that each specimen be accompanied with collection data, including date of collection, where found, and on what plant. Building on Simonds's efforts to educate farmers about insect pests, his successor in zoology, George F. Atkinson, was more active in this kind of outreach and published a number of papers to aid the agriculturists of the state.

An anonymous reporter (Anonymous, 1879c) praised Simonds for going round to visit the state's citizens and becoming acquainted with them, much like the ways of Elisha Mitchell. Simonds's visits provided an opportunity for the citizens to learn about the work of the university's professors and the benefits that they could gain.

Resignation from UNC

While at UNC, Simonds experienced periods of ill health, perhaps induced, in part, by his arduous workload, which "was too large for any one man, unassisted, to cover" (Simonds, [1908]a). In addition to his teaching responsibilities, he completed the course requirements for a doctoral degree through Syracuse University and served UNC as the librarian for a brief period (1877–1878) and as Secretary of the Faculty (September 1879–January 1881). According to Battle (1907–1912, 2: 105), Simonds suffered from pneumonia in May 1878. Between September and November 1878, Simonds was absent from the faculty meetings due to the illness, which also probably kept him from the classroom. An editor of the *University of North Carolina Magazine* noted that Simonds returned to teaching about November 1878 (Anonymous, 1878b); however, he looked rather thin and delicate (B., 1878). When his health took a turn for the worse in 1881, he decided to resign. In his letter of resignation to President Battle and the UNC trustees, dated May 5, 1881, he succinctly stated: "On account of continued ill health I hereby tender to you my resignation as Professor of Geology, Zoology and Botany to take effect at the close of the present collegiate year" [*i.e.*, spring 1881] (BOT 7: 411). He had actually ceased teaching early in the winter term. Thomas West Harris, who headed the newly established Medical School, took over the instruction of the physiology class after Simonds had delivered only eight of the twenty-six planned lectures (BOT 7: 473). Professor Carey Dempsey Grandy assumed responsibility for teaching mineralogy. Under Grandy's tutelage, "the greater part of [the students'] work consisted of laboratory practice, the object of which was to enable them to determine any ordinary mineral of interest, or value" (BOT 7: 472). Botany most likely was not taught that spring.

A disruptive classroom incident toward the close of 1880 undoubtedly also contributed to Simonds's decision to resign his professorship in his third year at UNC. Cornelia P. Spencer, Simonds's neighbor, related the following episode, which perhaps confirmed her initial impression of Simonds as too much a Yankee to teach successfully in the South. Upon entering his recitation class in early December 1880, Simonds was confronted with a room of noisy and disruptive students. He questioned whether they were little boys or gentlemen. Upon hearing his words, "certain of the southern bloods fired up at this remark as supposing he had said they were not 'gentlemen' (that name so dear to southern hearts & so little understood in truth by many who claim it), and when the lecture was over ... [three students] went up to [him] and one of them, [John Moseley Walker, Jr., a senior from Hickory, North Carolina, who was in Simonds's geology class] assailed him with most insulting profanity & giving him three minutes to take back his words, Walker was frantic & furious with rage—called Dr. S. a 'G___ d____ scoundrel, a d___ infernal hound of a Yankee—who only knew the insides of a grasshopper, etc. ... [Simonds] took back his words & gave up" (Spencer, 1880). Simonds went home and wrote out his resignation. He had lost face with his class, and he realized he could not remain at UNC. As a northerner, Simonds said (according to Spencer) that he did not "understand southern ways nor southern use of words." He noted that he was "not like you all & never shall be." Simonds felt as though he had failed to blend in with the students and local people.

This episode was the focus of a special meeting of the faculty held December 6, 1880, in the president's room (UNC Faculty 6: [480]–481). Walker appeared before the faculty (minus professors Mangum and Simonds). Although the faculty unanimously dismissed Walker from the university, he applied for readmission after writing a letter of apology, dated December 21, 1880, to the faculty (Walker, 1880). In his apology, he stated "that [he] had conceived [him]self grossly insulted and that [he] acted under that impression and upon the impulse of natural resentment caused thereby." Walker's application was backed by the recommendations of several influential citizens, including two university trustees. Simonds himself supported the student's reinstatement after receiving an apology from him. Nonetheless, Simonds noted that he "felt very much hurt at the insults he [*i.e.*, Walker] heaped upon [him]" (Simonds, 1880a). Walker subsequently resumed classes and earned the B.A. degree in 1881.

At their May 6, 1881 meeting, the faculty voiced its regret about Simonds's resignation. The members of the faculty paid him tribute by noting "his skill and attainments in his department, his uniform kindness and courtesy, his fidelity to his duties, and his success in the conduct and instruction of his classes" (UNC Faculty 6: 487).

Departure and Later Years

By late May 1881, Simonds and his wife had moved to Syracuse, New York. Simonds (1881a) wrote to President Battle that they were "rapidly gaining health and strength." Reflecting about their life in Chapel Hill, particularly about the Walker affair, Simonds admitted that "there [was] mingled with our life there, especially the last few months, much sorrow, but if this had not been we never would have known how many dear and generous friends we left behind." A couple of months later, he mentioned having a warm spot in his heart for his North Carolina friends (Simonds, 1881b). On November 20, 1881, Simonds and his wife left New York to live in San Jose, California, where his parents had taken up residency. Fourteen years later, Simonds (1895) reiterated in a letter to President Battle the good memories that he and his wife held and how they "always had, in [their] hearts, a tender regard for Chapel Hill and its people."

In California, their health continued to improve. Simonds taught science at San Jose High School for eighteen months and then became its principal (Simonds, [1908]a). He subsequently worked in a variety of positions: as lecturer of Economic Geology at Cornell University (1887); as Professor of Biology and Geology at the University of Arkansas (1887–1890); as Special Assistant to the Geological Survey of Arkansas (1887–1892); and as Associate Professor of Geology (1890–1895) and later Professor of Geology (1895–1936) at the University of Texas. At UT, where he worked for the remainder of his life, Simonds (Fig. 7.6) served as the chair of the Department of Geology from 1891 to 1921. In writing about the history of the geology department at the University of Texas at Austin, geologist Keith Young (1967) referred to Simonds's term as "the long régime of F.W. Simonds." At Austin, Simonds developed academic bonds with classical languages professor William J. Battle, President Kemp Battle's youngest son (Anonymous, 1940). When Simonds was a UNC faculty member, young Battle was "in knee trousers reading Greek at the age of six." (Benedict, 1916).

Simonds (Fig. 7.7) officially retired in 1936 and was placed on three-quarter time; in 1937 on one-half time, and he continued to serve the university until his death (Lockwood,

Fig. 7.6. Portrait of Frederic William Simonds, ca. 1903.

(From the *Cactus*; [Yearbook of the University of Texas at Austin], Published by the Athletic Association of the University of Texas, 1903, p. 22).

1941). During his so-called retirement, he continued to teach two classes: the physiography of North America and the physiography and geology of the American Southwest. At the University of Texas, he considered himself a teaching geologist, and "his career was one of a pioneer professor whose chief endeavor was to promote the study of the natural sciences, especially geology" (Anonymous, 1941a). He became an authority on the stratigraphic geology of the Southwest and of the physiography of North America, particularly of Texas and the Southwest.

At age eighty-seven, Frederic William Simonds died at his Austin, Texas, home on March 27, 1941 and was buried in Oakwood Cemetery, Austin, Texas (Fig. 7.8). At that time, he was the oldest member of the UT faculty. Additional biographical insight into Simonds's life can be gained from the literature (*Alcade*, 1915; Anonymous, 1941b; Anonymous, 1941c; Deussen, 1942; HTO) and archival sources (CU; UT).

Fig. 7.7. Portrait of Frederic William Simonds, undated.

(Courtesy of Department of Geological Sciences, Jackson School of Geosciences, conference room 2.120A, University of Texas, Austin. Original by University Studio; reproduction by Joseph Jaworski).

Legacy and Contributions

In recognition of his contributions to Arkansas geology, the University of Arkansas conferred on Simonds the D.Sc. in 1893. He was honored as one of fifty men and women who had contributed to the early development of the University of Texas at Austin (Eckhardt, [1979]). Simonds was a fellow of the American Association for the Advancement of Science and the Geological Society of America as well as an original fellow of the Texas Academy of Science (in which he served as president in 1899–1900 and as secretary for several years).

Fig. 7.8. Gravestone of Frederic William Simonds, Oakwood Cemetery, Austin, Texas, Section 2, Lot 2, June 19, 2005.

(Courtesy of Dale Flatt, Oakwood Cemetery).

He also held a membership in the Southwestern Geological Society. He was elected to Phi Beta Kappa, Sigma Xi (serving as president of the Texas Chapter in 1915–1916), Kappa Sigma, and Sigma Gamma Epsilon.

Simonds's artistic skills in drawing supplemented his teaching techniques in the classroom. At UNC, he was the first to substantively develop laboratory work in botany (as well as in geology and zoology). In the spring, his students collected and identified plants, which were preserved and housed in the university museum. Although his students were required to purchase textbooks, drawings, lectures, and laboratory exercises were his primary instructional tools. Students could also consult the plant specimens (including native woods and fossils) in the herbarium. He was an early proponent of educating farmers about insects harmful to crops. His instruction in natural history fostered a spirit of investigation among his students.

At UNC, Simonds served as librarian and later as the secretary of the faculty. His tenure at Chapel Hill was punctuated with periods of ill health and marred by a classroom incident that precipitated his decision to resign his professorship. He devoted the last forty-six years of his life to teaching at the University of Texas, where he inspired numerous students in geology and became an expert in that discipline. Fannie West Harris Pope (1915), one of his former students at UT, remembered Simonds's "beautiful enthusiasm in his work …; his perfect good nature and geniality; his kindly and helpful interest in his students; and above all, his sincerity of purpose."

References Cited

(References that conclude with page numbers within parentheses indicate sources from where quotes and pertinent information were obtained.)

Alcade, The (Austin, TX). 1915. Former students of the University of Texas at Austin published *The Alcade*. Volume 3, number 5, March 1915 features a tribute to Frederic William Simonds and contains the following articles: Reminiscences of a quarter of a century 1890–1915, by F.W. Simonds (pp. [435]–451); Frederic William Simonds, M.S., Ph.D., D.Sc., by H.W. Harper (pp. 451–458); Appreciations of former students (pp. 458–479); Dr. Frederic William Simonds, by T.U. Taylor (pp. 479–484)

Alderman, Edwin A. 1880. "Lectures on Physiology—Delivered by Prof. Simonds, 1880; taken by Ed. A. Alderman." (Edwin A. Alderman Papers, Southern Historical Collection 1869-z, Wilson Special Collections Library, University of North Carolina, Chapel Hill)

Alderman, William C. 1878. Letter of September 22, 1878 to [Belle]. (Alderman Family Papers #2994-z, Southern Historical Collection, Wilson Special Collections Library, University of North Carolina, Chapel Hill)

Anonymous. 1877a. The University. *Hillsborough Recorder* (Hillsborough, NC), new ser. 5(41): [3], Wednesday, September 12, 1877.

Anonymous. 1877b. The University. *Observer* (Raleigh) 1(39): [2], Tuesday, August 21, 1877.

Anonymous. 1877c. Meeting of the Society. *Raleigh News* (Raleigh) 12(4): [4], Thursday, October 18, 1877.

Anonymous. 1877d. Meeting of the Agricultural Society. *Raleigh News* (Raleigh) 12(42): [4], Friday, October 19, 1877.

Anonymous. 1877e. The University—Prof. Simonds and his work. *Observer* [Daily] (Raleigh) 2(165): [3], Friday, November 23, 1877.

Anonymous. [1878]a. University Museum—A valuable addition, p. 72 in Kemp P. Battle, *Scrapbook of Material Relating to the University of North Carolina*, 1869–1893, vol. 2. (Newspaper clipping without bibliographical data. North Carolina Collection, Wilson Special Collections Library, University of North Carolina, Chapel Hill)

Anonymous. 1878b. Professor Simonds. *North Carolina University Magazine* (University of North Carolina, Chapel Hill) 2(3): [129].

Anonymous. 1878c. Commencement exercises at the University of North Carolina, Chapel Hill, Thursday, June 5, 1878. *Observer* (Raleigh) 3(114): [3], Friday, June 6, 1878.

Anonymous. 1878d. [Department of Botany] in University items. *Weekly Ledger* (Chapel Hill, NC) 1(1): 2, Saturday, April 18, 1878.

Anonymous. [1879]a. The study of natural history—Mode of teaching at the University, p. 13 in Kemp P. Battle, *Scrapbook of Material Relating to the University of North Carolina, 1869–1893*, vol. 1. (Newspaper clipping without bibliographical data. North Carolina Collection, Wilson Special Collections Library, University of North Carolina, Chapel Hill)

Anonymous. 1879b. The University Normal School. The third term. *Weekly Ledger* (Chapel Hill, NC) 3(13): 1–2, Saturday, July 12, 1879.

Anonymous. 1879c. [Prof. Simonds] in Home matters. *Weekly Ledger* (Chapel Hill, NC) 3(2): 3, Saturday, April 26, 1879.

Anonymous. 1879d. The University–Normal School. *Chapel Hill Ledger* (Chapel Hill, NC) 3(14): 2, Saturday, July 19, 1879.

Anonymous. 1895. "Laurel Hill." *The Tar Heel* (University of North Carolina, Chapel Hill) 3(29): [3], May 9, 1895.

Anonymous. 1896. The Chapel Hill region. *Hellenian* 1896: 17–20. (p. 19)

Anonymous. 1940. Simonds observes 50 years' service. *Texan*, February 21, 1940.

Anonymous. 1941a. Oldest professor, Dr. Simonds, dies. *Texan*, March 28, 1941.

Anonymous. 1941b. Dr. F.W. Simonds, 87, noted as geologist. *New York Times* 90(30,380): 15, Saturday, March 29, 1941.

Anonymous. 1941c. Dr. F.W. Simonds dies: Texas U. geologist, 87. *New York Herald Tribune* 100(34, 467): 10, Saturday, March 29, 1941.

B., R.L. 1878. [Prof. Simonds] in University items. *Weekly Ledger* (Chapel Hill, NC) 2(6): 2, Saturday, November 30, 1878.

Baird, Spencer F. 1880a. Letterpress copy of March 30, 1880 to Frederic W. Simonds. (Smithsonian Institution Archives, Record Unit 53, National Museum, Correspondence and Memoranda, vol. 21, p. 374)

Baird, Spencer F. 1880b. Letterpress copy of May 10, 1880 to Frederic W. Simonds. (Smithsonian Institution Archives, Record Unit 53, National Museum, Correspondence and Memoranda, vol. 21, p. 687)

Battle, Kemp P. 1878a. *Report of Hon. Kemp P. Battle, President of the University of N.C. on the State of the University and the Normal School*. Daily and Weekly News Steam Book and Job Print, Raleigh. (pp. 4–5)

Battle, Kemp P. 1878b. [Broadside]. University of North Carolina, Chapel Hill, N.C., August 16th, 1878. (Box 9, folder 328, University Papers #40005, University Archives, Wilson Special Collections Library, University of North Carolina, Chapel Hill)

Battle, Kemp P. 1907–1912. History of the University of North Carolina. Printed by Edwards & Broughton, Raleigh. 2 vols. (Vol. 2, pp. 105, 639)

Benedict, H.Y. 1916. F.W. Simonds. *Daily Texan*, November 30, 1916.

Bessey, Charles E. 1880. *Botany for High Schools and Colleges*. (*American Science Series*). Henry Holt and Company, New York.

Bill. 1878. Letter from Chapel Hill. *Observer* (Raleigh) [Daily] 4(59): [2], Sunday, September 8, 1879.

BOT. Board of Trustees, University of North Carolina, Chapel Hill. *Minutes of the Board of Trustees*. (University Archives #40001, Wilson Special Collections Library, University of North Carolina, Chapel Hill) (Vol. 7, July 1868–March 1883; vol. 16, 1875–1881)

Chickering, J.W. 1880. A summer on Roan Mountain. *Botanical Gazette* 5: 144–148.

Coker, William C. 1919. The distribution of *Rhododendron catawbiense*, with remarks on a new form. *Journal of the Elisha Mitchell Scientific Society* 35: 74–82; plates 19–22.

Comstock, John Henry. 1879. *Report upon Cotton Insects*. Prepared under the direction of the Commissioner of Agriculture in pursuance of an act of Congress approved June 19, 1878. United States, Department of Agriculture, Government Printing Office, Washington, D.C.

Cornell University. 1875. *Programme of Exercises of the Seventh Annual Commencement of the Cornell University*. The University Press, Ithaca, NY.

CU. Cornell University. Frederoc [*sic*] W. Simonds *Scrapbook*. (Archives 37-5-187); Deceased Alumni Files (Archives #41-2-877, Box 118, folder on F.W. Simonds; Carl A. Kroch Library, Rare and Manuscript Collections, Cornell University)

CU-R. 1872. *The Cornell University Register. 1871–72*. At the University Press, Ithaca, NY. (p. 28)

CU-RS. *Cornell University Register of Studies*. (Archives #36/1/630, vol. B [1874–1880], pp. 90–91; Carl A. Kroch Library, Rare and Manuscript Collections, Cornell University)

Deussen, Alexander.1942. Memorial to Frederic William Simonds. *Proceedings Volume of the Geological Society of America, Annual Report for 1941*, pp. 193–200; Pl. 8 (portrait)

DMV. 1913. *Dudley Memorial Volume; Containing a paper by William Russel Dudley and appreciations and contributions in his memory by friends and colleagues*. (*Leland Stanford Junior University Publications, University Series*). Published by the University, Stanford University.

Eckhardt, Carl J. [1979]. *One Hundred Faithful: to The University of Texas at Austin*. [no statement of publisher or place]. (Composed of two parts: *Fifty Stars in the University Firmament* and *Fifty Who Loved and Served the University of Texas at Austin*.)

Faison, Henry E. 1877. Letter of September 9, 1877 to "Sister." (Box 3, folder 35, Henry Wilson Faison Papers #3789, Family Series, Southern Historical Collection, Wilson Special Collections Library, University of North Carolina, Chapel Hill)

Faison, Henry E. 1879. Letter of November 23, 1879 to "Pa." (Box 3, folder 35, Henry Wilson Faison Papers #3789, Family Series, Southern Historical Collection, Wilson Special Collections Library, University of North Carolina, Chapel Hill)

Fitzpatrick, Harry M. 1926. "An Historical Sketch of the Early Days of Botany in Cornell." Typed manuscript. (Copies at the Bailey Hortorium and Herbarium and the Library of the Hunt Institute for Botanical Documentation)

Gray, Asa. 1857. *First Lessons in Botany and Vegetable Physiology*. Ivison & Phinney and G. Putnam & Co., New York.

Gray, Asa. 1867. *Manual of the Botany of the Northern United States, Including the District East of the Mississippi and North of North Carolina and Tennessee, Arranged According to the Natural System*. 5th ed. Ivison, Phinney, Blakeman, & Co., New York; S.C. Griggs & Co., Chicago.

Gray, Asa. 1871. *Gray's Lessons in Botany and Vegetable Physiology: To which is added a copious glossary, or dictionary of botanical terms; Manual of the Botany of the Northern United States: Including the district east of the Mississippi and north of North Carolina and Tennessee: Arranged according to the natural system*. Ivison, Blakeman, Taylor, New York. (not seen)

Gray, Asa. 1876, © 1857. *Introduction to Structural and Systematic Botany and Vegetable Physiology*. Being a fifth and revised edition of *The Botanical Text-book*. Ivison, Blakeman, Taylor & Co., New York.

Gray, Asa. 1879. *Gray's Lessons in Botany and Vegetable Physiology*. Ivison, Blakeman, Taylor & Co., New York and Chicago. (This text was bound with Gray's *Manual* [5th edition, 8th issue] and served as an introduction to it.)

Harrington, Mark W. 1876. *The Analysis of Plants, Intended for Schools and Colleges and for the Independent Botanical Student*. Sheehan & Co., Ann Arbor, MI.

Haywood, Ernest. [1880]. ["Lectures on Botany, Delivered by Frederic W. Simonds, University of North Carolina Chapel Hill, spring 1880"]. 1 volume, containing 64 pages of handwritten notes, diagrams, and illustrations (many in color). (Folder 379, volume 94, Collection of Haywood Family, #1290, Southern Historical Collection, Wilson Special Collections Library, University of North Carolina, Chapel Hill)

Heitman, Rev. J.F. 1879. Mrs. Moon—Methodism in Chapel Hill—New Church—Commencement—Normal School—Railroad—Mission of Methodism. *Raleigh Christian Advocate* (Raleigh) 24(17): 2–3, Wednesday, April 30, 1879.

HTO. The Handbook of Texas Online. Frederic William Simonds. <http://www.tsha.utexas.edu/handbook/online/articles/view/SS/fsi19.html> (bibliography)

Hunt, T.S. 1873. The highlands of North Carolina. *Friends' Review* 27: 141–142.

Hutchins, J.R. 1878. Letter from J.R. Hutchins, Esq. Chapel Hill, April 26. *Observer* (Raleigh) 3(104): [2], Wednesday, May 1, 1878.

Johnson, Samuel William. 1879. *How Crops Feed: A treatise on the atmosphere and the soil as related to the nutrition of agricultural plants*. Orange Judd and Company, New York.

Johnson, Samuel William. 1880. *How Crops Grow. A treatise of the chemical composition, structure, and life of the plant, for all students of agriculture*. Orange Judd and Company, New York.

Knight, Edgar W. and Agatha Boyd Adams (editors). 1946. *The Graduate School: Research and publications*. Edited with a foreword by Edgar W. Knight and Agatha Boyd Adams. (*The University of North Carolina Sesquicentennial Publications*). University of North Carolina Press, Chapel Hill. (p. 8)

LeDuc, William G. 1878. Report of the Commissioner of Agriculture, pp. 5–95 in *Annual Report of the Commissioner of Agriculture for the Year 1878*. Government Printing Office, Washington, D.C. ("Botanical Division" report, p. 17)

Lockwood, Thelma. 1941. Letter of November 25, 1941 to R[obert] E. Coker. (Box 31, folder 454, Robert E. Coker Papers #3380, Southern Historical Collection, Wilson Special Collections Library, University of North Carolina, Chapel Hill)

Long, Augustus White. 1939. *Son of Carolina*. (*A Segment of the American Scene*). Duke University Press, Durham, NC. (pp. 95–96)

Love, James Lee. 1921. Diary (Vol. 6, James Lee Love Papers #4139, Southern Historical Collection, Wilson Special Collections Library, University of North Carolina, Chapel Hill) (pp. 45–48)

M., T. 1879. Prof. Simonds at Mebaneville. *Observer* (Raleigh) 5(96): [2], Tuesday, April 22, 1879.

McVaugh, Rogers, Michael R. McVaugh, and Mary Ayers. 1996. *Chapel Hill and Elisha Mitchell the Botanist*. The Botanical Garden Foundation, Chapel Hill, NC. (pp. 25, 81–82)

NHSR. Natural History Society Records. *Proceedings [of Meetings], 1876–1881*. (Archives #37-6-584, Carl A. Kroch Library, Rare and Manuscript Collections, Cornell University)

Oliver, Daniel. 1869. *Lessons in Elementary Botany*. 2nd ed. Macmillan and Co., London.

Pope, Fannie West Harris. 1915. [Letter of appreciation, Dallas, Texas, February 10, 1915], pp. 464–465 in Appreciations of former students. *Alcalde* 3(5): 458–479.

Porcher, Francis Peyre. 1863. *Resources of the Southern Fields and Forests, Medical, Economical, and Agricultural: Being also a medical botany of the Confederate States; With practical information on the useful properties of the trees, plants, and shrubs*. Prepared and published by order of the Surgeon-General, Richmond, Va. West and Johnston, Richmond, VA.

Prentiss, Albert N. 1871. On the methods of studying botany during the spring trimester. *Cornell Era* 3(22): 171, March 10, 1871.

Randall, William G. 1880. "Lectures on Zoology by Prof. Simonds." (Southern Historical Collection #3580-z, Wilson Special Collections Library, University of North Carolina, Chapel Hill)

Russel, William C. 1877. Letter of July 28, 1877 to George T. Winston. (Box 8, folder 302, University Papers #40005, University Archives, Wilson Special Collections Library, University of North Carolina, Chapel Hill)

Sachs, Julius. 1868, 1875. *Lehrbuch der Botanik*. Verlag von Wilhelm Engelmann, Leipzig. (In 1875 Alfred W. Bennett translated and annotated the book with the assistance of W.T. Thiselton Dyer under the title *Text-book of Botany*, published by Clarendon Press, Oxford, [England].)

Simonds, Frederic W. 1875. *The Geology of Ithaca, New York*. 8 pp. (Bachelor's thesis, Cornell University, Thesis Film 1875, #42, Olin Library, Cornell University)

Simonds, Frederic W. 1877a. Letter of August 14, 1877 to Kemp P. Battle. (Box 8, folder 302, University Papers #40005, University Archives, Wilson Special Collections Library, University of North Carolina, Chapel Hill)

Simonds, Frederic W. 1877b. The geology of Ithaca, New York, and the vicinity. *American Naturalist* 11: 49–51.

Simonds, Frederic W. 1878a. Entomological. By Prof. Simmonds [*sic*]. *Farmer and Mechanic* 1(11): 82, Thursday, January 3, 1878.

Simonds, Frederic W. 1878b. "Syllabus of a Course of Twenty-five Lectures on Physical Geography and Geology, Delivered at the University of North Carolina, Autumn Session, 1878." [no statement of publisher or place]. (North Carolina Collection, Wilson Special Collections Library, University of North Carolina, Chapel Hill)

Simonds, Frederic W. 1878c. "Synopsis of the Course of Instruction in General Zoology at the University of North Carolina, Autumn Session, 1878." [no statement of publisher or place]. (Inserted between pages 40–41 in Kemp P. Battle, *Scrapbook of North Carolina History and Biography, 1880–1894*, vol. 1. (North Carolina Collection, Wilson Special Collections Library, University of North Carolina, Chapel Hill)

Simonds, Frederic W. 1879a. On the habitat of *Rhododendron catawbiense*. *American Naturalist* 13: 777–778.

Simonds, Frederic W. 1879b. "Syllabus of a Course of Twenty-five Lectures on General Zoology, Delivered at the University of North Carolina, Autumn Session, 1879." [no statement of publisher or place]. (Inserted in lecture notebook of William G. Randall, Southern Historical Collection #3580-z, Wilson Special Collections Library, University of North Carolina, Chapel Hill)

Simonds, Frederic W. 1879c. Letter of September 8, 1879 to Asa Gray. (Gray Herbarium, Harvard University, Cambridge, Massachusetts). (Letter accompanied a specimen of *Rhododendron catawbiense* collected by Simonds on April 30, 1879.)

Simonds, Frederic W. 1880a. Letter of December 16, 1880 to Kemp P. Battle. (Box 11, folder 388, University Papers #40005, University Archives, Wilson Special Collections Library, University of North Carolina, Chapel Hill)

Simonds, Frederic W. 1880b. *University of North Carolina, Outline of Lectures on General Geology*. Edwards, Broughton & Co., Printers and Binders, Raleigh.

Simonds, Frederic W. 1880c. "[University of North Carolina], Schedule of Lectures on Physiology & Hygiene." 1 sheet pasted on the inside front cover of Edwin A. Alderman, "Lectures on Physiology—Delivered by Prof. Simonds; Taken by E.A. Alderman." (Notebook in the Edwin A. Alderman Papers, Southern Historical Collection #1869z, Wilson Special Collections Library, University of North Carolina, Chapel Hill)

Simonds, Frederic W. 1880d. Letter of June 21, 1880 to "The Commissioner of Agriculture, Washington, D.C." (Smithsonian Institution, Archives, Washington, DC, Record Unit 220, United States National Museum, Division of Plants, Correspondence, 1870–1893, Box 14)

Simonds, Frederic W. 1880e. Letter of June 30, 1880 to "The Commissioner of Agriculture, Washington, D.C." (Smithsonian Institution, Archives, Washington, DC, Record Unit 220, United States National Museum, Division of Plants, Correspondence, 1870–1893, Box 14)

Simonds, Frederic W. 1881a. Letter of May 28, 1881 to Kemp P. Battle. (Box 11, folder 399, University Papers #40005, University Archives, Wilson Special Collections Library, University of North Carolina, Chapel Hill)

Simonds, Frederic W. 1881b. Letter of July 16, 1881 to Kemp P. Battle. (Box 11, folder 404, University Papers #40005, University Archives, Wilson Special Collections Library, University of North Carolina, Chapel Hill)

Simonds, Frederic W. 1895. Letter of May 14, 1895 to Kemp P. Battle. (Center for American History, The University of Texas, Austin)

Simonds, Frederic W. 1905. Chapter VI. Vegetation, pp. 49–62 in Frederic William Simonds, *The Geography of Texas, Physical and Political*. Revised ed. Ginn and Company, Boston.

Simonds, Frederic W. [1908]a. "Frederic William Simonds." Manuscript (5 typed sheets). (Manuscript Collection 770, Reynolds/Thomas Research Papers, University Libraries, University of Arkansas, Fayetteville)

Simonds, Frederic W. 1908b. Chapter 26. Animals and plants: Their relations and distribution, pp. 280–287; Chapter 27, The distribution of useful plants, pp. 288–291 in *Maury-Simonds Physical Geography. [Being] Physical Geography*, by M.F. Maury. Revised and largely rewritten by Frederic William Simonds. American Book Company, New York.

S[lade], W[illiam] R[obert]. 1877. Chapel Hill, University of North Carolina, p. 35 in Kemp P. Battle, *Scrapbook of Material Relating to the University of North Carolina, 1869–1893*, vol. 2. (Newspaper clipping without bibliographical data. North Carolina Collection, Wilson Special Collections Library, University of North Carolina, Chapel Hill)

Spencer, Cornelia Phillips. 1876. Letter of January 16, 1876 to "daughter" June. (Box 1, folder 7, Cornelia Phillips Spencer Papers #683, Southern Historical Collection, Wilson Special Collections Library, University of North Carolina, Chapel Hill)

Spencer, Cornelia Phillips. 1877a. Letter of August 27, 1877 to Mrs. Swain. (Cornelia Phillips Spencer Papers, Personal Collection #7, North Carolina State Archives, Raleigh, North Carolina)

Spencer, Cornelia Phillips. 1877b. University Day. Chapel Hill, N.C., Oct. 13, 1877. *Observer* (Raleigh) [Daily] 2(133): [2], Wednesday, October 17, 1877.

Spencer, Cornelia Phillips. [1877]c. A letter from Chapel Hill. Typescript of article from the *Farmer and Mechanic*, [ca. fall 1877] in "Letters and Writings of Cornelia Phillips Spencer," transcript vol. 10. (Cornelia Phillips Spencer Papers #683, Southern Historical Collection, Wilson Special Collections Library, University of North Carolina, Chapel Hill)

Spencer, Cornelia Phillips. 1880. Letter of December 4, 1880 to "My dear & lovely daughter" [June]. (Box 1, folder 11, Cornelia Phillips Spencer Papers #683, Southern Historical Collection, Wilson Special Collections Library, University of North Carolina, Chapel Hill)

UNC Faculty. University of North Carolina, Chapel Hill Faculty. *Minutes of Faculty Meetings*. (University Archives #40106, Wilson Special Collections Library, University of North Carolina, Chapel Hill) (Vol. 6, 1856–1885)

UNC Student Records. *University of North Carolina Student Records and Faculty Reports*, Subgroup 2 (vol. 4, p. 34; vol. 11, p. 332). (University Affairs, Student Records and Faculty Reports, Subgroup 2: Miscellaneous Student Records, #40131 University Archives, Wilson Special Collections Library, University of North Carolina, Chapel Hill)

UT. University of Texas at Austin. Frederic William Simonds Papers (MS 70-1979 and vertical file, The Center for American History, University of Texas, Austin)

Vaughan, Victor C. 1877. Letter of April 22, 1877 to Kemp P. Battle. (Box 8, folder 294, University Papers #40005, University Archives, Wilson Special Collections Library, University of North Carolina, Chapel Hill)

Walker, John M., Jr. 1880. Letter of December 12, 1880 to the Faculty of the University of North Carolina. (Box 11, folder 387, University Papers #40005, University Archives, Wilson Special Collections Library, University of North Carolina, Chapel Hill) (The date of the letter is December 12, 1880 in the header; December 21, 1880 in the footer.)

Wood, Thomas F. (editor). 1879. [Announcement]. "Syllabus of a Course of Twenty-five Lectures on Human Physiology and Hygiene, Delivered at the University of North Carolina, Spring Session, 1879, by Frederic William Simonds, M.S. Professor of Geology, Zoology and Botany." *North Carolina Medical Journal* 3: 194.

Young, Keith. 1967. *History of Geology at the University of Texas*. Special issue for geology building dedication, November 16–17, 1967. (*Bureau of Economic Geology, Geological Circular*, 67-3). University of Texas, Austin.

Young, Wade Phillips. 1934. *A History of Agricultural Education in North Carolina*. A thesis submitted to the faculty of the University of North Carolina in partial fulfillment of the requirements for the degree of Doctor of Philosophy in the Department of Rural Social-Economics. Chapel Hill. (pp. 164–166)

Jos. A. Holmes

Chapter 8.

Joseph Austin Holmes (1859–1915)

Professor of Geology and Natural History (1881/1882–1890/1891)

Professor of Botany and Physiology, School of Medicine and Pharmacy (1881/1882–1884/1885, 1889/1890)

Professor of Geology, Botany, and Horticulture, College of Agriculture (1885/1886–1887/1888)

Professor of Geology and Mineralogy (1891/1892–1892/1893), absent on leave 1892/1893

State Geologist and Lecturer on the Geology of North Carolina (1893/1894–1903/1904)

State Geologist, Professor of Mining Geology (1904/1905, 1905/1906)

Early Years and Scientific Education

Born in the country village of Laurens, South Carolina, on November 23, 1859, Joseph Austin Holmes was the eighth of twelve children of Nancy Catherine Nickles and Zelotes Lee Holmes, a Presbyterian minister and teacher. His father spurned tradition by building a concrete, octagonal house in 1859. The first concrete structure in South Carolina and likely the first in the South (Irwin, 1963), the house now is listed on the National Register of Historic Places. Young Holmes attended Laurens Academy and later the Holmes Academy, both in his home village. He worked on the farm, gaining an interest in plants and agriculture. In addition to hikes in the woods, outdoor games, and horses, Holmes enjoyed reading books. His father's library, which was particularly strong in natural history, provided young Holmes "instruments for the study of natural sciences" (Manning, 1915). At age seventeen, he left home with a railway ticket and $15, heading directly for Cornell University (Sykes, [1916]). He began his studies there in January 1878 and worked his way through college.

At Cornell, Holmes pursued studies in agriculture, which allowed him to enroll in a number of scientific disciplines. According to archival records at Cornell, he chose courses in the following subjects: agricultural chemistry, agriculture, botany, chemistry, entomology, horticulture, math, and veterinary science. In botany, he took courses in systematics, histology, fungi, and "extra credit," which probably consisted of original research (CU-RS).

When Holmes enrolled at Cornell, the study of botany had been improved with new facilities and a new course of study. The spring of 1875 marked a significant milestone in the botanical teaching environment there. At this time, Sage College was completed. The new building provided ample space where the botanical lecture and recitation rooms, museum, and

Frontispiece Credits: (above) Joseph Austin Holmes, ca. 1890. (Harriet Morehead Berry Papers #2259, Southern Historical Collection, Wilson Special Collections Library, University of North Carolina, Chapel Hill). (below) Biological Laboratory, New East Building, University of North Carolina, Chapel Hill, ca. 1886. (North Carolina Collection Photographic Archives, Wilson Special Collections Library, University of North Carolina, Chapel Hill).

laboratories could be housed under one roof. Added features included greenhouses, from which instructors had a ready source of fresh plants for teaching purposes. In 1877, the course of study in botany had been considerably expanded. Besides the standard courses in vegetable physiology and systematic botany, several courses of lectures with accompanying laboratory sessions were added. The new offerings were on mosses and algae, ferns, and fungi (Anonymous, 1877). Students could also undertake additional lab work in a specialized area of botany. The teachers of botany were Albert N. Prentiss and William R. Dudley.

In about 1877, the botanical museum was enhanced with several gifts: a collection of specimens of foreign woods in sections from Professor [James] Johonnot, Brazilian woods from the Centennial Exhibition in Philadelphia, and ferns from South America. Students could consult "a very nearly complete local herbarium of mounted Phaenogamous plants and ferns, besides numerous collections in the lower groups" (Anonymous, 1877). During Holmes's senior year, courses of instruction in botany were offered in each of the following topics: physiological botany, Gramineae and Compositae, vegetable histology, vegetable physiology, systematic and applied botany, plant culture, higher cryptogams, fungi, and algae. Students could select courses to meet their academic needs and could also pursue advanced work, which entailed original investigations in a specialized area of botany (CU-R).

Other academic activities also cultivated Holmes's interest in science and agriculture. In the winter term of 1879, he joined the Natural History Society and subsequently served as vice-president (1879–1880). He was an active member of the organization, giving presentations on a variety of topics such as: phosphates and marl deposits in South Carolina, February 18, 1880; the life and scientific works of Humboldt, April 21, 1880; thermal springs, their distribution and nature, October 6, 1880; and the physical conditions under which prehistoric civilizations were developed, February 23, 1881 (NHSR). It is noteworthy that his presentations showed his early interest in geology, which subject would later become the focus of his career as State Geologist of North Carolina and then as an administrator of the United States Bureau of Mines.

The Greeley Agricultural Club (previously the Agricultural Society of Cornell) also provided Holmes with a stimulating environment in his chosen area of study. Established in 1872, the organization aimed to promote an interest in improving agricultural knowledge and to allow members to present their views on the subject. Joining the club in March 1879, Holmes became an active member, serving as recording secretary (1879, 1880), vice-president (1879–1880), and president (1880). He delivered several papers concerning southern agriculture, October 3, 1879; the life and works of Horace Greeley with extracts from his *Autobiography* and *What I Know about Farming*, March 12, 1880; apatite (a group of minerals containing calcium phosphate), December 2, 1880; and the cotton industry, March 11, 1881. Holmes took a keen interest in keeping abreast with relevant publications in agriculture. Toward that end, he recommended that the club develop a bibliographical "catalogue" of references on agriculture, for which members approved the purchase of catalog slips and a tin box for sorting and storing the bibliographical records (GACR).

By July 1879, Holmes was at the head of his class. In addition to his scholastic pursuits, he enjoyed playing football and baseball. Professor of Practical Mechanics John Edison Sweet thought Holmes was "a prodigy of integrity, industry and manliness" (Shepard, 1924). Holmes earned a Bachelor of Agriculture degree in 1881. At his commencement exercises, he delivered an essay on the "Agricultural Labor in the Southern States," which was received with "marked attention" (Anonymous, 1881a). His presentation was based on his bachelor's thesis of the same title (Holmes, 1881g). Additional biographical background on Holmes is recorded in the literature (AMC, [1915]; Anonymous, 1906; Anonymous, 1933; Battle, 1916; Cobb, 1910; M[cDonald], 1932; Pratt, [1916]; and Vinson, [1988]). Other relevant information is housed in archival collections (CU, SHC, SI).

Teaching at UNC

In advertising the vacant chair of geology and natural history, the university placed a notice in the journal *Nation* (Fig. 8.1). They also solicited names of qualified scientists from educators in the country. The pool of applicants grew. Letters of recommendation and inquiry housed in the University Papers at the University of North Carolina Archives reveal that at least fifteen men contended for the position. Among the applicants were William Trelease, who later became a noted botanist and first director of the Missouri Botanical Garden; Angelo Heilprin, Professor of Invertebrate Physiology at the Academy of Natural Sciences in Philadelphia; Thomas Nelson Dale from the United States Geological Survey Office, Newport, Rhode Island; J. Walter from the Museum of Comparative Zoology, Cambridge, Massachusetts; Henry Ward Turner, a Fellow at the Academy of Natural Sciences of Philadelphia; Professor John Foster Eastwood of Bethany College, West Virginia; and Joseph Austin Holmes, a recent graduate of Cornell University and the ultimate choice for the position.

PROFESSOR OF GEOLOGY AND NA-
tural History in the University of North Carolina. —A special meeting of the Trustees of the University of North Carolina will be held at the Executive Office, in Raleigh, on the 20th of September next, to elect a Professor of Geology and Natural History. Communications on the subject will be received by the President of the University, at Chapel Hill, N. C.

Fig. 8.1. Advertisement for Professor of Geology and Natural History in the University of North Carolina, 1881.

(*Nation*, Thursday, September 1, 1881, p. v).

In support of his application, Holmes's Cornell professors spoke highly of his scientific training. From the Botanical Laboratory at Cornell, Professor Albert Nelson Prentiss (1881) certified that Holmes had taken the full course of instruction in botany and that his "attainments as a student have been of a very high order." Assistant Professor of Botany William Russel Dudley (1881) stated that Holmes's "training ha[d] been a very practical one, having devoted considerable time to the natural sciences with a view to their bearing on the economic sciences, especially on agriculture." He went on, "He ha[d] pursued the laboratory work laid out in connection with lectures on botany with real care, fidelity and thoroughness." Entomologist John Henry Comstock (1881) reported that Holmes "ha[d] been an earnest and enthusiastic student" and that "he ha[d] a pleasing address and winning manner." He concluded: "Finally, as he [was] a Southern man and ambitious to advance the interests of the South, he [would] be more devoted to his work with you than a man from the North." Professor of Geology Samuel Gardner Williams (1881) gave additional supportive comments, saying that Holmes "[was] a young gentleman of excellent character, fine attainments, and of singularly pleasing manners." Even Cornell Registrar William Dexter Wilson (1881) wrote on behalf of Holmes and his superior scholarship. Addressing another aspect of Holmes's academic strengths, Professor of Agriculture Isaac Phillips Roberts (1881) wrote that Holmes had "shown very fine ability as a lecturer and teacher, in the addresses, and papers read before the college societies and local organizations. . . . He [was] unusually mature in thought and experience." Simon Gage (1881a), assistant professor of physiology and lecturer on microscopical technology, also noted Holmes's skill as a speaker. He was especially impressed by Holmes's lectures before the Natural History Society.

Several testimonials specifically addressed Holmes's devotion to the economic and scientific advancement of the South. Professor Gage (1881b) commented that Holmes had "the advantage of southern birth" and that his "heart [was] with the South and the southern people" (Gage, 1881a). Professor of German Bela Phillips MacKoon (1881) stated that Holmes's strong desire to better his beloved southern states was "an admirable passion with him—not an idle enthusiasm, but a steady, true inspiration of his life." When writing to President Battle about the faculty position, Holmes (1881a) explained that he attended Cornell to prepare himself "for studying and teaching the natural sciences in connection with their economic relations, with the hope that [he] might directly or indirectly, or in both ways, do something toward the development of the resources of our Southern states." He continued: "I am too thoroughly in sympathy with our southern people and too greatly interested in their prosperity to give up this purpose; and it is for this reason that I would desire, while not neglecting the theoretical part of the sciences, to devote considerable attention to them as applied sciences."

Commenting on his appointment and reflecting on the sentiments of the times, a Raleigh newspaper reporter stated: "In our judgment, it is best when practicable to employ North Carolinians, and next best to select our teachers from the South as being more likely to sympathize with the feelings of our boys and to know how to develop their best traits of character" (Anonymous, 1881b). When Holmes was hired, his salary, which should have been $2,000, was fixed initially at $1,500 per year (BOT 7: 491). Unfortunately, his employment at UNC followed the financial straits of 1879, when faculty salaries had been reduced. Holmes (1881c) acknowledged the difficult financial state of the university but accepted the temporary reduction in his salary with a proviso. He still wanted full appropriations for his department for purchasing books, as well as charts and models for illustrating his lectures. When the financial scene improved in July 1885, his yearly salary was increased to $2,000 (BOT 8: 170).

In accepting his appointment, Holmes (1881d) anticipated arriving at Chapel Hill around September 30, 1881. Even before his arrival, Holmes was thinking about the subjects he would teach at the university. In writing President Battle on September 2, 1881, he suggested "the addition of a course of lectures on economic geology and another on systematic and applied botany, including a study of the more important orders of plants from an economic standpoint, their position in the system of classification, their economic products, and the more important of the noxious weeds, parasitic fungi &c." (Holmes, 1881b). He also wanted "to conduct the work in such a way . . . as to create in [his] students an interest in their work, and a desire to experiment, and to investigate these subjects for themselves."

Upon his arrival in Chapel Hill, Holmes joined fourteen other faculty members who instructed 199 men. According to the UNC catalog of 1881–1882 (p. 38), tuition (including room and attendance) for the year was $85.00. During his tenure as professor of geology and natural history, the student enrollment changed little. Kemp P. Battle was university president until George Tayloe Winston became the successor to that office on July 2, 1891. During Battle's last years in office, student conduct had gotten "out-of-hand" and was marked by misbehavior in Chapel Hill and in the lecture halls, according to mathematics professor James Lee Love. Among the mischief were such pranks as nailing up the doors to professors' classrooms, greasing the blackboards, placing loaded wagons on top of the New West Building, and ringing the college bell for hours in the middle of the night (Love, 1921, vol. 7).

While Holmes served at UNC, the first legislative appropriation (of $5,000) to the university was made in 1881 and the original Memorial Hall was erected and later dedicated in 1885. In April 1886, the faculty forbade the playing of marbles between Old East, Old West, and South buildings during the mornings (UNC Faculty 7: 31). They also were concerned that hogs were a nuisance, and in February 1887 they detailed a servant to keep the animals off the campus grounds (UNC Faculty 7: 75). Chapel Hill's population was 831 in 1880, and a decade later it had increased to 1,017.

Concerning life in the dormitories in the early 1880s, James Lee Love (1921, vol. 5), who was then an undergraduate at UNC, gave a glimpse of his college days. In the cold months, Black men would make fires in the fireplaces in the early morning. They also brought buckets or pitchers of fresh water from the college well, and by forenoon they would return to sweep the rooms and to make the beds. Love noted that the sanitary conditions were rather rustic. Those few students who kept a slops jar pitched out the contents from their rooms. Most students did not keep such a jar, however, but just directly discarded their wastes. On the windowsill, many men kept a bottle, which served as a urinal, the contents of which they would dump onto the ground beneath. By August 1885, the Grounds Committee was charged with overseeing the emptying of slops and other wastes (UNC Faculty 7: 3). Further efforts to promote better sanitation were made at the end of 1886, when the faculty adopted a motion to place three urinals in convenient places on campus (UNC Faculty 7: 66). A committee recommended that the facilities be placed at the following locations: one abutted to the west end of New East, another abutted to the east end of New West, and another about thirty feet behind South Building (UNC Faculty 7: 67).

Outside of academics, students enjoyed various entertainments. One student recounted an evening of events in March 1884 that included exercises of the University Athletic Association, greased pig racing, baseball matches, and running (Grimes, 1884). The same night, students held a dance at Professor Hooper's house because the trustees would not let them use the ballroom (Smith Hall).

Upon moving to Chapel Hill, Holmes lived in the home of professor Charles Phillips and his wife Laura (Battle, 1916). Soon after, he shared bachelor's quarters with chemistry professor Francis P. Venable in the former Denison Olmsted house, which the university had purchased in 1822. Holmes lived on the upper floor, Venable on the lower. Venable (1916) affectionately related that occasionally various fauna (insects and reptiles) that Holmes had collected would wander down the stairs into his living area. When Venable married Sallie (also cited as Sally) Charlton Manning in November 1884, Holmes vacated the house (Bryant, 1999). On October 20, 1887, Holmes married Jeanie Isabella Sprunt of Wilmington, North Carolina; their nuptials took place in the First Presbyterian Church of Wilmington (Anonymous, 1887c). The bride was the daughter of Alexander Sprunt, one of the largest exporters of cotton in the South. In the same year, Holmes bought a parcel of land on Franklin Street in Chapel Hill and built a house (now 516 East Franklin Street) (Fig. 8.2).

Fig. 8.2. House of Joseph Austin Holmes Family, East Franklin Street, Chapel Hill, North Carolina, ca. 1894.

(Photo attributed to Walter Washington Foster in Battle Photo Album, North Carolina Collection, Wilson Special Collections Library, University of North Carolina, Chapel Hill).

At UNC, Holmes's instruction in botany spanned most of the 1880s—a period in which the professionalization of the discipline was further developed. Botany had earlier been viewed as "an effeminate and trifling pursuit" (Anonymous, 1848). Almira Hart (Lincoln) Phelps, a prolific nineteenth-century author of botanical textbooks, was instrumental in spreading knowledge of this discipline to American youth and adults. She presented her ideas about why botany was particularly suitable for females. Plants are beautiful and delicate, she argued, and the pursuit of their study provides exercise outside in the open air, contributing to health and cheerfulness (Lincoln, 1836). While enrolled in the botany class of Frederic Simonds, sophomore Henry Faison (1878) had questioned the place of botany among the sciences. He mentioned that botany "is not fully developed," and he continued: "The science applies only to wild flowers, therefore a Botanist would be out of place in a flower garden. All Botanists hold the theory that the more a flower is cultivated, the [more] abnormal it becomes."

When Holmes was teaching botany at UNC, Massachusetts physician J.F.A. Adams (1887) presented four reasons substantiating its importance: as a mental discipline, as a promoter of physical development, as a practical utility, and as a source of lifelong happiness. Two years later, North Carolina botanist Gerald McCarthy (1889) enumerated several additional benefits to be derived from botanical studies. He followed the dictum of Professor De Morgan of London University, who advised his students: "Whatever else you may do, some one subject you should thoroughly master for the purpose of giving the proper tone to the mind upon its regard to the use, province, and limits of knowledge in general." He believed that a systematic study of plants cultivated a love for form, symmetry, and color, which would incite a sense of aesthetics in students. Moreover, the study of plant life would provide students with the fundamentals of the phenomena of life, which in turn would prepare students to understand the more complex aspects of life, such as animals and human physiology. McCarthy also thought that plants could be studied more thoroughly than could minerals and animals, which it often required costly apparatus to investigate. Lastly, McCarthy noted that the study of plants would guarantee a virtually endless source of research and observation. Botany was loosed from the grips of natural history and taught as a distinct course. In America, the study of botany already had gained significant momentum in academia through the research of such botanists as Asa Gray of Harvard University and Daniel C. Eaton of Yale University (Ford, 1962).

Holmes was the first professor at Chapel Hill to build up a diverse curriculum in botany by offering several courses, improving laboratories, and promoting field studies. At UNC, botany indeed now offered practical utility, particularly in its application to agriculture. In the 1880s, classrooms were located in the dormitories. It is reported that Holmes lectured on botany, geology, mineralogy, and zoology on the eastern side of the first floor of New East (Love, 1945) (Fig. 8.3). However, he may have been teaching in the natural history room of Old East in the early 1880s. By 1897, he occupied the entire first floor of New East (Anonymous, 1897). As chair of the College of Natural History, Holmes was responsible for courses in the School of Geology and Mineralogy, the School of Zoology, and the School of Botany. (From 1885 to 1888, he also taught geology, botany, and horticulture to students matriculating in the College of Agriculture; and from 1881 to 1885 and in 1889, he taught botany and physiology to students in the School of Medicine and Pharmacy and in the Preparatory School of Medicine, respectively.) According to the university's catalog for 1881–1882 (p. 32), the School of Botany offered two courses, which focused heavily on agricultural concerns. The course descriptions for the School of Botany are quoted below.

SCHOOL OF BOTANY

PROF. HOLMES

The School of Botany is divided into two courses—a general course, including Physiological Botany, and an advanced course, including (a) Cryptogamic Botany, (b) Grasses and Forage Plants, (c) Vegetable Physiology.

Physiological Botany occupies three hours per week during the second half of the second term. The instruction is devoted to a study of the form, structure, and function of the parts or organs of the plant. Each student will be required to do work in plant analysis during the term, and at the close will hand in a collection of thirty mounted specimens.

The advanced course occupies two hours per week during one year. The instruction will consist of lectures and laboratory work.

Cryptogamic Botany, including a study with the microscope of the lower forms of plants, will be taken up during the first session. Special attention will be paid to rust, smut and other fungi parasitic upon field crops.

Grasses and forage plants will be studied during the first half of the second session. The time will be devoted to a consideration of the cereals and more important grasses, and valuable forage plants of other families.

Vegetable Physiology, or the life and growth of plants, will be taken up during the last half of the second term. Attention will be paid to the growth of the plant as effected by cultivation, different kinds of soil and fertilizers.

Text-books and works of reference: Gray's *Lessons and Manual of Botany* [combined edition, 1879a], Gray's *Structural Botany* [1879b], Bessey's *Botany* [1881], Johnson's *How Crops Grow* [1880] and *How Crops Feed* [1879], Chapman's *Southern Flora* [1865].[76]

Fig. 8.3. New East Building, University of North Carolina, Chapel Hill, 1886.

(North Carolina Collection Photographic Archives, Wilson Special Collections Library, University of North Carolina, Chapel Hill).

Holmes's instruction was intended to be practical. All of his scientific classes emphasized practical applications, particularly their influence on agriculture. Instruction involved lectures, recitations, lab work, textbooks, and field trips. During his first year, in addition

[76] Citations for these textbooks are included in the References Cited section of this chapter.

to a course in Physiological Botany (General Botany), Holmes taught "a course of lectures and laboratory work on economic Botany,[77] requiring two hours per week during one year" (BOT 7: 517). Among the first documented records of laboratory fees in natural history were those cited in letters of October 1882 (Holmes, 1882d, 1882e) and in mineralogy in a letter of January 1883 (Holmes, 1883a).

After Holmes completed one year of instruction at UNC, President Battle took note of his abilities and knowledge. He wrote of Holmes: "I am pleased to report that Prof. Holmes fulfilled all his promises made by the Cornell professors who endorsed him. A more energetic, practical, industrious, intelligent man, one more eager to learn in all possible ways everything useful in the department, I do not know" (BOT 7: 517). One year later, a campus magazine echoed these sentiments and added that Holmes "is quite young, but to find a man more devoted to his department, Natural Science, and more thorough in his work, would be a difficult undertaking. He keeps pace with the onward march of 'scientific research'" (Anonymous, 1883b).

In Holmes's second year of teaching, his zoology and botany class had fifty-nine students, whereas geology had eighteen, laboratory work twelve, and physical geography seven (BOT 7: 581). Students collected and mounted plants from their field excursions, and these were most likely placed in the growing herbarium. His botanical instruction must have been strenuous, as one student's recollection suggests: "Six hours a week in the Botany Laboratory, with examinations so near, is right hard on some of us; but Prof. Holmes, by his careful instruction, renders it not only instructive but very interesting" (Anonymous, 1883c). In 1882, a laboratory was fitted up for classes in mineralogy and in 1883, another was established, for zoology and botany. Holmes soon realized that improved and larger laboratory facilities were essential to practical work in zoology and botany.[78] He wanted to be "able to arrange during the present session [January 1883], a larger laboratory for the use of the class in Botany ... [and] a small expenditure for apparatus" (BOT 7: 582). He also wanted to acquire specimens, charts, and models as instructional aids in teaching. His efforts to acquire funding for teaching aids were sometimes successful, as when he procured $1,100 for purchasing books and apparatus (BOT 8: 414). He pointed out to the Board of Trustees, however: "that nearly every book, chart, and specimen that ha[d] been used in the [Natural History] Department (by students) since I came to the University ha[d] been procured at my own personal expense" (BOT 8: 371). In January 1886, Holmes identified the pressing needs in natural history: cases for housing insects and zoological specimens ($85), one compound microscope ($100), a set of botanical specimens ($25), and $50 for preparing specimens of wood and rocks before placing them in the museum (BOT 8: 206). In 1888, his wish list for botany consisted of a set of specimens of fungi and other vegetable parasites ($15), cases for collecting and storing plants ($40), apparatus for laboratory equipment ($50),

---

[77] According to botanist Gerald Wickens (2001), economic botany is defined as "the study of plants, fungi, algae, and bacteria that directly or indirectly, positively or adversely affect man, his livestock, and the maintenance of the environment. The effects may be domestic, commercial, environmental, or purely aesthetic; their use may belong to the past, the present or the future."

[78] By autumn of 1886, the Biological Laboratory was located on the first floor of the New East Building. UNC senior Stephen B. Weeks (1886a) noted that the seats in the former Natural History room were uncomfortable, stating that a person had "to change his position twenty or thirty times a minute to prevent being cramped to death. No wonder they get bored—the Professor would if he could have a little experience of this delightful sensation."

and models and charts for lectures ($50) (BOT 8: 372). In October 1888, Holmes (1888a) alerted the Board of Trustees to the need for additional microscopes for a special class of fifteen students preparing to be physicians. He further explained the need to purchase three microscopes by writing to William L. Saunders, Secretary of the Trustees. Holmes (1888b) believed "the use of microscope[s] in medicine ha[d] never before been so popular as now, nor the demand for it so great . . . [and that] the addition of two students would in tuition fees more than repay the present expenditure asked for." He also reminded Saunders that Davidson College was considering the establishment of a preparatory medical school and could possibly take these special students from UNC. In May 1892, the continued need for microscopes was backed by the Visiting Committee, which recommended six of them at a price of $240 for instruction in natural history (BOT 9: 161). In 1890, Holmes (1891a) personally purchased apparatus and specimens, which cost $300.

About 1885, a committee of the UNC Board of Trustees consisting of Kemp P. Battle, Charles M. Cooke, and Julian S. Carr ([1885]) asked the General Assembly for an increased appropriation. Among their concerns was that Professor Holmes's position, as the sole person to head his department and to teach geology, zoology, physiology, and botany, was stretched to its limit. The committee recommended that the practical aspect of his courses be strengthened. In geology, they suggested increasing the scope of the course to include agricultural applications (such as soils and marls) and additional laboratory and fieldwork. In botany, they mentioned that a practical course in agricultural botany and a short course in medical botany should be offered. In zoology, they recommended new courses of instruction in human and comparative physiology (including some lessons on domestic animals), breeding and feeding domestic animals, and a greater level of investigations of injurious and beneficial insects of crops. Among the new professorships the committee recommended was a position for an Assistant Professor of Economic Geology, Botany, and Entomology, with a salary of $1,500.

The university's requests were answered with a significant appropriation of $20,000 (up $15,000 from the previous year's allotment) from the General Assembly. The campus also received an additional $7,500 from the Morrill Land Grant Fund. The generous appropriation induced an excited fervor on campus, which culminated with a bonfire in the open space in front of South Building and the ringing of the college bell at the height of the blaze (Anonymous, 1885a). President Kemp P. Battle (BOT 8: 205) reported to the Board of Trustees with the hiring of Assistant Professor of Natural History George F. Atkinson, who took over teaching in zoology beginning in the fall term of 1885. Holmes could now devote additional time to instructing in geology and botany. The expanded program in botany that he developed is outlined in the UNC catalog of 1885–1886 (pp. 40–41) as quoted below.

II. BOTANY

Professor Holmes

UNDER-GRADUATE CLASSES.

The following courses are offered:

1. ELEMENTS OF BOTANY.—*Three lectures per week during the latter half of the second term*:

This is designed as an elementary course for the general student; and as an introductory course for students intending to do advanced work in Botany. Instruction is given by lectures and laboratory work. The lectures include a discussion of the important facts connected with the structure and the life and growth of plants. The time spent in the laboratory is devoted to a study of the structure of the plant and the determination of species.

2. ADVANCED BOTANY (ELECTIVE)—*Three hours per week during the year:*

Under this heading two classes have been organized. These alternate, the instruction in each being offered every two years.

(a) SYSTEMATIC AND APPLIED BOTANY:

The work in this class consists largely of the systematic study of the flowering plants of the Chapel Hill region, in the laboratory and the field. Students who desire to do so, may collect, identify and preserve specimens of plants at their homes during the summer vacation; and one hundred such specimens properly mounted and presented at the Botanical Laboratory at the beginning of the year, will count one hour of class work for that year. Instruction will be given in Applied Botany by lectures during both terms.

(b) AGRICULTURAL BOTANY:

The instruction is given by lectures and recitations, accompanied by work in the laboratory and field excursions. The course may be subdivided into the following:

(1) Cryptogamic Botany: including a study with aid of microscopes, of typical forms of the lower groups of plants. Special attention is paid to the fungoid diseases of agricultural plants. This work extends through the first term.

(2) Grasses and Forage plants, taken up during the first half of the second term. The instruction is devoted to a study of the common grasses, cereals and valuable forage plants of other families.

(3) Vegetable Physiology, or the life and growth of plants, is taken up during the latter half of the second term. Special attention is given to the growth of agricultural plants as affected by cultivation, different kinds of soils and fertilizers.

[3]. HORTICULTURE (ELECTIVE).—*Two lectures per week through the year*:

During the first term the instruction is devoted mainly to the subjects of forestry, orchards and vineyards.

During the second term the following subjects are treated of in the order of mention:

Native and tropical fruits; farm and garden soils, their improvement by drainage, manuring, &c; general management of the garden; cultivation of common vegetables and garden fruits; general management of flower gardens; some general principles in landscape gardening.

Text-books and books of reference.—Gray's *Lessons and Manual* of Botany [combined edition, 1879a], Gray's *Botanical Text-book* [i.e. *Structural Botany*], [1879b], Chapman's *Southern Flora* [1865], Sachs' *Text-book* [of *Botany*, 1875], Bessey's *Botany* [1881], and others of a special character.[79]

In offering an expanded curriculum in natural history, Holmes ([1885]) explained how the courses, particularly in the plant sciences, related to practical agriculture. Agricultural botany included studies in several pertinent areas: fungi parasitic on crops, grasses and forage plants and their cultivation and preservation, and the growth and cultivation of plants. Horticulture

[79] Citations for these textbooks are included in the References Cited section of this chapter.

featured studies on soils and methods for improving them through such methods as drainage and manuring; the management of orchards and forests with instruction on planting, grafting, and pruning; and the basic principles in cultivating garden and farm crops.

With an enlarged faculty, UNC could now offer its own graduates and those of other schools additional postgraduate courses of study (Anonymous, 1885c). In botany, three courses were now available: work in the field and laboratory on the trees, shrubs, and other flowering plants in the region of Chapel Hill; advanced work in the laboratory and field on any special group or groups of plants, particularly the grasses; and a more general course in economic and systematic botany that focused on the uses of plants.

Around this time, the faculty also offered a two-year program of study, providing a certificate of proficiency. This study program was aimed at those "young men, [who] from lack of money or time or previous preparation, are unable to complete one of the full four-year courses of study" (Battle, [1888]). These men could now gain an education and training in specialized subjects to assist them professionally or socially. Three course areas were offered: the Teachers' Course, the Business Course, and the Physician's or Pharmacist's Course. For the last cited course, physiology, zoology, and botany were included in the second year of study.

A Raleigh newspaper reporter (Anonymous, 1885d) noted how the new, reorganized curriculum at UNC prominently included the scientific disciplines, in contrast to earlier years in which Greek, Latin, English, and mathematics prevailed. The courses in science reflected "modern practical methods of instruction, [which were] adopted in theory and practice." The reporter was especially eager to learn about the teaching of botany at UNC, as it had been previously looked upon as a study of plants more suited to young women than young men. As noted above, the three botanical courses involved field observations, work in the laboratory (including the use of microscopes), and practical instruction. The intrigued reporter was amazed with his observations, noting, "the old time student of the University who took Dr. [Elisha] Mitchell's two weeks course in botany would never have dreamed that the study was to be expanded into one of so much practical importance." Laboratory and fieldwork honed the students' power of observation and interpretation as well as encouraging them to think and work instead of merely memorizing.

By the mid-1880s, the scientific laboratory experience was gaining prominence at UNC as a method of educating students. Different scientific departments acquired additional specimens and learning aids for illustration and instruction. The university was expanding its resources to promote scientific instruction and the use of new methodologies. Holmes became an active proponent of improving laboratory facilities in his disciplines of responsibility. He believed that the university should adopt laboratory practice "as far as possible in place of the old style text book system of instruction." He explained: "Seeing and examining an object gives a student a further understanding of it than does reading about it. All advanced institutions are adopting this plan" (BOT 8: 158).

Holmes thought that the Natural History Department was providing facilities and apparatus of a higher grade than ordinary colleges but that to maintain such excellence would require continual upgrades. A visiting committee, appointed by the Board of Trustees, met at the university on May 5, 1885 (Anonymous, 1885b). Their recommendations included improvements for the laboratories. Converting Gerrard Hall into lab facilities or possibly into quarters for the museum were options that were discussed. Another proposal considered building a facility for laboratories as an addition to Person Hall. The immediate needs could be met with purchases of instruments and equipment. In natural history, Holmes presented a detailed list of important purchases costing $1,500. By 1886, additions were made to the scientific laboratories, including a room on the first floor of New East (Weeks, 1886b) that was set aside for biological work (Frontispiece of chapter 8, lower image). According to the UNC catalogs of 1886 to 1891, the biological lab included a small herbarium, a collection of native woods, and vegetable products; beginning in 1889, it also included microscopes and other scientific apparatus. The "furnishing of these laboratories [was] of the best and superior advantages" (BOT 8: 269).

The enhancement of scientific laboratories in natural history at Chapel Hill garnered the attention of the media locally and internationally. One newspaper reporter took note that the university's "apparatus and improved appliances [had] been gradually added, until the equipment of the various laboratories compare[d] favorably with that of the more wealthy colleges." The laboratories were new features "in Southern institutions, dependence being commonly placed upon short courses of lectures, with no practical work to render the subject intelligible" (Alumnus, [1886]). Mention was also made of several German microscopes, constantly used by students. Another aspect of the teaching experience consisted of "frequent excursions [that] are made with biological classes about the surrounding country, collecting insects and plants, and studying their life-histories in their natural habitat" (Alumnus, [1886]). A note in the October 20, 1887, issue of *Nature* summarizes the "scientific revival" in the southeastern United States, including mention that UNC has a "small but modern Natural History Department" (Anonymous, 1887b).

The progress made in increasing the faculty and expanding the curriculum in botany (and other sciences) was short-lived. In 1885, the Legislature enacted the establishment of an Agricultural and Mechanical College in Raleigh and transferred to it the Land Grant allocation of $7,500 for instruction. The new institution was to be called the North Carolina College of Agriculture and Mechanic Arts and to be located west of and near the city of Raleigh on land given by prominent Raleigh businessman and philanthropist [Richard] Stanhope Pullen. To construct the campus "the Penitentiary authorities [were] commanded to furnish brick and laborers" (Anonymous, 1887a). The transfer of these funds was at the expense of faculty positions in special programs on UNC's campus, including those in pedagogics, ornithology, metallurgy, mining engineering, feeding and breeding of animals, and practical horticulture (Scales, 1888). George Atkinson was one of the newly appointed faculty members of 1885 who was let go in 1888. He subsequently accepted a position at the University of South Carolina. Professor of Chemistry Francis P. Venable (1888) summed up the feelings on campus by saying, "Several of us are quite blue about the University & its prospects. The last Legislature hit us quite a cruel blow."

Following Atkinson's departure, Holmes's teaching responsibilities were once again stretched to include all disciplines in natural history. Courses offered in botany were reduced to Elements of Botany and Systematics and Applied Botany. In an effort to assist Holmes, Joseph Volney Lewis, a student, was hired as an Assistant in Natural History (1888–1889) and then as an Instructor in Natural History (Fall of 1890). In the Board of Trustees minutes of February 20, 1890, it is reported that the Medical Department would be more efficient if it had "a more experienced teacher in the branches of Botany and Zoology, as assistant to Professor Holmes, in lieu of the two undergraduates now allowed him to help in his laboratories." The minutes continue: "A suitable man could be obtained for $500, which would only be a small advance on what we now pay, this advance would only be $180" (BOT 8: 452).

According to the trustees minutes of February 20, 1890, it was authorized that the Executive Committee should hire several instructors. In natural history, Victor Silas Bryant, an unnamed member of UNC, and Joseph Volney Lewis, were each paid $100 to assist those students planning to be physicians in the Biological Laboratory (BOT 8: 456). In the UNC catalog of 1889–1890 (p. 7), Lewis and Bryant are each listed as Instructor in Biology, the first time such a position is documented at the university. President Battle (1889) and the faculty had asked that "these teachers be called Instructors," a name that was being used at other institutions of higher education such as Cornell, Harvard, and Yale, among others. The minutes of the trustees meeting of February 11, 1891, record: "The department of Natural History has been embarrassed by the inability to employ an assistant for remuneration allowed. Mr. J.V. Lewis acting during the Fall Term, but declining to continue during the second Term" (BOT 8: 513). The university was still struggling to provide Holmes with a trained assistant who would earn a larger salary. Among the justifications for the position were that it would enable the provision of more thorough instruction in biology and microscopic technology for students who intended to become physicians.

This was a very demanding period in Holmes's university career. His numerous teaching responsibilities placed overwhelming pressure on him. In the fall of 1890, Holmes (1891a) taught six courses, requiring seventeen hours per week: General Geology and Mineralogy, 18 students; Advanced Geology, 4; Advanced Mineralogy, 4; Zoology, Physiology, and Botany, 40; Biology and Microscopic Technology, 20; and Advanced Botany, 4. In addition to heavy teaching loads during the fall and spring terms and geological fieldwork during many summers, Holmes lectured on the geology of North Carolina in the Summer School sessions in 1894–1896 and 1898–1901.

In January 1891, Holmes (1891a) reported to President Battle his ideas for bringing a new direction to the natural history department. On behalf of the students and the people of North Carolina, he proposed that two academic units should be formed from the present department. The state's physicians, he argued, required a more thorough and extensive instruction in biology and microscopic technology and its landowners needed more thorough and extensive instruction in geology and mineralogy as they applied to the agricultural and mineral resources of the state.

Holmes was intensely sincere in making this proposal. If the university failed to follow through on such a plan, he believed another institution would promote instruction in these areas of study. He thought the development of a Department of Geology and Mineralogy would not only support the populace of the state but would also bring academic strength to the university. Holmes wrote: "I have labored in vain, because as matters now stand, my time and energies are so fully taken up in teaching other matters, which I feel less competent to teach, and less deeply interested in, that I am unable to do what I see clearly can be and ought to be done for the work in Geology both at the university and throughout the State." A separate Department of Biology and Microscopic Technology could strengthen the curriculum for future doctors. Holmes's recommendations for change were placed before the trustees for their consideration and implementation. The creation of departments of biology and geology was favorably supported by the Committee of Visitation for the 1890/1891 academic year; however, the committee recommended postponing the election of an assistant professor of geology (BOT 9: 31).

From 1882 to 1891, Holmes taught botany and horticulture to nearly 335 students (Appendix 1). Of these students, nearly twenty-five took courses in horticulture and advanced botany. Among these young men, two became noted botanists: William Willard Ashe, who became particularly well known as a forester in the North Carolina Geological Survey and later in the U.S. Forest Service, and John Simcox Holmes (probably not directly related to Joseph Austin Holmes), who replaced Ashe as North Carolina's state forester.

The University Museum, Its Development and Service to Students and Citizens

In addition to his teaching responsibilities, Holmes was in charge of the University Museum, which served as an important instructive tool. His students could consult the herbarium of local plants and native woods. According to Battle (1907–1912, 2: 274), the natural history collections contained over 3,000 specimens of rocks, ores, minerals, and a growing assemblage of woods, as well as zoological and botanical specimens illustrating the local fauna and flora. Students were also expected to utilize specimens of a large collection of seeds, grains, and fibers from numerous countries in their experiments.

When the State University Railroad Company completed the rails leading to University Station, relatively close to campus, visitors to the museum increased. The rails enabled the state's citizens to visit the campus, whose isolation was mentioned by some as a way "to prevent the students from running away" (Anonymous, [1882]a). Concerning one of the first rides on the university branch, a newspaper reporter mentioned that "the train moved slow enough to give all an opportunity to enjoy the scenery and to let the little boys jump off now and then to gather apples and blackberries" (Anonymous, [1882]b). According to a newspaper reporter for the *North Carolina National*, the Union Sunday School of Company Shops and cooperatives employed by the Railroad Company—a group of nearly 300 people—rode the rails on June 17, 1882, to Chapel Hill (Anonymous, [1882]a). Among the spots of interest and instruction was the museum.[80]

Holmes was particularly concerned about the upkeep and development of the University Museum. While UNC was conducting its June 1882 commencement exercises, Holmes was visiting his South Carolina home, from where he wrote President Battle. Holmes (1882a) asked Battle that "whoever ha[d] charge of the Museum during the Normal School, [would] endeavor to have visitors keep 'hands off.'" Insight into Holmes's (1882c) active acquisition of natural history specimens for the UNC Museum is gained from a report that he wrote.

[80] Although James L. Love [1945] mentioned that the museum was cleared out of the second [*i.e.*, third] floor of Old East about 1883 as a place for teaching mineralogy by Professor Emile De Schweinitz, the UNC catalog of 1885–1886 mentions that the museum remained there.

During October 1882, donations of natural history specimens and cultural artifacts were received from the following people: from Professor Isaac P. Roberts of Ithaca, New York, twenty-one varieties of wheat (including stalks and grains); from Professor George H. Cook of New Brunswick, New Jersey, green sand marl; from Mr. R.D. Lacoe of Pittstown, Pennsylvania, thirty fossil plants from anthracite coalfields near Pittstown, [New Jersey], representing the more important orders of plants found there; from Mr. H.B. Short of Columbus County, North Carolina, two small alligators from Waccamaw Lake; from Master Kemp P. Battle, Jr., of Silerville, North Carolina, Indian arrowheads and spearheads from Anson County, North Carolina; from Mr. Lewis Battle of UNC, Indian arrowheads and moss from Edgecombe County, North Carolina; from Mr. L. Vann of UNC, hog caterpillars from Florida; from Mr. Thomas Radcliffe of UNC, fish, kingcrab, seaweeds, and the beak of a sawfish from Wilmington, North Carolina; and from Mr. J.C. Roberts of UNC, a rare fish from New Bern, North Carolina.

Holmes made a trip to Washington, D.C., in May 1883, in part to acquire specimens of rocks, fossils, and animals. At the conclusion of his second year of teaching, he reported that "the museum collections ha[d] increased somewhat in size and value during the [past] year and [were] beginning to render in practical teaching" (BOT 8: 23). Among the specimens added to the herbarium were most likely plants sent in two parcels from the USDA Botanical Division (Loring, 1883), for which Holmes (1883b) thanked Acting Commissioner of Agriculture Ezra Ayers Carman. Holmes mentioned that "the common plants and the grasses [would] be of considerable value to us and especially the latter [would] serve as a standard collection with which to compare other species." He would take "pleasure in rendering any service possible to [the] Botanical Division in the future in connection with the plants of this region." About this time, Holmes (1883e) requested a collection of wood specimens from George B. Loring, U.S. Commissioner of Agriculture.

In January 1887, Holmes noted that "the museum and collections in the different departments of science [were] receiving constant addition and prove valuable aids in illustrating the instruction" (BOT 8: 269). Holmes (1887) also referred to a donation from the USDA in which the university received about 500 specimens of grasses and other plants from Florida, New Mexico, and other areas of the "far West." This shipment of plants is likely the one for which Holmes (1886a) acknowledged receipt in June 1886. In the same letter, Holmes notified George Vasey about sending him specimens of *Senecio* that appeared to be *Senecio tomentosus*, although its leaves did not agree with published descriptions. In working on the systematics of the grasses, Holmes noted that he still needed two of Vasey's publications. During the 1887/1888 academic year, Holmes reported that the USDA presented the university with a collection of 1,000 plant specimens that were subsequently labeled and filed in the university herbarium (BOT 8: 343). Holmes also collected a large number of botanical specimens and several hundred specimens of rocks from various regions of the state. He likewise obtained specimens through exchange.

In the mid-1890s, two noteworthy collections of specimens were acquired: 200 specimens of wood representing economically important trees of the country from the American arboriculturist Charles Sprague Sargent (1880, 1884), who had been a special agent for the Tenth Census of the United States on Forestry, and numerous specimens of tropical fruits from Florida from the Atlanta Exposition (Anonymous, [1895]). The Cotton States and

International Exposition of 1895 (also known as the Atlanta Exposition) was originally conceived to exhibit the resources of the southern states and to promote trade with other countries, particularly with those of South America. Opening in Atlanta, Georgia, in Piedmont Park, on September 18, 1895, and concluding on December 31, 1895, the exposition featured over 3,619 displays from thirty-seven states and territories and thirteen foreign countries. It hosted over 779,560 visitors (Cooper, 1896). The North Carolina Department of Agriculture and the State Geological Survey sponsored an exhibit at the Atlanta Exposition. As State Geologist, Holmes helped in organizing the exhibit, which featured systematic collections of economically important minerals (*e.g.*, mica, kaolin, phosphate) and ores (*e.g.*, gold, silver, copper) as well as building stones. An inventory of the specimens displayed is housed in the archival collection of the North Carolina Geological Survey (NCGS). Holmes attended the event in mid-October 1895.

The trustees minutes of February 11, 1891, reflect a growing concern that the museum be properly housed (BOT 8: 531). If a new chapel were built, then Gerrard Hall could be converted for museum quarters "for the large and valuable collections." When Henry Van Peters Wilson became professor of the newly created Department of Biology in the fall of 1891, the biological facilities in New East included a laboratory, a lecture room, and a museum. Most likely, some of the natural history specimens from the University Museum were moved to this setting, while mineralogical and geological specimens were moved to the geological laboratory on the first floor of New East.

Botanical Investigations

Keeping current with the scientific literature and collecting specimens for preservation and instruction were important aspects of Holmes's scientific studies. While a student at Cornell University, he had begun frequently requesting publications from the Smithsonian Institution. His contact with the institution continued for at least the subsequent twenty years. His nearly fifty letters housed in the Institution's archives are concerned with obtaining publications and sending or receiving natural history specimens, among other topics. Of special note was his donation of Cretaceous invertebrates collected in Wilmington, North Carolina. His correspondence with scientists at the United States Department of Agriculture reveals distinct botanical research interests, especially with grasses, in his early years at Chapel Hill. Geological excursions and related research eclipsed Holmes's botanical investigations in later years. Holmes's interest in collecting scientific books continued during his tenure at UNC. On January 15, 1886, a committee consisting of professors Francis P. Venable, George F. Atkinson, Ralph H. Graves, and the librarian was constituted to consider the university's acquisition of a collection of books from Holmes's personal library (UNC Faculty 7: 17). Two weeks later, the committee recommended the purchase of the collection. Although a list of the books acquired has not been found in the UNC archives, volumes containing Holmes's signature occasionally show up in the UNC Science Library Annex.

Holmes had collected plants for scientific identification while an undergraduate at Cornell. Among his botanical collections from Ithaca, New York, were *Aralia nudicaulis* L., May 30, 1879, and *Tussilago farfara* L., May 17, 1887, of which specimens are preserved in the UNC Herbarium. From his correspondence, we further learn that Holmes (1881e) studied the grasses of Chapel Hill and environs soon after moving there. Holmes wanted to learn about the economic value of grasses. In October 1881, he wrote to agrostologist George Vasey at the USDA, asking for assistance in identifying a grass that he thought was a species of *Sporobolus* or *Vilfa*. Since Holmes's books and appliances for analysis had not yet been delivered to his new address, he was not able to confirm the identity of the grass. He offered to send Vasey grass specimens that were not already held at the USDA Herbarium. In thanking Vasey for identifying the grass *Sporobolus indicus*, Holmes (1881f) noted his goal of learning about the grasses and forage plants of North Carolina and the country at large. Nearly a year later, Holmes (1882b) wrote Vasey again to confirm the identity of a plant he found and tentatively identified as *Galium pilosum*. The plant had been collected in Warren County, North Carolina, where it was popularly called "snake weed," in reference to its reputed application in treating snake bites. Holmes (1886d) also sent Vasey a specimen of *Senecio* for identification, which was duly noted as *S. tomentosa* [*i.e., tomentosus*].

Holmes (1886b) pointed out to Vasey that he was attempting "to build up an herbarium with a small beginning," and he stated that any USDA duplicate specimens might be donated to UNC. He mentioned that he recently came across botanist [Gerald] McCarthy while botanizing in the eastern region of the state. Today, selected specimens of North Carolina plants collected by Holmes are archived at the UNC Herbarium, examples being *Lycopodium alopecuroides* (L.) Cranfill (Fig. 8.4) and *Woodwardia areolata*, both collected in Duplin County, North Carolina, October 10, 1885; and *Panax trifolius* from Orange County, North Carolina, April 25, 1888. Holmes wrote at least seven publications concerning botany and agriculture (Table 8.1). For several years, Holmes spent his vacation time conducting geological fieldwork, often taking along his advanced students (BOT 8: 525).

Table 8.1. A Bibliography of Agricultural, Botanical, and Natural History Publications by Joseph A. Holmes.

1883. Agricultural education in North Carolina, pp. 84–87 in *Proceedings of a Convention of Agriculturists Held at The Department of Agriculture, January 23, 24, 25, 26, 27, and 29, 1883* (second convention). (*Department of Agriculture. Miscellaneous. Special Report*, No. 2). Government Printing Office, Washington, D.C.

1884. Occurrence of *Abies canadensis* and *Pinus strobus* in central North Carolina. *Journal of the Elisha Mitchell Scientific Society* 1: 86–87.

*1885. The twisting of trees. *Journal of the Elisha Mitchell Scientific Society* 2: 94.

1885. *Taxodium* (Cypress) in North Carolina Quaternary. *Journal of the Elisha Mitchell Scientific Society* 2: 92–93.

1890. A study of plants in the garden and field. *North Carolina Teacher* 7: 474–479.

1891. Mineralogical, geological and agricultural surveys of South Carolina. *Journal of the Elisha Mitchell Scientific Society* 7: 89–117.

*1897. Brimley, H.H., J.A. Holmes, and T.K. Bruner. *Hand Book of the North Carolina State Museum, Giving Names and Localities of All Specimens*. Issued by the State Board of Agriculture, M.I. & J.C. Stewart Public Printers and Binders, Raleigh. Winston, NC. 273 pp.

*Not listed in Holmes's bibliography by Pratt (1916).

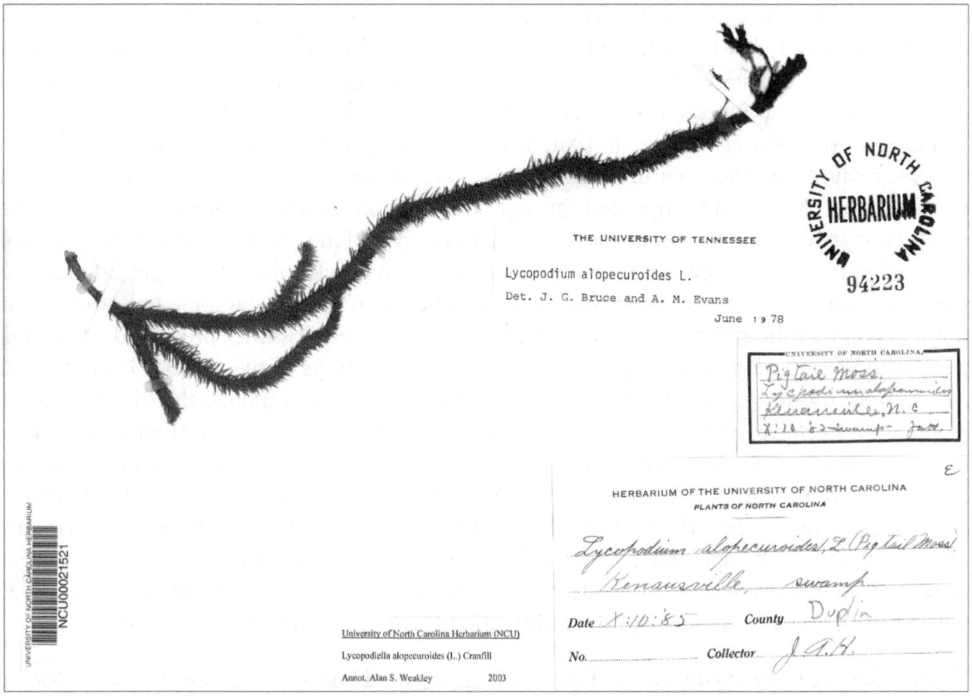

Fig. 8.4. Part of specimen sheet of *Lycopodium alopecuroides* L., collected by J.A. Holmes, swamp, Kenansville, Duplin County, North Carolina, October 10, 1885.

(Courtesy of University of North Carolina Herbarium, Chapel Hill [NCU]).

Holmes had strong interests in Indian antiquities and culture. While on a geological collecting trip in the fall of 1883, he also made a careful examination of an Indian burial mound at Kenansville, North Carolina (Holmes, 1883d). Apparently, he later sent specimens from the mounds to the Smithsonian Institution in exchange for their publications (Langley, 1890). Following a visit to the American West and Mexico in 1901, he reported how Indian women in New Mexico made bread, a sample of which he brought back to Raleigh. He noted that two kinds of corn were used, a blue variety and a white variety (Anonymous, 1901).

In the 1887/1888 academic year, Holmes mentioned that he had devoted considerable time to botanical and geological collecting. A two-week field excursion for his advanced geology class is well described in the *University Magazine* (Chapel Hill) (Anonymous, 1890). The class conducted original research on the geological formation in the "western" part of the state (Pilot Mountain and Mount Airy) and also visited the Egypt Coal Mine, Ore Hill, the Triassic formations of Rockingham County, and the Sam Christian Gold Mine.

The Elisha Mitchell Scientific Society

Holmes's active role in scientific organizations at Cornell probably influenced his interest in developing a learned scientific society at Chapel Hill. The introductory chapter of this book chronicles the formation of the Elisha Mitchell Scientific Society.

Holmes and future botanists at Chapel Hill provided guidance and leadership for the society. Initially, the society held meetings each month during the university's school year. At its first regular meeting on November 10, 1883, Holmes recounted the results of his observations and his readings on insectivorous plants, especially about the structure and habits of the Venus' flytrap, *Dionaea muscipula* (Anonymous, 1883a; Anonymous, 1883d). Most likely, his observations on these carnivorous plants were based on specimens he had collected in mid-June 1883 when he traveled through Wilmington, North Carolina. From there, he sent living specimens of these plants, and also pitcher plants, to President Battle, whom he asked to plant them "carefully in a moist place, with muck soil." He suggested planting some of each species near the spring in the northeastern part of campus and others near Battle's pond (Holmes, 1883c). During the next several years, Holmes presented additional papers (predominantly on geology) (Table 8.2) at the meetings of the society; his botanical papers were on the geographical distribution of North Carolina plants, the distribution of rhododendrons in North Carolina, and the flora of Angola Bay, North Carolina. Concerning the last topic, Holmes had spent the summer of 1884 conducting fieldwork on the plants and animals of the bay. At the October 18, 1884, meeting of the society, which included his paper on Angola Bay, he exhibited specimens of the plants collected on the trip (Anonymous, 1884a). Some of the specimens (Fig. 8.5) survive in the UNC Herbarium (NCU). In the society, Holmes served as vice-president (1883–1884, 1885–1886), as president (1892–1893, 1898–1899), and as a member of the Publication Committee (about 1885–1888).

TABLE 8.2. A CHRONOLOGICAL LIST OF TITLES OF PAPERS PRESENTED BY JOSEPH A. HOLMES BEFORE THE ELISHA MITCHELL SCIENTIFIC SOCIETY.

Title	Date
Insectivorous Plants	November 10, 1883
Southward Growth of Florida (An Address)	December 8, 1883
Volcanic Eruptions in the Straits of Sunda	February 9, 1884
*Observations on Movement of Water	February 9, 1884
Tornado of February 19, 1884	March 8, 1884
*Indian Mounds in Eastern North Carolina	April 12, 1884
*Geographical Distribution of Certain Plants in North Carolina	May 3, 1884
*Note on Indian Arrow Heads in Eastern North Carolina	May 3, 1884
Flora of Angola Bay	October 18, 1884
The Jetties at the Mouth of the Mississippi	February 7, 1885
A Petrified Human Body	April 25, 1885
Peculiar Animal and Plant-Life in Geology	April 25, 1885
Trap Rocks in Triassic Sandstone	April 25, 1885
Distribution of Rhododendrons in North Carolina	April 25, 1885
Character of Border-Rocks of N.C. Triassic	April 25, 1885
North Carolina Triassic	September 19, 1885

* Title only read.

TABLE 8.2. (CONTINUED)

Title	Date
Longevity of Frogs	September 19, 1885
The Condition of the North Carolina Indians	November 25, 1885
Note on the Folding of Rocks	November 25, 1885
History of the Geological Surveys of North Carolina	January 27, 1886
Work of Dr. Emmons in Connection with the North Carolina Geological Survey	February 17, 1886
The Coast Line at Cape Hatteras	March 17, 1886
Triassic Sandstone in North Carolina	May 19, 1886
Description and Exhibition of Core from Diamond Drill in the Coal Region of Pennsylvania	May 19, 1886
Subsistence along the Coast of New Jersey	November 1886
Report on the Recession of Niagara Falls	November 1886
Report on Earthquakes Occurring in Charleston, from 18th Century to Present Time	November 1886
Rainfall Statistics for North Carolina	January 25, 1887
Temperature Statistics for North Carolina	March 8, 1887
Average Elevation and Rainfall of North Carolina	April 12, 1887
The Mastodon in North Carolina	September 12, 1887
Triassic Traps	September 12, 1887
Geological Maps of Counties of South Carolina	September 12, 1887
Photographs of Formations of the Head of Chesapeake Bay	September 12, 1887
A Profile Map from Morrisville to Newbern	September 12, 1887
Report on the American Association for the Advancement of Science	September 12, 1887
A Cotton Plant Having Seeds and No Lint	October 11, 1887
Report on Progress in Geology	October 11, 1887
Note on the Origin of the Hudson River Palisades	November 8, 1887
Professor Kerr's Observations on Old Glacier Valleys in North Carolina	January 11, 1888
Professor Kerr on the Action of Frost on Superficial Layers of Soil	January 11, 1888
Biographical Sketch of Prof. W.C. Kerr	January 24, 1888
Triassic Rocks on New Hope Creek near Chapel Hill	March 13, 1888
Changes in the New Geological Map of North Carolina	April 11, 1888
Climatology of North Carolina	May 8, 1888
Report on Progress in Geology	January 8, 1889
An Account of an Interesting Fossil Found in the Neighborhood of Chapel Hill	February 12, 1889

* Title only read.

TABLE 8.2. (CONTINUED)

Title	Date
The Three Formations of the Atlantic Slope, with an Exhibition of Photographs and Specimens	March 12, 1889
Use of the Microscope in Geology and Mineralogy	April 2, 1889
On Arkose	September 17, 1889
Early Maps of the Carolinas	October 8, 1889
The Conglomerate and Pebble Beds of the Triassic and Potomac Formations in North Carolina	November 12, 1889
The Recent Geologic Formations on the Roanoke River	February 18, 1890
Exhibition of Some Minerals from the Mica Mines and of Some Fine Crystals	April 1, 1890
Exhibition of Specimens and Photographs	October 14, 1890
Problems of the Atlantic Coastal Plain	December 2, 1890
The Plan and Limitations of the North Carolina Geological Survey	February 9, 1892
Igneous Rock Formation in North Carolina	April 12, 1892
Work of the North Carolina Geological Survey	December 6, 1892
The Geographical Development of the Counties in Eastern and Middle North Carolina	October 10, 1893
The Clays of the Spout Springs Region	March 20, 1894
The Most Valuable Sources of Water Power in North Carolina	May 14, 1895
Experiments in Restocking Burned-over Areas with Long-leaf Pine	September 15, 1895
Underground Water Supplies of the Piedmont	January 14, 1896
Mica and Mica Deposits in North Carolina	October 13, 1896
Line of Exact Levels across the State	December 8, 1896
Mica Industry in the United States and the Important Relation of the North Carolina Mines	April 13, 1897
*Mica in the Rockies	November 23, 1897
The Volcanic Rocks of the North West	January 11, 1898
Artesian Well at Wilmington	March 8, 1898
Geological Conditions Favorable to Water Power Development in North Carolina	March 21, 1899
The Recent Geological Formations of the Mississippi Valley	May 14, 1901
Geology at the Meeting of the American Association for the Advancement of Science, Pittsburgh Meeting	October 14, 1902

* Title only read.

Fig. 8.5. Specimen sheet of *Kalmia angustifolia*, collected by J.A. Holmes, margin of swamp, Angola Bay, Pender County, North Carolina, July 22, 1884.

(Courtesy of University of North Carolina Herbarium, Chapel Hill [NCU]; scan of specimen by Stephen Seiberling, July 16, 2005).

Holmes encouraged his students to join the Mitchell Society as associate members. Three young men who pursued advanced studies in botany under Holmes's tutelage presented papers at the organization's meetings. On April 12, 1887, Leander Williams Lynch reported on the dates of flowering of nearly fifty plants growing in the region of Chapel Hill (EMSS 1: 47; Weeks, 1887). Gaston Battle presented a paper on the "pea-nut plant" on April 21, 1891 (EMSS 1: 61). He became a planter in Edgecombe County, North Carolina and was subsequently associated with Sinclair Oil in Atlanta, Georgia. After graduating from UNC in 1891, William W. Ashe, who had been an avid student of botany under the direction of Holmes, also presented papers at the society's monthly programs. Ashe prepared a paper on the longleaf pine and its struggle for survival, which Professor Venable read for him on October 10, 1894 (EMSS 1: 71). The paper was subsequently published (Ashe, 1894). Ashe also gave a presentation on the influence of high altitudes upon the growth of plants on November 23, 1897, which he illustrated with specimens from the Rockies (EMSS 1: 100).

Besides being a frequent presenter of talks at the Mitchell Society meetings, Holmes was also much sought in the community to give lectures. His focus was usually on the state's natural resources and geology or on his observations of the forests and geology of the American West, where he had traveled in the 1890s. His presentations were not only instructive but also captivating. They were illustrated by stereopticon views or lantern slides and Holmes also placed artifacts and natural history specimens on display. A reporter for the UNC student newspaper *The Tar Heel* enthusiastically recounted an especially entrancing lecture about a trip Holmes made to the western United States. A passage related that "one's hair had a tendency to stand up as he gave a vivid description of his lonely midnight ride through the great Oregon forest, entertained by the roarings of the mountain lion, the growls of 'old grizzly' and the barking of the wolves" (Anonymous, 1898).

At the request of and with the support of the North Carolina Board of Agriculture, Holmes took a two-month leave of absence in the fall of 1884 to collect and name mineralogical and botanical specimens for the North Carolina Exposition of 1884 in Raleigh and another showing in New Orleans (BOT 8: 121). Holmes (1884) told President Battle that "the University will get 5 specimens (duplicates) [of all the plants and flowers he collected],—& that without paying anything for them,—while the state gets but one." He would also make duplicate collections of wood specimens for the University Museum. While Holmes was on leave, Professor of Anatomy and Materia Medica Thomas West Harris (M.D.) took over the instruction of his classes. The exposition in Raleigh, held from October 1 to November 1, 1884, was an important promotional opportunity to display the state's natural resources and awakening industrial base, especially with an eye toward attracting northern investors and capital to set up new industries (Sumner, 1992). Holmes's name does not appear in lists of officers in a ninety-six-page guide to the exposition (Anonymous, 1884b). His efforts, however, certainly did not go unnoticed. A teacher from the New Bern graded schools noted that "the State exhibit of minerals and woods was undoubtedly the finest and most complete ever made in the South—the most durable and beautiful building and ornamental stones, the rarest and most precious gems, the most useful and valuable ores, the finest and most durable timbers—all are found in the bosoms of our native hills, and on our mountain slopes and fertile plains" (Ellis, 1884).

Holmes resigned the professorship in natural history in 1891. He was appointed the State Geologist that May, but he assumed his new responsibilities in June. Even before Holmes officially was given this appointment, the Trustees had known about the possibility. Their minutes of February 11, 1891, mention that it was possible that Prof. Holmes might be invited to take charge of the State Geologist position (BOT 8: 514). By retaining Holmes part-time as a lecturer in geology, the university hoped that their ongoing association with Holmes would enhance its reputation, secure specimens for the museum, and make instruction in geology more practical and of more evident value. In March 1891, Holmes made a trip to Washington, D.C., on survey matters. While away from campus, he (Holmes, 1891b) wrote President Battle, asking him to remind the botany class to examine a dozen wildflowers. From this letter, it is evident that he taught his final botany class just prior to resigning his position as professor of geology and natural history and becoming State Geologist. While serving as State Geologist, he continued to lecture at UNC as Professor of Geology and Mineralogy, from 1891 to 1893, taking a leave of absence in the academic year 1892/1893. He was Lecturer on the Geology of North Carolina from 1893 to 1904 and Professor of Mining Geology from 1904 to 1905. As State Geologist, Holmes focused the survey of North Carolina principally on improving and enlarging the mining industry of the state (Stuckey, 1965). In 1905, he relinquished his position of State Geologist to accept the post of geologist in charge of fuel investigations after the United States Geological Survey.

Contributions to Public Service

Holmes's concerns for conservation extended to the forests of North Carolina. At the Mitchell Society's meeting of May 14, 1895, he presented a paper about restocking burned-over areas with longleaf pine (EMSS 1). Four years earlier, he had suggested to American forester and conservationist Gifford Pinchot his thoughts about establishing a forest reservation in the Southern Appalachians (Smith, 1960). His interests in the state's forests continued while he served as the State Geologist of North Carolina. Testimonials to Holmes's efforts to conserve North Carolina's forests were given by forester William Willard Ashe, who commemorated his mentor by naming two trees after him: *Hicorius holmesia* (Ashe, 1896) (Fig. 8.6) and *Crataegus holmesiana* (Ashe, [1900]). According to a resolution of the North Carolina Forestry Association, Holmes was "the first to take up the investigation of the forestry problems of North Carolina and started the campaign of education looking toward the preservation and conservation of our forest areas" (Gibboney et al., 1916).

Although geological themes, mainly those of economic importance, were the focus of most of Holmes's nearly eighty publications, several of his early papers dealt with agricultural teaching in North Carolina; with *Abies canadensis*, *Pinus strobus*, and *Taxodium*; and with plants in gardens and fields (as noted in Table 8.1). He also served on the Executive Committee of the North Carolina Agricultural Society.

Holmes's horticultural interests were aimed at beautifying the campus. According to the minutes of the June 1, 1885, faculty meeting, he and Adolphus W. Mangum (Professor of Moral Philosophy, History, and English Literature) were chosen as "a committee on the campus" charged with planting shrubs and other plants (UNC Faculty 6: 600). In October 1885, the group was called the Grounds Committee. It would seek the counsel of the faculty on such matters as the advisability of planting shrubs on the campus (UNC Faculty 7: 10).

On February 22, 1880, one year prior to Holmes's arrival at UNC, the senior class initiated the custom of planting a tree on campus (S[pencer], 1880). The villagers, faculty, and students attended the event. At 4 P.M., the senior class emerged from the front door of South Building with class president Thomas H. Battle carrying the tree, a young sugar maple. The ceremony took place on the walkway transecting the north end of Old East and Old West. After the sapling was planted halfway between this walk and the well, the class orator, William B. Slade, spoke about the significance of the event and appealed to future classes to perpetuate the tradition of planting a tree on the campus grounds. Published documentation confirms that this practice continued through 1888 (Anonymous, 1886a; Battle, 1907–1912, 2: 391; Tim, 1882). Among the trees planted were a sugar maple in 1880, a water oak in 1882, a white pine in 1883, a tulip poplar in 1884, an oak in 1886, and a Norway spruce in 1888. A concrete marker (Fig. 8.7) commemorating the class tree of 1884 survives near the northeast corner of Old West.

In the *University Magazine* (Chapel Hill) for 1886–1887, it is recorded that Holmes assisted in the planting of trees on campus. His efforts included the care of the tree of the class of 1886, a young oak planted on the southeast corner of monument square, and the planting of pecan trees. Further insight into his efforts to maintain the landscape of the campus are found in archival records and publications. During the tree planting season of 1886, he was "seen with his elastic step hastening to beautify and adorn the Campus in any way possible." An anonymous account continued: "Some of the little trees [were] removed from the rows on Oida Avenue and planted at random in various parts of the campus" (Anonymous, 1886b). Holmes thought the scattered planting created a more naturalistic effect. Holmes was very observant of the condition of the trees and shrubs and continually advised workers about their care, including pruning. His recommendations for proper arboricultural care were very specific as noted in a memo of May 1887, in which he consulted with Mr. Patterson about trimming off several "bad looking dead [tree] limbs near the New W[est] building" and cutting down a partly dead oak near the Old Chapel Hall (Holmes, 1887). Recommending the removal of dead wood on crape myrtles along the walkway, he advised care in not cutting off new shoots growing from the roots. The minutes of the faculty meeting of January 31, 1890, reveal that Holmes was authorized to plant a hedge between the campus and the property of two adjoining hotels (UNC Faculty 7: 193). In the early 1900s, Holmes and Professor Joshua Gore unsuccessfully tried to drain a marshy site on the UNC campus, now Coker Arboretum. As a last resort, they sowed it with crimson clover and grass (Henderson, 1949).

Known to work into the early morning hours, much like one of his predecessors, Elisha Mitchell, Holmes also devoted himself to the cause of improving North Carolina's roads. The Office of Public Road Inquiries, U.S. Department of Agriculture, had appointed Holmes a special agent of its Southern Division. At the North Carolina Good Roads Convention, held in Raleigh on February 12 and 23, 1902, Holmes (1903) delivered a talk on "Road Building in North Carolina." He briefly reviewed the history of road improvement in the state and presented four features for planning the betterment of roads. One of his biographers (Sykes, 1916) considered Holmes "the father of the good roads movement in North Carolina." (Concerning Holmes's interests in improving the Chapel Hill landscape, Cornelia P. Spencer ([1894]) wondered if he had suggested "shaving off of that hill on Cameron Avenue.") He was active in the civic affairs of Chapel Hill, serving on the board of alderman. He also was interested in the sanitary conditions of the village and the grading and macadamizing of its streets.

NOTES ON THE HICKORIES OF THE UNITED STATES.

By W. W. ASHE.

Hicoria, Rafinesque.

The shag- or scaly-bark Hickories.—Bud scales 10 to 12, accrescent and persistent during leafing; nuts angled, white or mealy; leaflets 3 to 7; bark of trunk usually shaggy.

(*a*) Fruit globular; husks thick and splitting to the bottom and freeing the nut; nut white, sharply angled, taper-pointed and thin-shelled.

H. ovata (Mill.) Brit.—Fruit 1.5 inches in diameter; leaflets 5 to 7, large, tomentose. A large tree of bottoms and rich hillsides. Central New York to Michigan, south to Florida and Texas.

H. Holmesia, sp. nov.—Fruit about 1.25 inches in diameter; leaflets smaller than in H. ovata, ovate-lanceolate, smooth. New England.

H. Carolinæ-septentrionalis, sp. nov.—Fruit less than one inch in diameter; nut about .5 inch long; leaflets 3 to 5, oblong-lanceolate, smooth, glaucous beneath; bark exfoliating in broad sheets. A small tree growing on rocky hills. North Carolina to Georgia. Local.

(*b*) Fruit rarely globular; husks thin, not splitting to the base, and usually not freeing the nut; nuts more or less angled and pointed.

H. borealis, sp. nov.—Fruit flattened, .5 inch long; husk thin, not splitting; nut white, sharply angled, kernel very large; leaflets 3 to 5, oblong-lanceolate, smooth, not dotted beneath with resinous globules. Region of the Great Lakes.

H. odorata (Marsh.) Sarg.—Fruit subglobose, about 1 inch long; husk rather thin, splitting below the middle; nut ellipsoidal, obscurely angled, mealy; immature fruit sharply four-angled, the angles winged; leaflets large, ovate-lanceolate, smooth, beneath thickly dotted with resinous globules. Rich hillsides, New England to Minnesota and Kansas, and south to Texas and Florida. This is the pig-nut of the Central and Western States.

Var. villosa (Sarg.)*—Leaves, twigs and petioles somewhat pubescent, at least when young; husk rather thick; nut sharply angled; bark rough, not shaggy. Flinty hills near Allenton, Mo.

H. glabra hirsuta, var. nov.—Leaflets thicker; petioles and lower surface of leaves tomentose. North Carolina to Alabama.

H. pallida, sp. nov.—Resembling H. myristicæformis (Mx.) Britt. but the leaflets are usually 9, tomentose beneath as is the petiole; nut globose, not pointed, white; bark deep furrowed. Dry hills North Carolina.

Chapel Hill, N. C., Apr. 15, 1896.

* H. glabra villosa, Sargent Sylva. vol. vii. p. ~~148~~. 167.

Fig. 8.6. Circular on "Notes on the Hickories of the United States," by W.W. Ashe (1896).
(Courtesy of Botany Libraries, Harvard University, Cambridge, Massachusetts).

Fig. 8.7. Commemorative marker of the Class of 1884 tree (*Liriodendron tulipifera*), near northeast corner, Old West Building, University of North Carolina, Chapel Hill, March 2009.

(Courtesy of Brian Nalley, Photo Lab, Department of Biology, University of North Carolina, Chapel Hill).

Holmes's other scientific accomplishments and leadership deserve mention. In 1903, he was appointed Director, Department of Mines and Metallurgy of the Louisiana Purchase Exposition at the St. Louis World's Fair held in 1904. He was influential in getting a bill passed by Congress to found a marine biological laboratory at Beaufort, North Carolina, for the U.S. Fish Commission. He became the chief of the U.S. Geological Survey laboratories for testing fuels and structural materials at St. Louis (1905–1907) and then at Pittsburgh (beginning 1908). When the U.S. Bureau of Mines was established in 1910, he served as its first director, remaining in this position until his retirement in 1915.

Holmes was keenly interested in improving the safety of miners. He discovered that dust from bituminous coal was a greater danger to miners than was firedamp. For this work, he was greatly respected in the world of mining. In the summer of 1915, the Joseph A. Holmes Professorship of Safety and Efficiency Engineering was established at the Colorado School of Mines (Phillips, 1915). In 1916, the Joseph A. Holmes Safety Association was founded. The organization confers awards to individuals and companies responsible for reducing accidents and fatalities in the mining and related industries. A memorial tablet (Fig. 8.8) was erected in 1917 in the entrance rotunda of the Bureau of Mines Building, Pittsburgh, Pennsylvania, commemorating Holmes's contributions to the Bureau and his leadership in the movement for safety in mining. He was responsible for coining the well-known phrase "Safety first." In biographical records housed in the Cornell University Archives (CU-A), Holmes noted that the German, Belgian, Italian, and Japanese governments had also decorated him for his research on fuels and for other investigations.

Fig. 8.8. The Joseph Austin Holmes Memorial Tablet, rotunda entrance, Bureau of Mines Building, Pittsburgh, Pennsylvania.

(Box 5, folder 47, Joseph Austin Holmes Papers #3866, Southern Historical Collection, Wilson Special Collections Library, University of North Carolina, Chapel Hill).

Being the son of a minister and pursuing a career in science probably created a cause for Holmes's contemplation about religion versus science. In a newspaper article appearing in the *Presbyterian*, Holmes ([1886]c) discusses this topic. He supported scholarly pursuits in natural history and believed that there was not any special tendency toward materialism (atheism) among botanists, zoologists, geologists, chemists or physicians. He concluded by saying: "We should encourage the study of nature, not only for its own sake, but because we may hereby learn something more of God's works and God's laws. Can we not look outward through the natural, at least toward the supernatural?" In taking a stand in promoting scientific investigations, Holmes served society by challenging the skepticism of religious ideals and seeking scientific truths.

Legacy and Honors

In addition to his contributions as a professor of geology and natural history, a geologist, a conservationist, and an administrator, Holmes earned several honors. He was a fellow and charter member of the Geological Society of America and a fellow of the American Association for the Advancement of Science. He was a member of the Academies of Science in Washington, D.C., Raleigh, North Carolina, and St. Louis, Missouri; the American Forestry Association; the American Institute of Mining Engineers; the American Society for Testing Materials; and the American Society of Mechanical Engineers. In honorary societies, he was elected to Sigma Xi and Tau Beta Pi. Memberships in other organizations included the Cosmos Club (Washington, D.C.), the St. Louis Club, and the Engineer's Club (New York). A recipient of two honorary degrees in 1909, Holmes was awarded the LL.D. from UNC and the D.Sc. from the University of Pittsburgh. In conferring on Holmes the doctor of law degree at UNC, Dr. C. Alphonso Smith aptly described him as "A man of seasoned common sense, of winning personality, and of practical efficiency in all that he [undertook]" (Cobb, 1910).

Holmes and his wife Jeanie had two daughters (Jean Dalziel and Margaret Catherine) and two sons (Joseph Austin, Jr., and James Sprunt). Holmes's personal commitment to his wife was steadfast. While conducting field excursions and traveling to scientific meetings, he faithfully wrote letters to her about his professional endeavors and his caring thoughts of her. These letters and his diaries are housed in the Joseph Austin Holmes Papers in the UNC Southern Historical Collection. Holmes often overextended his energies on behalf of his scientific undertakings, working long days and making observations in the mines. When he retired in 1915, his health had been declining. In 1915, he entered a sanitarium in New Mexico and died on July 13 (at age fifty-five) from pulmonary tuberculosis in Denver, Colorado, according to his certificate of death. Joseph Holmes is buried in Rock Creek Cemetery, Washington, D.C. (Fig. 8.9).

Appearing on the North Carolina scene relatively soon after the Civil War and the reopening of the university in 1875, Holmes was a significant contributor to the development of the state's educational and industrial resources. During his tenure at UNC, there were three pivotal advancements in the scientific instruction of botany: the role of the laboratory as an adjunct to teaching and learning became more prominent, additional microscopes were acquired, and courses of instruction in botany with an application to agriculture were established. He also took his classes on field trips to collect natural history specimens (including plants for the herbarium) and to make observations. He increased the holdings of the University Museum

Fig. 8.9. Gravestone of Joseph Austin Holmes, Section M, lot 56, site 5, Rock Creek Cemetery, Washington, D.C., March 6, 2006.

(Courtesy of Kenneth J. Wurdack).

and avidly acquired literature in natural history. He actively acquired specimens for the university's growing herbarium, particularly through the Smithsonian Institution, to which he also sent plant specimens. Holmes was the first professor at the university to offer distinct courses in general botany and systematic botany as well as instruction in cryptogamic botany and horticulture. The economic and practical application of his courses was evident. Not until the arrival of William Chambers Coker in 1902 would botanical instruction again command such prominence in the Carolina curriculum. Holmes was probably the first to propose that the Department of Geology and Natural History be separated into two departments, Geology and Biology. He was one of the founding members of the Elisha Mitchell Scientific Society, whose establishment signaled the beginning of the professionalization of science in Chapel Hill and the South. A reporter for the *New York Times* considered Holmes a "martyr to miners," as he was overcome by ill health while seeking safety for miners (Anonymous, 1915).

A number of tributes were made in Holmes's honor at the 18th annual session of the American Mining Congress, held in San Francisco in September 1915. The testimonial of Samuel A. Taylor (AMC, [1915]), former president of the congress, summarized the sentiments of those assembled: "The name of Dr. Holmes will be remembered by all who knew him, as the embodiment of an earnest, conscientious man, performing with unstinted measure, any duty which came to his hand, and assisting in many works to which duty had no claim, but in which his ever present desire to serve his fellow man and do good to all, was the force behind his actions."

References Cited

(References that conclude with page numbers within parentheses indicate sources from where quotes and pertinent information were obtained.)

Adams, J.F.A. 1887. Is botany a suitable study for young men? *North Carolina Teacher* 4: 220–222.

Alumnus. [1886]. Practical and progressive. The attention that is given to the study of the sciences at the university, Chapel Hill, p. 21 in Kemp P. Battle, *Scrapbook of Material Relating to the University of North Carolina, 1869–1893*, vol. 8. (Newspaper clipping without complete bibliographical data. North Carolina Collection, Wilson Special Collections Library, University of North Carolina, Chapel Hill)

AMC. American Mining Congress. [1915]. *Joseph Austin Holmes 1859–1915.* Washington, DC, American Mining Congress. ("A record of tributes paid at the Memorial Exercises held at San Francisco, Tuesday, September 21, 1915, under the auspices of the American Mining Congress." Also includes tributes from correspondents.) (p. 80)

Anonymous. 1848. [Review]. *A Report on the Trees and Shrubs Growing Naturally in the Forests of Massachusetts* (1846). *North American Review* 66: 190–215. (p. 201)

Anonymous. 1877. The course in botany. *Cornell Era* 10(6): 4, October 27, 1877.

Anonymous. 1881a. Commencement week. *Ithaca Daily Journal* (Ithaca, NY) 18(140): [4], Thursday, June 16, 1881.

Anonymous. 1881b. The Professor of Geology and Natural History at the University. *News and Observer* (Raleigh) 18(166): [2], Thursday Morning, September 29, 1881.

Anonymous. [1882]a. Excursion to Chapel Hill, p. 13 in Kemp P. Battle, *Scrapbook of North Carolina History and Biography, 1880–1894*, vol. 3. (Newspaper clipping without complete bibliographical data. North Carolina Collection, Wilson Special Collections Library, University of North Carolina, Chapel Hill)

Anonymous. [1882]b. The excursion to Chapel Hill last Saturday—All aboard, p. 12 in Kemp P. Battle, *Scrapbook of North Carolina History and Biography, 1880–1894*, vol. 3. (Newspaper clipping without complete bibliographical data. North Carolina Collection, Wilson Special Collections Library, University of North Carolina, Chapel Hill)

Anonymous. 1883a. For the *Home and Democrat*. University of North Carolina. *Charlotte Home and Democrat* (Charlotte, NC), new series 13 (642): [2], Friday, November 16, 1883.

Anonymous. 1883b. [Professor Holmes], p. 54 in College Record, *University Monthly* (University of North Carolina, Chapel Hill) 2: 54.

Anonymous. 1883c. [Six hours a week in the Botany Laboratory]. *University Monthly* (University of North Carolina, Chapel Hill) 2: 184.

Anonymous. 1883d. University Day at the University. *Fayetteville Observer* (Fayetteville, NC) [Weekly], new series 1(42): [1], Thursday, November 22, 1883.

Anonymous. 1884a. [Elisha Mitchell Scientific Society meeting, held October 18· 1884]. *University Magazine* (University of North Carolina, Chapel Hill), new series 4: 87–88.

Anonymous. 1884b. *Visitors' Guide to the North Carolina State Exposition and City of Raleigh, with Description and Views of the State of North Carolina*. The Only Guide Book Sold on the Grounds. Authorized by the Exposition Executive Committee. Exposition Guide Publishing Company, Raleigh.

Anonymous. 1885a. [An additional appropriation for the University]. *University Magazine* (University of North Carolina, Chapel Hill), new series 4: 264–265.

Anonymous. 1885b. The University. *News and Observer* (Raleigh) 25(46): [2], Thursday, June 18, 1885.

Anonymous. 1885c. *University of North Carolina. Post-Graduate Courses of Study*. [no statement of publisher or place]. (p. 8)

Anonymous. 1885d. Branches of learning relating to agriculture at the University of North Carolina. *News and Observer* (Raleigh) 26(28): [2], Thursday, December 10, 1885.

Anonymous. 1886a. [The tree of the class of '86]. *University Magazine* (University of North Carolina, Chapel Hill), new series 5: 93.

Anonymous. 1886b. [Tree planting at UNC]. *University Magazine* (University of North Carolina, Chapel Hill), new series 5: 254.

Anonymous. 1887a. [Legislative news]. *Biblical Recorder* (Raleigh) 52(37): [2], Wednesday, March 9, 1887.

Anonymous. 1887b. [Note on a scientific revival in the southern states of America]. *Nature* 36: 587.

Anonymous. 1887c. The wedding bells. *Wilmington Messenger* (Wilmington, NC), Friday, October 21, 1887, p. 8.

Anonymous. 1890. University geological expedition. *University Magazine* (University of North Carolina, Chapel Hill), new series 10: 65–66.

Anonymous. [1895]. Contributions to the University Museum, etc. 2 pp. (Box 48, folder 1728, University Papers #40005, University Archives, Wilson Special Collections Library, University of North Carolina, Chapel Hill)

Anonymous. 1897. Department of Biology. *University Record* (University of North Carolina, Chapel Hill) 2(1): 7–8.

Anonymous. 1898. Prof. Holmes' lecture. *The Tar Heel* (University of North Carolina, Chapel Hill) 6(11): [1], January 18, 1898.

Anonymous. 1901. [Prof. J.A. Holmes commenting on Indian bread.] *News and Observer* (Raleigh) 50(107): 6, Tuesday, July 16, 1901.

Anonymous. 1906. Joseph Austin Holmes, p. 100 in *Prominent People of North Carolina, Brief Biographies of Leading People for Ready Reference Purposes*. Evening News Publishing Company, Asheville, NC.

Anonymous. 1915. J.A. Holmes dies—martyr to miners. *New York Times* 64(20,990): 9, Tuesday, July 14, 1915.

Anonymous. 1933. Holmes, Joseph Austin, pp. 104–105 in *The National Cyclopaedia of American Biography, Being the History of the United States*, vol. 23. James T. White & Company, New York. (portrait)

Ashe, W[illiam] W. 1894. The long leaf pine and its struggle for existence. *Journal of the Elisha Mitchell Scientific Society* 11: [1]–16.

Ashe, W[illiam] W. 1896. [Circular]: Notes on the Hickories of the United States. [The Author], Chapel Hill, NC. 1 p. (Based on a paper read before the Elisha Mitchell Scientific Society and distributed at the meeting held April 15, 1896. Copy in the Botany Libraries, Harvard University)

Ashe, W[illiam] W. [1900]. New East American species of *Crataegus*. (*Contributions from My Herbarium*. No. VI). *Journal of the Elisha Mitchell Scientific Society* 16: 70–79.

Battle, Kemp P. [1888]. Short Courses of Study. [University of North Carolina?], Chapel Hill, NC. (circular, 1 sheet)

Battle, Kemp P. 1889. Letter of September 13, 1889 to W.L. Saunders. (Box 18, folder 599, University Papers #40005, University Archives, Wilson Special Collections Library, University of North Carolina, Chapel Hill)

Battle, Kemp P. 1907–1912. *History of the University of North Carolina*. Printed by Edwards & Broughton, Raleigh. 2 vols. (Vol. 2, pp. 274, 391)

Battle, Kemp P. 1916. Dr. Joseph Austin Holmes at the University of North Carolina. *Journal of the Elisha Mitchell Scientific Society* 32: 20–23.

Battle, Kemp P., C.M. Cooke, and J.S. Carr (Committee). [1885]. *University Memorial*. Ashe & Gatling, State Printers and Binders, [Raleigh]. Pp. 3–14; pp. [1]–2: *Memorial of the Board of Trustees of the University of North Carolina*, by A.M. Scales, Governor. (Document No. 22, Session 1885; bound and collected with other documents in Executive and Legislative Documents of the State of North Carolina, Session 1885. P.M. Hale, State Printer and Binder, Raleigh)

Bessey, Charles E. 1881. *Botany for High Schools and Colleges*. 2nd ed., revised. (*American Science Series*). Henry Holt and Company, New York.

BOT. Board of Trustees, University of North Carolina, Chapel Hill. *Minutes of the Board of Trustees*. (University Archives #40001, Wilson Special Collections Library, University of North Carolina, Chapel Hill) (Vol. 7, July 1868–March 1883; vol. 8, June 1883–February 1891; vol. 9, April 1891–January 1898)

Bryant, Bernard Lee, Jr. 1999. *Occupants and Structures of Franklin Street, Chapel Hill, North Carolina at 5-Year Intervals, 1793–1998*. [The Author, Chapel Hill]. ("Xerox copy on 100% cotton paper of Mr. Bryant's original computer printout by J.D. Eyre, Chapel Hill Historical Society, 1999"). (p. 11)

Chapman, A.W. 1865. *Flora of the Southern United States: Containing abridged descriptions of the flowering plants and ferns of Tennessee, North and South Carolina, Georgia, Alabama, Mississippi, and Florida: Arranged according to the natural system*. The Ferns by Prof. Daniel C. Eaton. Ivison, Phinney & Co., New York.

Cobb, Collier. 1910. Joseph Austin Holmes. *Journal of the Elisha Mitchell Scientific Society* 26: 167–170. (Reprinted from the *Charlotte Observer*.)

Comstock, J. Henry. 1881. Letter to President [Kemp P. Battle] and Trustees of the University of North Carolina, August 25, 1881. (Box 11, folder 406, University Papers #40005, University Archives, Wilson Special Collections Library, University of North Carolina, Chapel Hill)

Cooper, Walter Gerald. 1896. *The Cotton States and International Exposition and South, Illustrated. Including the Official History of the Exposition,* by Walter G. Cooper. The Illustrator Company, Atlanta.

CU. Cornell University Archives, Carl A. Kroch Library, Division of Rare and Manuscript Collections, Ithaca, New York.

CU-A. Deceased Alumni Files. (Cornell University Archives #41-2-882, Carl A. Kroch Library, Division of Rare and Manuscript Collections)

CU-R. 1881. *Cornell University Register 1880–81*. Published by the University, Ithaca, NY. (p. 62)

CU-RS. *Cornell University Register of Studies*. (Cornell University Archives #36/1/630, vol. B [1874–1880], p. 648; Carl A. Kroch Library, Division of Rare and Manuscript Collections, Cornell University)

Dudley, William R. 1881. Letter of August 26, 1881 to President [Kemp P. Battle]. (Box 11, folder 406, University Papers #40005, University Archives, Wilson Special Collections Library, University of North Carolina, Chapel Hill)

Ellis, D.L. 1884. The Exposition as seen by a teacher. *North Carolina Teacher* 2: 206–208.

EMSS. Elisha Mitchell Scientific Society. *Records of the Elisha Mitchell Scientific Society*. (University Archives #40183, Wilson Special Collections Library, University of North Carolina, Chapel Hill). (Box 1 contains two volumes of minutes of the society's proceedings; volume 1 covers 1883 to 1940)

Faison, Henry E. 1878. Letter of April 14, 1878 to "Sister." (Box 3, folder 35, Henry Wilson Faison Papers #3789, Family Series, Southern Historical Collection, Wilson Special Collections Library, University of North Carolina, Chapel Hill)

Ford, Charles Edward. 1962. *Botany in the American Curriculum, 1642–1914*. A dissertation presented to the Graduate Board of Washington University in partial fulfillment of the requirements for the degree of Doctor of Education. Saint Louis. (Facsimile printed from a microfilm copy at UMI Dissertation Services, Ann Arbor, 2003)

GACR. *Greeley Agricultural Club Records* (including minutes of the Agricultural Society of Cornell University and the constitution and proceedings of the Greeley Agricultural Club). (Archives #37-6-2673, Carl A. Kroch Library, Division of Rare and Manuscript Collections, Cornell University)

Gage, Simon H. 1881a. Letter of August 25, 1881 to Kemp P. Battle and Trustees of the University of North Carolina. (Box 11, folder 406, University Papers #40005, University Archives, Wilson Special Collections Library, University of North Carolina, Chapel Hill)

Gage, Simon H. 1881b. Letter of September 7, 1881 to Kemp P. Battle. (Box 12, folder 407, University Papers #40005, University Archives, Wilson Special Collections Library, University of North Carolina, Chapel Hill)

Gibboney, James H., Robert R. Cotten and Joseph Hyde Pratt (Special Committee of the North Carolina Forestry Association). 1916. [Resolution on behalf of Joseph Austin Holmes], North Carolina Forestry Association. (Box 3, folder 29, Joseph Austin Holmes Papers #3866, Southern Historical Collection, Wilson Special Collections Library, University of North Carolina, Chapel Hill)

Gray, Asa. 1879a. *Gray's Lessons in Botany and Vegetable Physiology*. Ivison, Blakeman, Taylor & Co., New York and Chicago. (This text was bound with Gray's *Manual* [5th edition, 8th issue] and served as an introduction to it.)

Gray, Asa. 1879b. *Structural Botany, or Organography on the Basis of Morphology, To Which Is Added the Principles of Taxonomy and Phytography, and a Glossary of Botanical Terms*. Gray's *Botanical Text-book* (sixth edition) Vol. 1. Ivison, Blakeman, Taylor and Co., New York and Chicago.

Grimes, J. Bryan. 1884. Letter of March 1, 1884 to Miss Mannie. (Box 1, folder b, J. Bryan Grimes Papers, Collection 54, East Carolina Manuscript Collection, J.Y. Joyner Library, East Carolina University, Greenville, NC)

Henderson, Archibald. 1949. *The Campus of the First State University*. (*University of North Carolina Sesquicentennial Publications*). University of North Carolina Press, Chapel Hill. (p. 268)

Holmes, Joseph A. 1881a. Letter of August 25, 1881 to Kemp P. Battle. (Box 11, folder 406, University Papers #40005, University Archives, Wilson Special Collections Library, University of North Carolina, Chapel Hill)

Holmes, Joseph A. 1881b. Letter of September 2, 1881 to Kemp P. Battle. (Box 12, folder 407, University Papers #40005, University Archives, Wilson Special Collections Library, University of North Carolina, Chapel Hill)

Holmes, Joseph A. 1881c. Letter of September 5, 1881 to Kemp P. Battle. (Box 12, folder 407, University Papers #40005, University Archives, Wilson Special Collections Library, University of North Carolina, Chapel Hill)

Holmes, Joseph A. 1881d. Letter of September 23, 1881 to Kemp P. Battle. (Box 12, folder 408, University Papers #40005, University Archives, Wilson Special Collections Library, University of North Carolina, Chapel Hill)

Holmes, Joseph A. 1881e. Letter of October 31, 1881 to George Vasey. (Record Unit 220, United States National Museum, Division of Plants, 1870–1893, Records, Box 8, Folder 10, Smithsonian Institution Archives)

Holmes, Joseph A. 1881f. Letter of November 9, 1881 to George Vasey. (Record Unit 220, United States National Museum, Division of Plants, 1870–1893, Smithsonian Institution Archives)

Holmes, Joseph A. 1881g. *Agricultural Labor in the Southern States*. Thesis in Agriculture, Cornell University. (Thesis Film 1881, #38, Olin Library, Cornell University)

Holmes, Joseph A. 1882a. Letter of June 1, 1882 to Kemp P. Battle. (Box 12, folder 420, University Papers #40005, University Archives, Wilson Special Collections Library, University of North Carolina, Chapel Hill)

Holmes, Joseph A. 1882b. Letter of September 2, 1882 to George Vasey. (Record Unit 220, United States National Museum, Division of Plants, 1870–1893, Smithsonian Institution Archives)

Holmes, Joseph A. 1882c. [Report on] The Natural History Department of the University, October 20, 1882. (Box 12, folder 427, University Papers #40005, University Archives, Wilson Special Collections Library, University of North Carolina, Chapel Hill)

Holmes, Joseph A. 1882d. Letter of October 6, 1882 to Mr. Patterson. (Box 12, folder 427, University Papers #40005, University Archives, Wilson Special Collections Library, University of North Carolina, Chapel Hill)

Holmes, Joseph A. 1882e. Letter of October 6, 1882 to the University Bursar. (Box 12, folder 427, University Papers #40005, University Archives, Wilson Special Collections Library, University of North Carolina, Chapel Hill)

Holmes, Joseph A. 1883a. Letter of January 17, 1883 to the University Bursar. (Box 12, folder 431, University Papers #40005, University Archives, Wilson Special Collections Library, University of North Carolina, Chapel Hill)

Holmes, Joseph A. 1883b. Letter of May 22, 1883 to E.A. Carman. (Record Unit 220, United States National Museum, Division of Plants, 1870–1893, Smithsonian Institution Archives)

Holmes, Joseph A. 1883c. Postal card of June 14, 1883 to Kemp P. Battle. (Box 13, folder 440, University Papers #40005, University Archives, Wilson Special Collections Library, University of North Carolina, Chapel Hill)

Holmes, Joseph A. 1883d. Our Indian antiquities. Indian burial mounds in eastern North Carolina. The result of Prof. Jas. [*i.e.*, Jos.] A. Holmes' examination in Duplin County, p. 23 in Kemp P. Battle, *Scrapbook of North Carolina History and Biography, 1880–1894*, vol. 3. (North Carolina Collection, Wilson Special Collections Library, University of North Carolina, Chapel Hill)

Holmes, Joseph A. 1883e. Letter of May 22, 1883 to George B. Loring. (Record Unit 220, United States National Museum, Division of Plants, 1870–1893, Smithsonian Institution Archives)

Holmes, Joseph A. 1884. Letter of September 2, 1884 to President Kemp P. Battle. (Box 14, folder 480, University Papers #40005, University Archives, Wilson Special Collections, University of North Carolina, Chapel Hill)

Holmes, Joseph A. [1885]. [Studies in natural history that bear upon practical agriculture]. (Box 48, folder 1729, University Papers #40005, University Archives, Wilson Special Collections Library, University of North Carolina, Chapel Hill) (undated miscellaneous document)

Holmes, Joseph A. 1886a. Letter of June 18, 1886 to George Vasey. (Record Unit 220, United States National Museum, Division of Plants, 1870–1893, Smithsonian Institution Archives)

Holmes, Joseph A. 1886b. Letter of July 7, 1886 to George Vasey. (Record Unit 220, United States National Museum, Division of Plants, 1870–1893, Smithsonian Institution Archives)

Holmes, Joseph A. [1886]c. The study of God's work, p. 59 in Kemp P. Battle, *Scrapbook of North Carolina History and Biography, 1880–1894*, vol. 4, p. 59. (North Carolina Collection, Wilson Special Collections Library, University of North Carolina)

Holmes, Joseph A. 1886d. Letter of May 26, 1886 to George Vasey. (Record Unit 220, United States National Museum, Division of Plants, 1870–1893, Smithsonian Institution Archives)

Holmes, Joseph A. 1887. Letter of May 22, 1887 to Kemp P. Battle. (Box 16, folder 549, University Papers #40005, University Archives, Wilson Special Collections Library, University of North Carolina, Chapel Hill)

Holmes, Joseph A. 1888a. Letter of October 2, 1888 to the Board of Trustees, University of North Carolina. (Box 17, folder 572, University Papers #40005, University Archives, Wilson Special Collections Library, University of North Carolina, Chapel Hill)

Holmes, Joseph A. 1888b. Letter of October 2, 1888 to William L. Saunders, Secretary, Board of Trustees, University of North Carolina. (Box 17, folder 572, University Papers #40005, University Archives, Wilson Special Collections Library, University of North Carolina, Chapel Hill)

Holmes, Joseph A. 1891a. Report to Kemp P. Battle, January 24, 1891. (Box 19, folder 633, University Papers #40005, University Archives, Wilson Special Collections Library, University of North Carolina, Chapel Hill)

Holmes, Joseph A. 1891b. Letter of March 18, 1891 to Kemp P. Battle. (Box 19, folder 637, University Papers #40005, University Archives, Wilson Special Collections Library, University of North Carolina, Chapel Hill)

Holmes, Joseph A. 1903. *Proceedings of the North Carolina Good Roads Convention, Held at Raleigh, February 12 and 13, 1902*. (*United States Department of Agriculture, Office of Public Road Inquiries, Bulletin* no. 24). Government Printing Office, Washington.

Irwin, Betty W. 1963. Octagon-shaped house, first concrete structure in state. *Laurens Advertiser* (Laurens, SC) 78(43): Section 3: 1–2, Wednesday, April 24, 1963.

Johnson, Samuel William. 1879. *How Crops Feed: A treatise on the atmosphere and the soil as related to the nutrition of agricultural plants*. Orange Judd and Company, New York.

Johnson, Samuel William. 1880. *How Crops Grow: A treatise of the chemical composition, structure, and life of the plant, for all students of agriculture*. Orange Judd and Company, New York.

Langley, S[amuel] P. 1890. Letter of April 4, 1890 [to Joseph Holmes]. (Record Unit 112, vol. 48, folder 1, Smithsonian Institution Archives)

Lincoln, Almira H. 1836. *Familiar Lectures on Botany, Practical, Elementary and Physiological; With an appendix, containing descriptions of the plants of the United States and exotics, &c. for the use of seminaries and private students*. 5th ed., revised and enlarged. F.J. Huntington, New-York.

Loring, Geo. B. 1883. Report of the Commissioner of Agriculture, pp. 3–16 in *Report of the Commissioner of Agriculture for the Year 1883*. Government Printing Office, Washington, D.C.

Love, James Lee. 1921. [Diaries, Volumes 5 and 7]. (Box 5, folder 69; box 6, folder 71, James Lee Love Papers #4139, Southern Historical Collection, Wilson Special Collections Library, University of North Carolina, Chapel Hill). (Volume 5, p. 114; volume 7, p. 8)

Love, James Lee. 1945. *'Tis Sixty Years since; A Story of the University of North Carolina in the 1880s*. 2nd printing. University of North Carolina Press, Chapel Hill, NC. (p. [32])

MacKoon, B.P. 1881. Letter to whom it may concern, August 25, 1881. (Box 11, folder 406, University Papers #40005, University Archives, Wilson Special Collections Library, University of North Carolina, Chapel Hill)

Manning, Van H. 1915. Historical sketch and appreciation of Dr. Joseph A. Holmes, pp. 9–20 in *Joseph Austin Holmes 1859–1915*. American Mining Congress, Washington, D.C.

McCarthy, Gerald. 1889. Botany as a disciplinary study. *Journal of the Elisha Mitchell Scientific Society* 6: 33–38.

M[cDonald], P[hilip] B. 1932. Holmes, Joseph Austin, pp. 167–168 in Dumas Malone (editor), *Dictionary of American Biography*, vol. 9. Charles Scribner's Sons, New York.

NCGS. *North Carolina Geological Survey Records*. (Southern Historical Collection #2306, Wilson Special Collections Library, University of North Carolina, Chapel Hill). (Boxes 5 [folder 30 contains the inventory of the North Carolina exhibit at the Atlanta Exposition of 1895] and 6)

NHSR. *Natural History Society Records. Proceedings [of Meetings], 1876–1881*. (Archives #37-6-584, Carl A. Kroch Library, Division of Rare and Manuscript Collections, Cornell University)

Phillips, William B. 1915. Letter of August 12, 1915 to Mrs. Joseph A. Holmes. (Box 1, folder 2, Joseph Austin Holmes Papers #3866, Southern Historical Collection, Wilson Special Collections Library, University of North Carolina, Chapel Hill)

Pratt, Joseph Hyde. [1916]. Memorial sketch of Dr. Joseph Austin Holmes. *Journal of the Elisha Mitchell Scientific Society* 32: 1–15. (portrait and bibliography). ("Read at the annual meeting of the Geological Society of America, Washington, DC, December 28, 1915.")

Prentiss, Albert N. 1881. Letter of September 1, 1881 to Kemp P. Battle. (Box 12, folder 407, University Papers #40005, University Archives, Wilson Special Collections Library, University of North Carolina, Chapel Hill)

Roberts, I[saac] P. 1881. Letter of August 27, 1881 to Kemp P. Battle. (Box 11, folder 406, University Papers #40005, University Archives, Wilson Special Collections Library, University of North Carolina, Chapel Hill)

Sachs, Julius. 1875. *Text-book of Botany. Morphological and Physiological.* Translated and annotated by Alfred W. Bennett, assisted by W.T. Thiselton Dyer. At the Clarendon Press, Oxford, [England].

Sargent, Charles S. 1880. *A Catalogue of the Forest Trees of North America.* (*Department of the Interior, Tenth Census of the United States, Forestry*). Government Printing Office, Washington, D.C.

Sargent, Charles S. 1884. *Report on the Forests of North America (Exclusive of Mexico).* Government Printing Office, Washington, D.C.

Scales, Alfred M. 1888. *University of North Carolina. Special Announcement.* [no statement of publisher], Raleigh. (p. 1)

SHC. Southern Historical Collection. Joseph Austin Holmes Papers #3866, Wilson Special Collections Library, University of North Carolina, Chapel Hill. (includes numerous newspaper clippings)

Shepard, C.L. 1924. Letter of July 9, 1924 to Mrs. Donald Stanton. (Box 1, folder 2, Joseph Austin Holmes Papers #3866, Southern Historical Collection, Wilson Special Collections Library, University of North Carolina, Chapel Hill)

SI. Smithsonian Institution Archives, Washington, DC.

Smith, Charles Dennis. 1960. The Appalachian National Park Movement, 1885–1901. *North Carolina Historical Review* 37: 38–65. (p. 39)

S[pencer], C[ornelia] P. 1880. Tree planting. *Observer* (Raleigh) 6(223): [2], Sunday, March 21, 1880.

S[pencer], C[ornelia] P. [1894]. Letter of September 19, 1894 to Kemp P. Battle. (Box 7, folder 105, Battle Family Papers #3223-A, Southern Historical Collection, Wilson Special Collections Library, University of North Carolina, Chapel Hill)

Stuckey, Jasper Leonidas. 1965. *North Carolina: Its geology and mineral resources.* Department of Conservation and Development, Raleigh.

Sumner, Jim L. 1992. "Let us have a big fair": The North Carolina Exposition of 1884. *North Carolina Historical Review* 69: 57–81.

Sykes, R.H. [1916]. Joseph Austin Holmes. (Box 1, folder 1, Joseph Austin Holmes Papers #3866, Southern Historical Collection, Wilson Special Collections Library, University of North Carolina, Chapel Hill)

Tim. 1882. Class Day at Chapel Hill. *News and Observer* (Raleigh) 19(3): [2], Tuesday, April 4, 1882.

UNC Faculty. University of North Carolina, Chapel Hill. *Minutes of Faculty Meetings*. (University Archives #40106, Wilson Special Collections Library, University of North Carolina, Chapel Hill) (Vol. 6, 1856–1885; vol. 7, 1885–1901)

Venable, F.P. 1888. Letter of April 20, 1888 to [Thomas Fanning] Wood. (Personal Collection #1346, Box 2, Thomas Fanning Wood, State Archives of North Carolina)

Venable, F.P. 1916. Joseph Austin Holmes. *Journal of the Elisha Mitchell Scientific Society* 32: 16–19.

Vinson, Frank Bedingfield. [1988]. Holmes, Joseph Austin, pp. 177–178 in William S. Powell (editor), *Dictionary of North Carolina Biography*, vol. 3. University of North Carolina Press, Chapel Hill.

Weeks, Stephen B. 1886a. College record. *University Magazine* (University of North Carolina, Chapel Hill), new series 6: 27–28.

Weeks, Stephen B. 1886b. [Professor Atkinson], p. 65 in College record. *University Magazine* (University of North Carolina, Chapel Hill), new series 6: 64–70.

Weeks, Stephen B. 1887. Elisha Mitchell Scientific Society, pp. 421–422 in College Record. *University Magazine* (University of North Carolina, Chapel Hill), new series 6: 415–422.

Wickens, Gerald E. 2001. *Economic Botany; Principles and practices*. Kluwer Academic Publishers, Dordrecht. (p. 2)

Williams, S[amuel] G. 1881. Letter of August 25, 1881 to Kemp P. Battle. (Box 11, folder 406, University Papers #40005, University Archives, Wilson Special Collections Library, University of North Carolina, Chapel Hill)

Wilson, W.D. 1881. Letter of August 23, 1881 to Kemp P. Battle. (Box 11, folder 406, University Papers #40005, University Archives, Wilson Special Collections Library, University of North Carolina, Chapel Hill)

Chapter 9.

Henry Van Peters Wilson (1863–1939)

Professor of Biology (1891/1892–1903/1904), absent on leave 1902/1903
Professor of Zoology (1904/1905–1916/1917)
Kenan Professor of Zoology (1917/1918–1938/1939), absent on leave 1929/1930
Professor of Biology in Medical School
Professor of Biology in the School of Pharmacy

Youth, Education, and Early Career

Henry Van Peters Wilson, born February 16, 1863, in Baltimore, Maryland, was the son of Samuel Augustus and Sophia Ann (Stansbury) Wilson. As Wilson (1932a) noted, the name "van" in Van Peters was converted from a preposition to a name and subsequently capitalized, following a common American practice. Young Wilson was the second of six children. His father was a Methodist Episcopal Church, South clergyman, his mother a housewife. After completing lower classes in the public schools of Washington, D.C. and Baltimore, Wilson attended the Baltimore City College, a secondary school for boys, for three years. Established in 1839 as "The High School," the Baltimore City College is the third oldest public high school in the United States—those at Boston and Philadelphia being older. Its graduates include United States senators, noted authors, publishers, philanthropists, scholars, and one Nobel Prize recipient. There, several superb instructors (Charles C. Wight, Charles F. Raddatz, Stephen F. Norris, and Chapman Maupin) provided Wilson with strong foundations in English, Latin, German, and History.

Having scholarly promise, Wilson enrolled in 1880 at the Johns Hopkins University, which had been established four years earlier. It was founded on the principle of inquiry and research as developed in Germany, and it was the first graduate school in the United States. From its beginning, particular attention was directed to establishing a strong program in biology, including the physiology and morphology of animals and plants. Not only was the discipline significant as a science, but it was also closely allied to medicine, the study of which the university would eventually develop. Publishing research results was an important component of the Hopkins experience, as was ready access to biological literature, both journals and books (Anonymous, 1881). Another academic influence was the Baltimore Naturalists' Field Club, which was established in 1880 to promote field studies in natural science (Cowles, 1921). In applying for admission to Hopkins, Wilson desired "to obtain a liberal education in languages (Latin, German, and French), history, philosophy, and perhaps some branch of science" (JHU-A, no. 1). The Hopkins environment influenced Wilson and provided a model for the development of a biological curriculum at UNC nearly a decade later.

Frontispiece Credits: (above) Henry Van Peters Wilson, ca. 1902–1911. (below) Biology Laboratory, fourth floor of New East Building, University of North Carolina at Chapel Hill, ca. 1902. (Courtesy of North Carolina Collection Photographic Archives, Wilson Special Collections Library, University of North Carolina, Chapel Hill).

At the conclusion of his first year of studies at Hopkins, Wilson chose to study medicine. His academic mentors in biology included William Thompson Sedgwick (associate in biology), William Keith Brooks (comparative animal anatomist), and Henry Newell Martin (animal physiologist). While enrolled in a general biology course, an impressionable Wilson was influenced by his exposure to the methods and research investigations of the laboratory setting. Upon receiving the A.B. in 1883, he pursued summer work with Professor Brooks, studying marine biology at the seaside laboratory at Fort Wool, Hampton Roads, Virginia. Although a graduate student, he took additional undergraduate courses in biology the following year and served as a laboratory assistant.

Wilson then enrolled at the Medical School of the University of Maryland. After several weeks there, he decided to abandon this course of study, turning his scholarly pursuits to teaching and research. In November 1884, he became a private tutor in zoology to Milwaukee manufacturer Edmund Phelps Allis, Jr., who wanted to change his career to that discipline. In writing to Johns Hopkins University president Daniel C. Gilman about the tutorship, Wilson (1884) indicated his intentions of returning to Hopkins to pursue his doctoral studies.

In the fall of 1885, Wilson returned to Hopkins. He commenced doctoral studies in zoology under the direction of his former professor, William K. Brooks, who was a specialist on the embryonic development of marine invertebrates and had studied under Louis and Alexander Agassiz. Brooks was a Darwinian who used morphological studies as a means to understand the evolution of structure through animal embryology (Allen, 1978). As an academic mentor, Brooks's (Fig. 9.1) influence on young Wilson was of lasting significance. Like Brooks, Wilson became an evolutionist, a superb teacher who used original drawings on the blackboard, and an alumnus of the marine lab in Naples, Italy. Costello (1961) also mentions Wilson's preference for a simplicity of apparatus in the laboratory. No doubt this philosophy was nurtured at Hopkins, where a "distrust of fancy equipment" was evident (Shine and Wrobel, 1976). Nonetheless, Hopkins was an important center for biological research, and for the promotion of an experimental approach to science.

In Wilson's research environment were such notable students as Thomas Hunt Morgan, Ross Granville Harrison, and Edwin Grant Conklin (Maienschein, 1999). During the spring and summer of 1886, Wilson and classmate George Bowman Haldeman conducted their research projects at Beaufort, North Carolina, where Hopkins had first sponsored a marine zoology research station (Fig. 9.2) in 1880 under the direction of Brooks. Wilson studied the anatomy and histology of various medusae. In the fall of 1886, he returned to Baltimore, continuing his research. In that same year, he authored his first publication, in which he described the parasitic Cuninas of Beaufort (Wilson, 1886). The following year, he moved his work to Nassau, New Providence, Bahamas, where Professor Brooks set up a research lab during the spring. He returned to Nassau the following spring. There, he collected and reared larvae of the Chenille stone coral, of common occurrence in this region. Based on his work on the embryology of this coral, Wilson earned a Doctor of Philosophy degree in 1888. Wilson's (1888a) dissertation, "On the Development of *Manicina areolata*," published in the *Journal of Morphology* (2: 191–252, 1888) was embellished with seven lithographs (6 folded), which Wilson created. Wilson's classmate T. H. Morgan (1889) noted the dissertation and described the new facts that Wilson presented. During his last three years at Hopkins, Wilson was a graduate scholar (1885–1887) and a fellow (1887–1888).

Fig. 9.1. Portrait of William Keith Brooks, 1907.

(Courtesy of The Ferdinand Hamburger Archives, Sheridan Libraries, Johns Hopkins University).

While Wilson was completing his doctorate, he documented his various research interests in a letter to Hopkins president Daniel Coit Gilman. In the letter, Wilson (1888b) enumerated his reasons for wishing to return to study in the Bahamas on the Adam T. Bruce doctoral fellowship. Among them were the opportunities 1) to work out the life histories of marine animals, 2) to understand the breeding times of marine fauna, 3) to study in a warm climate where the marine life was numerous and varied and was likely to have breeding times throughout the year, and 4) to expand his investigations of coelenterates, which were well represented there, with the goal of determining the life history (cycle) of *Alcyormia* (sea fans and feathers). He also pointed out that his proposed studies would not be entirely focused on coelenterates. He also considered studying the embryology of a unique chiton found there and observing the sponges.

Wilson was awarded the fellowship, beginning June 1888. Classmate Charles Lincoln Edwards accompanied him to the Bahama Islands. Six months later, Wilson wrote to Gilman that he was stationed at Green Turtle, Bahamas. His primary work was collecting and raising embryological material, as well as making observations and notes on forms for investigation when he returned to Baltimore. One of the most significant results of his research would be a life history of a siliceous sponge. He also brought back to Hopkins ascidians, a tectibranch mollusk (*Aplysia*), a decapod crustacean (*Hippa*), and sea anemones (Wilson, 1889). Based on his work in the Bahamas, Wilson published several papers. His studies on sponges, which began in the Bahamas, would later occupy much of his future research, elucidating their embryology, classification, and physiology.

Fig. 9.2. The Seaside School of the Johns Hopkins University at Beaufort, North Carolina.

(From *Frank Leslie's Illustrated Newspaper*, November 20, 1880, p. 192; courtesy of North Carolina Collection, Wilson Special Collections Library, University of North Carolina, Chapel Hill).

Wilson's early professional experience began in May 1889, when he was appointed Scientific Assistant in the United States Fish Commission. He was assigned to head up the Fisheries Biological Laboratory at Woods Holl (now Woods Hole), Massachusetts. For the next two years, his investigations concentrated on vertebrate embryology, culminating in publications on the sea bass, *Serranus atrarius* (Wilson, 1890, 1891a), and then on the sponges. During the spring of 1891, Wilson began to search for an academic position.

Two letters of 1891 housed in the Johns Hopkins University archives shed light on Wilson's early efforts to gain an academic appointment. The letters, written to Gilman at Hopkins, indicate the genial association between these two men. In the spring of 1891, Wilson visited Rev. Dr. George Williamson Smith, president of Trinity College in Hartford, Connecticut. They discussed the newly established chair in biology there (Wilson, 1891b). Wilson did not get the position, most likely due to the fact that his academic background in geology was weak. He believed that Smith preferred to fill the vacancy with a biologist equally trained in geology.

Wilson later wrote about his appointment at Chapel Hill, mentioning that UNC president George Tayloe Winston had visited him in Baltimore toward the beginning of July. They discussed the new position in biology at UNC. Wilson thanked Gilman for recommending him to Winston (Wilson, 1891c).

Documentation concerning the nomination and election of a person to fill the newly created professorship in biology is sparse. Among the individuals interested in the position were Elmer E. Wolfe, whose professional background included positions as science teacher at Peddie Institute in Hightstown, New Jersey, and later as principal at Chapman Seminary, and Benjamin Franklin Koons, who had been principal (and would later be president) of the Storrs Agricultural College in Connecticut. We learn from the UNC Board of Trustees minutes of June 3, 1891, however, that the Committee of Visitation favored Thomas Hunt Morgan of the Johns Hopkins University to fill the chair of biology. Joseph A. Holmes had met Morgan, and his credentials had made a favorable impression. Morgan's testimonials so impressed the committee that it recommended his immediate election. Morgan, however, accepted a faculty appointment in biology at Bryn Mawr College. Thomas Hunt Morgan (1866–1945) became a pioneer geneticist who was named Nobel Laureate in 1933 for determining the role that chromosomes played in heredity.

Newly appointed UNC president George Tayloe Winston then suggested Morgan's college classmate, Henry Van Peters Wilson, whose subsequent election by the Executive Committee of the Board of Trustees was unanimous (BOT 9: 42). He was hired with a probation period of one year, after which time the trustees and the committee of election would assess his performance. A lifelong tenure at UNC would follow.

Wilson's starting date was September 1, 1891, and his yearly salary was set at $1,600 (BOT 9: 42) (Fig. 9.3), which was increased to $2,000 on September 1, 1895 (BOT 9: 462). In June 1894, the Executive Committee resolved that he "be allowed the use of his house rent free until his salary shall be raised, provided he keep it in ordinary use" (BOT 9: 375). In 1906, he built a house at 722 East Franklin Street, Chapel Hill (Fig. 9.4), where he lived until his death. UNC professor Samuel T. Emory next owned the house and subsequently UNC professor H. Garland Hershey became the third owner (Eyre, 2007). Although Wilson was educated and trained as a zoologist, his teaching assignment included botany until autumn 1902, when William Chambers Coker was hired as associate professor of botany.

In addition to his appointment to the UNC faculty, Wilson also would serve as associate Commissioner of Shell-fish, a position that gave him use of the commission steamer for collecting. At the same time that UNC filled its chair of Biology, Trinity College (then located in Randolph County, North Carolina, now Duke University in Durham) likewise hired its first biologist, John Moore Stedman (November 3, 1865–November 6, 1949). Stedman brought with him biological specimens, which he donated to the biological laboratory at Trinity (Anonymous, 1891c). He also became active in the Elisha Mitchell Scientific Society, joining in December 1892. Stedman's stay at Trinity was cut short when he accepted a professorship in biology at the experimental station of A & M College, in Auburn, Alabama in 1893.

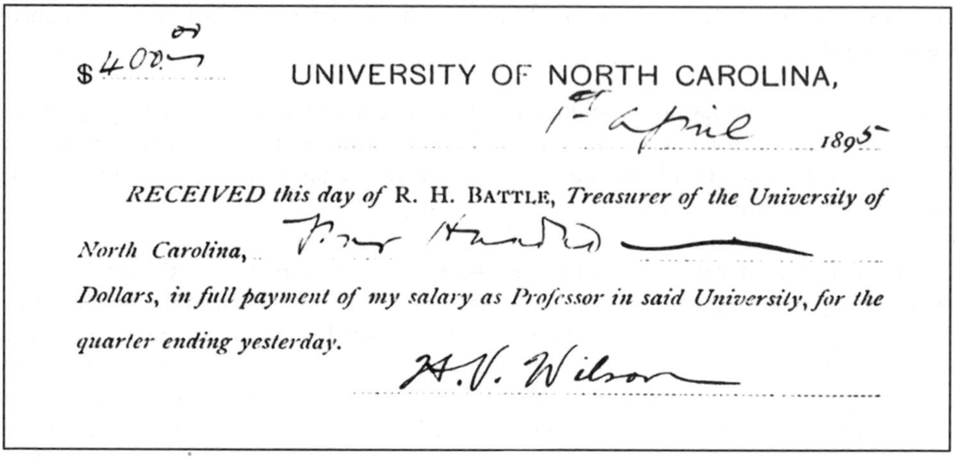

$400.00 UNIVERSITY OF NORTH CAROLINA,

1st April 1895

RECEIVED this day of R. H. Battle, *Treasurer of the University of North Carolina,* Four Hundred ———

Dollars, in full payment of my salary as Professor in said University, for the quarter ending yesterday.

H. V. Wilson

Fig. 9.3. Paycheck receipt of Henry Van Peters Wilson, April 1, 1895.

(Folder 106, Battle Family Papers #3223-A, Southern Historical Collection, Wilson Special Collections Library, University of North Carolina, Chapel Hill).

Fig. 9.4. House of Henry Van Peters Wilson, 722 East Franklin Street, Chapel Hill, North Carolina.

(Courtesy of Doug Eyre).

A Raleigh newspaper correspondent reported positively that Wilson was "a gentleman of fin[e] natural ability, of extraordinary culture, of large experience in biological work, and of national reputation as a scientist" (Anonymous, 1891a). In the same article, he described Wilson's strong endorsement by scientists at Johns Hopkins. Wilson was deemed "a genial, refined and cultivated gentleman, an enthusiastic worker and investigator, and a sympathetic, clear and forcible teacher." Of special note was the fact that he had a North Carolina connection, having conducted research in Beaufort.

A number of biographical papers provide further insight into Wilson's life (Beers, 1942; Coker, 1940a; Costello, 1961; House, 1964; Lewis, 1942; and MacNider, [1936]). Although archival papers concerning his professorship at UNC do not appear to have survived, there are several noteworthy documents housed at UNC (UNC-SHC) and Johns Hopkins (JHU-A).

The University in the 1890s

When Wilson began his professorship at UNC, he was one of two new faculty members to join the ranks of twelve others. (The other new member was Karl Pomeroy Harrington, Professor of Latin Language and Literature.) Wilson became chair of the Department of Biology, one of three new departments established at UNC, the others being History and Pedagogics. The population of the Chapel Hill village had reached slightly over 1,000 inhabitants. The summer of 1891 was a bad one because typhoid fever prevailed in the village, causing sickness and deaths (Spencer, 1853–1906). The university consisted of nine primary buildings (Fig. 9.5) and had 248 students. UNC medical student William Troy Turlington (1891) described the campus as a "beautiful place . . . covered with grass," which was kept mown. Willis Hinton Austin (1893), an undergraduate, mentioned the grove of nearly 25 or 30 acres of large trees. During rainy weather, however, the grounds were an "awful muddy place" (Turlington, 1892). The fall 1891 semester began as one of "the quietest . . . ever seen at the opening of the University" (Turlington, 1891). Even hazing was not obvious. The students released some of the exuberance of youth by building a bonfire at 11 P.M. in front of South Building.

Fig. 9.5. Photo of UNC campus, showing its nine main buildings, ca. 1893.

(From the UNC catalog of 1893–1894, plate inserted between pages 88 and 89).

According to the UNC catalog of 1891–1892, typical annual academic expenses were \$77.50 (\$60 tuition, \$10 registration, \$5 medical fee, and \$2.50 gymnasium). Students could choose from a limited number of unfurnished university rooms at no cost. Rooms were let out according to class level, with seniors receiving first choice, juniors next, and so forth. The majority of freshmen were placed in new campus buildings because the upperclassmen preferred rooms in the older buildings such as Old East (Graham, 1894). Space in South Building was available for a nominal cost (Patterson, 1891a) (Fig. 9.6).

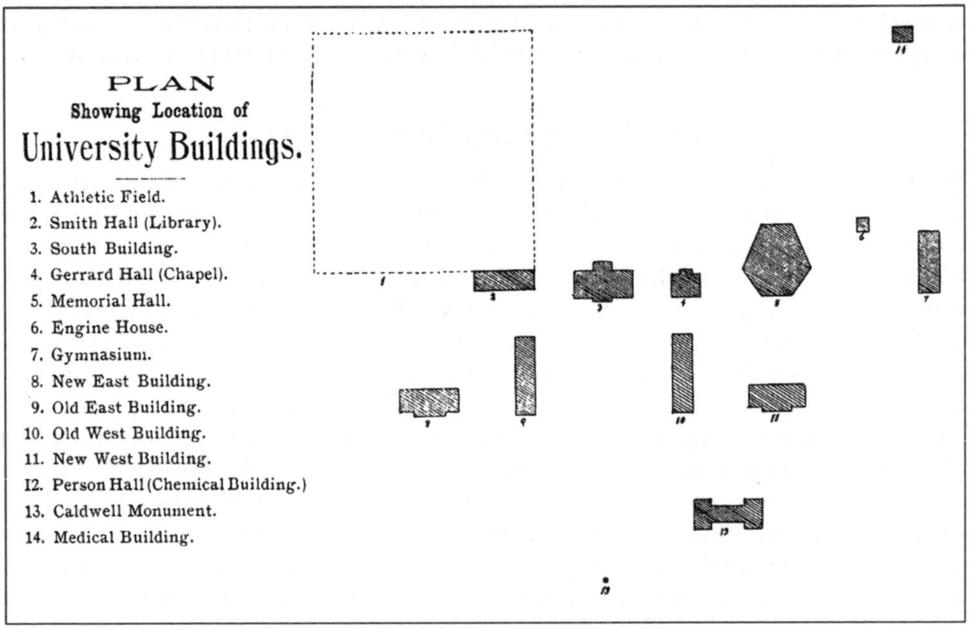

Fig. 9.6. Plan of UNC campus, 1892.

(From UNC catalog of 1892–1893, sheet inserted at end of text).

Additional expenses were board (at hotels or private houses) ranging from $8 to 13 each month, washing at $1 each month, a cord of wood for heating at $2, and books and supplies (Patterson, [1891]b). Around 1894, new books for a freshman cost nearly $24.00 (Graham, 1894). A student could keep the cost of books lower by buying used (secondhand) volumes, as a student did not "keep his books any longer than he [was] using" them (Graham, 1894). The UNC catalog of 1891–1892 also listed a small laboratory fee for students enrolling in scientific courses with labs. Total yearly costs would average between $200 and $300. Students such as UNC sophomore Eugene Berrien Graham (1893) thought the expenses greater than expected, but the university was nonetheless considered "the best place for a low student that can be found." Free tuition was available to the sons of ministers, candidates for the ministry, public school teachers planning to make teaching a profession, and men with physical disabilities. Graduates of colleges were also exempt from paying tuition. Scholarships and aid funds were available to qualifying students (Winston, 1892a).

On the same day that Wilson's appointment officially began, the UNC trustees requested a special meeting of the faculty to consider changing the system of grades. Professors William Cain and Karl P. Harrington constituted a Committee on Grading (UNC Faculty 7: 246). In their resulting report, they recommended a system of six grades (1 through 6) to indicate the student's standing in each course taken. The system would be based as follows (UNC Faculty 7: 250):

95% to 100%, inclusive required in grade 1
90% to 95%, including 90%, required in grade 2
80% to 90%, including 80%, required in grade 3
70% to 80%, including 70%, required in grade 4
50% to 70%, including 50%, required in grade 5
0% to 50%, including 0%, required in grade 6

Their report was adopted. Semi-annual grade reports were issued for all students and sent to their parents or guardians (Fig. 9.7). Students needed to maintain a grade of at least 4 to pass any class. Those receiving a grade of 6 in any class had to repeat the class (UNC Archives, no. 1).

In an editorial of the *North Carolina University Magazine*, it is mentioned that the undergraduates particularly criticized the new grading system (Anonymous, 1893a). Apparently, Harvard College had instituted the system several years earlier, considering that "the new method relaxe[d] the rigid lines which bound in the instructors of the past." The editorial concluded by saying: "It is well-nigh impossible to grade men justly on a scale of per cents."

Using this new grading system, the faculty documented its new guidelines for conferring scholastic honors on students receiving the bachelors degree. Those attaining a grade of 1 in five out of six of their classes graduated *summa cum laude*; those with a grade of 1 in half their classes or a 1 or 2 in five out of six of their classes graduated *magna cum laude*; and those with grades of 1 and 2 in half their courses, graduated *cum laude*. Undergraduates displaying meritorious scholastic performance were also bestowed special honors. Those receiving four out of grade 1 during an academic year received "Great Honor" and those attaining grade 1 in half or grade 1 or 2 in four out of five attained "Honor" (UNC Faculty 7: 277).

In 1897, the Committee on Courses recommended that the designation of honors (*cum laude*, etc.) for freshmen, sophomores, and juniors be abolished. The faculty approved the report (UNC Faculty 7: 400). In the session of 1901–1902, the faculty abolished the system of honors for graduating seniors (UNC Faculty 7: 483, 484, 498). Instead, these honors were replaced with special certificates.

The UNC Faculty (7: 299) minutes of November 25, 1893 document that candidates for the bachelor's degree were required to write either a graduating oration or a graduating (senior) thesis, the latter comprising not less than 2,500 words and not more than 5,000. Seniors selected a subject for their theses in consultation with the approval of a professor (UNC Faculty 7: 320–321). The selection had to be reported to the registrar by the end of January. A committee was charged to examine the theses and route them to appropriate faculty for their critical review (UNC Faculty 7: 305). From a manuscript version, students read their theses before the faculty. They then submitted their final, typewritten copies to the registrar by the middle of May (UNC Faculty 7: 403). The registrar had the theses bound and deposited in the university library. They are currently housed in the North Carolina Collection at UNC. The requirement of preparing senior theses concluded in 1912. At UNC, nearly twenty-nine extant theses written under the supervision of Wilson are housed in the collection (Table 9.1). Of these theses, all focus on zoology except for one that includes plants.

UNIVERSITY OF NORTH CAROLINA

TERM REPORT

Chapel Hill, N. C., Dec. 30, 1891.

During the term which closed Dec 22, *Mr.* T. J. Wilson Jr. *has made the following record in scholarship:*

[*Grade 1 denotes very high scholarship, grade 2 high scholarship, grade 3 very creditable scholarship, grade 4 satisfactory scholarship, grade 5 scholarship below the point required for graduation, grade 6 scholarship that is very unsatisfactory, suggesting the propriety of pursuing the same study another year.*]

STUDIES.	GRADES.	STUDIES.	GRADES.
Political Science		*Greek*	1.
History	1.	*Mathematics*	2.
Experimental Chemistry	1.	*Engineering*	
Industrial Chemistry		*Surveying*	
Agricultural Chemistry		*Drawing*	
Chemical Laboratory		*Philosophy*	
Physics		*Biology*	
Astronomy		*Physiology*	
Electrical Engineering		*Zoology and Botany*	
English	1.	*Histology*	
Essays and Orations		*Latin*	absent Ex. stood afterwards 2.
Anglo-Saxon		*Geology*	
French		*Mineralogy*	
German		*Physiography*	

Remarks

Mr. Wilson maintains his former high record for scholarship, conduct and character. Inter primos princeps.

Geo. T. Winston
President.

Fig. 9.7. Student term report, illustrating new grading system used at University of North Carolina, for term ending December 22, 1891.

(Folder 2, Thomas James Wilson Papers #2257, Southern Historical Collection, Wilson Special Collections Library, University of North Carolina, Chapel Hill).

TABLE 9.1. A LIST OF NAMES OF AUTHORS AND TITLES OF SENIOR BACHELOR'S THESES WRITTEN (1894–1912) UNDER THE DIRECTION OF HENRY VAN PETERS WILSON.

Student	Title	Year	Pages
James Robert Harris	*Nitrification*	1894	10
Charles Roberson	*Some Proofs of the Theory of Evolution—Taken from a Study of Embryology and Palaeontology*	1894	11
Daniel Rice Bryson	*Bacteriology. Growth and History of the Science Discoveries and Present Condition*	1896	13
Julius Alexander Caldwell, Jr.	*Mimicry in Animals and Plants—A Study in Involuntary Resemblances*	1899	[7]
Everett Augustine Lockett	*Functions of the Nervous System*	1899	14
Calvin Duvall Cowles	*The Geographical Distribution of Animals*	1901	5
Clarence Albert Shore	*Natural Selection As a Factor in Evolution*	1901	7
John William Turrentine	*On the Wide-spread Occurrence of Degeneration in the Animal Kingdom*	1901	9
Ivey Foreman Lewis	*The Migration of Birds*	1902	7
Greenville Ramsey Berkeley	*Degenerate Forms of Organisms* (Directed by J.E. Duerden)	1903	9
Ivey Foreman Lewis	*On Maturation in Ascaris megalocephala* (Directed by J.E. Duerden)	1903	9
Henry Grey Turner	*The Theory of Evolution*	1903	5
William Gray Amick	*Parthenogenesis*	1905	6
Walter Francis Cole	*The Geographical Distribution of Animals Explained by Evolution*	1905	6
Austin Flint Nichols	*Degeneration in the Crustacea*	1905	6
Theophilus Parker Cheshire	*The Direct Action of Habit and Environment on the Adult and Developing Organism*	1906	13
William Lawrence Grimes	*The Relation between Mind and Body*	1906	9
William Herbert Kibler	*Recent Researches into Possibility of Changing Protoplasmic Characteristics through the Application of Stimuli*	1906	12
Samuel Timothy Nicholson, Jr.	*Regeneration*	1906	11
John Carroll Wiggins	*Structure of Man—Viewed in the Light of Evolution*	1907	11
Robert Rufus Bridgers	*The Effects of the Environment on the Individual*	1908	14
Wortham Wyatt	*Insects as Disease Carriers*	1908	10
Cleaveland Fane Kirkpatick	*The Differences in the Structure of Animals as the Result of Their Environment*	1909	9

Table 9.1. (Continued)

Student	Title	Year	Pages
Robert Drane	*Protozoa and Disease*	1910	8
Samuel Coopersmith [previously Kuperschmidt]	*Oysters and the Oyster Problem*	1910	7
Orren Williams Hyman	*The Evolution of the Nervous System*	1910	6
John Manning Venable	*Some Points in Nervous Structure and Physiology*	1910	16
Lee Franklin Turlington	*Embryological Theories*	1910	8
Henry Wise Lyon	*Instinct and Intelligence in Birds*	1911	25
William White Falkener	*The Morphology of the Vertebrate Skeleton*	1912	24
David Reid Murchison	*The Doctrine of Heredity in Biology*	1912	8

Upon their arrival at UNC, new students were asked to present letters of recommendation or introduction at the president's office, situated on the first floor of South Building, where the bursar and registrar also had offices (Winston, 1892b). Here each student would receive a permit to be examined. Those who were admitted "on certificate" were required to present their certificates and were then exempted from taking entrance examinations. Entrance exams were given in mathematics (an hour long), Latin (two), Greek (one), English (one) and history (two) (Winston, 1892b). Students had the option of choosing a plan of coursework that required only an English exam. Those pursuing the Scientific Course were required to take examinations in English and mathematics (Winston, 1892a).

After passing their exams, the young men would remit their university fees to the bursar, who assigned campus rooms and issued a permit to register. The registrar then issued a card of registration, which served as a pass to use the gymnasium and library and to receive free medical services; it also served as proof of enrollment at the first classes of each term (Winston, 1892b). Dr. Richard H. Whitehead, Professor of Anatomy, Physiology and Materia Medica, examined each student, prescribing physical exercises to enhance good health and a sound body (Anonymous, 1891b). He later delivered a lecture before the students, giving practical advice on maintaining good health (Anonymous, 1891f). Compulsory prayers were held in the newly renovated chapel. By at least 1901, seniors and students in professional schools were exempted from compulsory attendance in the chapel (UNC Archives, no. 1). On the first day of classes in 1891, President Winston addressed the student body in Memorial Hall, the largest group for an opening day since 1860 (Anonymous, 1891d).

Development of Biology Department and Teaching

Under Wilson's guidance and dedication, research and instruction in biology became firmly established at UNC despite limited apparatus and a paucity of funds. He developed in his first two years a program consisting of six courses: Physiology, General Biology, Practical Biology, Vertebrate Embryology, Histology, and Animal Morphology. The objectives of the classes were to provide the knowledge of plants and animals fundamental for a liberal education, to prepare young men for careers as naturalists or teachers in biology, and to provide a background for men choosing medical careers. Concerning the last objective, Wilson mentioned that the new Department of Biology had an intimate connection with medicine because it offered courses to men preparing to study medicine. By helping to train future doctors, he argued, the department would benefit the state (BOT 9: 129).

Among Wilson's teaching responsibilities were two courses that included botanical instruction: General Biology and its associated class, Practical Biology (a laboratory experience). These courses are fully described in the UNC catalogs. In the former course, taught in both semesters (except only in the spring of the 1891–1892 school year), students were instructed on the fundamentals of zoology, botany, and comparative physiology, modeled after the "General Biology" course established at Hopkins. Wilson aimed to acquaint students with the general laws governing life "without a knowledge of which no intelligent understanding of the processes of nature (healthy or diseased) can be had" (BOT 9: 128). Students also learned how to dissect and how to use a microscope, and they gained a good foundation for future work in biology. Instruction consisted of lectures and laboratory work. In the latter part of the spring semester, studies of plant analysis (identification) were commenced. Textbooks for botany were Gray's *Lessons* (1881) and *Manual* (1889), as well as Prantl and Vines's *Botany* (1891). The last-cited title is an English translation of the textbook by noted German botanist Karl Prantl. It provided an abridged alternative to the famous *Lehrbuch der Botanik* by Julius von Sachs. Students were introduced to the morphology, anatomy, physiology, and classification of plants. (The UNC Library's copy of the Prantl book belonged to Wilson.) Wilson's other teaching aids were charts, models, and drawings that he made with color chalk on the blackboard.

Practical Biology was a laboratory experience that complemented, in part, General Biology. Emphasizing dissection and microscopic observation, it convened both semesters, with two-hour labs meeting twice a week. Lab notes and drawings were required. Wilson believed this experience prepared his students for research in any branch of biology. Students followed laboratory exercises that Wilson had written.

In an effort to meet the needs of students, Wilson wrote several guides for laboratory instruction, tailored to his teaching methods early in his career. He pointed out that other institutions were also developing such learning aids to increase the efficiency of their classes (BOT 9: 508). Wilson ([1895]a, [1895]b, [1895]c) developed guides in practical biology, histology, and elementary physiology. The first cited guide (*Practical Biology*) includes exercises in botany that provide an insight into Wilson's instruction in that section of coursework. Students were required to make notes and sketches of their observations.

In Practical Biology, the section of study on plants commenced with the lower forms of plants (as classified in Wilson's time). He singled out the molds *Mucor* and *Penicillium* for microscopic observation of hyphal cell walls and spores. The alga *Spirogyra* was studied next, followed by detailed work on *Nitella*. For instance, in observing the stems, students were to "cut (in potato or pith) and mount on slide, a transverse section through a node," then "observe the circle of smaller cells surrounding the central one," and "sketch" it (Wilson, [1895]a). *Pteris aquilina* L. was the representative fern; students particularly examined microscopically the rhizome, frond, and sporangia.

The botanical lessons concluded with the flowering plants. These investigations covered seeds (corn and bean), their structure, and germination. After studying cells and tissue of plant stems, leaves, and roots, the class examined flowers and their components. Wilson's approach included a tutorial style. In examining the fibro-vascular bundle of the bean, for example, students received the following instructions: "observe and sketch the following elements: 1. Pitted vessels—these are long tubes; do the pits ever become real clefts, extending entirely through cell walls? These vessels taper at the ends and are formed by the fusion of a row of cells. Can the septa between component cells be distinguished?"

Practical Biology sessions were held in the Biological Laboratory that had been developed in the old Philanthropic Society library on the top (fourth) floor of New East in 1891 (Anonymous, 1891e), having been moved from its prior quarters on the first floor of the same building (BOT 9: 125–126). At Hopkins, Wilson had studied and conducted research at a state-of-the-art biological facility, which was commodious, well lighted, and featured a library, museum, lecture rooms, photography lab, laboratories for teaching and research, among others. With the growing enrollment in biology courses, Wilson directed the enlargement of the UNC laboratory in the fall of 1894, securing space formerly occupied by the students on the fourth floor of New East (Anonymous, 1894a). According to the UNC catalog of 1894–1895 (p. 97), two adjacent rooms on the west end of the floor were merged (along with part of the hall) to form a biological lecture room seating 100 students (Winston, 1894). Other accommodations were developed from two rooms on the other end of the floor. According to the UNC catalog of 1894–1895, these included a smaller lab for advanced students in histology, embryology, and comparative anatomy, a private working room for Professor Wilson (Fig. 9.8), and a couple of storerooms. A year later, the enlargement of the old windows and the installation of several new ones greatly improved the area (BOT 9: 508). The Biological Laboratory had become a useful facility and was now well lighted. It had running water but lacked gas.

Concerning the supply of water to the laboratory, it is noteworthy that prior to 1893 the chemistry laboratory was unique in having running water for its operation. In the summer of 1893, however, water from a well was first pumped by steam into 15,000-gallon iron tanks situated in the attic of South Building. From there it was distributed to the labs of New East and to bathrooms in the basement of the University Library. As demand for running water increased, another well was dug. By 1899, the General Assembly had allotted funds to establish a waterworks facility to better serve the campus (Anonymous, 1899b).

In 1942, UNC zoologist Robert E. Coker (1942, pp. 5–6) provided a personal glimpse of the biology laboratory (Frontispiece of chapter 9, lower image). He noted a large room with a central space with tables and sinks, three alcoves on each side containing three or four tables in each, and the former library bookshelves now holding both dry and pickled specimens. A hallway entrance contained exhibition cases with specimens. A tank room for aquaria in which aquatic animals (and plants) were raised for student work was added about 1893.

Although the biology lab was developing nicely, Wilson recommended funds for the museum, which needed specimens, specimen jars, boxes, and cases. More pressing needs included additional microtomes, dissecting instruments, glassware, and charts and models (such as models from the Zeigler Studio). Wilson's request for added apparatus and scopes must have

Fig. 9.8. Portrait of Henry Van Peters Wilson, 1895.

(*News and Observer* [Raleigh, NC] 38(134): 21, Tuesday, July 9, 1895).

been successful, as we learn from the UNC catalog of 1893–1894 (p. 85) that the lab then had modern equipment, including a "large number [*i.e.*, nineteen] of Zeiss microscopes, dissecting microscopes, microtomes, paraffine bath, and incubator." Wilson reiterated a need for gas in the laboratory.

The academic year of 1893/1894 heralded the first biology courses listed under the heading "Graduate Department." They were also available to advanced undergraduates. These courses were confined to zoology, especially in areas of Wilson's specialty (the developmental biology of certain marine animals). In the fall semester, Wilson was initially assigned student assistants, the first being undergraduate Robert Palemon Jenkins. During the next nine years, twelve other students, both undergraduates and graduates, provided much needed help in the classroom and laboratory (Table 9.2). The university catalog for that academic year listed three courses for graduate students: Comparative Anatomy and Embryology of Vertebrates, Research in Zoology, and summer work in Marine Zoology.

Table 9.2. A List of Names of Student Assistants in the UNC Department of Biology, under the supervision of Henry Van Peters Wilson (1893–1903).*

	Undergraduate	Graduate
Robert Palemon Jenkins (Ph.B. 1896)	1893/1894	
Charles Roberson (B.S. 1894, C.Med. 1895; M.D. Long Island 1897)	1894–1895	
George Hughes Kirby (B.S. 1896, C.Med. 1897; M.D. Long Island 1899; Munich 1906)	1894–1896	1896–1897
Robert Ervin Coker (B.S. 1896, M.S. 1897; Ph.D. Johns Hopkins 1906)	1895–1896	1896–1897
Edward Jenner Wood (B.S. 1898, C.Med. 1899; Ph.D. Johns Hopkins 1906)	1897–1898	1898–1899
George Edgar Newby (C.Med. 1898; M.D. Jefferson 1900)	1897–1898	
Thomas Gilbert Pearson (B.S. Guilford 1897, UNC 1899; Harvard 1900)	1897–1898	
Albert Franklin Williams (A.B. 1897, C.Med. 1899; M.D. Maryland 1901)		1898–1899
William deBerniere MacNider (C.Med. 1901, M.D. 1903)	1899–1900	
Clarence Albert Shore (B.S. 1901, M.S. 1902; M.D. Johns Hopkins 1907)	1900–1901	
Dormond Steele Thompson (Ph.B. 1901)		1901–1902
Ivey Foreman Lewis (A.B. 1902, M.S. 1903; Ph.D. Johns Hopkins 1908)	1902–1903	

*Slightly modified from "Assistants in Biology," compiled by Robert E. Coker. Degrees without institutional designations are from UNC. (Box 31, folder 453, Robert E. Coker Papers #3380, Southern Historical Collection, Wilson Special Collections Library, University of North Carolina, Chapel Hill.) Ph.B. = Bachelor of Philosophy; C.Med. = Certificate in Medicine.

Among other resources of the Biological Laboratory was a basic (and growing) collection of important biological literature (monographs, reference works, and journals). Wilson noted that "these books [were] in constant use in the laboratory and [were] of much service to a number of students" (BOT 9: 410). After the study of specimens, Wilson considered the consultation of the original literature to be of utmost importance for students of biology. The development of the biology library is discussed in chapter 1.

In the UNC catalog of 1893–1894 (p. 85), we learn that "an extensive collection of the flowering plants of the state" was also maintained. In writing about "Biology at the University of North Carolina," Wilson (1899) mentioned that "an 'experimental garden' [gave] convenient opportunities for the study of cultivated plant forms and life-histories of many common insects." Additional details about this garden have not been found in the archival records. Wilson and his students added natural history specimens to the "museum" exhibits displayed in the lobby of the Biology Laboratory and in the lab itself. For example, in the summer of 1899, he, Professor Mensh of Ursinus College, and students Thomas Gilbert Pearson (later a noted ornithologist and conservationist) and Edward Jenner Wood (son of physician and botanist Thomas Fanning Wood and himself later an acclaimed physician) conducted marine zoological research at Beaufort. Their collection of nearly 150 specimens was added to the University Museum. Mr. Pearson prepared a similar lot of specimens for the museum in Raleigh (Anonymous, 1898).

Instruction in General Biology was first offered as a course incorporating Practical Biology during Wilson's fifth year at UNC. In the spring of 1896, Systematic Botany of the Flowering Plants debuted as a course. Its description is summarized in the UNC catalog of 1895–1896 (p. 36) as follows:

> Systematic Botany of the Flowering Plants. Reading, laboratory, and field work. Two hours a week (spring term).
>
> Representatives of the several orders of Phanerogams are studied, and each student is required to analyze, independently, a number of plants.

In the subsequent six years, the course was titled Botany. By spring of 1900, botany instruction focused on the structure and habits of selected cryptogams and flowering plants and was accompanied by laboratory work. It met three hours per week in the spring. Between 1897 and 1902 (except for 1900), Wilson taught the botany course in the spring semester. In his annual report for 1895–1896, Wilson did not note teaching botany. Cumulatively, around fourteen students took the course.

In 1897, the increased enrollment of students in biological courses (including those taking biology as an elective) placed a constraint on the limited space and personnel. Classes in General Biology were overcrowded, necessitating the division of the class, with thirty-six students working Thursday and Friday afternoons and the remaining students on Saturday morning, 9:45 to 1:30 P.M. (BOT 9: 689). The previous year, Wilson had suggested a plan that would provide additional lab stations for students. He wanted two old windows in the main lab replaced with a larger one to create more lighted areas for laboratory exercises. He also recommended that the roof of New East be raised to increase the pitch of rooms and make them less stuffy when the weather was hot (BOT 9: 591).

At the beginning of 1898, Wilson once again pointed out the overcrowding in his areas of instruction. The class in Elementary Physiology was so large that it was divided into eight sections (five sections meeting Monday and Friday under the direction of Albert Franklin Williams; three sections in another room on Monday and Wednesday under the supervision of George Edgar Newby.) Wilson even taught two classes at the same time. The freshman class was placed under the direction of "demonstrators" who were "too young and untrained to assume the actual charge of any of the other classes" (BOT 9: 691). To meet the increased enrollment in biology (especially in physiology and histology), Wilson requested ten new microscopes and "a considerable number" of dissecting instruments. The grim assessments of his teaching load reflected the strain that Wilson experienced.

Wilson's contribution to summer coursework was the creation in 1894 of the Summer School of Biology, which took place in a marine biological laboratory at Beaufort, North Carolina. The town, one of the oldest in the state, was a popular summer resort noted for its fisheries (Daniels, 1897). Located on Bogue Sound, it was a pleasant locale for spending summers and escaping the hot, humid weather of Chapel Hill. Wilson set up a laboratory in a large two-story house, bringing the necessary apparatus (microscopes, dissecting instruments, and books) from his university research facility (Anonymous, 1894b).

The first session of the Summer School of Biology opened on June 13, 1894, and continued for a month. Two courses were given: one for beginners, and one for advanced students. The former class focused on collecting and dissecting marine animals, the latter on the development of certain marine animals (Anonymous, 1894c). The primary objective of the session was to introduce students to the habits and structure of the Beaufort marine fauna, which was representative of the animals of the southeastern coast. There was a ten-dollar fee for the summer course. Wilson's academic mentor William K. Brooks was present at the first session.

The summer classes at Beaufort were dropped in 1895 from the UNC catalog, "doubtless because [Wilson] felt it distinctly preferable to devote the summers to research" (Coker, 1940a). Nonetheless, a limited number of students and faculty from UNC, and other academic institutions, continued to study there. Advanced investigations in subsequent years included embryology, the breeding of sponges, the sense organs of jellyfish, the action of various chemicals on undifferentiated cells (Anonymous, 1899a), and the life cycle of a species of barnacle. For instance, in 1899 twelve biologists studied at the lab, including seven from the Johns Hopkins University, one from Guilford College in North Carolina, three from UNC-Chapel Hill, one from Trinity College in North Carolina, one from the University of Alabama, and one from Ohio College (Anonymous, 1899a; Wolfe, 2000). In 1900, students and professionals from Columbia University, Hopkins, UNC-CH, and Trinity assembled at the lab. UNC President Francis Preston Venable (1901a) wrote to Wilson to acknowledge the success of the lab, adding that "it seem[ed] to be the old story about the South, it is hard to arouse interest."

On the occasion of the centennial of what was to become the Federal Biological Laboratory at Beaufort, Douglas A. Wolfe (2000) wrote a history of the facility. He noted that in 1899, Joseph A. Holmes recommended setting up a permanent lab facility there. Concurrently, his efforts were reinforced by those of Wilson. By May 1900, the U.S. Congress had authorized the founding of a U.S. Fish Commission marine biology lab at Beaufort. From 1899, when the United States Fish Commission first sponsored the marine laboratory at Beaufort, Wilson became its director, which post he held through the summer of 1901 (Wolfe, 2000). Subsequently, he spent most of his summers there conducting his marine research in zoology.

Having held the professorship of biology at UNC for a decade, Wilson alerted president Venable and the trustees that his workload was excessively heavy (BOT 10: 236–239). He emphasized that if the present work in biology was to be sustained, an adequate teaching staff was essential. He pointed out that other collegiate institutions with similar programs in biology were administered in two ways: either separate departments of Zoology, Comparative Anatomy, and Botany were established, or the Biology Department remained intact, administered by a department head who was supported by assistant professors and instructors. Wilson recommended the latter scenario.

Wilson further explained the trying situation under which he worked and even discussed the possibility of his resignation. In his direct manner of writing, he noted:

> My days and my evenings have been given up in attending to endless details necessitated by the modern laboratory methods of studying the various branches of Biology. My own personal investigations have, during the academic year, been practically to a standstill.—such publications as I have been able to send out represent work carried on during vacations. During these years, I have constantly been supported and encouraged by the conviction that this condition of affairs was but a temporary one and that better things would come. I have thus more than once refused to stand for other and easier places; and a year ago declined a chair to which the trustees of a southern college had elected me a chair. I may add to which were attached a salary of $3,000 and a new laboratory erected at the expense of $6,000. My interest in, and my attachment to, this University are very deep and I look back over my ten years of teaching and organizing there with real pleasure. I confess, however, that after so many years of strain, I have reached a point when I feel unable and unwilling to go on under the present conditions. If I am to remain in Chapel Hill, one of two things must be done; either the work of the department must be cut down, that is a number of electives (now sought for by some of our best and most earnest students who are either preparing themselves for the study of medicine or preparing themselves to teach Biology) must no longer be offered; or I must have more satisfactory aid than can be supplied by our system of student-assistants. The former alternative is one which I consider with the greater reluctance to go backwards and sacrifice courses which have been carefully planned to meet real needs, would be a hardship and disappointment to me, and would seriously curtail the opportunities of those classes of students above referred to.
>
> In considering the needs of the Department of Biology, the trustees should bear in mind the composite character of the department, the number of very dissimilar subjects (and classes).

While reminding the trustees about the diversity of subjects that were taught in the department, Wilson particularly emphasized the importance of the department in training medical students. Many medical schools, he continued, required superb training in various topics in biology. At UNC, the medical department prescribed several courses taught in its biology department, including Histology and Physiology. In fact, he proposed additional electives to support the medical department, such as a semester course on Microscopic Anatomy of Organs and one on experimental Comparative Physiology.

As a corollary to his request to build up the biological program, Wilson pointed out to the trustees the perennial need of an adequate budget for purchasing apparatus. He asked for $250.00. He reiterated the need to raise the pitch of the roof at each end of the laboratory to provide a less stuffy environment in the summer. He also reminded the trustees that the building should be supplied with gas.

President Venable (1901b) solidly endorsed Wilson's request to add an additional professor to strengthen the biology department. The trustees likewise recognized this need. At their January 16, 1902 meeting, they resolved that a professor of botany be elected after the next commencement at a salary of $1,000 (BOT 10: 302). In early June, the trustees selected botanist William Chambers Coker as an additional faculty member in the Department of Biology (BOT 10: 361). With Coker's appointment, botanical studies were nurtured; a Department of Botany was established in 1908, which earned national and international prominence.

When Wilson reached seventy-five years of age, he wrote to UNC president Frank Porter Graham, notifying him that he wanted to reduce his teaching load. He expressed a desire to continue instruction of three predominantly graduate-level courses: Morphology of Invertebrates, Advanced Embryology and Regeneration, and Research. He asked permission to discontinue instructing the class Zoology 42a (Development of Animals), an undergraduate class that required much of his time (Wilson, 1938b).

During his forty-seven-year tenure at UNC, Wilson saw many changes at the university and its environs. The small village more than tripled in number of inhabitants to 3,654 people, university enrollment had swelled thirteenfold to 3,207 students by the spring of 1939, the roster of faculty increased seventeenfold to nearly 235 professors (excluding instructors), and the campus physical plant increased nearly 750% to about eighty-two buildings, with running water and electricity now commonplace. From 1908 until 1922, Wilson was the Department of Zoology. Between then and his death in 1939, the department grew sixfold with the addition of Robert Ervin Coker in 1922, Charles Dale Beers in 1927, Donald Paul Costello in 1935, William Louis Engels in 1937, Irwin Clark Kitchin in 1939, and Robert Devore Boche in 1939. In 1897, women were admitted to UNC as postgraduates. (This topic is discussed in chapter 11.) The simpler ways of life at the university were giving way to an increasing number of regulations. By 1900, "all hunting and shooting on the Campus and land of the University east of the Campus [were] strictly forbidden" (UNC Faculty 7: 470). Robert House (1964) noted changes in the campus landscape, including the felling of trees and the loss of the natural springs due to either the incorporation of land into the town drainage system or the removal of trees.

Research

In research, Wilson's "work was marked by accuracy, keenness of observation, and quick appreciation of the significance of the observation, and readiness to put inferences to test in effective ways" (Anonymous, 1939b). During a fifty-three-year period, he published ninety-one papers, and among these, fourteen were of major significance, according to his biographer and UNC colleague Donald Costello (1961). These publications reflect the progression of Wilson's research interests. As already noted, his early scientific investigations focused on parasitic medusae and the development of a species of coral and sea bass. Subsequently, he devoted considerable attention in studies on the structure, embryology, and taxonomy of sponges, earning him recognition as an authority. About August 1891, Wilson received an offer from Alexander Agassiz to study the sponges collected on his Albatross trip. Between 1894 and 1904, Wilson worked up the taxonomy of these specimens. From studies in the early 1900s, Wilson attained worldwide reputation for his historically significant studies in embryology and developmental biology. First using sponges and later a species of soft-coral, he discovered that these organisms would reform their normal anatomies after total separation and random re-mixture of their cells. This discovery opened a new horizon in the thinking of biologists about embryonic development as being caused by properties and behaviors of individual cells. His last research elucidated the role of vacuoles in forming cell membranes (Wilson, 1938a).

During Wilson's years as professor, the preponderance of his research was conducted in Chapel Hill and at the Beaufort lab. During his tenure at Chapel Hill, he was granted two leaves of absence to study in Berlin (1902–1903) and Naples (1929–1930). His first leave, to Berlin, offered an opportunity to complete a report on the Agassiz collection of sponges, which work had been held in abeyance due to his heavy teaching load. UNC president Francis P. Venable (1901b) heartily approved of his requests for leaves as did the trustees who granted Wilson a year of research leave for the academic year 1902/1903 (BOT 10: 301–302). The Carnegie Institution not only financially sponsored his research, but also funded a substitute to teach his courses at UNC. James Edwin Duerden, whose name was proposed by Wilson (Venable, 1902), was hired as his substitute (BOT 10: 361). Like Wilson, Duerden was a doctoral graduate from the zoology labs of Hopkins, a Bruce Fellow, and an investigator of the development of corals (among other marine animals). He also was awarded a Carnegie Institution grant a year after Wilson was to study recent and fossil corals (Duerden, 1904).

During his leave of absence, Wilson conducted research on the morphology and classification of deep-sea sponges in Europe for fourteen months, from July 1902 to August 1903 (Wilson, 1905). He spent most of the year at the Zoologisches Institut in Berlin under the supervision of Franz E. Schulze (Wilson, 1932a). He had brought the Agassiz collection of sponges to the Institut, where he received assistance with the more difficult identifications from "Geheimrath" Schulze. The research setting provided an atmosphere free of teaching and administrative responsibilities. It also provided the services of a competent artist and photographer and access to an outstanding scientific library. Wilson also studied collections of sponges in the neighboring Museum für Naturkunde, which was headed by Professor Wilhelm Weltner. In order to examine type specimens, Wilson spent the summer of 1903 in other European institutions, including the British Museum of Natural History in London, the Rijks Museum in Leiden, and the Muséum d'Histoire Naturelle in Paris. A resulting monograph was published (Wilson, 1904), as well as an abstract (Wilson, 1905).

Nearly thirty years later the UNC trustees granted Wilson a second leave of absence to study the embryology of sponges at the Stazione Zoologica in Naples. The Kenan Foundation provided financial support for the leave during the academic year 1929/1930 (BOT 13: 453). Wilson's substitute, J. Manson Valentine, who stayed on an additional year after Wilson's return (Anonymous, 1930), later founded and managed a newly developed Museum of Natural History at UNC.

Wilson sailed for Naples from New York City on September 2, 1929. At the Naples Station, he had a working place (table) that was maintained by the Rockefeller Institute for Medical Research (Wilson, 1929). He investigated the metamorphosis of the larvae in monaxonid sponges and studied living larvae, which investigations were later continued at the Beaufort Laboratory (Anonymous, 1931). At the October 14, 1930, meeting of the Elisha Mitchell Scientific Society, he presented a paper on this work as well as an abstract (Wilson, 1932b).

In a letter to the newspaper editor of the *Chapel Hill Weekly*, Wilson (1930) described the setting of the Naples lab and some of the nearby spots of interest. He daily walked to the zoological station, "a large handsome building with a public aquarium on the ground floor" that attracted researchers from several countries. He was particularly impressed by the station's collection of journals. Wilson concluded his stay with an excursion to Camaldoni, an old monastery.

Another noteworthy research trip took Wilson to the Carnegie Institution Marine Biological Laboratory at the Dry Tortugas, where he conducted investigations on Loggerhead Key, about sixty to seventy miles west of Key West, Florida. He spent most of July 1925 there studying sponges (Anonymous, 1925). Wilson (1925) summarized his work on dissociated cells of Tortugas sponges in a report to the Carnegie Institution. Based on this research, Wilson (Anonymous, 1926) later presented exhibits of Tortugas sponges developed from artificially separated cells at the 1925 annual meetings—held in December—of the Carnegie Institution of Washington (Washington, DC) and the American Society of Zoologists (New Haven), as well as before the March 8th, 1926, meeting of the Elisha Mitchell Scientific Society (Chapel Hill), where he also presented an abstract.

Personality

Wilson was slight of frame, measuring 5 feet, 6 ½ inches tall, and never weighed more than 120 pounds (Wilson, 1932a). We learn about Wilson the man, his habits, temperament, and intellectual strengths, from accounts given by his associates and former students (Coker, 1940a; House, 1964; Wagstaff, [1950]), as well as others. Wilson was a quiet man and orderly in his work. He disliked noisy environments and administrative duties. He avoided publicity, preferring to dedicate himself to research to reveal scientific truths, rather than to seek fame. Possessing a great power of concentration and an ability to apply himself steadfastly, he was able to accomplish a great deal of work. His incisive thinking, clarity of speaking, and sharp wit were attributes that served him well when he expressed his views at faculty meetings, in the classroom, or elsewhere.

Despite his external appearance of sternness and austerity, he would display a genial side, particularly outside the lab. On occasions he invited small groups of students to his home, where they enjoyed his "few words of wisdom" and observed him puffing at his pipe (Eagle, 1939). Wilson was referred to as "Bugs." This nickname, however, was soon replaced by the then-famous moniker "Froggy" (also "Froggie"), which was affectionately assigned by students in his elementary physiology class (Barnett, 1939). Most likely the name was applied because the basic studies in zoology focused on experimentation with the frog (Ivey, 1968). Wilson didn't seem to resent this name (Lewis, 1942). When his students referred to his son as tadpole, however, he was not as accepting. Robert B. House (1964) recounted another version on the origin of Wilson's nickname: because Wilson particularly liked the frogs' vocalizations emanating from the ponds and marshes around Chapel Hill, he was called "Froggie."

Professor Wilson expected exacting standards of his students. Sloppiness and carelessness were not tolerated. Ivey Foreman Lewis (1942), one of Wilson's students at the beginning of the twentieth century who later was the Miller Professor of Biology and Agriculture at the University of Virginia, mentioned that Wilson imparted a sense of awe in his classes. Delivered with clarity and in rapid succession of facts, his lectures were supplemented with diagrams made in colored chalk on the blackboard. Lewis continued his account of Wilson's lectures, saying, "It was a real workout to take his notes, but when you got them, you had a full, concise, clear statement of the existing knowledge on the subject."

Robert B. House (1964) characterized Wilson's style in another light, reporting that Wilson would "put his thumbs in the armholes of his vest, walk up and down, and talk to us. Then he would stop in front-center, give a characteristic click of his tongue, clear his throat, and say, 'Take a little note.'"

Wilson's enthusiastic spirit of research and his standards of excellence were also guiding principles for his graduate students. Considering his overwhelming teaching commitments with undergraduate classes, it was remarkable that he devoted special attention in preparing twenty-nine graduate students (four Ph.D.s and twenty-five masters). Wilson believed strongly in conducting individual conferences with the graduate students, providing special scholastic guidance. After completing their regular subjects in zoology, graduate students who were assigned a subject commenced their work. Zoological topics were pursued, mostly in the fields of invertebrate morphology (anatomy, histology, embryology, and classification) and vertebrate embryology. The master's degree typically required a year of work and a thesis, the Ph.D. three years and a dissertation worthy of publication (Wilson, 1899).

Although dedicated to zoological investigations and teaching, Wilson also was knowledgeable and conversant on diverse subjects, including literature, classics, and arts. He never owned a car, preferring to walk from his home to the zoology laboratory. He worked steadily six and a half days each week, taking a respite on Sunday afternoons for a walk in nearby woodlands. After his wife Edith Theresa (Stickney), whom he had married on June 10, 1893 (Anonymous, 1893b), died in 1900, and his children moved on, Wilson acquired the life of a bachelor. In later years, he would often dine evenings at the Chapel Hill Coffee Shop, a landmark restaurant. There, he would eat alone or with friends.

Memberships and Honors

Soon after his arrival at UNC, Wilson joined his scientific colleagues in actively supporting the fledgling Elisha Mitchell Scientific Society. Elected to the society in September 1891, he became a dedicated member, giving papers, serving in leadership, and attending meetings. He delivered his first paper presentation, on phosphorescent bacilli, on September 15, 1891. He gave additional talks and published forty articles and abstracts in the *Journal of the Elisha Mitchell Scientific Society* during more than forty-seven years of membership. Virtually all of his publications in the *Journal* concerned zoological topics, among the exceptions being Wilson's (1897) "Notes on the natural history of the Wilmington region," reprinted from *Science* (n.s. 6[135]: 163–164, 1897). He briefly commented on the insectivorous plants growing there. In serving the society, Wilson was president twice (1905–1906; 1938–1939) and a member of the Publications Committee (1929–1937). As a faithful attendant at Society meetings, he was known as "a sympathetic and stimulating commentator upon the papers of others [and] as a general leader of discussion and as a personal exemplar of the spirit of research" (Coker et al., 1939).

Wilson held memberships in several other learned associations. Among them were the American Association for the Advancement of Science (fellow), the American Association of University Professors (charter member), the American Society of Naturalists (executive committee, 1914–1915), the American Society of Zoologists (formerly American Morphological Society, charter member, 1890; serving as vice-president eastern branch, 1908; president eastern branch and of the Society, 1911; executive committee, 1914–1916), the Boston Society of Natural History, and Société Linnéenne de Lyon. He also provided editorial assistance to the *Journal of Morphology*. When *Biological Abstracts* was first published in 1927, Wilson served as the editor of the section on "Porifera (Taxonomy, Morphology, etc.)" until the conclusion of 1930.

Wilson's original research and resulting discoveries earned him international acclaim. In recognizing his distinguished contributions to science, Wilson's colleagues at UNC elected him a Kenan Professor in 1917. Locally, besides devoting unabiding service to the Elisha Mitchell Scientific Society, he held active membership in the North Carolina Academy of Science. He was instrumental in the establishment of the latter organization and served as its president in 1912. On the national level, Wilson's scientific investigations were recognized in his election as a member of the National Academy of Sciences in 1927. He joined the ranks of only two other known southerners (from Missouri), as well as three of his former Hopkins classmates (E.G. Conklin [elected 1908], R.G. Harrison [elected 1913], and T.H. Morgan [elected 1909]). Wilson was also elected to the American Philosophical Society in 1932.

Wilson was honored with two lectureships. During 1915–1916, he was granted a Southern Exchange lectureship. At the Johns Hopkins University he delivered the Dohme lecture in 1936. First given in 1922, the Dohme lecture series commemorates Charles E. Dohme, one of the founders of Merck, a pharmaceutical company. The lectures were developed to foster a relationship between chemistry, medicine, and pharmacology.

In fraternal societies, Wilson was a member of Sigma Xi. He was one of eleven UNC professors who backed "A Petition for the Establishment of a Chapter of the Sigma Xi at the University of North Carolina" (Anonymous, 1915). Noted among their supporting documentation was the fact that in the graduate school of fifty students, twenty-four were pursuing scientific degrees. The charter of the North Carolina chapter was issued December 13, 1919; the exercises of installation and election of officers occurred May 26, 1920 (Mac Nider, [1936]).

Legacy

Henry Van Peters Wilson died of septicemia at age seventy-five on January 4, 1939 in Duke University Hospital. Ten days prior to his death, he had contracted a severe cold which eventually forced him to bed. On his last day, doctors gave him blood transfusions, which failed to help. Services were held at his residence the following Wednesday. Interment took place in the Old Chapel Hill Cemetery on the UNC campus (Anonymous, 1939a) (Fig. 9.9). He was survived by his three children: Edith Stedman (Mrs. Thorndike Saville), Eleanora Stansbury (Mrs. Howell Peacock), and Henry Van Peters Wilson, Jr.

Fig. 9.9. Gravestone of Henry Van Peters Wilson, Old Chapel Hill Cemetery, Chapel Hill, North Carolina, Section I, lot 008, April 2005.

(Courtesy of Brian Nalley, Photo Lab, Department of Biology, University of North Carolina, Chapel Hill).

Soon after Wilson's death, his friends and colleagues formed a committee to consider the establishment of a memorial to him at UNC. The first effort was the development of a memorial fund that raised nearly $4,600 from 155 people. From this fund, the committee commissioned Alice Kent Stoddard of Philadelphia to paint a portrait of Wilson (Battle, 1942) (Fig. 9.10). It currently hangs in a seminar room in the original wing of Wilson Hall, at UNC. With the remainder of the funds, the university set up an endowment to provide scholarships for students pursuing research in marine biology. The scholarship fit well with the intentions of Professor Wilson. One year earlier, he had personally provided funds to enable a graduate student to conduct research in marine zoology at the Beaufort lab. Wilson had intended to donate additional money in successive years (Battle, 1940).

Another tribute took place on November 20, 1941, when the new UNC zoology building, which had been completed and opened in 1940 (Coker, 1940b), was dedicated as the Wilson Zoological Laboratory (now called Wilson Hall). At 9 A.M. the facility was opened for tours, followed by a series of presentations presided over by UNC dean Robert B. House. According to the program of dedication, the talks and speakers were: "Origin and Development of Zoology," Robert E. Coker (1942); "The Wilson Memorial Fund," Kemp P. Battle (1942); "H.V. Wilson as I Knew Him," Ivey F. Lewis (1942); "H.V. Wilson's Scientific Contributions," C. Dale Beers (1942); and "Dedication and Naming of the Laboratory," UNC president Frank Porter Graham.

Wilson's children donated a collection of his books and pamphlets to the UNC library. Nearly 200 volumes of books and 7,000 pamphlets and reprints, useful in research and teaching, were added to the Zoology Library. The latter collection consisted of original scientific papers by Wilson's friends and correspondents throughout the world covering the period from 1875 until his death in early 1939. These were particularly rich in classical zoological monographs and papers on embryology and morphology (Anonymous, 1939c). Among the collection were sixty-nine papers on botany ([Coker], [1939]a).

After the UNC departments of botany and zoology merged to form a Department of Biology in 1982, Wilson's reprint collection of nearly 8,000 pamphlets (Coker, [1939]a), which had been interleaved with a larger collection of papers on bio-luminescence developed by E. Harvey Newton, was crowded out by other pressing needs for the space they occupied. Lawrence I. Gilbert, chairman of the UNC Biology Department, issued a memo on January 30, 1989, indicating that the collection would be disposed of by the following February. In an effort to preserve the collection from the trash heap, the UNC Biology Librarian contacted several scholars who might have shown an interest in the collection. In the end, William Provine, Professor in the Department of Ecology and Evolutionary Biology at Cornell University, accepted the entire lot of pamphlets, which were sent to Ithaca, New York, in fifty-nine cartons via Allied Van Lines movers.

During his years of research, Wilson had amassed numerous specimens of sponges and microscope slides. Without either a successor researcher of sponges or a desire to house this collection at UNC, Robert E. Coker (1939b), chairman of the Department of Zoology, contacted Waldo L. Schmitt at the United States National Museum about depositing the specimens there. Accepting a shipment of 209 lots of sponges on July 24, 1939, Schmitt

Fig. 9.10. Portrait of Henry Van Peters Wilson, painted by Alice Kent Stoddard, ca. 1942.

(Courtesy of Department of Biology, University of North Carolina, Chapel Hill).

(1939) considered it "a very wonderful one." The disposition of Wilson's sponge collection at the museum seemed most appropriate. Since first conducting investigations on sponges, Wilson had made numerous inquiries at the Smithsonian Institution. In their accession files are about fifty letters from him concerning specimens that he identified for the institution and specimens he donated. According to several additional inventories (Coker, 1939a), Wilson also left a collection of numerous other specimens of animals, some preserved in bottles and vials, others on slides. His negatives and prints were included in the inventory. The disposition of this collection is unknown.

Henry Van Peters Wilson devoted practically his entire professional career to the University of North Carolina in Chapel Hill, where he developed a one-man Department of Biology that later (1908) split into two departments, Botany and Zoology. He headed the latter for over three decades and taught until his death. Between 1897 and 1902, he developed a course in botany, which was clearly subordinate to his other academic teaching responsibilities, and it attracted a total of around fourteen students. His instruction in botany included the identification of species and their natural history as well as their structure. As Francis P. Venable developed chemistry on a research basis at the Carolina campus, so Wilson pursued the research endeavors in biology. Together these professors were early promoters at UNC of modern science based on the German model of scientific inquiry. A passionate researcher and teacher, Wilson also harbored a sincere love of the university and the village of Chapel Hill. In a tribute on the occasion of Wilson's seventy-fifth birthday, Rocky Mount attorney Kemp D. Battle (1939) wrote that Wilson was "the embodiment of the Scientific Spirit which seeks Truth always, without prejudices, without preconceptions, not caring where the research leads, but careful always that in the utmost detail the distinction be preserved between that which is known and that which is supposed."

References Cited

(References that conclude with page numbers within parentheses indicate sources from where quotes and pertinent information were obtained.)

Allen, Garland E. 1978. *Thomas Hunt Morgan; The man and his science*. Princeton University Press, Princeton, NJ.

Anonymous. 1881. Biology, 1876–81. *Johns Hopkins University Circulars* 9: 104–116.

Anonymous. 1891a. Dr. Henry V. Wilson elected Professor of Biology at the North Carolina University. *News and Observer* (Raleigh) 31(21): 4, Tuesday, July 28, 1891.

Anonymous. 1891b. Opening of the University. *News and Observer* (Raleigh) 31(44): [4], August 25, 1891.

Anonymous. 1891c. The chair of Biology at Trinity. *News and Observer* (Raleigh) 31(4): [4], Friday, August 21, 1891.

Anonymous. 1891d. Two hundred and twenty-five boys at the University the first day. *News and Observer* (Raleigh) 31(55): [4], Saturday, September 5, 1891.

Anonymous. 1891e. [The Biological Laboratory]. *North Carolina University Magazine*, new series 11: 37.

Anonymous. 1891f. [Dr. Whitehead's lecture]. *North Carolina University Magazine*, new series 10(1): 38.

Anonymous. 1893a. Editorials. *North Carolina University Magazine*, new series 12: [165].

Anonymous. 1893b. [Prof. Henry V. Wilson married]. *North Carolina University Magazine*, new series 13(1): 58.

Anonymous. 1894a. More room for Biology. *White and Blue* (University of North Carolina, Chapel Hill) 2(10): 1, November 16, 1894.

Anonymous. 1894b. Summer School of Biology. *White and Blue* (University of North Carolina, Chapel Hill) 2(1): 1, September 14, 1894.

Anonymous. 1894c. Summer School in Biology. *The Tar Heel* (University of North Carolina, Chapel Hill) 2(24): [3], May 4, 1894.

Anonymous. 1898. [Biological research at Beaufort, North Carolina], p. 14, (*University Record,* vol. 3, no. 1). University of North Carolina, [Chapel Hill].

Anonymous. 1899a. Elisha Mitchell Society. *The Tar Heel* (University of North Carolina, Chapel Hill) 8(4): [4], October 18, 1899.

Anonymous. 1899b. Water supply, pp. 36–37. (*University Record,* vol. 4, no. 3). University of North Carolina, [Chapel Hill].

Anonymous. 1915. *A Petition for the Establishment of a Chapter of the Sigma Xi at the University of North Carolina*. [no statement of publisher or place]. (North Carolina Collection, Wilson Special Collections Library, University of North Carolina, Chapel Hill)

Anonymous. 1925. Back from Tortugas. H.V.P. Wilson is home again after work in laboratories [and] Chapel Hill chaff. *Chapel Hill Weekly* (Chapel Hill, NC) 3(22): [1], Friday, July 31, 1925.

Anonymous. 1926. [Research conducted by H.V. Wilson], pp. 85–86; and Studies on sponge cells at the Tortugas (Abstract), p. 96 in Research in Progress July, 1925–July, 1926. (*The University of North Carolina Record,* no. 236). (*Graduate School Series,* no. 14). University of North Carolina Press, Chapel Hill.

Anonymous. 1930. New professors. *Chapel Hill Weekly* (Chapel Hill, NC) 8(15): [1], Friday, June 13, 1930.

Anonymous. 1931. [Research conducted by H.V. Wilson], p. 87 in Research in Progress, July, 1930–July, 1931. (*The University of North Carolina Record*, no. 276). (*Graduate School Series,* no. 24). University of North Carolina Press, Chapel Hill.

Anonymous. 1939a. Famed scientist of Carolina dies. *Durham Morning Herald* (Durham, NC), Section I, pp. [1], 2, Thursday, January 5, 1939. (portrait)

Anonymous. 1939b. Coker pays tribute to research achievements of friend and faculty associate, H.V. Wilson. *The Daily Tar Heel* (University of North Carolina, Chapel Hill) 47: [1], 2, Friday, January 6, 1939.

Anonymous. 1939c. [H.V. Wilson's Library]. (Box, 31, folder 439, Robert E. Coker Papers #3380, Southern Historical Collection, Wilson Special Collections Library, University of North Carolina, Chapel Hill)

Austin, Willis H. 1893. Letter of September 3, 1893 to Tommie [Lassiter]. (Folder 3, Thomas James Lassiter Papers #3852-z, Southern Historical Collection, Wilson Special Collections Library, University of North Carolina, Chapel Hill)

Barnett, Dewitt. 1939. Dr. "Froggy" Wilson always to remain a UNC tradition. *The Daily Tar Heel* (University of North Carolina, Chapel Hill) 47(74): [1]–2, Thursday, January 5, 1939.

Battle, Kemp D. 1939. [Letter]. Henry Van Peters Wilson. *Chapel Hill Weekly* (Chapel Hill, NC) 17(1): [1], Friday, January 6, 1939.

Battle, Kemp P. 1940. Memorial to Dr. Henry V. Wilson is planned by University alumni. *Alumni Review* (University of North Carolina, Chapel Hill) 28(4): 131, January 1940.

Battle, Kemp P. 1942. The Wilson Memorial Fund. *Journal of the Elisha Mitchell Scientific Society* 58: 8–9.

Beers, C.D. 1942. H.V. Wilson's scientific contributions. *Journal of the Elisha Mitchell Scientific Society* 58: 12–15.

BOT. Board of Trustees, University of North Carolina, Chapel Hill. *Minutes of the Board of Trustees*. (University Archives #40001, Wilson Special Collections Library, University of North Carolina, Chapel Hill) (Vol. 9, April 1891–January 1898; vol. 10, May 1898–April 1904; vol. 13, January 30, 1924–November 1930)

[Coker, Robert E.]. [1939]a. Inventories of research items of Henry V. Wilson. [Five inventories are identified: 1) "East Wall Shelves," 2 pp.; 2) "West Wall Shelves," 2 pp.; 3) "Reprints," 5 pp.; 4) "Dr. Wilson's Prints and Negatives," 3 pp.; and 5) "Slides of Dr. H.V. Wilson, Preserved Material, General Classification of Slides," 4 pp.] (Box 31, folder 439, Robert E. Coker Papers #3380, Southern Historical Collection, Wilson Special Collections Library, University of North Carolina, Chapel Hill)

Coker, Robert E. 1939b. Letter of July 10, 1939 to Waldo L. Schmitt. (Smithsonian Institution Archives, Accession #152962)

Coker, Robert E. 1940a. Henry Van Peters Wilson (1863–1939). *Year Book of the American Philosophical Society* 1939: 470–482. (pp. 473–474)

Coker, Robert E. 1940b. The new building for zoology, University of North Carolina. *Journal of the Elisha Mitchell Scientific Society* 56: 371–379 & Plates 20–22.

Coker, Robert E. 1942. Origin and development of the Department of Zoology. *Journal of the Elisha Mitchell Scientific Society* 58: 1–8.

Coker, Robert E., Archibald Henderson and Wm. deB. MacNider. 1939. Henry Van Peters Wilson. *Journal of the Elisha Mitchell Scientific Society* 55: 256–257.

Costello, Donald Paul. 1961. Henry Van Peters Wilson, February 16, 1863–January 4, 1939. *Biographical Memoirs, National Academy of Sciences of the United States of America* 35: [351]–383; portrait (plate between pages 350–[351])

Cowles, R.P. 1921. The Baltimore Naturalists' Field Club. *Alumni Magazine* (Johns Hopkins University), November 1921.

Daniels, C.C. 1897. Popular summer resort. Beaufort an ideal spot in which to spend the hot season. *News and Observer* (Raleigh) 42(100): 11, Sunday, June 27, 1897.

Duerden, James E. 1904. Grant No. 12. For investigation of recent and fossil corals. *Carnegie Institution of Washington, Year Book* 2(1903): xli–xlii.

Eagle, Watt W. 1939. Dr. Henry Van Peters Wilson. *Alumni Review* (University of North Carolina, Chapel Hill) 27 (February): 160.

Eyre, Doug. 2007. Historical notes: Illustrious UNC faculty had one house in common. *Chapel Hill News*, Section B (Sports), 84(13): B4, Sunday, April 15, 2007. (includes color photograph of house)

Graham, E[ugene] B[errien]. 1893. Letter of October 29, 1893 to Mr. T.J. Lassiter. (Folder 3, Thomas James Lassiter Papers #3852-z, Southern Historical Collection, Wilson Special Collections Library, University of North Carolina, Chapel Hill)

Graham, E[ugene] B[errien]. 1894. Letter of March 25, 1894 to Mr. T.J. Lassiter. (Folder 4, Thomas James Lassiter Papers #3852-z, Southern Historical Collection, Wilson Special Collections Library, University of North Carolina, Chapel Hill)

Gray, Asa. 1881. *Gray's Lessons in Botany and Vegetable Physiology*. Ivison, Blakeman, Taylor & Co., New York.

Gray, Asa. 1889. *Manual of the Botany of the Northern United States; Including the district east of the Mississippi, and north of North Carolina and Tennessee*. 6th ed. American Book Company, New York.

House, Robert B. 1964. Chapter 10. Woods, fields, streams, pp. 59–65; Chapter 20. Sturdy scientist, pp. 121–127 in *The Light That Shines; Chapel Hill—1912–1916*. University of North Carolina Press, Chapel Hill. (Personal reminiscences of H.V. Wilson)

Ivey, Pete. 1968. Basic secret was learned here. *Chapel Hill Weekly* (Chapel Hill, NC) 45(87): 2, Wednesday, October 30, 1968.

JHU-A. Johns Hopkins University Archives. No. 1. (Henry V.P. Wilson Student File, Record Group 13.010, Office of the Registrar, subgroup 1, series 2, box 221, Ferdinand Hamburger Archives, Johns Hopkins University, Baltimore)

Lewis, Ivey F. 1942. H.V. Wilson as a teacher. *Journal of the Elisha Mitchell Scientific Society* 58: 9–12.

MacNider, William deB. [1936]. North Carolina chapter, pp. 716–721 in Henry Baldwin Ward, Edward Ellery (compilers), with the assistance of chapter and club historians, *Sigma Xi Half Century Record and History 1886–1936*. Union College, Schenectady, NY.

Maienschein, Jane. 1999. Henry Van Peters Wilson, pp. 582–583 in John A. Garraty and Mark C. Carnes (general editors), *American National Biography*, vol. 23. Oxford University Press, New York, Oxford.

Morgan, T.H. 1889. Notice of Dr. H.V. Wilson's paper on the development of *Manicina areolata*. *Johns Hopkins University Circulars* 8(70): 39–40.

Nickers, T. Edward. 2002. For the birds. *Carolina Alumni Review* 91(6): 50–58.

Patterson, W.T. (bursar). 1891a. Broadside. *Prices of rooms in the South Building*. (Folder 1891, Announcements, bulletins, programs, etc., North Carolina Collection, Wilson Special Collections Library, University of North Carolina, Chapel Hill)

Patterson, W.T. (bursar) [1891]b. Brochure. *A Southern University*. [University of North Carolina, Chapel Hill]. (Folder 1891, Announcements, bulletins, programs, etc., North Carolina Collection, Wilson Special Collections Library, University of North Carolina, Chapel Hill)

Prantl, K. and S.H. Vines. 1891. *An Elementary Text-book of Botany*. From the German of Dr. K. Prantl. Edited by S.H. Vines. 5th ed. Swan Sonnenschein & Co., London.

Schmitt, Waldo L. 1939. Letter of August 3, 1939 to Mr. Commerford. (Smithsonian Institution Archives, Accession #152962)

Shine, Ian and Sylvia Wrobel. 1976. *Thomas Hunt Morgan, Pioneer of Genetics*. University Press of Kentucky, Lexington. (p. 20)

Spencer, Cornelia Phillips. 1853–1906. *Journal of Cornelia Phillips Spencer, 1853–1906*. (Box 14, folder 158, volume 2, Cornelia Phillips Spencer Papers #683, Southern Historical Collection, Wilson Special Collections Library, University of North Carolina, Chapel Hill). (pp. 181–182)

Turlington, W[illiam] T[roy]. 1891. Letter of September 16, 1891 to Mr. T.J. Lassiter. (Folder 1, Thomas James Lassiter Papers #3852-z, Southern Historical Collection, Wilson Special Collections Library, University of North Carolina, Chapel Hill)

Turlington, W[illiam] T[roy]. 1892. Letter of March 17, 1892 to Mr. T.J. Lassiter. (Folder 1, Thomas James Lassiter Papers #3852-z, Southern Historical Collection, Wilson Special Collections Library, University of North Carolina, Chapel Hill)

UNC Archives. University of North Carolina Archives, Chapel Hill. No. 1. "Regulations regarding students (1901)," University of North Carolina, Chapel Hill. (Box 21, folder 696, University Papers #40005, University Archives, Wilson Special Collections Library, University of North Carolina, Chapel Hill)

UNC Faculty. University of North Carolina, Chapel Hill. *Minutes of Faculty Meetings*. (University Archives #40106, Wilson Special Collections Library, University of North Carolina, Chapel Hill) (Vol. 7, 1885–1901)

UNC-SHC. University of North Carolina at Chapel Hill, Southern Historical Collection, Wilson Special Collections Library. Henry Van Peters Wilson Papers #3202, 4 items (including a letter from Alexander Agassiz [1903] and a copy of George B. O'Toole's The Case Against Evolution—New York, 1926, with numerous penciled annotations by Wilson supporting the theory of evolution). (Robert E. Coker Papers #3380, Southern Historical Collection, Wilson Special Collections Library, University of North Carolina, Chapel Hill, file folders 439–440)

Venable, Francis P. 1901a. Letter of July 3, 1901 to Henry Van Peters Wilson. (*President's Letter Copy Books*, vol. 7, University Papers, Series 3 #40005, University Archives, Wilson Special Collections Library, University of North Carolina, Chapel Hill)

Venable, Francis P. 1901b. Report of the president, pp. [13]–17. (*The Record, the University of North Carolina*, new series, vol. 1, no. 3). (*The University Bulletins*, no. 3). Published by the University, Chapel Hill. (p. 15)

Venable, Francis P. 1902. Letter of June 6, 1902 to J.E. Duerden. (*President's Letter Copy Books*, vol. 13, University Papers, Series 3 #40005, University Archives, Wilson Special Collections Library, University of North Carolina, Chapel Hill)

Wagstaff, Henry McGilbert. [1950]. *Impressions of Men and Movements at the University of North Carolina*. Edited with a prefatory note by Louis R. Wilson. University of North Carolina Press, Chapel Hill.

Wilson, Henry Van Peters. 1884. Letter of November 17, 1884 to President Daniel Coit Gilman. (Henry V.P. Wilson Student File, Record Group 13.010, Office of the Registrar, subgroup 1, series 2, box 221, Ferdinand Hamburger Archives, Johns Hopkins University, Baltimore)

Wilson, Henry Van Peters. 1886. The parasitic Cuninas of Beaufort. *Johns Hopkins University Circulars* 6(54): 45. (Abstract)

Wilson, Henry Van Peters. 1888a. Development of *Manicina areolata*. 126 pp., 6 plates. (A thesis for the degree of Ph.D., accepted by the Board of University Studies of the Johns Hopkins University, May 1888.) (Special Collections, Milton S. Eisenhower Library, Johns Hopkins University, Baltimore)

Wilson, Henry Van Peters. 1888b. Letter of June 5, 1888 to President Daniel Coit Gilman. (MS 1, Papers of Daniel Coit Gilman, series 1, box 54, Special Collections, Johns Hopkins University, Baltimore)

Wilson, Henry Van Peters. 1889. Letter of February 20, 1889 to President Daniel Coit Gilman. (Special Collections, Milton S. Eisenhower Library, Johns Hopkins University, Baltimore) (Also reprinted as Report of Dr. Henry V. Wilson as Bruce Fellow of the Johns Hopkins University, *Johns Hopkins University Circulars* 8([70]): 40. 1889)

Wilson, Henry Van Peters. 1890. On the development of the sea bass (*Serranus atrarius*). *Johns Hopkins University Circulars* 9(80): 56–59.

Wilson, Henry Van Peters. 1891a. The embryology of the sea-bass (*Serranus atrarius*). *Bulletin of the United States Fish Commission* 9: [209]–277.

Wilson, Henry Van Peters. 1891b. Letter of May 3, 1891 to Daniel Coit Gilman). (MS 1, Papers of Daniel Coit Gilman, series 1, box 54, Special Collections, Johns Hopkins University, Baltimore)

Wilson, Henry Van Peters. 1891c. Letter of August 1, 1891 to Daniel Coit Gilman. (MS 1, Papers of Daniel Coit Gilman, series 1, box 54, Special Collections, Johns Hopkins University, Baltimore)

Wilson, Henry Van Peters. [1895]a. *Laboratory Guide in Practical Biology*. [no statement of publisher or place]

Wilson, Henry Van Peters. [1895]b. *Directions for Laboratory Work in Histology*. [no statement of publisher or place]

Wilson, Henry Van Peters. [1895]c. *Laboratory Guide in Elementary Physiology*. [no statement of publisher or place]

Wilson, Henry Van Peters. 1897. Notes on the natural history of the Wilmington region. *Journal of the Elisha Mitchell Scientific Society* 14: [1]–4.

Wilson, Henry Van Peters. 1899. Biology at the University of North Carolina. *Gulf Fauna and Flora Bulletin* 1(1): [23]–25.

Wilson, Henry Van Peters. 1904. Reports on an exploration off the west coasts of Mexico, Central and South America, and off the Galapagos Islands, in charge of Alexander Agassiz, etc. XXX. The sponges. *Memoirs of the Museum of Comparative Zoology at Harvard College* 30(1): 1–164; 26 plates.

Wilson, Henry Van Peters. 1905. Grant No. 33. For morphology and classification of deep-sea sponges [and] Abstract of report. *Carnegie Institution of Washington Year Book* 3 (1904): 142–144.

Wilson, Henry Van Peters. 1925. Studies on dissociated sponge cells. *Carnegie Institution of Washington Year Book* 24 (1924–1925): 242–246.

Wilson, Henry Van Peters. 1929. Letter of July 8, 1929 to Harry Woodburn Chase. (Box 19, folder 1927–1936, Chancellors Records, Robert B. House series, University Archives #40019, Wilson Special Collections Library, University of North Carolina, Chapel Hill)

Wilson, Henry Van Peters. 1930. A letter from H.V. Wilson. *Chapel Hill Weekly* (Chapel Hill, NC) 8(7): 1–3, Friday, April 18, 1930.

Wilson, Henry Van Peters. 1932a. "H.V. Wilson—Biographical Sketch." 5 pp. (typewritten manuscript, dated February 22, 1932). (Archives, National Academy of Sciences, Washington, D.C.)

Wilson, Henry Van Peters. 1932b. [Abstract]. The larval metamorphosis in monaxonid sponges. *Journal of the Elisha Mitchell Scientific Society* 47: 27. (Abstract also published in Research in Progress, July, 1930–July, 1931. [*The University of North Carolina Record,* no. 276, p. 95]. [*Graduate School Series*, no. 24]. University of North Carolina Press, Chapel Hill)

Wilson, Henry Van Peters. 1938a. [Abstract]. Vacuolation a factor in the division of animal cells. *Science* 88: 435.

Wilson, Henry Van Peters. 1938b. Letter of March 17, 1938 to Frank Porter Graham. (Folder 440, Robert E. Coker Papers #3380, Southern Historical Collection, Wilson Special Collections Library, University of North Carolina, Chapel Hill)

Winston, [George T.]. 1892a. Hints to young men. *North Carolina Teacher Supplement*, April 1892. (Folder 1892, Announcements, bulletins, programs, etc., North Carolina Collection, Wilson Special Collections Library, University of North Carolina, Chapel Hill)

Winston, [George T.]. 1892b. Broadside. *Notice to New Students, August 29, 1892*. (Folder 1892, Announcements, bulletins, programs, etc., North Carolina Collection, Wilson Special Collections Library, University of North Carolina, Chapel Hill)

Winston, [George T.]. 1894. [The Department of Biology], p. 32 in The expansion of the University. *Alumni Quarterly* (University of North Carolina, Chapel Hill) 1: 31–39.

Wolfe, Douglas A. 2000. *A History of the Federal Biological Laboratory at Beaufort, North Carolina, 1899–1999*. U.S. National Oceanic and Atmospheric Administration, Washington, D.C.

C. P. Spencer

Ode to the Normal School

Let us sing of the Normal School
Where Nature, not Art, is the rule.
Where the Teacher is brought
Like a child to be taught
What it is that we call Education;
That not all the knowledge
He gains in a college,
Not the problems that vex,
Nor the laws that perplex,
Nor the strongest reliance,
On what be call "Science"
Are all he needs in his vocation.
But he learns that the Teacher,
As well as the preacher,
Must raise his thoughts higher
Than selfish desire
Of wealth, or of fame, or mere wordly well-doing.

That to hear the "well-done"
When his race he has run.
He must labor and wait, and "tho" faint, be pursuing."
"Twas with very much wondering,
And laughing and blundering,,
To the famous old Hill
We came with a will.
By a way most informal,
To look at "the Normal"
Not dreaming of what would befall.
And oh! it is past telling,
The reading and the spelling,
The grammar and the writing,
And the lectures we delight in,
And the kindness we met, above all.
Time would fail, should we tell,
Of the Campus, and the well,
Of the walks,
And the talks,
And the tuneful college bell.
What a treasure
Is the pleasure
That the six weeks have brought us;
Our hearts will ever turn
When our memories we turn,
To the thought of the lessons they taught us.

When each one of us became,
As a little child again,
And sat low at the feet of a master.
Our pulses will beat faster,
As we think of these long summer days;
When all the good and great,
Who adorn our native State,
Came to help, and to cheer, and to praise.

And now 'ere we go,
Let us pay the thanks we owe
To the College, and the President,
And every Chapel Hill resident,
For the kindness and the grace
That have so endeared the place.
Never was there such a LADD
As this Normal School has had,
To point them to their duty,
And show them all their beauty
Of a self-denying labor,
For the welfare of their neighbor.
Such instruction makes us glad,
Every lass must love a LADD,
And what true, and hearty gratitude,
We shall ever be OWEN
To him who has been showing
Us his notions
Of the oceans,
Of climate, dry and wet, and of longitude and Latitude.
In Professor A. McIver,
His quotients, and his fractions,
And other such distractions,
We are, each, a firm believer;
For, though he *teased* us much,
He *pleased* us much.
And tho' professor Winston
Kept our noses on the grin'-stone,
In a brave attempt to grind,
A bit of Latin in our mind,
Yet our thanks must be sent,
For we know 'twas kindly meant.
And as for Mr. Wilson,
We are sure that Madame Nilsson,—
Though the world is ringing
With her singing,—
Never draws,
More applause,
Than *our* Master's skillful rule
Merits from his grateful school.
Nowwhen all is said and one,
Here's Professor Tomlinson,
For such a FRIEND indeed
We have verily a need,
As many a kind glance will confess;
Yet with every disposition
To suggest
That a change in his condition
Would be best,
A lass!! Is all we can express.
And now, friends, Fare ye well!!
Our pen will never tell,
Of our hearts' true and lasting emotion;
Never more
As heretorfore,
Shall we rove,
Through the grove,—
But "in that higher School
Where Christ himself doth rule;"
And *there* we may believe
The faithful shall receive
The reward of his life-long devotion.

Chapter 10.

Summer Instruction and Science

Normal School (1877–1884)

The Normal School was established in 1877 at UNC to train North Carolina teachers, the first undertaking of its kind in the state (Leloudis, 1996). The efforts of many people brought this venture to fruition, particularly those of President Kemp Plummer Battle. On March 9, 1877, the North Carolina General Assembly ratified an act to establish the Normal School (Battle, 1878a). Although the school was founded for white males who were connected with UNC, it was open to North Carolina teachers and those intending to become teachers, including females. Publicity for the school included circulars (Fig. 10.1) that were distributed throughout North Carolina. Citizens of the state also learned about the fledgling school through articles published in the *Observer*, a Raleigh newspaper that outlined its purpose and instructional plans for that summer (Vance et al., 1877), and another in the *North Carolina Presbyterian* (Spencer, 1877b). Among the Normal School's objectives were teaching the latest and best methods of class management, inspiring interest, providing knowledge, and developing the minds of students as well as instructing on subjects that were being taught in the schools (Highsmith, 1923). Sessions of five to six weeks' duration were held during the summer vacations of the university. Professors and instructors, who came from UNC, other colleges, and city schools, were experts in their fields of specialty and were noted for their experience in the methodology of normal instruction.

State appropriations, as well as money from the Peabody Fund, helped finance the undertaking. The latter fund had been established in 1867 to assist in rebuilding the postbellum American South. The fund was named for George Peabody, who had donated over $2 million toward its establishment. Every effort was made to make attendance possible for both the rich and the poor. The Normal School provided educational enrichment to as many students as possible. Classes were tuition-free, and the pupils could either purchase their textbooks or acquire them through loan. Pupils had access to facilities for cooking and could live quite frugally. An anonymous newspaper reporter (Anonymous, 1878) noted that pupils could obtain board and lodging at the hotel or boardinghouses for $12.50 per month. Those scholars who brought their own bedding could subsist for $6 or $7 if they provided their own board. In 1877, "bread [was] brought up from the bakery in Durham every week—five cents a loaf. Fruit in Chapel Hill [was] abundant and cheap. Ladies [could] make their own tea and coffee, and provide their own food at astonishingly low rates, if they [chose] to make the experiment of 'camping out' for six weeks" (S[pencer], 1877c). Attendees—variously called normalites, normals, normaldies, normalists, or normaliters—also had unrestricted access to the university's facilities, including the museums, laboratories, and libraries. For those unable to pay basic expenses such as transportation costs, the Peabody Fund helped

Frontispiece Credits: Portrait of Cornelia Phillips Spencer. (Courtesy of North Carolina Collection Photographic Archives, Wilson Special Collections Library, University of North Carolina, Chapel Hill). "Ode to the Normal" School, by Cornelia P. Spencer. (From *Observer* [Raleigh, NC], Tuesday, August 14, 1877, p. [3]).

with their defrayal. The railroads and the Blackwater Line of steamboats provided students reduced-rate round-trip tickets to the region (Battle, 1878b).

UNIVERSITY NORMAL SCHOOL.

ANNOUNCEMENT FOR 1881.

The recent action of the General Assembly affords most gratifying evidence of the strong hold the University Normal School has taken on the public mind of the State. The enterprise will accordingly be prosecuted with renewed vigor and enthusiasm.

Teaching is not only a science, but an art. The Normal School will therefore be, as heretofore, under the charge of trained experts, who for years have made methods a study, and have had large practical experience in their working.

THE SCHEME OF INSTRUCTION WILL ENBRACE,

I. REGULAR INSTRUCTION FOR ALL STUDENTS, including

(1.) Daily Recitations and Lectures on Arithmetic, (mental and written,) English Grammar and Analysis, Orthography, Reading, Phonetics, and Geography.

(2.) Daily Lectures before the whole School, on School Organization and Discipline, Methods of Instruction, the relation of Teacher, Parent and Child, &c.

(3.) Daily Drill in Vocal Music.

II. SPECIAL INSTRUCTION FOR ADVANCED STUDENTS, in the following branches: Chemistry, including Laboratory Practice, Latin, (two classes,) Algebra, Book-Keeping and Penmanship.

III. Classes will be formed of pupils to be taught by the more advanced Normal Students, under the supervision of the Professors. In this connection will be exhibited the system and Methods of Graded Schools.

IV. A series of Lectures on the Geology of North Carolina, by Professor Kerr, the State Geologist.

V. Lectures by distinguished educators and specialists of North Carolina and other States.

A MODEL PRIMARY SCHOOL.

In order that the teachers of the State may see in practical operation a Primary School organized and conducted according to the most approved methods, such a school, composed of children from Chapel Hill and its vicinity, will be organized and conducted by a teacher of experience in the Public Schools of New York City.

ELOCUTION AND READING.

A distinguished teacher of Elocution and reading will give practical instruction in their arts, both to the entire School and to advanced classes.

THE FACULTY OF THE UNIVERSITY

will co-operate cordially, by occasional lectures and otherwise, in furthering the interests of the School.

THE TEACHERS' ASSOCIATION OF NORTH CAROLINA

will hold a regular meeting during the session.

THE NAMES OF THE PROFESSORS

composing the Faculty of the Normal School will be given to the public hereafter.

THE TIME OF OPENING.

The School will open on Thursday, the 16th of June, next, and will continue six weeks. It will be held in the University Recitation Rooms and Laboratories.

FREE TUITION.

There will be no charge for tuition. Dormitories in the University Buildings will be free to male students, who, however, must furnish their own bedding.

BOARD.

Table board can be had in private families at the rate of ten dollars per month. Arrangements for messing at a cheaper rate will be provided. Facilities for cooking will be afforded those bringing their own utensils. Last year some of the Normal students lived comfortably in this way at four dollars to five dollars per month.

TRANSPORTATION.

Half fare, in some cases less, will be granted by the Railroad Companies. *Students should apply for commutation tickets to the railroad agent at the station where they take the train. These commutation tickets must be obtained from the agents of each road travelled over. The ticket agent at the station, and not the conductor on the train, has authority to issue these tickets.*

It is hoped that the railroad will run to Chapel Hill by the opening of the School. If not, proper directions will be given.

TEXT BOOKS.

Facilities will be given to borrow or purchase such books as may be needed.

PECUNIARY AID.

Through the liberality of the Rev. Dr. Curry, Agent of the Peabody Fund, supplemented by the Board of Education, pecuniary aid will be given to those who are unable to pay their own expenses.

APPLICATIONS

for procuring pecuniary aid should be made on or before June 1st, next, to President Battle, at Chapel Hill, who will lay them before the Board. If the amount applied for does not exceed the fund at the disposal of the Board, all travelling expenses will be paid, so as to place those at a distance on an equal footing with those near at hand.

Applications should be accompanied by certificates from a Judge, a Chairman of the Board of County Commissioners, a Superior Court Clerk, or a member of the Legislature, vouching for the good character of the applicant and his inability to pay his own expenses.

In conclusion, the Board of Education assure the teachers of the State that the Normal School will fully respond to the increasing demand for better Education. It will exemplify by theory and practice, the best methods of instruction employed in Europe and America.

It is possible for any teacher in the State to attend the coming session of the Normal School at a cost not exceeding five dollars. Patriotism and self-interest alike call upon teachers everywhere to come and show their appreciation, both of their own profession and of the efforts that the State is making in behalf of Education.

THOS. J. JARVIS, *Governor,*
Ex-officio Chairman Board of Education.

J. C. SCARBOROUGH,
Sup't of Public Instruction
and Secretary Board of Education.

KEMP P. BATTLE,
President University of North Carolina.

Fig. 10.1. Circular for the 1881 Normal School, University of North Carolina, Chapel Hill.

(From Kemp P. Battle, *Scrapbook of Materials Relating to the University of North Carolina, 1869–1893,* vol. 3; circular inserted between pages 56 and 57. Courtesy of North Carolina Collection, Wilson Special Collections Library, University of North Carolina, Chapel Hill).

Several noteworthy sources of information about the school provide insight into this innovative venture. A major source of information about the Normal School is a series of more than 100 newspaper articles preserved in several scrapbooks pertaining to the history of the University of North Carolina. Compiled by President Battle (1869–1893) and currently housed in UNC's North Carolina Collection, these volumes are a rich resource for information on instructors and their instruction, as well as the events of the school's sessions. A noteworthy source of information about the first summer session is a series of newspaper articles by Cornelia P. Spencer, who described the classes and events comprehensively. Luly Thompson (1878), a pupil of the 1878 Normal School, kept class notes that described the nature of the lessons, including those offered in science.

The Normal School opened to students in the morning of July 3, 1877. In giving the opening address, Kemp Battle reinforced the primary purpose of the school by stating that "teachers are not born teachers"; he said that they must be trained, as other professionals are (Spencer, 1877d). Governor Zebulon Baird Vance (1877) was excited about the prospects of the school, saying, "I am full of hope, and do believe our old State is about to climb slowly upward." One of the highlights of the 1877 session was the instruction of a "new mode of teaching to read and spell—called for want of a better name, Phonetics" (S[pencer], 1877f).

On the night before the closing day of the first session, a copy of an "Ode to the Normal School" (Frontispiece of chapter 10) was left anonymously at President Battle's breakfast table (S[pencer], 1877a). Undoubtedly written by Cornelia Spencer, the verse charmingly praised the session. Commenting on the last class of the first session, Spencer (1877h) summed up the summer session, writing, "Well done." She added, "There never will be another Normal School like this one—for this was the *first*, and had all the fresh *first* flavor, the bloom of the grape, so to speak." Spencer's (1877g) favorable views of the new teaching enterprise emphasized its "grand success, from beginning to end." The school was, she wrote, "begun and carried on and ended with unbounded good will, and spirit, and sense" and it left "pupils and professors with a mutual respect and affection which time [would] not soon efface." In a letter to the editor of the *Observer*, a correspondent (Anonymous, 1877c) from Goldsboro, North Carolina, commended the educational enterprise. He considered it a "splendid success" and was encouraged that students gained new knowledge. He was particularly impressed by the enthusiastic spirit of the students, several of whom he had met. Barnas Sears (1877), administrator of the Peabody Fund, also praised the fledgling venture, congratulating President Battle "on the splendid success of [his] Normal School." The first session had attracted 235 teacher-pupils, of whom 128 were men and 107 women (Anonymous, [1877]b).

The students were kept busy. Their daily classes began at 9:00 A.M. and concluded at 6:00 P.M. (Anonymous, 1877c). At least twice a week, they attended special evening lectures, extending their days until 9:00 or 10:00 P.M. They endured the heat and drought of the height of summer (X., 1879a). Special lectures were also given during the day. John Wilson Lafferty (1883–1885), who attended the 1883 session, described a busy schedule for one week, which included six recitations each day and a special lecture every night. Students also attended debates that took place in the evening, one of which concerned the question "Should the sexes be educated together" (Lafferty, 1883–1885).

Entertainment included concerts and presentations of calisthenics performed by some of the normalites. Lucy Phillips Russell ([1957]), daughter of UNC professor of mathematics James Phillips, reminisced about the spirit of collegiality that was nurtured among the students by a weekly social gathering in the library, which was in Smith Hall. The students were expected to introduce themselves to one another.

When socials were given in the evenings, the Dialectic and Philanthropic halls and libraries were lighted up, and the ballroom (Fig. 10.2) in the University Library was "ablaze with a mellow light reflected from the gas reflector" (Anonymous, 1877a). Anothern person described a scene from an event early in the summer of 1879: "At night a social entertainment was held in the University Library. Books were thrown aside, pensive looks discarded, dignified instructors converted into youthful gallants, and the stern teacher actually changed, as the case might be, either into a talking beau or a laughing belle. There was no dancing, but a well-regulated, orderly 'walk-around.' The literary halls were thrown open, and here and there might be seen a few couples strolling stealthily from the hub-bub of the library into these quiet recesses" (P., 1879a). A couple of weeks later, one more social was held in the University Library on July 4. Again there was no dancing. Couples walked together, but "without the measured movements of the dance, and the boys took only the hands, but not the waists of the girls, in the style which Gov. Swain used to call 'as heretofore'" (X., 1879b). Concerts (Fig. 10.3) also provided another form of entertainment for the summer students. Spencer (1877g) noted that at an entertainment sponsored by the ladies of Chapel Hill in the library, as much flirting took place as if there had been true dancing.

Another diversion for the students was to stroll on the campus and in the nearby woodlands. Kemp Battle enjoyed giving tours around the campus region "to 'Roaring Fountain,' 'Black Tom's Den,' the 'Judge's Spring,' the 'Meeting of the Waters,' 'Piney Prospect,' and [the 'Dromgoole Stone,' to the moon rise over the 'Triassic Sea'"] (Russell, [1957]). John W. Lafferty (1883–1885) mentioned an evening visit to the campus cemetery, where he accompanied a young lady. There, they sat on the tombstones, enjoying a conversation "until the moon shone brightly." Lafferty noted visits to the University Museum as well as the Society Halls and their libraries. He also recorded in his diary that occasionally students collectively made trips to the surrounding countryside. One of these excursions included a visit to an iron mine, several miles from campus. They boarded an omnibus, which took them to the train depot, and traveled about a mile to their destination.

In 1881, a student from the lowlands of Pitt, North Carolina, described the campus:

> The natural scenery in and around this place fills me with ecstasy, and to my views, would afford painters an ample field in which they might feel proud to exhibit specimens. It is beyond the power of the pen to portray, one-tenth the beauty and grandeur [of the] surrounding hills. . . . The campus with the enclosure of rock-built walls, includes many acres of ground, delightfully shaded by the dense foliage of giant oaks and is covered entirely by nature's richest emerald carpet. The salubrity of this place is undisputed. The water is delightful to drink and when just drawn up is almost cool as ice (Normalite, 1881).

The school gained the admiration of North Carolinians. A correspondent (S., 1880) of the *Kinston Journal* (a newspaper that served the town of Kinston, North Carolina) praised the

Fig. 10.2. The ballroom in Smith Hall, University of North Carolina, Chapel Hill, ca. 1885.

(Stereograph Collection, image no. 43.68.9, negative 100-5; North Carolina Collection Photographic Archives, Wilson Special Collections Library, University of North Carolina, Chapel Hill).

school and thought it "among the greatest advancements toward education hitherto known." After the fourth year of conducting the "experimental" school, Battle (1881) also reported on its success and its good reputation, which was in all likelihood instrumental in the founding of similar summer normal schools in Virginia, South Carolina, and other places. The students derived benefits beyond the coursework and the opportunities for socialization. At the conclusion of the sessions, according to a reporter for the *Observer*, students could take examinations to earn public school certificates, which allowed them to teach in North Carolina (P., 1879c). The reporter further noted, "We hope that the teachers who obtain these certificates will be sought after, and will by their thoroughness and accuracy, so surpass those generally licensed by county examiners that all the public school teachers in the State will be brought to recognize the necessity of attending the Normal School."

Normal School pupils could also attend programs of the State Teachers' Association (also called the North Carolina Teachers' Association), which held its meetings at the university while the Normal School was in session. The pupils organized the association at the 1878

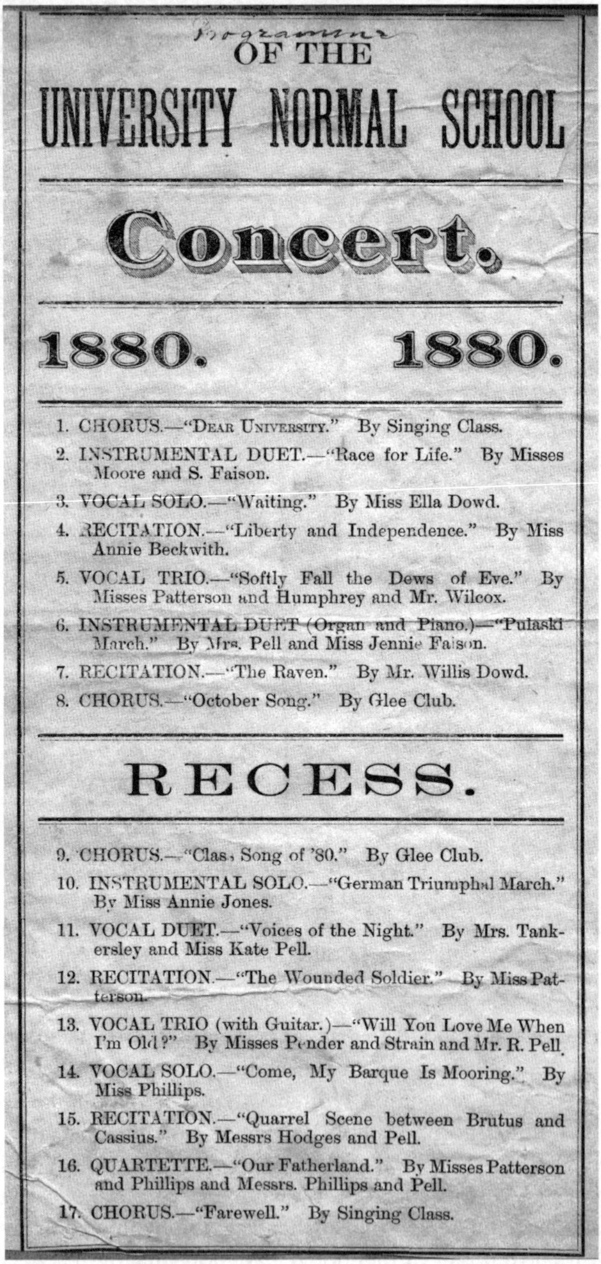

Programme OF THE

UNIVERSITY NORMAL SCHOOL

Concert.

1880. 1880.

1. CHORUS.—"Dear University." By Singing Class.
2. INSTRUMENTAL DUET.—"Race for Life." By Misses Moore and S. Faison.
3. VOCAL SOLO.—"Waiting." By Miss Ella Dowd.
4. RECITATION.—"Liberty and Independence." By Miss Annie Beckwith.
5. VOCAL TRIO.—"Softly Fall the Dews of Eve." By Misses Patterson and Humphrey and Mr. Wilcox.
6. INSTRUMENTAL DUET (Organ and Piano.)—"Pulaski March." By Mrs. Pell and Miss Jennie Faison.
7. RECITATION.—"The Raven." By Mr. Willis Dowd.
8. CHORUS.—"October Song." By Glee Club.

RECESS.

9. CHORUS.—"Class Song of '80." By Glee Club.
10. INSTRUMENTAL SOLO.—"German Triumphal March." By Miss Annie Jones.
11. VOCAL DUET.—"Voices of the Night." By Mrs. Tankersley and Miss Kate Pell.
12. RECITATION.—"The Wounded Soldier." By Miss Patterson.
13. VOCAL TRIO (with Guitar.)—"Will You Love Me When I'm Old?" By Misses Pender and Strain and Mr. R. Pell.
14. VOCAL SOLO.—"Come, My Barque Is Mooring." By Miss Phillips.
15. RECITATION.—"Quarrel Scene between Brutus and Cassius." By Messrs Hodges and Pell.
16. QUARTETTE.—"Our Fatherland." By Misses Patterson and Phillips and Messrs. Phillips and Pell.
17. CHORUS.—"Farewell." By Singing Class.

Fig. 10.3. [Programme] of the University Normal School Concert, 1880.

(From Folder 230, volume 73, p. 45, Cornelia Phillips Spencer Papers #683, Southern Historical Collection, Wilson Special Collections Library, University of North Carolina, Chapel Hill).

summer session (Battle, [1878]c). The association subsequently held annual meetings. In 1879, Battle gave the opening address, entitled "The Education of the Farmer" (Kinsey, 1879a). The organization provided a collective voice to promote the welfare of the profession, to foster the cooperation of teachers in aiding their interests, and to give their contributions to society public visibility (X., 1878). The pupils also worked toward the creation of county branches of the association throughout North Carolina. Thus the Normal School contributed to the teachers' overall betterment, offsetting the fact that teachers were "not sufficiently appreciated in North Carolina" and were paid poorly, as one contemporary noted (Anonymous, 1880a).

Science was not a prominent feature in the curriculum of the first Normal School in 1877. The superintendent of the school, John J. Ladd, remarked on the opening day that students would be given familiar talks and schoolroom experiences, among other lessons. Although he mentioned that science would not be among the topics set before the students (Spencer, 1877d), invited lecturers gave several special presentations with scientific themes. These included talks by UNC alumnus Fabius Haywood Busbee on the power of electricity; Dr. George Graham of Raleigh on the ear, including its structure and physiology; Dr. Richard Henry Lewis of Raleigh on the eye, illustrated by models, colored drawings, and diagrams on the blackboard; and Professor Alexander McIver on physiology.

During its eight-year existence, the Normal School (Table 10.1) did not offer courses in botany or natural history, but special lectures related to these topics were given. In 1877, Governor Vance lectured on the agricultural capacity of the country, President Battle talked on the relation of the university to the agricultural interests of the state, and State Geologist Washington Caruthers Kerr explained how climate determines the range of flora and fauna in a state; in 1878, Professor John Rennie Blake of Davidson College presented a talk on "Natural Science—Its Importance"; in 1879 and 1880, Battle delivered an address entitled "Education for Farmers"; in 1879, Julius L. Tomlinson, former professor of Santa Barbara University in California, presented a talk on California and its flora; in 1879, Professor Holt gave advice on why to study botany and described curious plants of North Carolina; UNC professor Frederic William Simonds gave five lectures in "Natural History" in 1879[81] and at least one lecture on "Curious Flowers" in 1880; and in 1881 and 1883, William Isaac Marshall of Fitchburg, Massachusetts, lectured on the "Big Trees of California" and "Yosemite Valley." Marshall's presentations in 1883 consisted of seven lectures delivered before the entire school during the first two weeks of the session. He illustrated the talks with dissolving views with the calcium light.[82] One student commented that the "pictures exhibited vividly to us the grand scenery in the Rocky Mountains" (Lafferty, 1883–1885). Some of Marshall's lectures are discussed in greater detail in the following pages. Other scientific subjects covered during the Normal School sessions, either in courses or by means of special lectures, were algebra, arithmetic, astronomy, chemistry, geology, geometry, geography, human physiology, mathematics, and natural philosophy (physics).

[81] Descriptions of Simonds's courses are given in chapter 7.

[82] The audience would see one image disappear and be quickly replaced by another image, a technique popularly known as dissolving views. Magic lanterns were the source of projection. Illumination was created by heating a piece of lime or calcium oxide, typically with a flame of combined oxygen and hydrogen. The resulting light or limelight was intense and white. When this kind of illumination was used for stage lighting, it became the basis for the phrase "in the limelight" (Robinson et al., 2001).

TABLE 10.1. A LIST OF DATES OF CLASSES FOR SUMMER INSTRUCTION, UNIVERSITY OF NORTH CAROLINA AT CHAPEL HILL, 1877–1884; 1894–1904.

Normal School	Summer School
1877: July 3–August 9	1894: July 2–July 28
1878: June 18–July 26	1895: June 25–July 26
1879: June 17–July 24	1896: June 21–July 19
1880: June 24–July 29	1897: June 22–July 23
1881: June 16–July 21	1898: June 21–July 13
1882: June 15–July 20	1899: June 20–July 15
1883: June 21–July 26	1900: June 12–[ca. July 7]
1884: June 17–July 17	1901: June 17–July 6
	1902: June 16–July 5
	1903: June 15–July 10
	1904: June 13–July 8

Newspaper accounts and Luly Thompson's classroom notes from 1878 (Thompson, 1878) describe these botanical and natural history lectures given at the Normal School in some detail. Among Thompson's teachers was Emily M. Coe, a specialist in the kindergarten grades, whose instruction imparted some knowledge of natural history. She believed that children should be led to study "Nature" by cultivating their perceptive faculties, stimulating thinking, and teaching them scientific terms. Teachers themselves, therefore, needed to learn about nature (animals, plants, and minerals). As part of a lesson, Coe's students had to write fifty names of things for each of the three kingdoms: animal, vegetable, and mineral.

Sidney H. Owens, principal of the Deshler Female Institute in Tuscumbia, Alabama, presented lectures on geography that were well attended and well liked. One of his lectures in July 1877 discussed the influence of water on climate and how climate in turn influenced the vegetation. He noted: "In Ireland the myrtle grows all the year round. Yet peaches and plums won't ripen. On the continent of Europe the myrtle won't always grow, yet peaches and plums will, because in Ireland, though it is warm and moist, yet the sun is not hot enough to ripen that fruit. Grapes, too, will not flourish in England, tho' [it is] warmer on the average than the Rhine where the vine does flourish" (S[pencer], 1877e).

Another noted lecturer of the 1879 summer session was Virginia transplant Jedediah Hotchkiss, who had been a topographical engineer during the Civil War. He later opened an engineering firm in the Shenandoah Valley and taught school in Staunton, Virginia. At the UNC Normal School, he presented a series of lectures during a week's period. He talked one hour during the day and again at night on the geography of different countries, including their topography, fauna, and flora. On Wednesday night, June 25, he spoke about the geography of North Carolina, highlighting its mountain ranges, coastline, water system, relief, peculiar flora, mineral resources, and soils (P., 1879b).

Toward the end of the 1879 session, Frederic W. Simonds gave a series of lectures on natural history. The subject and his style of delivery won the admiration of all in attendance. Joseph Kinsey (1879a), a North Carolina newspaper reporter for the *Kinston Journal*, was impressed by Simonds's lectures, writing: "It [was] truly astonishing what wonderful and strange things

[were] revealed to us by aid of the microscope in water collected from stagnent [*sic*] pools." Simonds also gave at least two lectures on the "Life and Habits of a Few Insects," in which he explained their anatomy, digestion, circulation, and respiration. "The interest manifested in his lectures was evident from the attention paid him," one source reported (Kinsey, 1879b). A reporter for the *Charlotte Democrat* highlighted Simonds's week of lectures on natural history, writing, "The most striking teachings of the week [were] Prof. Symond's [*sic*] daily lectures in the Chapel on the structure of plants, animals, and man" (X., 1879a). Of particular note was the skill and dexterity with which Simonds created illustrations on the blackboard using chalk of various colors and drawing with each hand in succession or both in tandem, all quickly and accurately. Even if Simonds had not uttered a word, the reporter claimed, his drawings would have "told a wondrous tale of the economies of bodily life." A more complete account of Simonds's lectures is given in chapter 7.

On July 11, 1879, Julius L. Tomlinson gave a presentation on California, stressing the wonders of the state, including its "unexampled flora, boundless farms, gigantic trees, and labrynthine caves" (P., 1879c). The excited listeners were captivated by his accounts of the large trees, many of which mentioned were named after distinguished Americans, presidents, and states of our country.

In a lecture of July 16, 1879, Professor Holt talked about the study of botany and then described several curious plants that grew in North Carolina (Anonymous, 1879). He claimed that children who became knowledgeable in the botany or ornithology of their native state would become "better known and remembered 100 years from now than any now prominent politician." He subsequently discussed several unique plants, such as sundews, Venus' flytraps, and pitcher plants.

In addressing his audience on agriculture at the third summer school, State Geologist Washington Caruthers Kerr displayed a map of North Carolina that he had shown at the Vienna Exhibition of 1873. Lettered in German, the large map showed in color the areas of the state where cotton and tobacco were raised as well as coal regions and mean annual temperatures in various locations. In reference to North Carolina's wide variety of climates, he mentioned that the state contained nineteen species of oak out of twenty-two known at that time to grow east of the Rocky Mountains (Anonymous, 1880b). (Kerr [1883] later enumerated the species of oaks occurring in North Carolina.) Kerr noted that the state was blessed with a climate and soil suitable for agriculture; he stimulated thinking among the pupils by asking them questions about these natural resources.

These special lectures expanded the knowledge of science and influenced the thinking of the Normal School pupils who enjoyed the scientific topics. One student praised Professor Carey Dempsey Grandy's lecture on the spectroscope, which was presented in July 1878 (Normal, 1878). The student wrote that Grandy "clearly and forcibly" explained about and placed on exhibit a spectroscope. Then in its infancy, the prism was employed to determine if a "substance under examination [was] a solid, liquid or a gas, and whether that gas [was] under a high or low pressure." The student further reported: "It gives us the cause of the changes of color of the leaves in autumn." In conclusion, Grandy told students that it was not possible to foresee the potential uses of this new method of research. Grandy was described as a "model student and professor."

Normal School students also chose scientific themes for special presentations, which they gave at commencement. In the closing exercises of the 1880 session, for example, Robert A. Southerland read his essay, "Natural History Should By [*i.e.*, Be] Taught in Our Common Schools" (UNC, 1880).

The Normal School was a prototype in the United States. Its claim of being the first school of its type in the country has gone undisputed. Historical documents, however, note that special summer schools had been held before 1877. Among these were summer courses in botany taught by botanist Asa Gray at Harvard in 1871 and 1872 (Love, 1944), a natural history school established by Louis Agassiz at Penikese in Buzzard's Bay, Massachusetts, in the early 1870s, and a biological program at Woods Hole, Massachusetts (Madry, 1925). In writing about the history of the Lawrence Scientific School in Harvard University, James Lee Love (1944) noted that Asa Gray's experiment of teaching courses in botany during the summers of 1871 and 1872 were the first summer courses to be taught in an American college or university. The Normal School at UNC ran for eight consecutive years and instructed 2,480 students (UNC, 1919). Except for the first session, classes typically began around the second or third week of June and continued into the third or fourth week of July. The school was suspended after the 1884 session due to the reassignment of state funds to establish summer schools in several other towns in 1885 (Highsmith, 1923). Also contributing to its closure was the withdrawal in 1884 of funds from the Peabody Fund.

Summer School (1894–1904)

The Normal School reopened as the "Summer School" in 1894. In his UNC presidential report of early 1898, Edwin Alderman noted that the school was "maintained without extra appropriation by the state" (BOT 9: 663). Over 120 teachers had enrolled in the first session (Anonymous, 1894a). Subsequent sessions offered advanced courses as well as general classes. Operating for the next eleven years, the school enrolled 1,541 students before it was suspended in 1904 (UNC, 1919). As with the Normal School, attendees had free admission to the university buildings (Fig. 10.4). Also retained was the emphasis on studies to enhance teachers' backgrounds and to educate those wishing to become teachers. The school was no longer an experimental enterprise, as it had become an established component of the university's educational mission (Alderman, 1894). The school was divided into two departments, the academic and the pedagogic. The former department addressed the educational needs of students who desired a university education and teachers who wanted to gain additional college training. The latter department aimed to establish the pedagogic principles of all disciplines taught in order "to develop with the pupil an orderly, consistent body of educational doctrine" and to demonstrate the scientific methodology of the teaching disciplines of the curriculum (Winston and Alderman, [1895]). Summer School students not affiliated with UNC could obtain certificates[83] that attested to their successful completion of coursework.

On many evenings, public lectures and entertainment were provided in the chapel and gymnasium, respectively. In 1894, talks included discussions of the Suez Canal, the German university, the Shakespeare-Bacon controversy, popular education, and the roads of Europe.

[83] From 1854 to 1915, teachers were certified by examination. Until 1897, each county was responsible for conducting exams and during the subsequent two decades some examinations were administered at the local level and others by the State Board of Examiners (Hillman, 1953).

Fig. 10.4. Scene of UNC campus facing north, ca. 1892. Though this picture precedes the opening date of the Summer School, it serves to show a view of campus as it would have appeared to students.

(North Carolina Collection Photographic Archives, Wilson Special Collections Library, University of North Carolina, Chapel Hill).

Evening entertainment featured music, germans (dancing), and special parties at professors' homes. The night before classes ended on July 28, 1894, UNC professor of modern languages Walter Dallam Toy hosted a "progressive euchre party" at his residence. He was a genial host, serving cream and cake. A prize was awarded to the winning female and male players, the former receiving a tray painted and produced on the peak of Mount Mitchell and the latter a pack of cards (Anonymous, 1894b).

Unlike the Normal School, the Summer School charged a modest fee (around $5.00) and offered a program stronger in the sciences. In 1894, life sciences was well represented in the curriculum, which offered not only botany but also marine zoology in the Summer School of Biology directed by UNC biology professor Henry Van Peters Wilson at Beaufort, North Carolina. The latter coursework, as well as coursework in the Summer School of Geology, could earn students regular university credit of three hours for one term (UNC Faculty 7: 302). Mathematics, physics, and chemistry were subsequently added to the register of summer courses. In 1897, the Summer School of Geology featured excursions in the regions of Chapel Hill, Sanford, Cumnock, Raleigh, and Cary (Anonymous, 1897). A promotional brochure for the sixth (1899) Summer School session mentioned that the science classes offered the previous year had been very popular (UNC, 1899). It noted that "The Scientific Department" would offer classes in elementary and advanced Sloyd[84] and physics taught by Professor William Charles Adam Hammel of the Maryland Normal School; in botany and zoology taught by Professor Robert Ervin Coker of Goldsboro, North Carolina; and in elementary chemistry taught by Professor Charles Baskerville of UNC. Classes in mathematics and geography were also scheduled for the session. Special evening lectures covered scientific

[84] Sloyd was a kind of skilled mechanical work, such as that used in wood carving. Training in sloyd included the practical use of the tools and materials and instruction in the design and interpretation of the plans and specifications.

topics as well. In botany, natural history, and biology, the following courses were given in the Summer School.

Course	Instructor	Year
Botany	Dixie Lee Bryant	1894
Marine Biology[85]	Henry Van Peters Wilson	1894
Botany	Austin Craig Apgar	1896
Nature Study	Wilbur Samuel Jackman	1897
Botany	Austin Craig Apgar	1898
Biology	Robert Ervin Coker	1899
Botany	George Francis Atkinson	1901
Biology	Clarence A. Shore	1901

The instructors were experts in their specialties. Chapters 11, 12, 13, and 14 of this book provide biographical background on instructors who taught botany as well as insight into their courses and methods of instruction.

Summer Term (1900–1901)

In 1900, the university first offered a "Summer Term," which consisted of twenty-one courses for which students could earn full credit. The instruction included four courses in chemistry. The rationales for setting up this program were several. It provided graduate and undergraduate courses in a variety of subjects and offered regular courses during the summer for those unable to attend at other times. The program also made good use of the university physical plant during an otherwise low-use time (Anonymous, 1900). Teachers and students who also enrolled in the Summer School could take courses in the Summer Term free of charge. Two years later, on the recommendation of the faculty (UNC Faculty 8:2) and President Venable, the trustees abolished the Summer Term (BOT 10: 301). No evidence has been found that botany was among the subjects taught at the Summer Term.

Conclusion

At UNC, the summer school sessions of about the last quarter of the nineteenth century provided two unique opportunities for the state's citizens: training to become teachers and democratizing education by allowing women enrollment during the summers. Special lectures at the Normal School introduced students to a variety of scientific subjects; however, the Summer School provided specialized courses in sciences, including botany and natural history, which covered disciplines then being taught in the public schools. For example, botany was a component of the curriculum by at least 1898 (SBE, 1898). Botanical studies were considered important in facilitating the powers of memory and observation early in the life of a child. Using fresh material from field excursions, teachers would stimulate a child's thinking through a series of questions. In fact, Wilbur S. Jackman, the father of the "Nature Study" movement, taught this method of teaching at the 1897 summer school at Chapel Hill.

[85] For coverage of this course, see chapter 9.

The summer sessions at Chapel Hill were very successful. They brought the university closer to the people and hired the foremost educators in the country, providing the best and most current methods of teaching as well as an opportunity to study specialized subjects.

References Cited

(References that conclude with page numbers within parentheses indicate sources from where quotes and pertinent information were obtained.)

Alderman, Edwin A. 1894. The University Summer Schools. *Alumni Quarterly* (University of North Carolina, Chapel Hill) 1(1): [17]–20.

Anonymous. 1877a. Normal School—One hundred and fifty students—Speeches by President Battle and Prof. Ladd—Handshaking and joking—Personal—Teachers taught, &c., p. 53 in Kemp P. Battle, *Scrapbook of Materials Relating to the University of North Carolina, 1869–1893*, vol. 2. (Newspaper clipping, without complete bibliographical information, from the *Tarboro Southerner* [Tarboro, NC], about July 9, 1877). (North Carolina Collection, Wilson Special Collections Library, University of North Carolina, Chapel Hill)

Anonymous. [1877]b. The University Normal School. Valuable statistics—List of teachers in attendance, together with their residence, p. 73 in Kemp P. Battle, *Scrapbook of Materials Relating to the University of North Carolina, 1869–1893*, vol. 2. (Newspaper clipping from the *News* (Raleigh) without bibliographical information). (North Carolina Collection, Wilson Special Collections Library, University of North Carolina, Chapel Hill)

Anonymous. 1877c. The State Normal School. *Observer* (Raleigh) 2(98): [2], Thursday, September 6, 1877.

Anonymous. 1878. The University Normal School. *Weekly Ledger* (Chapel Hill, NC) 1(9): 2–3, Saturday, June 15, 1878. ("From the *University Magazine* [University of North Carolina, Chapel Hill.]") (Cornelia Phillips Spencer most likely wrote the article.)

Anonymous. 1879. [Prof. Holt on curious plants]. *Weekly Ledger* (Chapel Hill, NC) 3(13): 1, Saturday, July 12, 1879.

Anonymous. 1880a. The University Normal School, p. 42 in Kemp P. Battle, *Scrapbook of Materials Relating to the University of North Carolina, 1869–1893*, vol. 1. (Newspaper clipping from the *News* [Raleigh], published about June 28, 1880). (North Carolina Collection, Wilson Special Collections Library, University of North Carolina, Chapel Hill)

Anonymous. 1880b. From Chapel Hill. *Kinston Journal* (Kinston, NC) 2(29): [2], Thursday, July 8, 1880.

Anonymous. 1894a. University Summer School, Chapel Hill, N.C., July 3. *News-Observer-Chronicle* (Raleigh) 36(141): [3], Wednesday, July 4, 1894.

Anonymous. 1894b. The Summer School, Chapel Hill, July 28. *News-Observer-Chronicle* (Raleigh) 37(7): [3], Sunday, July 29, 1894.

Anonymous. 1897. The Summer School of Geology. *University Record* (University of North Carolina, Chapel Hill) 1(4): 46.

Anonymous. 1900. The Summer Term. *University Record* (University of North Carolina, Chapel Hill) 5(3): 28.

Battle, Kemp P. (compiler). 1869–1893. *Scrapbook of Material Relating to the University of North Carolina, 1869–1893*. 8 volumes. ("Made up largely of newspaper clippings. The arrangement is approximately chronological, though in many cases date and name of the newspaper are lacking"—card catalog record) (North Carolina Collection, Wilson Special Collections Library, University of North Carolina, Chapel Hill)

Battle, Kemp P. 1878a. *Report of Hon. Kemp P. Battle, President of the University of N.C. on the State of the University and the Normal School*. Daily and Weekly News Steam Book and Job Print, Raleigh.

Battle, Kemp P. 1878b. [Report on the Normal School, University of North Carolina, Chapel Hill, Sept. 1, 1878], p. 8 in Kemp P. Battle, *Scrapbook of Materials Relating to the University of North Carolina, 1869–1893*, vol. 1. (Newspaper clipping without bibliographical information). (North Carolina Collection, Wilson Special Collections Library, University of North Carolina, Chapel Hill)

Battle, Kemp P. [1878]c. *Reports of the University Normal School of 1878*. [no statement of publisher or place]. (North Carolina Collection, Wilson Special Collections Library, University of North Carolina, Chapel Hill)

Battle, Kemp P. 1881. *Annual Report of the North Carolina University Normal School for 1880*. P.M. Hale and Edwards, Broughton & Co., State Printers and Binders, Raleigh.

BOT. Board of Trustees, University of North Carolina, Chapel Hill. *Minutes of the Board of Trustees*. (University Archives #40001, Wilson Special Collections Library, University of North Carolina, Chapel Hill) (Vol. 9, April 1891–January 1898; vol. 10, May 1898–April 1904)

Highsmith, Edwin M. 1923. The state normal school idea in North Carolina. *Meredith College Quarterly Bulletin*, series 17, no. 1, pp. 5–38. (p. 21)

Hillman, James E. 1953. The story of teacher education and certification in North Carolina. *Quarterly Review of Higher Education among Negroes* 21: 20–37. (p. 21)

Kerr, Washington C. [1883]. The oaks of North Carolina. *Weekly Star* (Wilmington, NC) 14(43): [3], Friday, August 24, 1883.

Kinsey, Joseph. 1879a. Chapel Hill Normal School. *Kinston Journal* (Kinston, NC) 1(30): [2], Thursday, July 10, 1879.

Kinsey, Joseph. 1879b. Chapel Hill Normal School, July 1, 1879. [continued]. *Kinston Journal* (Kinston, NC) 1(31): [2], Thursday, July 17, 1879.

Lafferty, John Wilson. 1883–1885. Diary of John Wilson Lafferty. (Miscellaneous Papers #517.57, Southern Historical Collection, Wilson Special Collections Library, University of North Carolina, Chapel Hill). (pp. [11–15])

Leloudis, James L. 1996. *Schooling the New South; Pedagogy, self, and society in North Carolina, 1880–1920*. (*The Fred Morrison Series in Southern Studies*). University of North Carolina Press, Chapel Hill. (p. 74)

Love, James Lee. 1944. *The Lawrence Scientific School in Harvard University 1847–1906*. [The Author], Burlington, NC.

Madry, Robert. 1925. University has pioneer record—Not only first State University but led way to summer sessions. *News and Observer* (Raleigh) 122(61): [1], Sunday, August 30, 1925.

Normal. 1878. [Letter to the editor, Chapel Hill, July 19th 1878]. *Hillsborough Recorder* (Hillsborough, NC), new series 6(35): [1], Wednesday, August 7, 1878.

A Normalite. 1881. Letter from Chapel Hill, p. 11 in Kemp P. Battle, *Scrapbook of Materials Relating to the University of North Carolina, 1869–1893*, vol. 3. (Newspaper clipping of a letter from "Chapel Hill, N.C., July 9th, 1881," without bibliographic information). (North Carolina Collection, Wilson Special Collections Library, University of North Carolina, Chapel Hill)

P., R.P. 1879a. The State Normal School, Chapel Hill, June 21, 1879. *Observer* (Raleigh) 5(154): [2], Wednesday, June 25, 1879.

P., R.P. 1879b. The State Normal School, Chapel Hill, N.C., June 26, 1879. *Observer* (Raleigh) 5(154): [3], Saturday, June 28, 1879.

P., R.P. 1879c. The State Normal School, Chapel Hill, July 17, 1879. *Observer* (Raleigh) 6(18): [3], Sunday, July 20, 1879.

Robinson, David, Stephen Herbert, and Richard Crangle (editors). 2001. *Encyclopaedia of the Magic Lantern*. The Magic Lantern Society, London. (pp. 89–91, 174)

Russell, Lucy Phillips. [1957]. The First Normalites, pp. 112–118 in *A Rare Pattern*. University of North Carolina Press, Chapel Hill.

S., R. 1880. The University Normal School. *Kinston Journal* (Kinston, NC) 2(26): [2], Thursday, June 17, 1880.

SBE. 1898. State Board of Examiners, North Carolina. *A Course of Study for Teachers in the Public Schools of North Carolina, together with Suggestions on Methods of Instruction*. Guy V. Barnes, Printer to the Council of State, Raleigh. (pp. 46–48)

Sears, B[arnas]. 1877. Letter of August 18, 1877 to Kemp P. Battle. (Box 6, folder 84, Battle Family Papers #3223-A, Southern Historical Collection, Wilson Special Collections Library, University of North Carolina, Chapel Hill)

S[pencer], C[ornelia] P. 1877a. Ode to the Normal School. *Observer* (Raleigh) 2(78): [3], Tuesday, August 14, 1877.

S[pencer], C[ornelia] P. 1877b. The University and Normal School. *North Carolina Presbyterian* (Wilmington, NC), new series, 10(493): [2], June 20, 1877.

S[pencer], C[ornelia] P. 1877c. University Normal School, Chapel Hill, June 22, 1877. *Observer* (Raleigh) 2(36): [2], Sunday, June 24, 1877.

S[pencer], C[ornelia] P. 1877d. University Normal School, Chapel Hill, July 3, 1877. *Observer* (Raleigh) 2(45): [2], Friday, July 6, 1877.

S[pencer], C[ornelia] P. 1877e. University Normal School. Sixteenth day. *Observer* (Raleigh) 2(58): [2], Saturday, July 21, 1877.

S[pencer], C[ornelia] P. 1877f. University Summer School. *North Carolina Presbyterian*, new series 10(500): [1], August 8, 1877.

S[pencer], C[ornelia] P. 1877g. University Normal School, Chapel Hill, Aug. 9, 1877. *Observer* (Raleigh) 2(76): [2], Saturday, August 11, 1877.

S[pencer], C[ornelia] P. 1877h. University Normal School. Resume of work—Closing exercises. *North Carolina Presbyterian*, new series 10(501): [2], August 15, 1877.

Thompson, Luly. 1878. Luly Thompson's lecture notes from the 1878 UNC Normal School in George Nicholas Thompson, "Diary of George Thompson, Leasburg, North Carolina, at University of North Carolina—1851—." (Southern Historical Collection #2367-z, Wilson Special Collections Library, University of North Carolina, Chapel Hill). (Luly Thompson's lecture notes from the 1878 UNC Normal School conclude the diary. There are two versions of the diary: the original, volume 1; a typescript, volume 2.)

UNC. 1880. University of North Carolina. *University Normal School of 1880, Closing Exercises.* [University of North Carolina, Chapel Hill.] (Folder 1741, University Papers #40005, University Archives, Wilson Special Collections Library, University of North Carolina, Chapel Hill)

UNC. 1899. "University of North Carolina Summer School, June 20–July 15, [1899]." University of North Carolina, Chapel Hill. (Announcements, bulletins, programs, etc, North Carolina Collection, Wilson Special Collections Library, University of North Carolina, Chapel Hill)

UNC. 1919. University of North Carolina. Announcement of the thirty-second session of the Summer School, June 24–August 8, 1919. *University of North Carolina Record*, Number [164]. Published by the University, Chapel Hill. (p. [7])

UNC Faculty. University of North Carolina, Chapel Hill Faculty. *Minutes of Faculty Meetings.* (University Archives #40106, Wilson Special Collections Library, University of North Carolina, Chapel Hill) (Vol. 7, 1885–1901; vol. 8, 1901–1919)

Vance, Z.B. 1877. Letter of July 6, 1877 to Cornelia P. Spencer. (Letters and Writings of Cornelia Phillips Spencer, Cornelia Phillips Spencer Papers #683, April 1876–August 1877, transcript vol. 9. Southern Historical Collection, Wilson Special Collections Library, University of North Carolina, Chapel Hill). (Photocopy of original letter in the State Archives of North Carolina, Cornelia Phillips Spencer Personal Collection, number 24)

Vance, Z.B., J.C. Scarborough and K.P. Battle. 1877. University Normal School. To the teachers of the State and those wishing to become teachers. *Observer* (Raleigh) 1(25): [3], Tuesday, May 22, 1877.

Winston, Geo. T. and Edwin A. Alderman. [1895]. *University of North Carolina. Chapel Hill, N.C.* (Broadside announcing the Summer School for 1895; North Carolina Collection, Wilson Special Collections Library, University of North Carolina, Chapel Hill)

X. 1878. The State Normal School at Chapel Hill. *Charlotte Democrat* (Charlotte, NC) 27(1343): [2], Friday, August 16, 1878.

X. 1879a. Correspondence of the *Charlotte Democrat*. University of North Carolina, Chapel Hill, July 12th, 1879. *Charlotte Democrat* (Charlotte, NC) 28(1390): [2], Friday, July 18, 1879.

X. 1879b. For the *Charlotte Democrat*. University of North Carolina, Chapel Hill, July 5th, 1879. *Charlotte Democrat* (Charlotte, NC) 28(1389): [2], Friday, July 11, 1879.

Dixie Lee Bryant

Chapter 11.

Dixie Lee Bryant (1862–1949)

Instructor of Physical Geography and Botany, Summer School (1894)

Early Years and Education

Born in Louisville, Kentucky, on January 7, 1862, Dixie Lee Bryant was the daughter of Butler and Martha Jane (Woolls) Bryant. Her father was a silversmith, jeweler, and keeper of a millinery; her mother a milliner and dressmaker. In the late 1860s, Bryant and her family moved to Columbia, Tennessee, where she grew up. There, she attended the Columbia Female Institute,[86] graduating around 1878. Bryant subsequently taught in a country school with combined grades in Culleoka, Tennessee, and in the first public school in Columbia. She then taught at Hamilton Female College, a finishing school for young ladies in Lexington, Kentucky. According to the annual catalogs of the college, Bryant was an instructor of history, art criticism, and English literature for the school years 1884–1885 through 1886–1887.

Up to this point, her education had not been too unusual for a fairly privileged American woman of the late nineteenth century. Bryant's education was directly influenced by the trends in women's education of the nineteenth century. Emma Hart Willard established the first American foundation for educating women, setting up the Troy Female Seminary in New York in 1821. Subsequently, other institutions were formed for women, including academies, seminaries, and colleges. Mount Holyoke Female Seminary (later Mount Holyoke College), which opened its doors in 1837, was a pioneer in the movement of higher education for women. It was particularly strong in teaching science (Rossiter, 1989). Between 1850 and 1880, several opportunities developed in offering higher education to women, including "denominational coeducational colleges, coordinate single-sex colleges (*i.e.*, a women's college associated with a men's college), secular coeducational public and private universities, and public single-sex vocational institutions" (Glazer-Raymo, 2002). Other factors also broadened the educational opportunities for women. The Morrill Act of 1862 emphasized the importance of higher education and promoted the founding of state colleges throughout the United States. The Morrill Act of 1890 furthered the cause by requiring federal funds to be equitably distributed between white and black campuses. When the Civil War ended, abolitionists (male and female) were concerned about the role of women in society. Suffragists linked efforts to give women voting rights to the goal of

[86] William B. Turner (1955) provides a short history of the Columbia Female Institute, which had its beginning in the early years of Columbia, Tennessee. Of particular interest is the fact that Dixie Lee's sister, Mary A. Bryant, was a teacher at the institute for nearly fifteen years (about 1875 to 1890), including several years as its principal. The institute closed in 1932, a casualty of the Depression, and was destroyed by a fire about 1959 (Duncan, 2002).

Frontispiece Credits: (above) Portrait of Dixie Lee Bryant, 1891. (Photographic Prints Collection, 104.9, Martha Blakeney Hodges Special Collections and Archives, University of North Carolina at Greensboro). (below) Photograph of Main Building, Science Laboratory, ca. 1893. (Photographic Prints Collection, 104.3.210.003, Martha Blakeney Hodges Special Collections and Archives, University of North Carolina at Greensboro).

giving them entry to higher education. Teachers, who were principally women by 1890, were also needed in public schools.

The establishment of women's colleges during the last half of the nineteenth century responded to a need to provide higher education to women, who were not admitted to most colleges (Harwarth, 2002). Among the colleges that later opened for the purpose of educating women were Vassar (1865), Wellesley (1870), Smith (1871), Radcliffe (1879), Bryn Mawr (1885), Women's College of Baltimore, now Goucher College (1885), and Barnard (1889) (Rossiter, 1989).

Oberlin College opened in 1833 and was the first coeducational college in the country. During the nineteenth century, a number of noted colleges opened their doors to women, including Cornell (1870), the University of Chicago (1892, its opening year), the University of Michigan (1870), and the University of Pennsylvania (1880) (Rossiter, 1989). By the beginning of the twentieth century, 61 percent of degree-granting colleges admitted women (Glazer-Raymo, 2002).

Bryant desired to enlarge her educational horizons in geology, but no southern university would accept women. She therefore sought an education in the North. Upon passing a qualifying exam, Bryant enrolled at the Massachusetts Institute of Technology in 1887 to study natural history (Bryant, 1905). MIT had admitted its first female student in 1870; by the mid-1890s, women made up nearly 6 percent of the student body (AMA, 1998–2005). In completing her degree, she wrote a thesis, "A Study of the Most Recent Geological History of the Tide-Water Region of Charles River," which is preserved in the archives at MIT. Bryant was the first student to earn a B.S. in geology at MIT after it began to be offered in 1890 (Shrock, 1977–1982). This was a remarkable accomplishment for a woman of the late nineteenth century (Bowles, 1967). She was among the earliest American women to obtain a college education in science and to develop a science curriculum at a southern college. Coordinate colleges were established at public universities at Virginia, North Carolina, South Carolina, and Georgia (McCandless, 1992).[87] Women were accepted into the regular program in 1897 provided they were postgraduates. This topic is discussed later in the chapter.

After Bryant's graduation, the New Hampshire State Normal School at Plymouth (now Plymouth State University) hired her to revise the curriculum in science (Bryant, undated). A year later, she became one of seven charter faculty members of the newly established Normal School in Greensboro, North Carolina. The Normal and Industrial School in Greensboro was established by the North Carolina General Assembly on February 18, 1891 (NC Laws, 1891). When the school opened its doors to students on October 5, 1892, it became "North Carolina's first state college for women" (Trelease, 2004). One of three applicants for the chair of Natural Sciences, Dixie Lee Bryant was elected to the position with a starting salary of $900.00 (BOD 1: 11). She mostly taught geology, biology, and physiography (physical geography). During her first semester of teaching in the fall of 1892, she commenced pioneering work in developing the science curriculum. After selling "the idea of lab teaching" to Greensboro Normal School president Charles D. McIver (Bryant, [1940]a), she developed laboratories

[87] Coordinate programs typically shared the facilities and faculties—but not the educational and social programs—of male colleges with female students.

in biology (Fig. 11.1), physics, and chemistry, the last mentioned laboratory being "the first such laboratory for women in the state" (Bowles, 1965). Although Bryant also developed lab facilities for physics, she never taught the subject (Bryant, [1940]a).

Fig. 11.1. Photograph of Main Building, Science Laboratory, late 1890s. Note Dixie Lee Bryant (standing center).

(Photographic Prints Collection, 104.3.210.005, Martha Blakeney Hodges Special Collections and Archives, University of North Carolina at Greensboro).

Bryant also furthered the welfare of women students. Her early students arrived at college thoroughly chaperoned. She initiated science courses for women and did away with the contemporary practice of not permitting women to answer and ask questions (Schuber, 1950). She developed academic skills in her students, such as teaching them how to take notes and make drawings during lectures and in laboratory sessions.

A number of publications on Bryant provide background on her life and her contributions in education (Bowles, 1967; Camp, 1955; Pearson, 1904; Shrock, 1977–1982; and Lawrimore, 2014). Archival records on Bryant that are housed at UNC-Greensboro (UNCG-A) contain correspondence, manuscripts, newspaper clippings, and photographs.

Teaching at the UNC Summer School

When UNC's Summer School reopened in 1894, Bryant joined a distinguished roster of faculty and instructors. She was selected to teach physiography and botany, becoming the first woman to teach botany at UNC. Superintended by Edwin A. Alderman, then UNC

Professor of History and Philosophy of Education, the school (Fig. 11.2) took place from July 2 to July 28. In the UNC catalog of 1893–1894 (p. 82), Bryant's courses are listed under the Department of Pedagogics, reflecting the fact that the focus of the educational program was on the school curriculum. According to the guidelines of the school, courses typically consisted of twenty lessons that involved lectures, the use of textbooks, parallel reading, and topical reports. Instructors were also expected to spend time conferring with the teachers and students.

A reporter for the *Raleigh News-Observer-Chronicle* gave a brief account of Bryant's botany course (H., 1894): "[Her] class of young girls [started] with the seed, the structure of which was gotten. By experiment the food found in the seed were determined. The movements of the germinating seed were observed, also what part of the seed became what part of the plant. The starch of the dry seed is turned into the sugar of the growing seed and thus the plant has power to feed itself." The reporter continued: "The plant was examined throughout its whole life and other plants compared with the type. In this way the girls laid for themselves the foundation for further scientific investigation and acquired the powers of intelligent analysis of flowers." Whether or not Bryant took her botany class into the field is unknown; however, her class in physical geography spent much of its class time on fieldwork.

Additional insight into Bryant's method of teaching botany may be inferred from information on how she presented the subject at the Normal School in Greensboro (Bryant, [undated]). There, she used Asa Gray's *School and Field Book of Botany* (1887), as well as one of Alphonso Woods's textbooks,[88] which also was used for identifying North Carolina plants. Fieldwork consisted of collecting plants, which were pressed to create a herbarium of twenty or thirty specimens. Lectures covered cells, cell structure, and flowerless plants. In explaining the cryptogams, Bryant used charts (measuring three feet by four feet) that student Bessie Battle had copied from Goebel's *Outlines of Classification and Special Morphology* (1887). In macroscopic laboratory work on plants, students discussed their observations in recitations and recorded their observations (including drawings) in notebooks (UNCG, 1898).

Bryant returned to the UNC campus in 1895 and 1897 as a student in the Summer School. Her name appears in a printed roster of students for the 1895 session, and her signature appears in the UNC Student Records for 1897, noting an arrival on June 23 (UNCA-2). In addition to enrolling in a class, another possible attraction to attend the 1897 Summer School may have been the presence of her sister, Mary A. Bryant, who was teaching English at the Summer School. Mary Bryant, then from the Ogontz School near Philadelphia, Pennsylvania, was a popular lecturer who returned to teach at UNC the subsequent summer. Dixie Lee gave a presentation on Saturday evening, July 3, in honor of Mary following a secret meeting of the attendees' literary societies held in the university chapel. The assembled group consisted of twenty-five to thirty young women who had attended the Normal School in Greensboro, North Carolina, and then gone on to teach in different schools of the state (Anonymous, 1897a; Anonymous, 1897e; Anonymous, 1897f).

[88] Bryant does not cite the title of the book by Alphonso Wood, who was a prolific writer of botanical textbooks in the latter half of the nineteenth century. She may have used his *Leaves and Flowers, or, Object Lessons in Botany. With a Flora* (A.S. Barnes & Company, New York and Chicago, 1888).

A SUMMER SCHOOL FOR TEACHERS

at the

University of North Carolina.

A Summer School for Teachers and Students who desire to strengthen themselves in knowledge and professional skill, will be held from July 2 to July 28, in the University buildings, at Chapel Hill, N. C.

The Tuition Fee is $5.00, which admits to all the instruction.* The Faculty includes Sixteen Professors, selected from the Faculties of the Universities of North Carolina, Texas and Louisiana, the State Normal and Industrial School for Women, and the Graded Schools of Wilmington, Raleigh and Goldsboro. **The following subjects will be taught:**

Latin,	**Greek,**	**French,**	**German,**	**English Language and Literature,**
Anglo-Saxon, Civics,		**History,**	**Pedagogics,**	**History of Education,**
Arithmetic,	**Geography,**	**Algebra,**	**Geometry,**	**Road Construction.**
Land Surveying,		**Botany,**	**Primary Reading,**	**Political Economy.**

Each Professor will devote a certain portion of time to conference with teachers concerning their special needs.

Excellent opportunities will be afforded for reading and private study, under intelligent guidance, in the University Library, comprising 35,000 volumes.

Methods of teaching will be discussed theoretically and illustrated practically by means of model classes.

Board and furnished room may be had for the month at the hotels for $15.00, more cheaply at clubs and private houses.

Reduced rates will be granted by all the railroads.

For circulars with full details, address

GEO. T. WINSTON, *President of the University.*

OR EDWIN A. ALDERMAN, *Superintendent of Summer School.*

CHAPEL HILL, N. C.

***The Schools of Biology (at Beaufort) and Geology (at King's Mountain) begin their sessions on June 14th, and the Tuition Fee in these courses is $10.00**

CHAPEL HILL, N. C.: PRINTED AT THE UNIVERSITY PRESS.

Fig. 11.2. Circular for "A Summer School for Teachers at the University of North Carolina," 1894.

(Announcements, Bulletins, and Programs, North Carolina Collection, Wilson Special Collections Library, University of North Carolina, Chapel Hill).

Bryant's association with UNC becomes more intriguing. In the UNC catalog of 1897–1898 (p. [81]), she is listed as a student in the Graduate School. Edwin A. Alderman, who was then UNC president, was a proponent of allowing women to enroll in postgraduate courses. At UNC, the Normal School (1877–1884) and Summer School (1894–1904) included both sexes. Alderman's advocacy for allowing women to take postgraduate courses was followed by a resolution made at the February 18, 1897, meeting of the Board of Trustees. It stated "that the Post Graduate courses at the University [would] be opened to women, under such regulations as the Faculty may prescribe" (BOT 9: 561).[89] In practice, women with at least a two-year college education could be admitted for postgraduate study. In Alderman's annual report, he mentioned that "five young women[90] have been enrolled in the University this [academic] year [*i.e.,* 1897–1898], one, a graduate of Massachusetts School of Technology [*i.e.,* Dixie Lee Bryant], in the graduate department" (BOT 9: 659). According to Pamela Dean ([1987]), the five women who enrolled in 1897 were Dixie Lee Bryant, Cecye Roanna Dodd, Mary McRae [*i.e.,* MacRae], Sallie Walker Stockard, and Lulie Watkins. A reporter for *The Tar Heel*, the university newspaper, noted that the opening session of the 1897–1898 academic year "mark[ed] the beginning of the history of the University as a co-educational Institution" (Anonymous, 1897b). The reporter also noted that Bryant was "pursuing an advanced course in Geology." In this article, the attendance of only four women was mentioned; Lulie Watkins was omitted.[91] A reporter for the *News and Observer* in Raleigh, North Carolina also mentioned that "four young women" had entered the university (Anonymous, 1897c). Yet if Bryant indeed registered for classes, it is doubtful that she completed them in her first semester at UNC, as her name does not appear in the student grade books for that period of time (UNCA-1). Furthermore, a newspaper reporter (Anonymous, 1897g) noted that in October 1897 Bryant "has gone to Greensboro to resume her work as Professor of Geology and Biology in the Normal and Industrial College."

Other evidence casts doubt that Bryant was among the original four or five women admitted to UNC in 1897. In 1922, the Women's Association of UNC compiled a twenty-nine-page pamphlet listing "every girl who ha[d] studied at the University since Mary MacRae registered twenty-five years ago" (UNC-WA, 1922). Bryant's name is not listed in this roster. Additionally, there is no mention of Bryant receiving a leave of absence from her teaching duties in Greensboro that would have permitted her to matriculate at UNC in the fall of 1897 (BOD, vol. 1). A letter from Bryant to Anna Maria Gove, who was her classmate at MIT and later the resident physician at the Greensboro Normal School, sheds additional light on this topic. Bryant ([1940]b) wrote of escorting the very first female student to UNC and of the student's unfortunate failure to graduate: "I chaperoned the first girl to appear on the grounds, to keep the boys from hazing her, also she did not make good in her studies, was perhaps totally unprepared, and left probably by request or advice, and they have taken her name off of their books." The result, she explained, was that "the University insists that Mary MacRae was the first [female] student." Perhaps it was Bryant's role as chaperone that caused her to be listed as among the first female students at the university.

[89] In its advertisements to attract prospective students, the university included information that graduate courses now were open to women (Anonymous, 1897d).

[90] UNC sophomore Needham Ward (1897) also mentioned that five ladies registered on campus, including one who enrolled in his mathematics class. He wrote: "It seem[ed] real odd to have a lady in the class with such a crowd of boys."

[91] An anonymous (1932) correspondent reported that Lulie Watkins arrived on campus in October 1897.

Later Years

At a meeting on May 21, 1901, the directors at the Greensboro school granted Bryant a three-year leave of absence from her teaching responsibilities to pursue graduate studies in Germany (BOD 2: 181). One of the stipulations made in granting the leave was that upon her return to the Normal School she would accept the same salary she was receiving when she left. Professor T. Gilbert Pearson of Guilford College filled the temporary vacancy (Anonymous, 1901a). Bryant set sail for Europe on June 13, 1901, accompanied by her colleague Miss Bertha Lee, who was traveling to Germany and Norway (Anonymous, 1901b).

In her first year overseas, Bryant studied microscopical petrography at the University of Heidelberg under the tutelage of Professor Harry Rosenbusch (Anonymous, 1949b). She then continued her mineralogical studies for two years under Professor of Mineralogy and Geology Hans Lenk (Fig. 11.3, left image) at the Friedrich Alexander University in Erlangen, where she also studied with professors Eilhard Ernst Gustav Wiedemann (Fig. 11.3, middle image), a physicist, and Hans Solereder (Fig. 11.3, right image),[92] a plant anatomist and director of the Erlangen botanical garden (Anonymous, 1949b; Bryant, 1905). Graduating *magna cum laude*, she earned a Ph.D. in June 1904 (Anonymous, 1904b)[93] and the distinction of being the first woman granted the doctorate at the Friedrich Alexander University (Anonymous, 1904a; Anonymous, 1904b). Her dissertation, "Beiträge zur Petrographie Spitzbergens" (Contributions to the Petrography of Spitzbergen) (Fig. 11.4), was published in 1905 (Bryant, 1905).

Fig. 11.3. Dixie Lee Bryant's professors in Germany. (left) Hans Lenk (Courtesy of Erlangen University Library, Portrait Collection, SUC 30029); (middle) Eilhard Ernst Gustav Wiedemann (Courtesy of Erlangen University Library, Portrait Collection, SUC 30028); and (right) Hans Solereder (Courtesy of Hunt Institute for Botanical Documentation, Carnegie Mellon University, Pittsburgh, Pennsylvania).

Returning to the Normal School in Greensboro in the fall of 1904, Bryant had the distinction of being the first person to serve there with a Ph.D. (Bowles, 1965). According to the terms of her leave of absence, her salary remained at $1,200. Nearly a year later, Bryant requested that her annual salary be set at $1,500. Although the Board of Directors thought the financial

[92] Cited as Colerader (Anonymous, 1949b).

[93] In this article, it is mentioned that Professor Roth in Munich calculated that then nearly 2,500 American women were furthering their education in Germany.

condition of the school precluded such an increase, President Charles McIver (1905) offered to reduce her teaching responsibilities by eliminating her work in the freshman class. Bryant apparently declined the offer and moved to Chicago, Illinois, where she began teaching science in the high schools of that city in 1905. Retiring in 1931, she relocated to Asheville, North Carolina. She remained active in civic and intellectual affairs (Camp, 1955) and was a member of Trinity Episcopal Church in Asheville. The nearly sixty letters that Bryant wrote

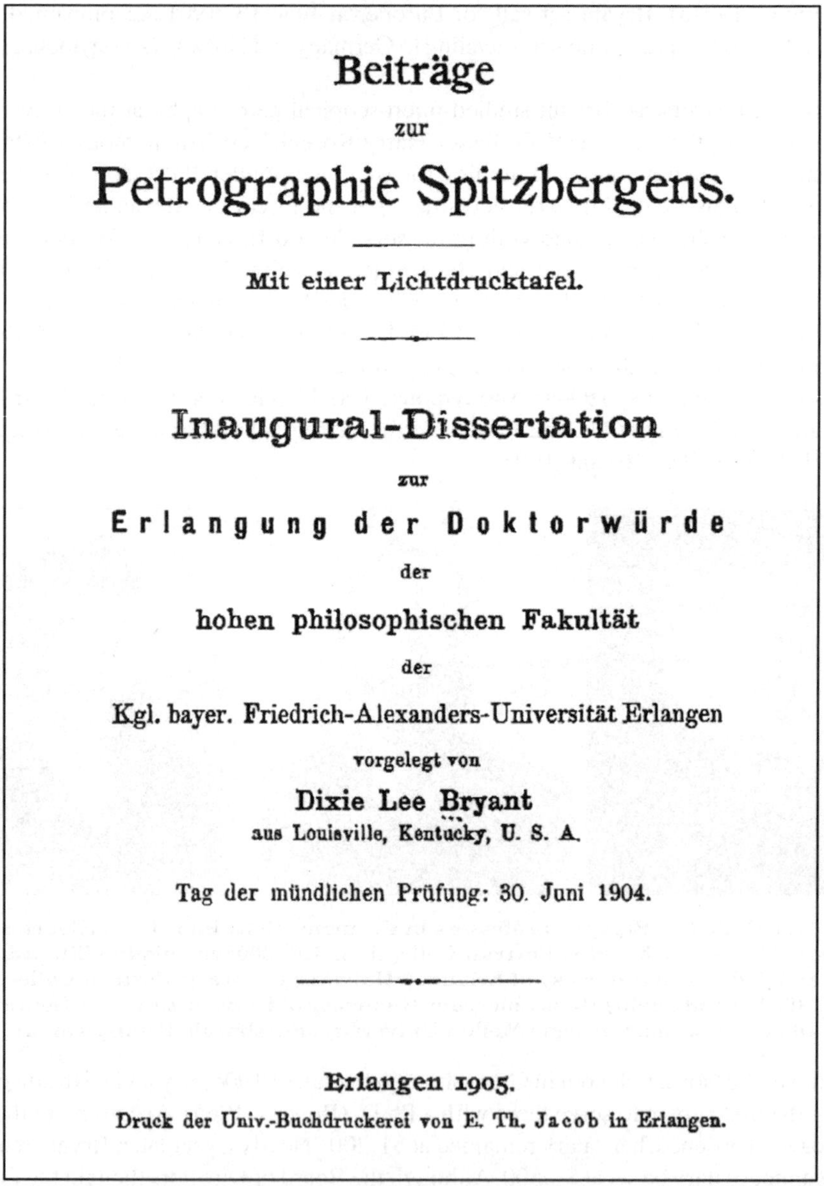

Beiträge
zur
Petrographie Spitzbergens.

Mit einer Lichtdrucktafel.

Inaugural-Dissertation
zur
Erlangung der Doktorwürde
der
hohen philosophischen Fakultät
der
Kgl. bayer. Friedrich-Alexanders-Universität Erlangen
vorgelegt von
Dixie Lee Bryant
aus Louisville, Kentucky, U. S. A.

Tag der mündlichen Prüfung: 30. Juni 1904.

Erlangen 1905.
Druck der Univ.-Buchdruckerei von E. Th. Jacob in Erlangen.

Fig. 11.4. Title page from *Beiträge zur Petrographie Spitzbergens,* doctoral dissertation by Dixie Lee Bryant (1905).

(Courtesy of Sterling Memorial Library, Yale University).

to her friend Anna M. Gove between 1932 and 1947 reveal that she engaged in reading and gardening, compiled her family genealogy, attended various organizational meetings, and participated in events of the American Association of University Women (Gove Collection, 1864–1952). During the early years of World War II, she visited the American Red Cross, where she made bandages for the soldiers of Fort Bragg, North Carolina (Bryant, 1941).

Legacy and Contributions

Earning several honors, Bryant was particularly proud of being elected to the American Association of University Women (then known as Collegiate Alumnae) in 1891. When the South Atlantic Regional Conference of the AAUW took place in Asheville in 1948, a tribute to Bryant was made. Two years later, local members of the organization designated a study grant as the Dixie Lee Bryant Fellowship, honoring not only her steadfast association with the organization but also her devotion to research and scholarship. She was also awarded honorary membership in the Gamma Chapter of Delta Kappa Gamma on November 8, 1947 (Camp, 1955).

On November 18, 1949, Dixie Lee Bryant, age 87, died in an Asheville, North Carolina, hospital after a prolonged illness (Anonymous, 1949a; Anonymous, 1949b). Having remained single her entire life, she was survived by two nieces in Texas, Mrs. H.P. Lowe of Dallas and Miss Selby Attwell of Weatherford, as well as a nephew, L.A. Bryon of Cape Girardeau, Nevada. She is buried in Riverside Cemetery (Fig. 11.5) in Asheville.

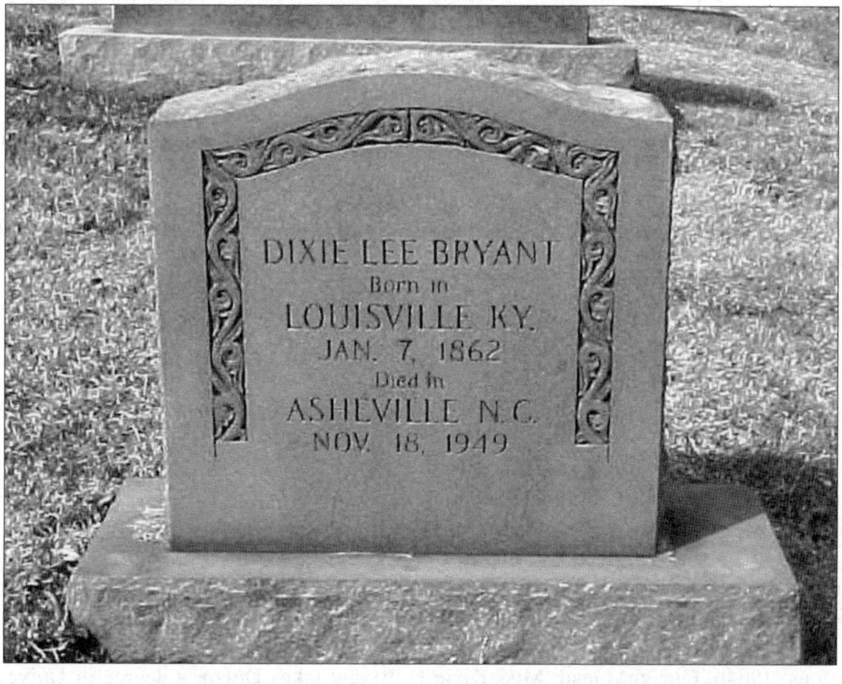

Fig. 11.5. Gravestone of Dixie Lee Bryant, Riverside Cemetery, Asheville, North Carolina, Section Y, Lot 2, Grave 75, February 20, 2006.

(Courtesy of David Olson, Riverside Cemetery).

In the field of education, Bryant was a trailblazer: the first person to obtain a baccalaureate in geology at MIT, the first woman to receive a doctorate at Friedrich Alexander University in Erlangen, Germany, and the first Ph.D. graduate to teach at the Normal School in Greensboro. In 1894, Bryant taught the first course devoted to botany in the UNC Summer School, and she was the first woman to teach botany at UNC. She established a science program with laboratories at the Normal School in Greensboro. When she was asked to comment about her life, Bryant modestly responded with a sweeping affirmation of all the women who had advanced the cause of women in higher education (Camp, 1955). She herself was a significant contributor to that movement.

References Cited

(References that conclude with page numbers within parentheses indicate sources from where quotes and pertinent information were obtained.)

AMA. Association of MIT Alumnae. 1998–2005. "Celebrating 125 years of women at MIT, 1873–1998. Technology Women." <http://72.14.207.104/search?q=cache:TM7BXf0gIpgJ:www.mit-amita.org/esr/practitioners.html+mit+women+admitted&hl=en&gl=us&ct=clnk&cd=9> (Accessed May 6, 2006)

Anonymous. 1897a. At Chapel Hill. *News and Observer* (Raleigh) 42(121): 4, Thursday, July 22, 1897.

Anonymous. 1897b. Co-Education. *The Tar Heel* (University of North Carolina, Chapel Hill) 6(1): [3]. September 24, 1897.

Anonymous. 1897c. Over 500 at University. *News and Observer* (Raleigh) 43(9): [1], Tuesday, September 14, 1897.

Anonymous. 1897d. The University. *News and Observer* (Raleigh) 42(112): 4, Sunday, July 11, 1897.

Anonymous. 1897e. University Summer School. *News and Observer* (Raleigh) 42(109): 4, Thursday, July 8, 1897.

Anonymous. 1897f. University Summer School. *News and Observer* (Raleigh) 42(123): 2, Saturday, July 24, 1897.

Anonymous. 1897g. [Miss Dixie Lee Bryant], p. 3 in Here and there. *The Tar Heel* (University of North Carolina, Chapel Hill) 6(4): [3], Friday, October 15, 1897.

Anonymous. 1901a. Prof. Pearson returns. *Greensboro Record* (Greensboro, NC), Tuesday, August 20, 1901.

Anonymous. 1901b. To build observation school. *Greensboro Telegram* (Greensboro, NC) 8(105): [5], Friday, June 7, 1901.

Anonymous. 1904a. Dixie Lee Bryant, Ph.D. *State Normal Magazine* 9(2): 30–31. [Managing editor, T. Gilbert Pearson]

Anonymous. 1904b. Our girls lead. Miss Dixie L. Bryant takes Doctor's degree in University of Bavaria. *News and Observer* (Raleigh) 55(113): 11, Sunday, July 17, 1904.

Anonymous. 1932. First women students scorned by resentful and jealous men. *The Daily Tar Heel* (University of North Carolina, Chapel Hill) 41(65): [1], [4], Sunday, December 11, 1932.

Anonymous. 1949a. Former teacher at college here is buried today. *Greensboro Record* (Greensboro, NC) 59(278): Section C: 8, Monday, November 21, 1949.

Anonymous. 1949b. Rites will be held Monday for Dr. Bryant. *Asheville Citizen-Times* (Asheville, NC) 20(12): Section C: 9, Sunday, November 20, 1949.

BOD. Board of Directors. *Proceedings, Board of Directors, State Normal & Industrial School, Greensboro, North Carolina*, vol. 1 (June 9, 1891–May 24, 1898) and vol. 2 (December 1, 1898–January 2, 1909). (University Archives and Manuscripts, Walter Clinton Jackson Library, University of North Carolina at Greensboro). (Vol. 1, p. 11; vol. 2, p. 181)

BOT. Board of Trustees, University of North Carolina, Chapel Hill. *Minutes of the Board of Trustees*. (University Archives #40001, Wilson Special Collections Library, University of North Carolina, Chapel Hill) (Vol. 9, April 1891–January 1898)

Bowles, Elisabeth Ann. 1965. *The University of North Carolina at Greensboro 1892–1931*. A thesis submitted to the faculty of the University of North Carolina at Chapel Hill in partial fulfillment of the requirements for the degree of Doctor of Education. (pp. 27–28)

Bowles, Elisabeth Ann. 1967. *A Good Beginning: The first four decades of the University of North Carolina at Greensboro*. The University of North Carolina Press, Chapel Hill. (Dixie Lee Bryant, p. 34)

Bryant, Dixie Lee. [undated]. "Reminiscences Concerning Early Curriculum." (Vertical file, University Archives and Manuscripts, Walter Clinton Jackson Library, The University of North Carolina at Greensboro)

Bryant, Dixie Lee. 1905. *Beiträge zur Petrographie Spitzbergens*. Inaugural-Dissertation zur Erlangung der Dokturwürde der hoheren philosophisches Fakultät der Kgl. Bayer. Friedrich-Alexanders-Universität. Druck der Univ.- Buchdruckerei von E. Th. Jacob, Erlangen, Germany. (vita, p. 40)

Bryant, Dixie Lee. [1940]a. Letter [of December 2, 1940] to Anna Gove. (Folder 1940, Anna Maria Gove Collection, 1864–1952, University Archives and Manuscripts, Walter Clinton Jackson Library, The University of North Carolina at Greensboro)

Bryant, Dixie Lee. [1940]b. Letter [of December 2, 1940] to Anna Gove. (Folder 1940, Anna Maria Gove Collection, 1864–1952, University Archives and Manuscripts, Walter Clinton Jackson Library, The University of North Carolina at Greensboro)

Bryant, Dixie Lee. 1941. Letter of September 3, 1941 to Anna Gove. (Folder 1941, Anna Maria Gove Collection, 1864–1952, University Archives and Manuscripts, Walter Clinton Jackson Library, The University of North Carolina at Greensboro)

Camp, Cordelia (editor). 1955. *Some Pioneer Women Teachers in North Carolina*. The Delta Kappa Gamma Society, North Carolina Organization, [no statement of place]. (pp. 41–42)

Dean, Pamela. [1987]. *Women on the Hill: A history of women at the University of North Carolina*. Division of Student Affairs, University of North Carolina at Chapel Hill.

Duncan, Bob. 2002. Electronic mail sent to William R. Burk, October 21, 2002. (Bob Duncan, Director, Maury County Archives, Columbia, Tennessee)

Glazer-Raymo, Judith. 2002. Women in higher education, pp. 699–704 in James F. Forest and Kevin Kinser (editors), *The Higher Education in the United States, an Encyclopedia*, vol. 2. ABC-CLIO, Santa Barbara, CA. (p. 699)

Goebel, K. 1887. *Outlines of Classification and Special Morphology of Plants. A new edition of Sachs' Text-Book of Botany, Book II*. Authorised English translation by Henry E.F. Garnsey and Isaac Bayley Balfour. At the Clarendon Press, Oxford.

Gove Collection. 1864–1952. Anna Maria Gove Collection, 1864–1952. (University Archives and Manuscripts, Walter Clinton Jackson Library, University of North Carolina at Greensboro)

Gray, Asa. 1887. *Gray's School and Field Book of Botany. Consisting of lessons in botany, and field, forest, and garden botany*, bound in one volume. Ivison, Blakeman, New York.

H. 1894. The Summer School. *News-Observer-Chronicle* (Raleigh) 37(6): [2], Saturday, July 28, 1894.

Harwarth, Irene. 2002. Women's colleges, pp. 705–709 in James F. Forest and Kevin Kinser (editors), *The Higher Education in the United States, an Encyclopedia*, vol. 2. ABC-CLIO, Santa Barbara, CA.

Lawrimore, Erin. 2014. Dixie Lee Bryant and scientific study at State Normal. *Spartan Stories*, Monday, March 31, 2014. <http://uncghistory.blogspot.com/2014/03/dixie-lee-bryant-and-scientific-study.html> (Accessed March 6, 2020)

McCandless, Amy Thompson. 1992. Part. I.—A brief history of the education of women in the South, pp. [1]–5 in Margaret Dittemore and Susan Tucker (compilers), *The Higher Education of Women in the South: An annotated bibliography*. With an introductory history by Amy Thompson McCandless. (*Archival and Bibliographic Series*, III). Newcomb College for Research on Women, Tulane University, New Orleans.

McIver, Charles D. 1905. Letter of June 10, 1905 to Dixie Lee Bryant. (Folder Br-By 1905, Charles Duncan McIver, General Correspondence, Special Collections, Walter Clinton Jackson Library, University of North Carolina, Greensboro)

NC Laws. 1891. *Laws and Resolutions of the State of North Carolina Passed by the General Assembly at Its Session of 1891, Begun and Held in the City of Raleigh on Wednesday, the Eighth Day of January, A.D. 1891*. Published by Authority; Josephus Daniels, State Printer and Binder, Raleigh. (pp. 126–128)

Pearson, T. Gilbert (editor). 1904. Dixie Lee Bryant, Ph.D. *State Normal Magazine* 3(2): 30–31.

Rossiter, Margaret W. 1989. *Women Scientists in America: Struggles and strategies to 1940*. Johns Hopkins University Press, Baltimore.

Schuber, Emily. 1950. The late Dr. Dixie Lee Bryant. *Asheville Citizen-Times* (Asheville, NC) 20(34): Section C: 1, Sunday, April 23, 1950.

Shrock, Robert Rakes. 1977–1982. *Geology at M.I.T. 1865–1965. A History of the First Hundred Years of Geology at Massachusetts Institute of Technology*. The MIT Press, Cambridge. (Vol. I. The Faculty and Supporting Staff; vol. II. Departmental Operations and Products). (Dixie Lee Bryant, vol. 2, pp. 396, 398–399; also 38, 449, 450, 493)

Trelease, Allen W. 2004. *Making North Carolina Literate, The University of North Carolina at Greensboro, from Normal School to Metropolitan University*. Carolina Academic Press, Durham, NC. (p. xiii)

Turner, William Bruce. 1955. *History of Maury County, Tennessee*. The Parthenon Press, Nashville, TN. (The Columbia Female Institute, pp. 131–133)

UNCA-1. University of North Carolina Archives. University Affairs, Student Records & Faculty Records #40131, Subgroup 1: Academic Records, vol. 5 (1882–1897), vol. 6 (1893–1902). (University Archives, Wilson Special Collections Library, University of North Carolina, Chapel Hill)

UNCA-2. University of North Carolina Archives, University Affairs, Student Records & Faculty Records #40131, Subgroup 2. Miscellaneous UNC Student Records (1875–1903, 1907), vol. 13, p. 25. (University Archives, Wilson Special Collections Library, University of North Carolina, Chapel Hill)

UNCG. University of North Carolina—Greensboro. 1898. *Sixth Annual Catalogue of the State Normal and Industrial College, Greensboro, N.C. 1897–8*. Presses of Edwards & Broughton, Raleigh. (pp. 34–35)

UNCG-A. University of North Carolina at Greensboro, University Archives and Manuscripts, Walter Clinton Jackson Library, Greensboro, North Carolina. (Includes Correspondence of Dixie Lee Bryant in the Charles Duncan McIver Records, 1855–1906 and in the Anna Maria Gove Collection, 1864–1952; and two documents by Dixie Lee Bryant: "Reminiscences Concerning Early Curriculum," (7 pp.) and "Early Days of the College," a photo-album.

UNC-WA. University of North Carolina, Women's Association. 1922. *Women Students in the University of North Carolina 1897–1922*. Published by the University of North Carolina Women's Association, [Chapel Hill, NC]

Ward, Needham. 1897. Letter of October 8, 1897 to Walter G. Ward (Box 1, folder c, #464 Walter G. Ward Papers, East Carolina Manuscript Collection, J.Y. Joyner Library, East Carolina University, Greenville, NC)

Austin C. Apgar

Chapter 12.

Austin Craig Apgar (1838–1908)

Instructor of Botany and Zoology (Summer 1896, 1898)

Early Years

Born on August 4, 1838, in Peapack, Somerset County, New Jersey, Austin Craig Apgar was the son of David and Hannah (Whitehead) Apgar. He attended local schools and in 1860 entered the State Normal School in Trenton, New Jersey, now the College of New Jersey. After graduating with honors in 1862, he taught in district schools and in the Farnum Preparatory School at Beverly, New Jersey. In 1866, he became a book salesman for Cowperthwait & Company, the publisher of several of his books, including *Apgar's Geographical Drawing Book* (1866), which he wrote with his brother Ellis A. Apgar, the senior author. The book underwent revisions and repeated printings and sold over one million copies (Anonymous, 1951). Further information on the life of Apgar has been published in the following sources: Anonymous, 1908b; Harshberger, 1899; Meeker, 1906; Stafleu and Mennega, 1992.

Career as a Science Educator

When he was first appointed to the faculty at the New Jersey State Normal and Model Schools in 1866, Austin Apgar instructed mathematics primarily, but by 1870 he had become Professor of Natural Science and Geography (Fromm, 1950). For his last nineteen years there, he also served as vice-principal. He devoted the remainder of his life to teaching at the Normal School and gained a favorable reputation for "his contagious enthusiasm as a naturalist and as a teacher" (Farley, 1908). He also served as head of the section on botany and birds at the State Museum of New Jersey in Trenton.

Apgar's philosophical approach to teaching was influenced by courses he took at the Andersen Summer School of Natural History at Penikese Island given by the famous naturalist Louis Agassiz in 1873 and by Alexander Agassiz (Louis's son) in 1874. Among his 1874 classmates was William K. Brooks, who later became Henry Van Peters Wilson's major professor at the Johns Hopkins University (Wilder, 1898).

Apgar encouraged students to develop their powers of observation and to follow Louis Agassiz's dictum: "Study nature, not books." In addition to emphasizing fieldwork, Apgar

Frontispiece Credits: (above) Portrait of Austin Craig Apgar. (From *New Jersey, a Historical, Commercial and Industrial Review*, by Ellis R. Meeker. [1906], p. 264). (below) State Normal School, Trenton, New Jersey, ca. 1907; Illustrated Postal Cards Co., New-York, Germany. (In the author's research files).

employed laboratory work to teach natural history. One scholar has explained, "He bitterly opposed the idea, then just emerging, of arranging botany from the lowest forms up. He argued that it is easier to study a buttercup than a moss, and that the Ranunculaceae was the proper beginning" (Anonymous, 1951).

Botanical Endeavors

Apgar's dedication and devotion to natural history, particularly to botany, is illustrated through his publications and his collaboration with botanists. He wrote five books and five articles on botany (Table 12.1) as well as several books on mollusks, birds, and geography. In his book on ornamental shrubs (Fig. 12.1), published in 1910, Apgar's artistic talents are apparent. Although Apgar died before publishing the book, his daughter Ada saw it through to completion. She inked in his 621 pencil illustrations. John Harshberger, Professor of Botany at the University of Pennsylvania and Apgar's good friend, critically edited the volume and compiled the glossary and index (Draycott, 1910).

Table 12.1. A List of Botanical Publications by Austin Craig Apgar.

1874. (with E[llis] A. Apgar) *Apgar's* [*i.e., Apgars'*] *Plant Analysis; Adapted to Gray's Botanies*. Ivison, Blakeman and Company, New York. [12] pp. & [122] forms. (Also published in 1875). (According to historian of botany Joseph Ewan [1969], the book "became the familiar work-book of botany classes.")

1887. A new variety of *Aralia nudicaulis*, L. *Bulletin of the Torrey Botanical Club* 14: 166–167.

1889. Extraordinary vitality in a girdled limb. *Journal of the Trenton Natural History Society* 2(1): 7–9; pl. 1.

1889. Study of plant life in our schools. *State of New Jersey, Sixteenth Annual Report of the State Board of Agriculture* 1888–89: 257–271; pl. I–II.

1891. *Pocket Key of Trees Both Wild and Cultivated of the Northern United States East of the Rocky Mountains*. Lucas & Company, Trenton, NJ.

1891. What shall be taught in country schools to educate farmers. *State of New Jersey, Eighteenth Annual Report of the State Board of Agriculture* 1890–91: 139–149.

1892. (with E[llis] A. Apgar) *Apgar's* [*i.e., Apgars'*] *New Plant Analysis, Adapted to All Botanies*. American Book Company, New York.

1892. *Trees of the Northern United States. Their Study, Description and Determination. For the Use of Schools and Private Students*. American Book Company, New York. (Also published in 1920)

1896. *How to Know Trees. Arbor Day Circular*. (Also published 1896 in State of New York, Department of Public Instruction, *Forty-second Annual Report of the State Superintendent, for the School Year Ending July 31, 1895*, vol. 2, pp. [1082]–1088, printed by Wynkoop Hallenbeck Crawford Co., Albany and New York)

1910. *Ornamental Shrubs of the United States (Hardy, Cultivated)*. American Book Company, New York. (Prepared for publication by Ada Apgar Draycott; glossary and index by John W. Harshberger)

ORNAMENTAL SHRUBS

OF THE UNITED STATES

(HARDY, CULTIVATED)

BY

AUSTIN CRAIG APGAR

AUTHOR OF "TREES OF THE NORTHERN UNITED STATES"
"BIRDS OF THE UNITED STATES," ETC.

FIG. 5.—Swamp Magnolia.

FIG. 50.—Alleghany Stuartia.

NEW YORK ·:· CINCINNATI ·:· CHICAGO
AMERICAN BOOK COMPANY

Fig. 12.1. Title page and Figures (inserted) from *Ornamental Shrubs*, by Austin Craig Apgar. (1910).

(From the author's library).

The Archives of the New York Botanical Garden (NYBG) holds a collection of twenty-six letters sent by Apgar between 1884 and 1889, to botanist Nathaniel Lord Britton. These letters reveal the level of assistance that he provided for Britton's *Catalogue of Plants Found in New Jersey* (1889). He sent Britton lists of plants and their localities as well as plant specimens from New Jersey. In return, Apgar asked his correspondent for help in identifying plants for the herbarium at the State Normal School, and he sent lists of desiderata for its collection. Botanical historian Joseph Ewan (1969) noted that Apgar founded the herbarium at the Normal School in Trenton, New Jersey. Pennell (1942–1943) mentioned that Apgar's "herbarium" (which may refer to his personal plant collection, not the State Normal School's) was donated to the State Museum of New Jersey. Its present location, however, is unknown.

Instructor at UNC's Summer Schools, 1896 and 1898

University administrators recruited leading educators in order to attract students to its summer school. They successfully hired Apgar to teach botany and zoology at the summer schools of 1896 and 1898. Billed as "the famous teacher of Natural Science" (UNC, 1896), he was the highest paid ($216.00) instructor for the 1896 session (Alderman, 1897). His credentials included thirty years of teaching science at the State Normal School in Trenton, New Jersey, as well as stints as an instructor at summer schools in Chicago, Illinois; Asbury Park, New Jersey; Glen Falls, New York; Fort Worth, Texas; Bedford City and Salem, Virginia; and institutions in Georgia, Massachusetts, Ohio, Pennsylvania, and Tennessee.

The 1896 UNC summer school started on June 23 and ended July 18. President George T. Winston acted as the superintendent while Professor Edwin A. Alderman, who was ill at the beginning of the school session, was recuperating in the Adirondacks. Alderman returned to Chapel Hill during the session, later assuming its superintendence (Anonymous, 1896a; Anonymous, 1896b). Mr. Marcus Cicero Stephens Noble was assistant superintendent. On the first day of classes of the third UNC summer school, Winston welcomed the teachers, who had assembled in the chapel. He spoke on the power, responsibilities, and opportunities of teachers. Professor Kemp P. Battle also addressed the group, providing glimpses of the early history of the university and describing the beauty of the campus (Anonymous, 1896a).

Apgar arrived in Chapel Hill accompanied by his wife and one of their daughters (Apgar, 1896). He was scheduled to teach botany and zoology courses, each consisting of fifteen lessons. The botany class was taught in the morning in a recitation room on the second floor of South Building, and it included practice as well as method. A Raleigh newspaper reporter noted that Apgar's first lecture was "charming." In fact, the popular botany class was so crowded that the reporter predicted it would be moved to larger quarters, probably in the campus chapel (Anonymous, 1896b). Topics studied were the growth of plants from seeds, trees of the Chapel Hill region, flowering plants, insect aid to seed formation, ferns, and the preparation of school collections of plants. Students were required to purchase and bring to the classroom a magnifying glass and three textbooks: Gray's *How Plants Grow* (1858), Gray's *Manual of Botany* (1890), and Apgar's *Trees of the Northern United States* (1892) (UNC, 1896).

Although not listed in the summer school circular as a required textbook, Apgar probably also used *Apgars' New Plant Analysis, Adapted to All Botanies*, which he wrote with his brother, Ellis A. Apgar (Apgar and Apgar, 1892). Much like Harrington's *Analysis of Plants* (1876) that Frederic Simonds used in teaching students how to identify plants, the Apgars' volume contained a section of botanical terms, but it was also highly illustrated to assist the students. The remainder of the volume consisted of blank forms for the students to complete in analyzing plants being identified. The authors provided several completed forms to serve as examples (Fig. 12.2).

16

ORDER OF DESCRIPTION.

(*Specimen Page* 2.)

Root..........	Primary, multiple, thickish, biennial.
Stem..........	Herbaceous, erect, branching, terete, grooved, hirsute, 2°.
Leaves........	*Radical* simple, 3 ribbed, oblanceolate, long-petiolate, obtuse, entire, hairy.
	Cauline few, alternate, sessile, rough-hairy.
Flowers.......	Terminal heads of about 10 ray flowers and many disk flowers; *receptacle* chaffy,
Calyx.........	Superior. ‖ with large, spatulate, hairy, purple bracts; *involucre* green, of one
Sepals........	No pappus. ‖ set of large, spreading, hairy scales.
Corolla.......	*Rays* bright orange, ligulate; *disk* dark purple, tubular.
Petals........	*Ray flowers* about 3 lobed; *disk flowers* 5 lobed, valvate.
Stamens......	*Ray flowers* none; *disk flowers* 5, syngenesious, epipetalous, short.
Anthers.......	Linear, 2, porous, oblong, rough, pointed.
Pistil.........	*Ray flowers* none; *disk flowers* 1; style long, 2 cleft.
Stigma........	2, slender, somewhat curved, stigmatic on inner side.
Ovary.........	Quadrangular-top shaped, inferior, 1 celled, basal.
Fruit.........	Oblong-rectangular, smooth, black, achene.
Seeds.........	1, oblong; embryo anatropous, without albumen.

Remarks.—A very rough, bristly-hairy plant, branching from the base, with large heads of bright orange ray flowers and dark purple disk flowers. Introduced from the west into eastern meadows, and spreading quite rapidly.

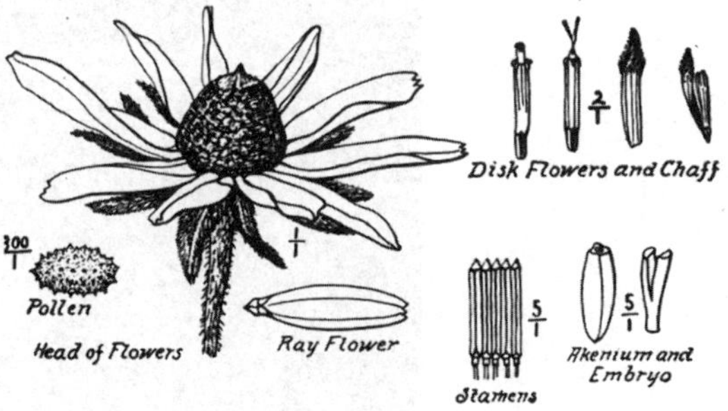

Classification.—Gamopetalous exogen. Compositæ.

Name.—Rudbeckia hirta, L. (Black-eyed Susan. Cone-flower.)

Locality and Date.—Ewing, N. J., August 6, 1892.

Fig. 12.2. A specimen page from *Apgar's* [*i.e., Apgars'*] *New Plant Analysis* (Apgar and Apgar, 1892, p. 16).

(From the author's library).

The fifth UNC summer school, superintended by Mr. Noble, opened June 21 and concluded on July 20, 1898. The UNC Summer School catalog of 1898 mentioned of Apgar, "Our teachers are fortunate in again having an opportunity of studying under this distinguished scientist" (UNC, 1898). Apgar's 1898 summer class in botany covered the same topics as his 1896 class had and was taught in the university chapel from 5:15 to 6:00 in the evening. His return to UNC elicited the following praise from a newspaper correspondent: "Mr. Apgar is so enthusiastic in his work that his pupils gain much of it" (Anonymous, 1898a). The faculty also gleaned new ideas from observing the leading teachers in the country. A series of evening lectures, two socials for the entire school, and a variety of smaller parties also contributed to the session's overall success (Anonymous, 1898b).

Honors and Legacy

On Tuesday evening, March 3, 1908, at age sixty-nine, Austin Apgar died suddenly of apoplexy at his home in Trenton. He was buried in Greenwood Cemetery in Hamilton, New Jersey (Fig. 12.3). Surviving him were his wife Maria Elizabeth Whitlock, whom he had married on December 27, 1866, and their three children, Ada, Walter, and Mrs. Alfred D. Cook. Apgar's colleagues recognized the breadth of his research and his scientific contributions. The Austin C. Apgar Society was established in his honor. Among the society's primary goals were stimulating interests and promoting amateur research in science and mathematics as well as fostering fellowship among students in these disciplines. The organization is still active but is now called the Biological Society. Through Apgar's efforts, the Mercer County Natural History Association was founded in 1905; after his death, it was named the Apgar Club in 1908. It dissolved around 1919 (Anonymous, 1932; Anonymous, 1935).

Fig. 12.3. Gravestone of Austin Craig Apgar, Section B-NW, Lot 59 North 1/3, Greenwood Cemetery, Hamilton, New Jersey.

(Courtesy of Leigh Miller).

A prominent scholar, Austin Craig Apgar probably helped to attract other high-caliber teachers to UNC's summer schools. In the 1890s, he was at the height of his teaching career in natural science and an avid proponent of laboratory and field studies. At Chapel Hill, he was a popular lecturer whose infectious and enthusiastic style promoted "liberal education [with] a truer regard for truth" (Anonymous, 1896b). He was a devoted student of botany not only as a science but also as "the Author of Science." Commenting on his death, one writer offered this epitaph: "The meanest flowers of the vale, //The simplest note that swells the gale, // The common sun, the air, the skies, //To him are opening paradise" (Anonymous, 1908a).

References Cited

(References that conclude with page numbers within parentheses indicate sources from where quotes and pertinent information were obtained.)

Alderman, Edwin A. 1897. Report of the President [1896], pp. [1]–2 in *Report of the President, Bursar and Treasurer of the University of North Carolina*. Document No. 24, Session 1897. Ordered to be Printed, M.I. & J.C. Stewart, Public Printers and Binder, [Winston, NC].

Anonymous. 1896a. Opening the Summer School—One hundred and twenty-five teachers at the university. *News and Observer* (Raleigh) 40(116): [1], Thursday, June 25, 1896.

Anonymous. 1896b. University Summer School closes the most successful session in its history. *News and Observer* (Raleigh) 40(136): 4, Sunday, July 19, 1896.

Anonymous. 1898a. University Summer School—Prof. Apgar and Prof. Joynes have arrived—a pleasant time. *News and Observer* (Raleigh) 44(105): 3, Tuesday, June 28, 1898.

Anonymous. 1898b. Summer School closes—The Summer School closes a successful and profitable season. *News and Observer* (Raleigh) 44(127): 3, Thursday, July 21, 1898.

Anonymous. 1908a. Austin C. Apgar dies suddenly from apoplexy. *Trenton Evening Times*, Wednesday, March 4, 1908, pp. 1, 6. (portrait and bibliography)

Anonymous. 1908b. [Tributes to Austin C. Apgar]. *New Jersey State Schools Signal* (Trenton, NJ). March 1908, pp. 85–89.

Anonymous. 1932. Austin C. Apgar Society. *The Seal of State Teachers College and State Normal School. Trenton, New Jersey*. 1932: 200.

Anonymous. 1935. Austin C. Apgar Society. *The Seal, State Teachers College. Trenton, New Jersey*. 1935: 101.

Anonymous. 1951. *Torrey Botanical Club Field Schedule, 1951*. Eighty-fifth Season, p. 3. (portrait on issue cover)

Apgar, Austin C. 1892. *Trees of the Northern United States. Their Study, Description and Determination. For the Use of Schools and Private Students*. American Book Company, New York. (also published in 1920)

Apgar, Austin C. 1896. Letter of July 15, 1896 to Mr. [Thomas W.] Harris. (Box 20, folder 657, University Papers #40005, University Archives, Wilson Special Collections Library, University of North Carolina, Chapel Hill)

Apgar, E.A. and A.C. Apgar. 1866. *Apgars' Geographical Drawing Book.* J.B. Cowperthwait, Philadelphia.

Apgar, E.A. and A.C. Apgar. 1892. *Apgar's* [*i.e., Apgars'*] *New Plant Analysis, Adapted to All Botanies.* American Book Company, New York.

Britton, Nathaniel L. 1889. Catalogue of Plants Found in New Jersey, pp. [25]–642 in *Geological Survey of New Jersey, Final Report of the State Geologist*, vol. 2. Mineralogy, Botany, Zoology. Printed by the John L. Murphy Publishing Company, Trenton, NJ.

Draycott, Ada Apgar. 1910. Editor's note, p. 6 in Austin Craig Apgar, *Ornamental Shrubs of the United States (Hardy, Cultivated).* American Book Company, New York.

Ewan, Joseph (editor). 1969. *A Short History of Botany in the United States.* With contributions by Chester A. Arnold, [and others]. Hafner Publishing Company, New York. (p. 11)

Farley, D.H. (chairman). 1908. In memoriam. 1838—Austin C. Apgar—1908, p. 87 in Anonymous, [Tributes to Austin C. Apgar]. *New Jersey State Schools Signal* (Trenton, NJ). March 1908, pp. 85–89.

Fromm, Glenn Everet. 1950. *A History of the New Jersey State Teachers College at Trenton, 1855–1950.* Ph.Ed. (School of Education), New York University. (Authorized facsimile, UMI Dissertation Services, 1999)

Gray, Asa. 1858. *How Plants Grow; A simple introduction to structural botany. With a popular flora, or an arrangement and description of common plants, both wild and cultivated. Botany for young people and common schools.* Ivison and Phinney, New York; S.C. Griggs & Co., Chicago.

Gray, Asa. 1890. *Manual of the Botany of the Northern United States, Including the district east of the Mississippi and north of North Carolina and Tennessee, Arranged according to the natural system.* 6th ed., Revised and extended westward to the 100th Meridian, by Sereno Watson … and John M. Coulter. Ivison, Blakeman, & Co., New York & Chicago.

Harrington, Mark W. 1876. *The Analysis of Plants, Intended for Schools and Colleges and for the Independent Botanical Student.* Sheehan & Co., Ann Arbor, MI.

Harshberger, John W. 1899. Austin Craig Apgar, p. 304 in *The Botanists of Philadelphia and Their Work.* Press of T.C. Davis & Sons, Philadelphia.

Meeker, Ellis R. (compiler). 1906. Austin Craig Apgar, p. 264 in *New Jersey, a Historical, Commercial and Industrial Review.* Commonwealth Publishing Co., Elizabeth, NJ. (portrait)

NYBG. New York Botanical Garden. LuEsther T. Mertz Library. 26 letters from Austin C. Apgar to Nathaniel L. Britton, dated 1884–1889.

Pennell, Francis W. 1942–1943. Botanical collectors of the Philadelphia local area (concluded). *Bartonia* 22: 10–31.

Stafleu, Frans A. and Erik A. Mennega. 1992. Apgar, Austin Craig, p. 135 in *Taxonomic Literature: A selective guide to botanical publications and collections with dates, commentaries and types, Supplement*, I (A–Ba). Koeltz Scientific Books, Koenigstein, Germany. (sources and bibliography)

UNC. University of North Carolina. 1896. *Summer School, June 23–July 17, 1896*. University of North Carolina, Chapel Hill.

UNC. University of North Carolina. 1898. *Courses of Instruction in the Summer School, June 21–July 19, 1898*. Published by the University, Chapel Hill.

Wilder, Burt G. 1898. Agassiz at Penikese. *American Naturalist* 32: 189–196.

Wilbur S. Jackman

1898

Chapter 13.

Wilbur Samuel Jackman (1855–1907)

Instructor of Nature Study (Summer 1897)

Early Years

Wilbur Samuel Jackman, born January 12, 1855, in Mechanicstown, Carroll County, Ohio, was the son of Barnard Conley and Ruth (Lilley) Jackman. When he was four or five years of age, his parents moved to Coal Center, outside the town limits of California, Pennsylvania, to live on a farm formerly owned by his father's family. Young Jackman attended the local schools of East Pike Run Township, now part of California, Pennsylvania, Washington County, and tended to farm chores. During these formative years, he developed an appreciation and knowledge of nature, which would influence his future career. He attended the South-Western State Normal School at California, Pennsylvania, now the California University of Pennsylvania, earning the M.E.D. and graduating first in his class in 1877. He then taught civil government, geography, and history there for the next three and one-half years. Jackman subsequently enrolled at Allegheny College, now Allegheny University, in Meadville, Pennsylvania. In about 1882, he transferred to Harvard College, where he earned the A.B. in June 1884 and wrote a thesis on the "Correlation of Science and History." Several sources provide biographical background on Jackman (Anonymous, 1907a; Anonymous, 1909; Bright, 1907; Butler, 1907; HU-A; J[udd], 1932; M[onroe], 1912; UCL).

Teaching

Jackman's early professional experience was as a teacher in charge of natural science at Central High School in Pittsburgh Pennsylvania, where he taught physiology and zoology one semester and botany the other (Jackman, 1887). In 1887, he published a pamphlet, "Syllabus of Fifty Lessons on Plants,"[94] used chiefly by his pupils. The following year, Jackman (1888) published a fifty-four-page booklet entitled "Practical Lessons on Plants." While in Pittsburgh, Jackman was active in the Iron City Microscopical Society as well as the Botanical Society of Western Pennsylvania, in which he served as corresponding secretary (Jackman, 1887).

As a successful educator, Jackman garnered the attention of Colonel Francis Wayland Parker, principal of the Cook County (Illinois) Normal School and a promoter of science education

[94] The author has not seen this publication.

Frontispiece Credits: (above) Wilbur Samuel Jackman, ca. 1900. (From "Wilbur S. Jackman," by Orville T. Bright, *Elementary School Teacher* 7: plate preceding p. 433). (below) Laboratory, University of Chicago Laboratory School, 1898. (Courtesy of University of Chicago Special Collections Research Center, Archival Photofiles Series IV: Student Activities Classes, Labs, Seminars, Lab School [Old], #24).

as an important component of the curriculum of the elementary school. Jackman accepted an invitation by Parker to take charge of the Department of Natural Science in the school beginning 1889. Jackman's assignment was a challenging one, being hampered, among other things, by inadequate apparatus. According to biographer Dennis M. Wint (1978), "Through Parker's influence, Jackman became involved in the reform of the elementary school curriculum through the addition of new content areas and the reduction of the formality common in schools of the period." In 1894, he also managed the Cook County Normal Summer School.

Teaching at the UNC Summer School

Wilbur S. Jackman was selected as an Instructor in Nature Study for the 1897 UNC Summer School, one of a faculty of twenty-four (Anonymous, 1897d). No doubt his presence on the roster of faculty attracted teachers to the session, which took place from June 22 to July 23, 1897. His course of instruction was Nature Study, and it consisted of lectures given on the first floor of South Building and of field excursions into the countryside around Chapel Hill (Anonymous, 1897e). Logan D. Howell (1897), superintendent of the Raleigh City Schools and teacher of Latin at the 1897 summer session, reported that Jackman's instruction took place only during the first three weeks of the session.

Clinton W. Toms, superintendent of the Summer School and at that time professor of pedagogy at UNC, welcomed the teachers who attended the school on the opening day of classes and encouraged them to take advantage of their educational opportunities (Anonymous, 1897a). On the same day, Jackman gave presentations on "Nature Work" in the morning and afternoon, elucidating "what it is and its place in education." According to a pamphlet describing the Summer School (Anonymous, 1897b), Jackman's course was composed of a series of lectures and conferences on natural science topics appropriate for classroom instruction of young students below high school age. Teachers were offered insight into the proper methods of instructing young students in natural science fieldwork and experimentation, skills in communicating the love of nature to children, and an understanding of the relationship of nature study to other disciplines and to the needs of the pupils' lives. Topics included the following (Anonymous, 1897b):

> Introductory: the scope and character of Nature Study.
>
> Painting—water color—and drawing as aids in developing pictures in nature. Illustrated by pupils' work.
>
> Number works as an aid in developing pictures of nature. Illustrated by practical problems.
>
> Reading and Writing as means of developing pictures of nature. Illustrated by primary reading lessons.
>
> The relation of History and Literature, including mythology to Nature Study. Illustrated by appropriate selections.
>
> The moral relations and value of Nature Study. The applications of the study to the practical affairs of life.

Jackman was an energetic man and an able organizer with a tendency toward being autocratic, according to teacher and historian Ida DePencier (1967). He augmented his teaching with the display of specimens of nature work made by his students in Chicago, which he had brought to Chapel Hill (Howell, 1897). He probably also distributed a syllabus on "Nature Study," which he used in summer schools from 1895 to 1898 (Jackman, 1899).

In teaching botany, Jackman believed that pupils should learn about plants in their natural habitats, receiving instruction about their distribution, growth and development; forms; habits; flowers; and fruit. His students also learned how plants interacted with other organisms and their landscape, as well as how they changed over time (Doris, 2002). He remained a strong advocate of taking his students on field trips, a feature of his teaching that he continued until his death. Concerning botanical fieldwork, Jackman (1894) stated, "Flowers when nodding and waving on their parent stem are far more beautiful than they are in vases or pressed in a herbarium." Even though some parents questioned whether field trips were too rigorous for children (DePencier, 1967), Jackman (1905) defended this aspect of his teaching, saying, "It is simply ridiculous to keep children confined in a schoolroom learning *about* a world while they must live *in* it."

In summary, the fourth Summer School was a resounding success, having attracted 185 teachers to the university. A sense of enthusiasm pervaded the session, and the evening lectures and receptions were also very popular. The attendees impressed the university librarian, who observed that the library was more heavily used than previously and that more books of a "higher order" had been read than during past sessions (Anonymous, 1897c).

Later Years and Legacy

Jackman returned to the Normal School in Chicago after teaching in Chapel Hill. Subsequently, he studied education in England, France, Germany, and Holland in the winter of 1899–1900 (Anonymous, 1939). In 1900, he assumed the deanship of the newly established Chicago Institute. When the institute was merged with the University of Chicago a year later, he became dean of the newly formed College of Education of the latter. In 1904, Jackman resigned this position to become principal of the University of Chicago's elementary school.

Jackman's years at the Cook County Normal School were accentuated by a steady flow of publications, including six textbooks and manuals (Table 13.1) (Fig. 13.1). His books and articles form a bridge between the early writers of children's literature and those of modern elementary science (Craig, 1957). While principal at the elementary school, he took on the editorship of the *Elementary School Teacher* and authored "*Part II. Nature Study*," published in *The Third Yearbook of the National Society for the Scientific Study of Education* (Jackman, 1904). According to noted elementary school science teacher Gerald S. Craig (1957), the book focused on the challenges of teaching science and was the first volume of the series of yearbooks to cover the subject.

TABLE 13.1. A SELECTED LIST OF TITLES OF BOOKS ON NATURE STUDY WRITTEN BY WILBUR S. JACKMAN FROM 1892 TO 1899.

1892. *Nature Study: For the common schools*. H. Holt and Co., New York.

1893. *Number Work in Nature Study,* Part 1. The Author; Press of La Monte-O'Donnell Co., Chicago.

1894. *Field Work in Nature Study*. A. Flanagan, Chicago.

1894. *Nature Study for the Common Schools*. H. Holt and Company, Chicago.

1896. *Nature Study and Related Subjects for the Common Schools*. The Author, Chicago.

1899. *Nature Study for Grammar Grades; A manual for teachers and pupils below the high school in the study of nature*. Macmillan Company, New York.

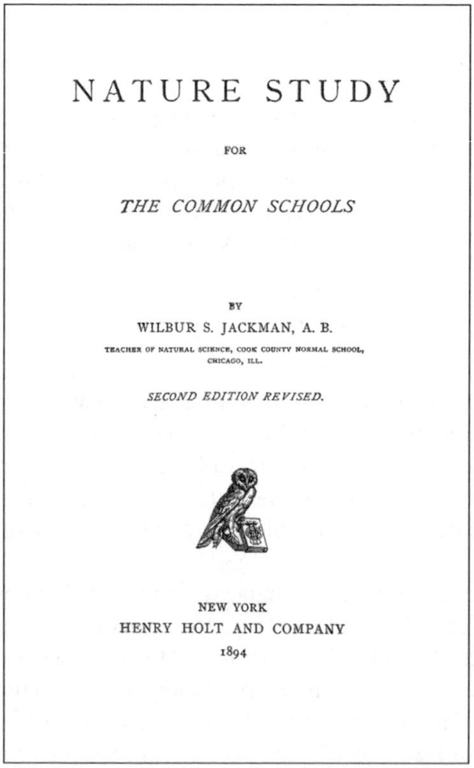
NATURE STUDY

FOR

THE COMMON SCHOOLS

BY

WILBUR S. JACKMAN, A. B.

TEACHER OF NATURAL SCIENCE, COOK COUNTY NORMAL SCHOOL, CHICAGO, ILL.

SECOND EDITION REVISED.

NEW YORK
HENRY HOLT AND COMPANY
1894

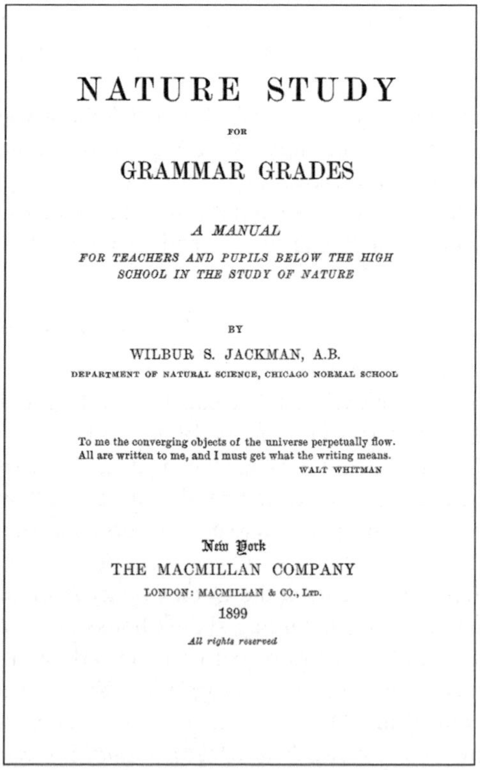
NATURE STUDY

FOR

GRAMMAR GRADES

A MANUAL

FOR TEACHERS AND PUPILS BELOW THE HIGH SCHOOL IN THE STUDY OF NATURE

BY

WILBUR S. JACKMAN, A.B.

DEPARTMENT OF NATURAL SCIENCE, CHICAGO NORMAL SCHOOL

To me the converging objects of the universe perpetually flow.
All are written to me, and I must get what the writing means.
WALT WHITMAN

New York
THE MACMILLAN COMPANY
LONDON: MACMILLAN & CO., LTD.
1899

All rights reserved

Fig. 13.1. Title pages of two books by Wilbur S. Jackman. (left) *Nature Study for the Common Schools* (1894). (right) *Nature Study for Grammar Grades* (1899).

(From the author's library).

Jackman's sudden death on January 28, 1907, due to pneumonia, startled the educational world. The following day, a private service was held at his home, and the next day a public memorial service was held at the Leon Mandel Assembly Hall at the University of Chicago. His remains were brought to California, Pennsylvania. Out of respect for Jackman, classes at his alma mater in California, Pennsylvania, were cancelled for the afternoon when a memorial service took place on January 31, 1907, in the campus chapel (Anonymous, 1907b). Surviving him at the time of his death were his wife, Ellen Amelia (Reis) Jackman, whom he had married on December 23, 1884, in Pittsburgh, Pennsylvania; two daughters, Ruth Reis, wife of Bradford Newcomb Stevens, and Louise Lilley, wife of Arthur A. Shillander; and a sister. He was buried in Highland Cemetery (Fig. 13.2) of his hometown.

Deemed the "father of modern elementary science" (Craig, 1957), Wilbur S. Jackman nationally pioneered the elementary education movement; he was particularly innovative in developing methods of nature study. Jackman and the nature study movement promoted self-expression and independent thinking in students (Doris, 2002). During his early professional years, Jackman taught courses in botany and was active in botanical associations. At UNC, he taught a summer course in Nature Study in 1897, through which he passed along his pedagogic philosophies to an enthusiastic group of teachers.

Jackman's alma mater in California, Pennsylvania, commemorated his name and his contributions to education by establishing in 1967 the Illustrious Californian Award. Renamed the W.S. Jackman Award of Distinction in 1983, it is given to one of the university's graduates who has attained professional acclaim on the regional, state, or national level. In a memorial tribute, Will S. Monroe (1907) noted: "Jackman was more than a great teacher and an active writer; he was a noble man, nobly planned. . . . He stood for all that was best in both his personal and professional life."

Fig. 13.2. Gravestone of Wilbur S. Jackman, lot 83 (west side of Highland Drive), Highland Cemetery, California, Pennsylvania, October 2004.

(Courtesy of Dana Sally).

References Cited

(References that conclude with page numbers within parentheses indicate sources from where quotes and pertinent information were obtained.)

Anonymous. 1897a. Summer Normal School. Over a hundred students in attendance at Wake Forest [*i.e.,* UNC], Chapel Hill, N.C., June 23. *News and Observer* (Raleigh) 42(98): 4, Friday, June 25, 1897.

Anonymous. 1897b. [*Summer School, University of North Carolina, Chapel Hill*]. (Pamphlet bound with *UNC catalogs* of 1891 to 1897, North Carolina Collection, Wilson Special Collections Library, University of North Carolina, Chapel Hill). (p. 8)

Anonymous. 1897c. University Summer School. The last week begun—Very successful—185 teachers registered—Discussions have been of great value. Chapel Hill, N.C., July 17. *News and Observer* (Raleigh) 42(118): [12], Sunday, July 18, 1897.

Anonymous. 1897d. University Summer School. The last week—The work done has been of the very best. Chapel Hill, N.C., July 23. *News and Observer* (Raleigh) 42(123): [2], Saturday, July 24, 1897.

Anonymous. 1897e. *University Summer School, June 22 to July 23, 1897, Schedule of Exercises.* [University of North Carolina, Chapel Hill]. (Folder 1897, Announcements, bulletins, programs, etc., North Carolina Collection, Wilson Special Collections Library, University of North Carolina, Chapel Hill)

Anonymous. 1907a. Dean of School expires suddenly. *Chicago Daily Tribune* 66(25): 4, Tuesday, January 29, 1907. (portrait)

Anonymous. 1907b. Wilbur S. Jackman. *Normal Review* (Southwestern State Normal School, California, PA) 17 (February): 10.

Anonymous. 1909. Wilbur Samuel Jackman. *Harvard College Class of 1884, Secretary's Report*, No. VII, pp. 127–128.

Anonymous. 1939. Jackman, Wilbur Samuel, p. 325 in *The National Cyclopaedia of American Biography, Being the History of the United States*, vol. 27. James T. White & Company, New York.

Bright, Orville T. 1907. Wilbur S. Jackman. *Elementary School Teacher* 7: 433–438. (portrait)

Butler, Nathaniel. 1907. Wilbur Samuel Jackman, a personal appreciation. *Elementary School Teacher* 7: 439–442.

Craig, Gerald S. 1957. Elementary school science in the past century. *Science Teacher* 24(1): 11–14, 37. (p. 13)

DePencier, Ida B. 1967. *The History of the Laboratory Schools; The University of Chicago 1896–1965*. Quadrangle Books, Chicago.

Doris, Ellen. 2002. *The Practice of Nature Study: What reformers imagined and what teachers did*. A thesis presented to the faculty of the Graduate School of Education of Harvard University in partial fulfillment of the requirements for the degree of Doctor of Philosophy. ("An authorized facsimile, made from the microfilm master copy of the original dissertation … published by UMI," 2002). (Chapter 4. The Landscape as Organism: Jackman's influence in Chicago, pp. 131–170)

Howell, Logan D. 1897. The Summer School. Prof. Howell's card saying the first two weeks are the most valuable, Chapel Hill, N.C., June 21. *News and Observer* (Raleigh) 42(95): [1], Tuesday, June 22, 1897.

HU-A. Harvard University Archives, Pusey Library, and Monroe C. Gutman Library, Cambridge, MA.

Jackman, Wilbur S. 1887. Wilbur Samuel Jackman. *Harvard College, Class of 1884, Secretary's Report* No. II, p. 43.

Jackman, Wilbur S. 1888. *Practical Lessons on Plants*. New England Publishing Co., Boston.

Jackman, Wilbur S. 1894. *Field Work in Nature Study*. 2nd ed., enlarged and illustrated. *A Hand-book for Teachers and Pupils. Through the Senses to the Soul*. A. Flanagan, Chicago.

Jackman, Wilbur S. 1899. Letter of February 20, 1899 to Professor Smith. (Harvard University Archives [UA III 15-88.10], Harvard University, Cambridge, Massachusetts)

Jackman, Wilbur S. 1904. *The Third Yearbook of the National Society for the Scientific Study of Education. Part II. Nature-Study*. Edited by Manfred J. Holmes. The University of Chicago Press, Chicago.

Jackman, Wilbur S. 1905. The University Elementary School, its general plan and course of study. *Elementary Science Teacher* 5: [585]–597. (p. 591) (Contributions by other teachers are presented on pages 597–665.)

J[udd], C[harles] H. 1932. Jackman, Wilbur Samuel, p. 525 in Dumas Malone (editor), *Dictionary of American Biography*, vol. 9. Charles Scribners Sons, New York.

Monroe, Will S. 1907. Wilbur S. Jackman: A tribute. *Journal of Education* 65: 128.

Monroe, Will S. 1912. Jackman, Wilbur Samuel, pp. 512–513 in Paul Monroe (editor), *A Cyclopedia of Education*, vol. 3. The Macmillan Company, New York.

UCL. University of Chicago Library, Special Collections Research Center, Chicago, IL. (Two letters written by Jackman in 1905 to University of Chicago President, William Rainey Harper; Presidents' Papers, 1889–1925, Box 41, folder 17)

Wint, Dennis M. 1978. Jackman, Wilbur Samuel, pp. 695–696 in John F. Ohles (editor), *Biographical Dictionary of American Educators*, vol. 2. Greenwood Press, Westport, CT.

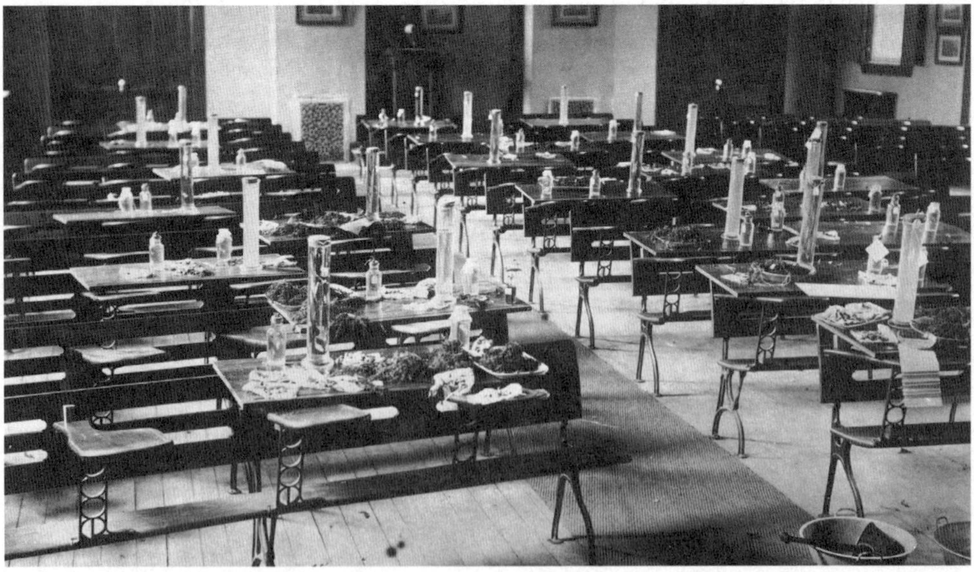

Chapter 14.

George Francis Atkinson (1854–1918)

Assistant Professor of Natural History (1885/1886–1886/1887)
Assistant Professor of Zoology and Entomology in the College of Agriculture (1886/1887)
Associate Professor of Natural History (1887/1888)
Associate Professor of Zoology and Entomology in the College of Agriculture (1887/1888)
Lecturer in Botany (Summer 1901)

Early Years

Born January 26, 1854, in Raisinville, Monroe County, Michigan, George Francis Atkinson was the son of Joseph and Josephine (Fish) Atkinson. Little is known about his boyhood years. In 1878, he entered Olivet College in Olivet, Michigan, where he took preparatory courses and continued his matriculation until his junior year. In January 1884, he transferred to Cornell University, where he petitioned for junior status. There, he came into contact with botanists William Russel Dudley and Albert Nelson Prentiss and zoologists John Henry Comstock and Simon Henry Gage. Atkinson's early interest in plants was evident in his enrollment in a number of courses in botany: physiology, systematics, woody plants, Compositae, fungi, and cryptogams (CU-RS). In addition to its pioneering program in botany, Cornell improved its facilities about 1883 by adding a conservatory which provided fresh plants for studying year round (Anonymous, 1883).

The Natural History Society also fostered Atkinson's scientific curiosity and after joining the organization in the spring of 1884, he quickly became an active member. He made several presentations including a talk on orchids and their modes of cross fertilization (Anonymous, 1884), a review of journals (Anonymous, 1885d), and a lecture on the Lemaneaceae (NHSR). The last mentioned topic was the focus of his bachelor's thesis, "Observations on the Development, Structure, and Alternate Generations of a Species of Fresh Water Alga of the Order Lemaneaceae" (Atkinson, 1885c) (Fig. 14.1). His thesis received honorable mention. In his subsequent work on *Lemanea*, which included specimens that he had collected in Chapel Hill (Atkinson, 1907) (Fig. 14.2), Atkinson ([1890]) developed this thesis into a published monograph.

Atkinson received the Ph.B. (Bachelor of Philosophy) in 1885. A number of publications provide biographical insight on Atkinson and his scientific writings: Anonymous, 1906; Anonymous, 1918; Cobb, 2015; CUP-GFA; Dirig, 2010; Farlow et al., 1919; Peattie, 1928; Stafleu and Cowan, 1976; Stafleu and Mennega, 1992; Thom, 1956; Tjossem, 1999; and Wright, 1953. Several archival sources (BE, CU, CUP, and UNC) house relevant manuscripts.

Frontispiece Credits: (above) George Francis Atkinson, ca. 1895. (image 15311). (below) Botany 1 Lecture room, Sage College, Cornell University, ca. 1891. (image 15312a). (Courtesy of Photograph Collection, Plant Pathology Herbarium, Cornell University).

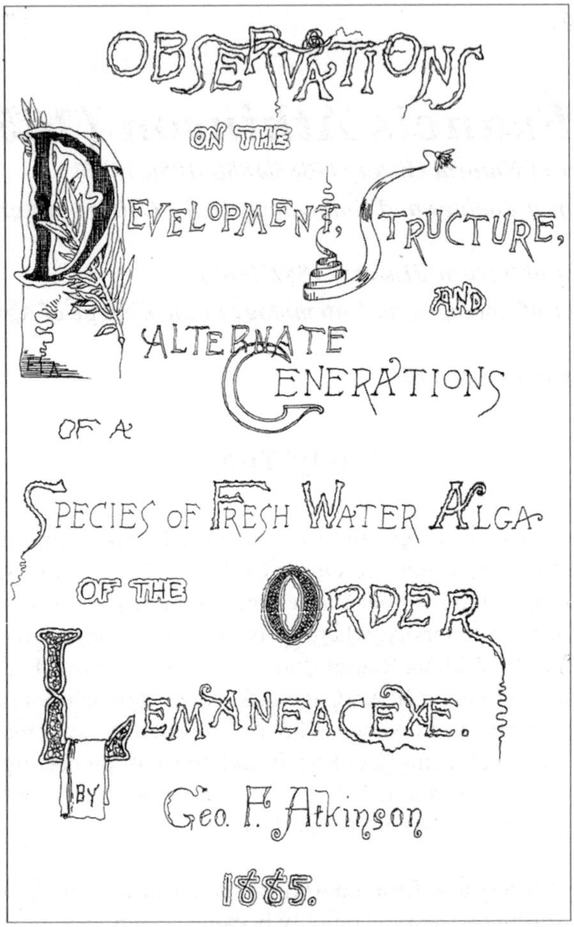

Fig. 14.1. Cover from *Observations on the Development, Structure, and Alternate Generations of a Species of Fresh Water Alga of the Order Lemaneaceae*, by George F. Atkinson (1885c).

(Courtesy of Olin Library, Cornell University).

Appointment and Teaching at UNC

George Francis Atkinson's candidacy for the position of assistant professor of natural history at UNC was strongly endorsed by both zoology professor Burt Green Wilder and President Andrew Dickson White of Cornell University, where Atkinson had specialized with "remarkable distinction" in zoology and botany as an undergraduate student (Anonymous, 1885b). Choosing him from a pool of over thirty applicants (UNC Archives), the UNC Board of Trustees unanimously approved Atkinson's appointment on July 1, 1885 and set his annual salary at $1,000 (BOT 8: 171), raising it to $1,200 in 1886 (BOT 8: 263).

In his letter accepting the UNC professorship, Atkinson (1885a) mentioned that he anticipated arriving in Chapel Hill around August 1, 1885, to begin his appointment. A reporter for a local newspaper was favorably impressed by Atkinson and noted that he was "brim full of energy and Yankee vim" (Anonymous, 1885a). In October 1885, Atkinson (1885b) wrote to his friend Comstock at Cornell that he had four students in entomology and was expecting a fifth. By the fall of 1886, he wrote that he had twelve students beginning entomology (Atkinson, 1886g). His workload became increasingly demanding. In the fall session of 1887, he had seventy-eight students taking courses in physiology (forty-five students), biology (ten), elementary entomology (nine), advanced entomology (five), and domestic animals (nine). The second term of that academic year was equally busy for Atkinson as he taught several classes: in zoology (thirty-seven students), biology (seventeen), elementary entomology (seven), advanced entomology (five), and domestic animals (eight) (BOT 8: 348–350). He supervised laboratory work in entomology, ornithology, and gross anatomy of animals and also supervised pre-med students. Students in the College of Agriculture took his courses in zoology and entomology. His courses in biology and entomology had a practical application and required laboratory work. Entomology also required fieldwork. Joking about his entomology classes, the editors of the *University Magazine* wrote, "Professor of Entomology is going to take his class out on a buggy ride this spring. (Cling! Clang!)" (Anonymous, 1887c). Atkinson's heavy schedule curtailed his aspirations of collecting specimens in the field during the fall 1887 session (Atkinson, 1887n). However, he had collected and prepared over 130 species of native birds, 70 species of butterflies, and over 200 species of other kinds of insects since his arrival in Chapel Hill.

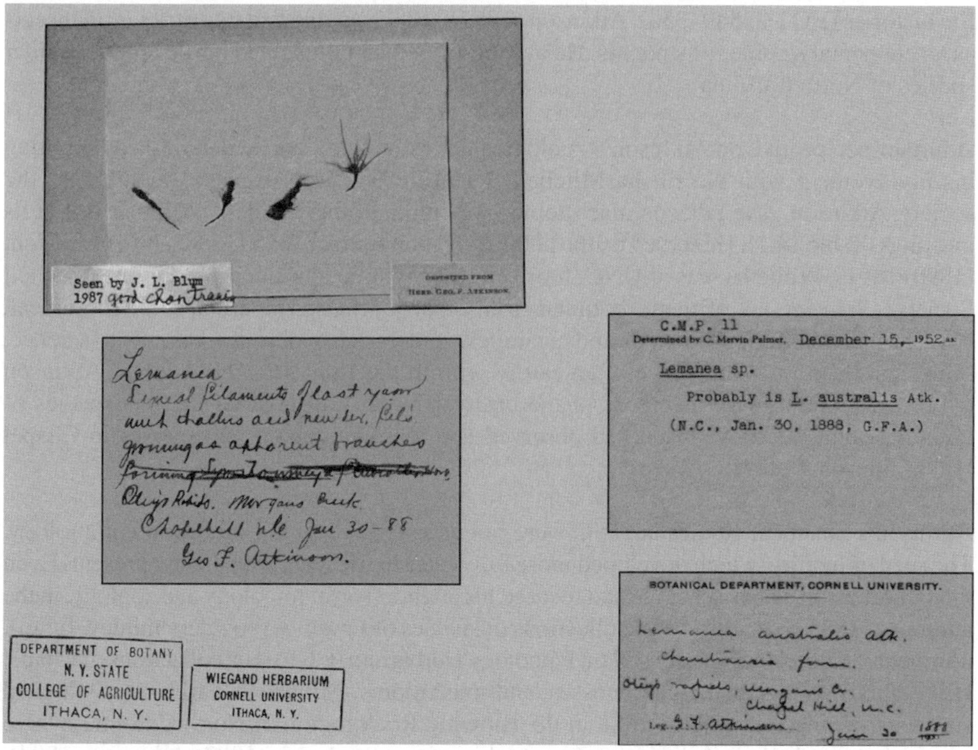

Fig. 14.2. *Lemanea australis* Atkinson, collected in Chapel Hill, North Carolina, January 30, 1888.

(Courtesy of Robert Dirig, Wiegand Herbarium, Cornell University).

Descriptions of Atkinson's courses of instruction are provided in the UNC catalog of 1885–1886 (pp. 42–43). Undergraduate classes offered were (1) Hygiene, (2) Physiology, (3) General Zoology, (4) Biological Laboratory, (5) Entomology, (6) Economic Entomology (elective), (7) Anatomy and Physiology of Domestic Animals, and (8) Feeding and Breeding of Animals. Students who wished to collect specimens for the University Museum during the summer could apply that work toward one of these courses. Postgraduate work included courses in (1) Animal Biology, Systematic Zoology, and Physiology and (2) Entomology, which covered systematic and economic aspects. Atkinson also taught "Special Courses of Instruction for Teachers" in Elementary Entomology and General Zoology (Battle et al., 1887). These classes began in February 1888 and lasted for three months.

Research and Publications at UNC

Beyond his university teaching, Atkinson conducted research and published articles. This work covered two broad areas: (1) the collection, identification, and description of animals, plants, and fungi, and (2) the practical application of his knowledge to assist farmers and horticulturists in controlling plant pests, both entomological and microbial.

Atkinson's entomological and zoological studies were diverse. Early in his career, he collected and preserved insects, many of which had been reared in the laboratory to determine their life histories (BOT 8: 349–350). Atkinson (1887a, 1887b) published two articles on insects as an elementary guide for students. He also described and illustrated new trap-door spider species of North Carolina.

A further perspective on Atkinson's zoological investigations can be determined by noting his involvement with the Elisha Mitchell Scientific Society. An active member of the society, Atkinson gave talks on many topics—including fungi, spiders, and insects—at its meetings (Table 14.1). He served on the publication committee (1885–1886) and as president (1890–1891). While he was at UNC, fourteen of his publications appeared in the Mitchell Society's *Journal*, six of them reprinted from other journals. The articles, all zoological, covered insects injurious to crops and cigarettes, spiders, a fish described as new to science, birds, and the phosphorescence of an earthworm. In the 1887 *Mitchell Journal*, Atkinson published a preliminary inventory of the birds of North Carolina. Of the 255 species of birds he enumerated, Atkinson had observed and identified about 120 species in Chapel Hill (Atkinson, [1887]o, 1887p).

Atkinson's botanical studies at UNC were not as extensive as his studies in entomology. The seed of inquiry, which developed more fully later in life, was nonetheless present. Even though Atkinson's busy schedule had focused his attention on entomology and zoology, in the autumn of 1885 he made several collections of species of *Erysiphe* (powdery mildew fungi), now held at Cornell University's Plant Pathology Herbarium (CUP). His collections of Chapel Hill plants also included algae, mosses, and mushrooms. Atkinson's earliest mycological surveys took place in August 1888 in the Blowing Rock region of North Carolina, where he also returned the following summer (Atkinson and Schrenk, 1892; Atkinson, 1894). These mycological excursions were probably undertaken at his own expense. Atkinson's published work (1891) described these fungi and others that he collected in South Carolina and Alabama. Mycology would later become his abiding focus of research. In the archives of Cornell's Plant Pathology Herbarium (CUP-GFA) is a small collection of watercolor paintings, probably made by Atkinson, of Chapel Hill mushrooms (Fig. 14.3). After leaving Chapel Hill and beginning to concentrate on botanical investigations, Atkinson continued to make extensive collections of fungi and plants in North Carolina.[95]

TABLE 14.1. A CHRONOLOGICAL LIST OF TITLES OF PAPERS PRESENTED BY GEORGE F. ATKINSON BEFORE THE ELISHA MITCHELL SCIENTIFIC SOCIETY.

Title	Date
Peach Tree Boring Beetle	October 17, 1885
Sketch of Dr. N. M. Hentz	November 25, 1885
Trap-door Spiders	November 25, 1885
Further Observations on the Trap-door Spider	January 27, 1886
Account of the Cigarette Beetle	February 17, 1886
A Nest of Ants with Domesticated Animals	March 17, 1886
Directions for Arranging a Natural History Cabinet	April 21, 1886
New Instances of Protective Resemblance in Insects	September 14, 1886
Report on a Recent Discovery in Biology	September 14, 1886
A New Glow-Worm	November 1886
Can an Air-bubble Function as an Organ of Respiration	January 25, 1887
A New Insect Epidemic	February 8, 1887
A Sarcoptid Mite	April 12, 1887
The New Glow Worm	April 12, 1887
Note on Apple Blight	May 10, 1887
[Preliminary List of the] Birds of Chapel Hill	May 10, 1887
Phosphorescence in an Earth-worm	September 12, 1887
A Peculiar Ant's Nest	September 12, 1887
Further Notes on The Glow-worm	September 12, 1887
A New Food Plant for *Blissus cucopterus* [*i.e.*, *leucopterus*]	November 8, 1887
Rust Mite on Cotton	November 8, 1887
A List of Birds Collected by Students at Chapel Hill, Spring of '77 [*i.e.*, 1887]	November 8, 1887
The Flight of Birds	May 8, 1888
Nematode Root Galls (by title only)	December 3, 1889
Some Erysipheae from Carolina and Alabama	December 2, 1890
Some Cercosporae from Alabama	December 8, 1891

[95] Accession ledgers of these collections, including his 1885 collections of *Erysiphe*, are housed at CUP. Volume 2 has entries for fungi: nos. 77–617, 639, Blowing Rock; nos. 626–632, Chapel Hill.—Volume 8, entries for plants and fungi: nos. 10001–12854, collected with Elias J. Durand, Blowing Rock (Wautauga and Caldwell Counties), August 19–September 22, 1901.—Volume 16, entries for fungi: nos. 25296–25359, vicinity of Hendersonville, July 23–31, 1918; no. 25360, Belmont, August 2, 1918; nos. 25361–25365, Belmont or Chapel Hill; nos. 25366–25408, Chapel Hill, August 5–10, 1918.—Volume 17, entries for fungi: nos. 25409–25442, Chapel Hill, August 12–14, 1918.

Fig. 14.3. Watercolor painting of an unidentified mushroom collected in Chapel Hill, North Carolina, ca. late 1880s.

(Courtesy of Plant Pathology Herbarium, Cornell University).

Atkinson's research endeavors also benefited the Chapel Hill community. Evidence of this is apparent in his participation in the Elisha Mitchell Society, discussed above, as well as in his donation of specimens to the University Museum. Around 1886, the museum's collection of insects was virtually destroyed by pests. Atkinson, however, had collected and preserved several thousand insects, representing nearly a thousand species. Most of the Chapel Hill specimens of birds collected by Atkinson and his students were also preserved and housed in the University Museum (Atkinson, [1887]o, 1887p). Currently, the UNC bird collection in the Department of Biology contains six skins collected by Atkinson in Chapel Hill between February 1886 and June 1887, including a Bachman's sparrow, a species no longer found near the town.

Atkinson's first known publication on a microbial plant pathogen was a well-researched article on the apple tree blight. Atkinson (1886b, 1887e) wrote this extensive scientific article while visiting Ithaca, New York, during the summer of 1886. The disease had infected trees in the orchards of Professor Adolphus W. Mangum and Mr. Martin in Chapel Hill and was generally prevalent throughout the state, according to Kemp P. Battle. From the twig samples that Atkinson examined, he determined that a microorganism, not an insect, was the cause of the blight. It is apparent that Atkinson kept current with scientific advancements in his field, as he pointed out that only "within the last year [was a bacterial cause] confirmed by experiments." He traced the history of apple tree blight (and its mode of action) caused by the bacterium *Micrococcus amylovorus* Burrill. A week later, the publication was reprinted from the *Raleigh News and Observer* in the *Progressive Farmer* (Atkinson, 1886c).

The farmers and horticulturists of North Carolina benefited from Atkinson's research in agricultural entomology and plant pathology. He disseminated current information about plant diseases, pests, and methods of controlling them through several local newspapers including the *News and Observer* in Raleigh, the *Bulletin* issued by the North Carolina Department of Agriculture in Raleigh, and the *Progressive Farmer*, a weekly agricultural newspaper published first in Winston and later (April 1887) in Raleigh. Interestingly, this work and these publications escaped the scrutiny of his biographers. It becomes clear that Atkinson's later work on plant pests and parasites was established through the pioneering research he undertook at UNC.

While at UNC, Atkinson devoted the preponderance of his research endeavors in educating the agriculturists and horticulturists of the state about injurious insects. (Plant pathology would become the focus of his research soon after leaving Chapel Hill.) Hitherto, he noted, there had been "no systematic efforts" to assist North Carolina's agriculturists in controlling the pests (Atkinson, 1887f). Therefore, he began studying insect habits and systematics and experimenting with remedies to control them. He asked the readers of the *Bulletin* issued by the North Carolina Department of Agriculture for their assistance in sending him insect specimens and diseased plants as well as illustrations showing the activity of the insects. With samples of diseased plants in his possession, Atkinson could determine if the causes of disease were insects or fungi. He also mentioned his special interest in parasitic fungi. Although he published reports in the *News and Observer* and the *Progressive Farmer*, the *Bulletin* became the primary medium in which he reported the results of his investigations and answered letters that he had received from farmers.

Atkinson devoted most of his entomological studies to insects that were injurious to crops and garden plants. Soon after arriving in Chapel Hill, he presented a paper on the peach tree-boring beetle at the October 17, 1885, meeting of the Elisha Mitchell Scientific Society. Although his conclusions about the insect were preliminary, he wanted to explain his research results for the benefit of society (Anonymous, 1885c).

Atkinson particularly focused on studying the chinch bug (*Blissus cucopterus* [*i.e.*, *leucopterus*]), which was a serious pest in North Carolina cornfields. In 1887, Professor Charles Phillips (1887) noted the insects were more numerous than ever, probably in the millions. Atkinson experimented with two types of sprays to destroy the insect on corn. He used a machine that he obtained from Washington, D.C., for applying the mixture. He reported success to his readers in controlling these insects with a mixture of common soap, water, and kerosene oil or with a solution of sulfo-tobacco soap. The chinch bug and its control were common themes in several of his articles (Atkinson, 1887g, 1887h, 1887i, 1888b). Atkinson believed that farmers should gain basic skills in identifying the "enemy" when it was first observed and then in controlling it before it destroyed their crops (Atkinson, 1887c). He also wrote about the corn billbug, or kazoo bug (*Sphenophorus sculptilis*); a mite that causes red rust on cotton; a fly (*Authoromyia brassicae*) that causes club-root in cabbage; the Colorado potato beetle (*Doryphora 10-lineata*); the apple tree tent caterpillar (*Clisiocampa americana*); the fall web worm (*Hyphantria textor*); and the harlequin cabbage bug (*Murgantia histrionica*) (Atkinson, 1886a, 1886f, 1887d, 1887g, 1887j, 1887k, 1887l, 1887m).

Another aspect of Atkinson's scientific investigations into the control of insect pests involved the use of fungi as a biological control; his interest in entomopathogenic fungi developed in Chapel Hill. At the Mitchell Society meeting of February 8, 1887, he reported on a species of *Entomophthora* that infected and killed soldier beetles, *Chauliognathus pennsylvanicus* DeGeer (Anonymous, 1887a; Anonymous, 1887b). He tried to culture the fungus in an effort to spread the disease to the injurious beetles. Although this experiment was unsuccessful, it was clearly a resourceful effort to promote the biological control of insects. He exhibited mounted specimens of goldenrod with the beetles attached and microscope slide mounts of the fungus. Atkinson also bred the species of insect parasites *Chiropachys* [*i.e., Cheiropachus*] *colon* Linnaeus and *Eupelmus* in his laboratory. He observed that the organisms were abundant on trees infected with scolytid tree-boring beetles and were very effective in keeping the spread of *Scolytus* in check (Atkinson, 1886i).

Atkinson became a knowledgeable resource from whom businesses and citizens sought assistance in the identification and control of insects; he was also often invited as a speaker. The February 1886 issue of the *University Magazine* reported on Atkinson's investigations of an infestation of the cigarette beetle (*Lasioderma serricorne* Fabricius) in tobacco warehouses in Durham, North Carolina. After the cigarettes were made, the beetles bored out, ruining the product (Anonymous, 1886a). Atkinson gave a talk on this subject at the February 17, 1886, meeting of the Elisha Mitchell Scientific Society (Anonymous, 1886b). In a related article, Atkinson (1886d) described the natural history of the beetle, which until then had not been fully determined, and he outlined several remedies for infestations.

Atkinson received specimens from residents of North Carolina not only for agricultural reasons but also for general identification. In late March 1886, Herbert Bemerton Battle, who was assistant chemist at the North Carolina Agricultural Experiment Station, sent Atkinson two kinds of insects that had formed swarms. Atkinson (1886h) described in detail the distinctive adaptive features and natural history of these insects. Of the first, "water scorpions"—more accurately giant water bugs belonging to the genus *Belostoma*—Atkinson explained to Battle that they were true "bugs" and were probably seeking more favorable ponds in which to live but had been attracted to the lights of the city. Out of their usual habitat, they died. The second group of insects were "water tigers" (*Dytiscus margicollis* LeConte), currently called predacious diving beetles. Their migratory fate, Atkinson explained, was similar to that of the giant water bugs.

On August 4, 1886, at the meeting of the North Carolina Horticultural Society (previously called the Fruit Growers of North Carolina) held in Fayetteville, North Carolina, Atkinson

(1886e) presented a paper on "The Relation of Insects to Horticulture." He prefaced his presentation with some comments about the benefits of insects but focused mostly on species that were injurious to fruit and fruit trees. His concluding remarks were about insecticides, of which he gave several examples, such as Paris green, London purple, and kerosene emulsions.

Although Atkinson's newspaper columns aided farmers, he recommended that the "tillers of soil, or those directly supervising the labor" should themselves acquire good schooling. He suggested a common school education followed by special training in agriculture, including training in the practical applications of the sciences such as chemistry, botany, and entomology (Atkinson, 1887c).

Career after Leaving North Carolina

Atkinson advanced in the academic ranks to associate professor of natural history in 1887 (BOT 8: 306), but his post was abolished when faculty positions were cut upon the establishment of the College of Agriculture in Raleigh and the loss of the Land Scrip fund ($125,000.00) at UNC. He wrote to Comstock at Cornell that his position would have to be vacated by the summer of 1888 (Atkinson, 1888a). President Battle deplored Atkinson's loss but gave him great praise as a scholar, teacher, and "a most accomplished expert in his department." Battle said: "No such intelligent and skillful work in the study of insects has ever been done within our limits. He had also the power of inspiring in his students" (BOT S-8: 381). Battle expressed additional praise for Atkinson in a letter of reference written to the administrators of the University of South Carolina, where Atkinson had applied for a professorship. Of Atkinson, Battle wrote: "He is remarkably skilled in conducting experiments, . . . and he has made original discoveries since his residence with us" (Anonymous, [1888]b). An editor of the *University Magazine* considered Atkinson's departure "a loss not only to the University, but to the State" because "he [had] made a fine impression upon the students who [regretted] his leaving" (Anonymous, 1888a). Twelve years after leaving Chapel Hill, Atkinson (1900a) still maintained "pleasant recollections of [his] work at the University."

In the fall of 1888, Atkinson commenced working as a professor of botany and zoology at the University of South Carolina, where he also served as *ex officio* entomologist to the Experiment Station. He was offered a salary of $1,500.00. In a letter of May 16, 1888, to Comstock, he mentioned that he would have "three fine rooms in a large Hall (which is being remodeled from a large and imposing building for a science Hall) for lecture room, laboratory & private office." In addition, he reported, "They give $1,000.00 to purchase apparatus & fit up tables & cabinets, & one or two hundred dollars for books" (Atkinson, 1888c). In the following year, however, he left South Carolina and took the position of biologist at the Polytechnic Institute and the Agricultural and Mechanical College of Alabama (now Auburn University) and in its Experiment Station. While in Alabama, Atkinson's mycological studies blossomed; he collected and identified fungi in Lee County with his students Benjamin Minge Duggar and Clifford Lewis Newman. Their resulting list of some six hundred species included sixty-one described as new to science (Davis, 1992).

Returning to his *alma mater* in Ithaca, New York, in 1892, Atkinson accepted a professorship in cryptogamic botany, replacing his former professor William R. Dudley. When Albert N.

Prentiss retired, Atkinson assumed leadership of the Botany Department in 1896 (Fitzpatrick, 1926a). He advanced in the academic ranks: assistant professor (1892–1893), associate professor (1893–1896), professor (1896–1917) and professor emeritus (1917–1918). At Cornell, his mycological studies, especially with the agarics, culminated in the publication of one of the country's foremost and earliest guides to the subject, *Studies of American Fungi: Mushrooms, edible, poisonous, etc*. (Atkinson, 1900b) (Fig. 14.4). It remained popular into the early 1960s (Atkinson, 1961). American mycologist Curtis Gates Lloyd (1900) hailed the book as "the beginning of a new era in the study of American Fungi." For numerous years, Atkinson (Fig. 14.5) headed the Cornell Botany Department and served the Cornell Agricultural Experiment Station.

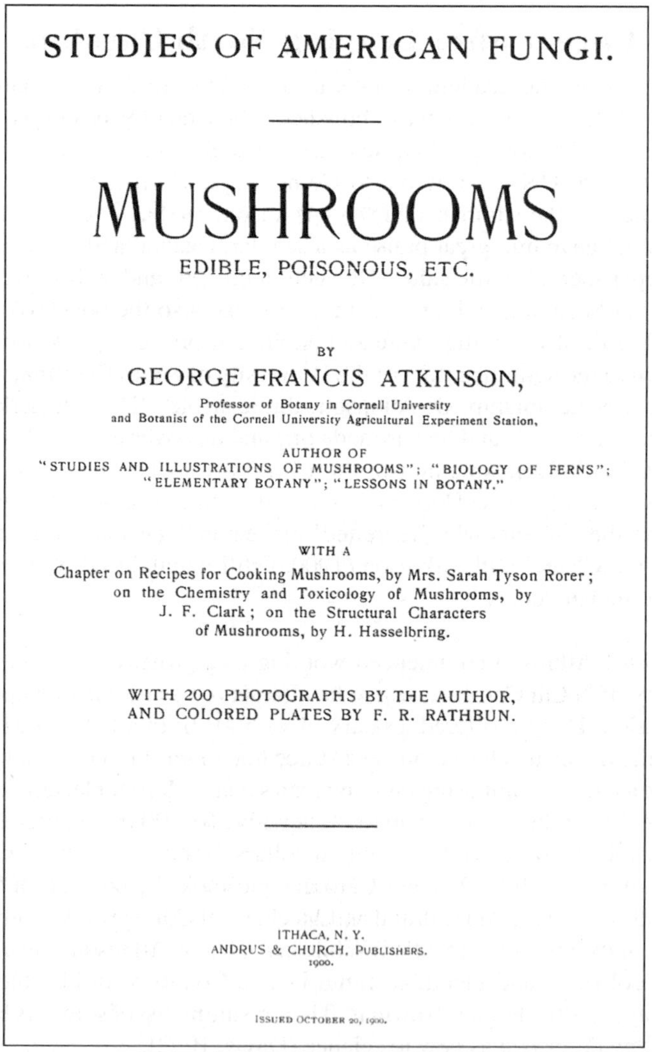
STUDIES OF AMERICAN FUNGI.

MUSHROOMS

EDIBLE, POISONOUS, ETC.

BY

GEORGE FRANCIS ATKINSON,

Professor of Botany in Cornell University
and Botanist of the Cornell University Agricultural Experiment Station,

AUTHOR OF
"STUDIES AND ILLUSTRATIONS OF MUSHROOMS"; "BIOLOGY OF FERNS";
"ELEMENTARY BOTANY"; "LESSONS IN BOTANY."

WITH A
Chapter on Recipes for Cooking Mushrooms, by Mrs. Sarah Tyson Rorer; on the Chemistry and Toxicology of Mushrooms, by J. F. Clark; on the Structural Characters of Mushrooms, by H. Hasselbring.

WITH 200 PHOTOGRAPHS BY THE AUTHOR,
AND COLORED PLATES BY F. R. RATHBUN.

ITHACA, N. Y.
ANDRUS & CHURCH, PUBLISHERS.
1900.

ISSUED OCTOBER 20, 1900.

Fig. 14.4. Title page from *Studies of American Fungi: Mushrooms, edible, poisonous, etc.*, by George F. Atkinson (1900).

(Courtesy of Walter R. Davis Library, University of North Carolina, Chapel Hill).

Fig. 14.5. George Francis Atkinson, 1892.

(Courtesy of Division of Rare and Manuscript Collections, Carl A. Kroch Library, Cornell University–Deceased Alumni Files, #41-2-877, Box 4 [Atki–Baco]).

Ongoing Links to the University of North Carolina and the State

Perhaps the prospect of collecting mushrooms in the South enticed Atkinson back to UNC to teach botany at its 1901 summer school. Concerning Atkinson's anticipated teaching in the summer of 1901, UNC President Francis Preston Venable (1901) commented that Atkinson's idea of taking his class on field excursions was "most admirable." According to information in course catalogs and announcements, his botany class was the first at UNC to place an emphasis on plant ecology; it was probably one of the first in the country to do so. Atkinson ([1904]) later delivered lectures on the relation of plants to environment in Cornell University's summer schools of 1903 and 1904. The UNC course entailed investigations of the following topics:

1. Plant Work: The parts of plants which work; the root, stem, leaf, flower. How plants obtain and use food and water.
2. Plant Behavior: The positions, forms, and movements of plants, or their parts, in securing favorable relations to their environment, for the purpose of work.
3. Plant Societies: The relations of plants among themselves, and to different kinds of territory. Congenial and uncongenial societies. Struggles among plants for territory, and for supremacy.

Atkinson may also have taught botany in the 1902 summer session. Although Kemp P. Battle (1907–1912, 2: 619) cited Atkinson as a summer lecturer in botany in 1902, no record of his teaching for that year can be found in the UNC faculty journal, the UNC catalog, or the trustees' minutes.

Between 1902 and 1918, Atkinson exchanged a series of letters with William Chambers Coker, the founder of UNC's Department of Botany, established in 1908. Nearly twenty letters are deposited in the University of North Carolina's Southern Historical Collection, over 110 in the archives at Cornell University, and one in the South Caroliniana Library at the University of South Carolina. Of special note is that Coker, upon his arrival in Chapel Hill in 1902 as the newly appointed associate professor of botany in the Department of Biology, carried with him a copy of Atkinson's (1900b) mushroom book (Couch and Matthews, 1954). One wonders if Atkinson's monograph inspired Coker's interests in mycology. No doubt exists, however, that after they began to correspond, Atkinson became a mentor to Coker, who subsequently sent numerous species of mushrooms for Atkinson's identification to Ithaca, New York. Among the specimens were those that Coker had collected on his Bahama expedition of 1903, which was sponsored by the Geographical Society of Baltimore. Atkinson determined the species of fungi for Coker's (1905) publication on the vegetation of the Bahama Islands.

A group of Atkinson's former students (Calvin Henry Kauffman, Charles Thom, Fred Carlton Stewart, Frederick Adolph Wolf, Henry Spencer Jackson, and Herbert Hice Whetzel) met informally during the December 1916 meeting of the American Association for the Advancement of Science in New York City. They discussed the possibility of publishing Atkinson's lifelong work on North American mushrooms, especially that on the agarics and clavarias (Whetzel, 1917b). Writing to over 125 research associates and former students of Atkinson's, Whetzel asked that they write Atkinson to acknowledge their support for this project (Whetzel, 1917a). The Board of Trustees relieved Atkinson of teaching and administrative responsibilities, so he could devote time to completing mycological fieldwork and preparing a manuscript. In the summer of 1917, Atkinson asked his mycological correspondents and former students for assistance in providing photographs and specimens of mushrooms (Whetzel, 1917a). Perhaps the grandest publishing venture in American mycology ever, it was envisioned as a work that would represent the agarics and clavarias of temperate North America, to be issued in ten volumes, each with approximately 300 pages and 150 plates of illustrations ([Atkinson, 1917]). Atkinson applied for grants to finance extensive fieldwork, which was necessary to complete this mycological venture.

With a grant received in 1918 from the United States Department of Agriculture, Atkinson commenced an expedition to collect, photograph, and study fleshy fungi. The field studies covered two collecting regions, the Southeast and the Pacific Northwest. Assisting him in the Southeast were Edwina Smiley and Dr. Gertrude Elizabeth Douglas, the latter working *pro bono*. Some of Dr. Douglas's letters pertaining to this collaborative work were published in *McIlvainea* (Haines, 1987). Atkinson (1918b) went to Florida, where he purchased a car and then spent "the first few days ... learning to drive." In late May, the collectors gathered mushrooms in Gainesville and made their way northward to Thomasville, Statesboro, and Augusta, Georgia; Columbia, South Carolina; and finally North Carolina. Arriving in Chapel Hill on August 5, Douglas wrote: "We are in the most adorable little town of 1,000 inhabitants, a place which has hardly changed since G. [*i.e.*, George Atkinson] was here. He's just reveling in it and has gone after Amanitas already." They were pleased to work in the comfort of Coker's laboratory. In Chapel Hill, Atkinson (1918b) had "a charming visit with Dr. Kemp Battle." He reminisced that when he was a professor at UNC, "many of our collections were made in a beautiful forest known as Battle Park; through this runs a winding stream along which are paths, the brush and small trees for the paths being cut by Doctor Battle himself." Battle's love for the woodlands was well known among the villagers. On a daily basis, he spent at least an hour or so not only clearing paths with his hatchet but also making bridges and seats and pointing out his favorite settings to friends and students (Anonymous, [1935]).

Atkinson (Fig. 14.6) then collected fungi in Hot Springs, Virginia, where he received a letter (dated August 21, 1918) from Coker (1918). Even though Coker had "scarcely had time to see" Atkinson in Chapel Hill, he had enjoyed having him at his house. Several days later, while still in Hot Springs, Atkinson (1918a) thanked his host for his hospitality and expressed his desire to return to Chapel Hill to collect mushrooms when a rainy period prevailed. After collecting fungi in Virginia, Atkinson returned in September to Ithaca for a brief stay. In writing to mycologist Herbert H. Whetzel about the mycological excursion, Atkinson (1918b) stated that it was quite successful, despite some dry weather.

Atkinson next traveled to the Pacific Northwest for additional collecting, working primarily by himself. In September, he passed through Seattle, Washington, and then spent several weeks collecting fungi on Mount Rainier (Frye, 1918). Securing a room in Tacoma, Atkinson made his last trip—to Spanaway Lake, nearly fifteen miles away. Overworked by his field collecting, Atkinson contracted influenza at age seventy-two. According to his certificate of death, he died of acute lobar pneumonia on November 14, 1918, in Tacoma, a victim of the great flu epidemic of 1918. University of Washington botanist Theodore Christian Frye related that Atkinson's last words consisted of a dictation of notes about his fungi to his nurse, who had been hired two days before his death. He was survived by his wife Elizabeth (Lizzie) S. Kerr from whom he was estranged (Thom, 1956). Kerr and Atkinson had been married on March 8, 1887, in Durham County, North Carolina. (Her father was Washington Caruthers Kerr, who had served as North Carolina State Geologist in 1864 and 1866–1882, and died on August 9, 1885.) Also surviving Atkinson was his son Francis Kerr. His daughters, Clara Packard Atkinson and Josephine Atkinson, predeceased him. Atkinson is buried in Covert, Michigan (Fig. 14.7), on the shores of Lake Michigan.

Honors and Legacy

Among Atkinson's honors were election to Phi Beta Kappa (1885) and Sigma Xi (1888). He was elected a fellow in the American Association for the Advancement of Science and the National Academy of Sciences (1918). In the AAAS, he served as secretary of the Botanical Section (1895) and vice-president (1896). Atkinson was a charter member of the Botanical Society of America, serving as its first president in 1907. He held memberships in the American Philosophical Society, the Mycological Society of America, the Society for Plant Morphology and Physiology, and the Torrey Botanical Club. At Ithaca, he was a member of the Natural History Society of Cornell University, which had been established in 1869 (Wright, 1954). For many years, he served on the editorial staff of the *Botanical*

Gazette, now the *International Journal of Plant Sciences*. He is commemorated in the names of one genus and twenty-five species and varieties of fungi (Table 14.2). In his memory, Atkinson's family gave his entire scientific library, notes, and plant collections to Cornell (Anonymous, 1920).

Atkinson's memory was honored at the International Congress of Plant Sciences, at Cornell University, Ithaca, New York, on August 20, 1926. For the first time, his students and friends gathered at an informal luncheon at Willard Straight Hall. Benjamin M. Duggar, Karl McKay Wiegand, Herbert S. Jackson, and Harry Philip Brown spoke on the occasion, and over forty people attended (Fitzpatrick, 1926b).

Fig. 14.6. George Francis Atkinson, ca. 1918. (image 15337).
(Courtesy of Photograph Collection, Plant Pathology Herbarium, Cornell University).

Fig. 14.7. Gravestone of George Francis Atkinson, Covert Cemetery, Covert, Michigan, June 2006.

(Courtesy of Miriam Downey).

Contributions to UNC

At UNC, Atkinson's instruction encompassed zoological aspects within the College of Natural History and the College of Agriculture. He also taught school hygiene to freshmen, and animal biology, systematic zoology, systematic physiology, and entomology to postgraduates. Attributable to his educational experience at Cornell University, fieldwork and experimental investigations in the laboratory were important components of his scientific endeavors, and he published numerous articles based on his research. Additionally, he devoted considerable time to collecting, classifying, and preserving birds and insects of the Chapel Hill area.

Atkinson's scientific studies also had a distinct economic and practical focus. He was a pioneer in the agricultural entomology and plant pathology of North Carolina; he conducted research on microbial and insect pests of economically important plants. He was the first to determine the life cycles and to experiment with ways of combating a number of species of insects. Atkinson's ardent and detailed observations were fundamental to his success as a scientist. His lectures in botany were the first at UNC—and among the first in the country—to focus on plant ecology.

TABLE 14.2. A LIST OF NAMES OF FUNGI COMMEMORATING GEORGE F. ATKINSON.

(ACCEPTED NAMES ARE IN BOLDFACE; SYNONYMS IN ITALICS, BASED ON THE PUBLISHED LITERATURE)

Ascomycetes

Atkinsonella Diehl, Agricultural Monographs 4: 48. 1950.
Atkinsonella hypoxylon (Peck) Diehl, Agricultural Monographs 4: 49. 1950.
Atkinsonella texensis (Diehl) Leuchtmann & Clay, Mycologia 81: 695. 1989. *Atkinsonella hypoxylon* var. *texensis* Diehl, Agricultural Monographs 4: 54. 1950.
Calonectria (Chiajaea) atkinsonii Rehm., Annales Mycologici 20: 178. 1904.
Cercospora atkinsonii Stev. & Solh., Mycologia 23: 388. 1931.
Meliola atkinsonii Hansford, Sydowia 10: 63. 1957.
Seynesia atkinsonii Stev. & Ryan, Bernice P. Bishop Museum Bulletin 19: 69. 1925.
Taphrina atkinsonii W. W. Ray, Mycologia 31: 66. 1939.

Basidiomycetes

Amanita atkinsoniana Coker, Journal of the Elisha Mitchell Scientific Society 33: 84. 1917. *Armillaria atkinsoniana* (Coker) Locq., Bulletin Trimestriel de la Société Mycologique de France 68: 167. 1952.
Boletus atkinsonii Peck, New York State Museum Bulletin 94 (Botany 8): 20. [1905].
Boletus atkinsonianus (Murrill) Coker & Beers, The Boletaceae of North Carolina. University of North Carolina Press, Chapel Hill. 49. 1943
Ceriomyces atkinsonianus Murrill, North American Flora 9: 144. 1910.
Conocybe atkinsonii Watling. in Roy Watling and Norma M. Gregory, Census Catalogue of World Members of the Bolbitiaceae. (*Bibliotheca Mycologica*, Bd. 82). J. Cramer, Vaduz, Liechtenstein. 95. 1981.
Corticium atkinsonii Burt, Annals of the Missouri Botanical Garden 13: 208. 1926.
Cortinarius atkinsonianus Kauffman, Bulletin of the Torrey Botanical Club 32: 324. 1905.
Galerina atkinsoniana A. H. Smith, Mycologia 45: 894. 1953.
Galerina atkinsoniana f. tetraspora Kühner, Bulletin Trimestriel de la Société Mycologique de France 88: 84. 1972.
Galerina atkinsoniana var. atkinsoniana A. H. Smith & Singer, A Monograph on the Genus *Galerina* Earle. Hafner Publishing Company, New York. 323. 1964.
Galerina atkinsoniana var. sphagnorum A. H. Smith. in A. H. Smith & R. Singer, A Monograph on the Genus *Galerina* Earle. Hafner Publishing Company, New York. 325. 1964.
Ganoderma atkinsonii Jahn, Kotlaba, and Pouzar, Westfälische Pilzbriefe 11: 98. 1980.
Kneiffia atkinsonii (Ell. & Ev.) Rick, Broteria, ser. 3, 2: 73. 1934. *Peniophora atkinsonii* Ell. & Ev., Proceedings of the Academy of Natural Sciences of Philadelphia. 1894. 324. (According to M. A. Donk. 1957. Taxon 6: 81, *Kneiffia* is impriorable).
Mycena atkinsoniana A. H. Smith, North American Species of *Mycena*, University of Michigan Press, Ann Arbor. 144. 1947.
Phlebia atkinsoniana W. B. Cooke, Mycologia 48: 393. 1956.
Puccinia atkinsoniana Dietel, Bulletin of the Cornell University (Science) 3(1): 19. 1897.
Puccinia atkinsoniana Dietel **var. bolleyana** (no author). in P. A. Saccardo, Sylloge Fungorum 17: 465. 1905.
Ramaria atkinsonii (Bres. in Atk.) R. H. Petersen, Persoonia 12: 227. 1984.
Lachnocladium atkinsonii Bres. in Atk., Journal of Mycology 8: 119. 1902.
Uredinopsis atkinsonii P. Magnus, Hedwigia 43: 123. 1904.

Conclusion

Atkinson's botanical fame and influence reached its zenith later, when he was employed at Cornell University. Most of his 150 publications appeared in botanical serials, including several foreign titles. He also wrote several textbooks on ferns as well as textbooks on general and college botany. In the words of Cornell mycologist H.H. Whetzel (1919), Atkinson was "for a generation one of the leaders of American botanical thought and activity. He was a master of the highest scientific ideals, unsparing in his criticisms, just and fair in his judgments, [and] generous with help and suggestions."

References Cited

(References that conclude with page numbers within parentheses indicate sources from where quotes and pertinent information were obtained.)

Anonymous. 1883. The botanical department. *Cornell Daily Sun* (Ithaca, NY) 3(78): [1–2], Friday, February 9, 1883.

Anonymous. 1884. The Natural History Society. *Cornell Daily Sun* (Ithaca, NY) 5(42): [2], Friday, November 14, 1884.

Anonymous. 1885a. Chapel Hill notes, p. 56 in Kemp P. Battle, *Scrapbook of Material Relating to the University of North Carolina, 1869–1893*, vol. 4. (Newspaper clipping without complete bibliographical data, possibly from *Durham Recorder*, around August 31, 1885. North Carolina Collection, Wilson Special Collections Library, University of North Carolina, Chapel Hill)

Anonymous. 1885b. The new Professors. *News and Observer* (Raleigh) 25(58): [2], Thursday, July 2, 1885.

Anonymous. 1885c. University news. *News and Observer* (Raleigh) 25(149): [2], Saturday, October 24, 1885.

Anonymous. 1885d. [Natural History Society]. *The Cornell Daily Sun* (Ithaca, NY) 5(109): [3], Thursday, April 16, 1885.

Anonymous. 1886a. [Prof. Atkinson in Durham, North Carolina]. *University Magazine* (University of North Carolina, Chapel Hill), new series 5: 151.

Anonymous. 1886b. The Mitchell Scientific Society. *University Magazine* (University of North Carolina, Chapel Hill), new series 5: 194.

Anonymous. 1887a. Elisha Mitchell Society. The February meeting. *News and Observer* (Raleigh) 28(68): [2], Thursday, February 17, 1887.

Anonymous. 1887b. A new insect epidemic. *University Magazine* (University of North Carolina, Chapel Hill), new series 6: 302.

Anonymous. 1887c. [Professor of Entomology]. *University Magazine* (University of North Carolina, Chapel Hill), new series 6: 233.

Anonymous. 1888a. [Prof. Atkinson]. *University Magazine* (University of North Carolina, Chapel Hill), new series 7: 243.

Anonymous. [1888]b. The new professors. A splendid team of intellectual athletes. Personal characteristics of the new men elected to the University faculty—the very power of scholarship—fresh blood in the educational veins of the State, p. 6 in Kemp P. Battle, *Scrapbook of North Carolina History and Biography, 1880–1894*, vol. 8. (Newspaper clipping without complete bibliographical data. North Carolina Collection, Wilson Special Collections Library, University of North Carolina, Chapel Hill)

Anonymous. 1906. Atkinson, George Francis. *National Cyclopaedia of American Biography* 13: 478–479.

Anonymous. 1918. Prof. Atkinson dies suddenly on west coast. *The Ithaca Journal* (Ithaca, NY), p. [1], Saturday, November 16, 1918. (portrait)

Anonymous. 1920. Prof. Atkinson's books presented to Cornell. *Ithaca Journal-News* (Ithaca, NY) 96(84): 3, Thursday, April 8, 1920.

Anonymous. [1935]. [University of North Carolina]. (Box 6, folder 166, University of North Carolina, Botany Department, Historical Collections #2053, Southern Historical Collection, Wilson Special Collections Library, University of North Carolina, Chapel Hill)

Atkinson, George F. 1885a. Letter of July 7, 1885 to UNC President Kemp P. Battle. (Box 15, folder 511, University Papers #40005, University Archives, Wilson Special Collections Library, University of North Carolina, Chapel Hill)

Atkinson, George F. 1885b. Letter of October 21, 1885 to John H. Comstock. (Division of Rare and Manuscript Collections, Carl A. Kroch Library, Cornell University, Ithaca, New York, George Francis Atkinson [Archives 21-29-103]; microfilm copy at Southern Historical Collection, Wilson Special Collections Library, University of North Carolina, Chapel Hill, George Francis Atkinson. Letters to John Henry Comstock [M-2959, A])

Atkinson, George F. 1885c. *Observations on the Development, Structure, and Alternate Generations of a Species of Fresh Water Alga of the Order Lemaneaceae*. Thesis completed for the requirements of the degree of Bachelor of Philosophy, Cornell University, Ithaca, New York. (Thesis Film, "T 1885 1," Cornell University Theses, 1884 #49–1885 #19, reel 20)

Atkinson, George F. 1886a. The apple tree tent caterpillar (*Clisiocampa americana*). *News and Observer* (Raleigh) 26(136): [2], Friday, April 23, 1886. Reprinted in *Progressive Farmer* (Winston, NC) 1(13): 2, May 5, 1886.

Atkinson, George F. 1886b. The apple blight (*Micrococcus amylovorus* Bur.). *News and Observer* (Raleigh) 27(33): [2], Wednesday, June 23, 1886.

Atkinson, George F. 1886c. The apple blight. The bacteria theory as explained by Prof. Atkinson. *Progressive Farmer* (Winston, NC) 1(20): 2, June 30, 1886.

Atkinson, George F. 1886d. The cigarette beetle (*Lasioderma serricorne*, Fab.). *Journal of the Elisha Mitchell Scientific Society* 3: 68–73.

Atkinson, George F. 1886e. The relation of insects to horticulture. *Report of the North Carolina State Horticultural Society* 1886: 11–14.

Atkinson, George F. 1886f. The fall web worm (*Hyphantria textor*, Harris). *Progressive Farmer* (Winston, NC) 1(31): 2, September 8, 1886.

Atkinson, George F. 1886g. Letter of October 14, 1886 to John H. Comstock. (Division of Rare and Manuscript Collections, Carl A. Kroch Library, Cornell University, Ithaca, New York, George Francis Atkinson [Archives 21-29-103]; microfilm copy at Southern Historical Collection, Wilson Special Collections Library, University of North Carolina, Chapel Hill, George Francis Atkinson. Letters to John Henry Comstock [M-2959, A])

Atkinson, George F. 1886h. Migration of water scorpions and water tigers. *News and Observer* (Raleigh) 26(116): [2], Tuesday, March 30, 1886.

Atkinson, George F. 1886i. Notes on the orchard scolytus (*Scolytus rugulosus* Ratz). *Journal of the Elisha Mitchell Scientific Society* 3: 74–75.

Atkinson, George F. 1887a. Elementary studies in insects. Part I. *North Carolina Teacher* 5: 119–124.

Atkinson, George F. 1887b. Elementary studies in insects. Concluded. *North Carolina Teacher* 5: 160–163.

Atkinson, George F. 1887c. Injurious insects. Value of an acquaintance with the common forms. *Progressive Farmer* (Winston, NC) 1(47): 2, January 5, 1887.

Atkinson, George F. 1887d. The harlequin cabbage bug. *Progressive Farmer* (Winston, NC) 2(4): 2, March 9, 1887.

Atkinson, George F. 1887e. The apple blight. *News and Observer* (Raleigh) 23(142): [2], Saturday, May 14, 1887.

Atkinson, George F. 1887f. An appeal to farmers, gardeners and the press. *Bulletin* (North Carolina Department of Agriculture) May 1887, pp. 4–5. (Also published in *Progressive Farmer* [Raleigh] 2(17): 4, June 16, 1887)

Atkinson, George F. 1887g. Injurious insects. *Bulletin* (North Carolina Department of Agriculture), June 1887, p. 4.

Atkinson, George F. 1887h. Successful experiments in destroying the chinch bug (*Blissus cucopterus* [*i.e., leucopterus*]) on corn. *Special Bulletin* (North Carolina Department of Agriculture), June 29, 1887. 1 sheet.

Atkinson, George F. 1887i. The chinch bug and how to get rid of it. *News and Observer* (Raleigh) 24(1): [4], Wednesday, June 29, 1887; also in the *Bulletin* (North Carolina Department of Agriculture), July 1887, p. 8. (These articles were reprinted from Atkinson's [1887h] special report on the chinch bug.)

Atkinson, George F. 1887j. The "red rust" on cotton. The discovery by Prof. Atkinson of the cause, and a remedy for its prevention. *Bulletin* (North Carolina Department of Agriculture), September 1887, pp. 7–8.

Atkinson, George F. 1887k. Injurious insects. *Bulletin* (North Carolina Department of Agriculture), November 1887, pp. 4–5.

Atkinson, George F. 1887l. The "club-root" in cabbage. *Bulletin* (North Carolina Department of Agriculture), November 1887, p. 4.

Atkinson, George F. 1887m. Injurious insects. *Bulletin* (North Carolina Department of Agriculture), November 1887, p. 4.

Atkinson, George F. 1887n. Letter of November 11, 1887 to John H. Comstock. (Division of Rare and Manuscript Collections, Carl A. Kroch Library, Cornell University, Ithaca, New York, George Francis Atkinson [Archives 21-29-103]; microfilm copy at Southern Historical Collection, Wilson Special Collections Library, University of North Carolina, Chapel Hill, George Francis Atkinson. Letters to John Henry Comstock [M-2959, A])

Atkinson, George F. [1887]o. Preliminary list of birds, p. 14 in Kemp P. Battle, *Scrapbook of Material Relating to the University of North Carolina, 1869–1893*, vol. 5. (Newspaper clipping without bibliographical citation. North Carolina Collection, Wilson Special Collections Library, University of North Carolina)

Atkinson, George F. 1887p. Preliminary catalogue of the birds of North Carolina, with notes on some of the species. *Journal of the Elisha Mitchell Scientific Society* 4(2): 44–84. (Pp. 85–87, "Appendix. Containing a list of birds which we may reasonably expect to yet take in the state.")

Atkinson, George F. 1888a. Letter of March 9, 1888 to John H. Comstock. (Division of Rare and Manuscript Collections, Carl A. Kroch Library, Cornell University, Ithaca, New York, George Francis Atkinson [Archives 21-29-103]; microfilm copy at Southern Historical Collection, Wilson Special Collections Library, University of North Carolina, Chapel Hill, George Francis Atkinson. Letters to John Henry Comstock [M-2959, A])

Atkinson, George F. 1888b. Destroy the bugs. Remedy for the "chinch bug" (*Blissus cucopterus* [*i.e., leucopterus*]) on corn. *Bulletin* (North Carolina Department of Agriculture), April 1888, p. 7.

Atkinson, George F. 1888c. Letter of May 16, 1888 to John H. Comstock. (Division of Rare and Manuscript Collections, Carl A. Kroch Library, Cornell University, Ithaca, New York, George Francis Atkinson [Archives 21-29-103]; microfilm copy at Southern Historical Collection, Wilson Special Collections Library, University of North Carolina, Chapel Hill, George Francis Atkinson. Letters to John Henry Comstock [M-2959, A])

Atkinson, George F. [1890]. Monograph of the Lemaneaceae of the United States. *Annals of Botany* 4: 177–229; plates 7–9.

Atkinson, George F. 1891. Some Erysipheae from Carolina and Alabama. *Journal of the Elisha Mitchell Scientific Society* 7: 61–74, 1890; plate 1.

Atkinson, George F. 1894. Additional note on the fungi of Blowing Rock, N.C. *Journal of the Elisha Mitchell Scientific Society* 10: 78.

Atkinson, George F. 1900a. Letter of April 2, 1900 to University of North Carolina President Edwin A. Alderman. (George Francis Atkinson Archives [21-29-103], Carl. A. Kroch Library, Division of Rare and Manuscript Collections, Cornell University)

Atkinson, George F. 1900b. *Studies of American Fungi. Mushrooms, Edible, Poisonous, etc.* With a chapter on recipes for cooking mushrooms, by Mrs. Sarah Tyson Rorer; on the chemistry and toxicology of mushrooms, by J.F. Clark; on the structural characters of mushrooms, by H. Hasselbring. With 200 photographs by the author, and colored plates by F.R. Rathbun. Andrus & Church, Ithaca, NY. (Later editions were: 2nd [1901, 1903]; 3rd [1911])

Atkinson, George F. [1904]. "Relation of Plants to Environment (or Plant Ecology). Outlines of a Course of Lectures Delivered in the Summer School of Cornell University 1903 and 1904." Published by the Author, Ithaca, NY.

Atkinson, George F. 1907. Letter of January 10, 1907 to F.L. Stevens. (George Francis Atkinson Archives [21-29-103], Carl. A. Kroch Library, Division of Rare and Manuscript Collections, Cornell University)

Atkinson, George F. [1917]. "Plan, Scope and Character of the Work." (George Francis Atkinson Archives [21-29-103], Carl. A. Kroch Library, Division of Rare and Manuscript Collections, Cornell University)

Atkinson, George F. 1918a. Letter of August 25, 1918 to William C. Coker. (Box 2, folder 115, William C. Coker Papers #3220, Southern Historical Collection, Wilson Special Collections Library, University of North Carolina, Chapel Hill)

Atkinson, George F. 1918b. Letter of September 7, 1918 to H.H. Whetzel. (Box 2, folder 116, William C. Coker Papers #3220, Southern Historical Collection, Wilson Special Collections Library, University of North Carolina, Chapel Hill)

Atkinson, George F. 1961. *Mushrooms, Edible, Poisonous, etc*. Hafner Publishing Company, New York. (Reprint of Atkinson's *Studies of American Fungi*, 2nd ed. [1901])

Atkinson, George F. and Hermann Schrenk. 1892. Some fungi of Blowing Rock, N.C. *Journal of the Elisha Mitchell Scientific Society* 9: 95–107.

Battle, Kemp P. 1907–1912. *History of the University of North Carolina*. Printed by Edwards & Broughton, Raleigh. (Vol. 2, p. 619)

Battle, Kemp P., Nelson B. Henry and F. P. Venable. 1887. [Broadside]. *Special Courses of Instruction for Teachers. University of North Carolina, Chapel Hill*. (North Carolina Collection, Wilson Special Collections Library, University of North Carolina, Chapel Hill)

BE. Biltmore Estates, Special Collections, Museum Services Department. George Atkinson, 1898–1904, Box Folder 2.0036, containing six letters from Atkinson to Carl Alwin Schenck, Director, Biltmore Forest School, Biltmore, North Carolina.

BOT. Board of Trustees, University of North Carolina, Chapel Hill. *Minutes of the Board of Trustees*. (University Archives #40001, Wilson Special Collections Library, University of North Carolina, Chapel Hill) (Vol. 8, June 1883–February 1891)

Cobb, Edward D. 2015. *150 Years of Botany at Cornell: A history of botany and plant biology*. Cornell University, Ithaca, NY. (p. 81)

Coker, William Chambers. 1905. Vegetation of the Bahama Islands, pp. [185]–270 & plates 1, 33–47 in George Burbank Shattuck (editor), *The Bahama Islands*. MacMillan Company, New York. ("Fungi," pp. 245–246)

Coker, William Chambers. 1918. Letter of August 21, 1918 to George F. Atkinson. (George Francis Atkinson Archives [21-29-103], Carl. A. Kroch Library, Division of Rare and Manuscript Collections, Cornell University)

Couch, John N. and Velma D. Matthews. 1954. William Chambers Coker. *Mycologia* 46: 372–383.

CU. Cornell University. Carl A. Kroch Library, Division of Rare and Manuscript Collections. George Francis Atkinson (Archives 21-29-103), contains extensive files of correspondence and other documents.

CU-RS. Cornell University Register of Studies. (Archives #36/1/630, vol. C [1880–1885], p. 188; Carl A. Kroch Library, Division of Rare and Manuscript Collections, Cornell University)

CUP. Cornell University, Plant Pathology Herbarium, Department of Plant Pathology and Plant-Microbe Biology.

CUP-GFA. Cornell University, Plant Pathology Herbarium, Department of Plant Pathology and Plant-Microbe Biology. George Francis Atkinson website, including links to several sites: the Atkinson Collection, Biography, Gallery, and Atkinson Bibliography. <http://www.plantpath.cornell.edu/CUPpages/Atkinson/index.html> (Accessed January 1, 2016)

Davis, Donald E. 1992. *History of the Department of Botany, Plant Pathology and Microbiology of Auburn University*. Assembled by Donald E. Davis. Edited by Robert T. Gudauskas and Bryan Truelove. [no statement of publisher], Auburn, Alabama. (p. 54)

Dirig, Robert. 2010. *George Francis Atkinson: Cornell's first mycologist*. Plant Pathology Herbarium (CUP), Cornell University, Ithaca, NY. Tri-fold brochure. (in author's files)

Farlow, W.G., Roland Thaxter and L.H. Bailey. 1919. George Francis Atkinson. *American Journal of Botany* 6: 301–302. (portrait; Publications of George Francis Atkinson, compiled by Harry M. Fitzpatrick, pp. 303–308)

Fitzpatrick, Harry M. 1926a. "An Historical Sketch of the Early Days of Botany in Cornell." Typed manuscript. (Copies at the Bailey Hortorium and Herbarium and the Library of the Hunt Institute for Botanical Documentation)

Fitzpatrick, Harry M. 1926b. In memory of George F. Atkinson. *Mycologia* 18: 291.

Frye, T.C. 1918. Letter of November 19, 1918 to H.H. Whetzel. (Box 2, folder 120, William C. Coker Papers #3220, Southern Historical Collection, Wilson Special Collections Library, University of North Carolina, Chapel Hill)

Haines, John (editor). 1987. The Georgian mushroom expedition. Letters of Gertrude E. Douglas, 1918. *McIlvainea* 8(1): 8–15.

Lloyd, C.G. 1900. Prof. Atkinson's book. *Mycological Notes* 6: 55–56.

NHSR. Natural History Society Records. *Proceedings [of Meetings], 1876–1881*. (Archives #37-6-584, Carl A. Kroch Library, Division of Rare and Manuscript Collections, Cornell University). (p. 121)

P[eattie], D[onald] C. 1928. Atkinson, George Francis, pp. 407–408 in Allen Johnson (editor), *Dictionary of American Biography*, vol. 1. Charles Scribner's Sons, New York.

Phillips, Charles. 1887. Letter of June 29, 1887. (The recipient's name is lacking.) (Box 1, folder 8, Charles Phillips Papers #2462, Southern Historical Collection, University of North Carolina, Chapel Hill)

Stafleu, Frans A. and Richard S. Cowan. 1976. Atkinson, George Francis, pp. 78–79 in *Taxonomic Literature* (2nd ed.), vol. I(A–G). Bohn, Scheltema & Holkema, Utrecht. (sources and publications)

Stafleu, Frans A. and Erik A. Mennega. 1992. Atkinson, George Francis, pp. 200–203 in *Taxonomic Literature*. Supplement, I(A–Ba). Koeltz Scientific Books, Koenigstein, Germany. (sources and publications)

Thom, Charles. 1956. George Francis Atkinson 1854–1918. *National Academy of Sciences of the United States of America, Biographical Memoirs* 29 (2d memoir), pp. 17–44.

Tjossem, Sara F. 1999. Atkinson, George Francis. *American National Biography* 1: 712–713.

UNC. University of North Carolina, Wilson Special Collections Library. George Francis Atkinson. Letters to John Henry Comstock (M-2959,A) and Letters to William C. Coker (#3220) in the Southern Historical Collection and eighteen letters in the University Papers, University Archives.

UNC Archives. University Papers #40005, University Archives, Wilson Special Collections Library, University of North Carolina, Chapel Hill. (Box 48, folder 1734)

Venable, Francis Preston. 1901. Letter of March 12, 1901 to George Francis Atkinson. (George Francis Atkinson Archives [21-29-103], Carl. A. Kroch Library, Division of Rare and Manuscript Collections, Cornell University)

Whetzel, H.H. 1917a. Letter of March 28, 1917 addressed to former students and friends of Prof. Atkinson. (George Francis Atkinson Archives [21-29-103], Carl. A. Kroch Library, Division of Rare and Manuscript Collections, Cornell University)

Whetzel, H.H. (chairman). 1917b. "Project for the Completion and Publication of Prof. G. F. Atkinson's Studies on the Fleshy Fungi of North America, Especially the Clavariaceae and Agaricaceae." (George Francis Atkinson Archives [21-29-103], Carl. A. Kroch Library, Division of Rare and Manuscript Collections, Cornell University)

Whetzel, H.H. 1919. George Francis Atkinson. *Botanical Gazette* 67: 366–368.

Wright, Albert Hazen. 1953. *Pre-Cornell and Early Cornell. IV. Biology at Cornell University 1868–1928*. (*Studies in History*, no. 18). [no statement of publisher], Ithaca, NY. ("Reprinted from *Bios*, Vol. XXIV, no. 3, October, 1953.")

Wright, Albert Hazen. 1954. *Pre-Cornell and Early Cornell. V. Natural History Society of Cornell University 1869–1899. The history of natural science in Cornell for the first thirty years*. (*Studies in History*, no. 19). [no statement of publisher], Ithaca, NY.

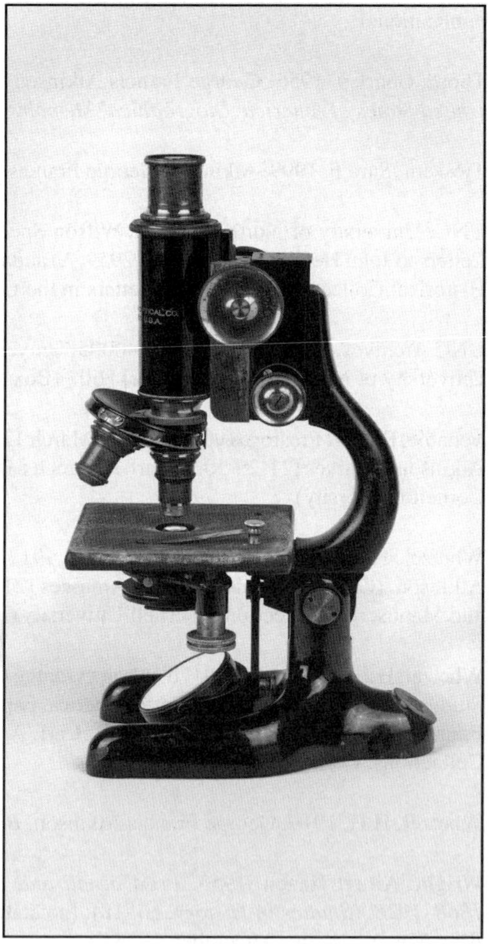

Chapter 15.

History of the Department of Botany at the University of North Carolina at Chapel Hill

Establishing the Department of Botany

At its meeting of January 16, 1902, the UNC Board of Trustees (BOT 10: 302) resolved to strengthen the Department of Biology by adding a new faculty position, namely an Associate Professor of Botany. Elected to the post on June 3, 1902 (BOT 10: 361), William Chambers Coker began his long-term career at UNC. By the time Coker joined the faculty at UNC, the role of research had taken hold and campus administrators had clearly strengthened the university's science laboratories to aid in this endeavor. Other significant developments were the growth of the Graduate School, the development of a comprehensive research library (including several departmental libraries), and the evolving role of the Elisha Mitchell Scientific Society in providing a forum for presenting talks and publishing scientific discoveries.

The Department of Botany was created in 1908 when the Department of Biology split into two departments: Botany and Zoology. Coker became the chair (and founder) of the new botanical unit. In mid-May of that year, the UNC Committee of Visitation recognized the increased responsibilities that Coker had assumed. It recommended that Coker be promoted to Professor of Botany and that his salary be increased from $1,250 to $1,750 (BOT 11: 180). UNC President Francis Venable confirmed that Coker was promoted to Professor of Botany in 1908 (BOT 11: 205).

Also in 1908 Davie Hall (Fig. 15.1) was completed to house the fledgling Departments of Botany and Zoology. The newly constructed Davie Hall, which was locally called the Biological Laboratory, was named in honor of William R. Davie, "Father of the University." A two-story structure located east of New East Building, the structure cost $35,000 and had a debt of $8,000 when it was completed (BOT 11: 207). Botanical facilities included an herbarium and a general botany laboratory on the first floor; and a biology library, a private office and laboratory for Coker, and a laboratory for advanced study on the second floor. A photographic studio was on the attic floor and a darkroom was in a basement room (Anonymous, 1908; Anonymous, [1909]). UNC administrators considered that the building would "increase [the] efficiency given in the instruction of zoology and botany" and

Frontispiece Credits: Instruments of research and botanical teaching model used in the UNC Botany Department. (right) Bausch and Lomb FS-7 school microscope, ca. 1920. (upper left) Model of a plant cell manufactured by Carolina Biological Supply Company, Burlington, NC, ca. 1960. (lower left) Eastman View No. 2 camera produced by the Eastman Kodak Company, ca. 1921. (Courtesy of Brian Nalley, Photo Lab, Department of Biology, University of North Carolina, Chapel Hill).

would aid "in keeping the university upon a high plane in comparison with other southern universities" (BOT 11: 207).

Fig. 15.1. Old Davie Hall, ca. 1935.

(Photo by Bayard Wooten; Courtesy of North Carolina Collection Gallery, Ck. 1347.4, Wilson Special Collections Library, University of North Carolina, Chapel Hill).

In fall 1925 the Department of Botany occupied a new wing added to Davie Hall,[96] which was a three-story, fireproof structure. The wing was built with a $50,000 appropriation from the North Carolina legislature. The main entrance on the west side of the "Botanical Wing," as it was called locally, opened into a museum that displayed botanical specimens. The first floor contained a student laboratory, a faculty office, and a faculty laboratory; another student laboratory and also the herbarium were on the second floor; and the third floor contained a mycological laboratory for advanced students, two private laboratories, and the botany library (Anonymous, 1925; Anonymous, 1926). A special feature of the new facility was the addition of modern equipment for the labs, including several new microscopes. Nearly two decades after its establishment, the Department of Botany had developed research programs in taxonomy and morphology of the lower plants, morphology and cytology of plants, and studies on the flora of North Carolina (Anonymous, 1925; Anonymous, [1927]). In 1940 when the Zoology Department moved out of old Davie Hall to its new quarters, Wilson Hall, Botany took over the entire building.

In anticipation of Zoology's move, UNC implemented a new building program for Davie Hall in August 1939 that included renovations. A series of new labs underneath the first floor were developed, and new workrooms were built. Even the North Wing of Davie Hall received attention with many rooms getting a fresh coat of paint. It was further reported in *The Daily Tar Heel*, the student-run newspaper, that "the pride of the rejuvenated Davie is their Herbarium, which contains about 85,000 specimens" (Anonymous, 1940).

Milestones of the Department of Botany are noted in Table 15.1. Background on the history of the department is recorded in several documents and publications as well as an exhibit on the UNC campus. UNC research assistant in botany Alma Holland Beers (1945a, 1946a) provides a brief background on the formation of the department as well as the educational and research role of the department during the chairmanship of William C. Coker, 1908–1944. Professor of Botany John N. Couch (1947, 1956b) prepared two noteworthy reports concerning the department. Botany chairperson Victor A. Greulach (1966) prepared a document titled "History, Status and Plans of the Department of Botany at the University of North Carolina at Chapel Hill" that covers a ten-year plan. The Wilson Special Collections Library at UNC holds two significant collections of resource material that the Botany Department prepared: annual reports and newsletters. Other archival sources include Department of Botany Records (UNC-A) and a collection of research material that William R. Burk (2015) compiled. In 2008 botany librarian William R. Burk and graduate assistant Tom Hailey mounted a historical exhibit in the lobby of the original wing of Wilson Hall on the UNC campus. The display, which is titled "From Organism to Molecule: A Centennial History of the Biology Department at the University of North Carolina at Chapel Hill," contains a component on the Botany Department. Coverage includes items used in teaching and research in botany, portraits of important personages, and a timeline of the department (Hailey, 2008).

[96] Background on the history of Davie Hall is noteworthy. As already mentioned, the original Davie Hall was built in 1908 and its "Botanical Wing" was constructed in 1925. In March 1966 demolition of the original Davie Hall began. The new structure, also called Davie Hall or New Davie Hall, was placed in front of the 1925 wing and formed a "T" with it. Steve Price (1966), a reporter with the student newspaper *The Daily Tar Heel*, disclosed that: "a portion of the [old Davie Hall] was left standing" and was remodeled and became part of the new building. New Davie Hall was completed around August 1967 and became the home of the Department of Psychology.

Table 15.1. Timeline: Department of Botany, University of North Carolina at Chapel Hill, 1908–1982.

1908	Botany and Zoology Departments founded and move into newly constructed Davie Hall (also called the Biological Laboratory); Herbarium founded.
1908–1944	William C. Coker, first chair. Disciplines of mycology and plant systematics excel under his administration.
1912	John A. McKay receives department's first M.A. in Botany.
1918	Alma L. (Holland) Beers, department's first woman botanist, becomes research assistant, retiring in 1951.
1920	William C. Coker becomes Kenan Professor of Botany; subsequently Kenan Research Professor of Botany in 1944.
1923	Henry R. Totten receives department's first Ph.D. in Botany.
1925	North Wing of Davie Hall is completed to house the Botany Department.
1927	Velma D. Matthews, first woman student to earn advanced degrees in Botany (M.A., 1927; Ph.D., 1930).
1930/1931	Department has three faculty members.
1940/1941	Department has four faculty members.
1943	John N. Couch elected a member of the National Academy of Sciences.
1944–1960	John N. Couch, second chair; maintains excellence in mycology and plant systematics; strengthens department in plant morphology/anatomy and physiology; sets up phycology.
1945	John N. Couch becomes Kenan Professor of Botany.
1948	Earlene Atchison is appointed first female botany faculty member.
1950/1951	Department has nine faculty members, including one professor emeritus and two instructors.
1952	UNC Trustees approve seventy-two acres of land for a botanical garden. Additional acreage is later added, and in 1961 the area is named the North Carolina Botanical Garden.
1953	William Chambers Coker Fellowship in Botany established to enable an advanced graduate student to devote full time to research.
1955	Flora of the Carolinas Project begins with the initial investigations of J. Edison Adams, Harry E. Ahles, C. Ritchie Bell, Albert E. Radford (chief investigator), and Henry R. Totten.
1956–1968	Flora of the Carolinas Project collects over 200,000 plants.
1960/1961	Department has nine faculty members, including two instructors.
1960–1972	Victor A. Greulach, third chair; strengthens disciplines of genetics and ecology; sets up horticulture; faculty size more than doubles.
1961	C. Ritchie Bell becomes first director of the North Carolina Botanical Garden, serving in that position into 1986.
1962	Bachelor of Arts degree with honors in Botany is first offered.
1963	Department moves into newly constructed Coker Hall.
1964	Curriculum in Genetics established as an interdepartmental program of study involving the Anatomy, Bacteriology, Biochemistry, Biostatistics, Botany, Pathology, Psychology, and Zoology Departments.

Table 15.1. (Continued)

1964/1965	Albert E. Radford serves as acting chair of Botany Department.
1965	Albert E. Radford and Harry E. Ahles establish the organization of the Vascular Flora of the Southeastern United States Project.
1966	Nature Trails, the first public offering of the North Carolina Botanical Garden; Botanical Garden Foundation Inc. created.
1968	*Manual of the Vascular Flora of the Carolinas* published.
1968	Curriculum in Marine Sciences established as interdepartmental program of study involving the Bacteriology,* Biochemistry,* Botany, Chemistry, Environmental Sciences and Engineering, Geology, Physics,* and Zoology Departments. (* Left the program in 1970.)
1969	Curriculum in Biology established as interdepartmental program of study involving the Anatomy, Bacteriology, Biochemistry, Botany, Parasitology, Physiology, and Zoology Departments.
1969	Master of Science degree in Botany is first offered.
1970/1971	Department has twenty-one faculty members, including four professors emeriti.
1970	(last half of summer) William J. Koch serves as acting chair of Botany Department.
1971	Curriculum in Ecology established as an interdepartmental program of study involving the Anthropology, Biomedical Engineering and Mathematics, Biostatistics, Botany, City and Regional Planning, Environmental Sciences and Engineering, Epidemiology, Geography, Sociology, and Zoology Departments; Professor of Zoology Alan Stiven is first chair. (In 2008 the curriculum merged with the undergraduate programs in Environmental Studies and Environmental Science administered by the Carolina Environmental Program to become the Curriculum for the Environment and Ecology, and later in 2018 the Environment, Ecology and Energy Program.)
1972–1982	Tom K, Scott, fourth chair; develops paleobotany and electron microscopy; Botanical Garden expands program.
1972/1973	Edward G. Barry serves as acting chair of Botany Department.
1974	Francis J. LeClair Award established to recognize the academic excellence of a senior in Botany (later Biology) with interests in the plant sciences.
1975/1976	Herbarium accessions its 500,000th specimen.
1976	The Totten Garden Center, named for Henry R. and Addie Totten, opens at the North Carolina Botanical Garden.
1976/1977	Bachelor of Science degree in Botany is first offered.
1979	Botany Library named the John N. Couch Library.
1979	Mrs. W.C. Coker Graduate Fellowship in Botany established to provide funds to worthy graduate students to attend the University of North Carolina at Chapel Hill.
1980	First volume of the *Vascular Flora of the Southeasten United States* is published.
1980/1981	Department has twenty-three faculty members, including four professors emeriti and one research professor.
1982	(January–June) Aristotle J. Domnas serves as acting chair of Botany Department.
1982	(July) Department merges with Zoology, forming a Department of Biology.

William Chambers Coker: First Chair (1908–1944)

William Chambers Coker (Fig. 15.2), born October 24, 1872 in Hartsville, South Carolina, was the son of Major James Lide and Susan Armstrong Stout Coker. Young Coker developed an early interest in nature. As was the custom of those days, he received his education at his home. He subsequently earned a bachelor's degree from the University of South Carolina in 1894. Following a banking position in Wilmington, North Carolina, Coker decided to attend the Johns Hopkins University in 1897. There he studied botany, earning his doctoral degree in 1901.

Fig. 15.2. William C. Coker, ca. 1945.

(Box 1, unlabeled folder of documents regarding W.C. Coker, William R. Burk Collection on University of North Carolina at Chapel Hill Botany and Zoology Faculty, University Archives #40478, Wilson Special Collections Library, University of North Carolina, Chapel Hill).

Coker joined the faculty at UNC in 1902 as associate professor of botany. His title subsequently became Kenan Professor of Botany (1920–1944) and Kenan Research Professor of Botany (1944–1945). Coker spent the remainder of his innovative and productive career at the University of North Carolina at Chapel Hill. He supervised twenty-eight graduate students, who earned thirty-six advanced degrees (M.A., 27; Ph.D., 9). Among Coker's numerous publications are the following books: *The Plant Life of Hartsville, SC*, 1912; *The Saprolegniaceae*, 1923a; *The Clavarias of the United States and Canada*, 1923b; *The Gasteromycetes of the Eastern United States and Canada* (with J.N. Couch), 1928; *Laboratory Guide in General Botany*, 4th ed. (with H.R. Totten), 1939; *The Boletaceae of North Carolina* (with A.H. Beers), 1943; *Trees of the Southeastern States*, 3rd ed. (with H.R. Totten), 1945; and *The Stipitate Hydnums of the Eastern United States* (with A.H. Beers), 1951. Coker stepped down as chair of the department effective July 1, 1944, and he officially retired as of September 1, 1945. UNC conferred an honorary Doctor of Science degree on Coker in 1947. Coker died June 27, 1953 and was buried in the Old Chapel Hill Cemetery on the UNC campus. Several biographers provide insight into his life and accomplishments (Couch and Matthews, 1954; Joslin, 2003; Methven, 2000; Totten and Couch, 1954; and Totten, Wager and Couch, 1956).

Coker first taught classes in fall 1902, presenting lectures and instructing laboratory sessions in General Biology. The following academic year (1903/1904) he taught General Botany, and in the spring he instructed on Lower Fungi, the first course devoted to this discipline to be taught at UNC. In 1905 he gave a course on Higher Fungi. During his long career at

UNC, he taught a wide range of classes in fungi, algae, and higher plants [97] and conducted research on various fields of botany, particularly mycology. Coker's interest in teaching mycology and conducting research in the taxonomy of fungi would clearly be influential in developing the department. Early in his tenure at UNC, Coker organized the Botany Journal Club for the purpose of reviewing and discussing botanical literature.

Some insight into the style of teaching botany during Coker's administration is revealed in an interview he had with a reporter (Anonymous, 1931) with the student-run newspaper, *The Daily Tar Heel*.

> The beginning classes in botany are all taught by professors who keep in mind the cultural as well as the professional needs of the students. The higher classes are all taught by a method that we might call in a way a personal report method, the individual student coming in personal daily contact with his professor and making frequent reports. These reports are especially emphasized as training in teaching and it is constantly kept before the student that one of his principal objects is to be able to present a subject clearly before an audience. We believe that this kind of training is fully as important as a preparation for teaching as any course in pedagogy.
>
> Almost immediately after the first year the student is given particular problems and individual attention. A number of our students have produced research results worthy of publication before graduation.

Coker had a profound impact on founding and developing the following educational and research resources at UNC: the Botany Library, the University Arboretum, the University Herbarium, and the department Photo Lab.

Botany Library

Chapter one of this book includes a discussion of the early development of a library collection in the Department of Biology.[98] Coker was instrumental in developing the collection, sometimes using his own funds to purchase rare books. Two decades following his arrival on campus in 1902, William Coker (1922) noted the library room "is completely filled and outgrown to such an extent that a large number of botanical books are now relegated to the laboratories." The collection of botanical and zoological books and journals remained as a single unit until fall 1925, when the Department of Botany moved into the "North Wing" of Davie Hall. The botanical library holdings were weeded out of the biology collection and moved to a library room on the third floor of the Botany Department's new quarters.

By 1935 the Botany Library occupied three separate rooms and required personnel to manage the collection and care for numerous scientific reprints. In the early years the departmental secretary/stenographer served as a part-time librarian. A number of people managed the

[97] According to UNC catalogs, Coker taught numerous courses, including the following: Structure and Development of the Fungi; General Biology; Lower Fungi; Higher Fungi; Introduction to Systematic Botany; Plant Morphology; General Botany; Special Morphology of Algae and Fungi; Special Morphology of the Fungi; Special Morphology of Mosses, Ferns and Higher Plants; Morphology of Seed Plants; Problems in Fungi; Structure and Physiology of Fungi; History of Botany; Field and Economic Botany; Plant Physiology; and Structure and Classification of Seed Plants.

[98] First established in 1891, the Department of Biology subsequently split into two departments (Botany and Zoology) in 1908. A new Department of Biology was formed when the Botany and Zoology departments merged in 1982.

library: Alma Holland (later Beers), Mrs. William Whatley Pierson, Mrs. Francis Ghigo, Mrs. W.L. Wilson, Virginia Andes, Louise J. Burks, Mrs. George Christenberry, Velma Matthews, Mrs. James Francis, and Marion Seiler. Despite its cramped quarters, the library was considered one of the most valuable botanical collections found in any southern university (Coker, 1935). When Coker stepped down as chair of the department in 1944, the Botany Library had developed an outstanding collection of mycological works, including a number of the earliest books on this subject. Other strong aspects of the library's holdings were old books of the herbal type, classical old works, numerous systematic works on plants of America, standard manuals, histories of botany, and runs of journals of the significant American titles as well as many foreign titles (Beers, 1945a).

Two decades later Professor of Botany John N. Couch (1956a) reported that the collection had grown to the extent that the responsibility of operating the library was too great for the part-time secretary/librarian. In fact, Couch noted the library room was "so small and overcrowded that frequently we do not have sufficient seats to accommodate the students." The university administration subsequently approved a part-time librarian, a post that Betty Zouck assumed in October 1957 (Couch, 1958). In the early 1960s Mr. R. Philip Hanes, Jr., of Winston-Salem, North Carolina, gave the library a magnificent collection of mycological books. The botany librarian position was made full-time around 1964, and under Betty Zouck's management the library became an integral and influential research and educational unit of the Botany Department. Zouck retired on June 30, 1978. John B. Darling served as the acting botany librarian until a new librarian was appointed.

William R. Burk became the next botany librarian, effective February 1, 1979. During his tenure several noteworthy events occurred. The Botany Library was dedicated as the John N. Couch Library on February 21, 1979 to honor Professor Couch, who was a lifelong advocate of the library. The library administratively (but not physically) merged with the Zoology Library to form the John N. Couch Biology Library in 1986, after which it was identified as the Botany Section of the John N. Couch Biology Library. The Botany Section library moved to the south wing of the Louis Round Wilson Library, where it merged with the staff and collections of the Zoology Section and Chemistry libraries in May 2008. Burk retired February 1, 2009, and David Romito became the next biology librarian.

In 2010, when a newly constructed Venable Hall opened its doors, the biology librarian and one technical staff member of the Biology and Chemistry Library moved into offices in the newly created Kenan Science Library on the ground floor of Venable Hall. The Biology and Chemistry book and journal collections and several technical staff members remained in the south wing portion of Wilson Library, which then became known as the Science Library Annex. The following year, in July 2011, the collections of the Geological Sciences Library and the Math/Physics Library merged with the other science collections in the Science Library Annex of Wilson Library.

Arboretum

In her biographical book on her "Uncle Will," Mary Coker Joslin (2003) notes that William C. Coker founded UNC's five-acre garden of woody plants in 1903. With an appropriation of ten dollars, Coker and a gardener began developing the land, which today is called Coker

Arboretum. Realizing that Americans did not place much emphasis on propagating native species, Coker decided to create an arboretum featuring woody specimens of the southeastern United States. Botany students learned about the plants in this setting and also examined specimens in the laboratory. The arboretum additionally served as an ornamental garden for the UNC campus. The arboretum became a unit of the North Carolina Botanical Garden in 1982. Information on the Coker Arboretum is available in publications by Diane G. Birkemo and J. Kenneth Moore (1990) and Daniel Stern (2004).

University Herbarium

Home to over 750,000 vascular plant specimens, 53,000 algae, 38,000 fungi, 6,000 lichens, 3,500 bryophytes and 10,000 plant fossils, the University Herbarium had its roots in humble beginnings in 1908. But even earlier, UNC professors assembled plant specimens for teaching purposes. As the herbarium grew through the decades, it expanded outreach in research and education. As mentioned earlier in this book, William H. Smith, Frederic W. Simonds, and Joseph A. Holmes required their students to collect and identify plants, which they then preserved as dried specimens. These specimens were placed in a cabinet housed in the University Museum on the third floor of Old East and called an herbarium. Around 1891, when Holmes stopped teaching botany, the Museum collections were dispersed to various locations on campus, but the fate of the nineteenth-century herbarium is unknown.

When the Department of Botany was established in 1908, it moved into newly constructed Davie Hall. The new building provided space to expand services, including the construction of herbarium cases, which initially housed a hundred or so plant specimens that William W. Ashe, a student under Joseph A. Holmes, had collected. During the next five decades, the herbarium quickly outgrew its space. Relief came in 1963, when the department moved into a new building, Coker Hall, where the herbarium occupied rooms on the fourth floor. By 1977, herbarium cases had overflowed into the halls of the building, a testament to sustained and even heightened interest in botany.

Guided by seven directors (William C. Coker, John N. Couch, Albert E. Radford, Jim R. Massey, Mark Chase, Peter S. White, and Alan S. Weakley) and run by a succession of dedicated curators and assistants, the herbarium grew and flourished during its first century. Originally administered by its home department (Botany, later Biology), the herbarium became a unit of the North Carolina Botanical Garden on June 18, 2000. The collection increased not only through specimens collected in the field by staff, faculty, students, and other researchers, but also through exchanges of specimens with other institutions, special acquisitions such as the personal herbaria of W.W. Ashe (20,000 specimens) and Thomas G. Harbison (12,000 specimens), and donations such as one from Dartmouth College (around 8,000 specimens). Selected milestones in the approximate growth of the collection are 100,000 specimens (1942), 250,000 (1964), 500,000 (1980), and currently about 810,000. The oldest specimen (a vascular plant) dates from 1827.

Building on its original objective of preserving plant specimens for research and teaching, the herbarium and its staff have adapted new technologies to expand their services. Some of these include loaning materials (as a digital copy or by mail), answering reference questions, and giving talks and teaching classes. The herbarium also provides online access to its collections and locally created research tools, which can be found at https://ncbg.unc.edu/research/unc-herbarium/specimen-portals/. Particularly noteworthy, the herbarium features online sites for viewing type specimens, a biographical resource on the collectors of the UNC Herbarium, and drafts of current director Alan Weakley's ongoing creation of *Flora of the Southern and Mid-Atlantic States*. Weakley (2020) posted his *Flora of the Southeastern United States* in late 2020.

The UNC herbarium is now the largest museum and research collection of southeastern plants in the world. It also houses field notebooks, maps, original illustrations, photographs, and slides. The herbarium has supported the publication of numerous books and research papers, including the landmark *Manual of the Vascular Flora of the Carolinas* by A.E. Radford, Harry E. Ahles, and C. Ritchie Bell ([1968]) as well as the *Flora of Virginia*, by Alan S. Weakley, J. Christopher Ludwig, and John F. Townsend (2012). Laurie S. Radford and Albert E. Radford (2000) and staff at the UNC Herbarium (NCU, 2019) provide additional background on the herbarium.

Photo Lab

Since its establishment in 1908, the Botany Department valued photographic facilities as a laboratory and service to support research and publishing. Under Coker's supervision the photo lab became an integral unit of the Botany Department. As already noted above, both Davie Hall and its subsequent "North Wing" housed a studio for taking photographs and a darkroom for their development. A room was also set up in the basement of Davie Hall for microphotography with adjoining darkrooms. These facilities improved by the 1940s as Couch ([1947]) noted, "There is a well-equipped dark room and stationary and portable cameras."

In 1960 the department hired artist Marion Seiler[99] as research associate in botany to complete art work for faculty and staff publications and to assist with their photographic work. When the department moved into newly constructed Coker Hall, there was a modern facility with two darkrooms, an area for photography, and tables for illustration and design. Seiler's position became department artist and photographer. She retired on June 30, 1977.

Anna E. Birkner, known as Betsy, subsequently filled Seiler's post. She took a leave of absence beginning February 1, 1980 but decided not to return to her position. In March 1980, Susan Sizemore (later Whitfield) was appointed as a temporary to fill Birkner's position and Larry Blanton was hired as a temporary photographer/darkroom manager. In June 1980, Sizemore became the permanent departmental scientific illustrator/photographer. With the advance of technology, her position later involved additional responsibilities as a computer-based design specialist. Whitfield's expertise and devotion to serving the Botany (later Biology) Department were recognized in 2006, when she became one of two winners of the 11th Robert R. Cornwell Unsung Hero Award. In February 2007, the Photo Lab of the Biology Department moved to rooms 203 and 211 in the newly renovated original wing of Wilson Hall. Whitfield retired effective January 1, 2020.

99 Marion Seiler served as the Botany Department's secretary/librarian from 1949 to 1954 and subsequently was a medical illustrator at the North Carolina Medical Hospital.

Professors of Botany[100] Joining the Department under Coker's Administration

Henry Roland Totten (1914–1917, 1919–1923, instructor; June 1923–June 1963, assistant professor to professor)[#]

Research Interests: fungi, ferns, higher plants (especially taxonomy of the woody plants of the southeastern United States), pharmacognosy.

Courses Taught: General Botany, Pharmacognosy, Dendrology, Plant Taxonomy, Morphology and Development of Seed Plants, Field and Economic Botany, Pharmaceutical Botany, Introduction to Systematic Botany, Morphology of the Fungi, Ferns, Advanced Macropharmacognosy, Advanced Taxonomy of Seed Plants, Field Botany, Local Flora, Structure of Seed Plants.

Selected Publications: *Laboratory Guide in General Botany*, 4th ed., 1939 (with W.C. Coker); *Trees of the Southeastern States*, 3rd ed., 1945 (with W.C. Coker); *Laboratory Guide in Pharmaceutical Botany and Pharmacognosy*, 7th ed., 1957; and numerous articles in the *Journal of the Elisha Mitchell Scientific Society*.

Graduate Students: M.A., 1; M.S., none known; Ph.D., 2.

John Nathaniel Couch (1922–1925, instructor; September 1925–October 1945, assistant professor to professor; November 1945–June 1967, Kenan Professor of Botany)[101]

Research Interests: water fungi, fungi parasitic in mosquito larvae, symbiosis between fungi and scale insects, *Septobasidium*, Actinomycetales, *Coelomomyces*.

Courses Taught: Special Morphology of the Fungi; Cryptogamic Botany; Structure and Development of the Fungi; Structure and Classification of Seed Plants; Structure and Physiology of Fungi; General Botany; Algae; Comparative Survey of the Non-Vascular Plants; Bryophytes; Fungi: Phycomycetes, Ascomycetes, Basidiomycetes; Problems in Fungi; Studies in the History of Botany; Elements of Plant Biology; Problems in Fungi; Advanced Work in Fungi; Oomycetes, Fungi Imperfecti; Advanced Mycology: Phycomycetes, Ascomycetes, Basidiomycetes, Fungi Imperfecti.

Selected Publications: *The Gasteromycetes of the Eastern United States and Canada*, 1928 (with W.C. Coker); *The Genus Septobasidium,* 1938; *The Genus Coelomomyces*, 1985 (with C.E. Bland); and numerous papers and abstracts.

Graduate Students: M.A., 27; M.S., 1; Ph.D., 16.

[#] Totten subsequently began a part-time continuance for one year.

[100] Apart from William C. Coker, thirty professors of botany served the Department of Botany under four chairpersons. Background on each professor provides the following information: years of tenure and ranks attained, primary areas of research, courses taught, selected publications, and graduate students advised. Information on a professor's period of service and a number of selected publications may extend past July 1982, when the Botany and Zoology Departments merged to form the Department of Biology. Seminars and graduate courses such as research in botany or special topics in botany are not included. Individuals who held instructorships in the Botany Department are not included in the following rosters of botany faculty members except for the following individuals who held instructorships (as hereby designated) and who advanced through the ranks of the department: H.R. Totten (1914–1917, 1919–1923), J.N. Couch (1922–1925), J.E. Adams (1935–1936), L.S. Olive (1942–1944), A.E. Radford (1946–1948), W.J. Koch (1954–1956), M.H. Hommersand (1959–1960), and C.J. Umphlett (1961–1962).

[101] Couch was appointed assistant professor of botany in 1925, but he took a leave of absence without pay from 1925/1926 to 1926/1927.

Joseph Edison Adams (1935–1936, instructor; September 1936–July 1969, assistant professor to professor)

Research Interests: plant anatomy, plant morphology, classification and phylogeny of flowering plants, wood anatomy.

Courses Taught: Botany, Advanced Taxonomy of Seed Plants, Advanced Micropharmacognosy, Plant Geography, Plant Anatomy, Plant Biology, Lower Plants, Morphology of Lower Plants, Comparative Morphology of Vascular Plants, Phylogeny and Classification of Flowering Plants, Pharmaceutical Botany and Microscopical Pharmacognosy, Advanced Work in Seed Plants, Pharmacognosy, Comparative Survey of Vascular Plants, Phylogeny and Classification of Flowering Plants, Advanced Pharmacognosy, Problems in the Taxonomy of Flowering Plants, Problems in the Morphology of Flowering Plants, Morphology of Seed Plants, Advanced Work in the Structure and Classification of Seed Plants, Structure of Seed Plants, Comparative Morphology of Lower Vascular Plants.

Selected Publications: *Laboratory Problems in Plant Science*, 1954 (with C.E. Wood, Jr. and V.A. Greulach); *Plants: An introduction to modern botany*, 3rd ed., 1976 (with V.A. Greulach).

Graduate Students: M.A., 11; M.S., 1; Ph.D., 13.

Earl Holland Newcomer (ca. May 1941–December 1947, assistant professor)

Research Interests: plant cytology and cytogenetics, mitochondria in plants, colchicine-induced polyploidy and its morphological and physiological consequences, forest genetics, history of botany.

Courses Taught: Cytology, Botany, Dendrology, Cytogenetics, Elements of Plant Biology.

Selected Publications: A colchicine-induced tetraploid cabbage, 1941; Induced parthenocarpy in *Ginkgo*, 1945; Concerning the duality of the mitochondria and the validity of the osmiophilic platelets in plants, 1946.

Graduate Students: M.A., 1; M.S., none known; Ph.D., 1.

John Nathaniel Couch: Second Chair (1944–1960)

Born October 12, 1896, in Prince Edward County, Virginia, John Nathaniel Couch (Fig. 15.3) was the son of John Henry and Sallie Terry Couch. In 1914 he attended Trinity College (now Duke University). He transferred in 1917 to the University of North Carolina at Chapel Hill, where he studied botany. At Carolina Couch earned three degrees in this discipline: B.A. (1919), M.A. (1922), and Ph.D. (1924) under the supervision of William C. Coker. Couch was appointed assistant professor of botany in 1925, but he took a leave of absence without pay from 1925/1926 to 1926/1927. He devoted most of his professional career to UNC, advancing through the academic ranks from assistant professor to professor. The National Academy of Sciences elected Couch a member in 1943. In 1945 his scholarship and outstanding scientific investigations were recognized with the conferral of a distinguished professorship, the Kenan Professor of Botany. UNC conferred an honorary Doctor of Science degree on Couch in 1973. He retired effective July 1, 1967 and died on December 16, 1986. He is buried in the Old Chapel Hill Cemetery on the UNC campus. Accounts of his life and

Fig. 15.3. John N. Couch.

(Box 2, folder: Couch, John N., William R. Burk Collection on University of North Carolina at Chapel Hill Botany and Zoology Faculty, University Archives #40478, Wilson Special Collections Library, University of North Carolina, Chapel Hill).

contributions to science, as well as works enumerating his writings, have been published (Burk, 2000; Burk and Bland, 1989; [Couch], 1956c; Shanor, 1968; and Szaniszlo, 1993).

Following in the footsteps of his mentor, William C. Coker, Couch assumed the leadership of the Department of Botany on July 1, 1944 and held this post until June 30, 1960. He played an instrumental role in promoting the establishment of the North Carolina Botanical Garden. In 1952 Couch requested that the Board of Trustees establish a botanical garden by developing a seventy-two-acre tract of land from the Mason Farm in Chapel Hill. Historical background on the garden provides insight into its development from its beginning to 2016 (NCBG Greenbriers, 2019).

Further insight into the Botany Department during the closing years of Couch's chairmanship is gained from a booklet about science on the UNC campus (Anonymous, [1958]). Several photographs depict laboratory scenes and other venues of botanical instruction.

Professors of Botany Joining the Department under Couch's Administration

Earlene Atchison (September 1948–June 1951, assistant professor to associate professor)[102] (Fig. 15.4)

Research Interests: interspecific hybridization among Leguminosae, plant cytogenetics and cytology.

[102] Earlene Atchison was the first female faculty member in the Botany Department. She received the A.B. (1941) from Huntington College and the M.A. (1943) from the University of Alabama. Under the supervision of botanist Orland E. White, she earned a Ph.D. from the University of Virginia in 1946. Atchison's (1946) dissertation entailed the "Chromosome Numbers in *Erythrina* L., *Crotalaria* L., *Gleditsia* L., *Mitchella* L., and the Myrtaceae." Her early experience was as a high school teacher in McCalla, Alabama. Prior to her appointment at UNC, she was a Fellow at the Rockefeller Foundation in Mexico City (1947–1948). Her UNC appointment was effective as of September 1, 1948 (Anonymous, 1948).

Earlene Atchison (Continued)

Courses Taught: Botany, Cytology, Cytogenetics, Cytotaxonomy.

Selected Publications: Chromosome numbers for two species of Mexican Commelinaceae, 1949 (with H.E. Moore and C.E. Wood, Jr.); Inhibition of growth and cell division in onion roots by maleic hydrazide, 1950 (with V.A. Greulach); Studies in the Leguminosae. VI; Chromosome numbers among tropical woody species, 1951; and additional papers on the topic of her expertise.

Graduate Students: M.A., 3; M.S., none known; Ph.D., none known.

Fig. 15.4. Earlene (Atchison) Rupert.
(Courtesy of Clemson University Libraries' Special Collections and Archives, Series 100, #74-0152).

Albert Ernest Radford (1946–1948, instructor; January 1949–May 1987, assistant professor to professor)

Research Interests: plant taxonomy, plant conservation, floristics, ecosystematics.

Courses Taught: General Botany, Local Flora, Dendrology, Cryptogamic Botany, Field Botany, Field Training in Natural Diversity, Lower Plants, Plant Ecology, Seed Plants, Taxonomy of Flowering Plants (later titles Taxonomy of Vascular Plants, Principles of Plant Systematics, Principles of Taxonomy), Plant Taxonomy Laboratory, Problems in the Taxonomy of Flowering Plants, Advanced Taxonomy, Plant Biology for Secondary School Teachers, Principles of Botany, Floristics, Plant Communities of the Carolinas, Plant Ecosystematics, Taxonomy of Special Groups of Vascular Plants, Vascular Plant Taxonomy Laboratory.

Selected Publications: *Laboratory Directions for Botany 41 [General Botany I]*, 1954; *Laboratory Direction and Exercises for General Botany*, 1961 (with C.R. Bell); *Atlas of the Vascular Flora of the Carolinas*, 1965 (with H.E. Ahles and C.R. Bell); *Manual of the Vascular Flora of the Carolinas*, [1968] (with H.E. Ahles and C.R Bell); *Vascular Plant Systematics*, [1974] (with others); *Natural Heritage,* [1981] (with others); *Fundamentals of Plant Systematics*, 1986 (with T.M. Barkley).

Graduate Students: M.A., 29; M.S., 13; Ph.D., 20.

Victor August Greulach (September 1949–July 1974, associate professor to professor)

Research Interests: plant growth, especially growth substances and photoperiodism; biological education.

Victor August Greulach (Continued)

Courses Taught: Botany, General Botany, Plant Biology, Plant Physiology, Plant Growth, Introductory Plant Physiology, Plant Water and Mineral Soil Relations, Plant Metabolism and Biochemistry, Physiology of the Fungi, Advanced Plant Physiology, Plant Biology for Secondary School Teachers, Problems in Plant Physiology, Plant Dynamics, Principles of Botany.

Selected Publications: *Laboratory Problems in Plant Science*, 1954 (with C.E. Wood, Jr. and J.E. Adams); *A Laboratory Manual for Elementary Plant Physiology*, 1955; *Botany Made Simple*, 1968; *Plant Function and Structure*, 1974; *Plants: An introduction to modern botany*, 3rd ed., 1976 (with J.E. Adams); and numerous articles in journals and books.

Graduate Students: M.A., 6; M.S., none known; Ph.D., 6.

Carroll Emory Wood, Jr. (September 1951–July 1954, assistant professor to associate professor)

Research Interests: plant systematics, particularly on the flora of the southeastern United States.

Courses Taught: Botany, Plant Biology, Cytology, Biosystematics, Cytogenetics, Advanced Morphology and Taxonomy, Cytotaxonomy, Problems in the Taxonomy of Flowering Plants, Problems in Cytology and Cytogenetics.

Selected Publications: *Laboratory Problems in Plant Science*, 1954 (with V.A. Greulach and J.E. Adams); and over fifty-nine papers, chiefly on the flora of the southeastern United States and written after his tenure at UNC.

Graduate Students: M.A., 1; M.S., none known; Ph.D., none known.

Clyde Ritchie Bell (September 1955–June 1990, assistant professor to professor)[103]

Research Interests: plant systematics, flora of southeastern United States, pollination biology, Sarraceniaceae, plant evolution and speciation.

Courses Taught: Economic Botany, General Botany, Cytology, Variation and Evolution in Plants, Biosystematics, Cytogenetics, Local Flora, Taxonomy of Vascular Plants, Field Botany, Plant Physiology and Morphogenesis.

Selected Publications: *Laboratory Direction and Exercises for General Botany,* 1961 (with A.E. Radford); *Atlas of the Vascular Flora of the Carolinas*, 1965 (with A.E. Radford and H.E. Ahles); *Plant Variation and Classification*, [1967]; *Manual of the Vascular Flora of the Carolinas*, [1968] (with A.E. Radford and H.E. Ahles); *Vascular Plant Systematics*, [1974] (with A.E. Radford and others); *Florida Wild Flowers and Roadside Plants*, 1982 (with B.J. Taylor); *Wild Flowers of North Carolina*, 2nd ed., 2005 (with W.S. Justice and A.H. Lindsey); numerous books and videocassettes on wild flowers, foliage, and fruits published by Laurel Hill Press; over 100 journal papers and abstracts.

Graduate Students: M.A., 13; M.S., 5; Ph.D., 15.

[103] "C. Ritchie Bell retired June 30, 1985 and received a five-year fixed term appointment beginning July 1, 1985" (Gilbert, 1986).

William Julian Koch (1954–1956, instructor; November 1956–June 1986, assistant professor to professor)

Research Interests: culture, comparative morphology, sexuality, and mobility of fungal reproductive cells; fungi parasitic on algae; plant diversity.

Courses Taught: General Botany, Plant Biology, Comparative Survey of the Non-Vascular Plants, Comparative Morphology of the Non-Vascular Plants, Fungi, Physiology of the Fungi, Algae, Bryophytes, Plant Dynamics, Fungi Laboratory, The Plant Kingdom, Taxonomy of Vascular Plants, Evolutionary Survey of Plants, Phycomycetes, Marine Mycology, Fungi and Algae, Plant Life, Taxonomy of Special Groups of Non-Vascular Plants.

Selected Publications: *Mycological Studies Honoring John N. Couch*, 1968; *Fungi in the Laboratory*, 2nd ed., [1972]; *Plants in the Laboratory*, [1973]; *Fungi*, 1976; *Plant Close-ups: Designs: Annotated photographs*, 2007 (also issued as a compact disk); numerous manuals adapted to his classroom teaching, including *Leaflets for Buds*, 1980; *Plants & Life*, 2nd ed., 1985a; and *In Touch with Plants*, 2nd ed., 1985b; and nearly fifty journal articles and abstracts.

Graduate Students: M.A., 5; M.S., 3; Ph.D., 2.

Max Hoyt Hommersand (1959–1960, instructor; July 1960–December 1997, assistant professor to professor)

Research Interests: morphology, systematics and phylogeny of marine algae, with an emphasis on red algae.

Courses Taught: Plant Biology, Algae, Photobiology of the Lower Plants, Plant Dynamics, Radioisotope Tracer Techniques, Plant Biochemistry, Photosynthesis, Marine Algae, Algal Physiology, Fungi and Algae, Marine Phycology, Algal Physiology Laboratory, Taxonomy of Special Groups of Nonvascular Plants.

Selected Publications: *The Morphology and Classification of Some Ceramiaceae and Rhodomelaceae*, 1963; *Ceramiales*, 1993 (with C.A. Maggs).

Graduate Students: M.A., 4; M.S., 7; Ph.D., 19.

Victor August Greulach: Third Chair (1960–1972)

Victor August Greulach (Fig. 15.5) was born December 6, 1906 in Convoy, Ohio. After receiving an A.B. (1929) at DePauw University, he pursued graduate work at The Ohio State University, earning an M.S. (1933) and Ph.D. (1940) in botany. He held various academic positions: instructor in biology at Muskingum College (1933–1935); assistant professor (1935–1940), associate professor (1940–1946), and chair of the Division of Biological Sciences (1944–1946) at the University of Houston; and subsequently associate professor (1946–1949) at the Agricultural and Mechanical College (now University) at College Station, Texas. He joined the Botany Department at UNC in September 1949 and became its chair on July 1, 1960. He took a leave of absence without pay in the academic year 1964/1965 to serve as the first executive director of the Commission on Undergraduate Education in Biological Sciences. During his absence, Albert E. Radford was the acting chair of the department. As of June 30, 1972, Greulach stepped down as chair, and he retired effective July 1, 1974. His chief research interest was the effect of growth substances and environmental factors on plant growth. He also had strong interests in biological education. Greulach died on December 31, 1984 and was buried in the Chapel of the Cross Episcopal Churchyard, Chapel Hill, North Carolina. Additional biographical information on Greulach has been published (Anonymous, 1960b; Anonymous, 1985a; Anonymous, 1985b; Anonymous, 1985c; [Greulach], 1972).

Fig. 15.5. Victor August Greulach.

(Image box 14, News Services of the University of North Carolina at Chapel Hill Records, 1924–2013, University Archives #40139, Wilson Special Collections Library, University of North Carolina, Chapel Hill).

Early in Greulach's administration, the Department of Botany focused in three areas of research, as described by UNC botany professors Victor Greulach, Albert Radford, and William Koch (1962) in an informative paper presented before the 571st meeting of the North Carolina Academy of Science on November 14, 1961. They noted that research and graduate instruction in botany at UNC were focused on mycology and algology, plant taxonomy, and plant physiology. Research in plant taxonomy was particularly strong. A group of plant taxonomists were involved in the preparation of a vascular flora of the Carolinas, with Albert E. Radford (Fig.15.6) as the principal investigator. The resulting landmark flora, *Manual of the Vascular Flora of the Carolinas*, appeared in 1968.

Under Greulach's early years of leadership the department experienced a period of noteworthy growth. The number of full-time faculty members nearly doubled. The number of undergraduate botany majors increased from four to twenty and graduate majors from fifteen to forty-eight. There was also an extensive reorganization of botany courses. Among the changes in the curriculum was the establishment in 1962 of an undergraduate thesis course designated as Botany 99. The course allowed students to take an A.B. in botany with honors. Four additional developments during Greulach's administration deserve mention: the establishment of the first botany electron microscope laboratory, the erection of a new building for the Botany Department, educational outreach in botany beyond UNC, and participation in curricula that promoted interdisciplinary coordination, all of which are discussed below.

Fig. 15.6. Albert E. Radford lecturing in the Plant Taxonomy Laboratory, Room 418 Coker Hall, UNC, 1960s.

(Photo by Laurie S. Radford. Courtesy of UNC Herbarium [NCU]).

First Botany Electron Microscope Laboratory

The acquisition of an electron microscope (Fig. 15.7) around 1961 further developed botanical research at UNC. Toward the end of January 1961, the Botany Department received an NSF grant of $24,200 that enabled the unit to purchase a Philips Electron Microscope-75 and complementing equipment such as a Porter-Blum ultramicrotome and a high vacuum evaporator. The microscope would be available to all the botanists in the department and would expand scientific investigations in various areas of botany. But the acquisition would particularly enhance research in the Laboratory for Studies in Mycology under the direction of John Couch and William Koch, who had procured the grant (Anonymous, 1961c; Anonymous, 1961d).

Soon after the microscope was installed, Couch and Koch began using it in their research. Koch continued his studies on the flagellar apparatus in various bacteria and fungi with the new instrument. Couch and UNC botanist Clyde Umphlett would soon employ it in their research on the fungus *Coelomomyces* (Anonymous, 1962). In 1963, when the Botany Department moved from Davie Hall to its new quarters, Coker Hall, the electron microscope was moved to the second floor of the building. The instrument broke down in 1975 (Anonymous, 1975a), but it was subsequently repaired (Brown, 2020).

Fig. 15.7. Electron microscope that UNC botanists used around 1962.

(Photographic Archives #22783, North Carolina Collection, Wilson Special Collections Library, University of North Carolina, Chapel Hill).

Just as Elisha Mitchell persevered to purchase the magnificent Vienna cabinet of minerals in 1842 (discussed in chapter 3), Couch and Koch devoted great effort to acquiring the electron microscope. Both purchases required significant sums of money and both used government assistance in acquiring them. In the case of the electron microscope, United States Representative from North Carolina Horace R. Kornegay assisted in getting U.S. House of Representatives Bill 3385 passed for duty-free entry of the microscope into the country. A decade later UNC botanist R. Malcolm Brown, Jr. further developed facilities for ultrastructural research in botany. That endeavor is discussed later in this chapter.

A New Botany Building

As early as 1944 UNC's long-range goals included plans for the construction of a modern, fireproof building to house the Botany Department (Anonymous, 1944). Twelve years later John N. Couch (1956b) reminded campus administrators about the need for a new building. He inserted a noteworthy two-page document, "Need for a New Building for Botany Department," into his departmental annual report of 1955/1956. He emphasized that the need for a fireproof building was urgent.[104] Among his additional reasons for a new physical plant, he noted that the present facilities for teaching elementary classes were outdated, graduate student rooms were cramped, Davie Hall could no longer adequately support research programs, and the Botany Library was filled to capacity and did not have enough seating room to accommodate the students. He recommended that the new building should be constructed close to the Zoology Department. University administrators approved funding for the building in spring of 1959, following the passage of a bond referendum that included $750,000 for the project (Anonymous, 1959a; Medlin, 1960). In the early 1960s Couch obtained a grant from the National Science Foundation that awarded $80,250 to support funding for the construction and furnishing of the new building. The grant was applied over a two-year period ([NSF], 1961). The National Institutes of Health also granted funds ($306,510) for the new building (Greulach, 1961).

A committee charged with planning the new structure consisted of C. Ritchie Bell (chair), Victor Greulach, and William Koch. Greulach's (1961) annual report of 1960/1961 notes that plans for the building were progressing. He itemized a number of features of the proposed structure.

> The plans (Fig. 15.8) for the new building call for an office-laboratory and a full research laboratory for each present and prospective staff member and for separate teaching laboratories for both graduate and undergraduate courses. Graduate students are to be provided with offices that can be used as laboratories for part of their research. All major equipment will be housed in the large research laboratories according to the needs in each discipline, and will be available for use by all staff members, their assistants, research fellows, and graduate students doing research.

Construction of the new Botany building, named Coker Hall in honor of William C. Coker, began on September 12, 1961. At the groundbreaking ceremony Victor Greulach turned the first shovelful and John Couch dug the second shovelful (Ivey, 1961). The department occupied its new quarters in March 1963. The nearly 64,000-square-foot building (Fig. 15.9) was constructed of reinforced concrete and was of Williamsburg architectural style. The south end of the building included a 2,500-square-foot greenhouse. A dedication celebration was held on May 30, 1964. The new building provided expanded and modern facilities for research and scientific investigations as well as adequate laboratories for teaching the discipline.

---

[104]Concerning the condition of old Davie Hall, a reporter with the university's newspaper noted that "the entire framework of Davie Hall is of heart pine, now dry and ripe for a spark. A valuable library and much equipment would be greatly endangered" (Anonymous, 1959b).

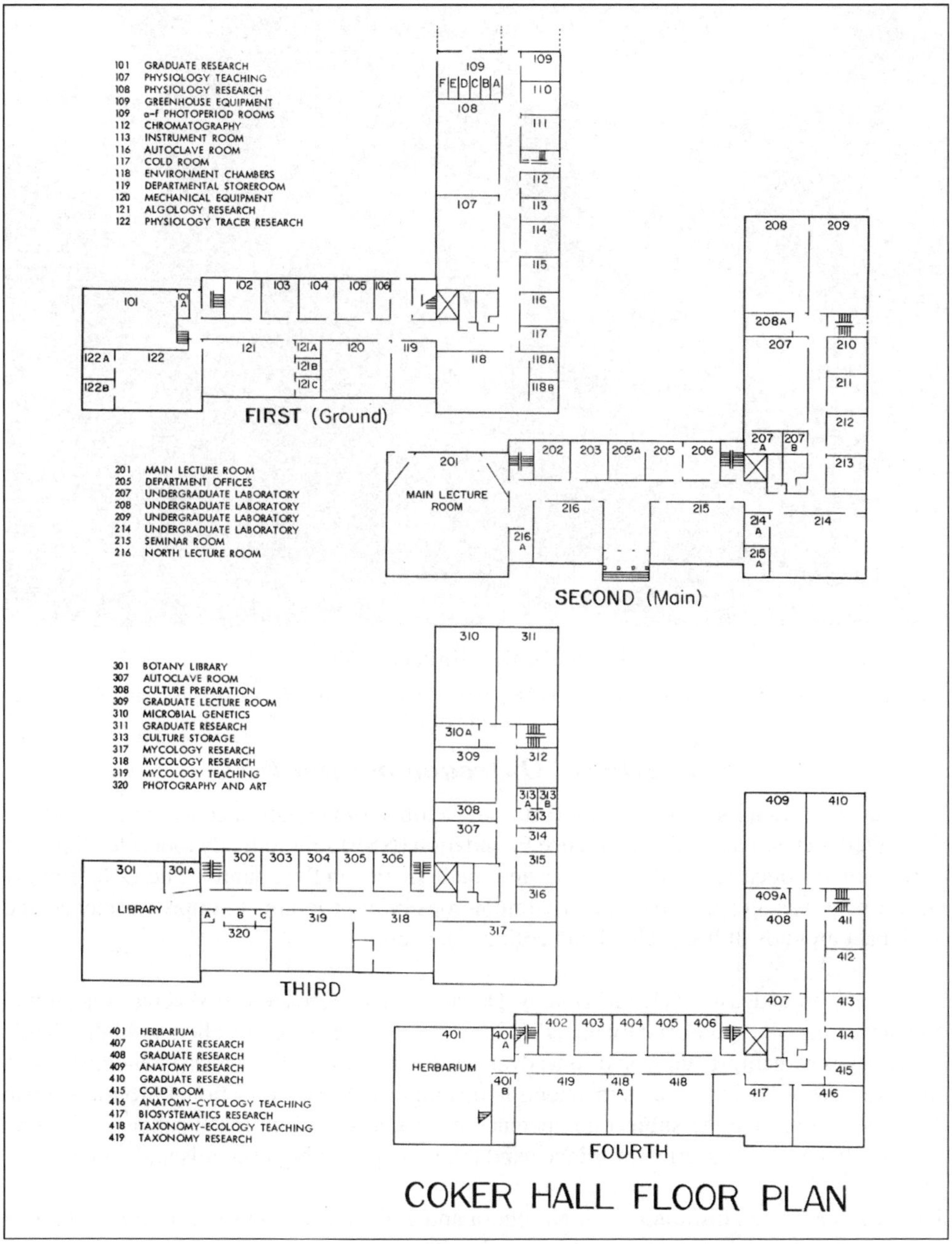

Fig. 15.8. Floor plans of Coker Hall. Report of the Self-study Committee, Department of Botany, University of North Carolina, Chapel Hill, April 9, 1973 (Hommersand, 1973), sheet inserted between pp. 25 and 26.

(Box 5, unlabeled folder, William R. Burk Collection on University of North Carolina at Chapel Hill Botany and Zoology Faculty, 1920–2010, University Archives #40478, Wilson Special Collections Library, University of North Carolina, Chapel Hill).

Fig. 15.9. Coker Hall, ca. 1980s.

(Courtesy of Susan Whitfield, Photo Lab, Department of Biology, University of North Carolina, Chapel Hill).

Educational Outreach beyond UNC

The launch of Sputnik by the Soviet Union in October 1957 spurred educational reform in the United States. The National Science Foundation (NSF) subsequently took a leading role to rejuvenate education in the sciences and mathematics in the country. The UNC Botany Department was an active participant in numerous NSF programs to improve science and math backgrounds of high school and college teachers.

From 1957 to at least 1971, the Botany Department participated in several educational outreach programs sponsored by the NSF. The programs provided teachers in high schools and colleges with information on recent advances in various scientific fields, including botany, and a review of basic concepts in scientific disciplines. These educational programs were designed to improve the subject competence of teachers, and ultimately their instruction would enhance the education of science and mathematics in high schools and colleges.

The NSF-sponsored institutes for high school and college teachers were of three principal kinds: summer, academic year, and in-service. The last-mentioned type entailed night or weekend classes during the school year; it is unknown if the UNC Botany Department participated in these. UNC botanist Victor A. Greulach and Howard University botanist Marie C. Taylor (1958) provide information on the institute programs. Beginning in 1957, Greulach and UNC mathematics professor Edward A. Cameron co-directed a series of summer institutes for high school science and mathematics teachers held on the UNC campus.

NSF Academic Year Institutes for High School Teachers

Archival sources documenting the Botany Department's role in the NSF Academic Year Institutes are scarce. The botany Department participated in the NSF Academic Year Institute from 1960 to around 1965. High school biology teachers enrolled in the program to earn the M.Ed. in science teaching. Over a period of six years, 104 teachers took part in this endeavor: 1960 (51 teachers), 1961 (17), 1962 (8), 1963 (11), 1964 (8), and 1965 (9).

NSF Summer Conferences in Botany for College Teachers

Under the aegis of the NSF Summer Institute for College Teachers, the Botany Department conducted seven NSF Summer Conferences in Botany (Fig. 15.10) from 1959 to 1965. Greulach directed the sessions except for those of 1962 and 1965, which were directed by

Fig. 15.10. Group photo of participants in the National Science Foundation Summer Conference in Botany, June 1963, held at the University of North Carolina, Chapel Hill. (Bottom row: left to right) Ruth E. Griffith, Elizabeth Anne Colley, Pauline D. Pages, Mildred McAusland Marengo, James Harvey Craft, James S. Fletcher, and Harry E. Williams; (2nd row) Sister Mary Baptist Cleary, Vida Grace Wood, Sarah A. Staples, Sister M. Gervasia Schreckenberg, Gordon Earle Geeseman, Charles Frederick Wolf, and Melvin L. Conrad; (3rd row) Sister Adrian Marie Hofstetter, Sister Casimir Czurles, Carlita L. Georgia, Agnes Rose Lisle, Pauline M. Longest, Elizabeth McGowan, and Lamar Johanson; (4th row) Alfred M. Wolfson, Ben Thomas Winn, Alfonso Roscoe Vick, John Randall Keller, Margaret Haskell, John B. Whitney, Jr., and William Canno Lenoir, Jr.; (5th row) Victor A. Greulach, James M. Craig, Walter Raymond Leonard, Edward G. Barry, Raymond Westra, N.E. Tolbert, Richard Kirk Speairs, Jr., Herald H. Hopkins, S. Lane Wilson, Max H. Hommersand, and Charles William Shrader.

(Photographic Archives #24447, North Carolina Collection, Wilson Special Collections Library, University of North Carolina, Chapel Hill).

C. Ritchie Bell. These sessions aimed to provide college teachers of general biology and general botany courses with a current knowledge of botany and research advancements in the discipline. They also presented important botanical concepts and principles that teachers may not have received in their training.

The UNC Botany Department publicized the conferences through notifications placed in several botanical and biological journals. It also prepared bifold brochures that described the learning experience. Each year that a conference was offered, the department mailed the brochures to colleges and universities in the eastern half of the country that offered botany courses or botanical topics in general biology courses. The conferences featured morning sessions of lectures and afternoon sessions of laboratory experiences, demonstrations, and discussions. Evening lectures were presented; and field trips, usually to areas of taxonomic and ecological interest, took place on weekends. Lecturers, experts in their field of specialty, came from not only UNC's Botany Department but also from renowned institutions throughout the East. Historical documents in UNC's archives (UNC-A-1) contain numerous records regarding the conferences. Information includes proposals, reports, brochures, and lists of participants. Table 15.2 provides information on the summer botany conferences.

A short description of the second NSF Summer Conference in Botany that took place at UNC on August 1–19, 1960 provides insight into the nature of the program and the participation of UNC botanists (Anonymous, 1960a). Two non-credit short courses were given, one in plant physiology, the other in plant evolution. Three professors from other universities lectured on the former topic, which included plant growth, the biochemistry and physiology of DNA, RNA and protein synthesis, and respiration. Three UNC botanists lectured on the latter topic: C. Ritchie Bell on cytogenetics and evolution, J. Edison Adams on the evolutionary significance of certain morphological features of vascular plants, and William J. Koch on the evolutionary significance of some recent mycological research. In addition, at least six lectures were presented in the evenings, one of which was given by John N. Couch. Albert E. Radford conducted a field trip for the class.

TABLE 15.2. NSF SUMMER BOTANY CONFERENCES HELD AT UNC, 1959–1965, PROVIDING TITLE OF CONFERENCE, DATE HELD, ORGANIZER, THEME, AND NUMBER OF PARTICIPANTS.

1959

A Botany Conference for College Teachers of General Botany and General Biology
July 27–August 14, 1959
Victor A. Greulach, Director
Plant physiology, plant anatomy and morphology with an emphasis on plant physiology
24 participants

1960

A Botany Conference for College Teachers of General Botany and General Biology
August 1–19, 1960
Victor A. Greulach, Director
Plant physiology, plant evolution
31 participants

TABLE 15.2. (CONTINUED)

1961

A Botany Conference for College Teachers of General Botany and General Biology
July 31–August 18, 1961
Victor A. Greulach, Director; C. Ritchie Bell, Associate Director
Plant physiology, plant cytology with particular reference to recent advances in these fields
30 participants

1962

A Conference for College Teachers of General Botany and General Biology
July 23–August 10, 1962
C. Ritchie Bell, Director
Plant physiology, plant morphology with particular reference to recent advances in these fields
17 participants

1963

A Plant Physiology and Biochemistry Conference for College Teachers (also titled NSF Summer Botany Conference)
June 10–28, 1963
Victor A. Greulach, Director
Plant physiology, plant biochemistry
33 participants

1964

A Botany Conference
June 29–July 17, 1964
Victor A. Greulach, Director
Nucleic acids and proteins, cytology and cytogenetics, photosynthesis and respiration, biosystematics, morphogenesis, and plant growth
29 participants

1965

A Botany Conference for College Teachers (also titled Conference on Botany: New Methods of Botanical Investigation and Classification)
June 14–July 2, 1965
C. Ritchie Bell, Director
New methods of botanical investigation and classification, including chemotaxonomy, breeding structure of plant populations, numerical taxonomy, and paleobotany
28 participants

NSF Summer Short Courses

Around 1959, the Botany Department first participated in an NSF program that offered college teachers of general botany, general zoology, and general biology short courses in special scientific fields, such as ecology and mycology. The department participated in the program for nearly ten consecutive years. Taking place in summers, the courses of three weeks' duration reviewed basic concepts and reviewed information on recent advances in the subjects. Some of the sessions were cooperatively taught with Duke University and North Carolina State University as well as scientists invited from other institutions. A typical course consisted of morning lectures and afternoon sessions in which demonstrations, discussions, and laboratory work occurred. On occasional evenings, instructors gave lectures, and on weekends field trips were taken. Examples of the short courses include the following:

1. Mycology Summer Short Course, June 9–27, 1968; directed by John N. Couch. Of the nine instructors, four were UNC botanists: John N. Couch, Aristotle J. Domnas, William J. Koch, and Lindsay S. Olive (Anonymous, 1968a).
2. Ecology Summer Short Course, June 10–July 5, 1968; directed by J. Frank McCormick.
3. Mycology Summer Short Course, June 9–27, 1969; directed by Lindsay S. Olive. Among the participating lecturers were four notable mycologists from outside institutions: Peter R. Ray, Melvin Fuller, John Raper, and Kenneth Raper.
4. Ecology Summer Short Course, June 7–July 3, 1970; directed by J. Frank McCormick.

NSF Summer Research Participation Programs for Undergraduates

Around 1959, the department began an NSF Summer Research Participation Program for Undergraduates. Serving as research assistants to various UNC professors, undergraduate students from other institutions participated in the experience. The program offered the students firsthand experience in conducting basic research. It also provided training in laboratory procedures and techniques (Adams, 1962). The Botany Department was involved in this program for at least six years. Professor C. Ritchie Bell directed most of the programs in the later 1960s.

The experiences of several college undergraduate students who served as summer research assistants in the Botany Department in 1962 provide noteworthy insight into the program. Virginia E. Bowers of Hopewell, Georgia, collaborated on a project in the laboratory of John Couch. She assisted with studies on the fungus *Coelomomyces*, which parasitizes mosquito larvae. In professor Max Hommersand's lab, Julius Hill of Murphy, North Carolina, collaborated in physiological investigations of two microscopic species of one-celled algae. John E. Ward of Mocksville, North Carolina, and Donald R. Farrar of Perryville, Missouri, assisted Ritchie Bell and Albert Radford in the Flora of the Carolinas Project. Farrar studied the size of pollen in a variety of weedy plants, and Ward investigated the effect of plant hormones on the growth of the plant *Coleus* (Adams, 1962).

NSF Summer Research Participation Programs for College Teachers

The Botany Department was involved in a research participation program for college teachers. The primary aim of the program was to provide an opportunity for college teachers to resume research that had been prevented because of a lack of lab facilities, space, or time. Teachers could engage in productive research under the guidance of competent investigators. The Botany Department conducted at least two such research participation programs on June 16–August 23, 1969 and June 14–August 21, 1971. Participants chose an area of research from several specialties in botany: mycology, phycology, plant biochemistry, genetics, ecology, plant taxonomy, and systematics (Anonymous, 1970). Victor Greulach (1970) directed the first program, and Professors Ed Barry, C. Ritchie Bell, Aristotle Domnas, Max H. Hommersand, H.T. Odum, Clifford Parks, and Clyde Umphlett collectively mentored eight participants. When the session concluded, the advisors unanimously noted that the program was valuable to them as well as to the participants.

Interdisciplinary Curricula

Toward the last years of Greulach's chairmanship, scientific disciplines on the UNC campus had increasingly overlapped in their areas of research. The Botany Department participated in at least four interdepartmental curricula. These were the Curriculum in Genetics (beginning in 1964 and later named Curriculum in Genetics and Molecular Biology), the Curriculum in Marine Sciences (1968), the Curriculum in Biology (1969), and the Curriculum in Ecology (1971).[105] The various departments that participated in the programs are cited in Table 15.1.

Planning for the Future

In 1965, UNC formed a committee to evaluate the life sciences curricula at its campus (Anonymous, 1965). In response to the committee's request for background on the status of the Botany Department's program, the botany professors prepared six informative reports. The accounts outlined the needs, such as faculty, staff, and facilities. They also described aspects of research programs and current strengths in the specialized disciplines. The reports regarding ecology and plant physiology/biochemistry noted the need for a program at the molecular level in the Botany Department.

In the Botany Department annual report of 1965/1966, Greulach (1966) appended a five-page document titled "History, Status and Plans of the Department of Botany at the University of North Carolina at Chapel Hill." The report outlined the plans and needs of the department for the ensuing decade. He stated that the department as of 1966 was still known for its outstanding strength in mycology and plant taxonomy. Efforts were also being focused on further developing several other disciplines: plant physiology and biochemistry, ecology, microbial genetics, and algology. He noted that a special emphasis would be placed in three areas: developmental biology, environmental biology, and systematic biology. Nonetheless, other disciplines, such as genetic, molecular, metabolic, and regulatory biology, would not be ignored.

In accomplishing these objectives, Greulach enumerated a number of needs:

1. A substantial increase in non-professional technical personnel, including research associates, research assistants, and technicians.
2. Completion of furnishing Coker Hall with casework and fixed equipment specified in the original plans (11 laboratories).
3. Six additional controlled environment plant growth chambers.
4. An additional unit to the Coker Hall greenhouses doubling the space currently available.
5. An addition of at least 25,000 square feet to Coker Hall.

[105]The Curriculum in Ecology merged in 2008 with the undergraduate programs in Environmental Studies and Environmental Science administered by the Carolina Environmental Program to become the Curriculum for the Environment and Ecology, and later in 2018 was transformed to become the Environment, Ecology and Energy Program.

Professors of Botany Joining the Department under Greulach's Administration

Edward Gail Barry (September 1962–December 1996, assistant professor to professor)

Research Interests: fungal genetics, microbial genetics, cytogenetics with special reference to *Neurospora*.

Courses Taught: General Botany, Genetics, Cytogenetics, Microbial Genetics, Genetics Laboratory, Genetics Systems.

Selected Publications: Diffuse diplotene stage of meiotic prophase in *Neurospora*, 1969; Strains of *Neurospora* collected from nature, 1976 (with D.D. Perkins and B.C. Turner); Reversal of a *Neurospora* translocation by crossing over involving displaced rDNA, and methylation of the rDNA segments that result from recombination, 1986 (with D.D. Perkins, R.L. Metzenberg, and N.B. Raju).

Graduate Students: M.A., 2; M.S., 1; Ph.D., 1.

Clyde Jefferson Umphlett (1961–1962, instructor; [September] 1962–August 1970, assistant professor to associate professor)

Research Interests: mycology.

Courses Taught: General Botany; Fungi, Algae, and Bryophytes; Ecology of Fungi.

Selected Publications: *General Botany Laboratory: Exercises and directions*, 1965; Cytological and morphological studies of a new species of *Phlyctochytrium*, 1967 (with L.W. Olson); Two new dentigerate species of *Phlyctochytrium* (Chytridiomycetes), 1969 (with W. J. Koch).

Graduate Students: M.A., 1; M.S., none known; Ph.D., 2.

Donald Benton Anderson (September 1963–May 1970, professor)[106]

Research Interests: plant physiology, history of botany.

Courses Taught: Development of Modern Concepts in Botany, General Botany.

Selected Publications: numerous papers and several textbooks on plant physiology written prior to his teaching in the UNC Botany Department.

Graduate Students: none known.

Walter Earl Loomis (September 1964–December 1965, visiting professor)

Research Interests: plant physiology, plant growth.

Courses Taught: Plant Physiology, Plant Growth.

[106] Greulach (1964) reported that Anderson "was a professor in the Department [of Botany] effective September 1, 1963, although he is not at present engaged in active duties in the Department." He was then Vice-President of the Consolidated UNC but taught a signature course on Development of Modern Concepts in Botany at UNC. Botanist James R. Troyer (1996) provides an insightful biography on Anderson. Anderson was named University Professor in 1966 with the privilege of teaching at the three then-existing branches of the Consolidated University in Chapel Hill, Greensboro, and Raleigh. He taught plant physiology and developed a new course on the history of ideas in biology. His course at UNC was clearly developed from the latter concept.

Walter Earl Loomis (Continued)

Selected Publications: *Botany*, 3rd ed., [1962] (with C.L. Wilson) and papers on plant growth.

Graduate Students: M.A., 1.

J. Franklin McCormick (September 1964–August 1974, assistant professor to professor)

Research Interests: plant ecology (ecosystems, populations), land use planning.

Courses Taught: Plant Ecology, Autoecology, Radiation Ecology, Environmental Biology, Ecology, Physiological Ecology, Ecology Laboratory, Plant Life.

Selected Publications: Differential radiation sensitivities of races of *Sedum pulchellum* Michx, 1964; Ecotypic differentiation in *Diamorpha cymosa*, 1964 (with R.B. Platt); Irradiation of natural vegetation, 1966 (with F.B. Golley).

Graduate Students: M.A., 7; M.S., 2; Ph.D., 9.

Aristotle Jordan Domnas (ca. April 1965–June 1994, assistant professor to professor)

Research Interests: biochemistry of fungi, fungal lipid chemistry and sterol enzymology, studies on *Lagenidium*.

Courses Taught: Physiology of the Fungi, Plant Biochemistry, Instrumental Methods in Botany, Cell Biology, Cell Biology Laboratory, Plant Physiology, Comparative and Evolutionary Biochemistry.

Selected Publications: The active transport and metabolism of purines in the yeast *Candida utilis*, 1959 (with A.H. Rousch and L.M. Questiaux); Biochemistry of mosquito infection, 1974 (with P.E. Giebel and T.M. McInnis); Sterol requirement for zoospore formation in the mosquito-parasitizing fungus *Lagenidium giganteum*, 1977 (with J.P. Srebro and B.F. Hicks).

Graduate Students: M.A., none known; M.S., 1; Ph.D., 10.

Jan Justus Kohlmeyer (ca. 1966–June 2001 [part-time July 1998–June 2001], assistant professor of botany to professor of botany)[107]

Research Interests: marine fungi, Ascomycetes.

Courses Taught: Marine Mycology, Taxonomy of Special Groups of Nonvascular Plants.

Selected Publications: *Icones Fungorum Maris*, 1964–1969 (with E. Kohlmeyer); *Synoptic Plates of Higher Marine Fungi*, 3rd ed., 1971 (with E. Kohlmeyer); *Marine Mycology*, 1979 (with E. Kohlmeyer).

Graduate Students: none known.

Friedrich Heinrich Helmut Lieth (ca. June 1967–December 1977, associate professor to professor)[108]

Research Interests: biometeorology, biogeography, plant ecology, global ecology.

[107] In early 1964 Kohlmeyer became an assistant professor in the UNC Institute of Fisheries Research (currently the Institute of Marine Sciences) and advanced through the academic ranks to professor. Around 1966 he also joined the UNC Botany Department, becoming an assistant professor of botany and advancing to professor in July 1974.

[108] Following Lieth's resignation as of January 1, 1978, he became adjunct professor of botany for five years.

Friedrich Heinrich Helmut Lieth (Continued)

Courses Taught: Geobotany, Community Ecology, Ecology Laboratory, Systems Ecology.

Selected Publications: *Klimadiagramm-Weltatlas*, 1967 (with H. Walter); *Phenology and Seasonality Modeling*, 1974; *Primary Productivity of the Biosphere*, [1975] (with R.H. Whittaker); *Patterns of Primary Production in the Biosphere*, 1978; *Contributions to the Knowledge of Flora and Vegetation in the Carolinas*, 1979–1981 (with E. Landolt and R.K. Peet); and a number of additional books written after his professorship at UNC.

Graduate Students: M.A., 4; M.S., 2; Ph.D., 5.

Clifford Robert Parks (August 1967–June 2003 [part-time, July 2000–June 2003], assistant professor to professor)

Research Interests: genetic analysis of North Temperate Zone disjunct species, variation patterns in isozyme and flavonoid extractives in wide-ranging tree species of the eastern United States, compatibility and breeding studies in the genus *Camellia*, molecular systematics of the genus *Camellia*, inheritances of cold resistances of *Camellia* species.

Courses Taught: Horticulture, Chemotaxonomy, Plant Genetics, General Botany, Genetics, Genetics Laboratory, Plant Life, Plants and Life.

Selected Publications: *Botany 11: Introductory botany laboratory manual*, 1979 (with P.G. Gensel and A. White); Woody plants commonly cultivated in central North Carolina, 1979; *Collected Species of the Genus Camellia*, 2005 (with Jiyin Gao and Yueqiang Du).

Graduate Students: M.A., 4; M.S., 10; Ph.D., 17.

Lindsay Shepherd Olive (1938–1942, laboratory instructor; 1942–1944, instructor; January–April 1968, professor; May 1968–June 1982, University Distinguished Professor)

Research Interests: cytology, genetics, morphology, taxonomy of fungi and mycetozoans.

Courses Taught: Mycetozoa, Ascomycetes, Basidiomycetes, Fungi and Algae, Biology of the Fungi, Taxonomy of Special Groups of Nonvascular Plants, Ascomycetes and Basidiomycetes, Mycetozoans and Lower Fungi, The Mycetozoans and Their Allies.

Selected Publications: *The Mycetozoans* (with the technical assistance of Carmen Stoianovitch), 1975; and over 120 articles in area of specialty.

Graduate Students: M.A., none known; M.S., 4; Ph.D., 8.

Richard Malcolm Brown, Jr. (September 1968–June 1982, associate professor to professor)[109]

Research Interests: plant cytology; plant cell biology; algae; fungi; cytological methods; ultrastructure and cytochemistry; freshwater, airborne and volcanic algae; cellulose and its biosynthesis.

[109] Although appointed to the Botany faculty on September 1, 1968, Brown received an initial year of absence to complete a post-doctoral fellowship at the Botanical Institute, University of Freiburg. He arrived at UNC in August 1969.

Richard Malcolm Brown, Jr. (Continued)

Courses Taught: Algae, Cytology, Fungi and Algae, Cytological Methods, Cell Biology, Cell Biology Laboratory, Cytological Methods: Instrumentation, Cytological Methods: Specimen Preparation and Methodology, Taxonomy of Special Groups of Nonvascular Plants.

Selected Publications: *Contributions in Phycology*, [1971] (with B.C. Parker); *Cellulose and Other Natural Polymer Systems*, 1982; and numerous articles as well as several additional books published after his departure from UNC.

Graduate Students: M.A., 5; M.S., 3; Ph.D., 8.

William Campbell Dickison (August 1969–November 1999, assistant professor to professor)

Research Interests: comparative anatomy and morphology of vascular plants, plant diversity, phylogeny and classification of flowering plants.

Courses Taught: Vascular Plants, Comparative Morphology of Lower Plants, Plant Anatomy, Phylogeny and Classification of Flowering Plants, Comparative Morphology of Vascular Plants, Paleobotany, Plant Diversity.

Selected Publications: *Vascular Plant Systematics*, [1974] (with A.E. Radford and others); *Contemporary Problems in Plant Anatomy*, 1984 (with R. A. White); *BIOL 152: Plant Anatomy*, [ca. 1990]; *Selected Readings for Botany 51: Plant Diversity*, [ca. 1992]; *Integrative Plant Anatomy*, 2000 (which was published posthumously); and nearly 95 articles.

Graduate Students: M.A., 5; M.S., 6; Ph.D., 4.

Tom Keck Scott (August 1969–June 2001 [part-time, July 1998–June 2001], associate professor to professor)

Research Interests: growth development and hormone translocation in higher plants, phytohormones, gravitational plant biology.

Courses Taught: Introductory Botany, Introductory Biology, Plant Physiology, Plant Growth and Development.

Selected Publications: *Plant Regulation and World Agriculture*, 1979b; *The Functions of Hormones from the Level of the Cell to the Whole Plant*, 1984; *Plant Gravitational and Space Research*, 1984 (with T.W. Halstead).

Graduate Students: M.A., 1; M.S., 3; Ph.D., 5.

Norton George Miller (September 1970–December 1974, visiting assistant professor to assistant professor)

Research Interests: Quaternary paleobotany and paleoecology, Tertiary and Quaternary history of the Bryophyta.

Courses Taught: Bryophytes, Plant Diversity, Paleobotany.

Selected Publications: nearly 100 books and articles, most of which were published outside his years at UNC.

Graduate Students: M.A., 1; M.S., 2; Ph.D., none known.

Tom Keck Scott: Fourth Chair (1972–1982)

Born August 4, 1931 in Saint Louis, Missouri, Tom Keck Scott (Fig. 15.11) is the son of George Drake and Mary Ann (Keck) Scott. Following his secondary school education, he earned an A.B. in Botany (1954) from Pomona College. Subsequently he was 1st Lieutenant, United States Army Reserve, Chemical Corps (1954–1956). Scott then pursued graduate studies at Stanford University, where he earned an M.A. (1959) and a Ph.D. (1961), both in biological sciences. After receiving his doctorate degree, he was a postdoctoral researcher at Princeton University from 1961 to 1963. He then served as a member of the biology faculty at Oberlin College, Ohio, where he was an assistant professor (1963–1967) and associate professor (1967–1969).

Fig. 15.11. Tom K. Scott.

(Image box 37, News Services of the University of North Carolina at Chapel Hill Records, 1924–2013, University Archives #40139, Wilson Special Collections Library, University of North Carolina, Chapel Hill).

Scott joined the faculty in the Botany Department at UNC in August 1969. Beginning July 1972, he became chair of the department and served in that position until the department merged with the Zoology Department in July 1982 to become the Department of Biology. He took a leave of absence during August 1972 and the 1972/1973 academic year to serve as a Fulbright-Hayes Senior Lecturer at Ege University in Izmir, and as a consultant, the Ministry of Agriculture, Ankara, both in Turkey. During his leave, Edward G. Barry was the acting chair of the Botany Department. Scott also took a leave the first half of 1982 to spend a semester at Harvard University. During his absence, Aristotle J. Domnas was the acting chair of the Botany Department. Scott's leadership extended beyond the Botany Department. He was chair of the UNC Biology Curriculum (1970–1975) and director of the UNC Office of Research Services (1985–1991). He was a Fellow in summer 1984 with the Japanese Society for the Promotion of Science, Osaka, Japan. Scott was also a consultant to NASA for over 15 years. In 1994 the agency chose Scott to coordinate ground and space experiments on cell biology (Anonymous, 1994). Beginning July 1998, Scott took a phased retirement plan, working part-time until the end of June 2001. At least three sources provide additional biographical background on Scott (Gensel, 2001; [Scott], 2000; Scott, [2003]).

In the early 1970s, botanical facilities included electron microscopes with accessory apparatus such as a freeze-etch device, controlled environmental plant growth chambers, constant temperature rooms, radioisotope equipment, and a well-equipped photographic laboratory (Anonymous, [1971]b). According to a self-study report of 1973, the department aimed to focus on three primary goals (Hommersand, 1973) as noted below.

1. To provide a basic understanding and appreciation of plants to liberal arts students and to students specializing in the Biological Sciences.
2. To provide an intense, broadly comprehensive program for undergraduate majors and graduate students who seek training in Botany as a profession.
3. To engage in scholarly research and contribute to the fund of knowledge in the plant sciences.

This same report noted several downward trends of the department during the five-year period, 1968–1973. The growth of the department did not keep pace with the overall growth of the university. Total course enrollments remained at a static level, while there was a decline in the General Botany course. This portrait of the state of the department foreshadowed its future merging with the UNC Department of Zoology.

Among the noteworthy accomplishments during Tom K. Scott's term as chair of the Botany Department were the establishment of a bachelor of science degree in botany program and the completion of the second botany electron microscope laboratory and its resulting research. Scott was instrumental in setting up the Botany Undergraduate Student Association. A number of landmark publications by faculty members were issued during this administration.

Undergraduate Developments

The Bachelor of Science Degree Program

As early as November 1973, the Curriculum Committee of the Botany Department recommended that the department should offer the B.S. degree in botany (UNC-A-2a). Faculty members of the department discussed the proposed requirements for the degree in October of the following year, but university approval of the proposal was stalled by administrative tie-ups (UNC-A-2b). UNC President William Friday subsequently approved the offering of the B.S. degree in May 1976 (Gaskin, 1976). The B.S. degree program was first listed in the campus *Undergraduate Bulletin* for the 1977/1978 school year (Anonymous, 1977b).

The B.S. degree in botany provided a more focused and broad-based program for students who planned to follow a professional career in botany than the B.A. degree in botany provided. Requirements in 1977 for the B.S. in botany included a foreign language, mathematics, and either a course in computer science or statistics as well as four botany courses (General Botany, Plant Diversity, Field Botany, Plant Physiology and Morphogenesis), and four botany electives, one each chosen from four of the five groups: Group I. Algae, Fungi, or Taxonomy of Vascular Plants; II. Plant Physiology, Plant Physiology Laboratory, or Cell Biology; III. Ecology or Geobotany; IV. Evolutionary Mechanisms or Genetics; and V. Comparative Morphology of Vascular Plants, Plant Anatomy or Cytology. Students also had to take Botany 92 (Senior Seminar).

Undergraduate botany students could also apply for honors work and earn a B.S. in botany with honors. An overall grade average of B was a prerequisite for admission into the honors program. During their senior year, honors students were required to enroll in the course titled Senior Honors Thesis. The course consisted of individual reading, study, research, and the preparation of a thesis. The original version of the honors thesis (also called honors essay) was archived in UNC's North Carolina Collection, now part of the Louis Round Wilson Special Collections Library. A second copy was deposited in the Botany Library. Honors students prepared at least nineteen theses for the B.A. and B.S. degrees with honors in the Botany Department, including two honors theses with a botanical theme in the Program in Interdisciplinary Studies (Table 15.3).

Table 15.3. Titles of Theses for B.A. and B.S. Degrees in Botany with Honors, 1962–1982, Alphabetically Arranged by Author's Name.

Name	Date	Title	Degree	Major Advisor
Blanton, Richard Lawrence	1977	Studies on a new protostelid (Eumycetozoa) with a compound nucleolus	B.S.	Lindsay S. Olive
Doares, Steven Humphrey	1982	A general characterization of tranposon mutants of *Agrobacterium tumefaciens*	B.S.	Ann G. Matthysse
Fambrough, Douglas McIntosh, Jr.	1963	Experiments on the photosynthetic assimilation of carbon dioxide and evolution of oxygen by *Chlorella pyrenoidosa*	B.A.	Max H. Hommersand
Hall, Susan Hewlett	1967	An anatomical study of the dehiscence of the fruit of *Hibiscus esculentus* L.	B.A.	Edward G. Barry
Hood, Betsy Ann	1978	A measurement of chemotaxis of *Agrobacterium tumefaciens*	B.A.	Ann G. Matthysse
Kronenberg, Meredith (Alden)	1971	Some characteristics of N-acetyl-[beta]-D-glucosaminidase activity in *Streptosporangium roseum* #499	B.A.	Aristotle J. Domnas
Leach, Norman Edward, II	1968	Effect of indole-3-acetic acid on growth and cell division in *Chlamydomonas reinhardtii*	B.A.	Max H. Hommersand
Lundeen, Carl Victor, Jr.	1965	Some aspects of acetate metabolism in *Chlamydomonas reinhardtii* Dang.	B.A.	n.a.
McCall, Claire Morris	1979	An investigation of the inheritance of seven morphological traits in *Camellia japonica*	B.A.	Clifford R. Parks
Michener, David Charles	1977	Some aspects of the reproductive biology of two introduced species of the genus *Mazus* (Scrophulariaceae) in the United States	B.A.	Jim Massey

TABLE 15.3. (CONTINUED)

Name	Date	Title	Degree	Major Advisor
Motten, Alexander Fewell	1970	Use of pigment absorbance ratios and community metabolism as indices of maturity in New Hope Creek	B.A.	J. Frank McCormick
Olson, Lauritz William	1966	A cytological and morphological study of an undescribed species of *Phlyctochytrium*	B.A.	Clyde J. Umphlett
Oxman, Linda Lee (Kerley)	1962	The vegetation of a coastal island—Shackleford Bank	B.A.	n.a.
Proctor, Alan Ray	1967	A highly mutable strain of *Neurospora crassa*	B.A.	Edward G. Barry
Robinson, Thomas Marshall, Jr.	1975	An annotated checklist of collected Norwegian marine and terrestrial flora	B.S. (Biology)*	C. Ritchie Bell
Trail, Frances	1981	The effects of V-A mycorrhizae on the growth of *Sporobolus* in a pine savanna	B.A.	Peter M. Vitousek
Waterman, Kristin Clover (Taylor)	1964	The comparative effects of NAA, IBA and IAA on the initiation of dormant root primordia and the growth of roots in the stem of the Lombardy poplar	B.A.	Victor A. Greulach
Weakley, Alan Stuart#	1978	The vegetation of Swift Creek swamp forest, Edgecombe County, North Carolina	B.A.	Albert E. Radford
Zucchino, Lawrence Ray#	1975	A phytosociological study of selected wetland habitats in the lower coastal plain of North Carolina	B.A.*	Helmut Lieth

*Honors essays in the Program in Interdisciplinary Studies, University of North Carolina, Chapel Hill

#Alan Weakley and Lawrence Zucchino each received a French Broad River Garden Club Scholarship in the 1970s. The French Broad River Garden Club of Asheville, North Carolina established the scholarship in 1965. It provided "$500.00 per year for an award to a student majoring in botany, with preferences given to students from certain Western North Carolina counties. Awards are based on merit, character, and financial need" (Anonymous, 1968b).

Botany Undergraduate Student Association (BUSA)[110]

The Botany Undergraduate Student Association was established in fall 1975 and was open to botany majors as well as biology majors. In 1977 a revised charter provided for elections of a chairperson who would administer a steering committee of five members. Elections took place each April.

[110] The following account of the Botanical Undergraduate Student Association was derived primarily from information in several annual reports written by Department of Botany Chairman Tom K. Scott (1977b, 1978,

The group sponsored various programs such as career development meetings held in association with the UNC Career Planning and Placement Office and speakers who gave career development presentations. The association promoted informal departmental interactions through such events as potluck suppers that included undergraduate and graduate students, faculty members and staff. It also created a bulletin board in the Botany Department and a folder in the Undergraduate Library that highlighted internships, graduate programs, and special programs. BUSA leadership was instrumental in organizing a Career Mart. The organization also took a stand on political issues, such as supporting the Natural Diversity Preservation Act.

At the botany faculty meeting of October 3, 1975, chairperson Tom Scott recommended that departmental undergraduates serve on faculty committees. The faculty "moved and seconded that undergraduate representatives be included on appropriate committees and invited to attend, with option of refusal, pertinent faculty meetings" (Anonymous, 1975b). Members of the association played an active role in departmental committees. In the academic year 1978/1979, six undergraduates served on five committees. Tom Scott (1977b) praised the involvement of BUSA's members. For example, he noted "that BUSA representative to the Curriculum Committee, Larry Blanton, was very influential in bringing about a review of the B.S. degree requirements," which were then being assessed by the committee.

In its fifth and sixth years, BUSA was relatively inactive and was not engaged in any special endeavors. Only one association member served on a departmental committee in the academic year 1980/1981. In the following year, there was no student representation on a departmental committee, but members met on several occasions to discuss the future of the botany program in light of the upcoming merger of the Botany and Zoology Departments.

Noteworthy Botanical Publications

Plant taxonomy has played a significant role in research and teaching at UNC since the establishment of the Botany Department in 1908. Botanists at the UNC Herbarium prepared several landmark publications as noted earlier in this chapter. In 1964, even while the *Manual of the Vascular Flora of the Carolinas* was being completed, botanists Albert Radford and Harry Ahles "actively proceeded with the organization of The Vascular Flora of [the] Southeastern United States project" (Radford, 1965). Two years later, Radford, C. Ritchie Bell, James W. Hardin, and Robert L. Wilbur (1967) compiled a *Contributors' Guide* that provided instructions on writing manuscripts. The National Science Foundation and the George R. Cooley Trust contributed to the funding of the project and its resulting volumes.

The primary goals of the flora project were to: provide a floristic survey of the forested region of the southeastern United States west to the prairie and north to the southern limit of the moraines, and prepare a multi-volume manual to identify the species of vascular plants of this region. Arthur Cronquist (1980), botanist at the New York Botanical Garden, authored Volume 1, titled *Asteraceae* and published in 1980. The volume was prepared under the guidance of an editorial board consisting of Albert E. Radford (chairperson), James W. Hardin, Jimmy R. Massey, and Earl L. Core; with Laurie S. Radford as executive editor.

1979a, 1980, 1981b, 1982).

Following the publication of the first volume of *The Vascular Flora of the Southeastern United States*, the management of the project changed. In January 1981, Jim Massey became the chair of the editorial board and the principal investigator of the project. He also prepared a newsletter, *Contributors' Communique*. Issued from May 1981 to May 1983, four editions of the newsletter informed contributors of progress and problems regarding the flora. Massey and Laurie Radford (1981) subsequently produced a revised *Contributors' Guide* in 1981. In the same year, botanist James Matthews (1981) published a list of nearly 200 plant family assignments of the flora along with their contributors.

The second volume of the flora was published in 1990. Duane Isely (1990), professor of botany at Iowa State University, authored *Leguminosae* (*Fabaceae*), which constituted Volume 3, Part 2 of the flora. The subsequent volumes of the flora were to consist of contributions of the remaining families of vascular plants to be prepared by more than 150 authors; however, the flora project has remained in abeyance. The UNC Herbarium houses numerous archival records regarding the flora. Among the items are nearly 240 folders, many of which contain contributors' typed manuscripts. In many aspects, botanist Alan Weakley's (2020) online *Flora of the Southeastern United States* is an extension of the multi-volume flora that Radford and Ahles had first envisioned in 1964.

Second Botany Electron Microscope Laboratory

Botany chair Victor Greulach conceived the creation of an outstanding electron microscope lab in the late 1960s. He noted that "the establishment of a first-class electron microscope laboratory [would] be under the direction of R. Malcolm Brown" (Greulach, 1969b). Upon Brown's arrival in Chapel Hill in August 1969, the new professor began planning the development of the lab in Coker Hall. Various archival records regarding the Botany Department mention that extensive efforts were made to procure funds to purchase a Hitachi transmission electron microscope (TEM). By 1971 several sources had committed financial support of the acquisition: University Institutional grants, $11,360; the Research Corporation, $19,000; and the UNC Botany Department equipment budget, $6,000. In addition to these funds, Brown obtained grants from the North Carolina Board of Science and Technology, $13,000, as well as the National Science Foundation, amount unknown (Scott, 1972). During Scott's administration, the EM lab further developed and became an active center of research.

Besides purchasing a new TEM, renovations of the physical plant were necessary. In fall 1969 Greulach (1971) submitted a request to obtain funding for the renovation of space in Coker Hall to develop a TEM facility and a phycological research lab. Originally the lab was to be developed on the first floor of Coker Hall, but a more commodious area became available on the third floor in Room 318, when UNC botanist Clyde Umphlett resigned his position in August 1970 and subsequently vacated the room. Greulach submitted revised plans in March 1971 that now focused on the space of Room 318 and that were based on plans prepared by Brown. It is noteworthy that the job description of the work and accompanying drawings of the floor plans for the renovations survive in the archival records of the UNC Provost.

The TEM facility and phycological lab were probably completed by mid-1971. Brown eventually hired a technical assistant, Richard Santos, to manage and operate the microscope. Brown served as chair of the Electron Microscope Committee and prepared occasional committee reports that were presented at Botany faculty meetings. Numerous botany graduate students were trained to become proficient in using the lab. Botany faculty advanced their scientific investigations through work accomplished in the lab. Examples include John Couch's research with fungal parasites of mosquitoes and his resulting publications and Lindsay S. Olive's (1975) book, *The Mycetozoans*, which includes EM illustrations.

In conjunction with the establishment of the TEM laboratory, Brown set up a related facility in another room of Coker Hall. The facility contained a Balzers BA 360 M freeze-etch apparatus designed to create a metal replica of a frozen sample by evaporation of platinic carbide (Fig. 15.12). Brown and other botany researchers used the instrument to prepare specimens for the transmission electron microscope. According to Candace Haigler (2021), Brown's former graduate student and currently professor of Crop Science and Plant Biology at the North Carolina State University, the use of the apparatus led to pioneering discoveries that were initially controversial in some scientific circles. But the interpretations of Brown and students, particularly Susette C. Mueller, were ultimately confirmed to be correct: intramembrane protein complexes called rosettes function to synthesize cellulose microfibrils in land plants (Mueller and Brown, 1980). When Brown transferred to the University of Texas at Austin, he successfully negotiated with UNC administrators to move the Balzers apparatus to his research facility in Texas.

Fig. 15.12. R. Malcolm Brown, Jr. using a Balzers BA 360 M freeze-etch apparatus, Coker Hall, Department of Botany, University of North Carolina, Chapel Hill, August 1971.

(Photographic Archives #33649, North Carolina Collection, Wilson Special Collections Library, University of North Carolina, Chapel Hill).

The TEM became a productive research tool as well as a significant teaching aid. Brown noted that the facility achieved distinction in academic circles and a reputation that brought acclaim to the University of North Carolina. Brown resigned his professorship in summer 1982 to accept a faculty position at the University of Texas at Austin. The fate of the TEM is not documented in archival records. A former professor in the UNC Department of Zoology, however, conjects that the microscope probably was transferred to a unit in the university's School of Medicine.

Professors of Botany Joining the Department under Scott's Administration

Jim Massey (August 1971–July 2000, UNC Herbarium curator; 1983–1990, UNC Herbarium director; January 1980–July 2000, adjunct associate professor)

Research Interests: vascular flora of the southeastern United States, monograph on Polygalaceae, pollination-reproductive biology, species biology of threatened and endangered species.

Courses Taught: Local Flora, Introduction to Plant Taxonomy.

Selected Publications: *Vascular Plant Systematics*, [1974] (with A.E. Radford and others); *Laboratory Guide for an Introduction to Plant Taxonomy*, 1977 (with D.K.S. Otte and R.D. Whetstone); *Natural Heritage*, [1981] (with A.E. Radford and others); *An Atlas and Illustrated Guide to the Threatened and Endangered Vascular Plants of the Mountains of North Carolina and Virginia*, 1983 (with other authors).

Graduate Students: M.A., 1; M.S., none known; Ph.D., 1.

Patricia Gabbey Gensel (July 1975–current, assistant professor to professor)

Research Interests: plant evolution, particularly as evidenced by the fossil record; morphology, diversification and evolution of early land vascular plants; studies on Devonian plants; fossil plants.

Courses Taught: General Botany, Paleobotany, Comparative Morphology of Vascular Plants, Palynology, Plant Diversity.

Selected Publications: *Botany 11: Introductory botany laboratory manual*, 1979 (with A. White and C.R. Parks); *Plant Life in the Devonian*, 1984 (with H.N. Andrews); *Plants Invade the Land*, 2001 (with D. Edwards).

Graduate Students: M.A., none known; M.S., 7; Ph.D., 2.

Ann Gale Matthysse (July 1975–current, assistant professor to professor)

Research Interests: bacterial plant pathogens; bacterial genetics; *Agrobacterium tumefaciens*; interactions of bacteria with plant surfaces, including *Agrobacterium tumefaciens*, *Escherichia coli* and *Salmonella enterica*.

Courses Taught: Plant Physiology, Plant Physiology and Morphogenesis, Molecular Biology of Plants and Their Interaction with Microorganisms, Plant Physiology Laboratory, Microbiology, Microbiology Laboratory, Bacterial Genetics.

Selected Publications: The presence of *Agrobacterium tumefaciens* plasmid DNA in crown gall tumour cells, 1976 (with A.J. Stump); Plasmid-dependent attachment of *Agrobacterium tumefaciens* to plant tissue culture cells, 1978 (with P.M. Wyman and K.V. Holmes); Binding of *Agrobacterium tumefaciens* to carrot protoplasts, 1982 (with K.V. Holmes and R.H.G. Gurlitz).

Graduate Students: M.A., 1; M.S., 5; Ph.D., 1.

Robert Krug Peet (July–December 1975, lecturer; January 1976–June 2018, assistant professor to professor; July 2018–current, non-salaried research professor)

Research Interests: plant ecology, biogeography, plant communities and populations, ecoinformatics.

Courses Taught: Ecology, Ecology and Population Biology, Ecology and Evolution, Plant Community Ecology, Population and Community Ecology, Ecological Plant Geography, Field Plant Geography.

Selected Publications: *Contributions to the Knowledge of Flora and Vegetation in the Carolinas*, 1979–1981 (with H. Lieth and E. Landolt); *Plant Community Ecology*, 1985; *Plant Succession*, 1992 (with D.C. Glenn-Lewin and T.T. Veblen).

Graduate Students:[111] M.A., 2; M.S., 19; Ph.D., 27.

Peter Morrison Vitousek (January 1980–December 1983; associate professor)

Research Interests: environmental biology.

Courses Taught: Ecology, Ecology Laboratory, Ecosystem Structure and Function.

Selected Publications: The effects of ragweed (*Ambrosia artemisiifolia* L.) on nutrient cycling in a 1st-year old-field, 1980 (with M.M. Foster and P.A. Randolph); A comparative analysis of potential nitrification and nitrate mobility in forest ecosystems, 1982 (with J.R. Gosz, C.C. Grier, J.M. Melillo, and W.A. Reiners); Anion fluxes in three Indiana forests, 1984; and numerous publications subsequent to his time at UNC.

Graduate Students: M.A., none known; M.S., 1; Ph.D., 3.

There were also appointments of associated faculty in the Department of Botany. Nearly thirty individuals served as adjunct faculty members with appointments in other UNC departments or in an outside institution, visiting faculty members, lecturers, and research faculty. Table 15.4 provides the names of these faculty members with the years of their service at UNC.

TABLE 15.4. NAMES OF ASSOCIATED FACULTY MEMBERS OF THE UNC DEPARTMENT OF BOTANY, WITH THEIR TITLES AND YEAR(S) SERVED.

Name	Title	Year(s)
Ammann, Harriet	Visiting Faculty	1980/1981
Camp, Pamela Jean	Visiting Lecturer	Spring 1982
Cleland, Charles F.	Visiting Assistant Professor	1974/1975
Davis, Graham	Visiting Professor	Summer 1964
Denison, William Clark	Visiting Assistant Professor	1958/1959
Dykstra, Michael Jack	Visiting Lecturer;	July 1978–June 30, 1979
	Visiting Professor	Summer 1979
Edgell, Cora-Jean	Visiting Faculty	1980/1981
Gaulden, Mary Esther	Visiting Associate Professor	Fall 1954
Haefner, James Welt	Visiting Professor	Fall 1979
Hellwig, Robert Lee	Visiting Assistant Professor	1976/1977

[111] Several of the degrees were awarded in the Curriculum in Ecology, and the Curriculum for the Environment and Ecology after the establishment of the Department of Biology in 1982.

TABLE 15.4. (CONTINUED)

Name	Title	Year(s)
Higginbotham, Kenneth Orland	Visiting Assistant Professor	1974/1975
Kuenzler, Edward Julian*	Professor	1969–1975
Mangelsdorf, Paul Christoph	Lecturer	February 1968–1989
McGinnis, Michael Randy	Adjunct Associate Professor	January 1981–1989
McNitt, Rand	Visiting Lecturer	1975/1976
McVaugh, Rogers	Research Professor	January 1981–2009
Misch, Margaret Segur	Lecturer	1974–December 1976
Odum, Howard Thomas**	Professor	1966/1967–1970
O'Kelley, Charles	Visiting Professor	Summer 1964
Plyler, Daniel Belk	Visiting Professor	Summer 1965
Ramseur, George Shuford	Visiting Associate Professor	1966/1967
Silliman, Frances Ernestine	Visiting Professor	Summers 1962, 1964, 1965
Sovocool, George Wayne#	Adjunct Associate Professor	July 1981–1986
Stewart, Dorothy	Lecturer	1969/1970
Wagener, Klaus P.	Visiting Professor	Spring 1976, July 1977–April 1978
Wall, Monroe E.	Visiting Professor; Adjunct Professor	1963/1964 1964/1965–1965/1966
Whitney, John B.	Visiting Professor	Summer 1964
Woods, William Joseph##	Assistant to Associate Professor	1965/1966 –1977/1978
Wyatt, Raymond Louis	Visiting Associate Professor	Summer 1964

*Joint appointment with the UNC Department of Environmental Sciences and Engineering.
**Joint appointment with the UNC Department of Zoology.
#Joint appointment with the Analytical Chemistry Branch, Environmental Protection Agency, Research Triangle Park, North Carolina.
##Joint appointment with the UNC Institute of Marine Sciences in the Graduate Faculty of the Botany Department

Botany and Zoology Departments Merge

As early as March 1969, the idea of merging the Botany and Zoology Departments into a Department of Biology had been discussed among faculty members. Botany chairperson Victor A. Greulach (1969a) noted in a letter to UNC Chancellor J. Carlyle Sitterson: "We have reliable, but unofficial, information to the effect that several members of our Department of Zoology are actively working for the establishment of a Department of Biology on this campus." Greulach stated that he and members of the botany faculty vigorously opposed the dissolution of the Botany Department. He believed that botany as a discipline would thrive only when it was set up as a separate unit. He further outlined reasons to maintain a distinct Botany Department at UNC.

On September 16, 1981, members of the Department of Botany learned that the deans of the Graduate School and of the School of Arts and Sciences had made a preliminary decision to recommend consolidation of the Botany and Zoology Departments into a Biological Sciences Department (Scott, 1981a). Following this notification, a Committee to Study the Creation of a Department of Biological Sciences (1981) issued a report that provided background for the merger.

The committee pointed out that the Botany and Zoology Departments had developed their greatest strengths in traditional areas of research, but since the early 1960s they had inadequately developed research in the nascent fields of molecular and cellular biology. Also there had been a decline in enrollment and majors in botany, and much of the teaching contribution of the botany faculty was through the undergraduate biology curriculum. The committee thought that the new department would foster growth and enhancement in the biological sciences and would provide an opportunity to establish new programs, while preserving existing programs that were noted for their excellence.

Many of the biologists, particularly the botanists, had strong reservations and opposition to the merger. Numerous personal letters and media reports document the atmosphere surrounding this administrative decision (Radford, 1981–1982). The effective date for the establishment of a Department of Biology was July 1, 1982. Lawrence I. Gilbert, then chair of the Zoology Department, became the chair of the new department and held this role of leadership through June 1992.

Following the creation of the Department of Biology in 1982, sixteen faculty members who were botanists or whose research included studies on plants joined the department. Table 15.5 provides a list of these faculty members along with their years of service.

TABLE 15.5. BOTANISTS AND PLANT BIOLOGISTS WHO JOINED THE FACULTY OF THE DEPARTMENT OF BIOLOGY AFTER THE BOTANY AND ZOOLOGY DEPARTMENTS MERGED IN JULY 1982.

Faculty Member	Years of Service
Bedinger, Patricia Anne	July 1987–June 1994
Chase, Mark Wayne#	July 1988–September 1992
Dangl, Jeffery Lee	January 1995–
Grant, Sarah Robina*	January 1995–December 2020
Jones, Alan Monte	July 1986–
Kieber, Joseph	July 1999–
Liljegren, Sarah Jean	July 2003–July 2011
Mitchell, Charles Edward	January 2005–
Nimchuk, Zachary Luke Dudman	July 2015–
Pedersen, Karl	July 1983–October 1986
Quatrano, Ralph Stephen	July1989–June 1998
Reed, Jason Wardwell	January 1995–
Smith, Laurie Garland	January 1995–December 1997
Vision, Todd Jeremy	July 2001–
Weakley, Alan Stuart	June 15, 2002–
White, Peter Sherwood	July 1986–December 2019

#Mark Chase took a leave of absence beginning October 1992 to assume a position at the Royal Botanic Gardens, Kew, England. He officially resigned his post at UNC as of June 30, 1994.
*Research Assistant Professor

Significant Landmarks

The legacy of the Department of Botany includes other significant accomplishments and contributions that have not already been discussed in the above text. The department was "ranked among those that are 'Distinguished or Strong' in the American Council on Education Survey of graduate programs" (Anonymous, [1971]b). The department played a significant role in the leadership and growth of the Elisha Mitchell Scientific Society. UNC botanists were also engaged in research and teaching at the Highlands Biological Station. Seminars played a significant role in broadening the educational outreach in botany. Faculty members were at the forefront of actively publishing the results of their scientific investigations, writing papers in scientific journals, and publishing seminal books in the areas of their expertise.

Administrative Background

In the earliest days of teaching at UNC, faculty members assumed duties that assisted campus administrators in easing their workloads. Examples of these (such as bursar and buildings supervisor) have been discussed in previous chapters of this book. In the twentieth century similar roles assisted the chairpersons of academic departments. Committees consisting of faculty members were formed to address specific duties. By 1950, the number of committees of the Botany Department had increased due to the diversification and growth of the unit. Faculty reported on the proceedings of the committees at departmental meetings. In the late 1960s, botany faculty members met at department meetings at least once a month. Botany faculty meeting minutes covering the years 1957, 1959, and 1967 to 1982 are housed in the UNC Archives (UNC-A-2; UNC-A-3). It is unknown if other Botany Department meeting minutes exist.

In at least the last decade of the department's existence, graduate and undergraduate students also served on various committees. These students occasionally attended faculty meetings, as noted by a faculty resolution of 1975. In fall of that year, the botany faculty directed that "undergraduate representatives be included on appropriate committees and invited to attend, with option of refusal, pertinent faculty meetings" (UNC-A-2c). There were at least fifteen committees in 1981, including areas of coverage such as the physical plant, curriculum, safety, faculty endowment, graduate admissions and appointments, the Coker Arboretum, and the Botany Library.

Chairperson Victor Greulach (1967) noted that "within the Department the faculty makes all final decisions regarding plans, policies, personnel selection, course and curriculum changes, and solutions of various types of problems." The role of the faculty on departmental committees and their administrative contributions and decisions led to a democratic style of operation for the Botany Department.

Graduate Program

Early Years

Formal efforts to establish a graduate program at UNC began in 1881. The first degrees of master of arts and doctor of philosophy were conferred in 1883. Two years later a circular announced the requirements for advanced degrees in a number of disciplines, including botany (Anonymous, [1927]). As early as 1885, prior to the establishment of the Botany Department, various botany courses were restricted to graduate students (Knight and Adams, 1946).

John A. McKay earned an M.A. in 1912, the first graduate degree to be conferred in the Botany Department. Henry R. Totten received the department's first doctor of philosophy degree in 1923. Velma D. Matthews was distinguished as the first woman student at UNC to earn advanced degrees in botany (M.A., 1927; Ph.D., 1930).

Cooperative Program in Graduate Study in Botany

Following the North Carolina General Assembly's ratification of a Consolidation Act of March 27, 1931 and the action of the new Board of Trustees on July 11, 1932, the consolidation of the three campuses at Chapel Hill, Greensboro, and Raleigh occurred. Among the aims of the consolidation was the elimination or reduction of overlapping administrative and academic functions (Wilson, 1964). The Graduate School for the Consolidated System was subsequently based at UNC. Technically all graduate students had to enroll through that university. "All doctoral degrees were awarded at the University of North Carolina, regardless of where they were earned, although State College [*i.e.*, North Carolina State College of Agriculture and Engineering (now North Carolina State University)] was permitted to develop graduate work in the fields allocated to it by the trustees" (Reagan, 1987). (The Graduate School at NCSU was re-established around 1972, and separate Boards of Trustees were re-established for the constituent institutions which then numbered sixteen.)

In concert with the consolidation of the three campuses, the Botany Departments at Chapel Hill and Raleigh coordinated the training of graduate students (Anonymous, 1933). Through an agreement of cooperation between the two botanical departments, graduate students could earn advanced degrees. John Couch played a significant role in contributing to the program. He taught mycology to students at State College from the 1930s to the early 1960s. His mycology courses were also listed in several of the graduate catalogs at State College. According to plant pathologist Don E. Ellis (1976), "Couch contributed much to the development of plant pathology at N.C. State." He also served as a visiting research professor at State in summer 1943 when he conducted investigations on *Cercospora* leafspot of peanut.

Thomas Earle Smith was the first Ph.D. graduate of State College (through this cooperative program), and he majored in plant pathology (Ellis, 1976). Smith (1938) completed his dissertation, *Host Range Studies with* Bacterium solanacearum *E.F.S.*, under the supervision of State College plant pathologist Robert F. Poole.

Another graduate student, Don Edwin Ellis, received his doctoral degree through the program. Ellis (1945) wrote his dissertation on the *Anthracnose of Dwarf Mistletoe Caused by* Septogloeum gillii *n. sp.* under the supervision of plant pathologist James H. Jensen at North Carolina State College. John N. Couch[112] provided mycological advice to Ellis and served on his advisory committee but was not Ellis's major advisor. It is noteworthy that Ellis is listed as a Ph.D. alumnus of UNC and NCSU. Based on the above information, Smith's and Ellis's dissertations are not included in the list of doctoral degrees in Appendix 15.1.

Summer Graduate Study in Botany

In summer of 1961, the Botany Department introduced an expanded program in summer graduate study in botany (Greulach, 1961). The program offered graduate courses in

[112] Ellis (1981) considered Mycology, which was taught by John N. Couch, his most important course.

botany that would lead to master's degrees through summer study. According to a brochure (Anonymous, [1961]a) the program was developed to meet the needs of the following groups of people.

1. High school teachers studying on NSF Summer Fellowships,
2. High school teachers attending the NSF Summer Institute, and the subsequent Second Summer term,
3. College teachers holding NSF Science Faculty Fellowships for Summer Study,
4. Other college or high school teachers of botany or biology who wish to increase their knowledge of botany, and
5. Graduate or undergraduate students in biological sciences.

Graduate students could complete a Master of Education in Botany in a period of four summers. The Master of Arts in Botany could possibly be earned in three summer sessions, but the requirement of a thesis for this degree would likely extend the time needed to complete the degree. The M.S. degree was subsequently included in this program after the Botany Department first offered the degree in 1969. Work toward these master's degrees had to be completed within a six-year period. The director of the Summer Session at UNC, Arnold K. King, approved the increased offering of summer graduate courses for a two-year trial period, but the program was probably offered through 1971. The course announcements for the Summer Session of 1961 included a note about the program: "The Department of Botany has established a sequential offering of its basic graduate courses in a two-summer cycle permitting the obtaining of master's degrees by summer study" (Anonymous, 1961b). This note appeared in subsequent summer session announcements through 1971. By inference of the information in the summer course announcements, it is probable that the program concluded with the 1971 session.

Master of Science Degree in Botany

At a faculty meeting of the UNC Botany Department held December 3, 1968, the botany professors discussed a proposal that the department should consider offering the M.S. degree in botany (UNC-A-2g). In a subsequent meeting held on January 7, 1969, the "faculty debated the advisability and conditions for offering a Master of Science degree in Botany. Dr. [J. Franklin] McCormick moved that the department offer the degree. The motion carried unanimously. The requirements for the degree are set by the graduate school" (UNC-A-2h). The Master of Science degree in botany was first offered in 1969.

In the following year, the UNC Botany Department prepared an eight-page document that provided graduate students in the Botany Department with convenient information about the requirements for earning graduate degrees (M.A., M.S., and Ph.D.) (UNC-A-2i). The requirements for the M.S. in botany were outlined as follows:

1. A bachelor's degree from a recognized institution.
2. A minimum residence period of two semesters (or equivalent part-time residence). All work credited toward the degree must be completed within a period of five years, but it is expected that it will be completed much sooner.
3. A reading knowledge of one modern foreign language (normally German, French, Russian or Spanish).

4. Application for admission to candidacy by specified deadline.
5. At least 30 semester hours of graduate credit at U.N.C.-C.H., including at least 4 of the graduate core courses or their equivalents, and preferably more. Two s[emester] h[our] 300 level courses are also required.
6. A written examination (lasting no more than one day) in the field of the major (general botany), to be taken when course work is completed or in its final stages. The examination will be prepared and graded by a committee of at least three faculty members appointed by the departmental Director of Graduate Studies.

Although the M.S. in botany did not require a thesis, numerous botany graduate students prepared a research paper based on their scientific investigations. The degree provided a broad, general degree, particularly appropriate for students who wished to teach in junior colleges or liberal arts colleges. Students could complete requirements for the M.S. in botany in two academic years, or possibly in one academic year and a summer session.

During a period of seventy-five years, the department developed an outstanding program for graduate students, who earned 341[113] advanced degrees in botany (Appendix 15.1). According to Greulach's (1961) annual report of 1960/1961, the Botany Department reached a level of "productivity, both as regards graduate degrees and research ... [and] ranks among the first 20, and in national prestige it probably ranks about the same or slightly higher." In 1966, the American Council on Education noted that UNC's botany graduate program was one of the top three programs from southern universities and among the sixteen best in the country (Couch, 1968). In 1967 the Botany Department prepared an attractive, two-color brochure as an aid in publicizing and attracting students to its programs (Anonymous, 1967). Botanist Ed Barry ([1967]) edited the document (Fig. 15.13). The ten-page promotional booklet includes information on research programs, advanced degree programs, financial assistance, and a list of the botany faculty with their specialties. Eight photographs illustrate various scenes of the department.

It is noteworthy to mention that UNC conferred honorary (Doctor of Science) degrees on several former students who were botany majors at Chapel Hill. Recipients of these degrees were: Harry Ardell Allard (B.S., 1905; D.Sc., 1948), John Nathaniel Couch (B.A., 1919; M.A., 1922; Ph.D., 1924; D.Sc., 1973), Ivey Foreman Lewis (B.A., 1902; M.S., 1903; D.Sc., 1947), and Kenneth Bryan Raper (B.A., 1929; D.Sc., 1961).

Botany Graduate Student Association (BGSA)

The Botany Graduate Student Association, established in the early 1970s, promoted communication among botany graduate students and served as a liaison with the botany faculty (Scott, 1981b). The association was composed of all matriculating botany graduate students. The organization was involved with numerous activities: planning an annual fall and spring picnic, discussing the undergraduate and graduate curricula, conducting new student orientations, developing exhibits in the Botany Department display case, serving on departmental committees, and sponsoring guest scientists as speakers. Members of the

[113] Graduate students earned an additional 107 doctoral and master's degrees under the supervision of professors of botany whose tenure covered years after the Departments of Botany and Zoology merged to form a Department of Biology in July 1982.

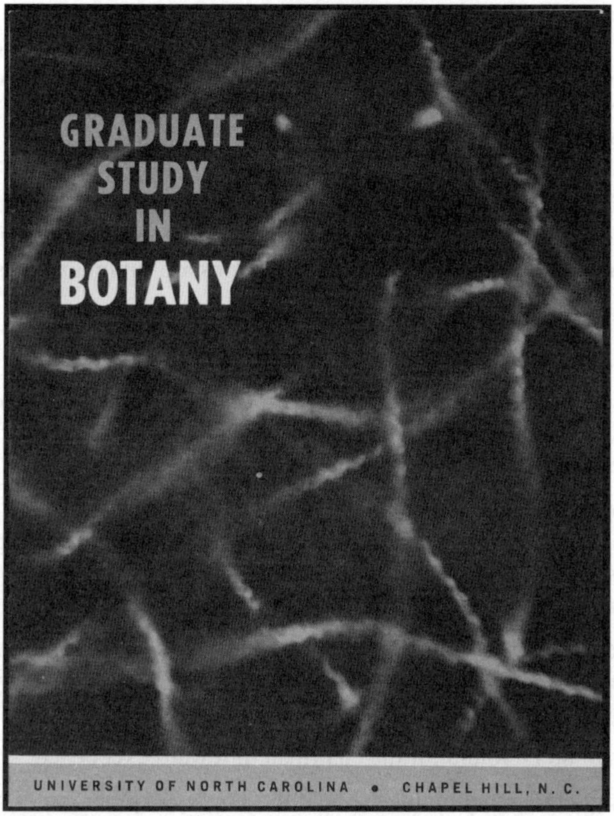

Fig. 15.13. Cover page of *Graduate Study in Botany* edited by Edward G. Barry ([1967]). (Courtesy of University Archives, Wilson Special Collections Library, University of North Carolina, Chapel Hill).

association actively engaged in discussions regarding the UNC administrators' plans in the early 1980s to merge the Botany and Zoology Departments (Scott, 1979a, 1980, 1982).

The Botany Department focused attention on developing its graduate programs. In 1977 the department prepared a *Procedural Handbook* aimed for new and returning graduate students. The handbook provided "an understanding of the programs, services and facilities [that were] available" (Anonymous, 1977a). A copy of this publication is appended to the Botany Department's annual report of 1976/1977.

In the previously cited self-study report of 1973, it was noted that "the lack of funds for the support of [the] graduate student [was] probably the most serious problem facing the Department at [that] time" (Hommersand, 1973). The department, however, offered major fields of interest and expertise in diverse areas of research: general botany, paleobotany, plant anatomy, physiology, biochemistry, mycology, phycology, cell biology, genetics, systematics, and ecology (Anonymous, 1980).

Graduate Program Review

In October 1980, a three-person committee conducted a review of the graduate program of the Botany Department. Members of the committee came from botany or biology departments at other institutions. An archival copy of the committee's report is housed with the Botany Department's meeting minutes (UNC-A-2e).

The committee members met with university administrators, department faculty, graduate students, and the chair of the unit. They commented that the communication they experienced was "open and frank." The results of their observations are summarized as follows:

1. Botany should remain a departmental entity.
2. Strength should be maintained in classical as well as modern areas.
3. At least one appointment arising from an imminent retirement in the classical areas should be in a modern area, such as plant molecular biology, photosynthesis, or some other aspect of plant biochemistry.
4. The Dean should appoint a committee immediately to conduct a search for a chairman to replace Tom Scott at the end of his term. It was urged that an appointment be made from outside. A person with strength in a modern area could also meet recommendation # 3.
5. Departmental in-fighting must be terminated. To this end, a strong new chairman would be extremely helpful. A student-faculty grievance committee or an ombudsman might assist in resolving some of the conflicts.
6. Some way should be found for the chairman to reward excellence.
7. A joint faculty-student committee should thoroughly explore issues such as course requirements, examination format, orientation to faculty research interests, requirements for seminar presentation, and length of time needed for degree completion.
8. Faculty members should examine their own graduate offerings critically, as there is such disparity in the levels of sophistication at which these courses are taught.
9. Every effort should be made to improve financing of graduate research and teaching assistantships.
10. The department must take a hard look at its roles within the university. Internal problems seem to have caused the faculty to lose sight of these roles.

The Elisha Mitchell Scientific Society

Members of the Department of Botany contributed significantly to the success of the Elisha Mitchell Scientific Society. Founded in 1883, the society provided a noteworthy forum for UNC scientists, and others, to share the results of their investigations at meetings and in the society's official organ, the *Journal of the Elisha Mitchell Scientific Society* (*JEMSS*). UNC botanists provided a significant editorial role in publishing the *JEMSS*. William C. Coker served as chairman of the editorial committee from 1902 to 1945. During his journalistic guidance, Coker's emphasis was mainly in the biological sciences of North Carolina and the Southeast with a botanical slant, particularly a mycological one (Wiggins, 1963). Other botanists who served as editors-in-chief were John N. Couch, 1946 to 1960; Victor A.

Greulach, 1966 to 1974; and William C. Dickison, 1975 to 1980. (Coker himself published twenty-five papers in the journal between 1924 and 1943, and John N. Couch authored at least forty publications in the journal.) These four UNC botanists provided editorial guidance to the journal for nearly seventy-four years of the journal's century of publication.

UNC botanists also held numerous other posts in the society,[114] such as president, vice-president, secretary/treasurer, and corresponding secretary. The botanists who held administrative posts are noted in Table 15.6.

Table 15.6. UNC Botanists Who Served in Positions in the Elisha Mitchell Scientific Society and Their Year(s) of Service.[115]

	President	Vice-President	Secretary/ Treasurer	Corresponding Secretary
William Chambers Coker	1919	1923		
Henry Roland Totten	1925	1924	1920, 1921, 1922, 1923	
John Nathaniel Couch	1937	1928		
Joseph Edison Adams	1945	1942	1938, 1939	
Albert Ernest Radford	1965, 1982	1958	1949, 1950	
Clyde Jefferson Umphlett			1967, 1968, 1969,1970	
William Campbell Dickison			1973, 1974	
William Julian Koch		1960, 1974, 1979, 1980 1981	1962, 1963	1964, 1965, 1966

Highlands Biological Station

Established in 1927 as the Highlands Museum Association, the present-day Highlands Biological Station (HBS) at Highlands, North Carolina has played an influential role in biological investigations and research in the southeastern U.S. UNC botanists contributed to the research and educational mission of the organization and held various leadership roles in the early years of its development. Emeritus Professor of English at Haverford College Ralph M. Sargent (1977) documented the first fifty years of the station's history. Today the HBS campus covers twenty-four acres, including four dormitory facilities that house up to fifty-two individuals, laboratories for teaching and research, classrooms, a nature center, and a botanical garden (Anonymous, 2021).

Author Mary C. Joslin (2003) noted the role that UNC botanist William C. Coker played in the development of the Highlands Biological Station. In 1928 Coker began his botanical investigations at the fledgling facility. His fieldwork, particularly on fungi, continued throughout his professional career. In 1933, when UNC joined a consortium of research organizations that supported the station (then called the Highlands Museum and Biological Laboratory), Coker became president of the biological station for the subsequent decade. He

[114] Several years cited are based on information published in the front cover material of the *JEMSS*.
[115] Years of service typically spanned an academic year. For instance, in the entry of H.R. Totten's presidency, he served during the school year 1925/1926.

was also director of the laboratory for eight years and had been instrumental in establishing a mountain laboratory at Highlands. His contributions to the institution were later recognized when a new wing to the laboratory facility was dedicated the Coker Laboratory in 1958 in his memory.

UNC botanists taught courses at and provided leadership to the organization. Information extracted from a list of officers of the station (Sargent, 1977) notes seven UNC-affiliated botanists who held administrative posts at the HBS (Table 15.7).

Table 15.7. UNC Botanists Who Held Offices at the Highlands Biological Station and Their Year(s) of Service.

Botanist	President	Vice-President	Year(s)
Ivey Foreman Lewis#	X		1930
William Chambers Coker		X	1931–1932
William Chambers Coker	X		1933–1943
John Nathaniel Couch		X	1945
Henry Roland Totten		X	1949–1954
Leland Shanor#		X	1958–1962
Albert Ernest Radford		X	1971
Albert Ernest Radford	X		1972–1975
Robert Edward Wyatt#	X		1999–2005

#Individuals whose names are followed by # studied botany at UNC and received the following degrees at Chapel Hill: Ivey F. Lewis (B.A., 1902; M.S., 1903), Leland Shanor (Ph.D., 1939), and Robert E. Wyatt (B.A., 1972).

There were other ways that UNC botanists participated in research, teaching, and the development of the station. Since 1928, professors, researchers, and students from the UNC Botany Department have engaged in educational and scientific endeavors at the HBS. Among these scientific investigators were nineteen individuals who wrote publications based wholly or partially on research completed there. Based on information from a bibliography of station publications (Bruce and Michener, 1977), Table 15.8 provides a list of these UNC botanists.

Table 15.8. UNC Botanists with Publications Based on Research Conducted at the Highlands Biological Station.

Harry Everett Ahles	William Julian Koch
John G. Baker	John Thomas Mullins
Lane Barksdale	Lindsay Shepherd Olive
Alma Holland Beers	Albert Ernest Radford
Clyde Ritchie Bell	George Shuford Ramseur
William Chambers Coker	Charles Leland Rodgers
John Nathaniel Couch	Leland Shanor
Elizabeth Kathleen Goldie-Smith	Henry Roland Totten
	Alma Joslyn Whiffen

UNC botanists conducted other noteworthy research at the biological station. Carroll E. Wood, Jr. collected specimens of *Robinia* for cytotaxonomic revision. In 1974 Albert E. Radford co-led an inventory of natural areas. His team of botanists included UNC botany students David E. Boufford, Lynn A. Schram, Diane E. Wickland, and Emily Wood. On two different surveys, Clifford R. Parks studied the elevational variation of trees in the Southern Appalachians. Lindsay S. Olive devoted extensive attention to developing the station's botanical garden. He served on and later chaired the HBS Garden Committee, oversaw the clearance of land for expanding the garden, and was instrumental in labeling plant specimens. In 1975, the Garden's executive director commended Olive for his steadfast contributions to the garden.

At the December 5, 1975 faculty meeting of the UNC Botany Department, Albert Radford reported that "the transferral of Highlands to the Consolidated University [of North Carolina] was approved unanimously; it now has to be processed through the Board of Governors and the State" (UNC-A-2f). Faculty members voiced their resounding approval of the transfer. Professor Ritchie Bell also expressed his hope that the station and the North Carolina Botanical Garden would collaborate on areas of common interest. The actual institutional transfer occurred on January 1, 1977 (Sargent, 1977).

After the UNC Botany and Zoology Departments merged to form a Department of Biology in mid-1982, UNC botanists continued their association with the Highlands Biological Station. Botanist Robert K. Peet served on the station's Board of Directors (1982–1984 and 1990–1993). Botanist Peter S. White, who was both a UNC professor of biology and the director of the North Carolina Botanical Garden, taught at HBS for over thirty years, served on several boards for the station, and spoke at two of their anniversary symposia (White, 2021). In 2018 the Highlands Biological Station became a unit of Western Carolina University, the westernmost institution of the University of North Carolina system.

Departmental Seminars

The earliest known information regarding the program of departmental seminars is recorded in the Botany Department annual report for the school year 1963/1964. Funds from an NDEA Title IV grant provided financial support for the initial program of twenty speakers who presented talks during that academic year. Guest lecturers from throughout the country proved to be "enlightening and stimulating" according to Botany Chair Victor A. Greulach (1964). The first year of seminars was a promising harbinger of an active program to follow for the ensuing two decades.

The initiation of the program coincided with the opening of the Botany Department's new building, Coker Hall. The department now had a commodious lecture hall where speakers could give seminars. Following the presentations, attendees met in an attractive Conference and Seminar Room, locally called the Coker Room. There, they interacted in conversations and enjoyed refreshments.

In the early 1970s, Paul C. Mangelsdorf, a lecturer in the department, and his wife Helen P. Mangelsdorf, made donations to support the seminar program (McCormick, 1973). In early 1975, the Mangeldorfs gave the department an additional donation ($1,000.00) to fund guest speakers (Dickison and Koch, 1975). Around 1977, the botany faculty resolved to establish the Paul C. and Helen P. Mangelsdorf Distinguished Lecture Series in Botany to recognize the donors' interest in scientific communication (UNC-A-2d). The first program was given in 1978 (Anonymous, 1978). Four renowned botanists representing various fields of botany presented lectures, seminars, and discussions. The program consisted of four sessions with the following guest botanists: Peter Albersheim, plant biochemist and plant physiologist (January); Jeremy Pickett-Heaps, plant cell biologist (March); Robert F. Thorne, plant taxonomist (early April); and Elizabeth Cutter, plant morphogenesist (late April). A copy of the program brochure is appended to Chairman Tom Scott's (1978) annual report of 1977/1978.

Although aimed primarily for the department's botany students and faculty members, seminars attracted attendees from other UNC departments and from nearby colleges and universities. Funding for the seminars came from various sources, the most prominent from lapsed salary funds. Proceeds from the Mangeldorfs' gifts selectively funded various seminars of the department's regular seminar sessions. The balance of the special fund was eventually spent down. Discretionary funds from the university administration also provided financial support. During periods when little or no lapsed funds were available, seminars were reduced (as in 1971/1972) or not given (as in 1972/1973). Speakers came from the United States or other countries with national or international reputation.

Some speakers were part of the department's interview process; others included UNC faculty (including botany professors), botany graduate students supporting their master of arts theses and their dissertations as well as honors students in botany supporting their honors theses. Seminars continued to be an integral component of the department's educational outreach in botany until 1982, when the department merged with the Zoology Department. During 1963 to 1982, thirteen known annual programs were given with over 300 seminar presentations.

The program was successful and instructive. Students and faculty members became acquainted with researchers in various fields of botany. These interactions provided a stimulus for their own research interests. Attendees were also exposed to botanical specialties not covered at UNC, which broadened their knowledge in botany. The exchange of research results and new ideas between faculty members and visitors also publicized UNC's botany program to people at other institutions of research and learning.

Publications

A significant indicator of research activity and scientific inquiry in academic units is reflected in the resulting publications of its faculty members. UNC research assistant in botany Alma Holland (later Beers) prepared several bibliographies concerning publications in science that UNC professors, graduate students, and teaching fellows published. Holland's (1934) first report covered the period from 1795 through 1934. She identified eight primary authors, who prepared nearly 160 abstracts, papers, and books in botany. In a continuation of this bibliography, Beers (1945b, 1946b) extended the coverage to 1944.

An additional compilation of the publications by UNC botany faculty and graduate students extended the coverage to 1966 (Anonymous, 1971a). This record shows a marked increase in the number of publications by UNC botanists. From 1902 to 1970 more than 737 published contributions appeared in six categories: research papers (463), abstracts (147), monographs (9), books and manuals (26), biological education (30), and miscellaneous publications (62).

In 2006 staff members of the UNC Botany Library issued a poster to commemorate the published contributions of Carolina botanists during the twentieth century (Burk and Appleton, 2006). The staff identified 2,300 publications by eighteen UNC botanists.[116] Over a period of at least twenty years, the library staff compiled the papers of each of these botanists, which they chronologically arranged and bound into nearly fifty volumes. These volumes are currently archived in the North Carolina Collection, Wilson Special Collections Library at the University of North Carolina at Chapel Hill.

Legacy

During seventy-five years of academic service, the Department of Botany was a distinguished unit at UNC. Its faculty members made numerous contributions in their fields of expertise through publications, guidance of over 300 graduate students, leadership roles in various scientific organizations, and advancements and discoveries in diverse areas of botany.

References Cited

(References that conclude with page numbers within parentheses indicate sources from where quotes and pertinent information were obtained.)

Adams, Martha. 1962. Professor praises work of summer lab assistants. *UNC News* (Chapel Hill, NC), Thursday, August 2, 1962, p. 7.

Anonymous. 1908. Davie Hall is completed. *The Tar Heel* (University of North Carolina, Chapel Hill) 17(4): 1, Thursday, October 8, 1908.

Anonymous. [1909]. The Biological Laboratory, pp. 161–162 in *The Catalogue, 1908–1909*. University of North Carolina, Chapel Hill.

Anonymous. 1925. Davie Hall has additional wing. *The Tar Heel* (University of North Carolina, Chapel Hill) 34(2): 2, Tuesday, September 22, 1925.

Anonymous. 1926. The Biological laboratories, pp. 33–34 in *The Catalogue, 1925–1926*. ([*University of North Carolina Record*, no. 232]). [University of North Carolina Press, Chapel Hill].

Anonymous. [1927]. The Graduate School, pp. 10–17; Department of Botany, pp. 35–36 in *The Graduate School Catalogue, [1927]*. (*University of North Carolina Record*, no. 244). (*Graduate School series*, no. 16). University of North Carolina, Chapel Hill.

[116] The following list provides the names of UNC botanists for whom the UNC Botany Library compiled collected papers: C.R. Bell, W.C. Coker, J.N. Couch, W.C. Dickison, A. Domnas, P.G. Gensel, V.A. Greulach, M.H. Hommersand, W.J. Koch, J.J. Kohlmeyer, J. Massey, L.S. Olive, C.R. Parks, A.E. Radford, T.K. Scott, and H.R. Totten, as well as R. Quatrano and P.S. White. The last two mentioned professors arrived after the Botany Department merged with the Zoology Department.

Anonymous. 1931. Resume of courses in six science departments ends student survey. *The Daily Tar Heel* (University of North Carolina, Chapel Hill) 40(68): 1, 4, Friday, December 11, 1931.

Anonymous. 1933. Example of the Botany Departments in coordination of work and consolidation of values, pp. 17–18 in *Report of the President*. (*University of North Carolina Record,* no. 287). University of North Carolina Press, Chapel Hill.

Anonymous. 1940. New labs have been built in basement of Davie Hall. *The Daily Tar Heel* (University of North Carolina, Chapel Hill) 49(30): 4, Tuesday, October 29, 1940.

Anonymous. 1944. Dean House announces plans for long-range expansion: Building program includes immediate and future needs. *The Daily Tar Heel* (University of North Carolina, Chapel Hill) 53(SW-9): 1, Tuesday, October 17, 1944.

Anonymous. 1948. Eleven new faculty appointments are approved by trustees' executive group. *The Daily Tar Heel* (University of North Carolina, Chapel Hill) 56(171): 1, 4, Thursday, May 20, 1948.

Anonymous. [1958]. *Science at Chapel Hill*. [University of North Carolina, Chapel Hill].

Anonymous. 1959a. All nine bond issues appear passed. *The Daily Tar Heel* (University of North Carolina, Chapel Hill) 68(35): 1, Wednesday, October 28, 1959.

Anonymous. 1959b. Botany Department seeking building; Oct. 27 bond election will decide. *The Daily Tar Heel* (University of North Carolina, Chapel Hill) 68(17, [i.e., 19]): 1, 3, Friday, October 9, 1959.

Anonymous. 1960a. Second summer botany conference for college teachers at the University of North Carolina. *Plant Science Bulletin* 6(2): 7.

Anonymous. 1960b. Greulach named head of Botony [*sic*] Department. *The Daily Tar Heel* (University of North Carolina, Chapel Hill) 78(172): 4, Sunday, May 22, 1960.

Anonymous. [1961]a. *Summer Graduate Study in Botany at the University of North Carolina: A sequential program of graduate courses in botany leading toward master's degrees through summer study*. University of North Carolina, Chapel Hill. (A copy of this bifold brochure is appended to Victor A. Greulach's *Department of Botany, University of North Carolina 1960–61 Annual Report*, p. 4.)

Anonymous. 1961b. *1961 Summer Session*. (*University of North Carolina Record,* no. 621). University of North Carolina Press, Chapel Hill. (pp. 27, 49)

Anonymous. 1961c. UNC Botany Dept. granted $24,200 to buy equipment. *Durham Morning Herald* 67(114): 4, January 22, 1961.

Anonymous. 1961d. Depts. of science to receive $36,000. *The Daily Tar Heel* (University of North Carolina, Chapel Hill) 69(88): 3, Friday, February 3, 1961.

Anonymous. 1962. Mycology. *Botany Newsletter UNC* 2: 7.

Anonymous. 1965. [Reports from the Department of Botany, University of North Carolina, Chapel Hill]. (Box 4, folder: Botany, Department of: General, 1965, Office of the Chancellor of the University of North Carolina, Chapel Hill: Paul Frederick Sharp Records, 1964–1965, Series 2. Academic Affairs, 1964–1966, University Archives #40021, Wilson Special Collections Library, University of North Carolina, Chapel Hill)

Anonymous. 1967. Department brochure. [Edited by E.G. Barry]. *Department of Botany News from Chapel Hill* (University of North Carolina, Chapel Hill) 1: 4.

Anonymous. 1968a. [NSF-supported summer short course in mycology for college teachers]. *Plant Science Bulletin* 14(4): 4.

Anonymous. 1968b. Restricted funds in designated fields of study: Botany, p. 450 in Benefactions: Fellowships and scholarships. *The General Catalogue Issue 1967–1968*. (*Record of the University of North Carolina at Chapel Hill*, no. 726). University of North Carolina Press, Chapel Hill.

Anonymous. 1970. Summer research participation in botany for college teachers at the University of North Carolina. *Plant Science Bulletin* 16(4): 9.

Anonymous. 1971a. Contributions from the Department of Botany, University of North Carolina at Chapel Hill, 1902 through 1970. *Department of Botany News from Chapel Hill* (University of North Carolina, Chapel Hill) 7(2): ii, 42, 1, 1, 6, 5.

Anonymous. [1971]b. *Graduate Study in Botany, University of North Carolina at Chapel Hill*. [Department of Botany, University of North Carolina, Chapel Hill]. (Box 4, folder: "Botany faculty in the Department of Biology ... Botany," William R. Burk Collection on University of North Carolina at Chapel Hill Botany and Zoology Faculty, 1920–2010, University Archives #40478, Wilson Special Collections Library, University of North Carolina, Chapel Hill) (Includes nine photographs)

Anonymous. 1975a. Minutes of the Botany faculty meeting, November 14, 1975. (Box 1, folder: Faculty meeting minutes, 1967–1989, Department of Botany of the University of North Carolina at Chapel Hill Records, 1922–1989, University Archives #40078, Wilson Special Collections Library, University of North Carolina, Chapel Hill)

Anonymous. 1975b. [Minutes of the] Botany faculty meeting, October 3, 1975. (Box 1, folder: Faculty meeting minutes, 1957, 1959, 1975, 1977, Department of Botany of the University of North Carolina at Chapel Hill Records, 1922–1989, University Archives #40078, Wilson Special Collections Library, University of North Carolina, Chapel Hill)

Anonymous. 1977a. *Procedural Handbook for Graduate Students, 1976–1977*. [Department of Botany, University of North Carolina, Chapel Hill] (A copy of the handbook is appended to the Botany Department *Annual Report 1976–1977* [Scott, 1977a]).

Anonymous. 1977b. Bachelor of Science in botany, pp. 38, 181 in *The Undergraduate Bulletin, 1977–1978 Issue*. (*Record of the University of North Carolina at Chapel Hill*, no. 854). University of North Carolina, Chapel Hill.

Anonymous. 1978. [brochure]. *Paul C. and Helen P. Mangelsdorf Distinguished Lecture Series in Botany*. Sponsored by Department of Botany, University of North Carolina at Chapel Hill, Appendix D in Tom K. Scott, *Department of Botany, University of North Carolina at Chapel Hill, Annual Report, April 12, 1978*. [Chapel Hill, NC].

Anonymous. 1980. *Evaluation of the Graduate Degree Programs in the Department of Botany, University of North Carolina, Chapel Hill, N.C.* [Department of Botany, University of North Carolina, Chapel Hill]. (Box 5, folder without label, William R. Burk Collection on University of North Carolina at Chapel Hill Botany and Zoology Faculty, 1920–2010, University Archives #40478, Wilson Special Collections Library, University of North Carolina, Chapel Hill)

Anonymous. 1985a. Former UNC department head dies. *Durham Morning Herald* 91(94): 6B, Wednesday, January 2, 1985.

Anonymous. 1985b. 3 outstanding faculty alumni die. *The Daily Tar Heel* (University of North Carolina, Chapel Hill) 93(99, [*i.e.*, 100]): 4, Wednesday, January 9, 1985.

Anonymous. 1985c. V.A. Greulach ex-professor at UNC dies. *News and Observer* (Raleigh) 239(3): 4C, Thursday, January 3, 1985.

Anonymous. 1994. Botanist named to direct NASA's cell biology in space experiments. *University Gazette* (University of North Carolina, Chapel Hill) 22(17): 6, September 7, 1994.

Anonymous. 2021. A Brief History of the Highlands Biological Station. <https:// highlandsbiological.org/hbs-research-history/> (Accessed January 17, 2021)

Atchison, Earlene. 1946. Chromosome Numbers in *Erythrina* L., *Crotalaria* L., *Gleditsia* L., *Mitchella* L., and the Myrtaceae. A dissertation presented to the Graduate Faculty of the University of Virginia in candidacy for the degree of Doctor of Philosophy. [University of Virginia, Charlottesville].

Atchison, Earlene. 1951. Studies in the Leguminosae. VI. Chromosome numbers among tropical woody species. *American Journal of Botany* 38: 538–546.

Atchison, Earlene, H.E. Moore, Jr. and C.E. Wood, Jr. 1949. Chromosome numbers for two species of Mexican Commelinaceae. *Science* 110: 41.

[Barry, Edward G.] (editor). [1967]. *Graduate Study in Botany*. University of North Carolina, Chapel Hill. (Copy is filed in the Office of the Provost of the University of North Carolina at Chapel Hill Records, 1917–2017. Series 2. Academic Affairs: College of Arts and Sciences: Institutes, Schools, Centers, Departments, and Curricula, 1937–1999, Box 65, Arts and Sciences, College of: Departments, Programs, and Curricula: Botany, Department of: 1954–1982, University Archives #40039, Wilson Special Collections Library, University of North Carolina, Chapel Hill)

Barry, Edward G. 1969. Diffuse diplotene stage of meiotic prophase in *Neurospora*. *Chromosoma* 26: 119–129.

Beers, Alma. 1945a. Botany, pp. 85–87 in Charles E. Rush (editor), *Library Resources of the University of North Carolina: A summary of facilities for study and research*. (*The University of North Carolina Sesquicentennial Publications*). University of North Carolina Press, Chapel Hill.

Beers, Alma. 1945b. Publications in the field of science from the University of North Carolina (1934–1944). *Journal of the Elisha Mitchell Scientific Society* 61: 301–375.

Beers, Alma. 1946a. Department of Botany, pp. [71]–73 in R.W. Bost, Chapter 3, Scientific Research Laboratories in Edgar W. Knight and Agatha Boyd Adams (editors), *The Graduate School Research and Publications*. (*The University of North Carolina Sesquicentennial Publications*). University of North Carolina Press, Chapel Hill.

Beers, Alma. 1946b. Science and mathematics, 1934–1944, pp. 387–430 in Edgar W. Knight and Agatha Boyd Adams (editors), *The Graduate School Research and Publications*. (*The University of North Carolina Sesquicentennial Publications*). University of North Carolina Press, Chapel Hill.

Bell, C. Ritchie. [1967]. *Plant Variation and Classification*. Wadsworth Publishing Company, Belmont, CA.

Bell, C. Ritchie and Bryan J. Taylor. 1982. *Florida Wild Flowers and Roadside Plants*. Laurel Hill Press, Chapel Hill.

Birkemo, Diane G. and J. Kenneth Moore. 1990. *Conservation Plan for the Coker Arboretum of the North Carolina Botanical Garden*. Botanical Garden Foundation, Chapel Hill.

BOT. Board of Trustees, University of North Carolina, Chapel Hill. *Minutes of the Board of Trustees*. (University Archives #40001, Wilson Special Collections Library, University of North Carolina, Chapel Hill) (Vol. 10, May 1898–April 1904; vol. 11, May 1904–September 1916)

Brown, R. Malcolm, Jr. 1982. *Cellulose and Other Natural Polymer Systems: Biogenesis, structure, and degradation*. Plenum Press, New York.

Brown, R. Malcom, Jr. 2020. Electronic mail of December 1, 2020 to William R. Burk. (Copy in author's files)

Bruce, Richard C. and David Michener (compilers). 1977. Bibliography of station publications, pp. 144–156 in Ralph M. Sargent, *Biology in the Blue Ridge: Fifty years of the Highlands Biological Station 1927–1977*. Highlands Biological Foundation, Inc., Highlands, NC.

Burk, William R. 2000. John Nathaniel Couch (1896–1986), his contributions to the *Journal of the Elisha Mitchell Scientific Society* and his scientific publications. *Journal of the Elisha Mitchell Scientific Society* 116: 297–306.

Burk, William R. 2015. William R. Burk Collection on University of North Carolina at Chapel Hill Botany and Zoology Faculty, 1920–2015. (University Archives #40478, Wilson Special Collections Library, University of North Carolina, Chapel Hill)

Burk, William R. and Betsy Appleton. 2006. [Poster]. "The Couch Biology Library Commemorates a Century of Carolina Botany." Botany Section, John N. Couch Library, University of North Carolina, Chapel Hill. (North Carolina Collection, Wilson Special Collections Library, University of North Carolina, Chapel Hill)

Burk, William R. and Charles E. Bland. 1989. John Nathaniel Couch (1896–1986). *Mycologia* 81: 181–189. (frontispiece portrait)

Coker, William C. 1912. *The Plant Life of Hartsville, SC*. State Co., Printers, Columbia, SC.

Coker, William C. 1922. Letter of October 14, 1922 to H.W. Chase. (Box, folder 1922–1929, *Department of Botany of the University of North Carolina at Chapel Hill Records, 1922–1989*, University of North Carolina, Chapel Hill, University Archives, #40078, Wilson Special Collections Library, University of North Carolina, Chapel Hill)

Coker, William C. 1923a. *The Saprolegniaceae: With notes on other water molds*. University of North Carolina Press, Chapel Hill.

Coker, William C. 1923b. *The Clavarias of the United States and Canada*. University of North Carolina Press, Chapel Hill.

Coker, William C. 1935. Letter of November 20, 1935 to R.B. House. (Box 2: 10, folder Botany, Dept. of, Annual Reports, 1935, [etc.], Academic Affairs, Dean of the College of Arts and Sciences Records, Series 2, Subseries 2, University Archives, #40076, Wilson Special Collections Library, University of North Carolina, Chapel Hill)

Coker, William C. and Alma H. Beers. 1943. *The Boletaceae of North Carolina*. University of North Carolina Press, Chapel Hill.

Coker, William C. and Alma H. Beers. 1951. *The Stipitate Hydnums of the Eastern United States*. University of North Carolina Press, Chapel Hill.

Coker, William C. and John N. Couch. 1928. *The Gasteromycetes of the Eastern United States and Canada*. University of North Carolina Press, Chapel Hill.

Coker, William C. and Henry R. Totten. 1939. *Laboratory Guide in General Botany*. 4th ed. Published by the Authors, Chapel Hill.

Coker, William C. and Henry R. Totten. 1945. *Trees of the Southeastern States, including Virginia, North Carolina, South Carolina, Tennessee, Georgia, and Northern Florida*. 3rd ed. University of North Carolina Press, Chapel Hill.

Committee to Study the Creation of a Department of Biological Sciences. 1981. *Report of the Committee to Study the Creation of a Department of Biological Sciences in the College of Arts and Sciences at the University of North Carolina at Chapel Hill, December 15, 1981*. (Box 5, folder without label, William R. Burk Collection on University of North Carolina at Chapel Hill Botany and Zoology Faculty, 1920–2010, University Archives #40478, Wilson Special Collections Library, University of North Carolina, Chapel Hill)

Couch, John N. 1938. *The Genus Septobasidium*. University of North Carolina Press, Chapel Hill.

Couch, John N. 1947. *Report of the Botany Department of the University of North Carolina on the Doctorate of Philosophy Curriculum, Facilities, Faculty, and Graduates for the Association of American Universities*. [Department of Botany, University of North Carolina, Chapel Hill]. (Box 4, folder without label, William R. Burk Collection on University of North Carolina at Chapel Hill Botany and Zoology Faculty, 1920–2010, University Archives #40478, Wilson Special Collections Library, University of North Carolina, Chapel Hill)

Couch, John N. 1956a. *Annual Report to the Chancellor, 1955–1956*. (Box 2: 10, folder, Botany, Dept. of: Annual Reports, 1935, etc., Academic Affairs, Dean of the College of Arts and Sciences Records, Series 2, Subseries 2, University Archives #40076, Wilson Special Collections Library, University of North Carolina, Chapel Hill)

Couch, John N. 1956b. [Appendix]: Need for a New Building for Botany Department, May 7, 1956. in *Annual Report to the Chancellor, 1955–1956*. (Box 2: 10, folder, Botany, Dept. of: Annual Reports, 1935, etc., Academic Affairs, Dean of the College of Arts and Sciences Records, Series 2, Subseries 2, University Archives #40076, Wilson Special Collections Library, University of North Carolina, Chapel Hill)

[Couch, John N.]. 1956c. John Nathaniel Couch, pp. [216]–219 in A.C. Howell, *The Kenan Professorships*. University of North Carolina Press, Chapel Hill.

Couch, John N. 1958. *Report to the Chancellor, [1957–1958]*. (Box 2: 10, folder, Botany, Dept. of: Annual Reports, 1935, etc., Academic Affairs, Dean of the College of Arts and Sciences Records, Series 2, Subseries 2, University Archives #40076, Wilson Special Collections Library, University of North Carolina, Chapel Hill)

Couch, John N. 1968. *Department of Botany, University of North Carolina, Annual Report 1967–1968*. (North Carolina Collection, Wilson Special Collections Library, University of North Carolina, Chapel Hill) (p. 1)

Couch, John N. and Charles E. Bland (editors). 1985. *The Genus Coelomomyces*. Academic Press, Orlando.

Couch, John N. and Velma D. Matthews. 1954. William Chambers Coker. *Mycologia* 46: 372–383.

Cronquist, Arthur. 1980. *Asteraceae*. (*The Vascular Flora of the Southeastern United States,* vol. 1). University of North Carolina Press, Chapel Hill.

Dickison, William C. [ca. 1990]. *BIOL 152: Plant anatomy*. Photocopy, Department of Biology, University of North Carolina, Chapel Hill.

Dickison, William C. [ca. 1992]. *Selected Readings for Botany 51, Plant Diversity*. Photocopy, Department of Biology, University of North Carolina, Chapel Hill.

Dickison, William C. 2000. *Integrative Plant Anatomy*. Harcourt/Academic Press, San Diego.

Dickison, [William C.] and [William J.] Koch (faculty recorders). 1975. Botany faculty meeting, February 14, 1975. (Box 1, folder: Faculty meeting minutes, 1957, 1959, 1975, 1977, Department of Botany of the University of North Carolina at Chapel Hill Records, 1922–1989, University Archives #40078, Wilson Special Collections Library, University of North Carolina, Chapel Hill)

Domnas, Aristotle [J.], P.E. Giebel and T.M. McInnis. 1974. Biochemistry of mosquito infection—Preliminary studies of biochemical change in *Culex pipiens-quinquepasciatus* following infection with *Lagenidium giganteum*. *Journal of Invertebrate Pathology* 24: 293–304.

Domnas, Aristotle J., James P. Srebro and Bruce F. Hicks. 1977. Sterol requirement for zoospore formation in the mosquito-parasitizing fungus *Lagenidium giganteum*. *Mycologia* 69: 875–886.

Ellis, Don Edwin. 1945. *Anthracnose of Dwarf Mistletoe Caused by* Septogloeum gillii *n. sp.* A thesis to the faculty of the University of North Carolina in partial fulfillment of the requirements for the degree of Doctor of Philosophy in Plant Pathology. [The Author], Raleigh.

Ellis, D[on] E[dwin]. 1976. *Plant Pathology in North Carolina 1776–1976*. School of Agriculture and Life Sciences, North Carolina State University, Raleigh. (pp. 19–20)

Ellis, Don Edwin. 1981. University of North Carolina at Chapel Hill General Alumni Association 1981 Questionnaire. (Held in the "Filing Room" of the Records Office of the UNC General Alumni Association, this document includes a handwritten note by Ellis regarding his doctoral education.)

Foster, Margaret Mead, Peter M. Vitousek and Polley Ann Randolph. 1980. The effects of ragweed (*Ambrosia artemisiifolia* L.) on nutrient cycling in a 1st-year old-field. *American Midland Naturalist* 103: 106–113.

Gao, Jiyin, Clifford R. Parks and Yueqiang Du. 2005. *Collected Species of the Genus Camellia: An illustrated outline*. Zhejiang ke xue ji shu chu ban she, [Hangzhou].

Gaskin, J.R. 1976. Letter of May 13, 1976 to Tom K. Scott. (Box 2: 9, folder including 1976, Academic Affairs, Dean of the College of Arts and Sciences Records, Series 2, Subseries 2, University Archives #40076, Wilson Special Collections Library, University of North Carolina, Chapel Hill)

Gensel, Patricia G. 2001. Dr. Tom K. Scott. (Box 4, folder: Scott, Dr. Tom K., William R. Burk Collection on University of North Carolina at Chapel Hill Botany and Zoology Faculty, 1920–2010, University Archives #40478, Wilson Special Collections Library, University of North Carolina, Chapel Hill)

Gensel, Patricia G., A. White and C.R. Parks. 1979. *Botany 11: Introductory botany laboratory manual*. [Revised ed]. Department of Botany, University of North Carolina, Chapel Hill.

Gensel, Patricia G. and Henry N. Andrews. 1984. *Plant Life in the Devonian*. Praeger, New York.

Gensel, Patricia G. and Dianne Edwards (editors). 2001. *Plants Invade the Land: Evolutionary and environmental perspectives*. (*Critical Moments and Perspectives in Earth History and Paleobiology Series*). Columbia University Press, New York.

Gilbert, Lawrence I. 1986. *Annual Report of the Department of Biology, [1985–1986]*. Department of Biology, University of North Carolina, Chapel Hill. (p. 55)

Glenn-Lewin, David C., Robert K. Peet and Thomas T. Veblen. (editors). 1992. *Plant Succession: Theory and prediction*. Chapman and Hall, London.

Greulach, Victor A. 1955. *A Laboratory Manual for Elementary Plant Physiology*. Burgess, Minneapolis, MN.

Greulach, Victor A. 1961. *Annual Report, 1960–1961*. [Department of Botany, University of North Carolina, Chapel Hill]. (pp. 4, 5, 14; p. 7 [second sequence of paging])

Greulach, Victor A, 1964. *Department of Botany, University of North Carolina, 1963–64 Annual Report*. [Chapel Hill, NC]. (pp. 1–2, 4–6)

Greulach, Victor A. 1966. History, Status and Plans of the Department of Botany at the University of North Carolina at Chapel Hill in Victor A. Greulach. *Department of Botany, University of North Carolina, 1965–1966 Annual Report*. (Box 2: 10, folder, Botany, Dept. of: Annual Reports, 1935, etc., Academic Affairs: Dean of the College of Arts and Sciences Records, University Archives #40076, Wilson Special Collections Library, University of North Carolina, Chapel Hill) (The five-page "History" is appended to the annual report.)

Greulach, Victor A. 1967. *Department of Botany, University of North Carolina, 1966–67 Annual Report*. [Chapel Hill, NC]. (p. 4)

Greulach, Victor A. 1968. *Botany Made Simple*. Doubleday, Garden City, NY.

Greulach, Victor A., 1969a. Letter of March 17, 1969 to J. Carlyle Sitterson. (Box 2: 9, folder: Botany, Dept. of: General, 1960; 1962; 1965–1969, Dean of the College of Arts and Sciences Records, University Archives #40076, Wilson Special Collections Library, University of North Carolina, Chapel Hill)

Greulach, Victor A. 1969b. *Department of Botany, University of North Carolina at Chapel Hill, 1968–1969 Annual Report*. [Chapel Hill, NC]. (p. 13)

Greulach, Victor A. 1970. *Department of Botany, University of North Carolina at Chapel Hill, Annual Report 1969–1970*. Chapel Hill, NC. (p. 20)

Greulach, Victor A. 1971. Letter of March 16, 1971 to UNC Dean Raymond H. Dawson. (Box 65 Arts and Sciences, College of: Departments, Programs and Curricula: Botany, Department of, 1954–1982, Folder: Botany January–June 1971, Office of the Provost of the University of North Carolina at Chapel Hill Records, 1917–2017, Series 2, Academic Affairs: College of Arts and Sciences: Institutes, Schools, Centers, Departments and Curricula 1937–1999, University Archives #40039, Wilson Special Collections Library, University of North Carolina, Chapel Hill)

[Greulach, Victor A.]. 1972. Greulach, Victor A(ugust), p. 2,319 in *American Men and Women of Science*, 12th ed., vol. 2. Jaques Cattell Press/R.R. Bowker, New York.

Greulach, Victor A. 1974. *Plant Function and Structure*. Macmillan, New York.

Greulach, Victor A. and J. Edison Adams. 1976. *Plants: An introduction to modern botany*. 3rd ed. Wiley, New York.

Greulach, Victor A. and Earlene Atchison. 1950. Inhibition of growth and cell division in onion roots by maleic hydrazide. *Bulletin of the Torrey Botanical Club* 77: 262–267.

Greulach, Victor A., A.E. Radford and William J. Koch. 1962. [Abstract]. Research activities in the Department of Botany. *Journal of the Elisha Mitchell Scientific Society* 78: [108].

Greulach, Victor A. and Marie C. Taylor. 1958. Botany in the National Science Foundation Institutes for High School Teachers. *Plant Science Bulletin* 4(6): 6–8.

Haigler, Candace. 2021. [Background regarding UNC's Balzers BA 360 M freeze-etch apparatus.] Electronic mail to William R. Burk, October 26, 2021.

Hailey, Tom. 2008. The Wilson Hall Exhibit: The making of an educational exhibit. A master's paper submitted to the faculty of the School of Information and Library Science of the University of North Carolina at Chapel Hill in partial fulfillment of the requirements for the degree of Master of Science in Library Science. University of North Carolina, Chapel Hill.

Halstead, Thora W. and Tom K. Scott (editors). 1984. *Plant Gravitational and Space Research: Report of a workshop held April 30–May 2, 1984 in Rossyln, Virginia*. (*Workshop Summaries*, 3). American Society of Plant Physiologists, Rockville, MD.

Holland, Alma. 1934. Publications in the field of science from the University of North Carolina (1795–1934). *Journal of the Elisha Mitchell Scientific Society* 50: 303–415.

Hommersand, Max H. 1963. *The Morphology and Classification of Some Ceramiaceae and Rhodomelaceae. (University of California Publications in Botany*, vol. 35, no. 2) University of California Press, Berkeley.

Hommersand, Max (chair). 1973. *Report of the Self-Study Committee. Department of Botany, University of North Carolina at Chapel Hill, April 9, 1973*. [Department of Botany, University of North Carolina, Chapel Hill]. (Box 5, folder without label, William R. Burk Collection on University of North Carolina at Chapel Hill Botany and Zoology Faculty, 1920–2010, University Archives #40478, Wilson Special Collections Library, University of North Carolina, Chapel Hill)

Isely, Duane. 1990. *Leguminosae (Fabaceae)*. (*The Vascular Flora of the Southeastern United States*, vol. 3, part 2). University of North Carolina Press, Chapel Hill.

Ivey, Pete. 1961. New weed building a pleasure to dig. *The Daily Tar Heel* (University of North Carolina, Chapel Hill) 69(12): 3, Tuesday, October 3, 1961.

Joslin, Mary Coker. 2003. *Essays on William Chambers Coker, Passionate Botanist*. With drawings by Sandra Brooks-Mathers. University of North Carolina at Chapel Hill Library and Botanical Garden Foundation, Chapel Hill, NC. (Chapter Six: Founder of the Arboretum, planner for the UNC campus, pp. 77–101)

Justice, William S., C. Ritchie Bell and Anne H. Lindsey. 2005. *Wild Flowers of North Carolina*. 2nd ed. University of North Carolina Press, Chapel Hill.

Knight, Edgar W. and Agatha Boyd Adams (editors). 1946. *The Graduate School, Research and Publications*. (*The University of North Carolina Sesquicentennial Publications*). University of North Carolina Press, Chapel Hill. (p. 8)

Koch, William J. (editor). 1968. *Mycological Studies Honoring John N. Couch*. University of North Carolina Press, Chapel Hill.

Koch, William J. [1972]. *Fungi in the Laboratory: A manual and text for studies of the culture, development, reproduction, cytology, genetics, collection, identification, and our general understanding of the major groups of fungi*. 2nd ed., completely revised and enlarged. University of North Carolina Student Stores, Chapel Hill.

Koch, William J. [1973]. *Plants in the Laboratory: A manual and text for studies of the culture, development, reproduction, cytology, genetics, collection, and identification of the major plant groups*. Macmillan, New York.

Koch, William J. 1976. *Fungi*. University of North Carolina Student Stores, Chapel Hill.

Koch, William J. 1980. *Leaflets for Buds: Plants & life leaflets*. Photocopy, [Department of Botany, University of North Carolina, Chapel Hill].

Koch, William J. 1985a. *Plants & Life*. 2nd ed. Photocopy, Department of Biology, University of North Carolina, Chapel Hill.

Koch, William J. 1985b. *In Touch with Plants*. 2nd ed. Photocopy, Department of Biology, University of North Carolina, Chapel Hill.

Koch, William J. 2007. *Plant Close-ups: Designs: Annotated photographs*. Wildot Press, Phoenix, AZ.

Kohlmeyer, Jan J. and Erika Kohlmeyer. 1964–1969. *Icones Fungorum Maris*. J. Cramer, Weinheim.

Kohlmeyer, Jan J. and Erika Kohlmeyer. 1971. *Synoptic Plates of Higher Marine Fungi: Identification guide to the marine environment*. 3rd ed. Stechert-Hafner Service Agency, New York.

Kohlmeyer, Jan J. and Erika Kohlmeyer. 1979. *Marine Mycology: The higher fungi*. Academic Press, New York.

Lieth, Helmut (editor). 1974. *Phenology and Seasonality Modeling*. Springer-Verlag, New York.

Lieth, Helmut (editor). 1978. *Patterns of Primary Production in the Biosphere*. Academic Press, New York.

Lieth, Helmut, E. Landolt and Robert K. Peet (editors). 1979–1981. *Contributions to the Knowledge of Flora and Vegetation in the Carolinas: Proceedings of the 16th International Phytogeographical Excursion (IPE), 1978, through the SE United States = Beiträge zur Kenntnis der Flora und Vegetation der Carolinas: Ergebnisse der 16. Internationalen Pflanzengeographischen Exkursion (IPE), 1978, durch die südöstlichen Vereinigten Staaten*. (Veröffentlichen des Geobotanischen Institutes der ETH, 68, 69, 77). Geobotanisches Institut der ETH, Stiftung Rübel, Zürich. 3 vols. (R.K. Peet was co-editor of volume 3.)

Lieth, Helmut and Robert H. Whittaker (editors). [1975]. *Primary Productivity of the Biosphere*. Springer-Verlag, New York.

Maggs, Christine A. and Max H. Hommersand. 1993. *Ceramiales* in *Seaweeds of the British Isles*, vol. 1, Part 3A (Rhodophyta). HMSO, London. (A collaborative project of the British Phycological Society and the British Museum [Natural History])

Massey, J.R. [and others]. 1983. *An Atlas and Illustrated Guide to the Threatened and Endangered Vascular Plants of the Mountains of North Carolina and Virginia*. (*General Technical Report*, SE-20). Illustrations by Susan Sizemore. Southeastern Forest Experiment Station, Asheville, NC.

Massey, J.R., D.K.S. Otte and R.D. Whetstone. 1977. *Laboratory Guide for an Introduction to Plant Taxonomy*. Burgess Publishing Company, Minneapolis.

Massey, Jim (chairperson, Editorial Board) and Laurie S. Radford (executive editor). 1981. *Revised Contributors' Guide for the Vascular Flora of the Southeastern United States*. Privately Published, [Chapel Hill, NC]. (A copy of the guide is housed in the archival files of the UNC Herbarium.)

Matthews, James F. 1981. Vascular flora of the southeast, pp. 159–161 in Index of current plant systematics and ecological research in the southeastern United States, 1981–1982. *ASB Bulletin* 28: 154–161.

Matthysse, Ann, Kathryn V. Holmes and Robin H.G. Gurlitz. 1982. Binding of *Agrobacterium tumefaciens* to carrot protoplasts. *Physiological Plant Pathology* 20: 27–33.

Matthysse, Ann G. and Anne J. Stump. 1976. The presence of *Agrobacterium tumefaciens* plasmid DNA in crown gall tumour cells. *Journal of General Microbiology* 95: 9–16.

Matthysse, Ann, Pat M. Wyman and Kathryn V. Holmes. 1978. Plasmid-dependent attachment of *Agrobacterium tumefaciens* to plant tissue culture cells. *Infection and Immunity* 22: 516–522.

McCormick, J. Frank[lin]. 1964. Differential radiation sensitivities of races of *Sedum pulchellum* Michx.: A useful method of plant identification. *Radiation Botany* 4: 247–251.

McCormick, J. Frank[lin]. (faculty secretary). 1973. Department of Botany, Faculty minutes, October 12, 1973. (Box 1, folder: Faculty meeting minutes, 1967–1982, Department of Botany of the University of North Carolina at Chapel Hill Records, 1922–1989, University Archives #40078, Wilson Special Collections Library, University of North Carolina, Chapel Hill)

McCormick, J. Frank[lin] and Frank B. Golley. 1966. Irradiation of natural vegetation. An experimental facility, procedures and dosimetry. *Health Physics* 12: 1467–1474.

McCormick, J. Frank[lin] and Robert B. Platt. 1964. Ecotypic differentiation in *Diamorpha cymosa*. *Botanical Gazette* 125: 271–279.

Medlin, Joe. 1960. Classroom triangle on tap for UNC. *The Daily Tar Heel* (University of North Carolina, Chapel Hill), Thursday, July 21 [*i.e.*, 28], 1960, p. 1. (Title at head of issue: *UNC News*, vol. 2, no. 8)

Methven, Andrew S. 2000. William Chambers Coker. *McIlvainea, Journal of American Amateur Mycology* 14(2): 27–33.

Mueller, Susette C. and R. Malcolm Brown, Jr. 1980. Evidence for an intramembrane component associated with a cellulose microfibril-synthesizing complex in higher plants. *Journal of Cell Biology* 84: 315–326.

NCBG Greenbriers (North Carolina Botanical Garden Greenbriers). 2019. *A Conservation Garden: The North Carolina Botanical Garden (1966–2016).* [North Carolina Botanical Garden, University of North Carolina, Chapel Hill].

NCU (University of North Carolina at Chapel Hill, Herbarium). 2019. Collectors and history of the UNC Herbarium. <http://herbarium.unc.edu/history.htm> (Accessed April 27, 2019)

[NSF (National Science Foundation)]. 1961. *The Eleventh Annual Report of the National Science Foundation*. Government Printing Office, Washington, DC. (Appendix D. Other Than Basic Research Grants, p. 251)

Newcomer, Earl H. 1941. A colchicine-induced tetraploid cabbage. *American Naturalist* 75: 62.

Newcomer, Earl H. 1945. Induced parthenocarpy in *Ginkgo*. *American Naturalist* 79: 186–187.

Newcomer, Earl H. 1946. Concerning the duality of the mitochondria and the validity of the osmiophilic platelets in plants. *American Journal of Botany* 33: 684–697.

Olive, Lindsay S. 1975. *The Mycetozoans*. With the technical assistance of Carmen Stoianovitch. Academic Press, New York.

Parker, Bruce C. and R. Malcolm Brown, Jr. (editors). [1971]. *Contributions in Phycology: A selection of papers in phycology written by former students of Harold C. Bold and dedicated to him in recognition of his more than 40 years of teaching excellence*. Students of Harold C. Bold, Lawrence, KS.

Parks, Clifford R. 1979. Woody plants commonly cultivated in central North Carolina. *Veröffentlichungen des Geobotanischen Institutes der ETH, Stiftung Rübel, Zürich* 68: 108–184.

Peet, Robert K. (editor). 1985. *Plant Community Ecology: Papers in honor of Robert H. Whittaker*. Junk, Boston.

Perkins, David D., Robert L. Metzenberg, Namboori B. Raju and Edward G. Barry. 1986. Reversal of a *Neurospora* translocation by crossing over involving displaced rDNA, and methylation of the rDNA segments that result from recombination. *Genetics* 114: 791–817.

Perkins, David D., Barbara C. Turner and Edward G. Barry. 1976. Strains of *Neurospora* collected from nature. *Evolution* 30: 281–313.

Price, Steve. 1966. New psychology building to open in September. *The Daily Tar Heel* (University of North Carolina, Chapel Hill) 74(61): [1], Tuesday, November 29, 1966.

Radford, Albert E. 1954. *Laboratory Directions for Botany 41 [General Botany I]*. University Book Exchange, Chapel Hill.

Radford, Albert E. (acting chairperson). 1965. *Department of Botany, University of North Carolina, 1964–65 Annual Report*. Chapel Hill, NC. (p. 4)

Radford, Albert E. (compiler). 1981–1982. [Notebook]. (Box 6, folders 1–3, William R. Burk Collection on University of North Carolina at Chapel Hill Botany and Zoology Faculty, 1920–2010, University Archives #40478, Wilson Special Collections Library, University of North Carolina, Chapel Hill)

Radford, Albert E. 1986. *Fundamentals of Plant Systematics*. With major contributions by T.M. Barkley. Harper and Row, New York.

Radford, Albert E., Harry E. Ahles and C. Ritchie Bell. 1965. *Atlas of the Vascular Flora of the Carolinas*. North Carolina Agricultural Experiment Station, Raleigh.

Radford, Albert E., Harry E. Ahles and C. Ritchie Bell. [1968]. *Manual of the Vascular Flora of the Carolinas*. University of North Carolina Press, Chapel Hill.

Radford, Albert E. and C. Ritchie Bell. 1961. *Laboratory Direction and Exercises for General Botany*. [Edwards Brothers, Ann Arbor, MI].

Radford, Albert E., C. Ritchie Bell, James W. Hardin and Robert L. Wilbur (compilers). 1967. *Contributors' Guide for the "Vascular Flora of the Southeastern United States."* [Department of Botany, University of North Carolina, Chapel Hill]. (Bibliographical information is based on a copy of the guide held at the Gray Herbarium, Harvard University.)

Radford, Albert E. [and others]. [1974]. *Vascular Plant Systematics*. Harper and Row, New York.

Radford, Albert E. [and others]. [1981]. *Natural Heritage: Classification, inventory, and information*. University of North Carolina Press, Chapel Hill.

Radford, Laurie S. and Albert E. Radford. 2000. *The Herbarium of the University of North Carolina 1908–2000: History and perspective*. With contributions from Peter S. White, Jimmy R. Massey and James F. Matthews. University of North Carolina Herbarium (NCU), North Carolina Botanical Garden, University of North Carolina at Chapel Hill.

Reagan, Alice Elizabeth. 1987. *North Carolina State University: A narrative history.* North Carolina State University Foundation and the North Carolina State University Alumni Association. (pp. 138–139)

Rousch, Allan H., Lillian M. Questiaux and Aristotle J. Domnas. 1959. The active transport and metabolism of purines in the yeast *Candida utilis*. *Journal of Cellular and Comparative Physiology* 54: 275–286.

Sargent, Ralph M. 1977. *Biology in the Blue Ridge: Fifty years of the Highlands Biological Station 1927–1977*. Highlands Biological Foundation, Inc., Highlands, NC. (pp. 138, 142–143)

Scott, Tom K. 1972. *Department of Botany, University of North Carolina at Chapel Hill, 1971–1972 Annual Report*. [Chapel Hill, NC]. (pp. 10–11)

Scott, Tom K. 1977a. Appendix E. *A Procedural Handbook for Graduate Students, 1976–1977* in Tom K. Scott, *Department of Botany, University of North Carolina at Chapel Hill Annual Report 1976–1977*. Department of Botany, University of North Carolina at Chapel Hill.

Scott, Tom K. 1977b. *Department of Botany, University of North Carolina at Chapel Hill, Annual Report 1976–1977*. Chapel Hill, NC. (pp. 10, 15, 16)

Scott, Tom K. 1978. *Department of Botany, University of North Carolina at Chapel Hill, Annual Report 1977–1978*. Chapel Hill, NC. (p. 6; Appendix D)

Scott, Tom K. 1979a. *Department of Botany, University of North Carolina at Chapel Hill, Annual Report 1978–1979*. Chapel Hill, NC. (pp. 7, 10)

Scott, Tom K. (editor). 1979b. *Plant Regulation and World Agriculture*. Plenum Press, New York.

Scott, Tom K. 1980. *Department of Botany, University of North Carolina at Chapel Hill, Annual Report 1979–1980*. Chapel Hill, NC. (pp. 10, 15)

Scott, Tom K. 1981a. Letter of November 3, 1981 to Dean Samuel R. Williamson, [et al.] in Albert E. Radford [Notebook]. (Box 6, folder 1, William R. Burk Collection on University of North Carolina at Chapel Hill Botany and Zoology Faculty, 1920–2010, University Archives #40478, Wilson Special Collections Library, University of North Carolina, Chapel Hill)

Scott, Tom K. 1981b. *Department of Botany, University of North Carolina at Chapel Hill, Annual Report 1980–1981*. Chapel Hill, NC. (pp. 9, 17,18)

Scott, Tom K. 1982. *Department of Botany, University of North Carolina at Chapel Hill, Annual Report 1981–1982*. Chapel Hill, NC. (pp. 9, 12, 13)

Scott, Tom K. (editor). 1984. *The Functions of Hormones from the Level of the Cell to the Whole Plant*. (*Encyclopedia of Plant Physiology, new series*, vol. 10). Springer-Verlag, Berlin, New York.

[Scott, Tom K.]. 2000. Scott, Tom Keck, p. 4,404 in *Marquis Who's Who*, 54th ed., vol. 2.

Scott, Tom K. [2003]. CV [Curriculum Vitae]. Prepared by T.K. Scott, Chapel Hill. (Copy in the research files of William R. Burk)

Shanor, L. 1968. The career of John Nathaniel Couch, pp. 1–7 in William J. Koch (editor), *Mycological Studies Honoring John N. Couch*. University of North Carolina Press, Chapel Hill.

Smith, Thomas Earle. 1938. *Host Range Studies with* Bacterium solanacearum *E.F.S.* A thesis submitted to the faculty of the University of North Carolina in partial fulfillment of the requirements for the degree of Doctor of Philosophy in the Department of Botany. [The Author], Chapel Hill, NC.

Stern, Daniel. 2004. *A Haven in the Heart of Chapel Hill: Artists celebrate the Coker Arboretum*. Botanical Garden Foundation, Chapel Hill, NC.

Szaniszlo, Paul J. 1993. John Nathaniel Couch, October 12, 1896–December 16, 1986. *Biographical Memoirs, National Academy of Sciences* 62: 38–64.

Totten, H.R. 1957. *Laboratory Guide in Pharmaceutical Botany and Pharmacognosy*. 7th ed. The Book Exchange, University of North Carolina, Chapel Hill.

Totten, H.R. and J.N. Couch. 1954. William Chambers Coker. *Journal of the Elisha Mitchell Scientific Society* 70: 116–118.

Totten, H.R., Paul W. Wager and John N. Couch. 1956. William Chambers Coker, pp. [108]–113 in A.C. Howell, *The Kenan Professorships*. University of North Carolina Press, Chapel Hill.

Troyer, James R. 1996. Donald Benton Anderson (1899–1994), scientist and educator. *Journal of the Elisha Mitchell Scientific Society* 111: 146–159.

Umphlett, Clyde J. 1965. *General Botany Laboratory: Exercises and directions*. [UNC Book Exchange for the] Department of Botany, University of North Carolina, Chapel Hill.

Umphlett, Clyde J. and William J. Koch. 1969. Two new dentigerate species of *Phlyctochytrium* (Chytridiomycetes). *Mycologia* 61: 1021–1030.

Umphlett, Clyde J. and Lauritz W. Olson. 1967. Cytological and morphological studies of a new species of *Phlyctochytrium*. *Mycologia* 59: 1085–1096.

UNC-A. [1922–1989]. *Department of Botany of the University of North Carolina at Chapel Hill Records, 1922–1989*. (University Archives #40078, Wilson Special Collections Library, University of North Carolina, Chapel Hill)

UNC-A-1. [1922–1989]. University of North Carolina at Chapel Hill, University Archives. *Department of Botany of the University of North Carolina at Chapel Hill Records, 1922–1989*. (Box 1, folder: General correspondence and memoranda, 1922–1971, University Archives #40078, Wilson Special Collections Library, University of North Carolina, Chapel Hill)

UNC-A-2. [1922–1989]. University of North Carolina at Chapel Hill, University Archives. *Department of Botany of the University of North Carolina at Chapel Hill Records, 1922–1989*. (Box 1, folder: Faculty meeting minutes, 1967–1982, University Archives #40078, Wilson Special Collections Library, University of North Carolina, Chapel Hill) (a. meeting of November 30, 1973; b. meeting of March 5, 1976; c. meeting of October 3, 1975; d. meeting of May 6, 1977; e. review follows meeting minutes of January 16, 1981; f. meeting of December 5, 1975; g. meeting of December 3, 1968; h. meeting of January 7, 1969; i. document follows meeting agenda of October 14, 1970)

UNC-A-3. [1922–1989]. University of North Carolina at Chapel Hill, University Archives. *Department of Botany of the University of North Carolina at Chapel Hill Records, 1922–1989*. (Box 1, folder: Faculty meeting minutes, 1957, 1959, 1975, University Archives #40078, Wilson Special Collections Library, University of North Carolina, Chapel Hill)

Vitousek, Peter M. 1984. Anion fluxes in three Indiana forests. *Oecologia* 61: 105–108.

Vitousek, Peter M., James R. Gosz, Charles C. Grier, Jerry M. Melillo and William A. Reiners. 1982. A comparative analysis of potential nitrification and nitrate mobility in forest ecosystems. *Ecological Monographs* 52: 155–177.

Walter, Heinrich and Helmut Lieth. 1967. *Klimadiagramm-Weltatlas*. G. Fischer, Jena.

Weakley, Alan S. 2020. *Flora of the Southeastern United States*. Produced from the Flora Manager database system by Michael T. Lee. University of North Carolina Herbarium (NCU), Chapel Hill; North Carolina Botanical Garden, University of North Carolina, Chapel Hill. Users can download a PDF of the flora after completing and submitting a form for its retrieval. See the following link: <https://ncbg.unc.edu/research/unc-herbarium/flora-request/>

Weakley, Alan S., J. Christopher Ludwig and John F. Townsend; Bland Crowder (editor). 2012. *Flora of Virginia*. Botanical Research Institute of Texas Press, Fort Worth, TX.

White, Peter S. 2021. Electronic mail sent to William R. Burk, January 22, 2021.

White, Richard A. and William C. Dickison (editors). 1984. *Contemporary Problems in Plant Anatomy*. Academic Press, Orlando.

Wiggins, Minnie Marguerite. 1963. *The Journal of the Elisha Mitchell Scientific Society, 1883–1963*. M.S. Thesis, Library Science, University of North Carolina, Chapel Hill.

Wilson, Carl L. and Walter E. Loomis. [1962]. *Botany*. 3rd ed. Holt, Rinehart and Winston, New York.

Wilson, Louis R. 1964. *The University of North Carolina under Consolidation, 1931–1963, History and Appraisal*. University of North Carolina, Consolidated Office. (pp. vii, 176–177)

Wood, C.E., Jr., Victor A. Greulach and J. Edison Adams. 1954. *Laboratory Problems in Plant Science*. University of North Carolina Book Exchange, Chapel Hill.

APPENDIX 15.1. NAMES OF GRADUATE STUDENTS WITH TITLES AND DATES OF THEIR DISSERTATIONS, THESES, AND GRADUATE PAPERS IN BOTANY COMPLETED AT THE UNIVERSITY OF NORTH CAROLINA AT CHAPEL HILL, INCLUDING THE NAMES OF THEIR MAJOR ADVISORS.

Compiled by
Jessica Kilfoil, William R. Burk, and Elizabeth A. Appleton,
with special assistance of Sarah Parnham
2006; revised and updated 2021

These lists provide bibliographical citations of titles of dissertations and theses written by graduate students in the Department of Botany at the University of North Carolina at Chapel Hill as well as names of major advisors. The lists also include titles of theses and dissertations in botany advised by professors of botany whose tenures extended past July 1982, when the Department of Biology was established. They also serve as a record of the print holdings of graduate theses and dissertations located in the UNC Science Library Annex. In compiling these lists, we update earlier efforts of Godfrey, Green, and Pierson (1947), and of the staff of the Humanities Division at the University of North Carolina Library (1960) in documenting graduate work at the University. Bibliographic information was verified for each title by looking at the volume itself. If the date of the dissertation or thesis differs from the date of the student's graduation, the latter appears in parentheses.

Appended to these lists is a selected list of Master of Science theses that students wrote for the Department of Botany and later the Department of Biology. The Department of Botany added the M.S. degree to its graduate program in 1969 (Jones, 1970). The department typically waived the requirement of writing a thesis for the M.S.; however, at least seventy students wrote papers in botany based on their research.

References Cited

(References that conclude with page numbers within parentheses indicate sources from where quotes and pertinent information were obtained.)

Godfrey, James L., Fletcher M. Green, and W.W. Pierson (editors). 1947. *The Graduate School Dissertations and Theses*. (*The University of North Carolina Sesquicentennial Publications*). University of North Carolina Press, Chapel Hill. (Botany, pp. 8–11)

Humanities Division, University of North Carolina Library (compiler). 1960. *The Graduate School Dissertations and Theses,* First Supplement: 1946–1959. (*University of North Carolina Library Studies,* no. 3). University of North Carolina Library, Chapel Hill. (Botany, pp. 8–13)

Jones, Lyle V. 1970. Advances in Graduate Education at the University of North Carolina at Chapel Hill 1960–61 to 1969–70. Office of the Dean, The Graduate School, University of North Carolina, Chapel Hill.

DOCTOR OF PHILOSOPHY LIST[117]

Name	Date	Title	Major Advisor
Aghajanian, John Gregory	1977	Light and electron microscope and cytochemical studies of morphogenesis in the fresh water red alga *Batrachospermum sirodotii* Skuja	Max H. Hommersand
Allen, Robert Bruce	1988	Latitudinal variation in southern Rocky Mountain forests	Robert K. Peet
Armentano, Thomas Vincent	1973	Population ecology and response to stress of *Aster macrophyllus* and *Cornus canadensis*	J. Frank McCormick
Ashworth, Ralph Page	1960	Investigations into the foliar anatomy and ontogeny of certain species of the genus *Ilex* L.	J.E. Adams
Baird, James Rex	1968 (1969)	A taxonomic revision of the plant family Myricac[e]ae of North America, north of Mexico	Albert E. Radford
Baker, Philip Carlton	1970	A systematic study of the genus *Vaccinium* L., subgenus *Polycodium* (Raf.) Sleumer, in the southeastern United States	C. Ritchie Bell
Baker-Brosh, Kathleen Faye	1996	The genetic consequences of self-thinning in two populations of loblolly pine (*Pinus taeda* L.)	Robert K. Peet
Bauer, Paul John	1986	Bumblebee pollination on a southern Appalachian grass bald and narcotic nectar in *Angelica triquinata* (Apiaceae)	C. Ritchie Bell
Beard, Luther Stanford	1963 (1964)	A taxonomic study of *Mimosa quadrivalvis* L. (*Schrankia* Willd. nom.cons.)	Albert E. Radford
Bennett, Bradley Charlton	1988	A comparison of life history traits in selected epiphytic and saxicolous species of *Tillandsia* (Bromeliaceae) from Florida and Peru	C. Ritchie Bell
Berkely, Edmund	1951	Morphological studies in the Celastraceae	J.E. Adams
Birk, Elaine Margaret	1983	Nitrogen availability, nitrogen cycling, and nitrogen use efficiency in loblolly pine stands at the Savannah River Plant, South Carolina	Peter M. Vitousek
Bland, Charles Everette	1968 (1969)	Structure and development in some genera of the Actinoplanaceae	John N. Couch
Blanton, Richard Lawrence	1981	The development of aerial sorocarps by the ciliated protozoan *Sorogena stoianovitchae*	Lindsay S. Olive
Boole, John Allen, Jr.	1955	Studies in the anatomy of the family Celastraceae	J.E. Adams
Bostick, Peter Edward	1966	A systematic study of the genus *Rhexia*	C. Ritchie Bell
Bowden, Sandra Taylor	1968	A study of the development of the gametophyte of *Polytrichum commune* L.	Max H. Hommersand
Bowerman, Linda Louise	1997	Ultrastructure of *Lagenidium caudatum*	Aristotle Domnas

[117] The dissertations of Thomas Earle Smith (1938) and Don Edwin Ellis (1945) are not included in this list of students who wrote dissertations. See the pertinent explanation about this exclusion on page 496.

DOCTOR OF PHILOSOPHY LIST (CONTINUED)

Name	Date	Title	Major Advisor
Bowmer, Richard Glenn	1960	The influence of various factors on the rate of translocation of fluorescein in *Gossypium hirsutum* L.	Victor A. Greulach
Box, Elgene Owen	1978	Ecoclimatic determination of terrestrial vegetation physiognomy	Helmut Lieth
Bozeman, John Russell	1971	A sociological and geographic study of the sand ridge vegetation in the coastal plain of Georgia	Albert E. Radford
Bradley, Ted Ray	1967 (1968)	Hybridization and variation in the species of *Triodanis* from North America	C. Ritchie Bell
Britt, Robert Franklin	1967	A revision of the genus *Hypoxis* in the United States and Canada	Albert E. Radford
Brown, Alissa Jane	2019	The occurrence of conspecific negative density dependence in temperate forest trees varies by species, plant traits, and environmental context	Robert K. Peet, Peter S. White
Brown, Rebecca Louise	2002	Biodiversity and exotic species invasion in southern Appalachian riparian plant communities	Robert K. Peet
Browne, Edward Tankard, Jr.	1956 (1957)	Morphological studies in *Aletris* L.	J.E. Adams
Burk, Carl John	1961	A floristic study of the Outer Banks of North Carolina	Albert E. Radford
Caddell, Gloria May	1989	Pollination, mating systems, and distribution of genetic variation within populations of *Camellia japonica* L. (Theaceae)	Clifford R. Parks
Cameron, Kenneth Michael	1996	Phylogenetic relationships of the vanilloid orchids: and integration of molecular, morphological, and anatomical data	William C. Dickison, Mark W. Chase
Campbell, Peter Hallock	1973	The phytoplankton of Gales Creek with emphasis on the taxonomy and ecology of estuarine phytoflagellates	Max H. Hommersand
Carson, Johnny Lee	1975	Ecological investigations of the soil algae and airborne microflora of the Hawaiian Islands	R. Malcolm Brown, Jr.
Christenberry, George Andrew	1940	A taxonomic study of the Mucorales of the southeastern United States	John N. Couch
Clark, Ross Carlton	1969	A distributional study of the woody plants of Alabama	Albert E. Radford
Clausz, John Clay	1969	Distribution of propagules of species of the Saprolegniaceae in the North Carolina Botanical Garden lake and some possible controlling factors	Clyde J. Umphlett
Coomans, Roy John	1986	A light microscope study of mitosis and vegetative development in Rhodophyta	Max H. Hommersand

Doctor of Philosophy List (Continued)

Name	Date	Title	Major Advisor
Corbin, Jeffrey Douglas	1997 (1998)	The effects of experimental nitrogen and phosphorus on a temperate deciduous forest ecosystem	Robert K. Peet
Corry, Patricia Mary	2006	Vegetation dynamics following grazing cessation on the Channel Islands, California	Robert K. Peet
Coskun, Fatih	2003	The phylogeny and biogeography of *Osmanthus, Cercis*, and *Tilia* based on its nuclear ribosomal, NDHF and TRNL-F chloroplast DNA sequence data	Clifford R. Parks
Couch, John Nathaniel	1924	Sexual reproduction and variability in the genus *Dictyuchus*	William C. Coker
Cox, Hiden Toy	1946 (1947)	The comparative anatomy of the Ericaceae	J.E. Adams
Culwell, Donald Edward	1970	A taxonomic study of the section *Hypericum* in the eastern United States	Albert E. Radford
Davis, Graham Johnson	1956	The effects of certain environmental and chemical factors on heterophylly in aquatic angiosperms	Victor A. Greulach
Dayton, Bruce Roy	1968	Accumulation and transfer of radiostrontium by young loblolly pines	J. Frank McCormick
Dean, David Devereux	1981	The extracellular proteolytic enzymes of *Lagenidium giganteum*	Aristotle Domnas
Deasey, Mary Clare	1982	Aspects of sorogenesis in the cellular slime mold *Fonticula alba*	Lindsay S. Olive
DeCoster, James Krause	1996	Impacts of tornados and hurricanes on the community structure and dynamics of North and South Carolina forests	Robert K. Peet
Dillon, Charles Ronald	1971	A comparative study of the primary productivity of estuarine phytoplankton and macrobenthic plants	J. Frank McCormick
Doubles, James Arthur, Jr.	1940	A taxonomic study of the Myxomycetes of North Carolina	William C. Coker, John N. Couch
Doyle, James Dennis	1990 (1991)	Systematics of the *Opuntia humifusa* complex	Jim Massey
Drapalik, Donald Joseph	1969 (1970)	A biosystematic study of the genus *Matelea* in the southeastern United States	C. Ritchie Bell
Duke, James Alan	1959 (1960)	The psammophytes of the Carolina fall-line sandhills	Albert E. Radford
Dykstra, Michael Jack	1975	*Sorodiplophrys stercorea*, a light and electron microscope investigation	Lindsay S. Olive
Edwards, Kathryn Louise	1974	Growth responses and their characteristics in corn root and coleoptile segments	Tom K. Scott

DOCTOR OF PHILOSOPHY LIST (CONTINUED)

Name	Date	Title	Major Advisor
Eierman, David Frederick	1985 (1986)	A lipoprotein from the oomycete fungus *Lagenidium giganteum*	Aristotle Domnas
Erdos, Gregory William	1971	Ultrastructure and development of *Endemosarca*: A new plasmodial parasite of ciliates	Lindsay S. Olive
Fredericq, Suzanne	1988	Developmental morphology and systematics of the Gracilariaceae (Rhodophyta)	Max H. Hommersand
Freshwater, David Wilson	1993	Molecular systematics of the Gelidiales based upon rbcL sequence analysis	Max H. Hommersand
Fridley, Jason Dale	2002	The influence of species diversity on ecosystem productivity in different environments: Experimental plant communities and theory	Robert K. Peet
Frost, Cecil Carlyse	2000	Studies in landscape fire ecology and presettlement vegetation of the southeastern United States	Robert K. Peet
Fulcher, William Ernest	1971	An anatomical and morphological study of *Batis maritima* L. with systematic implications	J.E. Adams
Gabrielson, Paul William	1980	Morphological studies of selected Northern Hemisphere genera in the family Solieriaceae, Gigartinales (Rhodophyta)	Max H. Hommersand
Gallagher, Patricia Ann	1992	Aspects of lipid metabolism in *L[agenidium] giganteum*	Aristotle Domnas
Gibson, Joan Reynolds	1987	The relationship of vegetation to diabase dikes and sills of the Gettysburg Basin, Pennsylvania	Albert E. Radford
Giebel, Peter Edward	1974 (1975)	Carbohydrate and amino acid metabolism in *Culex pipiens quinquefasciatus* infected with *Lagenidium giganteum*	Aristotle Domnas
Goldie-Smith, Elizabeth Kathleen	1953	Studies in the Plasmodiophoraceae	John N. Couch
Gramling, Joel Michael	2006	Understanding local and regional plant diversity: Species pools, species saturation and the multi-scalar effects of plant productivity	Robert K. Peet
Grant, Neil George	1972	The respiratory chain of *Chlorella protothecoides*	Max H. Hommersand
Graves, James Hazen	1995	Resource availability and the importance of herbs in forest dynamics	Robert K. Peet, Peter S. White
Gupton, Oscar Wilmont	1962 (1963)	An analysis of the taxonomic criteria as applied to the genus *Monotropa*	Albert E. Radford
Gurlitz, Robin Heath Greany	1983	Parameters of attachment of *Agrobacterium tumefaciens* to carrot tissue culture cells	Ann G. Matthysse
Haesloop, John Gamage	1960	The effect of gibberellic acid on cell division and cell elongation in stems of *Lycopersicum esculentum*	Victor A. Greulach

Doctor of Philosophy List (Continued)

Name	Date	Title	Major Advisor
Haigler, Candace Hope	1982	Alteration of cellulose assembly in *Acetobacter xylinum* by fluorescent brightening agents, direct dyes, and cellulose derivatives	R. Malcolm Brown, Jr.
Hansen, Gayle Ingrid	1976	The morphology and life history of *Cirrulicarpus carolinensis* sp. nov. (Kallymeniaceae, Cryptonemiales)	Max H. Hommersand
Hauk, Warren Douglas	1996	Phylogenetics of Ophioglossaceae: The evolutionary consequences of morphological reduction	Clifford R. Parks, Mark W. Chase
Hedberg, Mary Frances	1984	Investigation of the chloroplast genome in two marine algae, *Codium fragile* and *Monodus* sp.	Max H. Hommersand
Herr, John Mervin, Jr.	1957	Embryological studies in the genus *Ilex* L.	J.E. Adams
Higginbotham, Jeri Welsh	1987	Morphological and isoenzymic variation in Southern Appalachian *Betula alleghaniensis* Britt. and *B. lenta* L.	Clifford R. Parks
Hoey, Margaret Tevis	1990	Patterns of genetic divergence within the vicariad genus *Liquidambar* L. (Hamamelidaceae)	Clifford R. Parks
Horton, James Heathman	1961	A monograph of *Delopyrum* Small, *Dentoceras* Small, *Polygonella* Michx., and *Thysanella* Gray (Polygonaceae)	Albert E. Radford
Huck, Robin Bovard	1984	Systematics and evolution of *Dicerandra* (Labiatae)	Albert E. Radford
Huneycutt, Maeburn Bruce	1955 (1956)	Studies on the morphology, taxonomy, and physiology of some lower aquatic phycomycetes, especially keratinophilic forms	John N. Couch
James, Frederick Charles	1969	The woody flora of Virginia	Albert E. Radford
Jeffreys, Donald Bearss	1962 (1963)	The nature of growth inhibition in *Lycopersicon esculentum* plants treated with 5-fluorouracil and related compounds	Victor A. Greulach
Jensen, Douglas Paige	1997	Comparative morphology and evolution of the zosterphylls	Patricia G. Gensel
Johnson, Norma Grace	1989	*Bitelaria*, an early Devonian vascular plant, and its significance relative to the origin and radiation of early land plants	Patricia G. Gensel
Jones, Larry Hudson	1976	Transfer ribonucleic acid modification and its relationship to tumorous and non-tumorous plant growth	Tom K. Scott
Kann, Jacob	1997	Ecology and water quality dynamics of a shallow hypertrophic lake dominated by cyanobacteria (*Aphanizomenon flos-aquae*)	Robert K. Peet, Val H. Smith

DOCTOR OF PHILOSOPHY LIST (CONTINUED)

Name	Date	Title	Major Advisor
Katz, Barry Allan	1979	The ecology of pine litter decomposition: Phenology, horizontal and vertical distribution of fungi associated with loblolly pine needles (*Pinus taeda*) in the canopy and on the forest floor	Helmut Lieth
Kim, Won Kyum	1961	Influences of several plant growth substances on four species of algae	Victor A. Greulach
Knauss, Mark Ernest	1988	Reproductive development and systematics of some primitive families of Gigartinales (Rhodophyta)	Max H. Hommersand
Knox, Robert Gaylord	1987	Hypothesis testing in plant community ecology: Analysis of long-term experiments and regional vegetation data	Robert K. Peet
Koch, William Julian	1956	Studies in the Chytridiales, with special reference to the structure, movement and systematic significance of the swimming reproductive cell	John N. Couch
Kondo, Katsuhiko	1975	Cytological studies in cultivated species of *Camellia*	Clifford R. Parks
Lang, Andrew George	1936	Some observational and theoretical considerations of spermatogenesis in *Marsilea quadrifolia*, with special reference to the structure of the spermatozoid and the evolution of the nucleolus	John N. Couch
Lebrón, María Luz	1977	An autoecological study of *Palicourea riparia* Bentham (Rubiaceae): An ecologically important species in the recovery of a disturbed tropical rain forest in Puerto Rico	Tom K. Scott
Leisner, Robert Stanley	1958	Morphology and systematic anatomy of Empetraceae	J.E. Adams
Leonard, Richard Carlton	1968 (1969)	A physiological study of *Pilimelia terevasa* (Actinoplanaceae)	Aristotle Domnas
Liao, Lawrence Manzano	1995	A morphological study and taxonomic re-assessment of some genera and species of the Gracilariaceae (Rhodophyta, Gracilariales)	Max H. Hommersand
Lindsey, Anne Haskell	1979	Pollination and breeding system studies in *Thaspium* and *Zizia* (Apiaceae)	C. Ritchie Bell
Lipscomb, Harriett Ann	1968	An anatomical and morphological study of *Phryma leptostachya* L. with possible systematic implications	J.E. Adams
López Mata, Lauro	1994	Coexistence of *Quercus* and *Carya* in natural upland hardwood forests of the North Carolina Piedmont	Robert K. Peet

Doctor of Philosophy List (Continued)

Name	Date	Title	Major Advisor
Lugo, Ariel Emilio	1969	Energy, water and carbon budgets of a granite outcrop community	J. Frank McCormick
Marsho, Thomas Victor	1968	Simultaneous measurements of light-induced absorption changes and oxygen exchange in *Ulva lactuca*	Max H. Hommersand
Martin, William Wallace, III	1970	A morphological and cytological study of *Coelomomyces punctatus*	Clyde J. Umphlett
Mathen, Oravackal Mathai	1963	A taxonomic comparison of *Aureolaria*, *Agalinis* and *Tomanthera*	C. Ritchie Bell
Mathews, Andrew Clark	1939	The morphological and cytological development of the sporophylls and seed of *Juniperus virginianus* L.	William C. Coker
Matthews, Elizabeth Rainey	2011	Piedmont alluvial vegetation: Classification, geographic variation, and restoration	Robert K. Peet
Matthews, Velma Dare	1930	Studies on the genera *Nematosporangium* and *Pythium*	William C. Coker
McDougal, Karen McAlpin	1984	Patterns of variation in the foliar flavonoids of red oak, *Quercus rubra* L., and their relationships to migration and evolution in the species	Clifford R. Parks
McInnis, Thomas McLeod	1971	A physiological and biochemical investigation of the aquatic Phycomycete *Lagenidium* sp.; a facultative parasite of certain mosquito larvae	Aristotle Domnas
McLeod, Donald Evans	1988	Vegetation patterns, floristics and environmental relationships in the Black and Craggy Mountains of North Carolina	Albert E. Radford
McNitt, Rand Edwin	1973	Light and electron microscopy of *Phlyctochytrium irregulare* Koch	William J. Koch, Lindsay S. Olive
Mellinger, Clair Abram	1972	Ecological life cycle of *Viguiera porteri* and factors responsible for its endemism to granite outcrops of Georgia and Alabama	J. Frank McCormick
Meyer, Kenneth Albert	1972	The influence of nutrient availability and nutrient cycles upon ecosystem structure and stability	J. Frank McCormick
Michalski, Chester James	1971	Metabolic control, purification and properties of B-glucosidase from *Actinoplanes* sp. #1700	Aristotle Domnas
Miller, Charles Edward	1957	Morphology, cytology and life history of *Sorosphaera veronicae* and *Ligniera verrucosa*	John N. Couch
Miller, Gary Lee	1968	The influence of season on the radiation sensitivity of an old field community	J. Frank McCormick

DOCTOR OF PHILOSOPHY LIST (CONTINUED)

Name	Date	Title	Major Advisor
Millsaps, Vera	1936	The structure and development of the seed of *Paulownia tomentosa* Steud. and *Cynoglossum amabile* Stapf and Drumm	William C. Coker
Montezinos, David Lee	1977	Cell wall biogenesis on *Oocystis apiculata*	R. Malcolm Brown, Jr.
Morgan, Judy Tate	1966 (1967)	A taxonomic study of the genus *Boltonia* (Asteraceae)	Albert E. Radford
Morrill, Joy Frances	1976	Comparative morphology and taxonomy of some dorsiventral and some parasitic Rhodomelaceae	Max H. Hommersand
Mueller, Anna Maria	1974 (1978)	An evolutionary study of *Coreopsis* section Palmatae (Compositae)	C. Ritchie Bell
Mueller, Sabina Gertrud	1967 (1968)	The taxonomic significance of cuticular patterns and leaf morphology in populations of *Vaccinium* Ericaceae	C. Ritchie Bell
Mueller, Susette Carter	1979	Freeze-fracture and fast growth studies of seedling tissue: Cell wall synthesis, deposition and extension	R. Malcom Brown, Jr., Tom K. Scott
Mullins, John Thomas	1959 (1960)	Studies on the life history and development of *Pringsheimiella dioica* Couch	John N. Couch
Murphy, Peter George	1970	Ecological effects of acute beta radiation from simulated fallout particles upon a natural plant community	J. Frank McCormick
Musselman, Lytton John	1974	A comparative study of the structure and development of the haustorium in parasitic Scrophulariaceae	William C. Dickison
Nesom, Guy Lane	1980	A revision of the epappose species of *Erigeron* (Asteraeceae-Astereae)	C. Ritchie Bell
Nesom, Margaret Green	1973	Life cycle and ultrastructure of the cellular slime mold, *Copromyxa arborescens*	Lindsay S. Olive
Newell, Claire Leith	1997	Local and regional variation in the vegetation of the Southern Appalachian Mountains	Robert K. Peet
Nielsen, Peter Tryon	1964 (1965)	The uptake of chloride ions by *Chlorella pyrenoidosa* with special emphasis on photo-accumulation	Max H. Hommersand
Nifong, Timothy David	1998	An ecosystematic analysis of Carolina Bays in the Coastal Plain of the Carolinas	Albert E. Radford, Robert K. Peet
O'Connell, Jesse Elbert	1955	Systematic studies in *Sisyrinchium*	J.E. Adams
Olive, Lindsay Shepherd	1942	Cytology, morphology, and parasitism of *Thekopsora hydrangeae*	William C. Coker
O'Neil, Roger Alan	1983	Auxin effects on cell wall proteins of suspension culture carrot cells	Tom K. Scott

Doctor of Philosophy List (Continued)

Name	Date	Title	Major Advisor
Ott, Donald William	1973	Developmental cytology of *Vaucheria*	R. Malcolm Brown, Jr.
Ott, Jeffrey Eric	2010	Individualistic and phylogenetic perspectives on plant community patterns	Robert K. Peet
Owen, Joseph Donald	1973	The physiology of gametogenesis in *Chlamydomonas reinhardtii*	Max H. Hommersand
Palmquist, Kyle Ann	2014	Community assembly and vegetation patterns across space and time in the longleaf pine ecosystem	Robert K. Peet, Alan S. Weakley
Payne, Christopher Joseph	2018	Long-term temporal dynamics of the Duke Forest	Robert K. Peet
Plyler, Daniel Belk	1964	Investigations on the mode of action of maleic hydrazine	Victor A. Greulach
Powell, Martha Jane	1974	Developmental studies of the chytrid *Entophlyctis variabile* sp.n.: A light and electron microscopic investigation	William J. Koch
Prichard, Elmer Clyde	1954	Morphological studies in Rhamnaceae	J.E. Adams
Prince, Linda Marie	1998	Systematics of the Theoideae (Theaceae) and the resolution of morphological, anatomical, and molecular data	Clifford R. Parks
Qiu, Yin-Long	1993	Molecular divergence between Asian and North American species of *Magnolia* sect. *Rytidospermum* (Magnoliaceae)	Clifford R. Parks
Radford, Albert Ernest	1948	The vascular flora of the olivine deposits of North Carolina and Georgia	Henry Roland Totten
Ramseur, George Shuford	1958 (1959)	The vascular flora of high mountain communities of the southern Appalachians	Albert E. Radford
Reader, Richard Joel	1973	Leaf emergence, leaf coloration, and photosynthetic period-productivity models for the eastern deciduous forest biome	Helmut Lieth
Reilly, Lee Anne Jacobs	2010	A quantitative approach for defining rarity	Robert K. Peet
Rhyne, Charles Frederick	1973	Field and experimental studies on the systematics and ecology of *Ulva curvata* and *Ulva rotundata*	Max H. Hommersand
Rhyne, Clara Bonnie Carswell	1968	Some comparative studies on uptake and respiration of glucose and acetate in *Chlorella pyrenoidosa*	Max H. Hommersand
Risk, Patricia Louise Weigant	1993	High elevation heath communities in the Blue Ridge of North Carolina	Albert E. Radford

DOCTOR OF PHILOSOPHY LIST (CONTINUED)

Name	Date	Title	Major Advisor
Ritchie, Donald Dirk	1947	The cytology of the formation and structure of zoospores in *Allomyces*	Earl H. Newcomer
Rodgers, Charles Leland	1950 (1951)	The Umbelliferae of North Carolina and their distribution in the Southeast	Henry Roland Totten
Romanovicz, Dwight Keith	1975	Ultrastructural and cytochemical investigations of cell wall biogenesis in *Pleurochrysis scherffelii*	R. Malcolm Brown, Jr.
Rury, Phillip MacDonald	1982	Systematic anatomy of the Erythroxylaceae	William C. Dickison
Salgado, Teresa Terrazas	1994	Wood anatomy of the Anacardiaceae: Ecological and phylogenetic interpretation	William C. Dickison
Saterson, Kathryn Ann	1985	Nitrogen availability, primary production, and nutrient cycling during secondary succession in North Carolina piedmont forests	Peter M. Vitousek
Sears, Michael Novy	1970	A study of *Actinoplanes* and its urease	Aristotle Domnas
Sewell, Mitchell Martin	1992	Mating system and molecular variation in *Liriodendron tulipifera* L. (Magnoliaceae)	Clifford R. Parks
Shanor, Leland	1939	Studies in the genus *Olpidiopsis* (Cornu) Fischer	William C. Coker, John N. Couch
Sharitz, Rebecca Reyburn	1970	Population dynamics of two competing plant species	J. Frank McCormick
Sherwood, William Allen	1966 (1967)	Some physiological and genetical aspects of heterothalism in the water mold *Dictyuchus*	John N. Couch
Shew, Harry Wayne	1977	Genetics and cytology of *Neurospora intermedia*	Edward G. Barry
Silliman, Frances Ernestine	1957	*Chamaelirium luteum* (L.) Gray: A biological study	J.E. Adams
Slifkin, Miriam Kresses	1967 (1968)	Studies of somatic nuclear behavior in *Saprolegnia delica*	John N. Couch
Smith, Budd Elmon	1942	A study of the dicotyledonous flora of Darlington County, South Carolina	William C. Coker
Spiegel, Frederick William	1978	*Planoprotostelium aurantium*: A light and electron microscopic investigation of development in a simple eumycetozoan	Lindsay S. Olive
Spongberg, Stephen Alan	1971	A systematic and evolutionary study of North American Arctic and Alpine monocephalous species of *Erigeron* (Compositae)	C. Ritchie Bell
Stanford, Alice Margaret	1998	The biogeography and phylogeny of *Castanea*, *Fagus*, and *Juglans* based on matK and its sequence data	Clifford R. Parks

DOCTOR OF PHILOSOPHY LIST (CONTINUED)

Name	Date	Title	Major Advisor
Stevenson, John Court	1972	Evolutionary strategies and ecology of *Goodyera* and *Spiranthes* species (Orchidaceae)	C. Ritchie Bell
Szaniszlo, Paul Joseph	1967	Comparison of the cell-wall composition and intracellular pigmentation of some strains of Actinoplanaceae	John N. Couch
Taggart, John Bryant	1990	Inventory, classification, and preservation of coastal plain savannas in the Carolinas	Albert E. Radford
Tarasi, Dennis Darwin	2016	Community interactions and impacts of exotic plant species in the southeastern United States	Robert K. Peet
Thakor, Barbara Haren	1998	Phylogenetic relationships among the species of genus *Camellia* (Theaceae)	Clifford R. Parks
Totten, Henry Roland	1923	Description, life history, and mycorhizal character of *Rhizopogon parasiticus* Coker and Totten n. sp.	William C. Coker
Treiber, Miklos	1980	Biosystematics of the *Arisaema triphyllum* complex	C. Ritchie Bell
Triemer, Richard Ernest	1975	The ultrastructure of reproduction in *Chlamydomonas*	R. Malcolm Brown, Jr.
Umphlett, Clyde Jefferson	1961	Comparative studies in the genus *Coelomomyces* Keilin	John N. Couch
Walbridge, Mark Robert	1986	Phosphorous availability in acid organic coastal plain soils	Peter M. Vitousek
Walker, Joan L.	1985	Species diversity and production in pine-wiregrass savannas of the Green Swamp, North Carolina	Robert K. Peet
Warner, Steven Alan	1982	Biosynthesis and metabolism of sterols by pantonemic fungi	Aristotle Domnas
Wells, Elizabeth Fortson	1977	A taxonomic study of the genus *Heuchera* in eastern North America	Clifford R. Parks, Max H. Hommersand
Wendel, Jonathon Fox	1983	Electrophoretic analysis of genetic variation in wild and cultivated *Camellia japonica* L.	Clifford R. Parks
Wheeler, Brooke Ellen	2011	Species diversity of vegetation in the Carolinas: The influence and interaction of scale of observation, soil nutrients, and disturbance events	Robert K. Peet
Whetstone, Raymond David	1981	Vascular flora and vegetation of the Cumberland Plateau of Alabama: Including a computer-assisted spectral analysis and interpretive synthesis of the origin, migration, and evolution of the flora	Albert E. Radford

Doctor of Philosophy List (Continued)

Name	Date	Title	Major Advisor
Whiffen, Alma Joslyn	1941	Culture studies in the Chytridiales with relation to growth on synthetic media and decomposition of cellulose	John N. Couch
Whigham, Dennis Francis	1971	An ecological study of *Uvularia perfoliata* in the Blackwood division of the Duke Forest, North Carolina	Helmut Lieth
White, Alan Robert	1981	The characterization of the mode of action of the cellulase enzyme system of the fungus *Trichoderma reesei*	R. Malcolm Brown, Jr.
White, Jacqueline Michael	2014	The impacts of altered floodplain-hydrology of the lower Roanoke River on tree regeneration and floodplain forest composition	Robert K. Peet
Whitney, Kenneth Dean	1983	An ultrastructural study of sporocarp development in the mycetozoans *Schizoplasmodium cavostelioides* and *Cavostelium apophysatum*	Lindsay S. Olive
Wickland, Diane Elizabeth	1983	Vegetation patterns on derelict heavy metal mine sites in the North Carolina Piedmont	Helmut Lieth
Windler, Donald Richard	1970	Systematic studies in *Crotalaria sagittalis* L. and related species in North America (Leguminosae)	C. Ritchie Bell
Wiser, Susan Kathleen	1993	Vegetation of high-elevation rock outcrops of the Southern Appalachians: Composition, environmental relationships, and biogeography of communities and rare species	Peter S. White, Robert K. Peet
Wurdack, Kenneth John	2002	The molecular systematics and evolution of Euphorbiaceae *sensu lato*	Clifford R. Parks
Wyatt, Raymond Louis	1955 (1956)	Floral morphology and phylogeny of the Aristolochiaceae	J.E. Adams
Xi, Weimin	2006	Forest response to natural disturbance: Changes in structure and diversity of a North Carolina Piedmont forest in response to catastrophic wind events	Robert K. Peet
Xiao, Tiaojiang	2001	Phylogenetic relationships of the genus *Camellia* (Theaceae) based on the RNA polymerase II (RPB2) sequences	Clifford R. Parks
Yoshikawa, Norihiro	1999	Morphology and molecular analysis of *Camellia japonica* L. and closely related taxa from eastern China	Clifford R. Parks
Ziegler, Arthur William	1948	A comparative study of zygote germination in the Saprolegniaceae	John N. Couch

MASTER OF ARTS LIST

Name	Date	Title	Major Advisor
Aiuto, Russell	1963	Variation correlations in the *Phlox carolina* - *Phlox glaberrima* complex	C. Ritchie Bell
Allen, Ruth Ellis	1944	Four ascosporic species of *Aspergillus* found on tea leaves	John N. Couch
Ambrose, Lee Jeannie Howe	1977	A light and electron microscope study of the mucilage hairs of the liverwort *Pallavicinia lyellii* (Hook.) S.F. Gray	R. Malcolm Brown, Jr.
Bannister, Barbara Ann	1967	The ecological life cycle of *Euterpe globosa* Gaertn.	J. Frank McCormick
Barksdale, Walter Lane	1941	The pedicellate species of *Trillium* found in the southern Appalachians	J.E. Adams
Beard, Luther Stanford	1958	A floristic study along the Lower Deep River of North Carolina	Albert E. Radford
Beckham, Rebecca Joan	1947	The development of the ovule and female gametophyte of *Passiflora incarnata* L.	J.E. Adams
Bell, Clyde Ritchie	1948	A taxonomic study of the Sarraceniaceae of North America	J.E. Adams
Bernstein, Linda Beryl	1966	A biosystematic study of eight isolates of *Rhizophlyctis rosea* with emphasis on zoospore variability	William J. Koch
Bostick, Linda Roane	1966	Studies of the morphology and cytology of *Chytriomyces hyalinus* Karling	William J. Koch
Bostick, Peter Edward	1964	A geobotanical investigation of Chandler Mountain, St. Clair County, Alabama	Albert E. Radford
Bowden, Sandra Taylor	1964	Influence of some growth substances on selected mosses	Victor A. Greulach
Bozeman, John Russell	1964 (1965)	Floristic and edaphic studies of the Altamaha River sand ridge, Georgia	Albert E. Radford
Bradley, Ted Ray	1965 (1966)	Variation in mixed populations of *Specularia perfoliata* (L.) A. DC. and *Specularia biflora* (R. and P.) Fisch. and Mey. growing in North Carolina	C. Ritchie Bell
Britt, Robert Franklin	1960	The vascular flora of Robeson County, North Carolina	Albert E. Radford
Browne, Edward Tankard, Jr.	1950	Sporogenesis and chromosome morphology of *Lilium catesbaei* with taxonomic and ecological considerations	Earlene Atchison
Burk, Carl John	1959	A floristic study of a sandhill area on the North Carolina coastal plain	Albert E. Radford
Burton, Mary Gwendolyn	1940	Oogenesis and fertilization in *Pythiopsis intermedia*	William C. Coker, John N. Couch

Master of Arts List (Continued)

Name	Date	Title	Major Advisor
Carpenter, Carol Sue	1975	The morphology and relationships of *Oncotheca balansae*	William C. Dickison
Christenberry, George Andrew	1938	A study in the effect of light of various periods and wave lengths on the growth and asexual reproduction of *Choanephora cucurbitarum* (Berk. and Rav.) Thaxter	William C. Coker
Clark, Ross Carlton	1966	The vascular flora of the Fiery Gizzard gorges in south-central Tennessee	Albert E. Radford
Clausz, John Clay	1965 (1966)	Some factors affecting germination of oospores of *Achlya hypogyna*	William J. Koch
Colosi, Joseph Carmen	1975	Seed ecology of *Iva imbricata* Walter	Tom K. Scott
Compton, Sarah Winifred	1951	The effects of maleic hydrazide on growth and cell division of *Pisum sativum*	Earlene Atchison
Cooper, Kay McDonald	1978 (1979)	Callose deposit formation and the utilization of UDP-glucose by the root hairs of *Rhaphanus sativus*	R. Malcolm Brown, Jr.
Couch, John Nathaniel	1922	Spore formation and discharge in some genera of water molds	William C. Coker
Cox, Hiden Toy	1939	A new genus of the Rhizidiaceae	John N. Couch
Cumming, Fairman Preston	1969	An experimental design for the analysis of community structure	J. Frank McCormick
Damon, Carol Ann	1968	A linkage map of nineteen temperature sensitive mutants of *Neurospora crassa*	Edward G. Barry
Dayton, Bruce Roy	1965	A study of the vegetation of Iredell clay loam	Albert E. Radford
Dillon, Charles Ronald	1968	Distribution and production of the macrobenthic flora of a North Carolina estuary	J. Frank McCormick
Dorr, Laurence Joseph	1980	The reproductive biology and pollination ecology of *Zenobia* (Ericaceae)	C. Ritchie Bell
Doubles, James Arthur, Jr.	1938	Myxomycetes of the Chapel Hill area	William C. Coker
Duke, James Alan	1954 (1955)	Distribution and speciation of the genus *Ludwigia* in North Carolina	Albert E. Radford
Eliason, Nancy Blair	1930	Development of the seed in the morning glory (*Ipomaea* [*i.e., Ipomoea*] *purpurea*) and in dodder (*Cuscuta arvensis*)	William C. Coker
Farmer, Janis Marie	1982 (1983)	Pollination and fertilization effectiveness in *Asclepias syriaca* L. (Asclepiadaceae)	C. Ritchie Bell
Foust, Frances Katherine	1938	Studies on *Rozella* and *Pleolpidium*	John N. Couch

Master of Arts List (Continued)

Name	Date	Title	Major Advisor
Garwood, Marion Ethel	1942	*Rhipidium* and *Blastocladia* from North Carolina	John N. Couch
Gibson, Joan Reynolds	1969	The flora of Alder Run Bog, Tucker County, West Virginia	Albert E. Radford
Giebel, Kristen Phend	1976	Comparative wood anatomy and relationships of the Clethraceae	William C. Dickison
Griffith, Pamela Hirtle	1972	Evolutionary relationships within the southeastern yellow-flowered species of *Erythronium* (Liliaceae)	Clifford R. Parks
Gupton, Oscar Wilmont	1960	A vascular flora of the North Carolina sandhills wildlife management area	Albert E. Radford
Haesloop, John Gamage	1954	A comparative study of the protein metabolism of excised leaves of *Zea mays* and *Saintpaulia* sp.	Victor A. Greulach
Hansen, Gayle Ingrid	1976	The morphology and life history of *Cirrulicarpus carolinensis* sp. nov. (Kallymeniaceae, Cryptonemiales)	Max H. Hommersand
Harvey, James Vernon	1925	A study in the water molds and *Pythiums* occurring in the soils of Chapel Hill	William C. Coker
Hendricks, Caroline Frances	1942 (1943)	Observations on the development and cytology of *Blastocladiella cystogena*	John N. Couch
Henry, Nell Quinby	1932	The seed-development in *Cuscuta arvensis*	William C. Coker
Hewitt, Willard Chappell	1939	Seed development of *Lobelia amoena*	William C. Coker
Hinton, William Fred	1968	The taxonomic status of *Physalis lanceolata* in the Carolina Sandhills	C. Ritchie Bell
Hobbs, Jesse Harrison	1950	Studies in *Rhizidiomyces apophysatus* Zopf	John N. Couch
Horton, James Heathman	1957 (1958)	A vascular flora of Rowan County, North Carolina	Albert E. Radford
Houck, Duane Francis	1956	Flood plain flora of the Deep River Triassic Basin	Albert E. Radford
Huck, Robin Bovard	1979	Flora, vegetation and soils of the Bull Creek Watershed, Osceola County, Florida	Albert E. Radford
Huneycutt, Maeburn Bruce	1948 (1949)	Keratinophilic Phycomycetes. I. A new genus of the Saprolegniaceae	John N. Couch
Irwin, Julie English	1975	Structure of stump communities in a stream affected by thermal effluent	J. Frank McCormick
Jeffreys, Donald Bearss	1954 (1955)	Tropisms and responses to growth substances of *Coprinus sterquilinus*	John N. Couch, Victor A. Greulach

MASTER OF ARTS LIST (CONTINUED)

Name	Date	Title	Major Advisor
Johns, Robert Marvin	1952	A morphological study of *Gonapodya polymorpha*	John N. Couch
Johnson, Sidney Pickette	1952 (1953)	Effects of maleic hydrazide on the growth of *Avena* seedlings	Victor A. Greulach
Katz, Barry Allan	1973 (1974)	Phenology and enzyme activity of leaf litter inhabiting micro-fungi in the Blackwood division of the Duke Forest, North Carolina	Helmut Lieth
Kavookjian, Anahid Maleka-Anne	1976	Antigenic properties of the Golgi derived scales of *Pleurochrysis sherffelii*	R. Malcolm Brown, Jr.
Kessler, Peggy-Ann	1956	A floristic study of the Triassic sediments, Deep River Coal Field, North Carolina	Albert E. Radford
Kirk, Daniel Eddins	1950	The development of the ovule and megagametophyte in *Aralia chinensis* L.	J.E. Adams
Koch, William Julian	1950	A study on the motile cells of *Vaucheria*	John N. Couch
Kondo, Katsuhiko	1971	A comparison of variability in *Utricularia cornuta* Michx. and *Utricularia juncea* Vahl.	C. Ritchie Bell
Kowalczyk, Bruno Florian	1973	The pollination ecology of *Hedysarum alpinum* L. var. *americanum* (Mchx.) and *H. boreale* Nutt. var. *mackenzii* (Richards.) C. L. Hitchc. in the Kluane Lake area of the Yukon Territory, Canada	C. Ritchie Bell
Laing, Henrietta	1957	A spring and fall flora of Harnett County, North Carolina	Albert E. Radford
League, Elizabeth Anne	1954	Effects of temperature and light duration on reproduction in *Vaucheria sessilis*	Victor A. Greulach
Leitner, Jane Gage	1938	New species of *Achlya* and *Apodachlya*	William C. Coker
Lindsey, Anne Haskell	1975	Classification versus characterization in *Thaspium* and *Zizia* (Apiaceae)	C. Ritchie Bell
Longest, Pauline Moser	1947	The structure of the flagella in certain algae and fungi and its bearing on classification	John N. Couch
Marshall, Melissa Page	1977	A vascular flora of Bennington County, Vermont	Albert E. Radford
Mathews, Andrew Clark	1931	The seed-development in *Pinus palustris* (long leaf pine)	William C. Coker
Matthews, Velma Dare	1927	Studies on Saprolegniaceae and the genus *Pythium*	William C. Coker
McCurdy, Yolande de Boisferon	1975 (1976)	An ecosystematic analysis of the vegetation of Hanging Rock State Park, Danbury, North Carolina	Albert E. Radford

Master of Arts List (Continued)

Name	Date	Title	Major Advisor
McKay, John Archibald	1912	Cleistogamy	William C. Coker
McNeely, Jean Alexander	1960 (1961)	A floristic study of the Lake Norman area in Iredell and Catawba Counties	Albert E. Radford
Mehrhoff, Loyal Archie, III	1980	The reproductive biology of the genus *Isotria* (Orchidaceae) and the ecology of *Isotria medeoloides*	C. Ritchie Bell
Meyer, Kenneth Albert	1969	Seasonal fluctuation of phytoplankton composition, diversity and production in a freshwater lake	J. Frank McCormick
Michael, Jerry Lee	1969	The vascular flora of Bullhead Mountain, Alleghany County, North Carolina	Albert E. Radford
Miller, Charles Edward	1954	A new member of the aquatic Synchytriaceae	John N. Couch
Miller, Gary Lee	1965	The vegetation of the serpentine barrens in southern Lancaster County, Pennsylvania	Albert E. Radford
Morgan, Judy Tate	1962	A vascular flora of Morrow Mountain State Park, Stanly County, North Carolina	Albert E. Radford
Morton, Cynthia Mareen	1989	Comparative pollen morphology of the Styracaceae	William C. Dickison
Nesom, Guy Lane	1970	Variation in *Erigeron flagellaris* Gray and its presumed relatives endemic to southwestern Utah	C. Ritchie Bell
Nesom, Margaret Green	1969	Observations on the distribution of certain aquatic Phycomycetes in granite outcrop soil	Clyde J. Umphlett
O'Kelley, Joseph Charles	1948	A study of the effects of mistletoe on sugar maples	J.E. Adams
Olive, Lindsay Shepherd	1940	Studies in the genus *Gymnosporangium*	William C. Coker
Olson, Richard Kenneth	1975	The effects of simulated acid precipitation on nutrient cycling in granite outcrop ecosystems	J. Frank McCormick
Orr, Harold Eugene	1941	Gametogenesis and fertilization in *Pythium pulchrum*	John N. Couch
Otte, Deborah Kay Strady	1977 (1978)	The pollination ecology of *Hexastylis arifolia* (Michx.) Small var. *arifolia* and *H. minor* (Ashe) Blomquist (Aristolochiaceae) in the area of Chapel Hill, North Carolina	C. Ritchie Bell
Parker, Helen Moore	1946	Studies in the nutrition of some aquatic Myxomycetes	John N. Couch
Parrish, Mary Jo	1950	Effects of irradiation on meiosis and pollen grain development in a diploid hyacinth	Earlene Atchison

Master of Arts List (Continued)

Name	Date	Title	Major Advisor
Patterson, Paul Morrison	1927	Fertilization and oogenesis in *Achlya colorata*	William C. Coker
Patton, Ernest Gibbes	1950	Influence of flax root by-products on higher plants	J.E. Adams
Pendergrass, William Roper	1949	Studies on a plasmodiophoraceous parasite, *Octomyxa brevilegniae*	John N. Couch
Plyler, Daniel Belk	1962	The effects of maleic hydrazide and related compounds on plant growth	Victor A. Greulach
Qualls, Deborah Ann	1984	A revision of the new world species of *Lindernia allioni* (Scrophulariaceae)	Jim Massey
Radford, John Stewart	1971	Phenological delimitation of the growing season in North Carolina: Computer mapping and correlation of spring, fall and length of growing season data with environmental parameters	Helmut Lieth
Raper, John Robert	1936	Heterothallism and sterility in *Achlya* and observations on the cytology of *Achlya bisexualis*	William C. Coker, John N. Couch
Read, Mildred Conner	1950	An anatomical study of *Carpinus caroliniana* infected with *Septobasidium pseudopedicellatum*	J.E. Adams
Register, Thomas Eugene	1959 (1960)	Morphological variation in a new species of *Phlyctochytrium*	William J. Koch
Ritchie, Donald Dirk	1937	The morphology of the perithecium of *Sordaria fimicola* (Rob.) Ces. and De Not.	John N. Couch, William C. Coker
Rogers, Judith Lee	1965	Variation studies on three *Cuscuta* species of subsection Arvenses in the Southeast	C. Ritchie Bell
Rury, Phillip MacDonald	1976	Leaf venation patterns of the genus *Hibbertia* Andr. (Dilleniaceae) and their systematic and ecophyletic significance	William C. Dickison
Ryland, Alice Garnett	1947	A cytological study of the effects of colchicine, indole-3-acetic acid, potassium cyanide, and 2, 4-D on plant cells	Earl H. Newcomer
Sawyer, George Perry, Jr.	1968 (1969)	The vascular flora of William B. Umstead State Park, Wake County, North Carolina	Albert E. Radford
Schadel, William Edward	1978	Leaf anatomy and venation patterns of the Styracaceae	William C. Dickison
Scholz, Ruth	1934	Studies in the Leguminosae of North Carolina: Part I. Some diverse forms and peculiarities found in certain species of the family Leguminosae; Part II. The structure and development of the seed of *Strophostyles umbellata* (Muhl.) Britton	William C. Coker, Henry Roland Totten

MASTER OF ARTS LIST (CONTINUED)

Name	Date	Title	Major Advisor
Schue, Janette Ellen	1995	The effects of nutrient and light levels on the development of size inequality in monospecific stands of *Cosmos bipinnatus*	Robert K. Peet
Seaman, William Daniel	1969 (1970)	A study of the vascular flora of Mecklenburg County, Virginia	Albert E. Radford
Sears, Michael Novy	1966 (1967)	A floristic study of the limestone along Island Creek in Jones County, North Carolina	Albert E. Radford
Setzer, Elsie Graham	1941 (1942)	Sporogenesis in *Isoachlya anisospora* (de Bary) Coker	William C. Coker
Shanor, Leland	1937	Observations on the development and cytology of the sexual organs of *Thraustotheca clavata* (de Bary) Humph.	William C. Coker, John N. Couch
Sharp, Douglas Dwight	1974 (1975)	The phytosociology of the weed communities of Scotland County, North Carolina	Helmut Lieth
Sherwin, Helen Shedd	1944	Some wild yeasts found around Chapel Hill	John N. Couch
Shew, Harry Wayne	1974	Induced mitotic recombination in *Neurospora crassa*	Edward G. Barry
Silliman, Frances Ernestine	1955	A study of a colony of hybrid oaks among the white oak group	J.E. Adams
Slifkin, Miriam Kresses	1964 (1965)	Studies in the parasitism of *Olpidiopsis incrassata* on members of the Saprolegniaceae	John N. Couch
Smith, Budd Elmon	1934	A taxonomic and morphological study of the genus *Cuscuta*, dodders, in North Carolina	William C. Coker
Smith, Craig Allyn	1982	Cell wall biosynthesis by carrot protoplasts	R. Malcolm Brown, Jr.
Snider, Jerry Allen	1969 (1970)	A variation study of *Agalinis decemloba* (Greene) Pennell and *Agalinis obtusifolia* Raf. in the Carolinas	C. Ritchie Bell
Snyder, James Robert	1978 (1979)	Analysis of vegetation in the Croatan National Forest, North Carolina	Helmut Lieth
Sobel, Kathy Cochran	1974 (1975)	Systematics of three species of *Ranunculus* (Ranunculaceae)	Albert E. Radford
Spear, Ray William	1976	Late Quaternary vegetation change in southwestern New York State	Norton G. Miller
Stacy, Emily Mildred	1952	Studies in the ontogeny of the vessel	J.E. Adams
Stewart, Laurie Marguerite	1937	Studies in the life history of *Nitella hyalina* Agardh	William C. Coker
Stewart, Mary Josephine	1946	Studies on *Rhizophidium coronum* Hanson	John N. Couch

MASTER OF ARTS LIST (CONTINUED)

Name	Date	Title	Major Advisor
Stewart, Shelton Eugene	1958 (1959)	Early spring and summer vascular flora of Lee County, North Carolina	Albert E. Radford
Suiter, Karl Arthur	1984	Genetics of alloenzyme variation in *Gossypium arboreum* L. and *Gossypium herbaceum* L. (Malvaceae)	Clifford R. Parks
Suttie, Janet Louise	1982	A new model of native cellulose biogenesis to integrate structure and function in *Acetobacter xylinum*	R. Malcolm Brown, Jr.
Szaniszlo, Paul Joseph	1964	A study of the effects of light and temperature on oogonia formation in *Saprolegnia diclina* Humphrey	John N. Couch, Max H. Hommersand
Thorburn, Gary Warren	1980	Vertical crown distributions in hardwood stands of the North Carolina Piedmont	Robert K. Peet
Thrainsson, Sigurdur Armann	1986	Morphological studies on some members of the family Cystocloniaceae	Max H. Hommersand
Tingle, Constance Louise	1972	Some physiological aspects of oogonia formation in *Saprolegnia diclina*	William J. Koch
Titman, Paul Wilson	1948	Studies in the woody anatomy of the family Nyssaceae	J.E. Adams
Totten, Henry Roland	1914	I. Development of the seed in *Mayaca Aubleti*; II. Development of the seed in the cleistogamic flower of *Specularia perfoliata (With Mr. J. A. McKay)*	William C. Coker
Tucker, Gary Edward	1967	The vascular flora of Bluff Mountain, Ashe County, North Carolina	Albert E. Radford
Vardell, Mary Linda	1933	Seed development in *Spigelia marilandica* L.	William C. Coker
Ward, Mary Williams	1939	Studies in aquatic fungi: I. Observations on a new species of *Thraustotheca*; II. Observations on *Rhizophlyctis rosea* (de Bary and Woronin), Fischer	William C. Coker, John N. Couch
Ward, Rebecca	1930	Seed development in *Hibiscus esculentus*	William C. Coker
Weakley, Allison Elizabeth	2002	Evolutionary relationships within the genus *Philadelphus* (Hydrangeaceae): A molecular phylogenetic and biogeographic analysis	Clifford R. Parks
Wells, Edna Metz	1936	The structure and development of the seed in *Cimicifuga racemosa*	William C. Coker

Master of Arts List (Continued)

Name	Date	Title	Major Advisor
Wells, Elizabeth Fortson	1970 (1971)	A vascular flora of the Uwharrie Wildlife Management Area, Montgomery County, North Carolina	Albert E. Radford
Westra, Raymond Nelson	1965	Seed dormancy in *Uniola paniculata* L.	W.H. Loomis
Whiffen, Alma Joslyn	1939	The cytology of *Octomyxa achlyae*	John N. Couch
Whitlock, Patricia Anne	1960	Studies in the distribution and taxonomy of some members of the Actinoplanaceae of North Carolina	John N. Couch
Wilburn, Hampton Irene	1940	Seed and fruit development in *Smilax*	J.E. Adams
Williams, Delbert Erman	1974	A chemotaxonomic survey of four genomes of Gossypium	Clifford R. Parks
Wilson, Elizabeth Jean	1977 (1978)	A floristic study of the "savannahs" on pine plantations in the Croatan National Forest	Albert E. Radford
Wojcik, Violet Hicks	1968	Possible divergent evolution in *Helenium amarum* populations	J. Frank McCormick
Wyatt, Raymond Louis	1954	A biological and embryological study of four species of *Asarum* (Aristolochiaceae)	Carroll E. Wood, Jr.
Wyman, Patsy Marie	1977 (1978)	Attachment of *Agrobacterium* to plant tissue culture cells	Ann G. Matthysse
Ziegler, Arthur William	1941	Gametogenesis of *Achlya recurva* Cornu	William C. Coker, John N. Couch
Zollner, Jennifer Dow	1977	Variation in the developmental vegetative anatomy among the Gigartinaceae (Gigartinales, Rhodophyta)	Max H. Hommersand

Master of Science List

Name	Date	Title	Major Advisor
Adams, Nicholas Stephen	2012	A synthesis of rights-of-way native plant communities: Identifying their relevance to historical and contemporary Piedmont savannas	Robert K. Peet, Johnny Randall, Alan S. Weakley
Aune, Patricia	1989	Some aspects of epiplaytic and parasitic red algae	Max H. Hommersand
Avis, Peter Glen	1998	The effects of nitrogen and phosphorus additions on the mycorrhizal inoculum potential of an oak-maple forest in Mountain Lake, Virginia, USA	Robert K. Peet

MASTER OF SCIENCE LIST (CONTINUED)

Name	Date	Title	Major Advisor
Baker, William Lawrence	1980	Alpine vegetation of the Sangre de Cristo Mountains, New Mexico: Gradient analysis and classification	Robert K. Peet
Barnett-Laurence, Matthew S.	1992	On *Protobarinophyton ananier*	Patricia G. Gensel
Belcher, James	1985	Observations on a new genus of zosterophylls from the Devonian (Emsian) of New Brunswick	Patricia G. Gensel
Bennett, Walter Edward	1981	Not seen	Lindsay S. Olive
Binkley, Steven Wilson	1972	Not seen	William J. Koch
Bond, Harold Alonzo	1970	An ecological study of *Coelomomyces* in the Piedmont area of North Carolina	John N. Couch
Boufford, David Edward	1975 (1976)	Not seen	Albert E. Radford
Boyer, Marjorie Wright	1986	Floral phenology and reproductive systems in two species of *Oxypolis* (Umbelliferae)	C. Ritchie Bell
Brainerd, Sandra Lynn	2000	A morphological analysis of specimens of *Kavlangiophyton* from Gaspe Bay, New Brunswick, CAN	Patricia G. Gensel
Bryant, William Thomas	1974	Not seen	Albert E. Radford
Council, Orin Pete	1980	Changes in biomass and production of hardwood forests of the North Carolina Piedmont during secondary succession	Robert K. Peet
Courtney, Deborah Gotwalt	1970	Phenology in Duke Forest ecology	Helmut Lieth
Curtis, Marilou Barns	1978	Study of variation of *Betula lenta*	Clifford R. Parks
DiNome, Anthony Joseph	1980	Freeze fracture study of cell wall biogenesis in the desmids	R. Malcolm Brown, Jr.
Dopyera, Caroline E.J.	1994	Vegetative stem anatomy in a trimerophyte from the Emsian of New Brunswick	Patricia G. Gensel
Doyle, James Dennis	1981	The comparative water relations of three varieties of red maple (*Acer rubrum* L.)	Robert K. Peet
Dudley, Edith Louise	1973 (1974)	Not seen	Norton G. Miller
Ellis, Elizabeth Ann	1972	Not seen	R. Malcolm Brown, Jr.

MASTER OF SCIENCE LIST (CONTINUED)

Name	Date	Title	Major Advisor
Frantz, Vonda Lee	1984 (1985)	Reproductive biology of the Atlantic Coastal Plain endemic, *Lysimachia asperulaefolia* (Primulaceae)	C. Ritchie Bell
Gerbeth, Susan Kay	1980	The major flavonoids of *Betula lenta*	Clifford R. Parks
Gerschutz, Andrew David	2009	Changes in old-growth and second-growth eastern hemlock (*Tsuga canadensis*) communities in the Southern Appalachians after the introduction of the hemlock wooly adelgid (*Adelges tsugae*)	Robert K. Peet
Gilmour, Ronald Wayne	1996	Reevaulation of putative rhyniophytes from the Lower Devonian of Gaspe, Quebec	Patricia G. Gensel
Gleaves, Charles Taylor	1979	An investigation of hybridization between two birches in North Carolina	C. Ritchie Bell
Gramling, Andrea Elizabeth	2006	A conservation assessment of *Packera millefolium*, a Southern Appalachian endemic	Robert K. Peet
Hansen, Gayle Ingrid	1976	The morphology and life history of *Cirrulicarpus carolinensis* sp. nov. (Kallymeniaceae, Cryptonemiales)	Max H. Hommersand
Harris, Anne Fleishel	1975	Not seen	William C. Dickison
Higginbotham, Jeri Welsh	1982 (1983)	Pollen grain diameters of the Southern Appalachian birches	Clifford R. Parks
Hochman, Elissa Rachel	2004	Lower Roanoke River hydroperiods: Altered hydrology and implications for forest health and species response	Robert K. Peet
Hohn, Janet Elizabeth	1968	Not seen	J.E. Adams
Horn, James William	1997	Morphology and anatomy of the Sphaerosepalaceae and Diegodendraceae	William C. Dickison
Howard, Allen Richmond, III	1986	Genetic control of phosphoglucose isomerase and triose phosphate isomerase in *Acer rubrum* L.	Clifford R. Parks
Howell, Linda	1977	1) Stromatolites. 2) The anatomy of *Ophioglossum crotalophoroides*	Max H. Hommersand
Huffman, Anna K.	1978	Homology between plasmids from various *Agrobacterium* strains	Ann G. Matthysse
Israel, Kimberly Anne	2011	Vegetation change in the Duke Forest, 1977–2010	Robert K. Peet
Knight, Alan Lee	1976 (1979)	Host range and temperature requirements of *Culicinomyces clavosporus*	Lindsay S. Olive

MASTER OF SCIENCE LIST (CONTINUED)

Name	Date	Title	Major Advisor
Knott, Mark David	2001	Prescribed burning activity in remnant longleaf pine (*Pinus palustris*) ecosystems of North and South Carolina: Patterns, influencing factors, and policy recommendations	Robert K. Peet
Kolo (Hule), Lisa Pecquet	1993 (1994)	Hormone-responsive genes of *Agrobacterium tumefaciens*	Ann G. Matthysse
Kuruc, John Joseph	1973	Not seen	J. Frank McCormick
Lentz, Constance Bouricius	1977	Study of potassium uptake by normal and crown gall tissue culture cells of *Vinca rosea*	Ann G. Matthysse
Leonard, Steven Worth	1973	Not seen	Albert E. Radford
Lightfoot, Richard Thomas	1990 (1991)	Characterization of a genomic fragment of *Agrobacterium* DNA required for cellulose synthesis	Ann G. Matthysse
Lobdell, Ellen York	1974	Not seen	C. Ritchie Bell
Luxner, Laurie Ann	1977	Morphology and taxonomy of marine red algae	Max H. Hommersand
Martin, David Lance	1974	Not seen	Albert E. Radford
May, Mary Love	1981 (1982)	Vegetation seed reserve and environmental relationships in the sandhills, Sandhills Wildlife Management Area, Hoffman, North Carolina	Robert K. Peet
McChesney, Jasper	2010	Realizing the niche's breadth: Inferring ecological process with species generalism	Robert K. Peet
McDougal, Karen McAlpin	1979	An analysis of variation in leaf flavonoids of *Quercus rubra* over an altitudinal gradient	Clifford R. Parks
McGrady, Stephen Dwight	1987 (1988)	Observations on the bark of Cunoniaceae	William C. Dickison
Montgomery, Elizabeth	1974 (1975)	Not seen	Norton G. Miller
Moore, Julie	1976	Reproductive potential of the hemiparasite *Agalinis purpurea* (L.) Pennell	C. Ritchie Bell
Morgan, Sharon Webb	1980	Species General Information System: Species, population, habitat, and threat inventory; and species status summary for *Geum radiatum* Michaux	Albert E. Radford
Nelson, Mary Ann Pritchard	1975	Not seen	William C. Dickison
Overholser, Linda Ellen	1974	Not seen	Tom K. Scott

Master of Science List (Continued)

Name	Date	Title	Major Advisor
Parent, Josee Marie	1986	Flagellar sexual agglutins in *Chlamydomonas*	Max H. Hommersand
Pearce, Robert Elliott	1980	Polarographic studies of photosynthesis in *Elodea*	Max H. Hommersand
Plumblee, Barbara Zim Taylor	1979	Wood anatomy of *Rhododendron*	William C. Dickison
Pritchard, Mary Ann	1975	Not seen	unknown
Raisher, Eric Lawrence	1978 (1979)	Cytogenetics of *Neuospora crassa* meiotic behavior in translating strain AR 179	Edward G. Barry
Rao, Balakrishna	1971	Not seen	William J. Koch
Raubeson, Linda Ann	1986 (1987)	The conifer *Androvettia carolinensis* Berry from the Upper Cretaceous Black Creek Formation of North Carolina with comments on other elements of the flora	Patricia G. Gensel
Richards, Anne Marie	1983 (1984)	Herb distribution and microtopography in a northern hardwoods forest	Robert K. Peet
Rohrer, Joseph Raphael	1979 (1980)	The flora and vegetation of the Hanging Rock area, Avery and Watauga Counties, N.C.	Albert E. Radford
Roney, Gwen	1978	Apical dominance in seedlings shoots	Tom K. Scott
Sargent, Karen Gawthrop	1976 (1977)	Preliminary investigation into biological control of *Limnoria* using *Fusarium solani*	Aristotle Domnas
Saterson, Kathryn Ann	1982	Fine root biomass and nutrient cycling in *Aristida stricta* on a North Carolina coastal plain savannah	Peter M. Vitousek
Sather, James Dawson	1982	An ecosystematic analysis of selected areas of the Hot Springs Valley, Madison County, North Carolina	Albert E. Radford
Saxe, Bradford Eliott	1992 (1994)	Ribosomal DNA and plant phylogenetics—Its sequence divergence in *Platanus*	Clifford R. Parks
Schell, Wiley Alexander	1979	Technique for isolating fungi from blood specimens	Lindsay S. Olive
Schram, Lynn Alan	1975 (1976)	Not seen	Albert E. Radford
Scoville, Dorothy Allard	1981 (1982)	Herb and vine vegetation of single tree windthrow gaps in lowland forests of North Carolina	Robert K. Peet, Peter S. White
Seaton, Kathryn Paradiso	1981	A natural heritage study of Larkspur Ridge Natural Area, Halifax County, North Carolina	Clifford R. Parks

Master of Scienc List (Continued)

Name	Date	Title	Major Advisor
Senft, Amanda Ruth	2009	Species diversity patterns at ecotones	Robert K. Peet
Senior, Laura Bruns	1981	Study of the mycorrhizal organs of *Tipularia discolor*, the crippled crane fly orchid	William J. Koch
Seymour, Stephanie Denise	2011	Vegetation of non-alluvial wetlands of the southeastern Piedmont	Robert K.Peet, Alan S. Weakley
Sheftall, William Lowe	1980	A natural resources inventory of the research natural area Hitchiti Experimental Forest in Jones County, Georgia, Southeastern Forest Experiment Station, United States Forest Service	Albert E. Radford
Simpson, Michael George	1979	Anatomy and floral vasculature of *Lachnanthes* and *Lophiola* (Haemodoraceae)	William C. Dickison
Stiefel, Sonja Maria	1994	The food forest: The eastern deciduous forest as a model for a sustainable perennial polyculture	Robert K. Peet
Stuart, Charles Macaulay	1993	A preliminary analysis of internal transcribed sequence of ribosomal DNA from three *Camellia* and one *Gordonia* species	Clifford R. Parks, Mark W. Chase
Sullivan, Clare Moreen	1971	Not seen	Max H. Hommersand
Taverna, Kristin	2004	Mature hardwood forests of the central Piedmont of North Carolina: Landscape distribution and understory change	Robert K. Peet, Peter S. White
Thorburn, Patricia Lois Cooley	1978 (1979)	A vegetational analysis of the Cronn tract, Eastford, Connecticut	Albert E. Radford
Tingley, Carol Ann	1985 (1986)	An ecosystematic survey of floodplain vegetation at the mouth of the Roanoke River	Albert E. Radford
Trant, Charles Amon, Jr.	1983 (1984)	A new species of *Psilophyton* with a study of the variation of branching and its evolutionary significance	Patricia G. Gensel
Upson, James Robert, Jr.	1973	Not seen	J. Frank McCormick
Wang, Chi-Ann	1976	Not seen	Ann G. Matthysse
Wang, Yongan	1991	Manual of trees and shrubs distributed in warm temperate zones of China	Clifford R. Parks
Wendel, Jonathan Fox	1980	Genetic control of isozyme variation in *Camellia japonica* L. (Theaceae)	Clifford R. Parks
Wickland, Diane Elizabeth	1977	The relationship between the vegetation and soils of abandoned copper, lead and zinc mines in the Piedmont of North Carolina	Helmut Lieth

Master of Science List (Continued)

Name	Date	Title	Major Advisor
Williams, David Charles	1974 (1975)	Not seen	Albert E. Radford
Willis, Patricia Mary	1971	Not seen	R. Malcolm Brown, Jr.
Windham, Marcia Lee	1970	Not seen	Lindsay S. Olive
Wolf, Joan Walker	1980	Production and diversity in southeastern Coastal Plain savannas	Robert K. Peet
Wood, Emily Walker	1975 (1976)	Not seen	Albert E. Radford
Zhou, Shubo	1988 (1989)	Phytochrome-mediated swelling of etiolated-maize-leaf protoplasts and its possible biological significance	Tom K. Scott
Zucker, Ira Irwin	1975	Not seen	unknown

APPENDIX 1.
A LIST OF NAMES OF STUDENTS WHO ENROLLED IN COURSES IN BOTANY AND HORTICULTURE, INCLUDING YEAR AND TERM TAKEN, COURSE OR SUBJECT, AND GRADE RECEIVED, 1876–1904.[118]

Student	Year	Term	Course	Grade
ADAMS, WILLIAM GALES	1880	SPRING	BOTANY	63
ADAMS, WILLIAM JACKSON	1879	SPRING	BOTANY	90
ALBRITTON, JAMES ASHBY	1892	SPRING	BOTANY	83
ALDERMAN, EDWIN ANDERSON	1880	SPRING	BOTANY	96
ALDERMAN, WILLIAM COLIN	1878	SPRING	BOTANY	62
ALEXANDER, CHARLES TAYLOR	1882	SPRING	BOTANY	93
ALLARD, HARRY ARDELL	1903	SPRING	BOTANY	1
ALLARD, HARRY ARDELL	1903	FALL	BOTANY	2
ALLARD, HARRY ARDELL	1904	SPRING	BOTANY	2
ALLARD, HARRY ARDELL	1904	FALL	BOTANY	2
ALLEN, ALVIS WALDO	1879	SPRING	BOTANY	0
ALLEN, ALVIS WALDO	1880	SPRING	BOTANY	80
ALLEN, WILLIAM MARSHALL	1889	SPRING	BOTANY (BIOLOGY 4)	82
ALSTON, HOWARD	1890	SPRING	BOTANY	78
AMICK, THOMAS CICERO	1888	SPRING	BOTANY	94
ANDREWS, ALEXANDER BOYD, JR.	1891	SPRING	BOTANY	76
ANGIER, JOHN CICERO	1878	SPRING	BOTANY	77
ARMFIELD, EUGENE MOREHEAD	1888	FALL	HORTICULTURE	90
ARMFIELD, EUGENE MOREHEAD	1889	SPRING	HORTICULTURE	88
ASHE, WILLIAM WILLARD	1890	SPRING	ADVANCED BOTANY	NO GRADE
ASHE, WILLIAM WILLARD	1890	SPRING	BOTANY	99
ASHE, WILLIAM WILLARD	1890	FALL	ADVANCED BOTANY	98
ASHE, WILLIAM WILLARD	1891	SPRING	ADVANCED BOTANY	98
ASKEW, CHARLES THOMPSON	1876	SPRING	BOTANY	91
ATKINSON, HUGH HAMILTON	1893	SPRING	ZOOLOGY & BOTANY	6
ATKINSON, WADE HAMPTON	1886	SPRING	BOTANY	90
AVERY, JOHN MOREHEAD	1880	SPRING	BOTANY	96

[118] Notes to Appendix 1. Information in this appendix was obtained from Student Records & Faculty Reports, Series 1:1, Academic Records, University Archives #40131, Wilson Special Collections Library, University of North Carolina, Chapel Hill, vol. 4, Fall 1875–Spring 1885; vol. 5, Fall 1882–Spring 1897; vol. 6, Fall 1893–Fall 1902; and vol. 7, Fall 1898–Spring 1903. Some data regarding years were extrapolated from the records as accurately as possible.

The grades shown in the right-hand column reflect two systems of grading: grades represented in percentages and grades denoted 1–6. The latter system, which was implemented in 1891, is described in chapter 9, pages 350–351.

Appendix 1. (Continued) A List of Names of Students Who Enrolled in Courses in Botany and Horticulture

Student	Year	Term	Course	Grade
Aycock, Charles Brantley	1879	Spring	Botany	75
Bahnson, George William	1887	Fall	Horticulture	none cited
Baker, Thomas Atkinson	1884	Spring	Botany	80
Barlow, Thadeus Batts	1876	Spring	Botany	62
Barnard, Alfred Smith	1891	Spring	Botany	93
Batchelor, Francis Howard	1891	Spring	Botany	97
Batchelor, Oliver Douglas	1885	Spring	Botany	91
Battle, Gaston	1886	Fall	Horticulture	85
Battle, Gaston	1887	Spring	Botany	86
Battle, Gaston	1887	Spring	Horticulture	89
Battle, George Gordon	1882	Spring	Botany	97
Battle, Henry Lawrence	1880	Spring	Botany	86
Battle, Herbert Bemerton	1880	Spring	Botany	99
Battle, Kemp Plummer, Jr.	1877	Spring	Botany	approved
Battle, Kemp Plummer, Jr.	1878	Spring	Botany	94
Battle, Lewis Junius	1884	Spring	Botany	95
Battle, Thomas Hall	1878	Spring	Botany	99
Beall, Frank Harper	1889	Spring	Botany	85
Beall, John McCulloch	1884	Spring	Botany	45
Beaman, John Robinson, Jr.	1879	Spring	Botany	80
Bell, David	1877	Spring	Botany	approved
Bellamy, John Dillard	1889	Spring	Botany	98
Bellamy, John Dillard	1890	Spring	Advanced Botany	95
Bellamy, Spier Coffield	1880	Spring	Botany	n[o] exam
Bennett, John Tyler	1888	Spring	Botany	70
Benson, James William	1886	Spring	Botany	80
Benton, John Hogan	1879	Spring	Botany	0
Betts, William Archibald	1878	Spring	Botany	96
Blackman, Neill Robert	1898	Spring	Botany	2
Blankenship, James Moutton	1890	Spring	Botany	88
Blount, John Gray	1889	Spring	Botany	98
Bobbitt, James Henry	1882	Spring	Botany	77
Bond, Charles	1876	Spring	Botany	71
Borden, John Lemuel	1882	Spring	Botany	88

APPENDIX 1. (CONTINUED) A LIST OF NAMES OF STUDENTS WHO ENROLLED IN COURSES IN BOTANY AND HORTICULTURE

Student	Year	Term	Course	Grade
BOYDEN, VICTOR HUGO	1891	SPRING	BOTANY	70
BRASWELL, ARCHELAUS, JR.	1886	SPRING	BOTANY	82
BRASWELL, JAMES CRAIG	1889	SPRING	BOTANY	88
BRASWELL, JAMES CRAIG	1890	SPRING	ADVANCED BOTANY	96
BRASWELL, MACK CLAUDE	1880	SPRING	BOTANY	85
BRASWELL, MARK RUSSELL	1884	SPRING	BOTANY	78
BRIGHT, [FIRST NAME NOT CITED]	1897?	FALL	HORTICULTURE	78
BROOKS, JOHN	1884	SPRING	BOTANY	N[O] EXAM
BROOKS, THOMAS COOK	1877	SPRING	BOTANY	APPROVED
BROWN, ZENO	1880	SPRING	BOTANY	70
BRYAN, JAMES ALEXANDER	1884	SPRING	ADVANCED BOTANY	80
BRYAN, ROBERT THOMAS	1880	SPRING	BOTANY	98.5
BRYANT, VICTOR SILAS	1889	SPRING	BOTANY	92
BRYANT, VICTOR SILAS	1890	SPRING	ADVANCED BOTANY	98
BURCH, CHARLES HILL	1889	SPRING	BOTANY	70
BURNETT, JOSEPH HALSEY	1884	SPRING	BOTANY	N[O] EXAM
BURWELL, WILLIAM ARMISTEAD	1882	SPRING	BOTANY	88
BURWELL, ROBERT TURNBULL	1885	SPRING	BOTANY	88
BUTLER, GEORGE EDWIN	1888	SPRING	BOTANY	85
BUTLER, HENRY WILEY	1892	SPRING	ZOOLOGY & BOTANY	4
BYNUM, EDWARD TURNER	1876	SPRING	BOTANY	81
CALDWELL, ROBERT ERNEST	1876	SPRING	BOTANY	85
CARROLL, WILLIAM HOUSTON	1884	SPRING	BOTANY	90
CATES, CALEB GILMER	1886	SPRING	BOTANY	78
CHERRY, JESSE GLASGOW	1876	SPRING	BOTANY	68
CHERRY, TILMAN BROWN	1882	SPRING	BOTANY	N[O] EXAM
CLEGG, THOMAS DAVID	1885	SPRING	BOTANY	89
CLIFTON, LUNSFORD CLAYBRONE	1877	SPRING	BOTANY	APPROVED
CLINE, EDWARD BOST	1884	SPRING	BOTANY	75
CLINE, FRANK LEE	1879	SPRING	BOTANY	87
COBLE, ALBERT LUCIEN	1878	SPRING	BOTANY	97
COLLETT, WAITS AVERY	1892	SPRING	ZOOLOGY & BOTANY	4
COLLINS, PLATO	1890	SPRING	BOTANY	92
COLLINS, PLATO	1891	SPRING	ADVANCED BOTANY	70

Appendix 1. (Continued) A List of Names of Students Who Enrolled in Courses in Botany and Horticulture

Student	Year	Term	Course	Grade
Cooper, Thomas Jefferson	1890	Spring	Botany	4
Cornelius, William Henry	1885	Spring	Botany	70
Costner, Robert Edwin	1885	Spring	Botany	80
Covington, Charles Coleman	1876	Spring	Botany	95
Covington, Frank Leak	1888	Spring	Botany	99
Covington, Frank Leak	1889	Spring	Advanced Botany	95
Covington, Frank Leak	1889	Fall	Advanced Botany	90
Covington, Frank Leak	1890	Fall	Advanced Botany	approved
Covington, John Mallory	1889	Spring	Botany	78
Covington, Thomas Thomas	1880	Spring	Botany	90
Cox, Beriah Thaddeus	1885	Spring	Botany	77
Cox, Charles Gregory	1880	Spring	Botany	60
Cox, Joseph Brown	1886	Spring	Botany	72
Cox, Pierre Bayard	1884	Spring	Botany	90
Cox, Pierre Bayard	1885	Spring	Botany	77
Craig, Braxton	1882	Spring	Botany	n[o] exam
Craig, Braxton	1885	Spring	Botany	80
Craig, Locke	1878	Spring	Botany	97
Crowell, George Henry	1888	Spring	Botany	91
Crowell, James Lee	1886	Spring	Botany	82
Cuninggim, Jesse Lee	1890	Spring	Botany	96
Cuningham, John Somerville	1879	Spring	Botany	80
Currie, George Hendon	1889	Fall	Advanced Botany	85
Currie, George Hendon	1889	Spring	Botany	92
Currie, George Hendon	1890	Spring	Advanced Botany	89
Currie, Neill Alexander	1890	Spring	Botany	89
Curtis, Walter Makepeace	1887	Spring	Botany	89
Dale, William Jackson Beau	1885	Spring	Botany	85
Dale, William Jackson Beau	1887	Fall	Horticulture	96
Dale, William Jackson Beau	1888	Spring	Horticulture	93
Dancy, Frank Battle	1879	Spring	Botany	99
Darden, William Edward	1890	Spring	Botany	90
Davidson, Allen Turner, Jr.	1879	Spring	Botany	85
Davies, William Watkins, Jr.	1888	Spring	Botany	90

APPENDIX 1. (CONTINUED) A LIST OF NAMES OF STUDENTS WHO ENROLLED IN COURSES IN BOTANY AND HORTICULTURE

Student	Year	Term	Course	Grade
DAVIS, FREDERICK	1876	SPRING	BOTANY	98
DAVIS, HAYNE	1886	FALL	HORTICULTURE	96
DAVIS, HAYNE	1887	SPRING	HORTICULTURE	95
DAVIS, JAMES EDWARD BECTON	1888	SPRING	BOTANY	82
DAVIS, RICHARD BRYANT	1882	SPRING	BOTANY	86
DAVIS, ROBERT HENRY	1876	SPRING	BOTANY	91
DAVIS, SAMUEL LEE	1891	SPRING	BOTANY	PASSED
DAVIS, SAMUEL LEE	1893	SPRING	BOTANY "REPORTED BY HOLMES"	PASSED
DOCKERY, CLAUDIUS	1885	SPRING	ADVANCED BOTANY	70
DORTCH, WILLIAM THEOPHILUS	1882	SPRING	BOTANY	90
DOWD, JOSEPH CAREY	1878	SPRING	BOTANY	67
DUNSTON, WALTER SEATON	1884	SPRING	BOTANY	88
EASON, ROBERT RANSOM	1888	SPRING	BOTANY	90
EATON, LAFAYETTE BROWNE	1880	SPRING	BOTANY	70
EDGERTON, CHARLES NEWTON	1889	SPRING	BOTANY	70
EDMONSON, WILLIAM ELAM	1885	SPRING	BOTANY	78
EDWARDS, LUTHER BELL	1886	SPRING	BOTANY	70
ELLIS, ALEXANDER CASWELL	1891	SPRING	BOTANY	96
ELLIS, PEARSON	1882	SPRING	BOTANY	N[O] EXAM
ENGELHARD, EDWARD BENSON	1877	SPRING	BOTANY	APPROVED
ENGELHARD, JOHN COTTON	1885	SPRING	BOTANY	60
ERWIN, JOSEPH ERNEST	1885	SPRING	BOTANY	45
ERWIN, WILLIAM EDWIN	1880	SPRING	BOTANY	96
ESKRIDGE, THOMAS JOSEPH	1887	SPRING	BOTANY	88
EURE, MILLS ROBERTS	1888	SPRING	BOTANY	78
FAISON, HENRY ELIAS	1878	SPRING	BOTANY	85
FAISON, JAMES HICKS	1877	SPRING	BOTANY	APPROVED
FARMER, WILLIAM MOSES	1884	SPRING	BOTANY	N[O] EXAM
FARRAR, JOSEPH WILLIAM	1879	SPRING	BOTANY	89
FAUST, CHARLES GEORGE	1887	SPRING	BOTANY	70
FEARING, WOODSON BRADFORD	1876	SPRING	BOTANY	85
FEARINGTON, FREDERICK SHEPPERD	1889	SPRING	BOTANY	75
FERGUSON, HERBERT REEVES	1890	SPRING	BOTANY	NO GRADE RECORDED

APPENDIX 1. (CONTINUED) A LIST OF NAMES OF STUDENTS WHO ENROLLED IN COURSES IN BOTANY AND HORTICULTURE

Student	Year	Term	Course	Grade
Fields, Alpheus	1884	Spring	Botany	78
Fleming, John Martin	1890	Spring	Botany	90
Foreman, Ivey	1882	Spring	Botany	70
Foust, Julius Isaac	1886	Spring	Botany	91
Fuller, Claude Stokes	1890	Spring	Botany	92
Furgerson, Henry Beauregard	1879	Spring	Botany	75
Furgerson, MacMurray	1880	Spring	Botany	73
Gaither, James Frierson	1890	Spring	Botany	93
Gallaway, Charles Wilcher	1877	Spring	Botany	approved
Gatling, Bartholomew Moore	1891	Spring	Botany	80
Gattis, Fred Norwood	1889	Spring	Botany	89
Gattis, Samuel Mallett	1882	Spring	Botany	87
Gidney, Charles Chauncey	1885	Spring	Botany	73
Gill, Edwin James	1884	Spring	Botany	70
Gill, Edwin James	1885	Spring	Botany	75
Gill, Peter Hartwell	1890	Spring	Botany	60
Gill, Thomas Jeffries	1880	Spring	Botany	99
Gilliam, Henry Augustus	1887	Fall	Horticulture	65
Gilliam, Henry Augustus	1888	Spring	Botany	70
Gilliam, Henry Augustus	1888	Spring	Horticulture	75
Gilmer, Ellison Lindsay	1884	Fall	Horticulture	89
Gilmer, John Arthur	1891	Spring	Botany	87
Gilmer, John Lash	1892	Spring	Zoology & Botany	3
Goodman, Erastus Genair	1884	Spring	Botany	90
Graham, George Mordecai	1890	Spring	Botany	92
Graham, Paul Cameron	1890	Spring	Botany	95
Graham, William Alexander	1884	Spring	Botany	90
Grandy, Albert Sidney	1880	Spring	Botany	96
Grandy, Charles Taylor	1884	Spring	Botany	85
Grandy, Luther Bell	1883	Spring	Botany	80
Greene, Herbert Linwood	1884	Spring	Botany	77
Greenlee, Edward Thomas	1879	Spring	Botany	97
Greenlee, Robert Lee	1884	Spring	Botany	80
Gregory, Samuel Babcock	1890	Spring	Botany	85

APPENDIX 1. (CONTINUED) A LIST OF NAMES OF STUDENTS WHO ENROLLED IN COURSES IN BOTANY AND HORTICULTURE

Student	Year	Term	Course	Grade
GRISSOM, ROBERT GILLIAM	1884	SPRING	BOTANY	N[O] EXAM
GRISSOM, ROBERT GILLIAM	1885	SPRING	BOTANY	89
GUNTER, JOHN DOCTOR	1879	SPRING	BOTANY	75
GUYER, SAMUEL JACOB	1890	SPRING	BOTANY	85
HACKETT, RICHARD NATHANIEL	1884	SPRING	BOTANY	88
HAIGH, CHARLES THOMAS	1884	SPRING	BOTANY	78
HAIGH, DELAGNEL	1880	SPRING	BOTANY	88
HAMER, DOUGLAS	1891	SPRING	BOTANY	75
HAMLEN, HUBERT CHESLEY	1889	SPRING	BOTANY	83
HAMMOND, WALTER MONROE	CA. 1889	SPRING	BOTANY	70
HAMMOND, WALTER MONROE	CA. 1888	FALL	HORTICULTURE	88
HAMMOND, WALTER MONROE	CA. 1889	SPRING	HORTICULTURE	80
HAMPTON, DAVID A.	1880	SPRING	BOTANY	73
HARDING, FORDYCE CUNNINGHAM	1891	SPRING	BOTANY	86
HARGRAVE, ALFRED FRONTIS	1877	SPRING	BOTANY	APPROVED
HARGROVE, CHARLES BUDWOOD	1888	SPRING	BOTANY	84
HARPER, FRANCIS MARION	1885	SPRING	BOTANY	85
HARRELL, JAMES JOSEPH	1879	SPRING	BOTANY	0
HARRIS, ANDREW JACKSON	1882	SPRING	BOTANY	90
HARRIS, EUGENE LEWIS	1879	SPRING	BOTANY	99
HARRIS, HUNTER LEE	1886	SPRING	BOTANY	97
HARRIS, JAMES ROBERT	1887	SPRING	BOTANY	74
HARRISON, JOHN ADAMS, JR.	1879	SPRING	BOTANY	80
HAYWOOD, ERNEST	1880	SPRING	BOTANY	97
HEADEN, WILLIAM EDWARDS	1887	SPRING	BOTANY	81
HEDRICK, JAMES DAWSON	1886	SPRING	BOTANY	81
HENRY, WALTER RICHMOND	1878	SPRING	BOTANY	85
HERRING, JOSHUA JAMES, JR.	1886	SPRING	BOTANY	65
HESTER, BENNETT FRANKLIN	1877	SPRING	BOTANY	APPROVED
HESTER, JOSEPH FAIRFIELD	1891	SPRING	BOTANY	58
HESTER, ST. CLAIR	1884	SPRING	BOTANY	90
HICKS, JOHN WHITE	1879	SPRING	BOTANY	80
HILL, CHRISTOPHUS DUDLEY	1879	SPRING	BOTANY	95
HILL, EDWARD JOHN	1877	SPRING	BOTANY	APPROVED

Appendix 1. (Continued) A List of Names of Students Who Enrolled in Courses in Botany and Horticulture

Student	Year	Term	Course	Grade
Hill, Isham	1878	Spring	Botany	82
Hill, John Sprunt	1888	Spring	Botany	96
Hill, Robert Dudley	1888	Spring	Botany	60
Hines, Frank Gordon	1879	Spring	Botany	99
Hodges, Paul Arendel	1888	Spring	Botany	91
Holleman, Silas Alvin	1882	Spring	Botany	86
Holliday, Richard William	1889	Spring	Botany	74
Holmes, John Simcox	1886	Fall	Horticulture	91
Holmes, John Simcox	1887	Spring	Horticulture	90
Holmes, John Simcox	1887	Spring	Botany	94
Holmes, John Simcox	1887	Fall	Advanced Botany	90
Holmes, John Simcox	1888	Spring	Advanced Botany	94
Holshouser, Allan Rodolphus	1879	Spring	Botany	80
Holt, James Henry	1884	Spring	Botany	n[o] exam
Holt, Robert Lacy	1885	Spring	Botany	72
Holt, Robert Oscar	1879	Spring	Botany	99
Hoskins, Thomas Jones	1884	Spring	Botany	81
Huggins, William Sloan	1889	Spring	Botany	72
Hunter, Clifton Wheat	1876	Spring	Botany	75
Isler, Edwin Becton, Jr.	1892	Spring	Zoology & Botany	6
Jackson, Herbert Worth	1885	Spring	Botany	80
Jackson, Jonathan Worth	1880	Spring	Botany	90
Jackson, Max	1882	Spring	Botany	93
Jackson, Max	1884	Spring	Botany	80
James, Fernando Godfrey	1876	Spring	Botany	83
Jeffreys, John Osborne	1880	Spring	Botany	94
Jenkins, Joseph John	1884	Spring	Botany	70
Jimson, James Elliott	1889	Spring	Botany	87
John, Maxcy Luther	1884	Spring	Botany	80
John, Maxcy Luther	1885	Spring	Botany	70?
John, Roderick Belton	1879	Spring	Botany	98
Johnson, Jacob Coart	1884	Spring	Botany	80
Johnson, Jacob Coart	1885	Spring	Botany	82?
Johnston, Julius	1877	Spring	Botany	approved

APPENDIX 1. (CONTINUED) A LIST OF NAMES OF STUDENTS WHO ENROLLED IN COURSES IN BOTANY AND HORTICULTURE

Student	Year	Term	Course	Grade
JOHNSTON, RICHARD HALL	1891	SPRING	BOTANY	81
JOHNSTON, THOMAS JACKSON	1896–1897		BOTANY (BIOLOGY 4)	NO GRADE
JONES, KENNETH ANDERSON	1891	SPRING	BOTANY	86
JONES, LAWRENCE O'BRYEN BRANCH	1891	SPRING	BOTANY	88
JOYNER, JAMES YADKIN	1879	SPRING	BOTANY	85
JOYNER, JOHN PUGH	1879	SPRING	BOTANY	80
KAPP, JAMIE LYNN	1891	SPRING	BOTANY	83
KEECH, EDWARD JASPER	1889	SPRING	BOTANY	92
KENNEDY, DAVID SLOAN	1880	SPRING	BOTANY	75
KERNER, ROBAH BASCOM	1878	SPRING	BOTANY	65
KERNODLE, JAMES LOFTIN	1889	SPRING	BOTANY	82
KLUTZ, ADAM ALEXANDER	1882	SPRING	BOTANY	N[O] EXAM
KOONCE, FRANCIS DUVAL, JR.	1889	SPRING	BOTANY	84
KOONCE, WILLIAM MICHAEL	1885	SPRING	BOTANY	70
KOONTZ, ANDREW JACOB	1885	SPRING	BOTANY	65
KORNEGAY, DAVID ROBERT	1890	SPRING	BOTANY	72
LATHAM, HEBER AMOS LAMAR	1882	SPRING	BOTANY	90
LAWSON, ROBERT BAKER	1898	SPRING	BOTANY (BIOLOGY 4)	NO GRADE
LEACH, JAMES MADISON, JR.	1879	SPRING	BOTANY	80
LEHMAN, EUGENE WESLEY	1891	SPRING	BOTANY	85
LEWIS, HENRY WATSON	1885	SPRING	BOTANY	78
LEWIS, IVEY FOREMAN	1902–1903		BOTANY MINOR FOR MASTER'S DEGREE	
LEWIS, JOSEPH VOLNEY	1890	SPRING	BOTANY	97
LILLY, JOEL ALEXANDER	1890	SPRING	BOTANY	78
LITTLE, FRANK MILTON	1884	SPRING	BOTANY	70 SPECIAL EXAM
LLOYD, HENRY	1876	SPRING	BOTANY	71
LOCKETT, EVERETT AUGUSTINE	1899	SPRING	BOTANY (BIOLOGY 4)	3
LONDON, HENRY ADOLPHUS	1884	FALL	BOTANY	76
LONDON, JOHN HAUGHTON	1887	SPRING	BOTANY	52
LONG, AUGUSTUS WHITE	1882	SPRING	BOTANY	85
LONG, THOMAS HIRAM	1882	SPRING	BOTANY	81
LONG, VERNON WATSON	1885	SPRING	BOTANY	80

Appendix 1. (Continued) A List of Names of Students Who Enrolled in Courses in Botany and Horticulture

Student	Year	Term	Course	Grade
Love, Edgar	1889	Spring	Botany	94
Love, James Lee	1882	Spring	Botany	97
Lutterloh, Herbert	1878	Spring	Botany	92
Lynch, Leander Williams	1886	Spring	Botany	91
Lynch, Leander Williams	1887	Spring	Advanced Botany	95
Lynch, Leander Williams	1887	Fall	Advanced Botany	92
Lynch, Leander Williams	1887	Fall	Horticulture	94
Lynch, Leander Williams	1888	Spring	Horticulture	94
Lynch, Leander Williams	1888	Spring	Advanced Botany	94
Mangum, Charles Staples	1890	Spring	Botany	94
Manning, Frederick Haughton	1885	Botany	Botany	75
Manning, John Moore	1878	Spring	Botany	94
Manning, Pierre Beauregard	1884	Spring	Botany	95
Markham, Hugh Paris	1880	Spring	Botany	80
Massey, Charles Wesley	1884	Fall	Botany	84
Massey, Charles Wesley	1885	Spring	Botany	81
Mattocks, John Edwards	1892	Spring	Zoology & Botany	2
Mayhew, Thomas William	1880	Spring	Botany	70
McCorkle, George	1877	Spring	Botany	approved
McCullock, W.H.	1884	Spring	Botany	n[o] exam
McDonald, William H.	1885	Spring	Botany	84
McGehee, Lucius Polk	1884	Fall	Botany	99
McGehee, Lucius Polk	1884	Spring	Botany	90
McGuire, James, Jr.	1884	Spring	Botany	72
McIntyre, Alexander	1882	Spring	Botany	80
McIntyre, Peter	1882	Spring	Botany	85
McIver, Charles Duncan	1879	Spring	Botany	99
McIver, John Alton	1879	Spring	Botany	99
McIver, John Fenelon	1884	Fall	Botany	90
McIver, William Donald	1884	Spring	Botany	84
McKay, S. Prentys	1876	Spring	Botany	75
McKethan, Edwin Robeson	1890	Spring	Botany	88
McKinne, Frank Brothers	1892	Spring	Zoology & Botany	2
McKinnon, [first name not cited]	[1897]	Fall	Horticulture	75

Appendix 1. (Continued) A List of Names of Students Who Enrolled in Courses in Botany and Horticulture

Student	Year	Term	Course	Grade
McLeod, Gilbert	1879	Spring	Botany	90
McNeely, Thomas Chalmers	1884	Spring	Botany	50
McNeill, Charles Alexander	1878	Spring	Botany	82
McNeill, Neill Russell	1889	Spring	Botany	70
McNeill, William Hamilton	1883	Spring	Botany	70
MacNider, William deBerniere	1901	Spring	Botany	Passed
McRae, James Patterson	1879	Spring	Botany	0
McRae, Roderick	1879	Spring	Botany	98
Merritt, Leonidas Haywood	1889	Spring	Botany	91
Merritt, Rufus Grant	1876	Spring	Botany	83
Mewborn, William Edgar	1884	Fall	Botany	78
Mial, Leonidas LeMay	1879	Spring	Botany	98
Miller, Hugh Lee	1887	Spring	Botany	96
Miller, James Daniel	1882	Spring	Botany	99
Moore, Eugene Dick	1887	Spring	Botany	82
Moore, James	1877	Spring	Botany	approved
Moore, John Miller	1884	Spring	Botany	50
Moore, Thomas Lake	1888	Spring	Botany	88
Moore, Van Boddie	1877	Spring	Botany	approved
Morehead, John Motley	1884	Spring	Botany	70
Morehead, John Motley	1885	Spring	Advanced Botany	72
Morehead, John Motley	1888	Spring	Botany	91
Morris, Joseph Algenon	1884	Spring	Botany	80
Morris, Lewis Coleman	1891	Spring	Botany	86
Morton, William Zenas	1878	Spring	Botany	n[o] exam
Moye, Elbert Alfred, Jr.	1891	Spring	Botany	96
Murphy, John Gerald	1897	Spring	Botany	2
Neal, John William	1879	Spring	Botany	75
Nicholls, Joshua Robertson	1880	Spring	Botany	75
Nixon, Alfred	1879	Spring	Botany	93
Nixon, Henry Barber	1876	Spring	Botany	63
Nixon, James Whedbee	1876	Spring	Botany	58
Nixon, Robert Burns	1882	Spring	Botany	85
Norfleet, Pascal Paoli	1878	Spring	Botany	75

Appendix 1. (Continued) A List of Names of Students Who Enrolled in Courses in Botany and Horticulture

Student	Year	Term	Course	Grade
Odell, Oscar Charles	1884	Fall	Botany	93
Orr, Perry Leander	1880	Spring	Botany	77
Osborne, Thomas Samuel	1884	Spring	Botany	n[o] exam
Pailin, William White	1880	Spring	Botany	75
Palmer, Malvern Hill	1885	Spring	Botany	90
Palmer, Malvern Hill	1887	Fall	Horticulture	95
Palmer, Malvern Hill	1888	Spring	Horticulture	92
Parker, Haywood	1885	Spring	Botany	89
Parker, Romulus Bragg	1876	Spring	Botany	55
Patrick, George Lane	1884	Spring	Botany	81
Patrick, George Lane	1885	Fall	Horticulture	95
Patrick, George Lane	1886	Spring	Horticulture	90
Patterson, Andrew Henry	1888	Spring	Botany	100
Patterson, Gilbert Brown	1884	Spring	Botany	75
Patterson, John Legerwood	1892	Spring	Zoology & Botany	1
Patterson, Louis Morehead	1880	Spring	Botany	97
Patton, John Webster	1885	Spring	Botany	70
Peace, Leonidas Vassar	1880	Spring	Botany	70
Pearsall, Matthew James	1890	Spring	Botany	80
Pearson, Thomas Gilbert	1899	Spring	Botany (Biology 4)	4
Peebles, Henry Bruce	1880	Spring	Botany	88
Peele, William Joseph	1878	Spring	Botany	95
Pell, Edward Leigh	1879	Spring	Botany	0
Pell, Robert Paine	1878	Spring	Botany	n[o] exam
Pemberton, William David	1880	Spring	Botany	99
Pender, Herbert Ward	1880	Spring	Botany	n[o] exm
Person, Willie Mangum, Jr.	1885	Spring	Botany	78
Peschau, George Ludwig	1891	Spring	Botany	70
Peterson, Matthew Ransom	1884	Spring	Botany	90
Petty, George Edward	1889	Spring	Botany	certificate
Philbeck, Joseph Beattie	1885	Spring	Botany	87
Philips, James Jones	1890	Spring	Botany	92
Phillips, Alexander Lacy	1878	Spring	Botany	96
Phillips, John Leighton	1879	Spring	Botany	80

APPENDIX 1. (CONTINUED) A LIST OF NAMES OF STUDENTS WHO ENROLLED IN COURSES IN BOTANY AND HORTICULTURE

Student	Year	Term	Course	Grade
PHILIPS, WALTER EVERETT	1879	SPRING	BOTANY	87
POLLOCK, WILLIAM DURWARD	1882	SPRING	BOTANY	CONDITIONAL
POWELL, EDGAR	1879	SPRING	BOTANY	DISMISSED
PRIVOTT, HAYWOOD COLLINS	1885	SPRING	BOTANY	77
RADCLIFFE, THOMAS ROBERT	1880	SPRING	BOTANY	90
RANKIN, CHARLES ALEXANDER	1889	SPRING	BOTANY	PASSED
RANKIN, CHARLES ALEXANDER	1890	SPRING	ADVANCED BOTANY	90
RANSOM, JOSEPH EXUM	1878	SPRING	BOTANY	77
REID, REUBEN DAVID	1878	SPRING	BOTANY	80
RHEM, JOSEPH FRANKLIN	1889	SPRING	BOTANY	70
RHODES, WILLIAM HENRY	1884	SPRING	BOTANY	78
RICE, HENRY WILLIAM	1884	SPRING	BOTANY	88
RICHARDSON, CHARLES ELIJAH	1892	SPRING	ZOOLOGY & BOTANY	3
RICHARDSON, EDMUND ERNEST	1879	SPRING	BOTANY	0
RICHARDSON, SHAHANE	1878	SPRING	BOTANY	92
RICHMOND, ROMAN EUGENE	1887	SPRING	BOTANY	57
RICKS, WILLIAM BENJAMIN	1887	SPRING	BOTANY	91
ROBBINS, GASTON AHI	1878	SPRING	BOTANY	77
ROBERSON, WILLIAM STONE	1885	SPRING	BOTANY	89
ROBERTS, JAMES COLE	1882	SPRING	BOTANY	90
ROBERTSON, WILLIAM ROSS	1892	SPRING	ZOOLOGY & BOTANY	6
RODMAN, JOHN CROOM	1889	SPRING	BOTANY	80
RODMAN, WILLIAM BLOUNT, II	1880	SPRING	BOTANY	98
ROGERS, JUNIUS FOSTER	1880	SPRING	BOTANY	85
ROSS, EUGENE CLIFTON	1890	SPRING	BOTANY	89
ROSS, JOHN KIRKLAND	1897	SPRING	BOTANY	1
ROTHROCK, ROBAR BENJAMIN	1882	SPRING	BOTANY	75
ROUSE, NOAH JAMES	1879	SPRING	BOTANY	95
ROUSE, THOMAS RICHARD	1884	SPRING	BOTANY	80
ROWE, HOWARD MAY	1885	SPRING	BOTANY	70
RUFFIN, JAMES HIPKIN	1879	SPRING	BOTANY	95
RUFFIN, STERLING	1884	SPRING	BOTANY	75
RUFFIN, THOMAS, JR.	1880	SPRING	BOTANY	90
RUMBOUGH, HENRY THOMAS	1879	SPRING	BOTANY	80

Appendix 1. (Continued) A List of Names of Students Who Enrolled in Courses in Botany and Horticulture

Student	Year	Term	Course	Grade
Satterwhite, Solomon Green	1882	Spring	Botany	84
Saunders, Elijah Beattie	1884	Spring	Botany	75
Saunders, James Burrell	1884	Spring	Botany	78
Schenck, David	1885	Spring	Botany	30
Schenck, John Franklin	1885	Fall	Horticulture	89
Schenck, John Franklin	1886	Spring	Horticulture	92
Scott, James Edwin	1878	Spring	Botany	95
Scott, William Levy	1892	Spring	Zoology & Botany	4
Seawell, Aaron Ashley Flowers	1883	Spring	Botany	95
Self, William Augustus	1884	Spring	Botany	85
Sexton, Collin Hector	1880	Spring	Botany	50
Shaffner, Henry Fries	1885	Spring	Botany	93
Shaffner, William Francis	1887	Spring	Botany	95
Sharp, Hunter	1880	Spring	Botany	n[o] exam
Shelton, Charles Edward	1892	Spring	Botany	6
Sherrod, John Henry	1876	Spring	Botany	68
Sherrod, John Mayo	1890	Spring	Botany	92
Shields, Malcolm McGilvary	1884	Spring	Botany	75
Shore, Clarence Albert	1902–1903		Botany minor. for Ph.D	
Simmons, Albert Marchant	1885	Spring	Botany	90
Simmons, William Silas	1887	Spring	Botany	87
Sinclair, Neil Angus	1884	Spring	Botany	80
Skinner, Frederick Nash	1880	Spring	Botany	98
Slade, William Bonner	1879	Spring	Botany	98
Slocumb, John Charles	1882	Spring	Botany	77
Smith, Andrew Jones	1877	Spring	Botany	approved
Smith, Claudius Ferdinand	1884	Spring	Botany	78
Smith, Dulcedo	1890	Spring	Botany	78
Smith, George Worthington	1880	Spring	Botany	87
Smith, Robert Lee	1889	Spring	Botany	84
Snipes, William Seaton	1889	Spring	Botany	87
Spears, Henry Turner	1876	Spring	Botany	92
Spoon, William Luther	1888	Spring	Botany	96
Springs, Alva Connell	1878	Spring	Botany	n[o] exam

APPENDIX 1. (CONTINUED) A LIST OF NAMES OF STUDENTS WHO ENROLLED IN COURSES IN BOTANY AND HORTICULTURE

Student	Year	Term	Course	Grade
SPRINGS, LEROY	1879	SPRING	BOTANY	70
STAINBACK, DAVID CHANDLER	1876	SPRING	BOTANY	59
STARBUCK, HENRY REUBEN	1884	SPRING	BOTANY	78
STATON, JAMES RODERICK	1877	SPRING	BOTANY	APPROVED
STEDMAN, FRANK HAYWOOD	1879	SPRING	BOTANY	85
STEELE, EDWIN DOUGLAS	1876	SPRING	BOTANY	50
STEWART, WILLIAM BELLFIELD	1879	SPRING	BOTANY	85
STOKES, THOMAS DUDLEY	1879	SPRING	BOTANY	72
STRAYHORN, JOHN THOMAS	1880	SPRING	BOTANY	N[O] EXAM
STREET, NATHANIEL HEATH	1877	SPRING	BOTANY	APPROVED
STRICKLAND, EDWARD FOUNTAIN	1884	SPRING	BOTANY	75
STRONACH, ALEXANDER	1888	SPRING	BOTANY	94
STROWD, THOMAS WILSON	1888	SPRING	BOTANY	88
SULLIVAN, RUFUS BEAUREGARD	1885	SPRING	BOTANY	75
TAYLOR, ISAAC MONTROSE	1877	SPRING	BOTANY	APPROVED
TAYLOR, ISAAC MONTROSE	1878	SPRING	BOTANY	89
THOMAS, CHARLES RANDOLPH	1879	SPRING	BOTANY	92
THOMAS, JAMES	1884	SPRING	BOTANY	90
THOMPSON, SEYMOUR COLUMBUS	1888	SPRING	BOTANY	88
THORP, BENONI	1887	SPRING	BOTANY	89
TILLETT, HENRY AUGUSTUS	1880	SPRING	BOTANY	97
TILLEY, GEORGE HINTON VANCE	1887	SPRING	BOTANY	88
TOMS, CHARLES FRENCH	1890	SPRING	BOTANY	80
TOMS, CLINTON WHITE	1888	SPRING	BOTANY	87
TOY, THOMAS DALLAM	1891	SPRING	BOTANY	90
TRIPP, JOHN BRYANT	1886	SPRING	BOTANY	72
TROY, EDWARD	1882	SPRING	BOTANY	77
TROY, JOHN CLARK	1876	SPRING	BOTANY	84
TUCKER, WILLIAM RUFFIN	1885	SPRING	BOTANY	70
TURLINGTON, WILLIS EDGAR	1880	SPRING	BOTANY	60
TURRENTINE, JOHN WILLIAM	1901	SPRING	BOTANY	1
TURRENTINE, SAMUEL BRYANT	1882	SPRING	BOTANY	86
TYSON, BENJAMIN FRANKLIN	1887	FALL	HORTICULTURE	85
TYSON, BENJAMIN FRANKLIN	1888	SPRING	HORTICULTURE	80

Appendix 1. (Continued) A List of Names of Students Who Enrolled in Courses in Botany and Horticulture

Student	Year	Term	Course	Grade
Urquhart, Lewis Thompson	1889	Spring	Botany	74
Uzzell, John Randolph	1879	Spring	Botany	0
Uzzell, Kirby Smith	1884	Spring	Advanced Botany	75
Uzzell, Robert Lee	1884	Spring	Botany	78
Uzzell, Robert Lee	1885	Spring	Botany	85
Uzzell, Robert Lee	1887	Spring	Botany	74
Vann, Livingston	1882	Spring	Botany	95
Vaughan, Latimer Clark	1878	Spring	Botany	85
Vaughn, Robert Galloway	1888	Spring	Botany	89
Vines, Charles Carson	1876	Spring	Botany	93
Waldrop, Otis Theron	1886	Spring	Botany	75
Walker, John Moseley, Jr.	1879	Spring	Botany	97
Walker, Lucian Holmes	1879	Spring	Botany	99
Walser, Zeb	1882	Spring	Botany	90
Walser, Zenobian Ilmer	1891	Spring	Botany	89
Ward, Needham Erastus	1897	Spring	Botany	2
Warlick, Lee Martin	1882	Spring	Botany	92
Warren, William Young, Jr.	1892	Spring	Botany	82
Watkins, Henry Thomas	1877	Spring	Botany	approved
Webb, John	1879	Spring	Botany	70
Weeks, Stephen Beauregard	1884	Spring	Botany	90
Weill, Solomon Cohen	1885	Spring	Botany	96
West, Jesse Felix	1882	Spring	Botany	85
Whitaker, James Fletcher	1879	Spring	Botany	70
White, Benjamin Franklin	1882	Spring	Botany	n[o] exam
White, Benjamin Franklin	1884	Spring	Botany	95
White, Richard Street	1880	Spring	Botany	80
Whitfield, William Cobb	1880	Spring	Botany	60
Whitlock, Victor Emanuel	1891	Spring	Botany	90
Whitsett, George Walter	1880	Spring	Botany	84
Whitsett, William Thornton	1887	Spring	Botany	95
Wilkinson, William Stronach	1884	Spring	Botany	78
Wilkinson, William Stronach	1885	Spring	Botany	94
Willcox, Joseph Martin	1889	Spring	Botany	89

APPENDIX 1. (CONTINUED) A LIST OF NAMES OF STUDENTS WHO ENROLLED IN COURSES IN BOTANY AND HORTICULTURE

Student	Year	Term	Course	Grade
WILLIAMS, ALBERT SIDNEY	1888	SPRING	BOTANY	40
WILLIAMS, JOHN ROBERT	1889	SPRING	BOTANY	84
WILLIAMS, JOSEPH AUGUSTUS	1879	SPRING	BOTANY	0
WILLS, GEORGE STOCKTON	1888	SPRING	BOTANY	98
WILSON, ADOLPHUS ERWIN	1884	SPRING	BOTANY	50
WILSON, ALBERT ROBERT	1878	SPRING	BOTANY	100
WILSON, DELONZA TATE	1884	SPRING	BOTANY	81
WILSON, DELONZA TATE	1885	SPRING	BOTANY	91
WILSON, GEORGE GREGORY	1880	SPRING	BOTANY	97
WILSON, JOHN NESTOR	1879	SPRING	BOTANY	70
WILSON, NATHAN HUNT DANIEL	1884	SPRING	BOTANY	99
WILSON, STEPHEN PENDER	1884	SPRING	BOTANY	75
WILSON, WILLIAM ALBERT	1889	SPRING	BOTANY	89
WINBORNE, PETER PARKER	1890	SPRING	BOTANY	98
WINBORNE, ROBERT WARREN	1879	SPRING	BOTANY	90
WINDERS, JOHN WILLIAM	1884	SPRING	BOTANY	73
WINSTON, FRANCIS DONNELL	1877	SPRING	BOTANY	APPROVED
WINSTON, JOHN CAMM	1879	SPRING	BOTANY	75
WINSTON, ROBERT WATSON	1877	SPRING	BOTANY	APPROVED
WOOD, HENRY GILLIAM	1888	SPRING	BOTANY	80
WOOD, JULIEN	1882	SPRING	BOTANY	94
WOOD, WORD HARRIS	1892	SPRING	ZOOLOGY & BOTANY	5
WOODARD, LEONIDAS POLK	1884	SPRING	BOTANY	30
WOODARD, PAUL LEE	1887	SPRING	BOTANY	71
WOODARD, SIDNEY ALBERT	1882	SPRING	BOTANY	87
WOODLEY, DANIEL EDGAR	1886	SPRING	BOTANY	95
WOODLEY, DANIEL EDGAR	1886	FALL	ADVANCED BOTANY	NO GRADE
WOODSON, ROBERT SCOTT	1886	SPRING	BOTANY	95
WORTH, CHARLES WILLIAMS	1880	SPRING	BOTANY	95
WORTH, GEORGE CLARKSON	1889	SPRING	BOTANY	90
WRIGHT, CLEMENT GILLESPIE	1884	SPRING	BOTANY	71
WYCHE, BENJAMIN	1890	SPRING	BOTANY	90
WYCHE, RICHARD THOMAS	1890	SPRING	BOTANY	89
YARBOROUGH, FENNER	1876	SPRING	BOTANY	94
YATES, JOSEPH WALKER	1891	SPRING	BOTANY	91

SUBJECT INDEX (including names of individuals)

- **"UNC" = The University of North Carolina at Chapel Hill.**
- **Unless otherwise specified, all buildings, departments, libraries, museums, and faculty are those of UNC, and all organizations are those of North Carolina.**
- **A page number in bold indicates a drawing, photograph, or figure.**
- **The letter "t" after a page number indicates that information appears in a table.**
- **The letter "n" after a page number indicates that information appears in a footnote.**

B

C

D

E

H

I

Q

R

X

Y

Z

SCIENTIFIC AND COMMON NAMES INDEX

- **A page number in bold indicates a drawing, photograph, or figure.**
- **The letter "t" after a page number indicates that information appears in a table.**
- **The letter "n" after a page number indicates that information appears in a footnote.**
- **Scientific names were transcribed from the literature without updating the nomenclature but with correcting obvious misspellings.**
- **Various species of plants are selectively listed under terms shared by their common names. For instance, under the index entry of "grasses," names include species that are taxonomically unrelated but include the common name "grass."**

D

E

G

Colophon

PUTTING DOWN ROOTS: FOUNDATIONS OF BOTANY AT CAROLINA

Cover design by Linda Jacobson, Keeper, North Carolina Collection Gallery, Wilson Special Collections Library, University of North Carolina, Chapel Hill, USA.

Book layout by Ricki C. Herdendorf,
EcoSphere Associates, 4921 Detroit Road, Sheffield Village, Ohio 44054-2809, USA.

Titles are set in 11 to 19 pt. Century Schoolbook with 13.2 to 22.8 pt. leading.

The text is set in 9 and 10 pt. Times New Roman with 10.8 and 12 pt. leading used for captions, and appendices.

The index is set in 6.125 pt. Times New Roman with 7.35 pt. leading.

Final indexes created with SKY Index by Kelly E. Burch,
KEB Indexing Services, kelly@kebindexingservices.com.

This book was designed and produced on the iMac, using Adobe® InDesign®, CC 2019.

The cover is 300 gsm MGB C1S. Flex-binding, Smyth sewn, round back with head and tail bands, and hinged scored.

End papers are 150 gsm Woodfree white.

Text pages are acid free, 70 gsm New Classic paper white.

This book was printed and bound by Artpac, Seoul, South Korea.

Notes

Notes

Notes

Notes